Lecture Notes in Electrical Engineering   1045

The book series *Lecture Notes in Electrical Engineering* (LNEE) publishes the latest developments in Electrical Engineering—quickly, informally and in high quality. While original research reported in proceedings and monographs has traditionally formed the core of LNEE, we also encourage authors to submit books devoted to supporting student education and professional training in the various fields and applications areas of electrical engineering. The series cover classical and emerging topics concerning:

- Communication Engineering, Information Theory and Networks
- Electronics Engineering and Microelectronics
- Signal, Image and Speech Processing
- Wireless and Mobile Communication
- Circuits and Systems
- Energy Systems, Power Electronics and Electrical Machines
- Electro-optical Engineering
- Instrumentation Engineering
- Avionics Engineering
- Control Systems
- Internet-of-Things and Cybersecurity
- Biomedical Devices, MEMS and NEMS

For general information about this book series, comments or suggestions, please contact leontina.dicecco@springer.com.

To submit a proposal or request further information, please contact the Publishing Editor in your country:

China

Jasmine Dou, Editor (jasmine.dou@springer.com)

India, Japan, Rest of Asia

Swati Meherishi, Editorial Director (Swati.Meherishi@springer.com)

Southeast Asia, Australia, New Zealand

Ramesh Nath Premnath, Editor (ramesh.premnath@springernature.com)

USA, Canada

Michael Luby, Senior Editor (michael.luby@springer.com)

All other Countries

Leontina Di Cecco, Senior Editor (leontina.dicecco@springer.com)

** This series is indexed by EI Compendex and Scopus databases. **

Jason C. Hung · Jia-Wei Chang · Yan Pei
Editors

Innovative Computing Vol 2 - Emerging Topics in Future Internet

Proceedings of IC 2023

 Springer

Editors
Jason C. Hung
Department of Computer Science
and Information Engineering
National Taichung University of Science
and Technology
Taichung City, Taiwan

Jia-Wei Chang
Department of Computer Science
and Information Engineering
National Taichung University of Science
Taichung City, Taiwan

Yan Pei
Computer Science and Engineering
University of Aizu
Aizuwakamatsu, Fukushima, Japan

ISSN 1876-1100   ISSN 1876-1119 (electronic)
Lecture Notes in Electrical Engineering
ISBN 978-981-99-2289-5   ISBN 978-981-99-2287-1 (eBook)
https://doi.org/10.1007/978-981-99-2287-1

This Springer imprint is published by the registered company Springer Nature Singapore Pte Ltd.
The registered company address is: 152 Beach Road, #21-01/04 Gateway East, Singapore 189721, Singapore

Contents

The 6th International Conference on Innovative Computing (IC 2023)

The International Workshop on Big-Data, IoT, Cloud Computing Technologies and Applications (BICTA2023)

ICIC 2023

Computer Aided Simulation Experiment of Endogenous Microbial Oil Displacement

Wu Ze[✉]

Tertiary Oil Recovery Project Department, Seventh Oil Production Plant of Daqing Oilfield Limited Company, Daqing 16300, China
wz8186659@163.com

Abstract. Endogenous microbial oil recovery technology has attracted more attention because of its strong adaptability, good compatibility with reservoir and low cost. However, at present, the physical simulation experimental method of microbial oil displacement is quite different from the actual reservoir, and the evaluation method needs to be improved urgently. Therefore, this paper conducted a basic theoretical study on the influencing factors of endogenous microbial model experiment. The computer-aided simulation experiment of endogenous microbial oil displacement is a computer model that simulates the behavior of microorganisms in oil reservoirs. The purpose of this model is to study the effects of different conditions on microorganisms and petroleum hydrocarbons (oil). It can also be used to design remediation strategies for contaminated sites with high microbial activity and low dissolved organic carbon content.

Keywords: Computer-aided · Endogenous microorganisms · Oil displacement simulation

1 Introduction

Microbial oil recovery technology refers to the use of microbial metabolic activities and their metabolites to act on reservoir and reservoir fluids, So as to improve crude oil recovery[". Previous research and field practice have proved that microbial oil recovery technology is a cost-effective method to improve production and recovery. Microbial oil recovery technology has the advantages of low cost, strong adaptability, no damage to the reservoir and no pollution to the environment, and has a wide application prospect. The U.S. Department of energy has listed it as the fourth type of enhanced oil recovery technology after thermal oil recovery, miscible flooding and chemical flooding [1].

Different from chemical flooding, in the process of microbial EOR, microorganisms themselves carry out life activities, and their metabolites can also enhance oil recovery. Although microbial oil recovery technology has a history of decades, it is still difficult to quantitatively describe the detailed mechanism of microbial oil recovery technology due to the complexity of microbial life activities. Therefore, the basic research and evaluation of microbial oil recovery technology mainly rely on physical simulation means, and there

J. C. Hung et al. (Eds.): IC 2023, LNEE 1045, pp. 3–9, 2023.
https://doi.org/10.1007/978-981-99-2287-1_1

is still no mature and reliable numerical simulation software of microbial oil recovery [2].

The results of indoor physical simulation evaluation experiment show that both endogenous microbial oil displacement and exogenous microbial oil displacement have significant oil increase effects, and the enhanced oil recovery is between 5% and 15% under the indoor physical model conditions. However, the oil displacement tests carried out in many domestic oilfields show that the effect of microbial oil displacement field test is not significant. Through the analysis and comparison of the physical model of microbial oil displacement and field test literature, it can be seen that there are differences on many issues in the physical simulation of microbial oil displacement. For example, there is a prominent contradiction between the static growth and metabolism simulation of microorganisms in the physical model and the mine construction during the cultivation period, and there is no unified specification for the influencing factors such as core length, anaerobic environment and injection speed, etc. [3].

In conclusion, there are great differences between the physical model experiment of microbial oil displacement and the actual reservoir, and the evaluation method needs to be improved. The above factors should be fully considered, and the research on the physical simulation experiment technology of microbial oil displacement should be strengthened, so that the physical simulation evaluation results can guide the field application more accurately. Compared with the external microbial oil recovery technology, the internal microbial oil recovery technology has the advantages of strong adaptability, good compatibility with the reservoir, low cost and simpler construction [4]. Therefore, the research on the internal microbial oil displacement physical model evaluation technology is more urgent.

2   Related Work

2.1   Research Status of Microbial Oil Recovery Technology

From 1960s to 1990s, microbial oil recovery technology was booming. As the oil crisis in the 1970s hit the world economic development hard, the United States, the former Soviet Union, Canada and other countries have turned to the development of low-cost and high-efficiency oil recovery technology, and carried out a large number of theoretical research and field tests of microbial oil recovery technology. Key technologies such as microbial EOR mechanism, indoor evaluation method, field injection equipment, reservoir screening criteria, and microbial EOR numerical simulation have been comprehensively developed, and microbial huff and puff, wax removal and prevention technologies have been successfully applied in the oilfield. In 1967, hitzman of the United States proposed that because the oil layer is generally in an anaerobic environment, it is necessary to inject oxygen or air for microbial growth and metabolism when strange oxidizing microorganisms are used for oil recovery. In 1963, Kuznetsov et al. Found that some microorganisms released a large amount of methane in some oil and gas reservoirs, and speculated that hydrogen and carbon dioxide produced by bacterial metabolism could act to produce methane [5]. In China, Daqing Oilfield has mainly studied the use of microorganisms to judge the water absorption of oil layers. The results show that taking iron bacteria as indicator bacteria can qualitatively judge whether oil layers absorb water, and has been successfully applied to field tests. In 1966, Xinjiang Oilfield began to carry out research on microbial crude oil dewaxing; In 1986, the research work of microbial heavy oil dewaxing and methanol protein was carried out successively [6].

Since the 1990s, after decades of basic research, microbial single well huff and puff and microbial paraffin removal and control technology have stepped from the basic research stage to the large-scale field application stage. The research focus of this stage has shifted to microbial enhanced water drive technology, and the research content has also shifted from indoor and field qualitative research to numerical simulation quantitative research stage [7]. In the late 1980s and early 1990s, some foreign countries began to study the mathematical model and numerical simulation of microbial oil recovery. Knapp R.A., Zhang X., Islam M.R., Chang M.M. and others have successively put forward the research results of "microbial growth and migration model in porous media reservoir" and "mathematical model of microbial enhanced oil recovery". Sarkar A.K. and others published the research on "simulation of components of microbial enhanced oil recovery" at the international microbial enhanced oil recovery conference in 1994, pointing out that microbial enhanced oil recovery through the production of surfactant is the most potential development direction [8]. After entering the 1990s, China has also accelerated the research pace of microbial oil recovery technology, and introduced a variety of microbial products and microbial enhanced oil recovery technology from micro BAC company, NPC company in the United States and Casco company in Canada. CNPC has carried out field tests of various microbial oil recovery technologies in more than 1000 wells in Jilin, Dagang, Liaohe, Daqing, North China, Xinjiang and other oilfields, with a cumulative increase of more than 80000 tons [9].

2.2 Existing Problems

At present, microbial model and its application research are still in the development stage, and the main factors restricting its development are microbial oil recovery mechanism, the implementation scale of mine projects, the level of model development and the budget of projects. Due to the limitations of the above factors, microbial numerical simulation is far from as perfect as polymer flooding and chemical flooding numerical simulation. At this stage, the problems of microbial EOR numerical simulation are as follows:

(1) Only one microbial component is involved in the mathematical model of microbial oil displacement, that is, the reaction kinetic parameters of all microorganisms in the reservoir are the same, but from the indoor screening and field application of the endogenous microbial oil displacement nutrient system, it can be seen that all mathematical models of microbial oil displacement can not meet the simulation of the oil displacement process. Even if it is anaerobic activation, the kinetic parameters of anaerobic bacteria and methanogens are very different, Microbial components need to be reclassified [10].

(2) There are few studies on the formation process of endogenous microbial field and the adsorption law of microbial oil displacement. Most models only give the initial microbial concentration of suspended phase as a constant, and the initial concentration of adsorbed phase is not considered.

(3) The kinetic model of endogenous microbial reaction is not perfect. There are many studies on microbial growth kinetics, but few on product production kinetics and substrate consumption kinetics. The relationship between product production rate, substrate consumption rate and microbial growth rate is one-sided

(4) The structure design of model data body is quite different from that of commercial software, and has poor compatibility with other numerical simulation and geological modeling software.

(5) The solution of the difference equation involves a variety of solutions, and there is no demonstration of the reliability of the solution.

(6) There is no simulation application of the two-step activation process of endogenous microorganisms, nor a complete set of reaction kinetic parameters of the two-step activation process of endogenous microorganisms.

(7) The principle of microbial EOR mainly refers to the principle of chemical flooding EOR, which does not reflect the difference between endogenous microbial oil displacement process and chemical flooding.

3 Computer Aided Simulation Experiment of Endogenous Microbial Oil Displacement

3.1 Endogenous Microbial Field Model Components

The distribution of microorganisms in the reservoir depends on the structural and attribute characteristics of the reservoir, the degree of water injection development and the physicochemical characteristics of oil, gas and water. Mastering the distribution law of microbial communities in the reservoir is conducive to establishing an accurate endogenous microbial field and improving the accuracy of numerical simulation of endogenous microbial oil displacement. Generally, according to the characteristics of microbial nutrient consumption, metabolic pathway and oxygen demand in the reservoir, the endogenous microorganisms in the reservoir are classified as follows:

① Hob: Taking monsters as the only carbon source, the metabolic process is an aerobic process. The burning oxidizing bacteria are the main flora activated by endogenous microorganisms, and their activation systems are mainly phosphate and ammonium salts.

② Saprophytic bacteria (TGB): saprophytic bacteria are aerobic bacteria. Its growth and reproduction must be completed in an aerobic environment. It can use all kinds of sugary substances, decompose sugars, metabolize carbon dioxide, and change pH value.

③ Nitrate reducing bacteria (NRb): these bacteria can reduce nitrate to nitrite in anaerobic or low dissolved oxygen environment, and finally produce ammonia, nitrogen or coz. In addition, nitrate reducing bacteria can make better use of nutrients than sulfate reducing bacteria. Therefore, sulfate reducing bacteria can be inhibited.

④ Sulfate reducing bacteria (SRB): under anaerobic conditions, the bacteria can reduce the sulfate radical existing in formation water and injection water to produce reduced sulfur. Sulfur will combine with hydrogen to form HS, which will corrode various pipelines. It is generally considered as a harmful bacterium.

3.2 Mathematical Model of Endogenous Microbial Oil Recovery

Combined with Zhang Xu model, the mathematical model is cited. The model divides the microorganisms in each phase into three different components:

(1) Microbial flora with oxygen as the final electron acceptor (microorganism 1): including strange oxidizing bacteria and saprophytic bacteria;
(2) Flora that does not rely on oxygen for growth and reproduction (microorganism 2): fermentation bacteria;
(3) The main metabolite is methane, and the microbiological reaction rate parameters are significantly different from the first two components (microorganism 3): methanogens.

Considering the convection dispersion, adsorption desorption and precipitation of nutrients, microorganisms and their metabolites, combined with microbial reaction kinetics, the control equation of the biological field model:

$$\|e_{k+1}(t)\|_\lambda = \|C(t)\| \|\Delta x_{k+1}(t)\|_\lambda \tag{1}$$

To truly reflect the actual production situation, the indoor oil displacement experiment needs to adopt the "completely proportional" model, but due to the complexity of reservoir seepage and development process, it is impossible to simulate and reproduce these processes completely and truly. Therefore, it is necessary to deduce the similarity criteria of physical simulation according to the mathematical model to restrict the parameters in the development process, so that various factors affecting the physical model experiment can be agreed. The similarity principle requires that the similarity criteria between the physical model and the prototype must be completely equal, which can be satisfied for a simple physical system. However, for complex systems, it is difficult to meet all the similarity criteria. Sometimes, there are contradictions between the similarity criteria. Therefore, it is necessary to determine which similarity criteria are primary and which are secondary, which can be moderately relaxed. Due to the complexity of oilfield production and reservoir, it is impossible to design a simulation experiment with all the derived similarity criteria in proportion. Therefore, it is necessary to analyze and study the specific problems of the reservoir and select the similarity criteria that can be realized in the simulation and play a leading role in the recovery, that is, sensitivity analysis. Generally, there are analysis methods of numerical experiments and approximate modeling methods, as shown in Fig. 1.

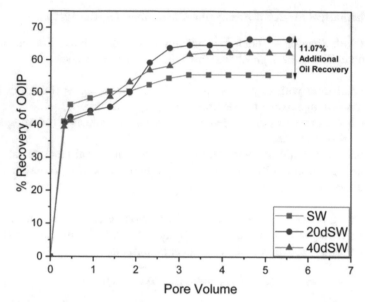

Fig. 1. Comparison of analytical methods of mathematical models of endogenous microbial oil recovery

4   Conclusion

Under the condition of medium and high permeability endogenous microbial oil displacement, whether gas injection or no gas injection, the EOR of quartz sand material is significantly higher than that of channel sand material, and the core materials with different wettability have a certain impact on EOR, among which the EOR of strong hydrophilic core is the largest, followed by neutral wetted core, and the EOR of weak hydrophilic core is the smallest. Finally, it is considered that the recovery rate range of water drive in channel sand cemented core with weak lipophilicity is closer to the actual reservoir, which can more truly simulate the pore structure of the reservoir and reflect the reservoir situation.

References

1. Zhao, J., Ying, F.: Research on the construction of virtual simulation experiment teaching center based on computer-aided civil engineering in colleges and universities. J. Phys. Conf. Ser. **1744**(3), 032115 (6p.) (2021)
2. Chen, L., Li, W., Li, Z.L., et al.: Experiment and simulation analysis of the pressure carrying capacity of ×80 pipe with metal loss defect on the girth weld. Mater. Sci. Forum **1035**, 813–818 (2021)
3. Yu, V.S., Popov, E.A.: Computer-aided simulation of high-dimensional event-continuous systems. J. Phys. Conf. Ser. **1791**(1), 012087 (5p.) (2021)
4. Langreck, J., Wong, H., Hernandez, A., et al.: Modeling and simulation of future capabilities with an automated computer-aided wargame. J. Defense. Model. Simul. Appl. Methodol. Technol. **18**(4), 407–416 (2021)

5. Moba, B., Akt, C., Rg, D.: Sustainable biorefinery process synthesis, design, and simulation: Systematic computer-aided methods and tools (2022)
6. Lewandowski, G.A., Klimczuk, T., et al.: Towards Computer-Aided Graphene Covered TiO2-Cu/(CuxOy) Composite Design for the Purpose of Photoinduced Hydrogen Evolution (2021)
7. Wang, Y., Nault, C., Givens, M., et al.: Computer-Aided Engineering Toolkit For Simulated Testing of Pressure-Controlling Component Designs, US20210342506A1 (2021)
8. Zhang, J., Lee, C., Farner, M.: Using computer-aided image processing to estimate chemical composition of igneous rocks: a potential tool for large-scale compositional mapping. Solid Earth Sci. 6(1), 21–26 (2021)
9. Ava, A., Vv, A., Sb, B., et al.: Computer aided cooling curve analysis (CACCA) of ADC-12 alloy (2021)
10. Oso, M., Regueira, A., Hospido, A., et al.: Fostering the valorization of organic wastes into carboxylates by a computer-aided design tool. Waste Manage. 142, 101–110 (2022)

Construction and Application of Intelligent Sensing Ability in Infrastructure Construction Site Based on Fish Swarm Algorithm

Cheng Zhang[1]([✉]), Yongguang Niu[2], Xuekai Zhang[1], Hao Zhang[3], and Weibin Lan[2]

[1] State Grid Shandong Electric Power Company, Jinan 250001, Shandong, China
63946017@qq.com
[2] Shandong Luruan Digital Technology Co., Ltd., Jinan 250000, Shandong, China
[3] Economic and Technology Research Institute, State Grid Shandong Electric Power Company, Jinan 250000, Shandong, China

Abstract. Intelligent sensing capability is the key technology of building intelligent infrastructure. It can be used in many fields, such as smart cities, smart buildings and smart agriculture. Intelligent sensing capability is an artificial intelligence (AI) technology that can sense the real-time environment. It is widely used in safety, environmental protection, medical and other fields. This paper will use the fish swarm algorithm to detect the fish swarm on the construction site based on the intelligent sensing ability. The application of fish swarm algorithm in intelligent sensing capability is to detect the construction site based on the information received from different sensors. The main objective of the project is to improve and enhance the performance of intelligent perception through the use of artificial intelligence technology. It will be used to detect and track any type of motion on the construction site, which will help reduce human errors and improve efficiency.

Keywords: Fish swarm algorithm · Infrastructure site · Perception

1 Introduction

As we all know, the new infrastructure includes information infrastructure, integration infrastructure and innovation infrastructure. It is an infrastructure system that provides services such as digital transformation, intelligent upgrading and integration innovation. However, in the face of such a huge system, many regions do not know where to start when developing new infrastructure.

In this regard, The "intelligent agent" released by Huawei on full connection 2020 puts forward a new idea, and all industries use "intelligent agent "To practice the new infrastructure as an entry point can further accelerate the implementation of the new infrastructure [1]. It is understood that the intelligent agent includes four layers of intelligent interaction, intelligent connection, intelligent hub and intelligent application. Based on the cloud and AI as the core, it will build an open intelligent system with three-dimensional perception, global coordination, accurate judgment and continuous

J. C. Hung et al. (Eds.): IC 2023, LNEE 1045, pp. 10–16, 2023.
https://doi.org/10.1007/978-981-99-2287-1_2

evolution, which can bring the whole scene intelligent experience for urban governance, enterprise production and resident life.

It can be seen that the emergence of intelligent agents, integrating various information technologies, can promote the construction and coordination of new infrastructure information infrastructure and reduce the difficulty of information infrastructure construction; From the perspective of application, as the technical reference framework for intelligent upgrading, the agent provides strong support for the construction of integration infrastructure and innovation infrastructure.

If the agent is Huawei's practice of new infrastructure, it is the reference framework for realizing the upgrading of government enterprise intelligence; The urban agent is a city level integrated intelligent collaborative system built by the collaborative innovation of multiple technologies such as connection, cloud [2], AI, computing and urban application, which can make the city feel, think, evolve and have temperature. In comparison, the "urban brain" often referred to in the industry in the past is based on the Internet, mainly focusing on the analysis and processing of urban data to realize the centralized management and monitoring of the city; while the "urban agent" is an integrated intelligent system, just like the "five senses", "hands and feet", "nerves", "trunk" and "brain" of the human body, so that the city can fully and real-time perceive the people, things, space and processes in the city, Through real-time data, we can timely find urban problems, study and judge the situation, prevent risks, and conduct real-time interaction [3]. Based on this, the research of this paper is the construction and application of intelligent sensing ability based on fish swarm algorithm.

2 Related Work

2.1 Fish Swarm Algorithm

The artificial fish swarm algorithm is that in a water area, fish can often find places with more nutrients by themselves or following other fish. Therefore, the place with the largest number of fish is generally the place with the most nutrients in the water area [4]. According to this feature, the artificial fish swarm algorithm simulates the foraging, clustering and tail chasing behavior of the fish swarm by constructing artificial fish to achieve optimization. Figure 1 below shows the iterative behavior flow of fish swarm algorithm.

Iterative behavior flow of fish swarm algorithm.

Artificial fish have the following typical behaviors:

(1) Foraging behavior: generally, fish swim freely in the water at random. When they find food, they will swim quickly in the direction of gradually increasing food.
(2) Swarm behavior: in order to ensure their own survival and avoid hazards, fish will naturally swarm in groups. There are three rules for Fish Swarm: separation rules: try to avoid overcrowding with neighboring partners; Alignment rules: try to be consistent with the average direction of neighboring partners; Cohesion rule: try to move towards the center of the neighboring partner [5].
(3) Tail chasing behavior: when one or several fish in a shoal find food, their neighboring partners will follow them to the food point quickly.

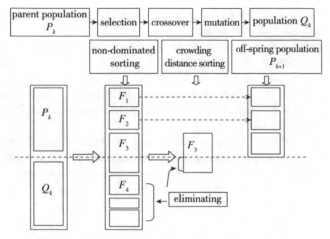

Fig. 1. Iterative behavior flow of fish swarm algorithm

(4) Random behavior: individual fish usually swim randomly in the water, in order to find food points or companions in a wider range.

$$O_{v,j} = h\left(\sum_{i=1}^{f_{k-1}} \sum_{u \in N[v]} w_{i,j,u,v} x_{u,i}\right), (j = 1, ..., f_k) \tag{1}$$

$$\max \sum_I \left[U^I(X^I) - C^I(X^I)\right] \tag{2}$$

Steps of implementing artificial fish swarm algorithm:

(1) Initialization settings, including population size n, initial position of each artificial fish, visual field of artificial fish, step size, crowding factor 5, and number of repetitions trynumber;

(2) Calculate the fitness value of each individual in the initial fish school, and give the best artificial fish state and its value to the bulletin board;

(3) Evaluate each individual and select the behaviors to be performed, including foraging pray, swarm, tail chasing follow and evaluation behavior Bulletin; (4) Implement the behavior of artificial fish, update themselves, and generate new fish schools;

(4) All individuals were evaluated. If an individual is superior to the bulletin board, the bulletin board is updated to that individual;

(5) When the optimal solution on the bulletin board reaches the satisfactory error range or reaches the upper limit of the number of iterations, the algorithm ends, otherwise, go to step 3.

2.2 Intelligent Sensing Hibernate Technology

Hibernate is a lightweight persistence layer solution. It is an open source ORM framework, i.e. object relational mapping framework 9 '. Hibernate encapsulates JDBC in a lightweight manner, and packages the relational database into an object-oriented model, so that developers can operate the relational database in an object-oriented manner, and only need to write simple HQL (hibernate query language) statements, thus greatly reducing the development time of manually writing SQL statements and processing JDBC. Hibernate advocates low intrusion design and fully adopts POJO programming.

There are many persistence layer solutions based on ORM framework, and Hibernate can stand out because hibernate has the following advantages compared with other ORM frameworks [6].

(1) Hibernate is completely free and open source.
(2) Hibernate is a lightweight framework that is non intrusive and avoids complex problems as much as possible.
(3) Active developers can have stable development.
(4) High scalability and open API. When the function is insufficient, it can be coded and expanded by itself.

Hibernate has five core interfaces: configuration, sessionfactory. Session, transaction and query. Hibernate accesses persistent objects and manages database transactions through these interfaces. The architecture of Hibernate is shown in Fig. 2.

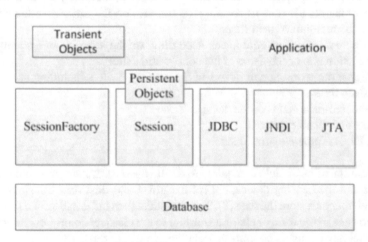

Fig. 2. Hibernate architecture

3 Construction and Application of Intelligent Perception Ability on Infrastructure Construction Site Based on Fish Swarm Algorithm

The construction site management function manages the construction projects undertaken by enterprises. A construction site is an engineering construction project. The commencement of a project needs to be added to the system and the relevant information of the construction site project can be configured. Then, the relevant information of the construction site can be modified and the completed construction site project can be deleted from the system. The module also needs to set up site personnel and internal law enforcement personnel at each construction site. In addition, when displaying the site list, you can display the completed site and uncompleted site as required.

The main function of the equipment management module is to manage the intelligent induction equipment installed on the construction site. One induction equipment is equipped with noise and dust sensors [7]. The functions of the device management module include:

(1) When new equipment is installed on the monitoring site, it is necessary to manually edit the basic information of the equipment: Province, city, district / county, construction site, installation location, detailed address and intelligent equipment coordinates (longitude, latitude and altitude).
(2) Manually enter the SIM card number of the device communication module, and the SIM card number uniquely identifies a device.
(3) Configure the p address and port number of the intelligent site monitoring system server in the area where the equipment belongs.
(4) Configure the equipment maintenance information, including the time of the last battery change, the time of cleaning the sensor, and the time when the equipment should be maintained next time.
(5) Inquire the induction equipment according to the district and county of the construction site or the name of the construction site.
(6) Configure the working parameters of the sensing device, including alarm interval and sleep time.
(7) Get the real-time status of the device.
(8) Enable and disable devices.
(9) Stop the equipment alarm function.

The data statistics module is required to define the data transmission mode between the server and the sensing device, collect the noise and dust data transmitted by the sensor, and filter and store the data [8]. The statistical data information can be displayed on the web page in the form of charts according to the district / county, construction site, specific intelligent sensing equipment, data type (noise or dust) and start and end time, so that relevant personnel can view the emission of noise and dust during construction.

4   System Design

The project construction enterprise will install a series of induction equipment at the project construction site, and then set up the site monitoring system, i.e. the data management server. The data exchange between the equipment and the server can be conducted in two ways. One is to send the configuration and response protocol and the sensor to the server through the GRPS wireless network using the TCP protocol. Second, the server sends the configuration command to the sensor device through the SMS cat. When using this method, it is unidirectional, and the sensor device does not need to send a response to the server. The server stores, statistics and analyzes the data sent by the sensor device. The user can easily view the real-time data information through the PC browser or the mobile phone mobile device installed with the intelligent site monitoring system application.

The project data, enterprise data and enterprise related personnel data of the system are based on the Enterprise GIS integrated business system, which is responsible for pushing the data of the project, enterprise and enterprise related personnel to the noise and dust monitoring system database [9]. The system can start working after obtaining these data. When a new intelligent sensing device is installed in a construction site project, the system will input the basic information of the device, such as installation location, SIM card number, etc., and store it in the database. Then, the system sends configuration information (server p address and port number) to the newly installed device, and configures the working frequency of the device (here, the working frequency refers to how often the intelligent sensing device performs data sampling) and the sampling frequency (the number of sampling times for each operation). Then the sensing device can start to work, collect data every once in a while and send the data to the server. After receiving the data, the server analyzes the data and stores the correct data into the database. According to the user's operation, the system makes statistical analysis on the data of a certain construction site in a specific time, and draws a chart on the front page for the user to view the real-time emission of noise and dust of the construction site project [10]. If the data transmitted by a construction site project exceeds the alarm threshold value of the construction site for several consecutive times, the system sends an alarm message to the on duty management personnel of the construction site through the SMS cat, and the management personnel can take relevant measures to deal with it, so as to ensure a good construction environment. If several alarm messages are sent to the site management personnel, and the noise or dust data still exceeds the standard, the alarm information will be sent to the law enforcement department for filing.

5 Conclusion

The construction of intelligent sensing capability on the infrastructure construction site based on fish swarm algorithm, the overall framework of the system, the main business process, and the database. Next, according to the overall design, each functional module is refined, and a good interactive communication interface is designed between each module, which greatly simplifies the system development complexity and development efficiency. During the development process, the strategy of "developing while testing" is implemented, which not only ensures the stability of the system, but also reduces the difficulty of positioning problems and the complexity of error accumulation in later testing. Finally, in the centralized test stage of the system, the function and performance of the system are comprehensively tested to ensure that the system has high performance while achieving the expected functional objectives.

References

1. Liu, Y., Chen, H., Liu, F.: Research and application of intelligent perception system for unmanned aerial vehicle inspection at construction site. Power System Protection and Control (2018)
2. Ling, Y., Fan, C.: Research on the design and application of intelligent emergency innovation service platform based on internet of things. China Comput. Commun. (2018)
3. Cheng, Y., Wang, J., Ji, S., et al.: Robust and secure data fusion algorithm based on intelligent sensing in wireless sensor networks. Wirel. Commun. Mob. Comput. **2020**, 1–14 (2020)
4. Jia, S.C., Yang, F.P.: Research on intelligent detection method of weak sensing signal based on artificial intelligence (2019)
5. Zhang, Z., Zhao, X., Chen, Z., et al.: Design of intelligent iot sensing module based on topology recognition. J. Phys. Conf. Ser. **1974**(1), 012011 (2021)
6. Yang, Y.: Design and application of intelligent agriculture service system with LoRa-based on wireless sensor network. In: 2020 International Conference on Computer Engineering and Application (ICCEA). IEEE (2020)
7. Liu, Y.L., Meng, X.J., Wu, Z.G., et al.: Design and application of intelligent sensing terminal for distribution transformer. In: 2020 IEEE 4th Conference on Energy Internet and Energy System Integration (EI2). IEEE (2020)
8. Xue, H., Wu, M., Zhang, Z., et al.: Intelligent diagnosis of mechanical faults of in-wheel motor based on improved artificial hydrocarbon networks. ISA Trans. **120**, 360–371 (2022)
9. Bai, S., Bao, F., Zhao, F., et al.: Development of an artificial fish swarm algorithm based on awireless sensor networks in a hydrodynamic background. **16**(5), 935–946 (2020)
10. Chen, Y.: Application of intelligent algorithm based on genetic algorithm and extreme learning machine to deformation prediction of foundation pit. Tunnel Construction (2018)

Construction Method and Typical Application of Data Analysis Service for Power Grid Enterprises Based on Data Middle Platform

Li Wenjuan[1], Liu Shi[1], Zhang Fan[1], Yang Zhi[1], Wang Honggang[1], Hu Xishuang[1], and Zhou Wenjin[2(✉)]

[1] Big Data Center of State Grid Corporation of China, Beijing 100052, China
[2] Beijing State Grid Xintong Accenture Information Technology Co., Ltd., Beijing 510030, China
kqx019@sina.com

Abstract. With the promotion of the construction of the power Internet of things of State Grid Corporation of China, the company's business is developing towards lean and digital, and the range of data demand is expanding. The company requires the construction of enterprise data center to provide various data analysis services, and support the big data application of important business efficiently and flexibly. The data analysis service is based on the unified construction of data center, It provides service retrieval and view to users in Web mode, and provides data analysis service to external business system and business platform through unified data analysis service in data platform to support rapid construction of business analysis scenario application.

Keywords: Data analysis Service · Data Center · Power Grid Enterprise · Power Data

1 Introduction

According to the company's strategic deployment, the data center of State Grid Corporation of China has been initially completed, and the data analysis service will be uniformly constructed through the data center to support cross department and cross level data sharing and analysis applications. In order to effectively manage the company's data assets and avoid the problems such as high threshold of data analysis service application, difficulty in data understanding and service acquisition, it is urgent to build enterprise level data analysis service construction standard and construction ability based on data platform, provide unified data analysis service for all specialties, improve data management level, and create a benign ecological environment for data assets [1]. At present, data center has become an indispensable technical support platform for power grid enterprises in production, operation, management and decision-making. With the continuous expansion of business, the scale of the enterprise is growing every year. In the process of operation, there are more and more failures, some of which even make system administrators unprepared [2]. After a fault occurs, it is often necessary to analyze the log to locate the problem. Log is an important tool for analyzing user behavior, monitoring system running status and locating system fault. And through log analysis can also find system vulnerabilities, optimize system design, help to quickly locate system faults, find the root cause of the problem [3].

However, at present, the development of the power industry is relatively backward, facing many serious problems, such as the old and lagging power grid construction, the lack of scientific basis for the allocation of power resources, and the sluggish infrastructure construction of the power market. Among them, the lack of reactive power supply is the main factor leading to power grid blackout, voltage fluctuation, harmonic distortion, serious loss and other problems, while power failure and power grid loss are the main factors restricting the growth of power grid enterprises. Therefore, through the local compensation of facts and other reactive power compensation equipment, it can reduce the network circulating current, reduce the network loss, improve the voltage and frequency stability, reduce the power grid failure probability, and improve the scale efficiency of power grid enterprises. In the future, with the gradual improvement of the power market and the introduction of new energy into the grid, power grid enterprises need to allocate reactive power resources more finely to improve the input-output efficiency of power grid enterprises [4].

In the era of electric power data, the resource allocation method of data aided decision-making is one of the necessary links in the upgrading and transformation of power grid enterprises. Power grid enterprises need to start from the demand of the power market, abandon the previous blind grid investment and construction methods, excavate the potential value information of power data, grasp the trend of regional industry and industry economic change, so as to form a refined power grid planning and construction strategy. In the rapid development and reform of the power industry, power grid enterprises need to study the operation efficiency optimization of power grid enterprises from the perspective of data, establish an effective incentive evaluation mechanism, accurately judge the power economic development trend, optimize the grid structure and resource allocation strategy, and drive the refined development of grid enterprises through data. This has important theoretical and practical significance for the stable supply of electric energy in the power grid industry and the steady improvement of the economic benefits of the power grid [5].

2 Metadata of Data Analysis Service

2.1 Classification of Data Analysis Services

Using the hybrid classification method, the data analysis services are classified in the following four aspects:

(1) Service form: classify the data analysis service according to different data analysis service forms;
(2) Construction unit: according to the different units of construction data analysis service, the data analysis service is classified;

(3) Business category: classify data analysis services according to different business categories involved in analysis services;

(4) Service release time: data analysis services are classified according to the year, month and day of the analysis service release [6].

Among them, there are three forms of data analysis services, which can be supplemented and improved according to needs.

Result dataset: it provides the query, subscription, push and other services of the result dataset, and supports the service consumers to quickly call the required result data through keyword, combined query and other forms.

Algorithm model: provide specific data analysis model and algorithm for service users to call according to their needs.

Analysis scene: provide accessible big data analysis scene page of finished product visualization.

2.2 Naming and Coding Rules of Data Analysis Service

The naming rules of data analysis service should meet the following requirements:

(1) A normalized full name of no more than 40 characters should be given;

(2) It should be named according to the format of "time + place + analysis object + analysis method", and the time and place can be default;

(3) Standard Chinese characters, numbers and English characters should be used instead of dialects, slang and obscure words;

(4) It should be unique and unambiguous within the scope of the unit;

(5) When the name of the self built data analysis service of each unit is the same or similar to that of the unified data analysis service of the headquarters, it shall be merged, renamed or revoked according to the specific rules and application conditions to ensure the uniqueness of the naming of the data analysis service.

Data analysis service code should include construction unit, service form, business category, creation time, serial number, etc.

Construction unit: representing the construction unit of data analysis service, adopting the company code in the human resource master data of State Grid Corporation of China, with a total of eight digits, including headquarters, provincial companies and directly affiliated units;

Service form: code according to the data analysis service form, with two digits in total;

Business category: code according to the first level business category and the second level business category;

Release time: the time when the analysis service is published, coded according to the data format of year, month and day;

Serial number: a six digit serial number generated automatically by the system when the service is published to ensure the uniqueness of the code [7].

2.3 Data Analysis Service Construction Method

(1) Service demand

The data analysis service requirements are managed in a unified way. The service demanders fill in the service requirements according to the demand reporting requirements, and enter the service construction link after being approved by the service manager.

The implementation method of data requirement reporting is to realize the management of new demand management, demand analysis summary, demand review management and other main functional components by expanding and enhancing development based on the current demand management situation of data analysis service.

New requirement: through enhanced development, the function of new requirement submission is realized to meet the requirement submission. After saving, the system will automatically record the information of the submitting personnel and the unique information code of the requirement number.

Requirement analysis and summary: the function of query, analysis and summary of requirements according to different dimensions is realized through enhanced development. Among them, the analysis function is to preliminarily screen and classify the submitted requirements, and supports the filling in of single or batch demand analysis opinions; the summary function is to classify and gather the qualified requirements according to the summary conditions.

Requirements review: the approval information can be filled in and returned for modification through enhanced development. The approval information should be filled in by category, recording the necessity, feasibility, risk and implementation details of the requirements in the requirement review; for the requirements that have not been approved, the historical records should be kept at the same time.

(2) Service Construction

The core of data analysis service construction includes result data set, algorithm model service and analysis scenario construction. Results the data set construction is based on the data platform sharing layer and analysis layer business data, based on the data platform technology component ability, through data processing, service development and other processes to complete the service construction; algorithm model service construction is based on the algorithm model component construction algorithm model service; business analysis scenario construction is based on the data platform self-service analysis component through the business After the business analysis scenario is formed by logical processing, the service is published [8].

2.4 Service Construction

Data analysis type: Based on the big data component of Alibaba cloud platform, access structured and collected measurement data. The structured data is output to online application area (ads, RDS, DRDS) after maxcompute data cleaning and summarizing, and the measurement data is stored in TS. Then according to different service types, the construction of services is completed.

Construction process of data analysis service: relying on the technical system of data middle platform, the analysis layer of data platform is constructed by using dimension model method to create public data layer and realize data exchange and sharing; open data application layer to meet personalized business application requirements, enrich data application of each business, and release data value; refine data service layer, construct public data service and support number According to the sharing of services and the rapid construction of data applications.

2.5 Service Launch

Data analysis service online is the process management of sharing and opening the data registered in the service API gateway. The data analysis service can be launched only after passing the security audit and confirming that it has been desensitized or does not contain sensitive information. Data analysis service users can query.

3 Typical Scenario Application of Data Analysis Service

Based on the support demand analysis of five special applications, including financial multidimensional, digital audit, material intelligent supply chain, integrated line loss and online state grid, the common demand characteristics are extracted from data demand (data type, demand frequency), data exchange, computing power and tools, data interaction mode, etc., and the application scenarios are classified as structured offline analysis and on the spot inspection Inquiry, acquisition, measurement and analysis, ad hoc query, real-time calculation and processing scenarios.

Ad hoc query scenario of structured offline analysis: the scenario focuses on structured data, and meets the ad hoc query requirements of applications after data access, data integration, analysis and calculation, scene display, etc. The off-line analysis and ad hoc query scenario of acquisition and measurement focuses on the acquisition and measurement data. After data access, analysis and calculation, data service and other links, it provides offline data query for applications to meet the ad hoc query needs of such applications. The real-time computing processing scenario is based on structured, measured and unstructured data, and provides real-time calculation results for applications after real-time data warehousing and real-time calculation.

4 Data Analysis of Bench Test

4.1 Functional Test Bug Distribution Analysis

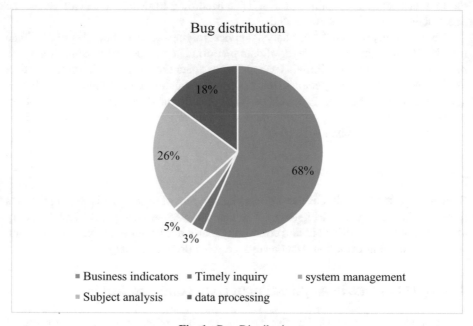

Fig. 1. Bug Distribution

A total of 387 test cases were designed in the functional test, of which 387 were effective cases, and 856 bugs were found. Among them, 96.9% were minor and general bugs, and there were no fatal bugs. All bugs have been fixed and closed. In the data accuracy test, 50 test cases were designed, 50 were effective cases, and 58 bugs were found. Among them, 100% were mild and general bugs, and there were no fatal bugs. All bugs have been fixed and closed. The function of the system meets the requirements of users (Fig. 1).

4.2 System Login Scenario Results and Analysis

Based on the different pressure of the design system, the compression performance of the platform is tested. According to the data in Table 1, the processing results of the platform system for concurrent login of users meet the test requirements, and there is no platform crash or obvious performance defects due to a large number of concurrent logins.

Table 1. System Login Scenario Results and Analysis Test

Scene	Execute Script	Number of Concurrent Users	Loading mode	Concurrency Srategy		Results of Enforcement		
				time interval	Synchronization Point Settings	Success	Fail	Average Response time
System login	System login	20	Start loading 10 people and load 2 people every 10 s	None	5	20	0	Satisfy
System login	System login	100	Start loading 20 people and load 5 people every 20 s	None	5	100	0	Satisfy
System login	System login	50	One time load	None	5	48	2	Satisfy
System login	System login	100	One time load	None	5	99	1	Satisfy
System login	System login	300	One time load	None	5	298	2	Satisfy
System login	System login	500	One time load	None	5	500	0	Satisfy

4.3   Performance Test of Log Analysis System

Through the analysis of the above test results, the log analysis system can meet the requirements of both performance and function (Fig. 2).

The former company analyzes the demand of operation and maintenance log. In terms of function, it realizes the fast and automatic analysis of operation and mainte- nance logs. It improves the efficiency of operation and maintenance personnel to deal with application system failure, and effectively ensures the efficient operation of the company's application system. In terms of performance, the system fully meets the existing daily log processing requirements. According to the growth rate of log in recent half a year, the computing power and storage capacity of the analysis system can meet the needs of log analysis and processing in the next two years.

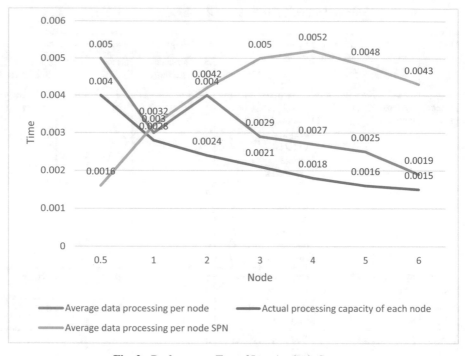

Fig. 2. Performance Test of Log Analysis System

5   Conclusion

The core value of data center is to precipitate common and reusable data assets. Through the rapid construction and iteration of data analysis application, it can realize multi-dimensional analysis and business exploration of business data, and simultaneously improve the intelligent ability of business application, so as to meet the requirements of providing agile and open data analysis and sharing for all disciplines, grass-roots units and external partners of SGCC The demand for service. With the continuous development of digital technology, the information level of power grid enterprises continues to promote, the power grid business continues to expand, data analysis services will also continue to improve with business expansion. The construction method proposed in this paper relies on the existing business status to make data analysis services have sustainable growth to face the analysis services generated by future business expansion. In the future, the construction of data analysis service still needs to be based on business needs, and the construction of data analysis service in this paper also needs to be continuously optimized and improved according to the actual work needs.

References

1. Lee, S.U., Soh, M., Ryu, V., et al.: Analysis of the health insurance review and assessment service data from 2011 to 2015. Int. J. Ment. Heal. Syst. **12**(1), 9–10 (2018)

2. Jing, L.I., Xiao-Hua, S., Yong-Ge, B., et al.: Research on data analysis and optimization in library self-service printing service— a case of shanghai Jiaotong university library. J. Acad. Libr. Inf. Sci. **11**(3), 19–21 (2016)
3. Walker, A.J., Curtis, H.J., Croker, R., et al.: Measuring the impact of an open online prescribing data analysis service on clinical practice: a cohort study in NHS England data (preprint). J. Med. Internet Res. **21**(1), 58–61 (2018)
4. Liu, L.G., ZLiu, H.X., Chen, W., Liu, Z.F.: Analysis of tripartite asymmetric evolutionary game among wind power enterprises, thermal power enterprises and power grid enterprises under new energy resources integrated. SCIENTIA SINICA Technologica **45**(12), 1297–1303 (2015). https://doi.org/10.1360/N092015-00244
5. Vergis, J.M., Vanderwall, D.E., Roberts, J.M., et al.: Unlocking the power of data. Lc Gc North America **33**(4), 1–2 (2015)
6. Badawi, O., Brennan, T., Celi, L.A., et al.: Metadata correction: making big data useful for health care: a summary of the inaugural MIT critical data conference. JMIR Med. Inform. **2**(2), 22–23 (2015)
7. de Montjoye, Y.-A., et al.: Unique in the shopping mall: on the reidentifiability of credit card metadata. Science **347**(6221), 536–539 (2015). https://doi.org/10.1126/science.1256297
8. Hong, S.D., Kim, J.T.: A construction of online luminance analysis service website. **12**(3), 45–48 (2017)
9. Zhang, B., Yu, L., Feng, Y., et al.: Application of workflow technology for big data analysis service. Appl. Sci. **8**(4), 1–5 (2018)

Construction of Computer-Aided Hierarchical Teaching Model for Higher Mathematics Courses

Jing Yang[✉]

Shandong Xiehe University Foundation Department, Jinan 250109, China
ckjbyyp@163.com

Abstract. The basic idea of this model is to use the hierarchical structure of mathematics to develop teaching methods that can be used in all higher mathematics courses. The main goal is to provide students with good mathematical knowledge and skills, which is crucial to their future study. The project consists of two parts: (1) establish a preliminary computer-assisted learning environment; (2) Formulation and evaluation of new teaching methods based on hierarchical teaching model. This paper introduces the construction of such a model, which includes not only the content, but also the teaching method. In addition, it also describes how to use this new teaching model as the basis for further research on computer-assisted learning methods in higher mathematics courses.

Keywords: Computer-aided · Advanced mathematics courses · Layered teaching model

1 Introduction

How to do layered teaching well in higher mathematics class? Hierarchical teaching is a teaching mode that faces all and teaches students in accordance with their aptitude. It emphasizes that "teachers' teaching should adapt to students' learning and" teach students in accordance with their aptitude. Today, the small series brings us effective teaching methods of higher mathematics.

Teaching objectives are the starting point and destination of classroom teaching, and play a role in regulating, guiding and controlling the whole teaching process. Teachers can carefully study the teaching materials before class, grasp the key and difficult points of the outline and teaching materials, consider the actual situation of students, and combine their own teaching to correctly formulate teaching goals at different levels, basic teaching goals and personal goals that everyone can achieve. In class teaching, I will let each student set a practical goal according to his own situation -- who I want to surpass, let each student find his own direction of effort, challenge him, and let the challenged child accept the challenge, and both sides obtain the learning goal at the same time.

The questions raised in the classroom should be carefully designed, and the questions raised by teachers should be similar to the students' thinking level, so that they can think

J. C. Hung et al. (Eds.): IC 2023, LNEE 1045, pp. 26–32, 2023.
https://doi.org/10.1007/978-981-99-2287-1_4

about it and solve it. At the same time, the questions raised should stimulate students' interest and thirst for knowledge, and also pave the way for the connection between new and old knowledge. In order to ensure that students at all levels have equal learning opportunities in the process of classroom questioning and make students at different levels think actively [1], I consciously divide it into upper, middle and lower levels when designing questions, among which basic questions such as review and basic questions are for group C students; Intermediate questions are for group B students; Difficult problems, such as those that can be solved by comparison, analysis and other thinking methods, are for group a students.

Homework can timely feed back the knowledge of students at different levels, reflect the teaching effect of a class, and achieve the purpose of initially consolidating knowledge. The different levels of homework are not only reflected in the amount of knowledge and thinking components, but also in the depth of knowledge and the level of thinking. Therefore, assignments should be carefully arranged, and assignments with different questions and difficulties should be designed for students at different levels. The homework of students at level C focuses on the memory and understanding of basic knowledge [2]. The homework of students at level B focuses on grasping concepts and general problem-solving methods. The homework of students at level a focuses on deepening the understanding of concepts and flexible and skilled application, considering the ideological methods and ability training of higher mathematics. In short, the amount and degree of homework should be based on the principle that every student can do his best to complete it, and try to meet the learning needs of students at different levels, so as to mobilize the learning enthusiasm of students at all levels, stimulate their learning hobbies, and mobilize the positive role of all students' non intellectual psychological factors. Based on this, this paper studies the construction of computer-aided hierarchical teaching model of higher mathematics curriculum.

2 Related Work

2.1 Computer Assisted Instruction

Computer aided instruction (CAI) is a variety of teaching activities carried out with the help of computer. It discusses the teaching content, arranges the teaching process, and carries out the methods and techniques of teaching training with students in the form of dialogue [3]. Cai provides students with a good personalized learning environment. The comprehensive application of computer technologies such as multimedia, hypertext, artificial intelligence and knowledge base overcomes the shortcomings of single and one-sided traditional teaching methods. Its use can effectively shorten the learning time, improve the teaching quality and efficiency, and achieve the optimal teaching objectives.

The main research contents of CAI technology include:

(1) Cai mode: at present, there are six kinds of teaching modes commonly used in the CAI System: (1) exercises: including arranging questions, comparing answers and registering scores, which are usually used as a supplement to normal teaching;

(2) Individual guidance: including teaching rules, evaluating students' understanding and providing application environment;

(3) Dialogue and consultation: also known as "Socrates" teaching mode, which allows students to have a relatively free "conversation" with computers;
(4) Game: create a competitive learning environment, and the content and process of the game are related to the teaching objectives;
(5) Simulation: use computers to simulate real phenomena (natural or man-made phenomena) and control them, such as simulating chemical or physical experiments and aircraft, vehicle and ship driving training;
(6) Problem solving: let students use rules and concepts in various ways to get the solution of the problem, which requires students not only to know the correct answer to the problem, but also to master the solution process. In the specific teaching process, according to the needs of teaching content expression and the requirements of teaching purpose, it is necessary to cross use these teaching modes in different contents of the same course or different teaching links[4].
(7) Making CAI Courseware: the core of CAI system engineering is courseware. It is compiled by the courseware designer with CAI writing tools or computer language according to the teaching requirements.
(8) Cai writing tools and environment: CA writing tools are the writing environment for course teachers to compile courseware. Good writing system and development tools are the key to improve the efficiency of CAI courseware development.

2.2   Advanced Mathematics Courses

As a basic discipline of natural science, mathematics not only plays a positive role in the progress of material civilization and human understanding of the world, but also has a very important impact on the development of human thinking. This impact will become more and more important and obvious with the advent of the information age.

The inquiry teaching mode of higher mathematics is actually a simulated scientific research activity, which includes three interrelated aspects: on the one hand, it is a learning environment centered on learning; The other is to provide students with necessary help and guidance to ensure that students can successfully discover scientific principles and concepts after scientific inquiry; The third aspect is the mathematical experiment link [5]. How to apply the explored principles to practice is the key to quality education. Today, when students' practical ability is vigorously advocated, it is of great help to carry out mathematical experiment courses so that students can experience the tool learning points and charm of basic mathematics in the experiment, so as to improve students' learning interest and exploration spirit.

The organic combination of higher mathematics and layered teaching has formed a teaching mode that meets the requirements of the times and has the characteristics of the times. The research on such problems is relatively few. This paper is about how to implement the construction of computer-aided hierarchical teaching model of higher mathematics, improve students' practical ability, and give an appropriate evaluation and analysis of the teaching effect of the hierarchical teaching model. The analysis is based on the statistical materials of the author's teaching practice in recent years, and the statistical materials are analyzed and sorted out by means of sampling survey and data analysis and processing, The corresponding evaluation index and evaluation model are given.

2.3   Computer Aided Hierarchical Teaching Model Construction of Higher Mathematics Courses

The construction of computer-aided hierarchical teaching model of higher mathematics courses refers to a teaching method and teaching concept that emphasizes the interaction between teachers and students and gives full play to students' subjective initiative and creativity by taking problem research as a means and comprehensively mastering and skillfully using the learned knowledge to solve practical problems [6]. The essence of inquiry learning based on problem solving is to cultivate students' problem awareness, habits of critical thinking Aiming at the ability to generate new knowledge and the quality of collaborative learning, we pay attention to the learners' subjective participation in the learning process, and highlight the problem-centered organization of the whole teaching and learning process. Figure 1 below shows the hierarchical teaching model framework.

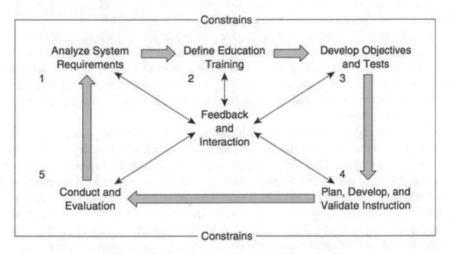

Fig. 1. Hierarchical teaching model framework

To design a computer-aided hierarchical teaching model of higher mathematics, we should grasp the two core concepts of "Computer-aided" and "hierarchical teaching". When applying computer-assisted instruction, we should fully consider the characteristics of computer-assisted instruction [7]. Therefore, when making computer-assisted instruction, we should grasp the micro principle and develop video. In terms of individual students, teaching content, learning objectives, after-school homework, evaluation and other aspects, we should consider the principle of hierarchical design, and formulate different teaching content, learning objectives, after-school homework and evaluation according to different levels of students. And we should also pay attention to the coherence of knowledge system while designing it hierarchically, that is, we should consider the systematic design. In addition, considering the current learning situation of college students, we should also pay attention to the interesting selection of computer-aided content when designing and developing the model, so as to attract students to study independently [8]. Finally, when designing the teaching implementation process, we should also pay attention to the interaction of students' activities. Only when students are more active in the construction of knowledge and experience, can students master the initiative of the classroom.

The introduction of computer-assisted instruction into layered teaching classroom greatly reduces the difficulty of teachers' layered teaching. Students at each level can learn differentiated knowledge in the same class, and after watching computer-aided video, they can carry out five links: hierarchical test, group discussion, teacher question answering, hierarchical homework and hierarchical evaluation. Students' differences are fully reflected in each link. Students with a good foundation will further expand the breadth and depth of teaching content, and increase the difficulty in tests and assignments, so that students with a good foundation can obtain more sense of achievement at the knowledge level [9]; For students with poor foundation, the breadth and depth of teaching content will be slightly simpler, and the difficulty of subsequent tests and assignments will also change. For students with poor foundation, it is more to let them enhance their self-confidence and interest in learning.

In addition, in the process of stratification, we should also pay attention to protecting students' self-esteem and self-confidence. Stratification does not divide students into three, six, nine grades and treat them differently, but for better teaching, which should be clearly told to students. And in the follow-up teaching process, teachers should also be consistent with their words and deeds, so that students can truly feel the equal treatment of teachers in emotion [10].

Generally speaking, Addie teaching design model tells us three aspects: what to learn, how to learn, and how to test the learning effect. In addition, the theory also has three advantages: first, systematic, integrating the five parts of analysis, design, development, implementation, and evaluation into a systematic process, avoiding the one sidedness of the development process and making it more complete; 2, Pertinence, according to the analysis stage of curriculum development, it avoids the blindness of curriculum resource development and meets the individual needs of each learner; 3, Supportability: ensure the quality of curriculum development through timely and effective evaluation of all links, combined with Addie teaching design model, as shown in Fig. 2.

Addie instructional design model is adopted as the model of computer-aided hierarchical instructional design, mainly because its advantages are more consistent with the requirements of instructional design expected by the author, which can make the effect of developed teaching resources better, and also ensure the orderly progress of quasi experimental research.

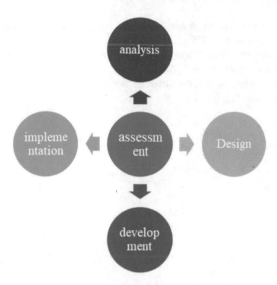

Fig. 2. Addie instructional design model

3   Conclusion

It is a new way to design computer learning materials to build a computer-aided hierarchical teaching model for higher mathematics courses. The purpose of this project is to develop an interactive system, which can be used as the basis for developing and implementing different types of multimedia textbooks (such as courses, exercises, tutorials, etc.). The main idea behind the development process is to create a graphical interface in which users can define mathematical topics and concepts. Provides a set of tools to facilitate the creation process.

References

1. Zhang, T.: Research on classification and training and hierarchical teaching model of higher vocational mathematics course based on professional group. Heilongjiang Sci. (2018)
2. Bao, H.F.: Exploration and practice on the hierarchical teaching model of higher mathematics. J. Jiamusi Vocational Inst. (2018)
3. Xu, X., Gong, Y., Xue, M., et al.: Reform and Practice of Classified and Hierarchical Teachingin Basic Computer Courses Based on "Internet +" for No-computer Major. China Modern Educational Equipment (2019)
4. Zhang, Liu, G., et al.: Goal-Driven "Five-in-one" Teaching Model Integrating Research and Practice into Mathematics Teaching Courses in Agricultural Universities (2018)
5. Ling, L.: University C J . Research on SPOC based online hierarchical teaching mode for computer basic courses. Computer Era (2018)
6. Zhi-Ping, Y.E.: Analysis on the Reform of Teaching Mode and Teaching Quality of Higher Mathematics——Taking the Northern University for Nationalities as an Example. Educ. Teach. Forum (2019)
7. Xiong, H.: On hierarchical teaching model for experiment of numerical analysis. Stud. College Math. (2019)

8. Chen, J., Qin, H.M.: On the hierarchical teaching of advanced mathematics in higher vocational colleges based on SPOC. J. Yangzhou College of Educ. (2019)
9. Chen, J., Mathematics SO: Thoughts on the innovation of teaching mode of higher mathematics. Educ. Teach. Forum (2018)
10. Ling, C.Y.: Research on teaching reform of advanced mathematics based on stratified teaching method. Heilongjiang Sci. (2018)

Construction of Folk Dance Resource Database Based on "Internet +"

Xinxiu Wang[✉]

Weifang Engineering Vocational College, Shandong 262500, China
15689175201@163.com

Abstract. In the 21st century, the Internet era is coming, and we are enjoying the great changes it has brought to our lives all the time. With the help of the Internet, all walks of life have realized the beautiful vision of resource sharing and cross-border integration. Dance, the body art, is also naturally bathed in the lucky spring breeze. Taking advantage of the development trend, insiders have responded to the banner of the times and moved towards the development avenue of "dance + Internet". In this paper, building a folk dance resource database based on "Internet +" is the first step to create an Internet-based dance resource project. The goal of the project is to provide a comprehensive and up-to-date collection of information on dance, choreography and related topics. The database will be built using the resources available at (www.folkdanceresource. www.folkdanceresour ce.org) and other sources, such as the world wide web, printed books, magazines, periodicals and newspapers, are listed under "references". However, it should be noted that there are some limitations in this regard.

Keywords: Folk dance · Internet + · Resource database

1 Introduction

China has entered the Internet era in the 21st century. It can be said that the Internet is "upgrading" traditional things at an unprecedented speed, and our lives have undergone earth shaking changes. As a new technology, it not only profoundly changes many aspects of human beings, but also changes the previous communication pattern as a new media, so that all communication content can be spread to specific audiences with a rapid trend and strong coverage. In today's daily life, we feel the convenience brought by the Internet everywhere. The Internet can break the restrictions of time and region, accept all kinds of information resources previously limited by time and region into the audience's own information base, and at the same time, it can also transmit the information of its own social circle to all parts of the world [1]. The global village of information is no longer a distant concept, but an accessible life circle. The Internet can obtain and reprint information by hundreds of millions of people through a large number of websites, so that it can spread to a larger field, so as to truly realize the dissemination role of mass media and become an extremely influential media [2]. Dance itself, as a kind of information, with the help of Internet media, makes it have a wider audience and return to the public.

As a historical breakthrough, Internet media has changed the master-slave relationship between the original communicator and the audience. In the face of the resources transmitted by the communicator, the discourse power has shifted from the original communicator to the audience. The audience can not passively accept the information resources as before, but enjoy full independent choice to find and obtain the content they want, Communicators who want their information to be more widely received and recognized must be consistent with the audience's search motivation to be more likely to achieve. Similarly, under the flood of Internet communication resources, whether resources can attract the attention of the audience has become the focus of attention of communicators [3]. Only the resources required by the audience can be widely disseminated. Therefore, in the Internet communication of dance, if the disseminator wants to get the attention of the audience, it must focus on the needs of the audience, so that the content of dance closely fits the needs of the audience, which is easy to make the content win the favor of the audience, So as to inject new impetus into the spread of dance [4]. Based on the "Internet +" folk dance resource database construction, this paper aims to study the supply and consumption status of online dance resources under the existing internet background.

2   Related Work

2.1   Database Concept

Going back to the source, the concept of database came into being in 1950, which is common in our daily life and learning. At first, its connotation was a shareable warehouse that organizes, stores and manages data according to the data structure. After the 1990s, data management changed from storing and managing data in the past to forming different forms of management according to the different needs of users. Compared with the traditional way that the database manages the paper version such as archives manually in the past, the document system, database system and advanced database stage evolved since then have realized the requirements of high efficiency and strict standards in people's production and life. These databases have visual interfaces, and the operations of adding, acquiring, segmenting and deleting data are more simple and convenient [5]. Through the Internet, data duplication can be effectively reduced and data sharing and real-time writing can be realized. Their systems are constantly updated in core technologies such as classification optimization and data model, making the databases independent, diverse, convenient and up-to-date. In terms of data analysis, it can know the whole leopard by random sampling of data, which reduces the management and operation cost of users. In addition, as long as the data is operated and saved properly, the database can run the diagnosis regularly through a series of algorithms and subroutines set by the program, find and correct the faults, and restore the damaged data according to the backup, so as to facilitate the centralized control of users and effectively ensure the safety of the database.

The "connecting everything" of modern advanced information technology in the "Internet +" era, as well as the "new normal" and "new kinetic energy" that can change the new sky derived from the connection and integration, have made people engaged in folk dance describe such a vision in their hearts: if they are in line with "Internet +" in the process of accelerating the construction of a modern folk dance service system, In terms of public services, we should carry out big data projects for the public, deeply explore folk dance service data, carry out application demonstrations of big data in the field of culture and education, and actively build a comprehensive cultural communication service platform supported by "Internet +" technologies such as big data and cloud computing, which should be able to seek more convenient and exquisite folk dance benefits for the majority of grass-roots people, and inject fresh blood into the construction of modern folk dance service system Add vitality and realize the great development and prosperity of our culture [6].

2.2 Construction Method and Significance of Folk Dance Art Resource Database

The construction of folk dance resources database mainly includes commercial database and self built collection resources. The construction methods of the database include the following aspects:

(1) Data collection

Although database construction itself is a kind of creative research work, the content of database is an integration of existing resources and services according to their own technological development ability and network technological ability. To build a thematic database, we must have a literature foundation for collecting professional data, which includes existing collection resources, authorized network resources, and other literature resources collection channels that comply with intellectual property rules. The construction of database is not only to classify documents or data collection, but also to provide readers with a relatively complete knowledge system or a series of research results. In selecting topics, the focus should be on "folk dance". We should rely on folk dance as the object of database construction to provide readers with literature reference [7].

(2) Give full play to professional advantages

Practice has proved that close cooperation with relevant disciplines and collections is an important guarantee for the smooth construction of the database. In order to make the database really meet the professional needs, we should fully estimate our own professional strength, give full play to the advantages of art professionals, and vigorously develop subject librarians. With the assistance of the University graphics Working Committee, we will make overall arrangements for the construction of self built databases in Colleges and universities to avoid duplication and waste. Regularly invite artists to communicate with the University Library Alliance and participate in the construction of art materials and documents in the library. The information contained in the thematic database should be consistent with the thematic direction of the database, and complete information should be collected for the theme of art and design [8]. Eliminate the inclusion of data with little relevance. For the same type of thematic database, try to ensure the specificity of the database structure standard, which can be convenient for producers and users at the same time.

(3) Standardization of production procedure

Quality control and database standardization. Data processing, data collection and data maintenance must be carried out in accordance with the relevant database standards at home and abroad, such as the international standard bibliographic description, the rules for standardizing the description of data items, the Chinese machine readable catalogue format, the Chinese Library Classification (Fourth Edition), etc., and the indexing and description must be carried out in strict accordance with the CNMARC standard description format to improve the construction of the database. It is convenient for producers to master database building skills by implementing specificity in the production process. Implementing programmed operation and simplifying complex work can improve the efficiency of database building [9]. The specificity of the production procedure can also monitor the quality of data records, divide the work in the input, indexing, classification and other work links, avoid mistakes, and ensure the accuracy of data, which is the basis for realizing the sharing of information resources in the future.

3   Construction of Folk Dance Resource Database Based on "INTernet +"

3.1   Functions of Folk Dance Resource Database

The fundamental purpose of folk dance resource database construction is to preserve and carry forward folk dance, so we should have its "preservation" and "promotion" functions.

First of all, the folk dance resource database should be able to collect folk traditional dance resources, including oral narration, recorded videos, and existing text, pictures, and videos; Using digital technology to sort it out and classify, and lay the foundation for the construction of dance resource database.

Secondly, the dance resources to be collected in the folk dance resource database are reprocessed. Through a variety of modern digital technologies, the dance resources are re recorded or modified to make them more vivid, lifelike and complete. At the same time, we should better preserve the dance resources that have been collected or processed, and digitize the dance resources as much as possible.

Finally, the best way to preserve culture, especially intangible culture, is to "carry forward", so that more people can master this culture and let more people preserve it. Therefore, the folk dance resource database should have the function of spreading folk dance, publicize traditional folk dance as much as possible, popularize basic dance knowledge, and let more people understand folk dance.

3.2   Design of Folk Dance Resource Database

Folk dance mainly includes four kinds of dances: Lantern Dance, shield dance, lion dance and tea picking song dance. One common feature of these four dances is that they all use props. Shield dance and lantern dance are more dependent on props. Therefore, the folk dance resource library can be realized through exhibition halls and databases. The exhibition hall is similar to a museum, including dance props and some text, pictures and audio, while the database is built on the network platform. Folk dance mainly includes four kinds: Lantern Dance, lion dance, tea picking dance and shield dance. Among them, lantern dance and lion dance are joyful dances for festivals and celebrations, tea picking dance is a life dance used to celebrate agricultural harvest, and shield dance is related to revolutionary politics [10].

Database Construction: 1. Database Planning and design stage (technical support and structural design); 2. Database production stage (using cloud computing and cloud storage to build the cloud); 3. Folk dance resources collection and processing stage (collect resources and process them through collecting styles, visiting old artists, querying materials, recording audio and video, etc.); 4. Database use and maintenance stage (emergency handling, IP protection and database updating).

The database operation structure includes two terminals and three platforms, as shown in Fig. 1: 1. The cloud is the database, which is used to store data; 2. The client is a software application, which is used to read data; 3. Platform 1 is responsible for editing refined resources and inputting data; 4. Platform 2 is responsible for extracting resources from the database and outputting data according to needs; 5. Platform 3 is responsible for searching market information and feeding it back to platform 1 to keep up with new data in time.

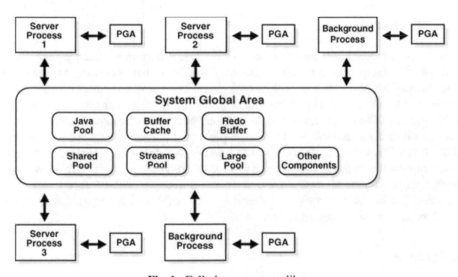

Fig. 1. Folk dance resource library

The design mainly includes three modules: resource collection and processing module, dance resource classification module, network platform management module and so on.

(1) Resource collection and processing module. This module is the "heart" of the whole folk dance resource database. The key point is the digital processing of dance resources, and the difficulty is the collection of a large number of dance original

resources. It is difficult to collect a large number of dance resources. Many kinds of folk dances are listed as provincial or even national intangible cultural heritage. Folk dance originates from the folk. Therefore, it is necessary to collect the original resources through a large number of data surveys and folk wind collection. While collecting dance resources, it is necessary to digitally process them to make them more convenient for preservation, use, and even secondary creation.

(2) Dance resource classification module. This module is the "skeleton" of the whole folk dance resource database. The module will use digital technology to classify, classify and screen folk dance according to the form of resources and dance resources. According to the form of resources, it is divided into: literature, pictures, audio, video, element animation; According to the type of dance, it can be divided into: shield dance, lantern dance, lion dance, tea picking song and dance, etc.

(3) Network platform management module. This module is the "mouth" and "hands and feet" of the whole folk dance resource database. This module includes three small modules: user management and information exchange, platform operation, and search engine. The module of user management and information exchange allows users to communicate with users. The operation of the platform is mainly aimed at the use and reading of dance resources, and the search engine is used to help users screen resources.

4 Conclusion

Folk dance originated from human labor and life. It is a mass dance activity created and performed by the people to express the cultural traditions, living customs and people's spiritual outlook of a nation or region. Folk dance has a strong regional character and carries a rich folk culture. The global economy and informatization have led to the rapid development of China's economy and society, and the people's living standards have also been significantly improved. With it, the people's growing demand for spiritual culture and a better life. From the perspective of "Internet + ", folk dance resource database, as an important part of the construction of modern folk dance service system, will use more scientific digital storage tools to protect and inherit traditional folk dance. It is expected that the construction of folk dance resource database can help the preservation and development of traditional dance.

References

1. Xin, W.: Thinking and rethinking the construction of chinese ethnic and folk dance featured educational(resource) database. J. Beijing Dance Acad. (2012)
2. Yang, H.: The Multi-dimensional extension of chinese folk dance teaching: the construction of internet resource collection of Anhui Hua Gu Deng Dance. J. Beijing Dance Acad. (2015)
3. Xiang-Kaiming, University Y: From cultural self-consciousness self-confidence to cultural self-reliance thinking on construction of folk dance classes. J. Xinjiang Arts Inst. (2017)
4. Yan, F.U., University G.: Construction of folk dance courses in Guangxi Universities. J. Beijing Dance Acad. (2018)
5. Liu, J.: Research on network education resource database construction based on intelligence recommendation. Int. Sym. Distrib. Comput. Appl. Bus. (2009)

6. Qin, Q.: Research on the construction of shared teaching resource database based on Web3.0. Educ. Teach. Forum (2015)
7. Gao, Y.J.: Study on the construction of teaching resource database of internet college psychology based on the theme. J. Hubei Corresp. Univ. (2017)
8. Chuan-Zhi, L.I.: Research on the construction and application of university education resource database based on 5G. China Electron. Educ. (2018)
9. Wang, W., Sun, H., Wang, J., et al.: Discussion on the construction and application of "Internet +" autonomous learning resource database for the practical training of traditional Chinese medicine. J. Changchun Univ. Chin. Med. (2017)
10. Zhang, L.: Construction of university students innovation and entrepreneurship resource database based on collaborative big data analysis. In: Jan, M.A., Khan, F. (eds.) Application of Big Data, Blockchain, and Internet of Things for Education Informatization. LNICSSITE, vol. 392, pp. 185–193. Springer, Cham (2021). https://doi.org/10.1007/978-3-030-87903-7_25

Construction of Network Teaching Platform of University Management Based on Internet

Yang Yang[✉]

College of Cultural Relics and Museums, Sichuan University of Culture and Arts,
Mianyang 621000, China
kekexili630@163.com

Abstract. Information technology and network technology are increasingly widely used in the field of teaching, and the idea of open teaching has become the core of modern teaching development. The teaching mode corresponding to the education system of using the Internet to carry out education and teaching has changed from teacher centered to student-centered. The system is a network teaching platform system using b/s structure, and the system is selected Net and SQL Server2005 are development environments. Starting from the characteristics of network teaching, through the investigation of network teaching platform, understand the business needs of the system. Using UML modeling technology, the system analysis is further elaborated.

Keywords: internet · College management · Network teaching

1 Introduction

With the increasing scale of construction, the number of students has increased sharply, and the teaching task has become more and more arduous. Online learning has long been an important supplementary content to teaching. Online learning involves all aspects of teaching management, which is more complicated, resulting in heavy work for the educational administration department. At present, the pilot online learning system has a relatively simple function, and most of it is used in students' basic information, course arrangement, course scheduling and other businesses, which cannot adapt to the current Internet + Online teaching system of cloud computing.

In January, 1999, the Ministry of education formulated the "opinions on the development of Modern Distance Education in China". In the same month, the State Council approved the action plan for revitalizing education in the 21st century proposed by the Ministry of education, which clearly proposed "implementing the modern distance education project, forming an open education network and building a lifelong education system" [1]. It is particularly noteworthy that on April 8, 2003, the Ministry of Education issued the document "the Ministry of education on starting the excellent courses of the teaching quality and teaching reform project in Colleges and universities" to promote the construction of excellent courses in Colleges and universities. In the "national excellent course evaluation index (2005)", as an important evaluation content, it establishes the

© The Author(s), under exclusive license to Springer Nature Singapore Pte Ltd. 2023
J. C. Hung et al. (Eds.): IC 2023, LNEE 1045, pp. 40–46, 2023.
https://doi.org/10.1007/978-981-99-2287-1_6

index items of the network teaching environment and defines the application of advanced teaching methods and modern educational technology means [2].

On February 10th, 2004, the Ministry of Education issued the "2003–2007 education revitalization action plan". The "plan" clearly puts forward "accelerating the construction of educational information infrastructure, educational information resources and talent training" and "comprehensively improving the application level of modern information technology in the educational system" in the column of "Implementing Educational Information Engineering" [3]. And further - it emphasizes "building a public service system for education informatization, building a network education public service platform shared by hardware and software", "strengthening the construction of campus network in Colleges and universities, creating a national education informatization application support platform" and "establishing a system of mutual communication between network learning and other learning forms, promoting the construction of digital campus in Colleges and universities, and promoting the development of network colleges". It can be seen that in a country like China, which mainly promotes practice with system guidance, the formulation of macro policies has laid a necessary foundation for the development of Network Education (the core component of Modern Distance Education) and the construction of resources and environment for network teaching (an important component of information-based teaching) [4].

2   Related Work

2.1   Internet+

The concept of "Internet +" can be traced back to November 2012, when Yu Yang, chairman of Analysys International, made a speech at the fifth mobile Internet Expo of Analysys. He pointed out that in the future, "Internet +" should be the product and service of our industry, which will be produced after combining with the multi screen, full network and cross platform user scenario we will see in the future. How to find the "Internet +" of your industry, is the problem that enterprises need to think about. In March, 2015, Ma Huateng, a deputy to the National People's Congress, submitted the proposal on promoting China's economic and social innovation and development driven by "Internet +". The proposal pointed out that "Internet +" refers to the use of Internet platforms and information technology to combine the Internet with all walks of life, including traditional industries, so as to create a new ecosystem in new fields[5]. In March 2015, Premier Li Keqiang first proposed the "Internet +" action plan in the government report at the third session of the 12th National People's Congress. The plan aims to promote the integration of mobile Internet, cloud computing, big data, Internet of things and modern manufacturing, promote the healthy development of e-commerce, industrial Internet and Internet finance, and guide Internet enterprises to expand the international market [6]. Although there is no clear definition of "Internet +" in this government report, we can further interpret it from the action plan, which points out that "Internet +" is a new business form of Internet development under Innovation 2.0, the evolution of Internet form driven by knowledge society Innovation 2.0, and the new form of economic and social development it has spawned. "Internet +" refers to "Internet + various traditional industries" in layman's terms, but this is not simply the sum of the two, but the use of information and communication technology and Internet platform to deeply integrate the Internet and traditional industries.

In the process of implementing the "Internet +" action plan, different fields have different understandings and understandings of "Internet +". In the field of education, the interpretation of "Internet +" mainly has the following two views: some scholars (such as Wang Qiaofeng's "analysis of the development of the" Internet + education "model", Zhao Guoqing's "Internet + education" opportunities, challenges and responses, Zeng Xiaojing, fan Bin's "reform of traditional education in the era of" Internet +") believe that" Internet +" That is, the network information means represented by the Internet, its impact on education is mainly to further promote the construction of educational informatization and provide technical support in the optimal allocation of educational resources; Some scholars (such as Liu Feng's future development trend of Internet + education, Jia mindI's survival law of Internet education, and Wang Guixiang's analysis and Research on the current situation and development trend of Internet Education) believe that the impact of "Internet +" on education is more manifested in the emergence of a new type of educational concept and educational model. In this process, "Internet +" is the source of the emergence of a new type of educational concept, And provide information technology support for its new teaching mode [7].

2.2   Basic Functions of Management Teaching

Function refers to the beneficial role played by things or methods. Usually, when people study the function of something, they often discuss it from two perspectives: explicit function and implicit function. Here, in terms of classroom teaching management, its explicit function is to ensure the effective completion of classroom teaching, while its implicit function can be extended to promote the healthy development of students' body and mind. Here, the author mainly discusses the explicit function of classroom teaching management, and analyzes it from three dimensions: adjustment function, control function and promotion function.

(1)   Adjustment function

Adjustment mainly refers to quantity or degree To adjust is to deal with the relationship between more than two kinds of people and things with mutual influence, in order to achieve a balanced state. In the process of classroom teaching, teachers not only need to integrate the scattered teaching resources effectively, but also need to sort them out and redistribute them to maximize their teaching efficiency and make every student get the best development as far as possible.

(2)   Control function

The meaning of control is to grasp that no arbitrary activity is beyond the scope, and the grasp of the development trend of things is mandatory, while regulation is not mandatory. In the whole process of classroom teaching, teachers should always grasp the dynamic direction of teaching and control the teaching progress in order to successfully complete the teaching plan. However, since the object of teaching activities is students with individual thoughts, various accidental phenomena and events often occur in teaching activities. At this time, teachers should use effective methods and means to control the development of the whole situation, minimize the impact of events, and try not to affect the overall teaching activities.

3   Construction of Internet-Based College Management Online Teaching Platform

3.1   The Demand of Network Teaching Practice for Network Teaching System Environment

The educational levels involved in online education in China show diverse forms, from basic education to higher education, from adult post vocational education to enterprise training, cadre training, from academic education to non academic education, and so on. Under such a wide range of network education forms, network teaching inevitably puts forward diversified requirements for the network teaching system environment. On the other hand, from the perspective of national long-term strategy, it is required to gradually build a lifelong education system based on modern distance education, realize a learning society, and meet the learning needs of all kinds of social members [8]. The characteristics of network teaching environment must correspond to the characteristics of network learners and adapt to the learning styles of different learners, so that network teaching can have a good teaching effect. Secondly, the requirements of different nature of teaching content for the network teaching system environment are also different, such as the conceptual, technical, operational or emotional teaching content needs to be arranged in an appropriate network teaching system environment.

Under the diversified needs of this network teaching system environment, the design and development of process oriented network teaching support system is a rich and complementary to the research of network teaching system environment in our country.

In addition, many online teaching support systems are widely used, and learners face a large number of networks The phenomenon that teaching resources are at a loss, and then lose interest in learning, so these support systems also reduce the effect and quality of online teaching to a certain extent. Although there are many factors affecting the effect and quality of online teaching, including system, social concept, evaluation mode, learner characteristics, etc., the impact from online teaching support system is unavoidable and particularly direct [9]. The author believes that there are at least the following problems in the widely used network teaching support system: (1) it rarely reflects the learning process. Although the teaching system provides the service of developing and managing online courses, online course resources are piled up independently and become reprints of textbooks, which separates learning resources from learning process. (2) The organization of online teaching activities is scattered and disordered, and the goal is missing. Students' learning time and progress are out of control, and teachers cannot effectively control and adjust students' learning process. Teaching quality assurance depends only on students' learning initiative and consciousness, which is obviously not enough under the current teaching system and evaluation mode in China. The so-called leading role of teachers only stays in good wishes. (3) The existing network teaching support system is not conducive to the use of process evaluation mode to comprehensively evaluate students.

Network course development system: refers to the system environment integrated by the tools for developing network courses, which provides complete functions for developing network courses. The relationship between the above three is shown in Fig. 1:

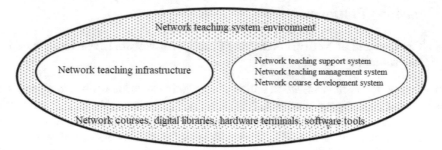

Fig. 1. Schematic diagram of network teaching system environment

3.2 Network Teaching Platform of University Management Based on Internet

The system adopts Net framework structure. In order to provide the maintainability and scalability of the system, the system is divided into three layers: user interface layer, business logic layer and data layer. Its hierarchical structure is shown in Fig. 2.

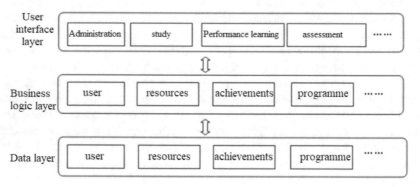

Fig. 2. System software architecture diagram

In the system software architecture diagram, the functions of each layer are described below.

(1) User interface layer

The user interface layer is used to provide the interaction between the operating user and the system. The function presentation layer is the interaction interface between the system operating user and the system, which can accept the input of the operating user, submit the input or selection of operation items on the client to the server, and display the processing results on the server, including the operation result prompt, data query results, etc. The function presentation layer is realized on the client side through HTML5, JavaScript and other technologies. In this system, the data interaction between the client side and the server side is realized, using JSON data format. The system adopts b/s structure, which is accessed by the client through the browser. Therefore, the static page design of the system completes the editing and modification of page layout and page elements (forms, menus, buttons, layers, etc.) under the web page design tool [10].

The static page design of the system is completed under the dreamer tool, which can complete the page design and provide the page editing function combining visualization and HTML code, which greatly facilitates the system page layout design. Some menus on the page are displayed through JavaScript code. In order to enhance the beautification effect of the system page, CSS style is adopted for some text boxes, buttons, menus and other element styles, These styles are editable in DreamEver. The system page layout adopts typical upper and lower parts. The upper part includes the system logo and function module menu, and the lower part is the function operation area. In the upper part, users can select the corresponding function menu, and in the lower part, users can input or display the corresponding processing results of the system.

(2) Business logic layer

The business logic layer is used to define the business processes and rules of the system. The definition of general business logic is stored in two ways: one is in the form of XML files, such as the system configuration file web In XML, another method is to store in the form of database table records. A database table is set in the system database table to store some configuration parameters of the system. This method based on the database record mode completes the setting of configuration information by reading and writing the data table records.

(3) Data interface layer.

This layer provides data support for the system, which realizes the corresponding operation of data with the database platform. The database of this system adopts SQL Server database, and defines the database access interface in the system data layer, including the instance object of the database, the authorized database role and the user password, etc. the system adopts JDBC interface to realize the operations of adding, deleting, modifying, etc. of the database, or complete the data query, statistics, etc. according to the demand, The data layer receives the database operation instructions transmitted by the business logic layer. In order to improve the efficiency of data operation, the database table is materialized in the data layer, the table fields are processed into the attributes of the entity class, and the corresponding methods are encapsulated in the class to complete the operation of the database.

4   Conclusion

The analysis, design and development of the system must be closely related to the actual application of the system. The function of the system cannot be separated from the application needs. It must realize the continuity of teaching and learning process, and provide an appropriate teaching environment for network teaching. However, the network teaching support system is only the physical environment for network teaching. It must be effectively integrated with teachers, students, well-designed teaching activities, and rich and high-quality teaching resources to jointly ensure the continuous coordination of the teaching process, the orderly organization of network teaching activities, the reasonable learning plan and progress of students, and finally improve the effect of network teaching.

References

1. Gu, L., Cai, M., Zhu, L., et al.: Enlightenment of China University MOOC on the construction of network teaching platform in colleges or Universities (2021)
2. Huang, W.: Study on the construction of total quality management based on internet educational teaching research projects (2020)
3. Shi, J.H., Chen, T.W., Wei, Y.Z.: Construction of MOOC teaching platform for supply chain management based on MOODLE system. J. Hebei Softw. Inst. (2020)
4. Huang, H., Chen, M.: Research on the construction of practice teaching system of business administration specialty based on artificial intelligence. J. Phys. Conf. Ser. **1648**(4), 042055 (9p.) (2020)
5. Li, S., Gao, X., Wang, W., et al.: Design of smart laboratory management system based on cloud computing and internet of things technology. J. Phys: Conf. Ser. **1549**, 022107 (2020)
6. Haiyang, M.A., Lejun, L.I., Xiaomei, L., et al.: Teaching design of tourism management based on information-based teaching method: a case study of selection of hotel construction site. Asian Agric. Res. **13**, 55–61 (2021)
7. Ren, Y.: Teaching Design of "fundamentals of power system engineering" based on curriculum ideology and politics. Open Access Libr. J. **9**(5), 7 (2022)
8. Chung, S.C., Wang, S.C., Liao, Y.T.: Construction and Application on the Web-based Network of Panorama for School Building Engineering (2020)
9. Wang, X., Ma, Y.: Construction of Practice Teaching System in Big Data Application and Management Major based on Information Technology (2021)

Construction of Teaching Evaluation Model of Computer Aided Pedagogy

Xiaohang Dong$^{(\boxtimes)}$ and Hui Li

School of Education Science, Harbin Normal University, Harbin 150000, China
xiaohang_1984428@163.com

Abstract. The computer aided instruction (CAPE) system should be able to collect data from students, teachers and parents. The collected data should be analyzed and then used in the decision-making process. The analysis should include both quantitative and qualitative aspects. There is a need to develop clear guidelines for the design of such a system. As this is a new field in educational research, it will take us some time to find an ideal model to describe how such a system works. The computer-aided teaching evaluation model (tepcap) is a teaching framework for designing and implementing a computer-based learning system. It was developed by Professor John J. Meister of the University of Illinois at Urbana Champaign in the 1990s. It is an extension of his early computer-aided instruction (CAI) work. Since then, the tepcap framework has been widely used, especially in higher education and distance education environments.

Keywords: Education major · Computer-aided · Teaching evaluation

1 Introduction

Since the founding of new China, pedagogy has been developing continuously. Development is a dynamic process. The development of pedagogy in New China has three meanings. First, from the perspective of development history, since the founding of the people's Republic of China, the Pedagogy Specialty has realized a process of change from scratch; Second, in terms of the development process, pedagogy has realized a quantitative change process from weak to strong, from fluctuation to stability; Third, in terms of the current situation and future trend of development, pedagogy has realized a qualitative change process from decentralization to integration and from edge to focus [1]. Pedagogy Major is the basic unit of education in Colleges and universities, which is based on the division of pedagogy and social occupation, aims to cultivate professional talents with solid educational theories, takes pedagogy and psychology as the main subject courses, and takes teachers and students as the main body of education. Its connotation is mainly reflected in the following aspects:

First, the Pedagogy Specialty is divided based on the pedagogy discipline and social professional division of labor. The division of majors is based on pedagogy and social professional division of labor. From the social point of view, the establishment of Pedagogy Specialty is to meet the training that people need to receive when they are engaged

J. C. Hung et al. (Eds.): IC 2023, LNEE 1045, pp. 47–52, 2023.
https://doi.org/10.1007/978-981-99-2287-1_7

in education related occupations [2]; From the perspective of higher education, the establishment of Pedagogy Specialty is to undertake the function of cultivating pedagogical talents for pedagogy discipline, which constitutes the knowledge base of Pedagogy Specialty. The social demand for talents in the field of education has promoted the emergence and development of pedagogy; Pedagogy provides knowledge and theory support for Pedagogy Specialty. Pedagogy Specialty is the manifestation of the cultivation of pedagogical talents. The trained pedagogical talents promote the development of pedagogy, meet the needs of society and promote the development of society [3].

Second, the education major aims to cultivate professional talents with solid educational theories. Colleges and universities cultivate high-level talents. As the most basic teaching carrier of higher education, majors undertake the task of cultivating high-level talents. As a platform and manifestation to realize the function of cultivating talents in pedagogy, pedagogy is aimed at cultivating professional talents with solid educational theories. On the one hand, the knowledge that pedagogy majors learn is advanced knowledge, which is different from the educational knowledge learned in primary and secondary education; On the other hand, the talents trained by the Pedagogy Specialty are specialized and have solid educational theories [4].

Third, pedagogy is a major course with pedagogy and psychology as its main subjects. The talent training of pedagogy can not be separated from the support of curriculum. The core and main course of pedagogy is pedagogy and psychology.

2 Related Work

2.1 Components of Pedagogy

From the perspective of the constituent elements of pedagogy, there are mainly the training objectives of pedagogy, the courses of pedagogy and the main body of education, that is, the people in pedagogy. No matter what changes happen in pedagogy, these three elements are indispensable.

First, the training objective of pedagogy plays a guiding and normative role in the whole educational practice. At the same time, the degree of realization of the training objectives is also one of the test criteria for the classroom teaching effect. The training objectives determine the direction of future talent development. They are required to be the decision-making standard and assessment basis for pointing out the future development direction, improving teaching efficiency and cultivating talents. They need to be operable and practical, keep pace with the times and constantly adapt to social needs [5]. There is a close relationship between the professional training objectives and the training objectives of colleges and universities. The training objectives of the specialty specify the talents with basic quality and basic ability to be trained in different fields of the specialty, which is the embodiment of the training objectives of colleges and universities in different levels of higher education and different specialized education. The training objectives of colleges and universities can be reflected through the professional training objectives.

Since the birth of pedagogy, the training goal of pedagogy has a relatively stable position. However, with the changes of the times and our understanding of education, the training objectives of pedagogy have also been differentiated. We can discuss it as five orientations [6]. First, to train special talents for high-level education, after graduation, they can engage in education teaching, education research or administrative management in primary and secondary schools, education research departments, education adminis-trative departments, etc. This kind of target orientation is actually the leading orientation of the training objectives of pedagogy in many normal universities. Second, train rele-vant teachers for primary school teachers or secondary school comprehensive activity courses. In the process of development, some colleges and universities have directly positioned the training objectives of pedagogy in the direction of primary education, and trained teachers for primary school teachers. Third, it is committed to training teach-ers of psychological counseling and mental health education in primary and secondary schools. Fourth, train educators from educational companies, news media, educational institutions or publishing units [7]. Fifthly, continue to pursue further studies and train master of education and doctor of education talents. For example, East China Normal University has cancelled the establishment of the undergraduate major of pedagogy and shifted the training focus to the training of master of pedagogy and doctor of pedagogy.

2.2 Computer Aided Teaching Analysis

In the traditional teaching mode, teachers usually use words, language, actions and other expressions to impart knowledge to students. Sometimes, for the sake of intuition, teachers will demonstrate and explain the contents to be explained with the help of some physical objects, models, teaching instruments, illustrations and other auxiliary teaching tools. These teaching aids are of course indispensable in teaching, but they also have their limitations: some of the described objects are too large or too small, such as the movement of celestial bodies or the thermal movement of molecules, which cannot be displayed in real objects; Some motion processes develop and change very rapidly, such as wave propagation process and electromagnetic wave process. Models and illustrations are not competent or even impossible to realize; There are also some processes, such as the biological growth process [8]. Although the means of video and film can reflect the process of movement change, it cannot be controlled by the teachers arbitrarily, such as changing the sequence before and after, changing the speed of change, or changing the display screen. Computer aided instruction just makes up for the deficiencies of the teaching aids mentioned above. It makes use of its interactive function to call out the multimedia information stored in the computer at will for teachers' selection. The text and sound are both rich, so that the content difficult to be expressed in traditional teaching is vividly and intuitively displayed in front of the students, so that the students' attention to learning is more concentrated, their understanding of the phenomenon or process is more profound, and their memory is more solid. The advantages of CAI can be summarized in the following three aspects.

① Computer aided instruction is helpful to highlight the key points, break through the difficult points, and turn abstract into intuitive. Multimedia is used to make some abstract, microscopic and complex dynamic change processes real and visualized, turn abstraction into intuition, turn difficulty into easy, and highlight the key points.
② Computer aided instruction is helpful to increase the classroom capacity and improve the classroom efficiency. Computer aided instruction (CAI) teaches students knowledge,

trains students' skills, and improves students' abilities, which is conducive to cultivating students' scientific methods and attitudes, so as to achieve the educational objectives in the emotional field and complete the task of quality education in an all-round way. If we simply explain in class, it is difficult to attract students. We can demonstrate some teaching computers that can be demonstrated by computers to students, which is intuitive and can deepen students' understanding of knowledge, saving time and improving classroom efficiency.

③ Computer aided instruction is helpful to optimize the classroom structure and strengthen some important contents in the teaching materials. Teachers should explain before class, explain and guide students to observe the phenomenon through demonstration, and finally summarize the conclusion.

The classification framework of CAI mode is shown in Fig. 1.

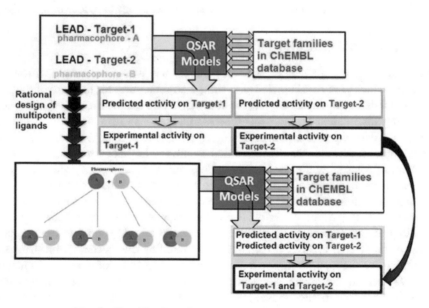

Fig. 1. Classification of computer aided instruction system

3   Construction of Teaching Evaluation Model of Computer Aided Pedagogy

3.1   Principles for Establishing Teaching Evaluation Indicators of Pedagogy in Colleges and Universities

According to the requirements of the guiding outline of physical education courses in national colleges and universities, the aim is to strengthen the educational evaluation concept of continuous progress between teachers and students, apply modern sociology, pedagogy, physical education, mathematics and other theories, and start with various conditions and related aspects of pedagogy teaching.

(1) The principle of objectivity.

The construction of the evaluation system at the present stage is based on various evaluation theories and the actual situation of ordinary colleges and universities in Kunming as the starting point. It comprehensively and objectively analyzes many existing factors, so as to make each component of the evaluation system objective and truly reflect the characteristics that can effectively promote the teaching effect. Especially in the process of evaluation, it is required to be objective, fair and reasonable, and can make a fair A fair and realistic judgment [9].

(2) Feasibility principle.

The indicators in the teaching evaluation system of the Pedagogy Specialty in ordinary colleges and universities should conform to the characteristics of the physical education discipline and the physical and mental development of students. Before formulating the evaluation objectives and index system, the current situation of the Pedagogy Specialty Teaching should be systematically investigated and analyzed, and the evaluation system should be built on the basis of in-depth understanding, so as to find out the existing problems and shortcomings and affirm the existing advantages.

(3) scientific principle.

The overall evaluation index system must be complete so that the evaluation system can fully reflect the requirements of the evaluation objectives. When screening indicators, we should follow the educational law and make the indicators independent. The indicators at the same level should not have causality, but should be level, and there should be no overlapping relationship between aggregation and complement [10].

(4) Principle of comparability.

The indicators in the evaluation system reflect the common attributes of the evaluation objects and should be measurable. That is to say, the indicators should be defined in specific and operable language. Through the use of evaluation methods, it can be observed and understood to obtain clear results. The setting of evaluation indicators should be as simple as possible and should be comparable.

3.2 Teaching evaluation model of Pedagogy

The analytic hierarchy process is used to calculate the weight coefficient of the teaching evaluation index set of Pedagogy Major in ordinary colleges and universities. The steps are as follows:

(1) Build an evaluation system. Through a large number of investigations and studies, the indicators are obtained, the indicators are sorted out, and the relationship between the indicators is analyzed.

(2) The hierarchical structure model is established. According to the survey results obtained from expert questionnaires and interviews, the indicators are divided into multiple levels in a pyramid shape according to the importance;

(3) And a corresponding judgment matrix is constructed. The pedagogical teaching experts of ordinary colleges and universities compare all pedagogical teaching evaluation indexes of ordinary colleges and universities at the same level in pairs, judge the importance of them according to the bid winning degree of analytic hierarchy process, and form a judgment matrix.

$$A = \begin{bmatrix} a_{11} & a_{12} & ... & a_{1n} \\ a_{21} & a_{22} & ... & a_{2n} \\ ... & ... & ... & ... \\ a_{n1} & a_{n2} & ... & a_{nm} \end{bmatrix} \tag{1}$$

4 Conclusion

In the practice of the application of information technology in education and teaching, we deeply feel that the excellent computer-aided teaching software originates from its contribution to the transformation of students' learning mechanism and its grasp of the characteristics of the application of information technology. Scientific evaluation can help teachers effectively apply information technology in specific pedagogy teaching. However, with the rapid development of information technology itself, there will be many new models of computer-aided instruction, and the evaluation technology of computer-aided instruction will also develop accordingly. We believe that the future computer-aided teaching evaluation technology will have three development directions.

References

1. Fu, Y.: An Analysis of the Construction of Teaching Evaluation Model Under the Framework of Web: Taking Educational Psychology as an Example (2021)
2. Xia, X., Yan, J.: Construction of music teaching evaluation model based on weighted Nave Bayes. Scientific Programming (2021)
3. Zheng, H., Miao, J.: Construction of college english teaching model in the era of artificial intelligence. J. Phys: Conf. Ser. **1852**(3), 032017 (2021)
4. Zhang, D., Chen, R., Yuan, Y.: Construction and application of comprehensive evaluation model of "golden classroom." J. Phys: Conf. Ser. **1880**(1), 012032 (2021)
5. Huang, J.: Research on the integrated teaching model of open education and vocational education. Open Access Libr. J. **8**(11), 10 (2021)
6. Wu, Y., Lu, P.: Comparative analysis and evaluation of bridge construction risk with multiple intelligent algorithms. Math. Probl. Eng. **2022** (2022)
7. Jing, Y., Mingfang, Z., Yafang, C.: Evaluation model of college english education effect based on big data analysis. J. Inf. Knowl. Manage. **21**(03) (2022)
8. Zhang, J.: Construction and exploration of virtual simulation experimental teaching platform for network security and computer technology. J. Phys. Conf. Ser. JPhCS (2022)
9. Han, Y.: The Construction of Chinese Online Resource Database and Teaching Implementation in Higher Vocational Education (2021)
10. Peng, P.F., Gong, L., Liu, Y.: Construction of teaching effectiveness evaluation system for command information system major based on kirkpatrick model. DEStech Trans. Econ. Bus. Manage. 2021(eeim)

Corpus Translator's Style in the Era of Big Data Under Data Mining Algorithm

Bing Chen[(✉)]

Chengdu Polytechnic, Chengdu 610000, Sichuan, China
applechen123@126.com

Abstract. The research of corpus translation style in the era of big data under data mining algorithm is to use big data to analyze the characteristics and changes of translation style. This study aims to achieve the following goals: to analyze how translation styles have changed in different periods from ancient times to the present; Study how translation style changes according to various factors such as age, gender and major; Investigate whether there are any differences between the translations of professional translators and non professional translators.. In this article, we will introduce two problems in dealing with big data: (1) big data analysis and (2) translation quality evaluation. We will discuss how to solve these problems by using machine learning algorithms such as neural networks and deep learning models.

Keywords: Data mining · Corpus · Translation style

1 Introduction

"Corpus refers to the real corpus collected and stored in the computer on a large scale according to the specific purpose of language research using computer technology and certain linguistic principles. These corpora are marked to a certain extent, easy to retrieve, and can be applied to descriptive research and empirical research." The application of u corpus provides a lot of data support for the study of translator's style, which makes up for the shortcomings of traditional translation studies [1]. The research method of corpus has also become a rookie in translation studies. However, by searching the existing corpus based research on translator style, it is found that this kind of research mostly focuses on the Chinese translation of English works and the English translation of Chinese works, and there is little research on the translation of other languages.

Corpus not only brings new methods to traditional translation studies, but also provides new research ideas. With the extensive use of corpus, the study of translator's style with English text as the research object is emerging in endlessly, constantly adding a new research perspective to corpus translatology. However, it is rare to use corpus to analyze the translator's style of English texts. "Corpus translatology refers to the study of systematically analyzing the essence, process and phenomenon of translation based on corpus, with real bilingual corpus or translation corpus as the research object, data statistics and theoretical analysis as the research methods, and linguistic, literary and cultural theories

J. C. Hung et al. (Eds.): IC 2023, LNEE 1045, pp. 53–58, 2023.
https://doi.org/10.1007/978-981-99-2287-1_8

and translatology theories." Although corpus translatology did not appear early, it has developed rapidly and achieved a number of excellent research results, which can be observed from two aspects: methodology and case study. The characteristics of English texts, word form reduction and part of speech tagging of English texts involved in this study provide some methodological references for the study of corpus translatology, and also enrich the research results of translator style with English novels as the research object to a certain extent [2]. Based on this, our research is about the style of corpus translators in the era of big data under the data mining algorithm.

2   Related Work

2.1   Data Mining Text Classification

There are many ways to extract text features. The teacher of this course focuses on TF-IDF and chi square verification. Let's first look at the calculation method of if-idf:

Term frequency (TF) refers to the frequency of a given word in the file.

Inverse document frequency (IDF). The IDF of a specific word can be obtained by dividing the total number of files by the number of files containing the word, and then taking the logarithm of the obtained quotient [3].

IDF is a measure of the general importance of a word. TF-IDF value is the product of TF value and IDF value.

TF-IDF comprehensively represents the importance of the word in the document and the document differentiation. However, it is not enough to use tf-df to judge whether a feature has discrimination in text classification. It does not consider the distribution of feature words among classifications. If a feature word is evenly distributed among various classes, such a word has little contribution to classification; However, if a feature word is concentrated in a certain class and hardly appears in other classes, such a word can well represent the characteristics of this class, and TF-IDF cannot distinguish between the two cases. The distribution of feature words in class internal documents is not considered. In documents within a class, if the feature words are evenly distributed among them, this feature word can well represent the characteristics of this class [4]. If it appears only in a few documents and does not appear in other documents of this class, it is obvious that such feature words cannot represent the characteristics of this class. Figure 1 below shows the data mining text extraction code.

In this paper, there are not too many strict requirements for words. This data mining experiment requires taking nouns, and all stop words are non nouns, so the operation of taking nouns only also removes all stop words. The problem of stop words mentioned below has been solved here. Similar operations are carried out on several corpora, turning each corpus into a noun [5]. If all nouns form a set without repeating elements (W1, W2, W3.. WN), this set is a dictionary. For each corpus, the values are 0 and 1 respectively according to whether the words in the corpus exist in the dictionary. So far, the work of converting corpus into word vector has been completed.

```
1   def getUrls(url):
2       req = requests.get(url).text
3       if (req == None):
4           return
5       bf = BeautifulSoup(req, 'html.parser')
6       div_bf = bf.find('div', attrs={'class': 'content_list'})
7       div_a = div_bf.find_all('div', attrs={'class': 'dd_bt'})
8       urltxt = open(b'F:\data\url.txt', 'a', encoding='UTF-8')
9       for div in div_a:
10          link = div.find('a').get("href")
11          urltxt.write(link+'\n')
12      urltxt.close()
```

Fig. 1. Data mining text extraction code

2.2 Analysis of English Noun Structure

An English noun phrase can be composed of three parts: head, premodifier and postmodifier, and its structural sequence can be expressed in the following four forms:

1) Headword (noun phrase containing only the headword, such as noun phrase "China")
2) Prepositional modifier + headword (a noun phrase consisting of a headword and its prepositional modifier, such as a noun

   Phrase "the right procedure")

3) Headword + Post modifier (a noun phrase consisting of a headword and its post modifier, such as a noun

   Phrase "corruption aplenty")

4) Prepositional modifier + head word + Post modifier (composed of head word and its prepositional modifier and post modifier

   Noun phrases formed by words, such as the noun phrase "the talkative man in the center of theroom")

The head word is an indispensable part of any noun phrase and the core of the noun phrase; In front of it All modifiers of (or on the left) are collectively referred to as pre modifiers, and all modifiers after them are collectively referred to as post modifiers; in English grammar rules, the head word of a noun phrase can be acted as by a noun, pronoun or a nominalized word. Because nouns often represent new information in the language, and pronouns are a review of the previous old information, noun phrases with the head word as nouns are shorter than nouns with the head word as pronouns At the same time, relevant studies have proved that in English texts, the number of nouns is the largest and far greater than the number of pronouns, which may mean that the number of noun phrases with nouns as the center word is more than that with pronouns as the center word [6]. Therefore, in this study, we mainly identify noun phrases with nouns as the center word.

3 Research on Corpus Translator's Style in the Era of Big Data Under Data Mining Algorithm

In recent years, in addition to the traditional translation study, which takes the translated language as the main research object, the study of translator's style has gradually moved from invisibility to dominance under the framework of traditional translation theory. The translation and the translator have gradually got rid of the shackles of the source text and become one of the focuses of translation research. The study of translator's style is also developing towards the direction of independent and individual research. From the perspective of grammatical rules, the head word is the core of noun phrases and an indispensable part of any noun; In addition, there are relatively few parts of speech of words that can act as the head word, especially in this study, the part of speech of phrases whose head word is a noun is more specific. Therefore, we use the method of identifying the head word first and then its modifier for noun phrase recognition.

"Translation language features can be roughly divided into two categories: the generality of translation language and the language features of specific language on the translated text. The former refers to the universality and regular characteristics of the translated text, such as simplification, manifesting, normalization, implicit, etc. these characteristics are not limited by the source language and target language, and are closely related to the translation process itself [7]. The latter refers to the language features formed by the differences between specific source language and target language, which are mainly manifested in It refers to the characteristics of the translated text at the lexical and syntactic levels. "

After the recognition module completes the recognition of noun phrases of a sentence, the software will submit the sentence, the recognized noun phrases and their head words to the verification and judgment module, which will verify and judge the reliability and integrity of these noun phrases based on the corpus. The reliability judgment here mainly refers to determining the recognition of a noun phrase according to the frequency of the identified noun phrase and its colligation in the corpus as the main parameter: the integrity judgment refers to combining the two adjacent noun phrases with high reliability and the words between them into a new noun phrase in the order from left to right, Then by judging the reliability of this new noun phrase, we can determine the integrity of the original two adjacent noun phrases. If the reliability of the newly merged noun phrase is low, it indicates that the integrity of the original two noun phrases is high, and all the information of the newly merged noun phrase is discarded [8]; If the reliability of the newly merged noun phrase is high, it indicates that the integrity of the original two noun phrases is low and should be discarded, and the newly merged noun phrase should be used to replace the original two noun phrases; Press this cycle until the integrity of all noun phrases in the sentence is judged.

4 Simulation Analysis

It refers to the establishment of corpora based on real translation corpora according to specific research objectives (including monolingual comparable corpuses, bilingual / multilingual parallel corpuses, translation corpuses, etc.). Based on the electronic text of the corpus and computer statistics, it describes all kinds of translation phenomena in a large or specific range, analyzes and explains translation phenomena on the basis of full description, or verifies various hypotheses about translation [9]. In essence, corpus translation is an interdisciplinary product of the combination of descriptive translation studies and corpus linguistics. The research on corpus translator style in the era of big data under data mining algorithm is the effective application of natural language processing technology to web documents. This method first linearizes the source code of the reconstructed web page, and preliminarily filters the noise of the web page, then filters and clusters the text blocks by using the methods of classification and clustering to get the text paragraphs of the web page, and finally absorbs the pseudo noise paragraphs to get the text of the web page [10]. This method does not need to build a tree for web pages, and has the characteristics of fast and accurate. However, due to the use of text classification, clustering, data mining and so on, there is a certain complexity. As shown in Fig. 2 below, the character code of the captured text is as follows.

```
1   head = requests.head(url)
2   req = requests.get(url)
3   req.encoding = 'GB2312'
4   bf = BeautifulSoup(req.text, 'html.parser')
5   div = bf.find('div', attrs={'class': 'content'})
6   h1 = div.find('h1')
7   head = re.sub(r'\s+', '', h1.get_text())
8   out = open(filepath + '.txt', 'w', encoding='GB2312', errors='ignore')
9   out.write(head+'\n')
10  timediv = div.find('div', attrs={'class': 'left-t'})
11  time = timediv.get_text().replace(" ", "")[0:16]
12  out.write(time)
13  p = div.find('div', attrs={'class': 'left_zw'}).find_all('p', text=True)
14  for ptext in p:
15      out.write('\n'+ptext.text);
16  out.close()
```

Fig. 2. Grab character feature code

5   Conclusion

The research on the translation style of corpus in the era of big data under data mining algorithm is a research aimed at analyzing and studying the translation quality of corpus. The main purpose of this study is to find out the difference between the translation quality of human translators and machine translators, which will help to improve the machine translation system. This research has another goal: to improve our understanding of the characteristics, characteristics and problems of machine translation.

Acknowledgments. 1.Research on the Construction and Development of "Cross-Border E-Commerce English" Loose-Leaf Textbook Based on Corpus (WYJZW-2021-2006).

2."Big Data Application Research Center for Higher Vocational Foreign Language Education" (19kypt06), a scientific research platform project of Chengdu Polytechnic.

References

1. Han, H., Jiang, Y., Yuan, X.: Corpus-based study of translator's style in the big data era. Foreign Lang. Educ. (2019)
2. Wang, Q.: Readjustment of translator's status in the era of big data. J. Panzhihua Univ. (2018)
3. A Brief Analysis of the Translator's Subjectivity in the Process of Com puter Aided Translation. **23**, 2 (2019)
4. Zhen, C., Jiang, C.: Overview of data mining in the era of big data. Int. Core J. Eng. **5**(10), 136–139 (2019)
5. Xinmin, Z., Qiang, N.: A comparative study of the translator's style——a corpus-based case study of lianghuiwang. **2**, 10 (2018)
6. Singh, K.K., Kushwaha, V.: Smart Wireless Network Algorithm in the Era of Big Data (2021)
7. Xiao, Z., Wang, Y.: The model of translator's information literacy in the new era. In: Proceedings of the 2019 4th International Conference on Modern Management, Education Technology and Social Science (MMETSS 2019) (2019)
8. Bazylev, V.N.: G.E. von Spilcker is the translator of the "Satires" of Antioh Cantemir (translation experience in linguistic and cultural context of the era) (2019)
9. Yang, Y.: On the embodiment of the literary and translator's subjectivity in the translation of literary works——taking the three translation versions of jane eyre as examples. J. Heihe Univ. (2019)
10. Wang, S.: The evolving paratexts in the C-E translation of Tang poems in the era of fragmentation reading. J. Xi'an Int. Stud. Univ. (2018)

Cost Optimization Technology of Construction Engineering Based on Genetic Algorithm

Wenen Li[✉]

China Gansu International Economic and Technical Cooperation Co. Ltd., Lanzhou 730050, Gansu, China
fanbenbao@163.com

Abstract. In recent years, with the rapid development of computer information engineering technology and the continuous improvement of the management level of construction projects, it is possible to achieve multi-objective quantitative optimization of construction projects. The main content of project objective quantitative management is the comprehensive optimization of construction period and cost. Relatively reasonable determination of construction period can bring benefits to the project to a certain extent. This study is based on the genetic algorithm of construction cost optimization technology. The purpose of this study is to find out whether genetic algorithm can be used to design the optimal structure and solve the construction engineering problems. The main purpose of this paper is to study the possibility of using genetic algorithm to optimize the cost, time and labor of construction projects. It also aims to find out whether there are other methods to replace the use of genetic algorithms, so as to not only reduce but also minimize the cost, time and labor involved in the construction project. In addition, this work will attempt to.

Keywords: architectural engineering · Genetic algorithm · Cost optimization

1 Introduction

The construction industry has a large volume and a pillar position in the national economy. It is an important carrier and material production department for the development of the national economy. According to incomplete statistics, the total output value of the construction industry in 2015 exceeded 18 trillion yuan, and its added value reached 4.6 trillion yuan, accounting for nearly the proportion of the secondary industry. On the other hand, the construction industry is closely connected with other industries, which can drive the development of more than 50 related industries such as building materials, metallurgy, petroleum, forestry and machinery. The development of the construction industry has important strategic significance in expanding domestic demand, promoting employment and stimulating economic growth [1].

Usually, project management has three major objectives, namely, cost, construction period and quality. However, in fact, most of them use construction period and cost as the main objectives for management. On the premise of ensuring quality, reducing construction period and cost means more economic benefits. Therefore, when people choose

J. C. Hung et al. (Eds.): IC 2023, LNEE 1045, pp. 59–65, 2023.
https://doi.org/10.1007/978-981-99-2287-1_9

bidding documents or construction schemes, the design with short project duration and lower cost will often be given priority. In fact, the relationship between construction period and cost is a unity of opposites, and there is no feasible solution [2]. It can achieve the two goals of short time and low cost at the same time. Reducing the cost will inevitably lead to the extension of construction period, and compressing the construction period to a certain extent will increase the cost input. In order to solve this problem, the time cost tradeoff method is proposed, which is also the main method to solve the time cost multi-objective optimization problem [3].

After many years of research, the multi-objective optimization problem of construction period cost has made great progress in theoretical research. However, due to the development of science and technology, the expansion of project scale and other reasons, the traditional optimization model of construction period cost can not meet the actual needs. In recent years, with the further improvement of modern computer technology, computer-aided computing has become the mainstream, and more and more artificial intelligence search algorithms have been applied to the optimization field, which provides a new direction for the development of duration cost optimization.

2 Related Work

2.1 Principle of Duration Cost Optimization

By analyzing previous research documents, it is not difficult to find that many documents seldom consider the time value of funds when studying the relationship between the duration and the cost of construction projects, and only analyze the relationship between the direct cost and the duration of the project, ignoring the relationship between the indirect cost and the duration of the project. Moreover, most studies assume that the direct cost and the duration of the project are linear, However, this assumption is not consistent with the specific project implementation process. With the shortening of the activity duration, the comprehensive cost of each project activity and its duration show a non monotonic nonlinear function relationship. For example, the shorter the construction period of the project, the greater the labor intensity of the project [4]. The exchange of resources for time will inevitably consume a lot of human and material resources. The direct cost increases, but the indirect cost decreases. Therefore, the duration and the comprehensive cost are not simple linear relations. To sum up, from the perspective of the contractor, the shorter the time to complete the project, the faster the funds will be withdrawn. Considering the time value of the funds, the faster the loan will be returned to the bank, the lower the interest, and the higher the income will be obtained. Therefore, it can more accurately reflect the real income of the contractor in the project construction, and is more conducive to the contractor to control its construction period and cost [5].

The total cost of construction projects is mainly composed of direct costs and indirect costs. The direct cost is calculated according to each work in the project. The shorter the execution time of each work, the more the direct cost of the work increases.

Indirect costs are apportioned comprehensively according to the whole project, that is, the shorter the construction period, the less the cost. Since the direct cost is inversely proportional to the construction period, and the indirect cost is directly proportional to the construction period, the total cost including the two must have a lowest point, which is the best point for time cost optimization. As shown in Fig. 1, the X point in the figure is the best point.

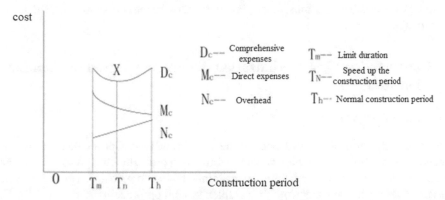

Fig. 1. Construction cost relationship

2.2 Cost Optimization Principle

The optimization principle of duration cost optimization is to minimize the project duration while reducing the project cost to the maximum extent, and improve or not affect the project quality. It is to sacrifice a certain number of certain objectives to achieve the overall gain or meet a specific demand. In short, the owner or the contractor can obtain the best arrangement and combination of the project subprojects by reasonably allocating the duration, construction cost or changing the construction sequence of the subprojects within a reasonable time and cost range through some optimization method. Under the same construction period, the total cost of any other sub project arrangement is higher than the optimization result, or under the same cost, the total construction period of any other sub project arrangement is longer than the optimization result [6].

From many research results of multi-objective optimization, we can see that there are many methods that can be used to solve the problem of multi-objective optimization. However, for the construction period cost optimization problem of building engineering construction, the most common method is to simplify the complexity and combine the multi-objective problems through certain variables or existing relationships, such as linear combination and objective weight ranking, so that the mathematical method can be used to solve the problem, that is, the optimization can be realized by solving the objective function. Therefore, the design of objective function is the basic premise of the whole optimization process [7].

The process of optimizing the objective function is the process that the duration cost optimization scheme gradually approaches the best scheme with shorter duration and lower cost. Under certain constraints, to select the best scheme from multiple schemes, it is necessary to use the optimization principle to continuously improve the original scheme according to the target until the most satisfactory scheme is found by the owner and the contractor. The duration cost optimization is to find a satisfactory solution according to the above steps [8]. After the operation, a group of non mainstream solutions will be obtained: under the same constraint conditions, this group of solutions has obvious advantages over other solutions, that is, Pareto front (solution set composed of all Pareto solutions).

3 Construction Cost Optimization Technology Based on Genetic Algorithm

3.1 Genetic Algorithm

The genetic algorithm does not depend on the domain and type of problems, and provides a general framework for complex system optimization problems (the greatest advantage of the genetic algorithm is that it is not constrained by the domain and type of problems, and provides a general framework for complex system optimization problems)[9]. The calculation steps are as follows:

(1) First, the individual expression m and the solution space are given, that is, the decision variables and various constraints are determined.
(2) Define the type of objective function and describe or quantify it in mathematical form.
(3) The individual gene set M and the search space, i.e., the coding gene of the chromosome representing the feasible solution, are defined.
(4) To clarify the decoding method, that is, to clarify the conversion relationship and method from individual genotype m to individual phenotype M.
(5) The quantitative evaluation method of individual fitness was defined and the regenerated individuals were selected.
(6) The specific operation methods of selection operation and crossover operation are specified. In this paper, the crossover probability is determined according to the following formula, and new individuals are generated through crossover.
(7) To clarify the specific operation method of mutation operation, this paper determines the mutation probability according to the following formula, and generates new individuals through mutation.

$$\begin{cases} \max, f(X) \\ s.t. X \in R \\ \quad R \subseteq U \end{cases} \tag{1}$$

(8) Specify the relevant operating parameters of the genetic algorithm, and generate a new generation of population through crossover and mutation, and return to step 2. See Fig. 2 for the specific algorithm diagram.

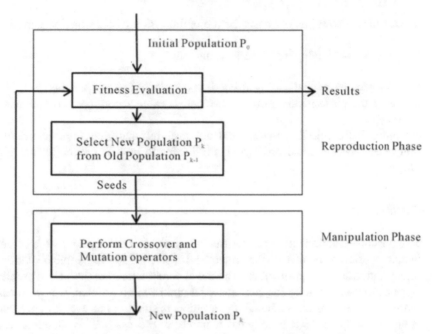

Fig. 2. Schematic diagram of genetic algorithm operation

3.2 Construction Cost Optimization Based on Genetic Algorithm

After the model is established, we need to solve it, that is, to find the optimal construction period and cost. In this paper, the genetic algorithm described above is used to solve the problem.

First, we need to set initial parameters, including genetic parameters and project parameters. Among them, genetic parameters include individual coding string length L, population size o, crossover probability p, mutation probability P and iteration algebra T. when binary coding is adopted, the selection of individual coding string length L is related to the accuracy of problem solving. When floating-point coding is used, it is the same as the number of decision variables and the population size O It refers to the number of individuals in the population. On the one hand, the value of Q should not be too large to reduce the operation efficiency of genetic algorithm; on the other hand, it should not be too small to reduce the diversity of the population. Therefore, the value range of Q is generally recommended to be 20 ~ 100. Project parameters: project duration T, cost C, project scale, logical relationship between processes, etc. [10].

Secondly, the initial population is randomly generated; The essence of genetic algorithm is the process of searching the optimal solution under the given solution set and the set constraints, so it needs to give an initial population. The individuals in the initial group shall be randomly generated by a random function according to the initial parameters, so as to obtain the initial group that does not contain any subjective intention of the decision maker. Therefore, the model needs to adopt the method of uniform sampling, and randomly generate n individuals in the problem solution set to form the initial set Q. through the following steps, the optimal resource allocation set of the process is finally generated, which also forms the basis for the optimization of construction period and cost.

General process of individual fitness evaluation:

(1) Decoding the individual coding string to obtain the individual phenotype.
(2) Calculate the corresponding objective function value according to the individual phenotype.
(3) According to the objective function and the constraint conditions, the objective function value is converted according to the following formula to obtain the individual fitness.

4 Conclusion

The comprehensive optimization of construction project duration and cost is of great significance to effectively manage the project and ensure the smooth implementation of the project. Duration cost optimization is a kind of combinatorial optimization problem. The biggest problem faced by the combinatorial optimization problem is how to solve the combinatorial explosion. Although the swallow is small, it has everything. Even a very small and simple project usually has hundreds of processes. This situation makes the duration cost optimization of engineering projects extremely complex and difficult. For this reason, it has attracted the research enthusiasm of many scholars at home and abroad. At present, from the perspective of time cost optimization technology, although enumeration method can obtain the optimal solution or satisfactory solution, its modeling data is difficult to collect, modeling is difficult and the amount of calculation is huge, which restricts its application in time cost optimization of large-scale projects. Due to the lack of corresponding mathematical rigor, it is difficult to ensure that the solution obtained by heuristic algorithm is the optimal solution, so its application in engineering projects is also greatly limited.

References

1. Zhang, Q.: Research on the construction schedule and cost optimization of grid structure based on bim and genetic algorithm. J. Phys. Conf. Ser. **1744**(2), 022065 (5p.) (2021)
2. Xia, F., Wang, P.: Research on power line communication technology of energy internet based on blockchain. IOP Conf. Ser. Earth Environ. Sci. **714**(4), 042085 (8p.) (2021)
3. Gong, Q., Zhan, J., Wang, P., et al.: Research on design, construction and maintenance technology of advanced marine research ship. IOP Conf. Ser. Earth Environ. Sci. **714**(2), 022061 (7p.) (2021)
4. Yu, Y.: Application and Optimization of Enterprise Financial Sharing Service Center Based on Cloud Computing (2021)
5. Kapse, S., Narasimhan, S., Thapa, R.: Descriptors and graphical construction for in silico design of efficient and selective single atom catalysts for the eNRR (2022)
6. Zhigang, Z., Zhimei, L., Haimiao, M., et al. Review of research on fireworks algorithm (2022)
7. Sustainability: Exploratory Research on the Use of Blockchain Technology in Recruitment (2022)

8. Qian, Z., Yang, X., Xu, Z., et al.: Research on key construction technology of building engineering under the background of big data. J. Phys. Conf. Ser. **1802**(3), 032003 (9p.) (2021)

9. Wang, Y., Liu, Z., Cai, C., et al.: Research on the optimization method of integrated energy system operation with multi-subject game. Energy. **245** (2022)

10. Karn, S.K., Dahal, K.R.: Cost optimization in building construction projects with special reference to kathmandu valley of Nepal. Am. J. Sci. Eng. Technol. **10**, 1 (2021)

Covariance Estimation and Algorithm Implementation of Hedge Fund Distribution Replication Model Based on AHP Algorithm

Bang Geng[✉]

Harbin University of Commerce, Haerbin 150028, Heilongjiang, China
gohnbrown@163.com

Abstract. Covariance estimation of hedge fund distribution replication model can help people identify relevant financial risks, so as to prevent and control them. In order to achieve this purpose, based on the hedge fund distribution replication model, this paper discusses the feasibility of AHP algorithm in model covariance estimation, and then introduces the algorithm implementation method. The research shows that AHP algorithm has obvious advantages in the covariance estimation of hedge fund distribution replication model, and accurate results can be obtained under reasonable application.

Keywords: AHP algorithm · Hedge fund distribution replication model · Estimation of covariance

1 Introduction

Hedge fund is a kind of fund different from the traditional mutual economy. It has the characteristics of high profit and high risk, so it attracts many "activist" investors. With the help of these investors, hedge fund began to be popular around the world. This background, the effects of high risk hedge funds increasingly obvious, and the types of hedge funds, so the influence of the high risk also increases gradually, so the people began to think of the problems of how to should hedge fund risk, thus was born the hedge fund distribution replication model, the main function of the model by means of covariance estimates for a hedge fund project, evaluating the risk so that investors can make accurate decisions according to their own situation. Although this can't completely eliminate risk, it can at least reduce the influence of risk and make the characteristic of high profit of hedge funds more prominent, which is conducive to the development of hedge funds. Because hedge funds distribution replication model is put forward by means of covariance estimation results, so the accuracy of the estimation results are very important, to ensure the accuracy, put forward a lot of fields related to covariance estimation algorithms, including AHP algorithm has been widely attention, think the results of this algorithm can well improve the accuracy, therefore, in order to understand the application method of AHP algorithm in the covariance estimation of hedge fund distribution replication model, this paper will carry out relevant research.

© The Author(s), under exclusive license to Springer Nature Singapore Pte Ltd. 2023
J. C. Hung et al. (Eds.): IC 2023, LNEE 1045, pp. 66–73, 2023.
https://doi.org/10.1007/978-981-99-2287-1_10

2   Basic Concept of AHP Algorithm

In real life or work, people often encounter a problem with multiple decision-making objectives. For example, in financial investment, people have to make joint decisions on the level of risk and profit of investment projects [1–3]. This kind of problem is called multi-decision objective problem. Concept in multiple decision objective problems early, it was found that there is almost no good method to provide help to the person, then solve these problems mainly rely on experience and subjective idea, but it also caused the attention of the related fields, and then through the study put forward some special algorithms for such problem, the AHP algorithm which is one of the representative algorithms. The core idea of AHP algorithm is: Will much decision goal is regarded as a system, namely the decomposition problem of all the decision goal, each decision target as a hierarchy, and find the relationship between the level, again according to the relationship between combination, of which each level contains the different index system, can adopt the qualitative index fuzzy quantification method for its calculation, get any level of weight, the hierarchy is sorted according to the weight, so that the importance degree of each decision-making objective in the problem can be known [4]. The decision-making focuses on the target with the highest importance degree for consideration, and finally obtains the optimal result. Figure 1 is the topology of the AHP algorithm.

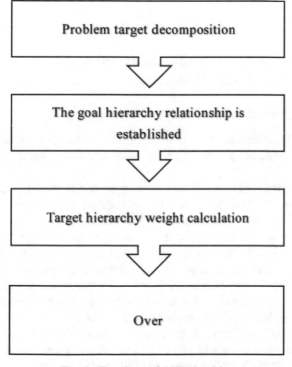

Fig. 1. Topology of AHP algorithm

It can be seen that the AHP algorithm is a kind of algorithm that builds the hierarchical model after the problem is decomposed, and then builds the judgment matrix of the model to solve the matrix eigenvector [5–7]. The results can show the correlation degree between different elements in different levels and an element in the previous level, so as to establish the relationship and distinguish the weight. These characteristics make AHP algorithm have a wide range of applications, basically any problem with hierarchical staggered evaluation index, but the target value is not qualitative index fuzzy quantitative description can be solved by AHP algorithm, this is due to the different functions of AHP algorithm, Table 1 is the functional characteristics of AHP algorithm in different problems.

Table 1. Functional characteristics of AHP algorithm in different problems

Type of problem	Functional features
Super multilevel problem	Build simplified models to improve efficiency
Some target values are not quantitatively described	Other target values of the system are described by fuzzy quantitative qualitative indicators to ensure the whole-body coordination of the target values
Staggered evaluation index relationship	Establish accurate index relation model and improve the transparency of results

3 Feasibility and Implementation Method of AHP Algorithm in Model Covariance Estimation

3.1 Feasibility Analysis

In order to verify the feasibility of AHP algorithm in covariance estimation of hedge fund distribution replication model, its advantages and disadvantages will be discussed below.

Advantages, AHP algorithm has very strong systemic, through analysis, comparison, judgment, comprehensive thinking ways of making decisions, the process does not give a combination of the relationship between, any combination of the factors will directly or reference effects on the final results, at the same time because of the fuzzy quantitative, so the final result has definite numerical value characteristic, It is very clear, so the AHP algorithm can give accurate and clear results in the covariance estimation of hedge fund distribution replication model. Compared with other algorithms, AHP algorithm is more concise in the covariance estimation of hedge fund distribution replication model, that is, AHP algorithm is not a pure mathematical method, and relevant theories tend to define it as an analytical method that combines qualitative and quantitative logics to calculate problems systematically. In short, is the thinking process of people thinking about the problem of mathematics, systematization, only in the final simple mathematical calculation, so that the form of the result is more accurate, so the AHP algorithm is more concise, use, operation difficulty is relatively small [8–10]. Finally, in the covariance estimation of hedge fund distribution replication model, other algorithms generally need a large amount of data support, but the AHP algorithm needs less data and generally only contains basic data information, such as the transaction amount of the fund, etc.

In terms of disadvantages, many modern fields will choose intelligent algorithms when solving problems, such algorithms can constantly give workers the choice of new solutions in decision-making, while AHP algorithm does not have this function, can only give the optimal results on the basis of existing data, so it may not be able to meet people's special needs. The AHP algorithm does not need too much data in the covariance estimation of the hedge fund distribution replication model, which makes the algorithm can give the answer quickly, but also leads to the lack of credibility of the answer given by the AHP algorithm. The AHP algorithm contains all the factors involved in the problem and the relationship between the combination of factors, so if faced with complex problems, there will be too many indicators, resulting in the weight is difficult to define the problem.

Combined with the advantages and disadvantages of AHP algorithm, first of all, the advantages of the algorithm also indicate the function of the algorithm, which indicates that it can be used in the decision of complex problems, and the covariance estimation of hedge fund distribution replication model is a typical complex problem, so the AHP algorithm has a good application value in the covariance estimation of hedge fund distribution replication model. Second hedge fund distribution copying model covariance estimation, will not affect mostly the drawback of AHP algorithm: first, the hedge fund distribution copying model covariance estimation, people do not need to get support of the new project, you just need to know details risk according to the covariance, therefore AHP algorithm cannot provide defect will not affect some of the new project; Secondly, AHP algorithm answer low credibility of defects can be eliminated by deduction method, namely AHP algorithm's answer itself although do not have too much credibility, but because the answer contains the whole development process, risk risk development process, so as long as the observation to determine whether a process is the same as the process description of the answer, if it is the same, it means that the results are still credible. Different representative processes have changed. Re-calculate and repeat this process to avoid the defect of insufficient credibility of the answer. Third, there are many improved AHP algorithms in modern times. The biggest difference between these algorithms and the classical AHP algorithm is that they can simplify the model. Therefore, the use of such algorithms can avoid the problem that the index is too complex and the weight is difficult to define. It can be seen that although the AHP algorithm has some disadvantages, they can be avoided by other methods, and the advantages represent that the algorithm is required by the covariance estimation of the model. Therefore, the AHP algorithm has good feasibility in the covariance estimation of the hedge fund distribution replication model.

3.2 Implementation Method

As for the implementation method of AHP algorithm in the covariance estimation of hedge fund distribution replication model, it is mainly divided into two steps, namely, the establishment of hedge fund distribution replication model and the estimation of AHP covariance. The specific contents are as follows.

3.2.1 Establishment of Hedge Fund Distribution Replication Model

In the establishment of hedge fund distribution replication model, a basic dynamic portfolio model should be established first, as shown in Formula (1).

$$\begin{cases} \pi_t = \sigma'(t)^{-1}\varphi_t \\ \varphi_t = \dfrac{\theta_t}{H_t}E\big[H_Tf'(L_T)\big] \end{cases} \tag{1}$$

where t is the investment time, H is a random vector, where the combination with t represents the state price density process, T is the change node of investment time, f'(L_T) is the end return, and the remaining elements are determined functions of t.

On the basis of formula (1), the establishment of hedge fund distribution replication model can be further divided into three steps, as shown in Fig. 2.

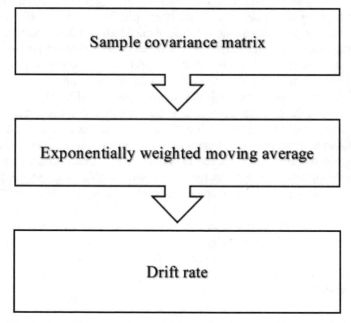

Fig. 2. Steps of establishing hedge fund distribution replication model

Built the sample covariance matrix, the definition of risk assets in a moment of the day logarithm yield, confirm the assets from one moment to any other moment, logarithm yield vector, and then by counting yield variance calculation, the results represent the risk, according to the known risk assets which belong to the risk assets, aimed at this kind of risk to establish covariance matrix, see formula (2).

$$\sum{}_t = \left(\sum{}_t^{ij}\right) n \times n \tag{2}$$

where i is the asset, j is the daily log rate of return, and n is the change vector of the daily log rate of return at two time points.

Second for index weighted moving average operation, first set up an assumed time series, and then take the index weighted moving average method to calculate the mean value of the sequence, the average according to the analysis of characteristics of time series data, if features preset standards, means you can continue, in contrast to change the time sequence, repeat the process.

Finally, the drift rate is confirmed, and the volatility of the hedge fund project can be calculated on the basis of formula (2). According to the daily logarithmic return series of the volatility risk asset and the assumed time series, the index weighted moving average method is used to calculate again, and the result is the drift rate of the project investment amount on the dynamic basis.

3.2.2 AHP Covariance Estimation

Combined with the hedge fund distribution replication model, covariance can be estimated in two directions: factor and contraction, but only factor direction is suitable for AHP algorithm, so this paper only considers the hedge fund distribution replication model in factor direction.

The hedge fund distribution replication model in the factor direction is characterized by a covariance definition structure, so the data dimension is less. The multi-level model can be decomposed into several single-factor models, and then the AHP algorithm is used to estimate the covariance. Applied to confirm the factor loading vector and the factor of intercept vector in this dimension, the residual vector, so that the full fa variance diagonal principle, after the completion of the combination of asset market index returns, assumes that the residuals are independent and normal distribution, can get covariance matrix, see formula (3), in the type into risk assets return, return time length can be estimated results.

$$\sum{}_t^{(1)} = \sigma_{market}^2 \beta\beta' + \Delta \tag{3}$$

where market is the return of asset market indicators.

3.3 Filter Establishment

Above plan basic function can let the AHP algorithm, its output hedge fund distribution replication model covariance estimation results, but must pay attention to the noise problem in actual applications, the large scale of financial data, and numerous sources, this leads to the financial data itself has no quality guarantee, which are likely to have a lot of data noise problem, For example, some complex data have irregular structure, which will have a certain impact on the result of AHP algorithm covariance estimation. Facing this problem, it is necessary to build a filter for denoising before data mining. In this paper, the mean filter is selected, and the denoising process of the filter is shown in Fig. 3.

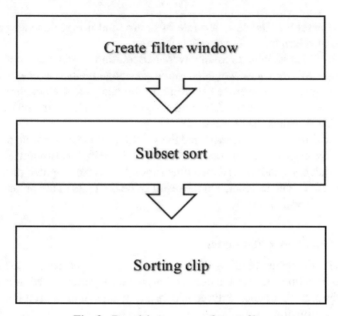

Fig. 3. Denoising process of mean filter

Combined with Fig. 3, the mean filter is first used for data cleaning, and several filtering Windows can be obtained in the process. Each window will have size differences according to the magnitude of internal data. Therefore, the subset sorting of filtering Windows is carried out according to the size differences to obtain the total sorting. According to the measurement results of the total ranking, the smallest and largest order is removed to complete the ranking pruning, and the noise input is also removed in synchronization, which can ensure the accuracy of the final results.

4   Conclusion

In conclusion, the characteristics of high profit, high risk of hedge funds that it is a "double-edged sword", but if we can reduce the incidence of risk, you can pass this kind of financial product brings more benefits, and the AHP algorithm could copy for hedge fund distribution model is used to estimate covariance, the results can help people to identify risks, convenient rule, this point is reached, it shows that AHP algorithm is a feasible method for covariance estimation and has good application value.

References

1. An, T.N., Le, T.N., Quyen, H.A., et al.: Application of AHP algorithm to coordinate multiple load shedding factors in the microgrid. IETE J. Res. 1–13 (2021)
2. Lin, Y., Lin, F., Huang, D., et al.: Voltage sag severity analysis based on improved FP-Growth algorithm and AHP algorithm. J. Phys. Conf. Ser. **1732**(1), 012088(8p.) (2021)
3. Zhang, N., Chai, R., Zheng, J.: Economic evaluation of rock oil project based on AHP algorithm. Arabian J. Geosci. **14**(9), 779 (2021)
4. Ibrahim, M.R., Suseno, J.E., Surarso, B.: Emergency service search using ant colony optimization algorithm and AHP-TOPSIS method. J. Phys. Conf. Ser. **1943**(1), 012104(7p.) (2021)
5. Yuan, Z., Wang, J., et al.: An MTRC-AHP compensation algorithm for Bi-ISAR imaging of space targets. IEEE Sens. J. **20**(5), 2356–2367 (2019)
6. Li, Y., Wu, L., Han, Q., et al.: Estimation of remote sensing based ecological index along the Grand Canal based on PCA-AHP-TOPSIS methodology. Ecol. Indic. **122**(2), 107214 (2021)
7. Meng, F., David, S.: Analysis of the optimal time to withdraw investments from hedge funds with alternative fee structures. IMA J. Manage. Math. **2021**(2), 2 (2021)
8. Vrjitoru, E.S., Boscoianu, M., Boscoianu, E.C.: Aspects regarding a new methodology for active portfolio management of hedge funds alternative in emerging markets-the case of Romanian capital market in the actual context of post-crisis recovery. In: International Conference Knowledge-Based Organization. Walter de Gruyter GmbH (2021)
9. Zhang, Y.J., Wu, Y.B.: The time-varying spillover effect between WTI crude oil futures returns and hedge funds. Int. Rev. Econ. Financ. **61**(MAY), 156–169 (2019)
10. Li, Y., Holland, A.S., Kazemi H.B.: Duration of poor performance and risk shifting by hedge fund managers. Global Financ. Journal,2019,40:35–47

Cross Border B2B E-Commerce Website Service Quality Evaluation System

Guo Mei[✉]

Hubei Communication Technical College, Hubei 430079, China
tingym2022@163.com

Abstract. With the popularity of the Internet, it plays an increasingly important role in people's lives. More and more people use the Internet to conduct online shopping, sales and other activities. Therefore, B2B e-commerce enterprises have shown a trend of vigorous development in China, and become one of the important fields of Internet information services in China. It has a considerable scale of transactions and users, and it continues to develop at a high speed. This paper studies the service quality evaluation system of cross-border B2B e-commerce websites, which can be used to evaluate the service quality of cross-border B2B e-commerce websites. The evaluation is based on a set of parameters, including user experience, product information and shopping experience. The results will be displayed in a table with corresponding scores for each parameter. Cross border B2B e-commerce websites are more and more popular with users because of their convenience and cost-effectiveness.

Keywords: Electronic Commerce · B2B · Website services · Quality evaluation

1 Introduction

In 2015, the scale of China's cross-border e-commerce export transactions accounted for 83.2% of the total scale of cross-border e-commerce imports and exports, and imports only accounted for 16.8%. From the perspective of the transaction structure of cross-border e-commerce, China's cross-border e-commerce has always been export-oriented. In order to better develop cross-border e-commerce, the state has issued many favorable policies in recent years. At the end of 2012, the state designated Hangzhou, Zhengzhou, Ningbo, Chongqing and Shanghai as the first batch of cross-border e-commerce pilot cities in China. From 2013 to 2014, 9 cities including Shenzhen, Suzhou and Qingdao were successively selected as the second batch of pilot cities for cross-border e-commerce in China [1]. In 2016, the State Council set up a new batch of cross-border e-commerce comprehensive pilot zones in 12 cities including Tianjin, Shanghai and Chongqing. In addition to the policies of pilot and pilot areas, China has also issued policies on cross-border e-commerce related payment, supervision and promotion of its healthy and rapid development.

At present, the development of cross-border e-commerce in China has ushered in a golden era. In addition to the export-oriented feature, cross-border e-commerce also

© The Author(s), under exclusive license to Springer Nature Singapore Pte Ltd. 2023
J. C. Hung et al. (Eds.): IC 2023, LNEE 1045, pp. 74–80, 2023.
https://doi.org/10.1007/978-981-99-2287-1_11

has another feature of taking B2B mode as the main body. According to the data of the national e-commerce research center, the total amount of cross-border c-commerce transactions in China in 2015 was 5.4 trillion yuan, of which the cross-border e-commerce B2B mode accounted for 88.5%, which has an absolute advantage. Compared with the cross-border e-commerce B2B mode, the development of the cross-border e-commerce B2C mode is a game of interests between the two countries. B2B mode is the focus of cross-border e-commerce development, because it meets the needs of stable growth of China's foreign trade, meets the needs of China's structural adjustment, and is also conducive to reducing the cost of China's supervision and improving the efficiency of goods clearance [2]. As shown in Fig. 1, the B2B mode of cross-border e-commerce plays an important role in the transformation of China's foreign trade.

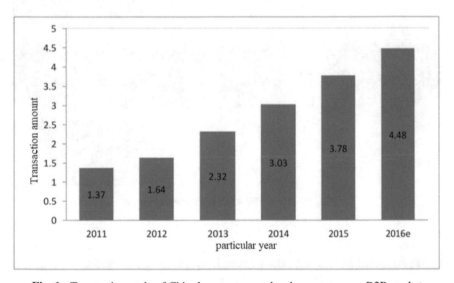

Fig. 1. Transaction scale of China's export cross-border e-commerce B2B market

2 Related Work

2.1 Meaning of B2B E-Commerce Enterprises

E-Commerce and enterprises are inseparable, but B2B e-commerce enterprises are different from e-commerce. Enterprises that have business operation behaviors among themselves are B2B e-commerce enterprises. The focus here is on the combination of network application and enterprises. As long as these two points are met, they can be included in the category of B2B e-commerce enterprises. Therefore, it includes the following two categories. First, the pure network enterprises that operate basically based on the network can not only provide services and products unique to B2B e-commerce, but also provide traditional services and products. Second, the network combines and complements the existing enterprise management behavior. Such enterprises usually conduct e-commerce activities on the premise of their original business activities [3]. The basic purpose of launching e-commerce is to improve and promote the original business behavior. Therefore, such e-commerce enterprises include the improvement of enterprise management and organizational functions through the network, so that the enterprises can operate more effectively and promote the indirect improvement of enterprise benefits [4]; Enterprises that provide services and products through the network, so as to improve the economic benefits of enterprises, as shown in Fig. 2.

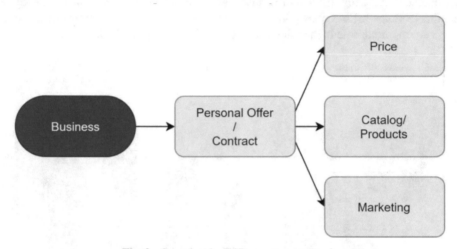

Fig. 2. Cross border B2B e-commerce mode

2.2   Development Status of Cross-Border E-Commerce B2B Mode

According to the data of the e-commerce research center, the cross-border e-commerce transaction volume of China in 2015 was 5.4 trillion yuan, an increase of 28.6% over 2014. Among them, the turnover of cross-border e-commerce B2B mode accounts for 88.5%, which has an absolute advantage. From the perspective of import and export structure, the scale of China's cross-border e-commerce B2B export transactions in 2015 was 3.78 trillion yuan, accounting for 79% of the scale of cross-border e-commerce B2B transactions. It can be seen that the B2B mode of cross-border e-commerce in China is mainly export. From 2008 to 2015, the growth of China's foreign trade slowed down, and even showed a negative growth [5]. At this time, the scale of cross-border e-commerce transactions continued to grow at a high speed. The development of cross-border e-commerce provided a way for small, medium and micro enterprises to "go global" and is expected to become a new growth point of China's foreign trade [6].

The geographical location of cross-border e-commerce B2B platform sellers is mainly concentrated in the economically developed coastal areas. With their superior geographical location and strong economic strength, they account for more than 70% of China's cross-border e-commerce export volume. The transaction volume in the central region grew rapidly. In 2015, the annual growth rate of cross-border e-commerce was

78.47%, making it the fastest growing region of cross-border e-commerce in China. The penetration rate of cross-border e commerce in the western region is low, and there is a large development space.

Developed countries such as Europe and America are mature trading partners of China's cross-border e-commerce B2B model. In addition, emerging market countries such as Brazil and Russia have begun to realize the benefits brought by cross-border e-commerce [7]. The governments of these countries have begun to make adjustments and improvements in customs, logistics and taxation.

In 2015, the most popular products in the European and American markets were mainly sports and outdoor products, while the ASEAN region mainly favored 3C, health and beauty products and clothing. In 2015, among the products exported by China's cross-border e-commerce, the products with the fastest demand growth were outdoor sports products, with an annual growth rate of 113%, ranking the first in the growth rate. China's high-quality, low-cost wedding dress products well show the brand image of our products [8].

3 Challenges Faced by Cross-Border E-Commerce B2B Mode

(1) Trust issues

The biggest challenge facing cross-border e-commerce is the trust between the buyer and the seller. The B2B mode of cross-border e-commerce has not completed the closed-loop transaction, and the final analysis lies in the trust issue. The buyer did not trust to send the payment to the seller when it did not receive the goods, and the seller was also worried that it would not receive the payment after the goods were transported to the buyer's location. Therefore, the cross-border e-commerce B2B platform has not completed the online transaction closed loop. Some people say that it is impossible to directly transfer the payment to the third-party payment platform like Taobao, and then transfer the payment to the seller's account after the user and the buyer receive the goods? Think about it carefully. International trade is not like domestic trade [9]. It has a large amount of capital, needs to settle foreign exchange, takes a long time to logistics and complicated transaction procedures. If the funds are transferred to the third-party payment platform and then transferred to the seller's account after the buyer receives the goods, it will delay a long time and affect the seller's working capital. Therefore, how to solve the trust problem is the key to realize the rapid development of cross-border e-commerce B2B mode.

(2) Payment issue

China's cross-border third-party payment is not mature enough. The country has not yet issued effective laws and regulations and lacks effective supervision. It takes a long time to review the information of cross-border payment, which indirectly increases the risk of payment. In the process of transaction data transmission, payment information may be lost due to information failure or system crash. There are also some behaviors of illegally stealing payment accounts and information through the Internet. Therefore, China's cross-border payment faces greater risks.

(3) Logistics problems

The construction of China's cross-border logistics system is not very reasonable, the infrastructure is relatively imperfect, and the degree of informatization is not high. Most of them need to be completed manually. The logistics form adopted by the B2 big B mode of cross-border e-commerce is similar to that of traditional foreign trade. However, the current foreign trade shows a trend of small orders. The goods of one enterprise may not be enough for one standard container. If the goods are delivered by the enterprise, the cost will be high. In the information age, the traceability of cross-border e-commerce B2B logistics is poor. How logistics should adapt to the current status of small and multi batch orders and how to track them from time to time is a challenge for cross-border e-commerce B2 big B mode. In addition, for B2 small B, the transportation mode selected is generally international small package, express, overseas warehouse, etc. International small package transportation takes a long time, the cost of international express delivery is too high, and the construction of overseas warehouses requires a lot of money. How to improve the logistics of China's cross-border e-commerce B2B mode is a problem that needs to be solved at present.

4 Cross Border B2B E-Commerce Website Service Quality Evaluation System

4.1 Establishment Principles of Indicator System

(1) Scientific principle

The correct theory needs to effectively guide practice and reflect the actual situation, and must be based on science. The scientific principle emphasizes the effective combination of theory and practice, and uses scientific research methods and tools to study specific problems. The establishment of a reasonable and effective indicator system needs to be guided by a scientific theoretical basis. At the same time, we should grasp the essence of the evaluation according to the actual characteristics of the evaluation object and carry out the indicator construction on the basis of theory and practice. The establishment of the service quality evaluation index system of e-commerce websites needs to be based on the service quality theory of the past scholars and combined with the characteristics of e-commerce websites [10].

(2) Comprehensive principle

The principle of comprehensiveness is to consider all aspects of things comprehensively and comprehensively from all angles of the research object. The service quality evaluation index system of e-commerce websites needs to comprehensively reflect the service quality of e-commerce websites. It needs to consider not only the dimensions that affect the service quality, but also the integrity of the index system and the correlation between various indicators. In the process of establishment, it is necessary to start from the whole process of obtaining business services, and comprehensively consider the three stages before, during and after obtaining services.

(3) Hierarchy principle

The hierarchy principle can ensure the connection and mutual relationship of all levels of the indicator system, and ensure the perfection and scientificity of the evaluation results. The establishment of the evaluation index system of service quality of e-commerce websites should be divided into different levels. The evaluation index system of service quality should be designed in different levels to ensure that each evaluation index is accurately and clearly defined, and that each evaluation index included in the same level does not overlap and is independent of each other.

4.2 Construction of Evaluation Index System

To establish the service quality evaluation model of tourism e-commerce website, it is necessary to determine the main impact on the service quality of tourism e-commerce website, analyze the main dimensions, and determine the content of its sub dimensions. In the process of establishment, we need to analyze the relevant theories based on service quality and the characteristics of tourism e-commerce websites.

(1) Primary dimension determination

Based on the summary of previous model studies such as sevqual, e-sevqual and self-service technology, and combined with the characteristics of tourism e-commerce website services, this paper proposes seven dimensions that affect its service quality: reliability, responsiveness, ease of use, security, empathy, trust and information quality. These seven dimensions borrow the service quality evaluation dimensions in SERVQUAL, E-S-QUAL, etc., but their meanings have also changed in the context of tourism e-commerce.

(2) Determination of evaluation index system

According to the analysis results of the previous section, the evaluation index system of tourism website service quality is established. KK_ Reliability, XY_ Responsiveness, YY_ Ease of use, AQ_ Safety, YQ_ Empathy and xxzl_ Information quality, XR_ Trust is the main dimension affecting the service quality of tourism e-commerce websites, and 21 indicators are the corresponding sub dimensions.

5 Conclusion

China's cross-border e-commerce B2B model is still in the early stage of development, and there are problems in credit, payment, logistics, customs clearance, foreign exchange settlement and sales, tax refund, financing, talent security and other aspects. Through the analysis of Alibaba international station, we understand that the "credit guarantee system" launched by Alibaba can solve the credit, payment and financing problems to a certain extent, but not many small and medium-sized enterprises participate in it. At the same time, the "one access" one-stop comprehensive foreign trade service launched by Alibaba can help small and medium-sized enterprises to complete one-stop customs clearance services and solve the problems of logistics, customs clearance, foreign exchange settlement and sales and tax refund. However, it is still in the promotion period. At present, under the pressure of foreign trade transformation, government policy support, big data assistance and the promotion of the "the Belt and Road" strategy, efforts should be made to promote the development of cross-border e-commerce B2B mode from the three aspects of enterprises, government and society.

References

1. Chen, L., Pan, T., Zhao, B.: Design of Evaluation System for Competitive Cooperation Capabilities of Cross-Border E-Commerce Comprehensive Pilot Area (2021). https://doi.org/10.1007/978-981-33-6141-6_31
2. Liu, C., Liu, R.: Application of BP neural network in cross-border e-commerce web pages quality evaluation. J. Phys. Conf. Ser. **1774**(1), 012015 (2021)
3. Sorwar, G., Hoque, R., Islam, S.: Assessing Cross-Border E-Commerce Success: A Cross-Country Analysis (2021)
4. Sun, P., Gu, L.: Optimization of cross-border e-commerce logistics supervision system based on internet of things technology. Complexity (2021)
5. Zhang, X., Liu, S.: Action mechanism and model of cross-border e-commerce green supply chain based on customer behavior. Math. Probl. Eng. **2021**(3), 1–11 (2021)
6. Zhang, H., Wang, L.: The service quality evaluation of agricultural e-commerce based on interval-valued intuitionistic fuzzy GRA method. J. Math. (2022)
7. Daniati, T., Roostika, R.R.: The effect of website design quality, e-service quality, and brand image on e-satisfaction and eloyalty of e-commerce customers. Int. J. Res. Sci. Innov. (10) (2021)
8. Luo, Y., Gui, X.: Logistics Efficiency Evaluation Model of Small and Medium Sized Cross Border E-Commerce Enterprises (2022)
9. Zhang, F., Yang, Y.: Trust model simulation of cross border e-commerce based on machine learning and Bayesian network. J. Ambient Intell. Humaniz. Comput. (11) (2021)
10. Luo, Y., Bai, Y.: Evaluation Model Construction and Empirical Analysis of Rural E-Commerce Logistics Service Quality (2021)

Cultural Heritage Network Courses in the Information Environment

Fang Lu[(✉)]

Xi'an Academy of Fine Arts, Xi'an 710065, Shaanxi, China
lufangxafa@163.com

Abstract. With the continuous penetration and integration of the Internet in the field of education, the availability of educational resources has been continuously improved, which has provided strong support for strengthening the construction of educational system resources. In the stage of rapid development of Internet technology, the coverage of wireless networks has continued to increase, and a large number of new Internet products have been born, which has provided a good boost for the networked development of educational products, and thus derived the Internet + education network online learning model. Cultural heritage is an important accumulation and basic source of contemporary cultural development and innovation in our country. In order to inherit cultural heritage and develop cultural undertakings, this paper has launched a cultural heritage online course learning system based on a Web platform. Any group who wants to understand national culture can enter platform for learning. With the support of information technology such as Web server and the Internet, the online courses of cultural heritage are developed, which can track students' learning records, and teachers can send students' learning information through text messages. Through the evaluation of the cultural heritage online curriculum design, some deficiencies in the curriculum design can be improved, and the curriculum design content such as cultural heritage teaching resources and teaching objectives can be more in line with the actual teaching.

Keywords: Information Technology · Cultural Heritage · Online Course Design · Web Platform

1 Introduction

The development of the Internet and information technology has changed the way people learn. The rapid economic development has also accelerated the pace of cultural progress, but due to the increasing abundance of modern culture and the continuous strengthening of its influence, it has caused a huge impact on the spread of national culture. Therefore, it is necessary to use modern technology to build a cultural heritage protection mechanism and scientifically manage national culture, so that cultural heritage can also be effectively guaranteed in the information environment.

So far, many scholars have carried out research on the design and development of online courses for cultural heritage in the information environment, and have achieved

J. C. Hung et al. (Eds.): IC 2023, LNEE 1045, pp. 81–88, 2023.
https://doi.org/10.1007/978-981-99-2287-1_12

extraordinary results. For example, Blackboard, WebCT, Moodle and Sky Classroom are the most widely used environments for online teaching. Some scholars use the Moodle platform to develop English online courses, and students can learn interactively through this platform [1]. Taking quasi-physical education in this province as an example, a scholar explores the inheritance status and problems of quasi-sports culture in physical education in primary and secondary schools, and believes that the setting of educational goals, changes in teaching methods, compilation of textbooks, development of school-based courses, and strengthening of cooperation inside and outside the school should be carried out. In order to promote the inheritance effect of sports cultural heritage in school physical education [2]. Some researchers have integrated the costume features and cultural features of ethnic minorities, and digitized them with the help of computer technology, and then promoted the ethnic culture through the platform, popularized the ethnic cultural characteristics to the general public, and made the heritage culture into the field of vision of more people [3]. Some scholars take advantage of UML to design a network platform for the digital inheritance and protection of minority culture through object-oriented design methods in software engineering [4]. Although many scholars have used a lot of modern information technology in the development of cultural heritage online courses, the cultural heritage online courses still need to strengthen the curriculum design form and collect more cultural heritage resources in order to attract students to learn national culture.

This paper first introduces the concept of online courses, and then analyzes the design process of cultural heritage online courses. Then build a cultural heritage network course system based on Web platform, and support students and teachers to learn and teach network courses through Web server. Finally, the online course design of cultural heritage in the platform is evaluated through the evaluation of teachers and students, and the content of course design is optimized.

2   Cultural Heritage Online Course

2.1   Online Courses

Online courses are a new type of courses produced with the development of distance education. The main features are network, interactive, and synchronous learning and asynchronous learning. Therefore, higher requirements are placed on the autonomy of learners. Without the guidance of traditional classroom teachers, learners can set their own learning pace according to their own learning styles, participate in various learning activities according to specific learning goals, and finally complete learning tasks [5, 6].

2.2   Cultural Heritage Online Course Design Process

2.2.1   Identify Teaching Objects and Goals

In the process of designing the teaching process of online courses, it is necessary to first determine the teaching objects and teaching objectives, formulate a cultural communication plan according to the characteristics of the national culture, and use a variety of communication methods to convey cultural ideas to users. Teaching can also gradually arouse users' interest in national culture, so as to obtain better teaching effect [7]. The cultural heritage curriculum platform based on the network information environment should carry forward, inherit and develop national culture. Therefore, most of the courses are designed to attract users' attention through ethnic cultural pictures, videos, etc., and interactive programs are set up for flexible teaching. The development of this cultural heritage course has a wide range of users, so the course arrangement of the entire cultural resources online learning will not be too difficult and complicated. The goal of this course is to display cultural heritage information, so that ethnic culture lovers can learn Heritage culture so as to increase the spread and inheritance of the entire heritage culture [8, 9].

2.2.2 Clarify the Teaching Mode and Highlight the Teaching Characteristics

In the process of online course teaching, it is necessary to establish a unique teaching style such as cultural learning methods, provide personalized courses according to the needs of different users, and understand a variety of ethnic cultures, so that users can learn in a relaxed environment and achieve teaching purposes. Online course teaching must have multiple functions. In addition to basic text display, it is best to set some animation, sound effects, small video and other functions through information technology [10].

2.2.3 Teaching System Structure and Module Design

In the process of carrying out online course teaching, special teaching modules should be established according to the learning styles of different students. The contents of the modules are inherited and echo each other. In this way, a learning system is created, which provides students with a variety of learning experiences and improves their comprehensive learning ability [11]. Therefore, the design of the cultural heritage online course can determine multiple teaching modules, and the modules are related to each other, so as to master and learn the teaching content of the online course one by one.

2.2.4 Integrate Material Information

Create a material web page, on which students can search for the course resources they need to find based on keywords. For example, according to the unique characteristics of a certain ethnic group's cultural costumes, they can search for the text introduction, picture display, and historical sources of this type of costume. Wait. A jump function can also be set on the webpage, that is, students can jump to the corresponding material by clicking on the relevant picture or video, which is convenient for students to find relevant cultural heritage information and knowledge, realizes the effective integration of teaching resources, and effectively solves the problem of collecting materials. In the process of data, due to no purpose and no strategy, the quality of the material is affected [12].

2.3 Scoring Feature Extraction

This article evaluates the design of the cultural heritage online course and adopts a scoring system. The scoring method is that each user performs scoring feature extraction on the satisfaction of the online course design according to their own scoring standards.

$$M = \frac{\sum_{k=1}^{n} ru(k)}{n} \tag{1}$$

$$P = \{p_1, p_2, ..., p_n\} \tag{2}$$

Among them, M is a new feature, P represents the set of items evaluated by user u, n is the total number of items evaluated by the user, ru(k) represents the user's rating on item k, and $k \in P$.

3 Design of Cultural Heritage Online Course System Based on Web Platform

3.1 Online Course Development Platform - Web Platform

The Web server includes various components, and its main task is to create a friendly human-machine interface, so that users can enter the platform and form a good interactive mode with the platform. Web server plays a very important role in system design, and its popularity is gradually expanding. It is an important consideration in application design. Web frameworks have their own particularities, so this is a quick and easy framework to use when designing web courses.

3.2 Design of Cultural Heritage Resource Library

In order to better digitally protect and disseminate national cultural heritage in the information age, this paper establishes a cultural heritage resource database. The cultural resources included in the database mainly include national costumes, customs, songs and dances, transportation, food, housing and other cultures and resources. The presentation forms include pictures, text, video, and audio.

3.3 Support Services Architecture

3.3.1 Student Support Services Module

Due to the rich resources of this online course, in addition to the text chapters, there are many extracurricular tutoring resources. In order to enable students to enter the learning state after logging in to the online course, and not spend time on aimless web browsing because they can't find the learning focus, the course learning guidance is provided to the students. The prompt of the course guide is "according to the system setting time, the current learning scope is: which chapters, specific chapter titles, please pay attention to arrange your learning progress reasonably", which appears on the homepage of the course. Part of the content of the guidance prompt, "which chapters, the specific chapter titles" changes automatically with time, and the change is based on the course schedule, which is arranged by the teacher at the beginning of each semester. The specific initialization is done by the system administrator.

3.3.2 Teacher Support Module

Due to the different time and asynchronous characteristics of online teaching, it is impossible to meet the needs of students by displaying important notices or teaching guidance only through web pages. Mobile learning just provides us with a new way of thinking. Timely and effective push to students, in order to provide good support for students' online learning. Considering the large number of students, it is impossible for teachers to send text messages through mobile phones one by one. In order to solve this kind of problem, this support service has designed a mode of sending text messages through the course platform, that is, in the online course, use the computer keyboard to input the content of notifications or teaching guidance, and use the Select the object to receive the message with the mouse, and click the Send button to complete the message sending.

4 Learning Research and Experimentation

4.1 Cultural Heritage Online Course Resources

Data mining has always been a research hotspot of scholars. Data mining technology is to mine hidden information from massive information. Data mining can be applied to decision support systems to provide support for user decision-making. In this paper, data mining technology is used for cultural heritage resource mining. Fuzzy theory is a common form of data mining. The Gaussian triangular membership function in the membership function of fuzzy theory is as follows:

$$u(x) \begin{cases} 1 - \frac{|x-w|}{\varepsilon}, & |x - w| \leq \varepsilon \\ 0, & |x - w| > \varepsilon \end{cases} \tag{3}$$

Among them, w and ε are the center and width of the fuzzy set, respectively, and u(x) represents the membership function.

$$v(x) = \exp\left(-\frac{(x - y)^2}{l^2}\right) \tag{4}$$

where y and l represent the center and width of the fuzzy set, and v(x) represents the Gaussian triangular membership function.

The cultural heritage resource library plays a central role in resource sharing, providing users with various ethnic cultural resources for users to use and learn. The cultural heritage resource library includes three contents, as shown in Fig. 1.

Cultural Browsing and Keyword Retrieval: As a portal website, the main function of Cultural Resource Bank is to provide search and retrieval pages. Users can find a lot of cultural heritage resource information by entering cultural heritage keywords on the course platform. This operation is also very simple. It is affected by the situation that we usually use the search bar. There is no technical level, and it is supported by information technology and network. The search efficiency is very high.

Thematic display: Different cultural learning topics can be made for very special cultural elements, cultural origins or story backgrounds in national culture to guide

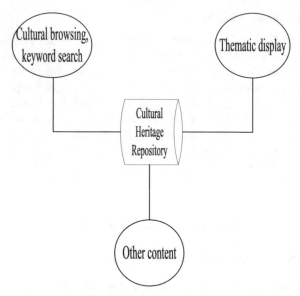

Fig. 1. The use of information technology for teachers

learners to learn better. For example, as the most characteristic symbol of each ethnic group, clothing is displayed on the portal website in the form of clothing culture topics.

Other content: Considering the validity and sufficiency of the data in the repository, in addition to providing portal services, it is also necessary to provide system introduction, system dynamics and other related systems.

4.2   Cultural Heritage Online Course Assessment

The last link in the design of the teaching mode of the online course of cultural heritage based on the Web platform is evaluation. Through assessment, feedback can be used to modify specific teaching links.

Take teacher self-assessment as an example. Teachers' self-evaluation method evaluates the "excellent" standard content in five aspects: curriculum design, curriculum resources, curriculum interaction, curriculum interface and curriculum evaluation. Teachers will judge and score according to the content of their own curriculum design. Excellent is 5 points, good is 3–4 points, and medium is 1–2 points. In the process of self-assessment, teachers can discover the areas that need improvement in the above five areas, so as to revise and improve the curriculum as soon as possible. A questionnaire was given to 30 teachers who participated in the design of the cultural heritage online course for course self-evaluation. The evaluation results are shown in Table 1. From the results in the figure, it is not difficult to see that the design of course interaction and course interface needs to be strengthened.

Take student evaluations as an example. A questionnaire was sent to 30 online learners who participated in cultural heritage learning, and they gave feedback on the course design stage, course development stage and course implementation stage respectively.

Table 1. Teacher self-assessment results

	5	3–4	1–2
Course Design	12	14	4
Course Resources	24	3	3
Course interaction	6	19	5
Course interface	5	21	4
Course evaluation	10	17	3

If the proportion of excellent is more than 70%, it can be defaulted as no modification, otherwise, the modification of this part needs to be considered. The evaluation results of the course design stage are shown in Fig. 2.

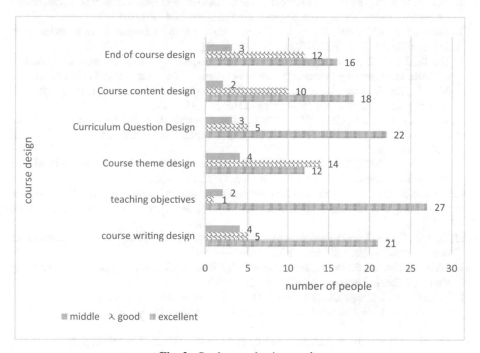

Fig. 2. Student evaluation results

Analyzing the results in Fig. 2, it is not difficult to find that the student users are not satisfied with the theme design of the cultural heritage course, but other aspects are acceptable evaluations. Therefore, according to the feedback information, the designer needs to revise the cultural heritage theme design part.

5 Conclusion

This paper uses the Web development platform to design the cultural heritage online course, and hopes to contribute to the modern inheritance, protection and innovative development of cultural heritage by promoting the cultural heritage online course to the majority of users. In the process of designing cultural heritage online courses, we investigated the evaluation results of students and teachers on curriculum design, and found that when designing cultural heritage online courses, the design of course interaction, course interface and course theme still needs to be optimized.

References

1. Brosius, A., Elsas, E.V., Vreese, C.D.: Trust in the European Union: effects of the information environment. Eur. J. Commun. **34**(1), 57–73 (2019)
2. Bekele, M.K., Pierdicca, R., Frontoni, E., et al.: A survey of augmented, virtual, and mixed reality for cultural heritage. J. Comput. Cult. Herit. **11**(2), 1–36 (2018)
3. Faulds, D.J., Raju, P.S., Rosen, E.: The new information environment: an interview with Emanuel Rosen. Bus. Horiz. **62**(3), 337–342 (2019)
4. Haislip, J.Z., Richardson, V.J.: The effect of CEO IT expertise on the information environment: evidence from earnings forecasts and announcements. J. Inf. Syst. **32**(2), 71–94 (2018)
5. May, L., Mei, B., Heap, R., et al.: Rapid development studio: an intensive, iterative approach to designing online learning. SAGE Open **11**(3), 77–94 (2021)
6. Chin, K.Y., Lee, K.F., Chen, Y.L.: Using an interactive ubiquitous learning system to enhance authentic learning experiences in a cultural heritage course. Interact. Learn. Environ. **26**(1–4), 444–459 (2018)
7. Nafziger, J.: The UNESCO convention on the protection of the underwater cultural heritage: its growing influence. J. Marit. Law Commer. **49**(3), 371–400 (2018)
8. Alsuwaida, N.: Online courses in art and design during the coronavirus (COVID-19) pandemic: teaching reflections from a first-time online instructor. SAGE Open **12**(1), 59–66 (2022)
9. Shin, S., Cheon, J.: Assuring student satisfaction of online education: a search for core course design elements. Int. J. E Learn. **18**(2), 147–164 (2019)
10. Gaston, T., Lynch, S.: Does using a course design framework better engage our online nursing students? Teach. Learn. Nurs. **14**(1), 69–71 (2019)
11. Monteiro, E.P., Gomide, H.P., Remor, E.: Massive open online course for Brazilian healthcare providers working with substance use disorders: curriculum design. BMC Med. Educ. **20**(1), 1–10 (2020)
12. Castel, A., Gutfreund, P., Cabane, B., et al.: Stability of fluid ultrathin polymer films in contact with solvent-loaded gels for cultural heritage. Langmuir **36**(42), 12607–12619 (2020)

Data Analysis Method of English Education Based on Improved Deep Learning Algorithm

Jie Wu[✉]

English Department, School of Foreign Languages, Dalian Polytechnic University,
Dalian 116033, Liaoning, China
wujie@dlpu.edu.cn

Abstract. In the reform of college foreign language education, it is often one of the focuses of discussion whether the college curriculum can make students fully prepared to cope with the challenges of learning, future life and career. This is mainly because some students still have low learning efficiency at the stage of higher education, rely on mechanical memory and recitation, and lack high-level thinking ability, especially the ability to solve practical problems. Deep learning algorithm is the latest research trend of English education data analysis. It has been widely used in computer vision, speech recognition and text processing. Deep learning algorithm is an artificial intelligence (AI) technology, which uses neural networks to achieve advanced cognition. It can be divided into two types: 1. Deep convolution network 2. Deep recursive network in this paper, we will mainly focus on using deep convolution network to analyze English data, and provide a method to improve the accuracy of its results by using its improved deep learning algorithm.

Keywords: English education · Deep learning · Data analysis

1 Introduction

Since the Ministry of Education promulgated the College English curriculum teaching requirements in 2002, the reform of College English education has aroused heated discussions. Scholars have conducted extensive discussions on the increasingly prominent problems in teaching objectives and teaching methods. The low teaching efficiency has gradually become a dilemma faced by College English education, which is mainly caused by the traditional teaching mode of Teacher centered and simple teaching of language knowledge. Under its influence, students often adopt traditional learning methods such as memorization and recitation and exercises to learn English, which is not conducive to the internalization and transfer of language knowledge [1]. Therefore, although the traditional learning methods have made some achievements, it is difficult to achieve more efficient English learning today. In order to solve this problem, researchers began to pay attention to the role of "deep learning" in improving learning efficiency, and regarded it as one of the indicators to measure students' meaningful learning. Deep learning originates from Marton and saljo's research on the learning process and learning strategies

© The Author(s), under exclusive license to Springer Nature Singapore Pte Ltd. 2023
J. C. Hung et al. (Eds.): IC 2023, LNEE 1045, pp. 89–95, 2023.
https://doi.org/10.1007/978-981-99-2287-1_13

of college students in 1976. It is proposed that when using shallow learning strategies, students should pay attention to the retelling of original materials and the solution of surface problems, while when using deep learning methods, they should pay more attention to the integration and rational solution of themes and viewpoints [2].

Domestic research on deep learning started relatively late. The earliest discussion on deep learning appeared in the article "promoting students' deep learning" published by Li Jiahou in 2005. From 2005 to 2014, deep learning gradually became a hot research field [3]. The research focuses mainly on education and teaching application, computer field, strategy research and technical support, Among them, the research on deep learning methods is the main focus of researchers, namely, strategy research, teaching mode, environmental design and model design (fan Yaqin, Wang Binghao, Wang Wei, & Tang yewei, 2015). For the front-line educators, it is their primary concern to design reasonable teaching programs and adopt appropriate teaching methods to promote students' deep learning, which is particularly important to localize the deep learning theory and improve teaching practice [4].

2 Related Work

2.1 Deep Learning

The research on students' learning methods began with Marton and saljo (1976), who explored the two methods of students' reading academic articles and defined them as deep learning and shallow learning. The former tries to find meaning in the problems studied and critically links them with other experiences and thoughts; The latter relies on rote learning and treats the learning content separately. Biggs initially divided learning methods into three categories: shallow learning, deep learning and achievement learning. Shallow learning is based on the external motivation of students' learning. The main goal is to complete the learning tasks as effortlessly as possible. Therefore, the characteristics of learning are to concentrate on important topics and accurately reproduce them; Deep learning is based on internal motivation and curiosity. Students connect the content with meaningful situations or prior knowledge, and then conduct theoretical derivation to obtain expanded knowledge [5]; Achievement learning is affected by competition and self-achievement, and the main goal is to obtain high scores. Although deep learning and shallow learning are mutually exclusive concepts, achievement learning can be related to them and only depends on the learning environment; For this reason, in the follow-up studies, researchers often focus on both deep learning and shallow learning, while omitting achievement learning. When exploring the characteristics of shallow learning and deep learning, the researchers found that deep learning reflects the learners' investment in understanding the content of materials, which is reflected in the use of a variety of strategies, such as extensive reading, integration of resources, peer discussion, and practical application of knowledge. While shallow learning focuses on the characteristics or signs of information [6]. The goal of learning is to avoid failure, rather than mastering key concepts and understanding their relationship with other information and the application of this information in other situations. In short, when students use deep learning methods, they are often internally driven and try to understand the learning content.

Correspondingly, when they use shallow learning methods, they are usually character-ized by rote learning and aiming at passing the examination. Both deep learning and shallow learning can be regarded as the combination of students' learning motivation and accompanying learning activities.

2.2 Deep Learning Model

According to the "3P" teaching and learning model, the mechanism of learning mecha-nism can be divided into three stages: presage, process and product. The learning process and results can be predicted by students' personal factors (prior knowledge and preferred learning methods) and teaching environment (teaching objectives, evaluation methods, environment, teaching and institutional procedures, etc.); The specific learning methods used in the learning activities also affect the learning results and other factors before learning [7]; The learning results include quantitative results based on facts and skills and qualitative results based on knowledge structure and transfer, which reflect the con-text based learning methods and also interact with the first two processes, as shown in Fig. 1.

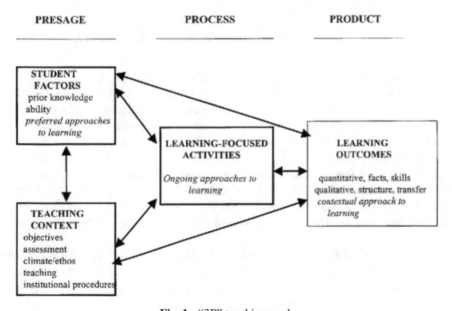

Fig. 1. "3P" teaching mode

3 Improved Deep Learning English Education Data Analysis Scheme

In the process of medical education, a large amount of teaching data will be generated. In order to optimize the education program, it is necessary to analyze the teaching data to improve the educational effect of the teaching program. In this study, the improved deep learning algorithm is used for processing, the artificial intelligence algorithm is used to optimize the education information data, and the education status is analyzed through the data characteristics of different levels. The purpose analysis scheme of English education and training is shown in Fig. 2.

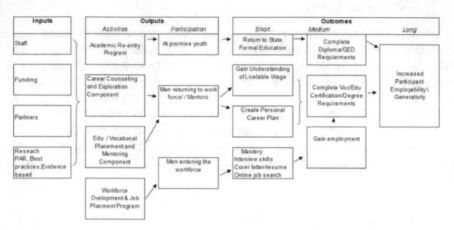

Fig. 2. Analysis on the purpose of English education and training

Combined with Fig. 1, this study analyzes the development mode of social enterprises, the life style required by people and the talent demand of higher vocational education, so as to improve the talent education and training ability and improve the talent output efficiency and quality. Optimize the education and training objectives according to the government's policies on talent cultivation, design the training programs according to the types, levels and functional requirements of skills needs, adopt different knowledge structures, ability structures and relevant moral character optimization training programs, and put forward different requirements at the national level, regional level and institutional level to achieve multi-level and multi-faceted training programs for talents, And then complete the teaching plan of the enterprise's demand for various talents [8].

In order to realize the multi-level training program for medical education, it is necessary to analyze the education data. In this study, we use classroom teaching response data, experiment execution data, homework and examination results data to analyze the students' learning level data, and analyze the job requirements of employees in relevant enterprises. We use back propagation (BP) algorithm to train the data, and use distributed algorithm (DA) to optimize the BP neural network algorithm, so as to improve the efficiency of algorithm searching for the optimal value, And prevent the algorithm from reaching the local optimal value.

4   Improved Deep Learning Model

In this study, the improved neural network algorithm is used to analyze medical education data, output English education quality information, so as to evaluate the teaching quality, and complete the teaching courses and education programs required by the students through the long-term and short-term interest bias setting sequence recommendation program [9].

The key technologies of this study are analyzed below. Firstly, the Da Optimized BP neural network algorithm model is constructed. Because the data analysis speed of the neural network algorithm is slow and the analysis ability is poor, it may be trapped in the local minimum value. Therefore, this research optimizes the neural network algorithm by using distributed algorithms and uses meta heuristics to optimize the algorithm. The algorithm solves global and local data by imitating the colony behavior of dragonflies, and optimizes the data through five different behavior modes of dragonflies: separation, alignment, gathering, foraging and avoiding enemies [10]. The separation behavior in the distributed algorithm is as follows (L):

$$S_i = \sum_{j=1}^{n} w_{ji} x_j - \theta_i \tag{1}$$

where, s represents the degree of separation between Dragonfly I and surrounding dragonflies. In this study, dragonflies represent medical education data information elements; 10. It represents the coordinates of dragonfly J in space and can display the distribution position of medical education data information elements in a spatial range. There are n dragonflies in the algorithm, that is, the medical education data information unit is n. by calculating the sum of the distances between Dragonfly I and each Dragonfly around, that is, the sum of the distances between the medical education data information element and the surrounding data information element, the separation degree of dragonflies can be obtained, that is, the association degree between the output medical education data information elements. The alignment speed between dragonflies is the convergence speed between information elements of medical education data.

It can effectively improve the optimal solution in the information element solution algorithm of medical education data. The operation flow diagram of the neural network algorithm is shown in Fig. 3.

In Fig. 3, the optimized neural network algorithm designed in this study uses Da algorithm to search for the global optimal solution for the weights of the neural network algorithm, and finally obtains the optimal initialization weights. The steps are as follows: first, the initial parameters are set, and the fitness function value of each medical education data information element is calculated. The position of the medical education data information element is the direct source of the students' demand for English data information. The position with the worst fitness is designed to be the position where the medical education data information element is interfered by external information (such as noise, clutter, data use environment, etc.). Based on this, the line vector of the neural network algorithm is established, and the weight coefficient is updated through the European distance function, By using the iterative function to update the position and step size of the medical education data information element, the search for the optimal weight of the neural network algorithm is realized through the continuous iterative optimization process.

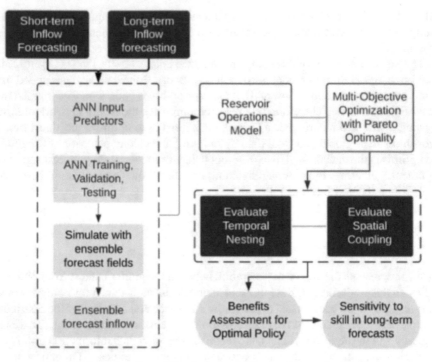

Fig. 3. Operation flow of Da Optimized BP neural network algorithm

5 Conclusion

In the research on the analysis method of medical education data, this research designs teaching plans for different needs by taking advantage of different demands of different posts for knowledge and skills, trains the collected teaching data by using neural network algorithm, solves the teaching quality value, and optimizes the neural network algorithm by using distributed algorithm. This study analyzes the calculation results of the algorithm and the actual degree, and studies the operation efficiency of the algorithm under different frameworks.

References

1. Liang, M.: Optimization of quantitative financial data analysis system based on deep learning. Complexity **2021**(1), 1–11 (2021)
2. Duan, X.X., Duan, P.: Research on Pattern Recognition Method of Online English Education Based on Feature Self Learning (2021)
3. Zhang, H., Cao, H., Liu, Z., et al.: Research on multi-angle face detection method based on improved YOLOV2 algorithm. J. Phys. Conf. Ser. **1848**(1), 012024 (7pp) (2021)
4. Hu, L.: Research on english achievement analysis based on improved CARMA algorithm. Comput. Intell. Neurosci. (2022)
5. Lu, H., Li, M., Zhang, Y.: Research on optimization method of computer network service quality based on feature matching algorithm. J. Phys. Conf. Ser. **1982**(1), 012005 (2021)

6. Ruan, Y.: Design of intelligent recognition english translation model based on deep learning. J. Math. (2022)
7. Liu, H., Ko, Y.C.: Cross-media intelligent perception and retrieval analysis application technology based on deep learning education. Int. J. Pattern Recognit. Artif. Intell. (2021)
8. Liu, W., Kou, F., Huang, H.: Research on Deep Learning Algorithm in Cultural and Creative Product Design. Hindawi Limited (2021)
9. Jiang, Y., Li, C., Zhang, Y., et al.: Data-driven method based on deep learning algorithm for detecting fat, oil, and grease (FOG) of sewer networks in urban commercial areas. Water Res. **207**, 117797 (2021)
10. Zhou, J., Hu, L., Jiang, Y., et al.: A correlation analysis between SNPs and ROIs of Alzheimer's disease based on deep learning. Biomed. Res. Int. **2021**, 1–13 (2021)

Design and Research of Airfield Navaid Lighting Monitoring System Based on Data Mining Algorithm

Xiaoshuo Zhao[(✉)]

Civil Aviation Transportation School, Sanya Aviation and Tourism College, Sanya 572000, Hainan, China
258365548@qq.com

Abstract. Navaid lighting system is an important visual navaid to ensure the flight safety of aircraft, and plays a key role in the process of aircraft approach and landing. The normal operation of the navaid lighting system is directly related to the safety of aircraft takeoff and landing, so it must always work normally. With the rapid development of civil aviation industry, the original manual inspection method of navaid lighting can not meet the requirements. It has become the basic requirement of the navaid lighting system to automatically monitor the navaid lighting equipment and improve the reliability of the navaid lighting system. The design and research of airfield navaid lighting monitoring system based on data mining algorithm is a project involving the design, development and implementation of airfield navaid lighting monitoring system. For the sake of safety, NAV light monitoring system will be designed to monitor all lights in the airport. It can also detect any abnormal light patterns or light pattern changes, which may indicate a possible problem with one or more navigation lights. This helps the airport authorities to take immediate action before any problem occurs with one or more navigation lights, thus preventing accidents. Design includes designing a software.

Keywords: Airfield navaid lighting · Data mining algorithm · monitoring system

1 Introduction

Airport navaid lighting, poor visibility or navigation signal engineering facilities on the runway at night to provide visual guidance; In the past two decades, the airport navaid lighting system has generally adopted two ways of power supply: first, it is connected by parallel power lines, that is, the power lines are directly placed in the middle or both sides of the airport runway, and the airport navaid lighting is connected to two lines on the power supply, one is a high-voltage line, the other is a zero voltage line [1]. This wiring method is cheap and easy to increase or reduce the load; The second type is the series power mode, that is, no matter how many lights are connected in series, the navigation lights; In order to ensure that the open circuit lamp will cause minimum damage to the system, a bypass device is added and an isolation transformer is used; In order to supply

J. C. Hung et al. (Eds.): IC 2023, LNEE 1045, pp. 96–102, 2023.
https://doi.org/10.1007/978-981-99-2287-1_14

power to the series circuit and in different circuits, a constant current dimmer (CCR) with constant current source must be added to ensure that the cable line passes through the same current [2]. The cost of modern airport navaid lighting system has increased a lot because of the addition of a large number of isolation transformers for navaid lighting (the number of isolation transformers is the same as that of navaid lighting), and the high price of constant current dimmers, However, the addition of these two devices ensures that the brightness of all bulbs on the same line is the same, and the light brightness can be adjusted from 1 to 5 levels according to different visibility conditions; The function of constant current dimmer (CCR) is to provide a stable signal source and can feed back according to the actual output current and voltage, and can quickly stabilize at a fixed current level. It exists as the "heart" of the aviation lighting system. It and other parts of the aviation lighting system, such as independent transformer room, diesel engine room, lighting cable, light box, isolation transformer, All kinds of navaid lamps together form the airfield navaid lighting system [3].

The constant current dimmer is the stable power supply of the airport navaid lighting system. It not only has stable output current, but also has adjustable output current. The constant current dimmer is a sign of the technical level of the whole airport navaid lighting system. A constant current dimmer with stable output, adjustable light level, rapid and accurate response plays an indispensable role in the safety and reliability of the whole airport navaid lighting system; In addition, the amount of navigation light cables used in the airport is nearly 46 km, 876 cable heads are made, 438 isolation transformers of various models and 438 navigation lights of various models, with a total power of 35300w [4]. Therefore, the cable heads, isolation transformers and navigation lights are introduced in detail; Based on foreign experience and taking a military standby airport in the east of Beijing as an example, combined with the actual construction situation of airport navaid lighting in China, this paper has achieved a stable and reliable navaid lighting system. The brightness of many navaid lights is consistent and adjustable, providing safe and reliable visual guidance to aircraft pilots during night flight and low runway visibility.

2 Related Work

2.1 Brief Introduction of Airport Visual Navaid Lighting System

During flight, the pilot can control the aircraft in two ways: using autopilot or manual control. Manual control is divided into two methods: one is to refer to the instrument panel, and the flight command instrument system makes some judgments for the pilot; The other method is to fully refer to the external world and make all judgments by using visual references. The latter method is based on sufficient visibility and clear horizon, which is called visual flight.

In visual flight, the most difficult task for piloting an aircraft is to judge the approach to the runway and the subsequent landing maneuver. At this time, the pilot must carefully control the speed and constantly make three-dimensional adjustments to track the correct course [5]. In order to achieve a smooth grounding, the speed and descent rate must be reduced at the same time in the "leveling" operation, so that the aircraft wheels just touch the runway when the wing stalls or is about to stall. After grounding, it is necessary to estimate the length of the remaining runway, so it is necessary to get the prompt of runway exit position in advance. When leaving the runway, the aircraft must be parked in the apron correctly through the taxiway.

The research shows that the average time required for the driver to change from external visual reference to instrument, and then from instrument to external reference is 2.5 s. Since the high-performance aircraft will travel 150 m during this period, the visual aids should provide the maximum possible guidance and information under possible conditions, so that the pilot does not need to check his instruments when moving forward.

The requirements for navaid lighting include configuration, color, candelas, and coverage, which are referred to as four "C" for short. Configuration and color can provide important information for dynamic 3D positioning. The configuration provides guidance information, and the color tells the driver his position in the system. For example, runway edge lights, runway threshold lights and runway end lights belong to different types of configurations [6]. They are used to describe the location of the runway and provide guidance information for pilots. The pilot can judge the position of the aircraft in the system by observing the light configuration, color and color changes, and take measures to control the flight attitude. Candela and effective range refer to the characteristics of light that are very important for the normal play of the role of configuration and color.

The airfield navaid lighting system consists of five parts, including lamps (bulbs), isolation transformers, cables, step-up transformers and dimmers. The schematic diagram is shown in Fig. 1.

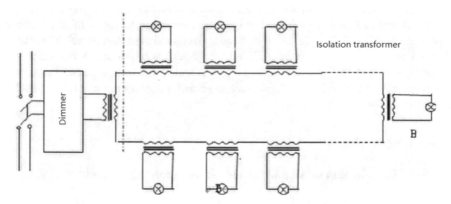

Fig. 1. Composition of airport navaid lighting system

2.2 Navigation Light Patrol Monitoring System

General medium-sized airports have more than ten circuits, including runway edge lights, runway center line lights, end lights, approach lights, slope lights, sequence lights, with a total of thousands of lights. Large airports have tens of thousands of lights. As the navaid lights are related to the safety of aircraft taking off and landing, the airport department requires the airport to inspect every day, and timely replace and repair the bulbs that are not bright or dim.

The navaid lights are distributed on the runway, taxiway and extension lines at both ends of the runway. There are many and wide points. It takes about 1–2 h for all small airports and 2–3 h for large airports. During the patrol inspection, the navaid light needs to be on all the time, which consumes a lot of power. It can be seen that manual patrol inspection requires a lot of manpower and financial resources.

Manual inspection of navaid lights is not only an economic loss, but also a conflict with aircraft flight. In the busy airport, there is an endless stream of flights from morning to night. There is no break time between them. The staff have no opportunity to patrol the runway. They can only proceed after the night flight. It is cold in winter and hot in summer. The staff work very hard, and most of the patrol inspection bulbs are not bright or dim. In addition, there are potential safety hazards in the navaid lighting. In the case of heavy fog or poor visibility, the navaid lighting needs to turn on the high light level [7]. The bulb forms a high temperature in the brightest state for a long time, and is subject to the shock wave of aircraft roar. The bulb is most likely to break the core or cause the lamp to go out or dim. The flight accumulation caused by bad weather increases the continuous time of turning on the lights. During this time period, the light bulb is easy to break, and the staff can not go on the runway for patrol inspection. There may be problems with the navaid lights, but they can not be found in time, thus affecting the visual effect of the navaid lights and flight safety. This hidden danger is an inconvenient problem to be solved in the manual patrol inspection of the navaid lights at present.

The application of intelligent monitoring system can free the staff from the heavy inspection work, and more importantly, it can find out the potential safety hazards in time to ensure flight safety, which is of great significance to flight safety.

3 Design of Airport Navaid Lighting Monitoring System Based on Data Mining Algorithm

3.1 Data Mining Algorithm

Data mining technology refers to the process of mining useful patterns or knowledge from a large amount of data. Many people call data mining knowledge discovery in database (KDD), while others regard data mining as a step of knowledge discovery. The process of knowledge discovery mainly includes data cleaning, data integration, data selection, data transformation, data mining, pattern evaluation and knowledge representation, as shown in Fig. 2.

In Fig. 2, the knowledge discovery process can be summarized into three main steps, namely, data preprocessing, data mining, knowledge evaluation and representation. Data preprocessing mainly includes data cleaning, data integration, data selection and transformation [8]. Data cleaning mainly refers to deleting irrelevant data, duplicate data, smoothing noise data in the original data set, filtering out data irrelevant to the mining topic, and processing missing values, outliers, etc. Data integration refers to merging data from different data sources into a unified format. After data cleaning and data integration, the dirty data with noise from different data sources will be transformed into intermediate data to be further processed, and then the intermediate data will be transformed into the target data required by data mining after selection and transformation. Data selection is to select the features needed for data mining from the integrated data. Data transformation mainly refers to the standardized processing of data and the transformation of data into appropriate forms to facilitate the needs of data mining. Data cleaning and integration, data selection and transformation together constitute the data preprocessing process [9]. Data preprocessing is an important step of data mining. It is the key to the success of data mining. After obtaining the target data through data preprocessing, we can use the relevant data mining algorithms to mine the potential useful patterns in the data [10]. It should be noted that not all the patterns found by data mining algorithms are valid. We also need to evaluate them according to the actual situation. The effective patterns after evaluation are called knowledge. Because the knowledge obtained by data mining is often abstract, we often need to visualize the relevant knowledge in order to facilitate understanding.

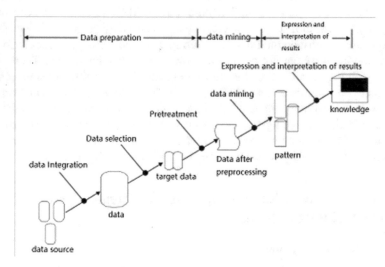

Fig. 2.   General process of data mining in monitoring system

3.2   Navigation Light Patrol Monitoring System

The main functions of the system include:

(1)   Complete the detection of normal, open circuit, dark bulb in each lamp position and water ingress in the lamp barrel enclosing the isolation transformer.
(2)   provide fault warning information, and display the lamp position distribution diagram of each dimming circuit and other relevant information.
(3)   store the fault light position information, and support the query and printing of the fault light position information. At the same time, it also needs to store the work operation records.
(4)   The tower computer sends the patrol command to the lighting station management computer to realize the unattended working mode of the lighting station; The

lighting station management computer requests the tower computer for weather information, flight information, etc.

According to the functions to be completed by the system, it can be determined that the navaid lighting patrol monitoring system is mainly composed of four parts, namely, the tower computer, the lighting station management computer, the main control unit and the light position detection unit.

The working process of the system is as follows:

(1) The tower computer directly sends the command to the lighting station management computer, and the lighting station management computer issues the lighting on and patrol inspection commands; Or after receiving the telephone notification from the tower, the staff of the lighting station will operate the management computer of the lighting station to issue the lighting on and inspection commands, which will be sent to the corresponding main control unit.

(2) After receiving the command from the lighting station management computer, the main control unit sends the light level setting command to the specified dimmer.

(3) The dimmer sets the light level.

(4) After the circuit is stable, the lamp position detection unit detects the lamp state, and transmits the detection results to the main control unit using the power line carrier time sharing.

(5) The main control unit transmits the received information to the lighting station management computer, which processes, displays and stores the data.

(6) If necessary, the tower computer can view the light position patrol information in the light station management computer.

4 Conclusion

It is an inevitable trend to install and use navigational lighting patrol monitoring system in airports. In this paper, the design of airport navaid lighting monitoring system based on data mining algorithm realizes the automation of inspection and monitoring of navaid lighting system. The system can monitor the operation of lighting equipment in real time, realize the intelligent monitoring of the whole navaid lighting system, and ensure the safety of aircraft approach and landing. The use of two-way power frequency communication technology eliminates the need to re lay communication lines and saves human and material resources. The monitoring computer can reasonably arrange the light maintenance time and improve the maintenance efficiency. The use of the navigation light patrol monitoring system is of great significance to improve the reliability of the lighting system, enhance the ground support capability of the airport, reduce the malignant accidents caused by the navigation light, and ensure flight safety.

References

1. Zhang, L., Wang, Z., Xiu, X., et al.: Design and research of infusion monitoring system based on OneNET platform. IOP Conf. Ser. Earth Environ. Sci. **440**(5), 052061 (4pp) (2020)

2. Wang, J., Cheng, H., Yuan, S., et al.: Design and Research of the Intelligent Aquaculture Monitoring System Based on the Internet of Things (2020)
3. Sudjoko, R.I., Hartono, Hariyadi, S., et al.: Design and simulation of airfield lighting system using 8 luminaire in airfield lighting laboratory at politeknik penerbangan surabaya. J. Phys. Conf. Ser. **1845**(1), 012034 (8pp) (2021)
4. Sun, J.-R., Cheng, X.-N.: Design and research of agricultural environmental monitoring system based on wireless sensor. In: 2020 5th International Conference on Automation, Control and Robotics Engineering (CACRE) (2020)
5. Hao, H., Chi, X., He, J., et al.: Design and research of CAS-CIG for earth system models. Earth Space Sci. **7**(7) (2020)
6. Dris, M., Ramli, M., Khusaini, N.S., et al.: Design and development of insole monitoring system for runner. Appl. Mech. Mater. **899**, 103–113 (2020)
7. Zhang, S.L., Deng, J.H., Lu, H.L.: Research and design of Laboratory environmental monitoring system based on stm32 (2020)
8. Liu, X.Y., Wu, D.L., Hou, J.: Design and analysis of a scheme for the naval gun test shell entering the bore (2021)
9. Ogle, S.M., Butterbach-Bahl, K., Cardenas, L., et al.: From research to policy: optimizing the design of a national monitoring system to mitigate soil nitrous oxide emissions. Curr. Opin. Environ. Sustain. **47**, 28–36 (2020)
10. Zheng, L., Xiao, C., Chen, F., et al.: Design and research of a smart monitoring system for 2019-nCoV infection-contact isolated people based on blockchain and internet of things technology (2020)

Design and Research of Chinese Painting Authenticity Identification System Based on Image Recognition

Weitong Chen[✉], Yawei Yu, and Ping Zhu

Yunnan Normal University, Kunming 650504, Yunnan, China
cwt33366096@163.com

Abstract. The design of the system is based on image recognition technology. The main function of the system is to identify the authenticity of Chinese paintings through image analysis. The system can be used for the authentication and identification of ancient paintings in museums, galleries or private collections. It will also help collectors obtain more information about their works by comparing them with other works at different times and places. The project is designed as a multi-layer method, using statistical pattern recognition, neural network, fuzzy logic and other methods to identify real Chinese paintings according to their images. The most common way to verify ancient Chinese paintings is to compare them with other similar works and then determine whether they are true. However, there are some problems in this traditional way: first, it is difficult to find similar works that have been lost over time; Secondly, due to changes in materials and technology, there may be some differences between the original and its copies. In order to solve these problems, we need a new authentication system based on image recognition technology, which can accurately identify the originality of artworks from images.

Keywords: Image recognition · Identification of authenticity · Pictures · System development

1 Introduction

To determine the authenticity of a work of art is actually an extremely complicated task. In the absence of a direct and powerful determination of the elements of a work, such as the certification certificate or the detailed introduction in the list or catalogue, art historians and experts have to act as detectives.

First of all, art experts should analyze the paintings and confirm the authenticity of the paintings based on the author's historical and biographical knowledge. Use a pair of "good eyes" to determine the style of art with known knowledge [1]. This skill needs long-term training. First give an instruction of painting, and then rely on science and technology and use scientific tools to deeply understand.

At a deeper level, for example, X-ray ° can be used to detect whether there are traces of the beginning of the painting under the surface layer of the painting. The radioactivity measurement ° of lead in the oil painting is very strong evidence. It may be found that the paintings are wrong with the times, which is inconsistent with the assumed age of the paintings. The typical performance is that the pigments used were invented after that time, that is, the imitations are not authentic.

We can also analyze according to the thread structure of the canvas (bump, repeat in each part of the canvas, whether it is an old feature, hand woven or woven frame), the brush used (the shape of the brush, the length or width of the brush), or compare it with other works of the same author [2]. The location of the artist's signature is very important. Observe whether it is the same era as the painting. Based on this, the design of Chinese painting authenticity identification system based on image recognition is studied.

2 Related Work

2.1 Traditional Chinese Painting Identification Method

Since ancient times, China has been a country with a long and rich culture and art. In the long river of history and culture, the art of calligraphy and painting can be regarded as a very dazzling part. Chinese calligraphy and painting began 10000 years ago. In the Neolithic Age of China, painting technology appeared on pottery.

Calligraphy and painting include calligraphy and calligraphy and painting. Of course, some calligraphy and painting are combined. For example, literati paintings after the Song Dynasty are typical works of art that combine calligraphy and painting. With the continuous improvement of people's knowledge and appreciation of ancient calligraphy and painting, more and more people like to collect some famous calligraphy and painting [3]. With the increase of the collection, the price becomes higher and higher, and a lot of counterfeits, that is, imitations, are produced. So how to distinguish the authenticity of famous calligraphy and painting?

1. It can be distinguished from the contemporary flavor of famous calligraphy and painting. The times of each famous artist are different. The economic, cultural, technological and other aspects are different in different times. The calligraphy and paintings created by famous artists in this period have a very strong flavor of the times. We can see the flavor of the same era from his paintings without clothes. If the flavor of the times is different, it may be a fake.
2. Distinguish from the personality style of famous calligraphy and painting. The personalities and styles shown by the calligraphy and paintings of famous masters in each period are very different. The calligraphy and paintings in the Wei and Jin Dynasties show a very willful and liberal attitude, which is what we often call the demeanor of the Wei and Jin Dynasties. The paintings in the prosperous Tang Dynasty are very brilliant and rich. Heroic. The style of each period after that was very different [4]. There are too many subdivisions and too complex, which requires time to understand
3. Distinguish from the brush and ink of the calligraphy and painting. The ink used by ancient people has changed a lot. The composition of the ink is very different from that of the present ink. Even if the imitation is made with the ink at that time, there is still a big difference. It is possible to distinguish the true from the false by using modern technology.

4. Comprehensive comparison of calligraphy and painting. Generally, calligraphy and painting are signed and sealed by the artist. Comparative analysis can be conducted from this aspect. In addition, two calligraphy and paintings of the same artist can be compared. One of the authentic works can be compared with the uncertain one, and the specific characteristics and styles can be analyzed to determine the authenticity.

In short, the identification of calligraphy and painting has existed since ancient times. Since the beginning of counterfeiting, the identification has also come into being. If you want to really identify the authenticity of calligraphy and painting, you'd better ask a special person to identify it [5].

2.2 Image Shape Features

(1) Point feature extraction: the properties of feature points are determined by each pixel. G, and B components in the color image. Point feature extraction is to extract salient points in the image, such as house corners and circles.
(2) Line feature extraction: line features include "edge" and "dotted line" of the image. "Edge" generally refers to the boundaries between different regions, such as the sudden change of gray level and texture performance. "Line" refers to an edge with very small width and the same image characteristics, which constitutes the middle area. Two edges with small distance can form a line. The line feature can reflect the properties of its small neighborhood, such as the strength, direction, density, average value and variance of the line and edge.
(3) Edge feature extraction: the binary image edge processing is to find the mutation positions of the pixel gray value, and then set the pixel value to "1" at these positions, and set the remaining pixel value to "0" to determine the boundary target. Edge feature extraction of image is usually realized by operators, such as Sobel, Prewitt operator, Kirsch, Laplace and so on [6]. These operators are based on an image region (3×3) Multiplied by the edge intensity at the center of the region. First, the edge intensities of all pixels in the image are obtained, and then the pixel points whose edge intensities are greater than this point are extracted. The pixel value is set to "1", and the remaining assigned pixel value is "0". Then the edge of the image region is detected according to the direction statistics, and the shape feature of the histogram is obtained.

3 Design and Research of Chinese Painting Authenticity Identification System Based on Image Recognition

The image recognition method is used to analyze the artistic style of Chinese painting. Firstly, the elements representing the artistic style of Chinese painting are extracted, and the artistic style characteristics of different painters are extracted from the artistic style elements, so that the computer can classify and identify the Chinese painting images and identify the authenticity.

The steps of extracting artistic style from Chinese painting images are as follows:

(1) The overall Chinese painting image is regarded as a local area, and the statistical characteristics and structural characteristics are calculated.
(2) Calculate the pixel points and characteristics of Chinese painting that can express the artistic style of Chinese painting in some local areas, and make statistics on the local areas to obtain the application of ink method and geometric characteristics.
(3) Brush movement and stroke feature extraction in Chinese painting, extracting stroke movement features, edge and line strength, direction and statistical value in local neighborhood.

After obtaining the features of all pixels, the segmentation of Chinese painting image is divided into two stages. In the first stage, the algorithm constructs an HMM to describe the characteristics of the pixels in the first row of the image. In the second stage, the linear time adaptive dynamic programming method is used to process the features of all pixels in the image line by line [6]. When new pixels are processed, the parameters of the HMM are updated. In addition, if the existing state in the HMM cannot transmit the feature vector with an acceptable probability, a new state is generated in the HMM to describe the feature vector of the new pixel. After the dynamic planning process is completed, the label is assigned to the pixels in the image.

This method uses a set of convolution operators to process Chinese painting images, and obtains feature vectors for each pixel in the image. Then, the feature vectors are processed step by step to construct an adaptive HMM. An adaptive Viterbi algorithm is used to determine the state to which each pixel is assigned [7]. If the feature vector of the new pixel is not generated in the existing state in the HMM with the maximum possibility, the parameters of the HMM will be updated when a new pixel is processed and a new state is created and included in the HMM.

4   Simulation Analysis

Traditional Chinese painting expresses the shape of objects through lines, so it needs to draw the outline and three-dimensional feeling of objects with the help of lines. The lines of Chinese painting are usually long and thin brush strokes, which are the basic elements of Chinese painting and also the expression methods of Chinese painting. The brushwork can reflect the painter's brushwork skills, such as the direction and strength of brushwork. Brush strokes play a key role in the artistic style of traditional Chinese painting. Therefore, the extraction of detailed style features is mainly to extract the brush stroke information in the whole traditional Chinese painting image, so as to capture the brush stroke characteristics that can best show and highlight the artist's local details in the painting [8].

The local detail feature extraction algorithm adopted in this section is as follows:

$$\min f(x_1, x_2, ..., x_n) = \frac{1}{m} \sum_{i=1}^{m} \left[Y_i^0 - Y_i \right]^2 \tag{1}$$

The next step is to divide the local area. There are many ways to divide the area, such as the quadtree method, ROI region of interest method, fixed size segmentation, etc. However, this paper uses the simple and convenient segmentation method of fixed module area, which has a good recognition effect. The size of the local area is usually defined as 128×128, 64×64 or 32×32. In order to obtain a sufficient number of local detail features and reduce the amount of calculation, experiments have proved that a 64 × The 64 pixel image block is used to find the local area in the image that most reflects the detailed pen operation characteristics:

$$K(\Delta d)\Delta d = \Delta f = f_{ext,n+1} - f_{int}(d_{n+1}) \tag{2}$$

After locating the region, extracting the gray histogram of local features is not simply extracting the shape, color and other features after edge detection [9]. It is necessary to make the local details fully useful. As shown in Fig. 1 below.

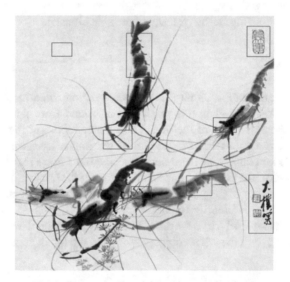

Fig. 1. Extracting local feature gray histogram

Each tested Chinese painting image contained three different regions, including white matter, gray matter, and CSF. The segmentation results were compared with real images, and the segmentation accuracy of these three substances was evaluated based on sensitivity and specificity. Sensitivity is the percentage of pixels in a specific area correctly identified in the segmentation result. Specificity is the percentage of correctly identified pixels in the corresponding region provided by the segmentation result [10]. Accuracy was assessed by calculating the average of sensitivity and specificity. As shown in Fig. 2 below.

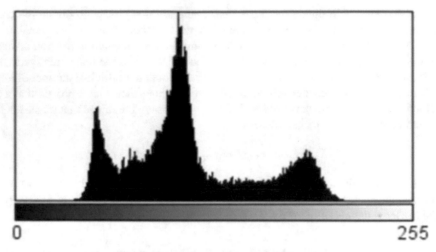

Fig. 2. Image feature extraction histogram

5 Conclusion

The design and research of the Chinese painting authenticity identification system based on image recognition studied in this paper can be found from the above description that the overall characteristics of the image can better represent the characteristics of the object layout, pen strength, ink use, etc. of the image from a macro perspective. In addition, the local features of the image describe the feature information in a small neighborhood, such as the strength, shape and contour of the brush movement, which can reflect the characteristics of the painter's brush movement. It has been proved by experiments that when experts identify the authenticity of calligraphy and painting works, they usually put the art works in a far place first, and appreciate the whole picture from the overall perspective of the whole object layout and the blank area.

References

1. Peng, W., Lin, Q.: Research on plot authenticity identification method based on PSO-MLP. J. Phys. Conf. Ser. **1486**, 042003 (2020)
2. Li, Y., Pi, X., Xu, Y.: Reliability of peanut oil authenticity identification based on fatty acids and establishment of adulteration analysis model. IOP Conf. Ser. Earth Environ. Sci. **512**(1), 012063 (11pp) (2020)
3. Chen, B., Liu, D., Jiang, T., et al.: A triboelectric closed-loop sensing system for authenticity identification of paper-based artworks. Adv. Mater. Technol. **5**, 2000194 (2020)
4. Pan, F.: A Food Traceability System Based on Blockchain and Radio Frequency Identification Technologies. 电脑和通信(英文) (2020)
5. Wang, Y., Sun, Q., Rong, D., et al.: Image source identification using convolutional neural networks in IoT environment. Wirel. Commun. Mob. Comput. (2021)
6. Zhi, Y., Ouyang, H., Xue, M., et al.: A research on driver nodes identification in Chinese interbank networks: based on the controllability theory of complex network (2022)

7. Zhu, B., Zhang, Y., Zhao, J., et al.: Accurate pressure control based on driver braking intention identification for a novel integrated braking system. SAE Int. J. Adv. Curr. Pract. Mob. (3–4) (2021)

8. Qu, J., Chu, Z., Chen, C., et al.: Design of urban rain-waterlogging monitoring system based on video image recognition. Microcontrollers Embed. Syst. (2020)

9. Qu, J.Y., Li, Z., Su, J.R., et al.: Development and validation of an automatic image-recognition endoscopic report generation system: a multicenter study. Clin. Transl. Gastroenterol. **12**(1), e00282 (2020)

10. Wadhwa, S.S., De Horta, A.Y., Filipovic, M.D., et al.: Simplified Method for the Identification of Low Mass Ratio Contact Binary Systems that are Potential Red Nova Progenitors (2022)

Design of Disinfection System for Medical Inspection Instruments Based on Intelligent Algorithm

Xiaoyong Huang and Haiyan Shi[✉]

Medical College of Yan'an University, Yan'an 716000, Shaanxi, China
shyhxy1005@163.com

Abstract. The design of medical examination instrument disinfection system is the most important. It is a process involving selection, formulation and implementation of plans to ensure that there is no pollution in the area where medical examination instruments are used. It must be able to keep the instrument clean, safe and hygienic, which can be achieved by implementing cleaning, disinfection and sterilization processes. The design should also conform to national standards. The design of medical inspection instrument disinfection system based on intelligent algorithm is the process of designing a disinfection system for medical inspection instruments. This is a very important task because it involves many factors, such as product specifications and characteristics required by customers. Type of disinfectant and its concentration, temperature, time, etc. Characteristics of the environment in which the product is used (such as room temperature and humidity). Cleaning frequency, water supply rate and other operating conditions.

Keywords: Intelligent algorithm · Medical examination · Appliances · Disinfection system

1 Introduction

Medical examination is a science that tests materials taken from the human body in microbiology, immunology, biochemistry, genetics, hematology, biophysics, cytology and other aspects, so as to provide information for the prevention, diagnosis, treatment of human diseases and evaluation of human health.

At present, in the process of medical inspection, a large number of inspection instruments need to be used. In order to ensure the hygiene of inspection instruments and the accuracy of test results, medical staff need to regularly place inspection instruments in disinfection devices for disinfection. However, existing disinfection devices usually place inspection instruments on a platform and spray disinfection water from the top of inspection instruments toward inspection instruments [1]. However, inspection instruments are usually glassware, which is cylindrical, If the opening of the inspection instrument is upward and the disinfection water is sprayed from the top, a large amount of the waste liquid of the disinfection water will be stored in the inspection instrument after the disinfection is completed, and the laboratory personnel still need to pour it out manually in the future. If the opening of the inspection instrument is downward, the interior of the inspection instrument will not be disinfected, resulting in incomplete disinfection.

In the prior art, the disinfection method used to disinfect the medical instruments used in the laboratory is to put all the instruments in the disinfection box or disinfection box and soak them in the disinfectant [2]. The medical instruments are directly immersed in the disinfectant and placed in a disorderly manner. Because of the disorderly placement of the instruments, it is very inconvenient to take them, which brings unnecessary trouble to the staff. Moreover, it is easy to get the disinfectant on the hands during the process of taking out the instruments, Causing corrosion damage to the hands of medical personnel; In addition, if medical devices after detoxification are not used immediately, secondary pollution will be caused [3].

During the use of some existing disinfection and cleaning devices, they need to be disinfected in the disinfection device first, then taken to the cleaning device for cleaning, and finally dried. The disinfection and cleaning process is cumbersome, which increases the workload of staff to take instruments back and forth. At the same time, because of the high frequency of use, the commonly used medical devices in the inspection department are simply soaked in disinfectant or wiped with alcohol, It is easy to disinfect incompletely, and it is easy to cause cross infection when reused, which increases the treatment pain of patients, affects the accuracy of test data, and reduces the disinfection efficiency of appliances.

2 Related Work

2.1 Overview of Intelligent Algorithms

The concept of artificial intelligence was put forward at a conference. One of the important topics of this conference is whether machines can accurately describe and simulate all human learning processes. The term "artificial intelligence" has been widely used in social life, but scholars have interpreted it differently from different angles, and there is no unified answer. Some scholars have defined it from the perspective of the research content of artificial intelligence, while r John McCarthy and Wu Handong have interpreted it from the purpose of artificial intelligence research [4]. The former believes that artificial intelligence is to let computers learn and imitate human behavior, while the latter believes that artificial intelligence is not only to imitate human behavior, but also to extend human behavior and expand human intelligence@ Many scholars have different understandings of artificial intelligence, but the difference between artificial intelligence and human beings is the consensus of scholars.

Algorithm is one of the core elements of artificial intelligence. The term "algorithm" itself has multiple meanings. Its concept originated from a systematic work on algebra written by Persian mathematicians in the 9th century. The concept of algorithm is a historical category. With the continuous deepening and progress of subject research, it can be simply understood as programming operation through formula and obtaining results [5]. In the field of artificial intelligence, algorithm mainly refers to a computing method that uses a set of clearly defined steps in computer science to achieve a certain goal and solve a certain problem. It is a coding program that converts input data into desired output results based on the operation mode specified by the code.

The algorithm itself has unique properties. First of all, the operation of the algorithm is complex, opaque and difficult to interpret; Secondly, through the continuous development of the algorithm, the algorithm even began to have a certain degree of autonomy; Finally, as far as the current legal norms are concerned, the legal status of artificial intelligence and algorithms is not clear. At the same time, based on the above two points, algorithms are also difficult to be accountable. As far as the algorithm of artificial intelligence is concerned, at present, the artificial intelligence of China is still in the weak artificial intelligence stage. At this stage, the algorithm of artificial intelligence has the characteristics of technical defects, data dependence and action limitation [6]. Medical artificial intelligence algorithm is a branch of artificial intelligence algorithm research, but because the application of medical artificial intelligence is likely to involve people's basic rights such as life and health, this paper believes that its regulation should be different from that of artificial intelligence algorithms in other fields.

2.2 Characteristics of Medical Examination

(1) Comprehensive. Medical testing is to test and analyze various human body test samples collected through a series of chemical, physical and biological professional knowledge, use scientific methods of various disciplines, and with the assistance of test equipment and reagents, so as to provide clinical test results and assist clinicians in disease diagnosis. It is necessary to combine scientific medical theoretical knowledge, operate according to the inspection technical specifications, ensure that the inspection technicians have correct inspection methods, and that the laboratory quality standards meet the quality control requirements. At the same time, it is required to have multiple requirements on the sampling requirements of the inspection objects (such as some items requiring fasting blood sampling, sampling posture), sample storage requirements, and inspection time. Finally, accurate data and inspection reports are obtained, It also provides clinicians with the results and explanations of inspection and analysis [7]. Through medical inspection, it can provide reliable clinical basis for clinic, such as disease prevention, treatment detection, treatment effect evaluation, and finding the cause of disease. Therefore, comprehensiveness is an important feature of medical examination.

(2) Complexity. Medical examination is a scientific experimental research work, which requires scientific rigor. With accurate examination results, it provides important evidence for clinical diagnosis, treatment and monitoring. There are also many factors that affect the test results, such as whether the test reagents are stored properly, whether the operation of the test instruments is standardized, whether the quality control system is perfect, whether the sample collection meets the standards, and the proficiency of the technical operators will have a great impact on the test results. Therefore, no link in the inspection can be neglected. The inspection staff must strictly investigate the influencing factors of each link, standardize the inspection process, ensure the accuracy of the inspection results, provide accurate inspection

results for the clinic, and avoid clinical diagnosis errors caused by incorrect inspection results, miss the best treatment opportunity, and cause medical disputes [8]. Therefore, the medical examination work has certain complexity.

(3) Professionalism. In the process of medical laboratory experiments, professional technical personnel, precision instruments and professional technology are indispensable and important parts. Technical operators need to undergo professional learning and training, have practical experience in standardized operation, and constantly learn new technology and operation knowledge. The quality control system of the laboratory also needs to be supported by professional ISO systems, such as is015189, cap and other quality management standards, to ensure the inspection quality of the laboratory. Testing instruments and equipment requires sophisticated scientific facilities and environmental requirements.

3 Disinfection System Characteristics of Medical Inspection Instruments

An intelligent algorithm based disinfection system for medical inspection instruments, comprising:

(1) A sterilization bin, an ultraviolet lamp is arranged on the top wall of the sterilization bin, which is used to irradiate ultraviolet rays into the sterilization bin to sterilize the instruments in the sterilization bin by ultraviolet rays, and a first visual collector is arranged on one side of the ultraviolet lamp to collect image information of the instruments; A heating wire is arranged on the inner side wall of the sterilization chamber to increase the temperature in the sterilization chamber to sterilize the appliance at a high temperature, and a first temperature detector is arranged on the side wall adjacent to the side wall where the heating wire is located to detect the temperature in the sterilization chamber; A first conveyor belt is also arranged at the lower side of the sterilization bin to load and move the instruments to be sterilized, and a disinfectant recovery tank is arranged at the bottom of the first conveyor belt to collect the disinfectant;

(2) A disinfectant storage tank, which is arranged on the top of the disinfectant bin, is used to store disinfectant. The disinfectant storage tank is externally connected with a connecting pipe [9]. The other end of the connecting pipe vertically penetrates the top wall of the disinfectant bin and is provided with a nozzle at the end of the connecting pipe, which is used to spray disinfectant into the disinfectant bin to disinfect the appliance;

(3) A drying bin, which is arranged side by side with the sterilization bin, for drying the sterilized instruments output from the sterilization bin, a side wall adjacent to the drying bin and the sterilization bin is provided with a conveying hole, and the conveying hole is at the same level with the output end of the first conveyor belt, so that the first conveyor belt conveys the sterilized instruments to the drying bin; A second vision collector is arranged on the inner top wall of the drying bin to collect image information of the sterilized appliance surface; A fan is arranged on the top of the drying bin, the fan is externally connected with a drying pipe, and the drying

pipe vertically penetrates the top wall of the drying bin to cooperate with the fan to deliver air to the drying bin, and a heater is arranged at the end of the drying pipe to heat the air delivered by the fan; A second conveyor belt is arranged at the lower side of the drying bin, and the input end of the second conveyor belt is at the same level with the conveying hole, for receiving the sterilized instruments output by the first conveyor belt, and a fan is arranged at the bottom of the second conveyor belt to dry the sterilized instruments in the drying bin;

(4) The central control unit is respectively connected with the ultraviolet lamp, the first visual collector, the heating wire, the first temperature detector, the first conveyor belt, the spray head, the second visual collector, the fan, the heater and the fan, so as to adjust the operating parameters of the corresponding components to corresponding values when the system disinfects or dries the appliances;

(5) When the system disinfects the appliance, the central control unit starts the ultraviolet lamp and controls the spray nozzle to spray the disinfectant into the disinfection chamber. When the time length of spraying the disinfectant reaches T0, the central control processor controls the first visual collector to collect the average distribution density p of the disinfectant droplets on the surface of the appliance and compare P with the preset droplet distribution density standard P0,

If $P \geq P0$. The central control unit determines that the spraying measurement of the disinfectant reaches the standard, the central control unit controls the nozzle to stop spraying the disinfectant, controls the UV lamp to continuously irradiate, and starts the heating wire to sterilize the appliance at high temperature when the total irradiation time reaches the preset value;

If $P < P0$, the central control unit determines that the spraying measurement of disinfectant does not meet the standard, and the central control unit controls the nozzle to perform secondary spraying and controls the first visual collector to collect the diameter of disinfectant droplets on the surface of the appliance, calculate the average diameter r of disinfectant droplets on the surface of the appliance Correct P0 to P0 'according to R, and re count the average distribution density p' of disinfectant droplets on the surface of the appliance when the duration of secondary spraying reaches T0, compare p 'with P0' and determine whether the spraying measurement of disinfectant meets the standard according to the comparison result. If the central control unit determines that the spraying measurement of disinfectant after secondary spraying still does not meet the standard, The central control unit controls the nozzle to start and compares the average distribution density of disinfectant droplets on the surface of the appliance with P0 'when the spraying time reaches t0 until the average distribution density of disinfectant droplets on the surface of the appliance is greater than or equal to P0'.

4 Design of Disinfection System for Medical Inspection Instruments Based on Intelligent Algorithm

The intelligent algorithm based disinfection system for medical inspection instruments according to the requirements, which is characterized in that when the central control unit determines that the spraying measurement of disinfectant does not meet the standard and corrects the preset droplet distribution density standard P0, the central control unit controls the image collector to collect image information of the surface of the instrument and statistics the average droplet diameter r of the surface of the instrument according to the image information, Compare R with the preset average radius pre stored in the central control unit and select the corresponding and distribution correction coefficient to correct Po according to the comparison result [10];

The central control unit is provided with a first preset average radius R1, a second preset average radius R2, and a first preset distribution density standard correction coefficient α 1 and the second preset distribution density standard correction coefficient A2, wherein R1 < r2,0 < A2 < A1 < 1;

When R $\leq$ R1, the central control unit does not correct P0;

When R1 $\leq$ R $\leq$ R2, the central control unit selects A1 to correct P0; When r > R2, the central control unit selects A2 to correct P0;

When the central control unit selects AI to correct P0, set I = 1,2, and the corrected preset droplet distribution density standard is marked as P0 ', and set P0' = P0 $\times$ ai.

Specifically, when the central control unit determines that there is an appliance to be disinfected in the sterilization bin 1, the central control unit controls the first visual collector 12 to collect appliance image information, and the central control unit analyzes the contour characteristics of the appliance to determine the type of the appliance. If the central control unit determines that the type of the appliance is a container or an appliance, the central control unit determines whether the spraying measurement of the disinfectant meets the standard by detecting the droplet distribution density on the surface of the appliance; If the central control unit determines that the type of the appliance is clothing, the central control unit controls the spray head 21 to pre spray the disinfectant and sets the spraying amount of the disinfectant to 0.3 $\times$ VO, after spraying, the central control unit controls the first visual detector to collect image information of the surface of the appliance, as shown in Fig. 1.

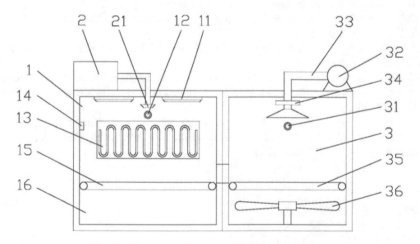

Fig. 1. Structural diagram of disinfection system

5 Conclusion

In this paper, by using independent compartments to disinfect and dry the appliances respectively, incomplete disinfection can be avoided, so that secondary infection or cross infection can be effectively avoided when using the appliances, and the disinfection efficiency of the system described in this paper against the appliances can be effectively improved. At the same time, this paper determines whether the spray nozzle evenly sprays disinfectant on the surface of the appliances by using the average distribution density of disinfectant droplets, So as to quickly determine whether the disinfection of the surface of the appliance is completed and further improve the disinfection efficiency of the system described herein for the appliance.

References

1. Gao, B., Ge, R., Zhang, D., et al.: Design and implementation of quality and safety traceability system for reusable medical devices disinfection based on RFID technology. Chin. J. Med. Instrum. **45**(2), 167–171 (2021)
2. Shi, M.: Design of intelligent planning system for tourist scenic route based on ant colony algorithm. Int. J. Ind. Syst. Eng. IJISE **39**(3), 377–393 (2021)
3. Li, J., Tu, Y., Lu, B.: Design of intelligent early warning robot system for warehouse autonomous patrol based on scale estimation algorithm of probability theory. IOP Conf. Ser. Mater. Sci. Eng. **1179**(1), 012002 (13pp) (2021)
4. Xia, K., Li, X., Li, X., et al.: Intelligent anti-epidemic mask based on KF and ECF fusion algorithm. Electron. Lett. **57**, 724–726 (2021)
5. Ma, H., Zhao, C., Zhai, C., et al.: Design and field experiment of precise control and monitoring system for a solid fumigant sterilizer based on IoT. Comput. Electron. Agric. **189**, 106387 (2021)
6. Heibeyn, J., Knig, N., Domnik, N., et al.: Design and evaluation of a novel instrument gripper for handling of surgical instruments. Curr. Dir. Biomed. Eng. **7**(1), 1–5 (2021)
7. Cui, G.: Design of Intelligent Recognition English Translation Model Based on Feature Extraction Algorithm (2021)
8. Zeng, J.: Design and implementation of intelligent EOD system based on six-rotor UAV. Drones **5**, 146 (2021)
9. Gong, S., Kumar, R., Kumutha, D.: Design of lighting intelligent control system based on OpenCV image processing technology. Int. J. Uncertain. Fuzziness Knowl. Based Syst. **29**, 119–139 (2021)
10. Cheng, Q., Wang, S., Fang, X.: Intelligent design technology of automobile inspection tool based on 3D MBD model intelligent retrieval. Proc. Inst. Mech. Eng. Part D J. Automobile Eng. **235**(10–11), 2917–2927 (2021)

Design of English Teaching Resource Management System Based on Collaborative Recommendation

Yu Jie[✉]

Hunan Sany Polytechnic College, Changsha 410011, Hunan, China
dominic_71268@163.com

Abstract. With the development of modern network and the fast and convenient characteristics based on network, the construction of network education resources has become the focus of information construction in Colleges and universities. However, due to the lack of unified resource description standards and construction standards, most of the resource database construction is still in an isolated and decentralized state, and it is difficult to achieve sharing and reuse. Therefore, by deeply analyzing the metadata standard of educational resources, studying XML and its related technologies, the author has carried out the "design of English teaching resource management system based on collaborative recommendation", which is a tool for managing English teacher resources. It can be used by teachers, students and school management departments. The main function of the software is to ensure that all resources are in place before the start of the course and to track their status after each course. Detrsc also allows teachers to create different types of reports according to their needs: from a simple list containing names and descriptions to more complex reports, such as activity reports or performance reports.

Keywords: Collaborative recommendation · Resource management · English teaching resources

1 Introduction

With the rapid development of science and technology and the continuous progress of computer technology, network technology and multimedia technology, the traditional teaching mode can no longer meet the needs of modern, rich and interactive teaching activities. More and more teachers begin to use the network to obtain English teaching resources, so that network education is crossing the boundaries of time and space and becoming the darling of the new era, Make the construction of network education resources become the key project of education information construction [1].

At present, colleges and universities have begun to attach great importance to the construction of educational informatization, and have spent a lot of money on the hardware construction of campus network. However, the construction of educational informatization is not only a simple construction of modern hardware facilities, but also includes

J. C. Hung et al. (Eds.): IC 2023, LNEE 1045, pp. 117–124, 2023.
https://doi.org/10.1007/978-981-99-2287-1_17

the construction of teaching software and material resource library. Without the support of teaching software and material resource library, the hardware can only be a simple decoration, which is difficult to play its strong supporting role in teaching [2]. So far, some schools have built their own teaching resource management software, but due to the lack of unified construction standards and description standards, it is difficult to share and reuse resources. Therefore, the urgent task is to establish a standard, rich and practical teaching resource database in Colleges and universities, so as to integrate the scattered network teaching resources, realize the sharing of resources in a large range and the exchange between heterogeneous systems, and improve the utilization of teaching resources [3].

The construction of English teaching resources is a long-term and arduous basic project, the key project of campus network construction, and the top priority of information construction in Colleges and universities. It can fundamentally solve the contradiction of "there are roads but no cars, and there are cars but no goods" in the process of information. This paper is a case study of teaching resource management software proposed under this background - the design of English teaching resource management system based on collaborative recommendation. Through specific analysis, research and implementation, this paper provides system support for the construction of standardized resource database.

The construction of educational resources is the foundation and key to the realization of educational informatization. According to the standards related to the description of educational resources in China's online education technology standard system, combined with celts-3.1 and CELTS-41 specifications, this topic uses the learning object metadata standard (LOM) to describe the media type teaching resources, and proposes to design a set of "design of English teaching resource management system based on collaborative recommendation", Users can realize the standardization of resource storage, retrieval, uploading, evaluation and other applications by performing relevant operations in the interface as required [4]. At the same time, it studies the use of XML technology to bind resources, which solves the problems of scattered and isolated teaching resources, non-standard description, incomplete information, inaccurate search, imperfect evaluation, and the need for a large number of repeated construction. Through the research and use of XSLT, it realizes the mutual transformation between XML and HTML, thus laying a good foundation for the large-scale sharing and interaction of teaching resources. In addition, a good interface is designed in the system, which can well complete the expansion of metadata elements in the teaching resource library [5]. In a word, the research, design and development of this system will not only help to fully understand, widely promote and apply standards, but also further promote the future development of online education in China and the construction of standard resource database, and accelerate the pace of educational informatization.

2 Related Work

2.1 Research Status of Resource Collaborative Filtering Recommendation

Personalized recommendation systems is an information filtering system proposed to reduce the additional labor cost in the process of obtaining information. General information filtering system is also called Resource Recommendation System in general. It can not only recommend potential information, service or product information and other resources that users may need according to users' preferences, interests, behaviors and needs, but also integrate the recommendation system with the operation structure of enterprise e-commerce, so as to tap many potential benefits for enterprises [6]. For example, through the information recommendation system, businesses analyze and judge users' interests according to users' past purchase or browsing records, and make recommendations based on this, so as to stimulate users' consumption and increase sales opportunities. In this process, businesses not only consolidate their contacts with existing users, but also vigorously attract new users.

Collaborative filtering technology is a way to obtain recommendation rules, which was first proposed by Goldberg and other scholars in a research report in 1992. They study how to filter e-mails that are useful to users. The premise is that users should directly identify what kind of information they do not want to get, that is, users need to participate in the filtering. Another pioneering work of collaborative filtering is movie-lens, a web-based research-based movie recommendation system built by researchers at the University of Minnesota in 1996. The system finds the similar users of the target users through the users' ratings of some movies, and then recommends the ratings of other movies through the similar users. After that, collaborative filtering algorithm has been a great success and widely used [7]. For example, the famous collaborative filtering systems abroad include GroupLens, an online news filtering system, which allows users to collaborate to find the content they are interested in from a large number of Usenet news: jester, a joke recommendation system, finds similar users based on users' comments and generates recommendations based on the comments of similar users; Online bookstore amazon COM is to recommend books after analyzing users' hobbies and finding similar users according to their browsing and purchasing information.

At present, almost all large-scale e-commerce systems such as Amazon and eBay use collaborative filtering recommendation technology to varying degrees. At the same time, this recommendation technology has also been widely used in all walks of life. The recommended objects include books, audio-visual products, web pages, articles, news, learning resources, etc. the specific application fields of the resource recommendation system include e-commerce, digital library, network education, news recommendation and search engine.

2.2 Performance Requirement Analysis of Teaching Resource Management System

Performance requirements specify the timing constraints or capacity constraints that the system must meet, usually including the corresponding time, information rate, main memory capacity, disk capacity, security and other requirements of the system. For this project, in terms of performance, it can not only realize users' browsing, downloading and other operations; The system should also strictly check the data entered by the user to eliminate human errors as far as possible; Flexible and fast information query and safe data storage; At the same time, the system is stable, safe and reliable. The operation of the background is strictly restricted. Users without permission will not be able to log in to the system through any channel to view any information and data of the system, ensuring the tightness and security of the system [8]. The system adopts the functions of backup database and restore, which can backup the database immediately. When the system fails, the system database can be restored after troubleshooting, so that the original data will not be lost, which greatly strengthens the security of the teaching resource system.

Apache is the world's most used web server software. It can run on almost all widely used computer platforms. Because of its cross platform and security, it is one of the most popular web server-side software. As an open source software, Apache has its own unique advantages. It can work with most of the current mainstream servers, and has a very high running efficiency. Apache and Tomcat can be integrated so that Apache handles static HTML and Tomcat handles servlets. There are also many users who use Apache, which is rich in information and convenient for communication [9].

Through analysis, the system adopts b/s development mode, which can not be limited to the client software. As long as the browser is installed, the system can be accessed. The c/s development mode is not adopted because the latter is limited to installing client software to connect to the server for communication. This mode has great limitations and is not easy to maintain and modify in the future. Therefore, this paper adopts the b/s development mode and uses Apache to publish the teaching resource database management system, so that the system has greater flexibility.

3   Collaborative Recommendation Technology

3.1   Rule Based Filtering Recommendation Method

The rule-based filtering recommendation method is to establish a special utility function f (U, I) based on statistical data or user profiles. This data is usually obtained at the user registration stage and is static. Based on this information, some if then rules are established to select relevant items for specific user groups for recommendation. The recommendation process is shown in Fig. 1. However, this method relies on the pre classification of users and projects. For example, online marketing often divides users by gender and age, and then provides different products and services for these users.

The relationship between users and projects (also known as filtering rules) is usually defined by domain experts, so the recommendation quality of this method completely depends on the quality of the rule base. Therefore, this method usually has some serious maintenance problems, such as the need for experts to formulate rules, difficult verification of effectiveness, and so on. In addition, user registration information is often a subjective description of personal interests, which often has deviation and lack of update over time. Therefore, this method's reliance on user registration information will lead to deviation in recommendations.

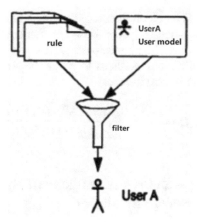

Fig. 1. Rule based filtering recommendation mode

3.2 Content Based Filtering Recommendation Method

Content-based filtering recommendation is to discover and filter resources by comparing the similarity between user requirements and project descriptions. Its concept is mainly derived from the method of "information retrieval (IR)". The content-based filtering recommendation system compares user needs and project information to determine what projects users may be interested in. For example, a news recommendation system uses text classification to extract important terms from news, and matches these terms with user needs and interests to select appropriate news recommendations to users. The recommendation process is shown in Fig. 2.

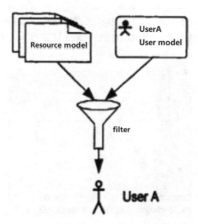

Fig. 2. Content based filtering recommendation mode

In filtering recommendation, utility function f (U, I) is an important concept, which estimates the user U's preference for item I, usually expressed as a numerical value. From a formal point of view, the estimation of the utility function f (U, I) of content-based filtering recommendation is based on the user U's evaluation of I's similar items I. For example, a Book Recommendation System Based on content filtering is to first analyze the books that users have scored high in the past. Only books similar to these books can be recommended. There are many methods to judge the similarity, among which the cosine similarity algorithm is most used:

$$P(i,j) = \overline{R}_i \frac{\sum_{n \in Ni} sim(i, N) \cdot (R_{n,j} - \overline{R}_N)}{\sum_{n \in Ni} sim(i, N)} \tag{1}$$

4 English Teaching Resource Management System Based on Collaborative Recommendation

4.1 Overall System Framework Design

In order to achieve the above objectives, the system uses the combination of file management system and relational database to store and manage teaching resources. The file system is used to store different teaching resources and their description metadata, namely XML documents. The database is used to store the key information that can completely describe the resources, namely the core metadata information. The system consists of six functional modules (taking the administrator interface as an example): user management module (including adding administrator, modifying password and ordinary user authorization), resource editing module (including importing metadata resources and exporting resource metadata), resource management module (including registering metadata resources, modifying and deleting resource metadata), resource retrieval module (including simple retrieval and composite retrieval), Resource evaluation module (including adding resource evaluation and modifying resource evaluation) and LOM element set introduction module (including LOM element set introduction and adding new elements) [10]. The system structure diagram is shown in Fig. 3.

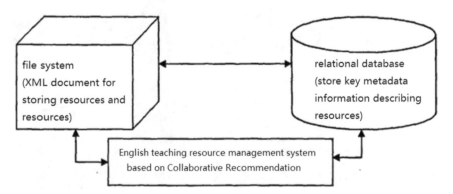

Fig. 3. System structure of English teaching resources

4.2 Resource Editing Module

With the development of network and network education, how to bring the ready-made resources on the network into the resource database, improve the utilization of resources, or publish resources to the network to realize resource sharing, has become one of the design goals of the resource management system, that is, "import metadata resources" and "export resource metadata". On this basis, considering that the document format on the network is still mostly HTML, this system designs and implements the conversion function from XML document to HTML document in the system by studying the XML conversion technology.

(1) Import metadata resource sub module: by checking the validity of metadata of external resources, legal resource metadata is included in the resource database, and the resource and resource description information, i.e. XML documents, are stored in the file system. When using this module to import resources, the system user must first select the type of resources to be verified and the corresponding XML document, and then click the "verify document" button. The system will detect whether the XML file structure of external resources is well structured. If it is well structured, the data legitimacy will be judged according to the DTD file or schema file it locates, and then the legal metadata resources will be imported into the system. This ensures the standardization and unity of resource description documents and database description information, and helps to widely share resources.

(2) Export resource metadata sub module: use XSLT to transform XML documents into HTML documents for display and saving. Before using this module to export resources, the system user needs to select the XML document to be converted and the corresponding style sheet, and then click the corresponding button. The system will automatically call the XSLT processor to scan the entire selected XML document, and use the XSLT style sheet to convert the XML document into an HTML document for display or saving. At present, html is still the mainstream markup language on the network. After the conversion from XML to HTML is realized by designing this module, it is more conducive to expand the sharing of resources uploaded to the network. This function is another innovation of this research mentioned above.

5 Conclusion

With the rapid development of network education, the theme of educational information-ization is to give full play to the advantages of modern information technology, build, manage and make good use of educational resources. However, at present, due to the lack of overall planning standards, unified resource construction standards and description standards, the resources of colleges and universities can not be quickly circulated and information can not be shared. Therefore, how to convert a large number of non-standard teaching resources into standardized resources, so as to maximize the use of existing resources and realize the sharing of resources in a wide range, is a topic worthy of study. Although this paper puts forward some opinions on the standardization construction of teaching resources in theory and develops a system to realize this standardization, due to the limited time and energy, in this paper, I only describe the media materials in teaching resources in XML, and do not involve other types of teaching resources.

References

1. Yan, Q.: Design of teaching video resource management system in colleges and universities based on microtechnology. Secur. Commun. Netw. (2021)
2. Yang, Z., Feng, B.: Design of key data integration system for interactive english teaching based on the internet of things. Int. J. Continuing Eng. Educ. Life-Long Learn. **31**(1), 1 (2021)
3. Wang, J.: Research on college english teaching system based on computer big data. J. Phys. Conf. Ser. **1865**(4), 042141 (10pp) (2021)
4. Zhu, M.: Research on english teaching model with computer aid. In: CIPAE 2021: 2021 2nd International Conference on Computers, Information Processing and Advanced Education (2021)
5. Hendy, N.T.: The effectiveness of technology delivered instruction in teaching human resource management. Int. J. Manag. Educ. **19**(2), 100479 (2021)
6. Xie, X.: Design of english course in middle school based on multiple teaching methods. Open Access Libr. J. **9**(3), 10 (2022)
7. Musa, R.J., Ejovi, O.M., Oghenerhovweya, F.O.: Rethinking the Design of English Language Teaching Online Using the Flipped Classroom Approach. Social Science Electronic Publishing (2021)
8. Zhu, M., Su, C.: The Curriculum Design of SPOC-based Online and Offline Blended Teaching Model of English Linguistics in Flipped Classroom (2022)
9. Nie, A.: Design of english interactive teaching system based on association rules algorithm. Secur. Commun. Netw. (2021)
10. Lin, H., Wei, Y.: Design and implementation of college English multimedia aided teaching resources. Int. J. Electr. Eng. Educ. 002072092098351 (2021)

Design of Financial Modeling System Based on Decision Tree Analysis

Shihui Du[✉] and Xiaochen Guo

Weifang Engineering Vocational College, Weifang 262500, China
gjx1107752384@163.com

Abstract. The design of financial modeling system is based on decision tree analysis. The main objective of this analysis is to determine the best portfolio of investors and choose the best investment strategy. The process includes determining which factors are important for making decisions about investment strategies, how investors should make these decisions, what criteria they should use to choose different investment strategies, and how they will ultimately apply the strategies they choose. It also helps them understand why some investments perform better than others, so it enables them to identify which factors contribute most to this performance. The decision-making process includes three steps: 1. Data analysis: this step involves collecting data from different sources, such as the company's annual report or other documents, and then analyzing them using statistical techniques. 2. Decision making: according to the results obtained in step 1, you can take appropriate actions for next year's performance through this method. 3. Implementation of decision making: at this stage, you must take appropriate measures (such as setting new strategies or changing existing strategies) to implement the decision.

Keywords: Decision tree · Financial modeling · Systems analysis

1 Introduction

The quality of the company's financial situation must be the focus of attention of the enterprise itself, investors and creditors. A well run and financially healthy company can not only improve its reputation in the market, but also broaden its financing channels. On the contrary, any financial problems of the enterprise will restrict the long-term development ability of the enterprise and even directly lead to the bankruptcy of the enterprise [1]. Therefore, early detection of the company's financial crisis signal enables the company's operators to take effective measures to improve the business operation and prevent the crisis in the embryonic stage of the financial crisis.

The financial situation of the company will also affect the credit scale of the bank to the company. As the bond market in China is still underdeveloped, the debt financing of the company mainly comes from the loans of commercial banks. In order to control risks, commercial banks, as creditors, will reduce the scale of credit to companies with poor financial conditions. Reducing the credit scale of commercial banks will inevitably aggravate the financial difficulties of enterprises themselves. Therefore, the company's

J. C. Hung et al. (Eds.): IC 2023, LNEE 1045, pp. 125–130, 2023.
https://doi.org/10.1007/978-981-99-2287-1_18

managers must focus on the company's own financial risks and respond in a timely manner. Once the company encounters special treatment, it will have a great negative impact on the company's subsequent operation [2]. Among them, computer financial management includes financial modeling, which can effectively provide decision-making suggestions for managers by establishing financial data related models and optimization. Based on this, this paper studies the design of financial modeling system based on decision tree analysis.

2   Related Work

2.1   Decision Tree Modeling

First, to put it simply, the decision tree is a "if judgment tree structure", which follows the divide and rule strategy. Each node is a feature. The key to dividing data according to the feature is to find out how to select the feature as the node and divide the data, and how to define the loss function to evaluate the result. Each decision in the decision-making process is a "test" of a certain attribute, and the final decision result corresponds to the final decision result. In general, a decision tree includes a root node, several internal nodes and several leaf nodes. It is easy to know:

Each non leaf node represents a feature attribute test.

Each branch represents the output of this characteristic attribute in a certain value range.. Each leaf node stores a category.

The sample set contained in each node is divided into child nodes through attribute test, and the root node contains the full set of samples.

The construction of a decision tree is a recursive process. There are three situations that will lead to recursive returns: (1) all the samples contained in the current node belong to the same category. At this time, the node is directly marked as a leaf node and set as the corresponding category; (2) If the current attribute set is empty or all samples have the same value on all attributes and cannot be divided, the node is marked as a leaf node and its category is set as the category with the largest number of samples in the node; (3) The sample set contained in the current node is empty and cannot be divided. At this time, the node is also marked as a leaf node and its category is set as the category with the largest number of samples in the parent node [3]. The following Fig. 1 shows the flow of the decision tree algorithm.

It can be seen that the key to decision tree learning is how to select the partition attributes, and different partition attributes will lead to different branch structures, thus affecting the performance of the whole decision tree. The goal of attribute division is to make the divided child nodes as "pure" as possible, that is, they belong to the same category. According to the improvement of specific methods for quantifying purity, there are three kinds of decision tree algorithms: ID3, C4.5 and cart (most commonly used). Information entropy, gain rate and Gini index are used as measurement methods respectively.

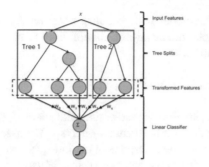

Fig. 1. Decision tree algorithm flow

Pruning is the main means for the decision tree algorithm to deal with overfitting. There are two pruning strategies:

Pre pruning: in the process of construction, it is first evaluated before considering whether to branch. Pre pruning makes many branches of the decision tree cut off, thus greatly reducing the training time cost and reducing the risk of over fitting. On the other hand, since pruning also cuts off the branches of the current node and the subsequent child nodes, the nature of "greedy" pre pruning prevents the expansion of branches, and to a certain extent, brings the risk of under fitting.

Post pruning: after constructing a complete decision tree, evaluate the necessity of branching from bottom to top. Post pruning usually retains more branches, so the performance of the decision tree adopting the post pruning strategy is often better than that of the pre pruning [4]. However, it traverses all nodes from bottom to top and calculates the performance. The training time cost is greatly improved compared with that of the pre pruning.

For continuous value processing: for continuous value attributes, it is not feasible if each value is taken as a branch, so it needs to be discretized. The commonly used method is dichotomy. The basic idea is: given the sample set D and the continuous attribute d, the dichotomy tries to find a partition point t to place the sample set D in the attribute α It is divided into $< T$ and $> t$.

2.2 Financial Analysis

Financial analysis is an economic management activity based on accounting, statement data and other relevant data, which uses a series of special analysis techniques and methods to analyze and evaluate the profitability, operating capacity, solvency and growth capacity of enterprises and other economic organizations in the past and present related financing activities, investment activities, operating activities and distribution activities.

It is an economic application discipline that provides accurate information or basis for investors, creditors, operators and other organizations or individuals concerned with enterprises to understand the past of enterprises, evaluate the current situation of enterprises, predict the future of enterprises and make correct decisions [5].

The financial analysis model is to establish some models through some financial indicators to make complex financial information easy to understand. There are three main financial analysis models:

1. Enterprise strategy analysis model:

On the basis of clarifying the purpose of financial analysis, enterprise strategic analysis is the starting point of enterprise financial analysis. The essence of strategic analysis R is to clarify the position of the enterprise in the industry and the competitive strategy that should be adopted through the analysis of the industry in which the enterprise is located or the industry that the enterprise intends to enter, so as to weigh the benefits and risks, understand and master the development potential of the enterprise, especially the potential in enterprise value creation or profit.

2. Accounting analysis model:

The essence of accounting analysis is to clarify the connotation and quality of accounting information, that is, to reveal its actual meaning from the surface of accounting data. The analysis includes not only the analysis of the connotation of each accounting statement and related accounting subjects, but also the analysis of accounting principles and policy changes, accounting method selection and changes, accounting quality and changes, etc.

3. Financial statement analysis model:

The analysis of financial statements is based on the financial statements, adopts scientific evaluation standards and applicable analysis methods, follows standard analysis procedures, and makes judgment, evaluation and prediction on the business situation and performance of the enterprise by comparing and analyzing the financial situation, operating results, cash flow and other important indicators of the enterprise [6].

3   Design of Financial Modeling System Based on Decision Tree Analysis

The most important thing about the design of financial modeling system based on decision tree analysis method studied in this paper is that because of this tool, the data architecture of the whole company can become standardized, and the next step is to build the big data platform of the enterprise. Moreover, it is written in Java and supports secondary development. It is an excel like designer. It is very easy to start, whether it is it or business: editing SQL optimization and data set reuse are all small cases, greatly reducing the threshold of report development. In the aspect of data security, which is most concerned in the enterprise, it supports multiple people to develop the same set of reports at the same time, and prevents editing conflicts through the template locking function [7]; Ensure data security through data analysis and permission control.

Whether in the database or in the DW/Bi design, it is necessary to do dimension modeling. However, due to different performance and objectives, the rules for dimension modeling are different. What I am talking about here is mainly dimension modeling in the DW/Bi design. The following is an example of the process of constructing a decision tree. The information gain is used to find the field with the largest amount of information in the data table, and a node of the financial analysis decision tree is established. Then, each branch of the tree is established according to the different values of the field, and then the lower node and branch of the tree are repeatedly established in each branch. Taking the traffic cost data set as an example, the classification decision tree of "overall traffic cost exceeding standard level" is constructed [8]. Classification label: Class P It refers to the overall over standard level of transportation expenses = "1", and category n refers to the overall over standard level of transportation expenses = "0". Similarly, a small amount of data is selected here to establish a branch of the tree and highlight the classification process.

4   Simulation Analysis

The financial data analysis model is established according to the decision tree ID3 algorithm, and the financial data set is tested. The prediction accuracy can be reached, and whether it can be reported as the budget of the next year. These problems are also completed through the decision tree algorithm.

In the process of financial data analysis, two-thirds of all data sets are used as data training sets, and the prediction model is established according to the decision tree ID3 classification algorithm to generate a decision tree and evaluate the accuracy of the established model. Then one-third of the whole data set is used as the data test set to test the accuracy of the model algorithm on the experimental platform. Data_ The pretreatment() function is mainly used to discretize the samples with continuous test attributes. Because ID3 decision tree algorithm cannot classify continuous attributes. If classification is enforced, each continuous value of an attribute will be treated as an attribute category [9]. In this case, the tree will be very large and the efficiency of building the tree will be very low. Even if the tree is constructed, because of the overfitting of continuous attributes, the accuracy of classification of unknown samples using the tree will be very poor. The test item code is shown in Fig. 2 below.

When different sample numbers are selected, when the two accuracy rates are in a stable or rising state, it means that this data set is well adapted to this kind of algorithm, and can be considered to provide auxiliary decision-making information for the actual financial work; Otherwise, it means that this data set is not suitable for this algorithm, can not be predicted by classification analysis algorithm, and can not guide the work. There are many factors affecting the evaluation results, such as the number of samples in the training set and the test set, the number of samples in a single tree, and the selection of algorithm parameters. With the increase of the number of samples, the decision tree structure will become more complex, and the accuracy may fluctuate [10]. Therefore, it is necessary to conduct multiple tests to determine the classification performance.

```
t0=clock;

[E correct_test correct_train

  error_train_num]=M_ID3(handles.train,handles.test,clas,attribute_kind,long,thet);

correct_test_num=sum(correct_test,2)*100;

set(handles.edit5,'string',correct_test_num/(handles.len_test));

set(handles.edit6,'string',1-error_train_num/handles.len_train);

set(handles.edit9,'string',length(E));

axes(handles.axes1)

plot(correct_train)
```

Fig. 2. Test item code

5   Conclusion

The design of financial modeling system is based on decision tree analysis. The main purpose of this type of analysis is to let users understand the relationship between different data sets and the relationship between them. This process first identifies all possible combinations or permutations of the variables in question and then proceeds through a series of steps designed to help you determine which combination is best suited to your particular situation. Decision trees are used as part of many different types of analysis, including those related to marketing, finance, sales forecasting, and customer relationship management.

References

1. Zhang, W.: Research on English score analysis system based on improved decision tree algorithm and fuzzy set. J. Intell. Fuzzy Syst. **39**(4), 5673–5685 (2020)
2. Zhang, X.Z.: Analysis and design of tourism CRM system based on decision tree. Mod. Comput. (2018)
3. Ming, Q., Li, R.: Analysis and Design of Personalized Learning System Based on Decision Tree Technology (2020)
4. Ming, Q.: The analysis and design of the personalized learning system based on decision tree. Microcomput. Appl. (2018)
5. Geng, X., Yang, D.: Intelligent Prediction Mathematical Model of Industrial Financial Fraud Based on Data Mining. Hindawi Limited (2021)
6. Jin, M., Wang, H., Zhang, Q., Luo, C.: Financial management and decision based on decision tree algorithm. Wireless Pers. Commun. **102**(4), 2869–2884 (2018). https://doi.org/10.1007/s11277-018-5312-6
7. Le, Y., Zizong, T., Panpan, W.U., et al.: Visual modeling of rice root growth based on B-spline curve (2022)
8. Huang, Z., Sun, Y., Gan, L., et al.: Durability Analysis of Building Exterior Thermal Insulation System in Hot Summer and Cold Winter Area Based on ANSYS (2022)
9. Varghese, V., Krishnan, V., Kumar, G.S.: Evaluating pedicle-screw instrumentation using decision-tree analysis based on pullout strength. Asian Spine J. **12**(4), 611–621 (2018)
10. Huang, Z., Liang, Y.: Research of data mining and web technology in university discipline construction decision support system based on MVC model. Library Hi Tech (2019)

Design of Information Intelligent Management System for Nursing Training Room Based on Big Data Environment

Chan Tang[⊠]

Chengdu Polytechnic, Chengdu 610000, Sichuan, China
464867984@qq.com

Abstract. With the development of economy and the strong support of governments at all levels for the in-depth integration of production, learning and research in Colleges and universities, the scientific research cooperation projects of colleges and universities show a sharp upward trend. In this case, the rapidly rising data brings great pressure on the processing efficiency of the scientific research management system of colleges and universities. The design of nursing training room information intelligent management system based on big data environment is to improve teaching efficiency by using modern technology and Informatics, such as artificial intelligence, machine learning, deep learning and other new technologies. The design also includes the improvement of the teaching content system. In this case, computer-based teaching materials are used. This will provide more effective content for students' learning and make students' learning process more convenient and efficient. By using different teaching methods at one time, or dividing them into several parts according to the level of students, so as to promote students' understanding of knowledge.

Keywords: promotion of information technology · big data · Nursing training room · intelligent management

1 Introduction

Since the 21st century, China has paid unprecedented attention to vocational education. The difference between vocational school teaching and other types of school teaching lies in its special emphasis on practical teaching, which is determined by the specific training objectives of vocational education. Practical teaching is an important part of vocational education teaching system. It is an important way to cultivate students' professional technology application ability and ability to analyze and solve problems [1]. It is an indispensable link in the process of vocational education. The implementation of practical teaching requires the construction of training bases. China's governments at all levels from the central to local governments and vocational schools attach great importance to the construction of training bases. At the implementation level, the construction of various types of vocational education training bases in various regions is also booming, which is the basic condition for doing a good job of vocational education. However, the development of Vocational Education in China is unbalanced [2]. For example, nursing education has always been subordinate to medical education and has not been paid attention to by vocational education. The reform of in-service education curriculum has been a slow step. However, with the cultivation of nursing professionals included in one of the national skills shortage talent cultivation, the theory of vocational education curriculum has gradually attracted attention in the field of nursing education.

The information construction of colleges and universities in China has made phased achievements. However, the information construction of colleges and universities in China is still in the exploratory stage [3]. There are some problems in policy guidance, fund planning, implementation and use, which requires colleges and universities to seek new ways in continuous attempts. With the popularization and application of cloud computing, big data, Internet of things, mobile Internet, social network and other concepts and technologies, as well as the implementation of the national "Internet+" action plan, the field of university informatization has also entered a new stage of development. As an important part of the social innovation ability of colleges and universities, scientific research informatization ability plays an increasingly important role in the information construction of colleges and universities [4]. At present, the difficulty of scientific research management in Colleges and universities is increasing, and the management process is becoming more and more complex. Only combining with traditional management methods will not meet the needs of scientific research operation in Colleges and universities in the information age.

With the development of science and technology, information technology is more and more closely related to modern life, and information resources have become one of the most important resources in the new era. From the perspective of education, the information construction of colleges and universities is the inevitable product of the development of science and technology. Information construction in Colleges and universities is the practical need for colleges and universities to realize educational modernization [5]. It is also the actual requirement for building a modern education system and cultivating modern talents who meet the requirements of the times. Using information technology to carry out modern management can greatly improve the efficiency of campus management, provide decision support for university managers, reduce the workload for managers, and promote more scientific and reasonable campus management.

2 Related Work

2.1 Nursing Training Room

"Nursing is a profession" is put forward from the perspective of nursing development, because the traditional nursing work is limited to simply being the assistant of doctors, the training of nursing practitioners is also trained in the form of short-term training, and the school running form is also hospital run nursing school, so it reflects more on the level of profession. Due to the continuous development of nursing, nursing has gradually developed from a profession or a simple technology to a profession, It should be said that this is the result of social development and progress. This change has the same occurrence in all fields [6].

Specifically, specialty has three meanings: one is the discipline of higher education, the other is the division of department specialty, and the third is specialized in a certain job or occupation. Therefore, nursing specialty can also have the following understandings. The first level is a discipline in higher medical education. From this point of view, higher education emphasizes nursing as an independent discipline, so it pays more attention to the construction of its discipline system. The second meaning is the different division of labor in medical and health undertakings. The third meaning is specialized in nursing work. The professional concept here refers to the specialization requirements brought by social division of labor, job classification and occupational differentiation [7]. In this sense, the occupation is more biased towards things, and the specialty is more biased towards "people"; Occupation focuses on the normative requirements of things to people, while specialty focuses on the knowledge, skills and quality requirements required for people to do things well. Generally speaking, a career is what you want to do, and a major is what you want to do. Professionalization and specialization are the most important and basic management tasks for all current work fields.

The goal of nursing vocational education is to cultivate applied talents engaged in nursing work in medical and health institutions. Therefore, it is far from enough to only have practical teaching links such as theoretical verification, demonstration and simple mechanical decentralized practice [8]. It must emphasize that in a certain real situation, after mastering individual skills, the implementation of overall nursing activities for nursing objects, namely "practical training", can cultivate qualified nursing talents required for professional posts. "Practical training" mainly refers to the process of training and improving people's practical operation ability and enabling trainees to acquire technical skills through on-the-spot experience, practical operation and practical exercise. "Training" has a clear orientation: that is, the purpose of training is to improve the technical skills of trainees, the content of training is to focus on practical operation skills, the means of training is to focus on physical operation and simulation, and the function of training is to reflect whether to promote employment.

2.2 Analysis of Targeted Problems in Training Room

The training room information management system is an information management tool that combines the information technology with the database as the core and the management needs of the training room. With the continuous appearance of the disadvantages of the original manual training room management, the training room information management system has greatly improved the efficiency of training room management and use, and specifically solved the following problems:

(1) Low efficiency of training room management

The basic information collection and management of the training room is still in the stage of manual operation or document management. The efficiency of query and management is very low, and the error rate is quite high, so it is difficult to save the data. When the evaluation and inspection or data reporting is required, it is necessary to work overtime to catch up with the materials, which not only makes the already heavy management task more arduous, but also inevitably leads to the omission of everything and the loss of a lot of information. In order to solve this problem, there is an urgent need for a training room management system to manage information such as experimental projects, personnel, rooms, equipment, training room construction, log documents and so on.

(2)   It is difficult to count the equipment and consumables in the training room

Training equipment and low value durable goods have not well established an open query information base, which is not conducive to understanding the status of these equipment and the maintenance of these experimental equipment; The collection, borrowing, repair and scrapping of instruments and equipment are still in the manual processing stage, and the processing process is cumbersome and prone to mistakes, resulting in the loss of equipment; The management of experimental consumables also has great subjective randomness, which is easy to cause the waste of consumables [9]. The statistical process of instrument and equipment information is complex, takes up a lot of working time, and the consumption of consumables can not be well counted.

(3)   The management of archives in the training room is relatively backward

All kinds of materials in the training room are not kept in a single page, and the file function is insufficient. All kinds of materials and files in the training room, such as construction materials, technical materials, management materials, teaching materials and other important file information, are not digitally saved. All kinds of paper materials are scattered in each training room, which are easy to be damaged and lost, and are not conducive to statistical data and future utilization.

3   Design of Information Intelligent Management System in Nursing Training Room

3.1   Java EB Software Development Technology

Java EE (Java Platform Enterprise Edition) is an enterprise application standard platform launched by Sun company for enterprises. In terms of actual development, Java EE can be regarded as a development framework or an industry standard.

Generally, the architecture of the program is divided into C/S (client/server) architecture and B/S (Browser/server) architecture. The schematic diagram of C/S architecture and BS architecture is shown in Fig. 1. Among them, C/S architecture, i.e. client/server architecture, is to install the client on the computer of local users, which is more convenient to call local hardware resources. It is suitable for various tool application software, audio and video software, various large-scale game software, various system development software and other application software. The B/S architecture, i.e. browser/server architecture, realizes the user interface through the browser. Under this architecture, the main thing logic is realized on the server. Therefore, it simplifies the workload of the client and reduces the workload of the local computer, that is, the system does not need to follow the new client, which greatly reduces the cost of users [10]. Java EE is mainly used for web development platform based on B/S architecture. It can be said that Java EE is one of the most widely used parts of Java.

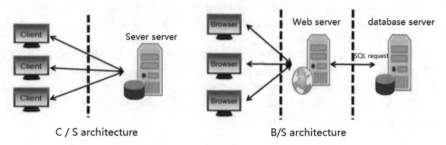

C / S architecture   B/S architecture

Fig. 1. Schematic diagram of C/S architecture and BS architecture

3.2 Overall System Architecture Design

Based on the big data environment, the information-based intelligent management system of nursing training room adopts B/S architecture in the overall system architecture, and the system development uses Java EE tools. These are relatively mature technologies. The overall architecture of the system is shown in Fig. 2.

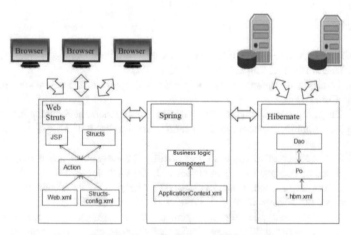

Fig. 2. Overall system architecture

The web layer of the information intelligent management system in the nursing training room is realized by JSP and struts, with the JSTL basic class library as the core, and the common class object undertakes the model function in MVC. Each action class in struts undertakes the controller function in MVC. Various operation requests are sent to their action modules through HTTP and processed logically. For example, in the docking of information between science and technology system and financial system, the push out information table to the financial system will be processed through fundallot action and finally stored in the database server through the persistence layer.

The business layer of the system uses the spring framework for the overall association of business components, analyzes the web layer and persistence layer, and unifies various management things. Using the spring framework can greatly enhance the compatibility and scalability of the system.

The persistence layer of the system will persist the data of the database through the Hibernate framework, map the database tables in Oracle 10g into data operation classes, simplify the operation of the database into the operation of classes, greatly simplify the coding work of Dao layer, and realize the addition, deletion, modification and query of data through a small number of SQL statements.

4 Conclusion

With the higher and higher requirements of management informatization of nursing training room in Colleges and universities, the functions provided by the system can be further optimized. The system also has some defects, which need to be improved in details, and the needs of users are not 100% satisfied. Relevant functions will continue to be improved in the future work. The intelligent review of scientific research contract text can have a more intelligent solution, which is limited to working time and energy. It has not been fully developed during the study period, and will continue to be developed in the future work.

References

1. Li, Q., Chen, Y.: Application of intelligent nursing information system in emergency nursing management. J. Healthc. Eng. **2021**(13), 1–13 (2021)
2. Fang, S.: Design and implementation of intelligent property management system. J. Beijing Polytech. Coll. (2020)
3. Feng, L., Zhao, J.: Research on the construction of intelligent management platform of garden landscape environment system based on remote sensing images. Arab. J. Geosci. **14**, 1–19 (2021)
4. Li, F., Gao, W.: Research on the design of intelligent energy efficiency management system for ships based on computer big data platform. J. Phys. Conf. Ser. **1744**(2), 022026 (2021)
5. Song, G., Zuo, T.H., Zhao, L.L.: Design and realization of intelligent vehicle based on embedded system. J. Phys. Conf. Ser. **1754**(1), 012080 (4pp) (2021)
6. Armyanova, M.: Possibilities for intelligent cities development with design patterns. In: International Scientific and Practical Conference "Construction Entrepreneurship and Real Property". University of Economics – Varna (2021)
7. Huang, Y., Chen, X., Xiao, X., et al.: Design of intelligent meter life cycle management system based on RFID (2020)
8. Kim, H., Choi, J.: Intelligent access control design for security context awareness in smart grid. Sustainability **13**, 4124 (2021)
9. Mumtaz, Z., Ilyas, Z., Sohaib, A., et al.: Design and implementation of user-friendly and low-cost multiple-application system for smart city using microcontrollers (2020)
10. Li, H.B.: Modeling method of tax management system based on artificial intelligence. Int. J. Artif. Intell. Tools, **29**(7n08), 2040023 (2020)

Design of Mental Health Platform for Adolescent Group Based on Random Forest Algorithm

Haiyang Ding[✉] and Qixuan Sun

Southwest University, Chongqing 404100, China
sean25@foxmail.com

Abstract. Because the mental health level of Chinese adolescents is generally poor, it is necessary to design a mental health platform to help understand the mental health status of adolescents and provide decision support for psychological intervention. This paper mainly introduces the status quo of adolescent group psychology, and then carries on the platform design based on random forest algorithm. Through the research, this platform can analyze the mental health status of the adolescent group under the effect of random forest algorithm, and can play a role in the early warning and evaluation of the adolescent mental health intervention.

Keywords: Random forest algorithm · The youth group · Mental Health Platform

1 Introduction

Youth group is in the important stage in psychological development, the stage of their psychological changeable, vulnerable, so many teenagers have emerged due to the impact of external forces in mental health problems, the present study for teenagers psychological activities and future personal development has seriously affected, even will cause a distortion of the young people personality construction, Therefore, attention should be paid to the mental health of adolescents. From this point of view, in order to avoid and correct the mental health problems of adolescents, people need to understand the specific causes and development of the problems before making accurate decisions for intervention. During this period, the time of the formation of mental health problems should be taken into account, so that accurate and punctual intervention can be carried out. To achieve this purpose, the related group psychological health think teenagers can design a platform, through this platform to solve related problems, help do accurately and on time intervention, and the key lies in the platform design algorithms, the algorithm can make the platform independent analysis of adolescent mental health status, timely provide accurate results, so that human decision-making, the random forest algorithm is proposed in this field. How to design the platform based on the algorithm is the main problem at present, which needs to be studied.

J. C. Hung et al. (Eds.): IC 2023, LNEE 1045, pp. 137–143, 2023.
https://doi.org/10.1007/978-981-99-2287-1_20

2   Mental Health Status of Adolescents

At present, the youth community mental health level of comprehensive is uneven, the overall is low, the characteristics of the overall point of view, the vast majority of teens are more or less have some unhealthy psychology, common types are shown in Table 1, at the same time, a lot of young people is more than one mental health problems, and higher degree, so the overall level is low.

Table 1. Common types of unhealthy psychology among adolescents

The target	Type
Youth group	The negative psychological
	Sensitive psychological
	Radical psychological

In terms of causes, the unhealthy psychology of adolescents is mainly caused by two factors: One is the pressure that teenagers age span is larger, contains a group of students and graduates, but all of these groups to study, live, work, such as pressure, the pressure is there for a long time, thus affects their psychological, such as students under pressure from their studies becomes hot-tempered, or pleasure, on the other hand, the graduates may become impulsive, blind and overly sensitive, which indicates that pressure is one of the main factors leading to the unhealthy psychology of teenagers. Second is the message that modern society is more open, the height of the combined with the network popularization, the youth group almost daily access to the vast amounts of information, and these information in some bad information, affected by such information, the ideas of the youth group is likely to change bad, indirectly lead to their psychological unhealthy, for example, some students had psychological is very positive and optimistic, however, under the influence of negative information from the outside world, she changed her attitude towards life and gradually became negative. She said, It is better to be negative because a positive attitude will not succeed. In addition, there are other influencing factors, which will not be further described in this paper [1–3].

Influence on unhealthy psychological for youth groups in each life stage has the serious influence, negative psychology, for example, the psychological causes people not active, so suppose teenagers at the learning stage, affected by the psychological youth academic decline, and difficult to ascend, this phenomenon may make it harder for teenagers to become the social needed talents, the future employment will be indirectly affected, but also lead to the distortion of the outlook on life and development of young people, and even lead to the idea of "suicide" [4–6]. It can be seen that unhealthy psychological influence is numerous and significant, so people should pay attention to it.

3 Design of Mental Health Platform for Adolescent Groups

3.1 Platform Framework Design

Figure 1 shows the basic framework of this platform.

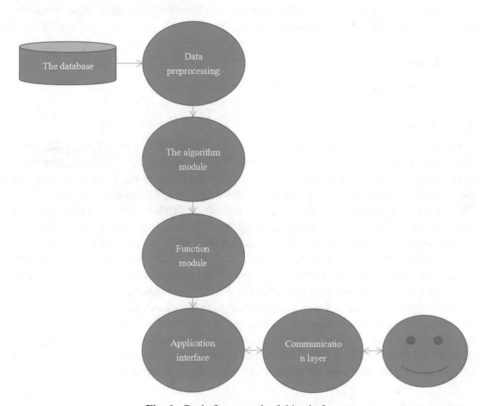

Fig. 1. Basic framework of this platform

Combined with Fig. 1, this platform can be divided into the database, data pretreatment, algorithm module, function module, application interface, communication layer six plate, through database receiving teenager groups in the operation of the mental health of data, and complete the data classification, and then for all kinds of data preprocessing, respectively, to remove the low quality data, after the pretreatment of data into the algorithm module, mental health model, model calculation, get the results of the analysis of mental health, such as negative and sensitive unhealthy psychology can be divided into two categories, data through data training, every appear negative psychological characteristic data, just vote for the negative psychological category, and so on, the final results, with the most votes for teenagers both concrete existence of the unhealthy psychology, The whole analysis process will be presented in the form of a report, from which the causes can be known. The result report will be sent to the application interface and displayed together with the function module. The function module mainly integrates the application functions of the system, including data scheduling and report review, etc. [7–10]. All the functions are displayed on the application interface. The application interface is not only responsible for displaying the calculation results and functions, but also serves as the receiving and sending end of the communication layer, enabling users to enter the application interface through the communication layer, use relevant functions and view corresponding results. The communication layer is mainly responsible for establishing communication channels.

3.2 Platform Implementation

The implementation method of this platform is as follows.

3.2.1 Database

Database, this paper mainly considers the two types of database, the physical database, the virtual database, which the former is mainly based on the physical server, create the data storage space, the space of the specific capacity depends on the resources of the physical server array size scale, the larger the magnitude, the greater the capacity, so the physical database is a database storage capacity is limited, the advantages are high safety level and strong realizability of functions. The disadvantages are high cost and tedious expansion operation. Which is based on the virtual server, the data stored in the virtual network space, space is infinite capacity, but for security reasons, it is necessary to limit the space, if has the capacity to operate again can, expansion and advantage of low cost, simple operation, strong function can be realized, the disadvantage is that low level of security (protection) can be performed, and therefore need frequent expansion. According to the characteristics of the two types of databases, because there are many types and causes of mental health of teenagers, and the number of teenagers is huge, the demand for data storage is high. The virtual database can better meet the demand, and this paper mainly selects the cloud database which is more representative.

3.2.2 Data Preprocessing

The realization of data preprocessing is to choose the preprocessing tools according to the possible quality problems of data. Through simple statistics, this paper learned that there are two types of possible quality problems in the mental health data of adolescents, as shown in Table 2.

Table 2. Common data quality problems

The data type	Describe
Repeat classes	Refers to data with a repetition rate of 100%
Incomplete class	Data that does not conform to a standard format

According to Table 2, data deduplication and data comparison tools are mainly selected in this paper, among which the former can identify each byte of data and establish a byte group. The repetition rate between data can be determined by comparing byte groups. If the repetition rate reaches 100%, either item will be deleted, and the repeated data with the later input sequence will be preferentially selected. The latter mainly compares the data according to the standard format. If the comparison result shows that there is a blank plate after the data is filled in the standard format, it means that the data is incomplete and cannot be input.

3.2.3 Algorithm Module

Firstly, according to the mental health analysis needs of young people, the selection of algorithms includes random forest algorithm and ID3 algorithm. Two algorithms, by contrast, are the function of data analysis and adolescent mental health, but the ID3 algorithm of information gain to take more flawed, this one defect on behalf of ID3 algorithm in practical application of universality and may ignore some elements, such as some rare mental health problems, or is some special causes mental health problems, etc., These performances will make the system unable to fully serve every youth group, indicating that the system performance is limited, while random forest algorithm does not have such defects, so this system chooses random forest algorithm.

Second is a kind of random forest algorithm based on the integration method of machine learning algorithms, able to build multiple decision trees around data set first, and then will be more than the decision tree integration, show that the algorithm is based on decision tree, each decision tree is the equivalent of a classifier, the same types of data, so each input sample into N, N classification tree as a result, based on these results, the classification voting method is adopted to make a judgment, in which the category with the most votes is the final output. On this basis, as for the implementation method of random forest algorithm module, this paper mainly uses Java language for module programming, and establishes the algorithm model, as shown in Formula (1).

$$H(X) = - \sum_{i=1}^{n} p_i \log P_i \tag{1}$$

where: n represents n different discrete values of X, P_i represents the probability that X takes the value of i ($i \in n$), and log base 2 or e.

3.2.4 Function Modules

The relevant functions of the function module include data scheduling, report review, etc. The main development technology of these functions is the same, that is, the Java language is used for programming, but the difference lies in the development port. First is data scheduling function, the function is mainly from the database of scheduling algorithm data needed for the module operation, for operations, so the port is the functional development database, in combination with cloud database, this paper chose the corresponding tools in the cloud resources environment, port development is studied by using tools, gives the port corresponding privileges, make the database open to the data scheduling function, users can also get the function of scheduling data through this function. Followed by the report of access function, the function of main results, and the transmission of the report to the application interface, so in addition to using programming technology for development, mainly in the application interface port development, so that when the report generation is in standby state, if the user starts the function, can get the corresponding feedback report.

3.2.5 Application Interface

The application interface mainly uses UI development technology, which is developed on the Web page. In addition to the exquisite layout of the interface, it also takes into account the functional operation logic. Taking the data scheduling function as an example, the user should log in first before selecting the function for use. Therefore, the UI design should put the "login" button first and the "function" button second.

3.2.6 Communication Layer

The communication layer is between the application interface and the user side, enabling them to interact and connect. Considering that the users facing the platform may be a large number of teenagers or related groups, the phenomenon of facing a large number of user requests at the same time will occur in practical applications, which will lead to communication congestion and even lead to data loss. In order to avoid such phenomenon, this paper selects communication protocols in the design of communication layer, and the selected items are shown in Table 3.

Table 3. Communication protocol selection items

The name of the protocol	Introduce
TCP	Large public network
Ethernet	Semi-open LAN
VPN	Private network

The first is TCP. As a large public network, TCP can support multiple people online and receive user requests synchronously, so it basically meets the communication requirements of this platform. However, TCP is relatively open and its security level is insufficient, which means that it is necessary to build a security protection system when using TCP, which will increase the cost and difficulty of platform implementation. Followed by Ethernet, its security is higher, can also support a certain number of communication, but as a half open LAN, normal only for a specific network within the scope of the user to provide services, to the foreign service of network users, requires the administrator to open access, and user identity verification, identification and so on, its operation is complicated, the convenience of the platform system. The last is VPN, which only supports one-to-one communication, but multiple one-to-one communication links can be established to support multi-person access. This method has a high cost, but the actual effect is the best. After comparing with each other, this paper chooses TCP, and set up a firewall system to ensure security.

4   Conclusion

In conclusion, random forest algorithm can well analyze the mental health problems of adolescents, so it is feasible and practical value to build a platform around this algorithm. Relying on the platform, people can understand the specific types of mental health problems of young people and learn the causes, so as to make accurate decisions, intervene, avoid and correct unhealthy psychology.

Acknowledgements. This work is supported by Project S202210635003 supported by Chongqing Municipal Training Program Of Innovation and Entrepreneurship for Undergraduates.

References

1. Saadoon, Y.A., Abdulamir, R.H.: Improved random forest algorithm performance for big data. J. Phys. Conf. Ser. **1897**(1), 012071 (13pp) (2021)
2. Zhang, X., Xu, J., Chen, Y., et al.: Coastal wetland classification with GF-3 polarimetric SAR imagery by using object-oriented random forest algorithm. Sensors **21**(10), 3395 (2021)
3. Kim, J.Y., Lee, M., Min, K.L., et al.: Development of random forest algorithm based prediction model of Alzheimer's disease using neurodegeneration pattern. Psychiatry Invest. **18**(1), 69–79 (2021)
4. Ali, M.S., Islam, M.K., Haque, J., et al.: Alzheimer's disease detection using m-random forest algorithm with optimum features extraction. In: 2021 1st International Conference on Artificial Intelligence and Data Analytics (CAIDA) (2021)
5. Papineni, S., Reddy, A.M., Yarlagadda, S., et al.: An extensive analytical approach on human resources using random forest algorithm. Int. J. Eng. Trends Technol. **69**(5), 119–127 (2021)
6. Lei, G., Su, S., Liao, W.: Classification of credit card holders based on random forest algorithm. In: ICMLSC 2021: 2021 The 5th International Conference on Machine Learning and Soft Computing (2021)
7. Bao, S., Pan, H., Zheng, W., et al.: Multicenter analysis and a rapid screening model to predict early novel coronavirus pneumonia using a random forest algorithm. Medicine **100**(24), e26279 (2021)
8. Ghorbanian, A., Zaghian, S., Asiyabi, R.M., et al.: Mangrove ecosystem mapping using Sentinel-1 and Sentinel-2 satellite images and random forest algorithm in Google earth engine. Remote Sens. **13**(13), 2565 (2021)
9. Dong, X., Meng, Z., Wang, Y., et al.: Monitoring spatiotemporal changes of impervious surfaces in Beijing city using random forest algorithm and textural features. Remote Sens. **13**(1), 153 (2021)
10. Nurwarsito, H., Nadhif, M.F.: DDoS attack early detection and mitigation system on SDN using random forest algorithm and Ryu framework. In: 2021 8th International Conference on Computer and Communication Engineering (ICCCE) (2021)

Design of Tour Guide Course Reform System Based on Virtual Simulation Resources

Hongyan Li[✉]

Shandong Institute of Commerce and Technology, Jinan 250103, Shandong, China
lihongyan1020@163.com

Abstract. The research on the system design of tour guide course reform based on virtual simulation resources is a research on the effect and potential of traditional teaching methods reform in tour guide course. This study aims to understand how to improve students' learning ability through the use of virtual simulation tools, which are widely used in educational institutions around the world. The key factors affecting students' learning ability include: (1) their own level of understanding; (2) Its interest level; (3) Incentive level; (4) Their knowledge acquisition ability, etc.

Keywords: Virtual simulation teaching · Tour guide course · reform in education

1 Introduction

With the development of information technology, the traditional tourism market operation mode has been greatly impacted, which is the inevitable requirement of China's transformation from traditional tourism industry to information-based tourism industry. At the same time, it also puts forward new requirements for the practitioners in the tourism industry. Under this background, it is not only required that tour guides master rich and professional tourism knowledge, but also required to have the necessary information-based ability, and become compound tourism talents with both professional ability and information-based quality. The traditional teaching methods are no longer suitable for the demand of tourism talents under the background of virtual tourism [1]. As the main training base of tourism talents, tourism schools should be based on the training of knowledge and technology composite tourism talents in the process of integration of information technology and tourism industry, and apply information technology such as virtual reality technology to teaching, which can not only provide a more real learning environment for students' learning, It is also conducive to the cultivation of students' information literacy. Therefore, tourism colleges and universities should pay attention to the training of tourism talents, cultivate tourism talents who adapt to the transformation, and strive to occupy a strong position in the international tourism industry.

As the resource production of the virtual simulation training course needs a certain amount of computer programming technology support, for most teachers of the design specialty, they are more specialized in the professional teaching field, do not know much about the implementation technology of the virtual simulation training course, and it is difficult to spare a lot of time for technical learning. When making and implementing virtual simulation training course resources, many professional teachers often have no way to start. Virtual reality technology is regarded as an important technological progress in the application of basic education in the next 2–3 years [2]. Therefore, the use of virtual reality technology to carry out immersive learning has attracted great attention in the educational field in recent years. At the same time, virtual reality technology has developed rapidly and gradually entered the field of education, giving birth to a new teaching mode, providing students with multi-path and strong interactive experience. The integration of real learning environment and virtual learning environment provides new ideas for solving problems in education and teaching. Based on this, this paper studies the design and research of tour guide curriculum reform system based on virtual simulation resources.

2 Related Work

2.1 Virtual Simulation Technology

Virtual reality technology, also known as "immersive multimedia", is an interactive simulation technology that can create a virtual learning environment, but it is obviously different from traditional multimedia technology. First, in terms of technical support, compared with traditional multimedia technology, virtual reality technology relies on a number of technical support, such as computer graphics and sensing technology, which greatly improves the user's sensory experience. Secondly, the media richness, social presence and self openness of virtual reality technology in social experience are significantly stronger than ordinary social media.

Virtual reality technology is a real or near real three-dimensional virtual environment constructed by computer programs. Learners can use it to enter the virtual environment and build a reasonable understanding of the real world through its interaction with the virtual environment. Virtual situations rely on reality and ultimately return to reality. Learners interact with virtual situations in order to better understand the real world. The application of virtual reality technology in tour guide teaching is mainly reflected in two aspects: immersion and existence. The technical attribute of virtual reality technology is immersion, which is a stimulating experience for learners to feel their environment and interact with it. The sense of existence is a kind of psychological feeling, which is the core of virtual reality technology. Immersion and sense of existence are the symbols to measure whether the virtual reality technology is mature and whether the experience is real, especially the application of virtual tour guide teaching [3]. The following Fig. 1 shows the characteristics of the virtual simulation technology.

Practical training teaching refers to simulating the actual working environment and using the actual cases of real work projects for teaching. Practical training teaching can be divided into narrow sense and broad sense. In a narrow sense, practical teaching refers to the practical teaching in various professional courses, and the teaching time is included in the teaching time of the course, and the teaching is usually conducted in the training room of the school. In a broad sense, practical teaching refers to all practical teaching contents of a major, including practical training in the course, on-the-spot investigation, holiday social practice, on-the-job practice and graduation practice.

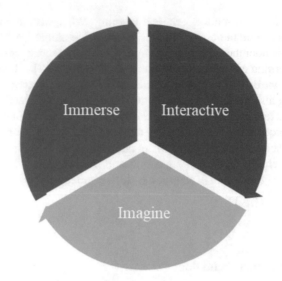

Fig. 1. Characteristics of virtual simulation technology

The "virtual" of virtual training refers to the simulation environment designed under the support of VR technology. Virtual training is to simulate some specific situations by using such simulation environment, and learners can conduct simulation and repetitive training teaching in such virtual simulation environment, So as to gain experience or form meaning [4]. The difference between virtual training and traditional field training is mainly that the objects operated by the former in the training are virtual objects or virtual models, and the objects operated in the real field training are physical objects.

2.2 Problems in Tour Guide Course

The current teaching situation of the course is understood from the aspects of learning interest, teaching methods, teaching conditions and suggestions for teaching improvement. The interviewed teachers have repeatedly mentioned the words "dull atmosphere, poor effect, low efficiency and low enthusiasm", which indicates that the teaching effect is poor under the current teaching mode. In terms of teaching methods, the interviewed teachers all said that teaching is mainly conducted in the form of teaching, In terms of the current teaching condition survey, the interviewees said that the school has the teaching conditions to carry out virtual practical training. However, due to the existence of "no use" or "bad use", traditional teaching is still carried out in the virtual teaching environment. Teachers use mobile learning devices as tools for courseware display and data upload. Teachers only use wechat, QQ and other commonly used social software to assist teaching in teaching, The professional virtual teaching software is basically not used. In the aspect of using the virtual and real learning environment to carry out teaching, the interviewees expressed that they were very willing to use this method to carry out teaching, and also hoped to have professional personnel to carry out training or conduct lectures to carry out training.

1. The training room of L school is well constructed and has the conditions for virtual training. However, since the teachers do not know how to carry out the training and teaching activities under the virtual and real learning environment, although the teachers use the virtual learning environment, they still only stay in the relatively shallow application of courseware display, resource upload and so on Promoting the development of students' thinking also needs to be realized by further designing teaching programs [5].
2. The interviewed teachers said that the current teaching is only limited to the classroom, and they hope to expand the students' learning space with the help of the virtual and real learning environment, so as to achieve better results in pre class preview and post class expansion learning.

3   Research on the Design of Tour Guide Course Reform System Based on Virtual Simulation Resources

The research mainly focuses on the in class training, which is the training teaching completed in the school's training room within the teaching time of the course according to the requirements of the training outline of a professional course. For example, the "simulated tour guide" course of the tour guide specialty is generally conducted in the simulated tour guide training room. The in class training of tour guides is an integral part of the classroom teaching of tour guides, and is a teaching activity that visualizes professional knowledge and practices operational skills.

The virtual learning environment in this study refers to the guide business training platform, the learning pass platform and the Xiaolu guide app. The guide business training platform and the Xiaolu guide app provide support for students' training simulation and business training. The learning pass platform is the general teaching platform of the school, including the computer end and mobile learning end, to facilitate the development of classroom teaching activities. With the support of virtual learning environment and the development of learning activities in real classroom teaching, a virtual learning environment is formed [6]. The virtual learning environment will be described in detail below.

1. Guide business training platform
The guide business training platform applies 3D real scene simulation, through the shooting of real pictures, and uses virtual reality technology to achieve a highly simulated scene simulation effect of the scenic spot, so that students can experience the scene guide environment in the virtual platform, enable students to make more scientific, reasonable and effective service operations through the platform, improve students' skills, and make students' training more attractive, interesting and practical [7], In addition, the platform also provides rich case resources and video resources.

The guide business training platform mainly provides students with guide business training process training and scenic spot simulation training. The platform arranges training links according to the work processes of local escort guides, full escort guides, outbound leader guides and scenic spot guides, which conforms to the actual employment workflow of tour guides. As shown in Fig. 2, there are different types of guide training modes. The platform takes the current work scope of tour guides, i.e. the four major businesses of local escort, full escort, tour leader and scenic spot announcer, as the core of the system, takes the business process of their work as the main line of training, and takes the rich and varied training cases, i.e. the tour guide's "travel plan", as the source of the whole training operation. The detailed business process settings, highly simulated simulated tour guide environment, emergency settings, and various types and scales of training plans in the platform allow students to experience the actual environment of tour guide tour guide from simple to complex in an all-round way, and enable students to personally experience various links and operating processes of tour guide work under the support of the platform, so as to obtain a real tour guide experience from the virtual training [8], And in the process of virtual training, learn the basic professional skills required in the service of tour guides, so as to meet the practical needs of tour guide teaching.

Fig. 2. Different types of guide training modes

4 Design of Learning Activities in Virtual Environment

This study analyzes the elements of the learning activity system according to the activity theory, including the design elements such as the activity task, the activity subject, the learning objective and the learning evaluation. The activity theory also emphasizes the hierarchical structure of the activity. When designing the learning activity, the activity, behavior and operation are designed serially, and the activity process is also analyzed in combination with the experience learning theory, So as to ensure the orderly progress of the learning process. In essence, experiential learning under the virtual reality fusion environment is a situation based and process learning method [9]. The virtual reality fusion learning environment can effectively support the design of learning situations. At the same time, teaching in the virtual reality fusion environment enriches the form of learning activities. Therefore, the learning situation under the virtual reality fusion environment is also analyzed.

To sum up, this study will analyze the design elements such as activity tasks, activity subjects, learning objectives, activity processes, learning situations and learning evaluation.

The experiential learning process supported by the virtual reality integration environment is composed of several learning activities. In order to achieve the relevant teaching objectives with different emphasis, the tasks to be completed in the teaching activities are also different. According to the characteristics of vocational education, i.e. employment oriented, the author reconstructs the task of learning activities in the form of projects, so that the setting of activity tasks can meet the needs of students going to work in the future [10]. At the same time, it is necessary to define the activity objectives, determine what activity tasks are given to students to achieve the learning objectives according to the learning objectives, and then analyze the sub links and composition of activities,

Make the learners' learning behavior work towards the goal to achieve the requirements of the activity task, and constantly adjust and correct their behavior in the process of completing the task, so as to achieve the predetermined learning goal.

5 Conclusion

Using the method of case study, this paper studies the design of tour guide course reform system based on virtual simulation resources. The purpose of this study is to find out whether there are any problems in the curriculum reform of tour guides and how to make it more effective. In addition, this study is divided into two parts: the first part focuses on finding the problems in the course reform of tour guides; The second part focuses on finding solutions to these problems.

Acknowledgements. Project from: Shandong Institute of Commerce and Technology, D211.

References

1. Xu, X., Zhang, B., Wu, X., et al.: Reform countermeasures of course system of environmental specialty based on virtual simulation. In: 2020 6th International Conference on Social Science and Higher Education (ICSSHE 2020) (2020)
2. Zhang, B., Zhao, J.: The reform of chemistry technology course system based on simulation technology (2020)
3. Xie, Y.: Design and implementation of virtual simulation system based on basketball teaching and training. In: CIPAE 2020: 2020 International Conference on Computers, Information Processing and Advanced Education (2020)
4. Fu, Q., Lv, J.: Optimal design of information elements in virtual reality system based on TOPSIS. J. Phys: Conf. Ser. **1654**, 012073 (2020)
5. Yang, Q., Zhang, L., Zhang, J., et al.: System simulation and policy optimization of China's coal production capacity deviation in terms of the economy, environment, and energy security. Resour. Policy **74**, 102314 (2021)
6. Zhao, L.: Design of digital circuit experiment course based on FPGA. World J. Eng. Technol. **9**(2), 346–356 (2021)
7. Dong, H., Zhang, X., Kong, D., et al.: Design and implementation of intelligent tour guide system in large scenic area based on fog computing. In: 2020 International Conference on Big Data, Artificial Intelligence and Internet of Things Engineering (ICBAIE) (2020)
8. Hu, Y., Sun, W., Liu, X., et al.: Tourism demonstration system for large-scale museums based on 3D virtual simulation technology. Electr. Libr. (2020). ahead-of-print(ahead-of-print)
9. Chen, L., Deng, Z., Wang, Y., et al.; Design of interferometer direction finding experimental teaching system based on LabVIEW. Exp. Technol. Manage. (2020)
10. Yuan, Y.B., Wang, J.J., Lin, M.H., et al.: Building virtual simulation teaching platform based on electronic standardized patient. Sheng li xue bao: [Acta physiologica Sinica] **72**(6), 730–736 (2020)

Design of Vocabulary Query System in Computer Aided English Translation Teaching

Yuyun Su[✉]

Hunan College of Foreign Studies, Hunan 410200, China
576520120@qq.com

Abstract. Application of Bayesian network in higher vocational english practical ability test.

Mining knowledge from massive data for decision support, analysis and prediction has become a new demand for information systems, but the technology of data processing and data extraction is scarce. Bayesian network, which originated from Bayesian statistics, represents the probability distribution and causal relationship of objects with its unique uncertain knowledge expression form, rich probability expression ability, incremental learning method of comprehensive prior knowledge and other characteristics. It has become one of the most eye-catching focuses of current data mining methods and an important knowledge discovery method in the field of knowledge discovery. This paper is based on the application of Bayesian network in Higher Vocational English practical ability test. Practical ability test is a part of foreign students' English proficiency test. It aims to determine the level of English required by students to receive higher vocational education in China. The practical ability test consists of four parts: listening, speaking, reading and writing. Each part has three sub parts: listening comprehension (LC), oral comprehension (SC) and reading comprehension (RC).

Keywords: Vocational English · Bayesian network · Application ability test

1 Introduction

From ancient times to the present, there are many kinds of examinations. The original intention or basic purpose of the examination we mentioned here is to use various ways to investigate students' mastery of knowledge, so as to feed back these information to teachers or students, so as to timely and appropriately adjust the next step of teaching activities and further improve the teaching quality and efficiency. Because of the powerful function of the computer, people try to use the computer to achieve this goal, so there are a variety of computer examination systems [1].

In order to be practical, the computer examination system needs to have a lot of domain knowledge, have the ability to organize reasonable examination content, provide convenient user access and friendly user interface. Moreover, with the development of the system, intelligence becomes more and more important in the computer examination

J. C. Hung et al. (Eds.): IC 2023, LNEE 1045, pp. 150–156, 2023.
https://doi.org/10.1007/978-981-99-2287-1_22

system. Therefore, the computer intelligent examination system is produced. It should be able to provide targeted and adaptive examination services according to the different conditions of each user [2]. In order to achieve this goal, the student model has been added to the computer examination system as an important part. Student model is a mathematical model that uses the feedback information of students to estimate the current knowledge level of students. The student model makes the computer examination system into an intelligent computer examination system, and gradually becomes the core and key part of the whole system.

The development of the current network has further provided a new way for users to access the higher vocational English application ability examination system. The network-based Higher Vocational English application ability test system has become a development trend of the test system. This kind of long-distance examination system, which provides services on the Internet, has higher requirements for the personalization and intelligence of the system, and the role of student model in the system has been further strengthened [3].

Therefore, in the implementation of Higher Vocational English application ability test system, the student model is the key to whether the system can be personalized and intelligent. The main purpose of the student model in the system is to use the computer to simulate the students' mastery of knowledge, so as to provide a reasonable estimate of the students' mastery of knowledge for the decision-making part of the system. At present, there are many methods to build student models. They can provide more accurate estimates of students' abilities in some specific fields [4]. At the same time, they are more or less limited in the scope of application, human factors are too large. In this paper, we try to use Bayesian network, a powerful tool in the field of uncertainty reasoning in the field of artificial intelligence, to realize the student model, and then realize an Internet-based Higher Vocational English application ability examination system with general significance.

2 Related Work

2.1 Definition of Bayesian Network

Bayesian network is also known as belief network, probability network, causal network, knowledgemap, etc. Its detailed definition is:

Let $V = \{x1, X2,..., xn\}$ be n random variables in the range u, then the Bayesian network in the range u is defined as BN (BS, BP), where BS is a directed acyclic graph t (DAG) defined on V, V is the node set of the directed acyclic graph factory, and e is the edge set of T. If there is a directed edge from node x to node x, it is called X, which is the parent node of X, and X, which is the child node of X. Note that all parent nodes are R,. And bp = {p (XT, [0.1]|x ∈ V)}. For each node in V, a set of conditional probability distribution functions P (x, [0,1]) is defined.

Bayesian network is a combination of probability theory and graph theory. It is mainly composed of two parts, corresponding to the qualitative description and quantitative description of the problem field respectively:

(1) one part is directed acyclic graph (DAG), which is usually called Bayesian network structure. Dag is composed of several nodes and directed edges connecting these nodes. The nodes represent random variables in the problem domain, and each node corresponds to a variable. The definition of variables can be phenomena, components, states or attributes of interest in the problem, which have certain physical and practical significance [5]. The directed edge between the connected nodes represents the dependency or causal relationship between the nodes, the arrow of the connected edge represents the directionality of the impact of the causal relationship (from the parent node to the child node), and the default connection between the nodes indicates that the variables corresponding to the nodes are conditionally independent. As shown in Fig. 1 (a).

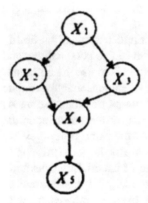

	X_2, X_3	$X_2, \neg X_3$	$\neg X_2, X_3$	$X_2, \neg X_3$
X_4	0.4	0.9	0.5	0.3
$\neg X_4$	0.6	0.1	0.5	0.7

(a) Directed acyclic graph   (b) Conditional probability table of node X4

Fig. 1. Bayesian network

(2) The other part is the local probability distribution set that reflects the correlation between variables, that is, the probability parameter, which is usually called the conditional probability table (CPT). As shown in Fig. 1 (b), the probability value represents the correlation strength or confidence between the child node and its parent node, and the node probability without a parent node is its prior probability. Bayesian network structure is the result of abstracting data instances. It is a macro description of the problem domain. The probability parameter is an accurate expression of the correlation strength between variables (nodes), which belongs to the quantitative description.

2.2   Application of Bayesian Network

In recent years, the concept of Bayesian decision theory and its uncertainty representation and computing technology have been integrated into the mainstream of uncertainty processing in artificial intelligence. Its fields include computer vision, natural-languageprocessing, robot navigation, planning, machine learning, and building and analyzing software systems using Bayesian technology. In addition, Bayesian network and Bayesian technology have also been widely used in data mining, and become an important knowledge discovery method in the field of knowledge discovery [6].

In 1982, pearl began to use Bayesian networks for probabilistic reasoning in artificial intelligence, and proposed message passing algorithms for tree networks and multi tree networks respectively. Since then, Bayesian network has been more and more applied to many expert systems. A typical example is Pathfinder, which is used to help diagnose and analyze the symptoms of "lymph nodes"; Another example is the cpcsbn telemedicine system, which has 448 nodes and 908 edges, which is superior to the world's major telemedicine diagnosis methods.

In the field of commercial application, a group of companies represented by Microsoft have applied Bayesian network to their own products [7]. In 1995, Microsoft launched the first expert system based on Bayesian network, a website for infant health care - Microsoft on parent (www.onparenting.msn.com), so that parents can diagnose their infants' diseases online. Microsoft's windows operating system and office series have been integrated with Bayesian network technology in many aspects.

Generally speaking, at present, the application of Bayesian network is mainly manifested in the following aspects:

(1) Diagnose. Find out the cause of the fault according to the fault characteristics, and carry out real-time monitoring and fault prevention according to the frequent faults or the existing state of the system. For example, the troubleshooting in Microsoft Windows software can help users solve the software and hardware problems they encounter; Industrial fault diagnosis (such as auxillary turbine diagnosis of General Electric Corporation of the United States) and aerospace fault diagnosis (such as diagnosis of spaceshuttle propulsion systems jointly developed by NASA and pockwell) [8].
(2) Expert system. Provide expert level reasoning, simulate human intelligence, and solve practical problems in the professional field. For example, the application of Bayesian networks in medicine.
(3) Planning. According to the causal probability reasoning, the probability of various events is predicted. For a given goal - saving money or time, the planning of a project is obtained.
(4) Learning. Help with learning. Help beginners quickly master the causal relationship and laws of events.
(5) Sorting. Bayesian network is used for cluster analysis and classification. It has important applications in data mining and pattern recognition.

At present, the application fields of Bayesian methods and technologies continue to expand, such as data mining based on probabilistic causality, intelligent information retrieval, real-time decision support system, speech recognition and handwriting recognition, and modeling of intelligent agent and MAS.

3   Application of Bayesian Network in Higher Vocational English Application Ability Test

3.1   Collect and Describe Data

Collecting and describing data is a very important part of the whole data mining work. The data of this paper comes from the admission files of a certain class of students in Anhui Vocational College of industry and Commerce and the score information table of English band B test. Data preprocessing is the basis for building a good prediction model. It should not only ensure the correctness and consistency of data values, but also ensure that these values record the same thing in the same way. In this paper, there are mainly differences in subjects and scores from different provinces and regions. The surname and source of students can be extracted directly, while the scores will be processed in the form of grading according to different provinces [9]. The disciplines will be classified according to the nature of different disciplines. Grade B test scores are directly extracted through the score information mark. In this paper, there are 2994 examples, 5 conditional attributes and 1 Category attribute in the experimental data. All values are discrete values. The Category attribute is English level B: excellent, pass and fail. In order to ensure the accuracy and robustness of the model, two-thirds of the data are randomly selected as the training set and the remaining one-third as the test set.

3.2   Prediction Model of English Practical Ability Test Based on Bayesian Network Classification Model

According to our experience and knowledge, the Bayesian network structure diagram needs to be modified in order to be better applied to the actual classification model. The relationship between conditional attribute gender (a) and Category attribute English level B has not been learned. However, according to our actual experience and knowledge, in higher vocational colleges, there is a certain relationship between English performance and gender [10]. Therefore, according to the network diagram learned in Fig. 1, we have constructed the Bayesian network prediction model shown in Fig. 2 for the prediction of English application ability test.

For the prediction model shown in Fig. 2, we can judge the accuracy of the model through accuracy evaluation. The main method is to extract the data in the test set according to the classification model for testing. The design idea is to extract the information of the test student from the test set, and then extract the data matching the student from the calculated results for operation to obtain the test results. The test accuracy is obtained by comparing with the actual test. For the prediction model of Anhui Vocational College of industry and Commerce students' English application ability test, if the calculated accuracy is greater than or close to 90%, we believe that the established model can classify and predict unknown samples. Otherwise, the random sample data shall be re sampled and re modeled until the accuracy reaches the standard.

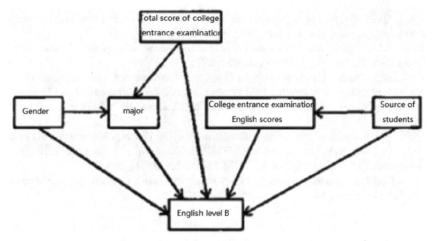

Fig. 2. Prediction model of English practical ability test based on Bayesian network

4 Conclusion

Bayesian network is a tool to help people apply probability and statistics to complex fields, carry out uncertainty reasoning and data analysis. It provides a natural method to express causality, and is used to find the potential relationship between data. Bayesian network uses graphical method to describe the relationship between data, with clear semantics and strong comprehensibility. In recent years, Bayesian network has been widely used in the field of uncertain reasoning and knowledge discovery because of its intuitive, easy to combine prior knowledge, easy to understand and so on. It has become one of the most eye-catching focus of many data mining methods.

Acknowledgements. "Research on Business English Teaching Standards of Higher Vocational Colleges under the Background of Modular Teaching" (Project No.: ZJBZ2021084) "Research on the Diagnosis and Improvement of Mixed College English Teaching in the Post-epidemic Era" (Project No.: XSP21YBC127).

References

1. Lai, F.: Design of vocabulary query system in computer aided english translation teaching (2021)
2. Trushnikov, V.E., Baranova, N.N., Grishin, M.V.: Development of methods and models of computer-aided design of security system against information threats for aviation-instrument-making. IOP Conf. Ser.: Mater. Sci. Eng. **760**(1), 012061 (2020)
3. Aljadani, A.S.: The influence of frequency on the acquisition and textbooks design of second language vocabulary (2020)
4. Li, L., Wei, H.: Design of the upper computer system for the electrical testing platform of shearer. Sci-tech Innov. Prod. (2020)
5. Wang, J., Ondago, O.: Optimization of computer-assisted vocabulary assessment system in international chinese language teaching. Comput.-Aided Des. Appl. **18**(S4), 106–117 (2021)

6. Wang, Q., Zhang, S., Liu, W.: Design and simulation of computer aided chinese vocabulary evaluation system. Comput.-Aided Des. Appl. **18**(S3), 1–11 (2020)
7. Xuan, L.: Computer aided optimization of picking system in logistics distribution center. Comput.-Aided Des. Appl. **19**(S4), 46–55 (2021)
8. Venkatanarasimhan, L., Chowdhury, A., Sen, C.: A vocabulary of function features for computer aided modeling of thermal-fluid systems. In: ASME 2020 International Design Engineering Technical Conferences and Computers and Information in Engineering Conference (2020)
9. Qi, B., et al.: Design of the computer management system for reproductive toxicity tests. In: Hung, J.C., Yen, N.Y., Chang, J.W. (eds.) FC 2019. LNEE, vol. 551, pp. 1565–1570. Springer, Singapore (2020). https://doi.org/10.1007/978-981-15-3250-4_205
10. Li, L., Cao, L.: Semantic analysis of literary vocabulary based on microsystem and computer aided deep research. Mob. Inf. Syst. **2021**, 1–13 (2021)

Development and Design of English Micro Reading Website for Network Technology Specialty

Changzhen Ju[✉]

Wuhan Donghu University, Wuhan 430212, Hubei, China
changzhenju@163.com

Abstract. English reading has always been a difficult problem for most Chinese Internet industry practitioners and professional students. Aiming at this problem, this paper explores how to use information technology to assist the majority of network technology practitioners and professional students to quickly and effectively improve their professional English reading ability, so that readers can easily and happily carry out reading training and gradually and solidly improve their professional English reading ability. Through the work of several stages such as demand analysis, algorithm research, system design and coding, the algorithm design and system architecture design are carried out around the three core problems of English new word statistics and analysis, intelligent collection of reading materials and the application of Ebbinghaus memory curve theory in English reading, and the web system is developed with the help of Ruby on rails development framework, With the help of mobile Internet and smart phone technology, we can improve the usability of web system and improve the reading efficiency of readers' professional English.

Keywords: Major in network technology · Website · English micro reading

1 Introduction

The project comes from a specific problem in the field of quality training of network technology specialty - the cultivation of professional English reading ability. In the network maintenance, operation and management technicians and network professional teaching projects, technicians or students often have to contact English software tools with strong professionalism, English operation interface and English technical materials and information related to reading, which are used to solve technical problems or learn professional knowledge [1]. The reading ability of professional English directly affects the working effect and the improvement speed of professional level of technicians, Affect students' learning effect and the cultivation of learning interest. Due to the lack of this ability, which affects the improvement of professional ability, many students and practitioners finally give up relevant professional work, which also indirectly affects the development of the industry.

J. C. Hung et al. (Eds.): IC 2023, LNEE 1045, pp. 157–163, 2023.
https://doi.org/10.1007/978-981-99-2287-1_23

Although many professional English words have been accumulated through the above methods, which indirectly improves the reading and understanding ability of professional English materials and information, because the overall English vocabulary is low, it is still unable to read and understand large professional documents smoothly [2]. In addition, due to the uneven reading level, it is difficult to quickly and accurately collect English reading materials that are suitable for the level of readers, fit the interests and concerns of readers, and have strong practicability and professionalism, so as to help readers indirectly improve their vocabulary and reading level. In addition, with the fast pace of modern life and work and the explosion of information, it is difficult to find large blocks of time. Even if there are suitable English materials, it is difficult for many people to sit down and read carefully and improve their level [3].

Based on the above situation, the research work of this project focuses on the goal of finding effective solutions. By exploring the use of fragmented time (such as elevator, bus, etc.), with the help of smart phones and other mobile terminal devices, intelligently sort out English materials suitable for readers' vocabulary level, so as to gradually improve readers' professional English reading level [4]. Taking exploration as the starting point and research work as the carrier, in order to realize a set of application system, based on the familiar English vocabulary of readers, intelligently collect short English reading materials suitable for readers' English level, so as to gradually improve their professional English reading level. Therefore, there is the topic of this paper, which aims to realize this idea and help more practitioners, network technology teachers and network technology students to improve their professional English reading ability and professional vocabulary.

2   Related Work

2.1   Target Analysis of English Micro Reading Website

To achieve the above-mentioned core objectives, we need to make an in-depth analysis from four aspects: the reader's professional English foundation, the arrangement of reading materials, the way of reading and the evaluation of reading effect, find out the core problems that the software system needs to solve, analyze the dependence between the main elements, summarize the general process of the whole reading, and design a clear system architecture to complete the detailed design of the software system [5].

To improve the professional English reading ability of readers, and then turn it into the reading ability that can solve practical problems, we must first take the reader's actual professional English foundation as the starting point: on the one hand, we should understand the reader's vocabulary and the mastery of English grammar, which is the basis of subsequent reading training and short text content screening; On the other hand, we also need to have a necessary understanding of readers' interests, educational background, professional foundation and concerns, so as to better meet the needs of readers and improve the matching degree, so as to reduce reading resistance [6].

In the process of sorting out reading materials, in addition to the efforts of the reader himself, there should also be the role of reading assistant to help the reader improve reading efficiency and reduce reading difficulty and resistance. The role of reading assistant can be realized by the final software system, and can also be played by the readers' stakeholders (relatives, friends, teachers, etc.), so as to improve the reading efficiency and reduce the reading difficulty and resistance. For different types of readers (practitioners and students in school), we should absorb different groups to play this role and participate in the collation and recommendation of readers' reading materials.

In the selection of reading methods, we should integrate the social background, objective reality, the author's personality characteristics and human characteristics, as well as the current situation of scientific and Technological Development for analysis: since entering the 21st century, people's pace of life and work has been accelerating, and the increasing sense of pressure has forced everyone to make every effort in life and work, and the role of economic interests lies in the media Human beings themselves and real life are constantly enlarged, which makes people generally in an impetuous living state [7]; As a result, more and more people began to lament that it is difficult to have time to read carefully, not to mention reading books or reading plans and materials at work. There is only one reason - time is urgent. Therefore, people have to try their best to occupy time and complete various tasks, ignore or even give up the methods of improving through self-study and training, and unilaterally rely on external forces to save themselves and help themselves pursue more interests.

2.2 Readers' Information Management

First of all, we should solve the problem of recording the basic situation of each reader and provide the basis for subsequent reading material push and personalized reading plan. The specific solutions are as follows:

(1) Perfect user basic files

First of all, for readers, in addition to recording the basic information such as learners' age, gender and learning experience, they should also record their interests and concerns in network technology as the basic conditions for screening reading documents in the future. Users become system users through self-service registration, and other information registered and registered is recorded in the database at the back of the system [8]. All kinds of information are uniformly stored in the form of data tables, which can facilitate developers to quickly add and delete form words in the in-depth application of the system. In addition, the relevant technical requirements for subsequent problem solving should be taken into account in the design of system architecture.

(2) Management of user basic files

Established a perfect basic file, but also enable the system to facilitate the self-service management of information and modify information after ordinary users log in to the system. Counselors should be able to easily read the basic files of ordinary users and have a more comprehensive understanding of the situation of ordinary users [9]. On this basis, the design of the system architecture should leave enough space to ensure that in the future development, special functional modules or components can be used to sort, analyze and extract information into keywords, so as to facilitate the system to retrieve information through search engine and database, and lay a foundation for the push of reading materials and user communication.

3   Development and Design of English Micro Reading Website

3.1   Selection of Development Aids

If you want to do a good job, you must first use its tools. Excellent development auxiliary tools can greatly improve the development efficiency, improve the operation stability of the system and reduce the maintenance cost of the system.

(1)   Git code version control tool

Programmers need code version control and management tools in the development process. Software tools with this function can help programmers manage source code more flexibly and effectively, especially the source code written under the guidance of different stages and ideas. This also includes code merging problems that often occur in the process of cooperation with the results of the development team.

Before the emergence of GIT, CVS (Concurrent Versions System) and SVN (subversion) were widely used by programmers. Compared with git, their functions and characteristics are more complex to get started. For large-scale distributed development such as Linux, some are not flexible, and their speed and management are cumbersome. Therefore, Linus Torvalds, the author of Linux kernel, has specially developed git as a tool to replace CVs and SVN, so that distributed developers can share and manage code conveniently and flexibly. The specific content can be deeply understood through the search engine. Limited to space, this paper will not be expanded [10].

(2)   Code editing and debugging tool VIM

The development of Ruby on rails advocates the use of tools similar to text editors for code writing and debugging, and Textmate, the exclusive code editing and debugging tool of Mac OS platform, is the first recommended by the majority of Ruby on rails programmers. Therefore, with the support of the hard work of programmers in the open source community, VIM, a widely used code editing tool under the Linux/Unix platform, can well support the development of Ruby on rails and has most of the features of Textmate, so as to ensure that the development of Ruby on rails under the Linux/Unix platform will not be affected.

VIM software has both command-line version and graphic version, but most of the plug-ins of the two versions are common. With these plug-ins, developers can use VIM to complete code editing and even operation and debugging of almost all programming languages. For the development of Ruby on rails, there are special VIM quick development plug-ins that can be used. For example, use G to switch between different rails project files. There are also supporting code highlighting and code completion. Some functions are realized as shown in Fig. 1. In the figure, the left small window shows the directory structure of the programming part of the project, the right window shows the contents of the view file, the middle two are the contents of the controller and routing configuration file respectively, and the floating window in the middle is the shortcut window of the code completion function.

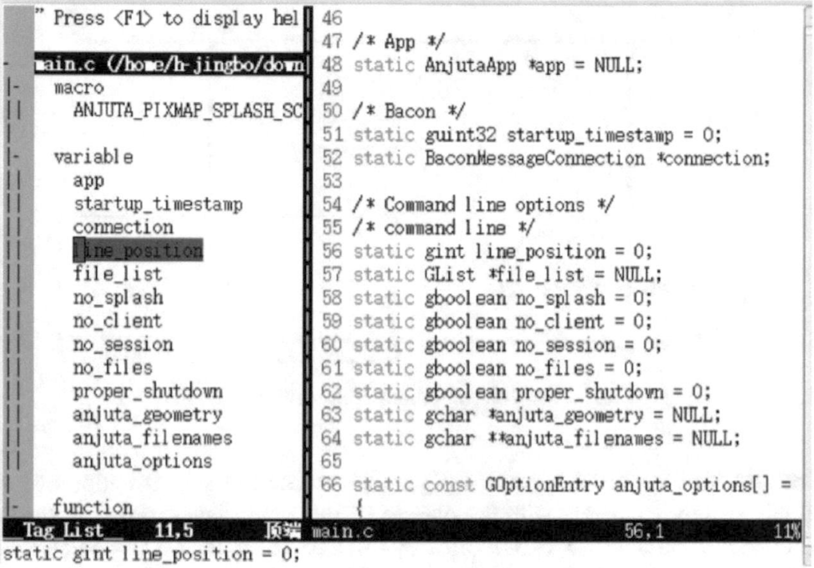

```
" Press <F1> to display hel   46
                              47  /* App */
main.c (/home/h-jingbo/down   48  static AnjutaApp *app = NULL;
|- macro                      49
| |   ANJUTA_PIXMAP_SPLASH_SC  50  /* Bacon */
|                             51  static guint32 startup_timestamp = 0;
|- variable                   52  static BaconMessageConnection *connection;
| |   app                     53
| |   startup_timestamp       54  /* Command line options */
| |   connection              55  /* command line */
| |   line_position           56  static gint line_position = 0;
| |   file_list               57  static GList *file_list = NULL;
| |   no_splash               58  static gboolean no_splash = 0;
| |   no_client               59  static gboolean no_client = 0;
| |   no_session              60  static gboolean no_session = 0;
| |   no_files                61  static gboolean no_files = 0;
| |   proper_shutdown         62  static gboolean proper_shutdown = 0;
| |   anjuta_geometry         63  static gchar *anjuta_geometry = NULL;
| |   anjuta_filenames        64  static gchar **anjuta_filenames = NULL;
| |   anjuta_options          65
|                             66  static const GOptionEntry anjuta_options[] =
|- function                       {
 Tag_List   11,5      顶端  main.c                      56,1           11%
static gint line_position = 0;
```

Fig. 1. VIM code editing interface

3.2 System Architecture

Based on the above technical route selection, the application architecture of the project will be determined as the hybrid structure of BS and CS, as shown in Fig. 2. The service end takes the web system as the core, and the B end interacts with the web system through the web browser; The C-end accesses the web system through the app application to achieve the goal of reading training by using fragment time.

Because the design of rails framework strictly follows the MVC architecture, the system architecture of web system will strictly follow the MVC architecture, and organically combine the background data, functional logic and front-end interface through the logical association of model, controller and view, so as to ensure that the subsequent development and function expansion can be faster and faster on the premise of reducing the complexity of the system It is flexible without affecting the stability of the original system.

Because ruby is a thorough object-oriented programming language, the MVC architecture of rails also makes full use of this feature. With the help of the design concept of rails framework, based on the above requirements, this project tries to abstract relevant things and data into model classes as tables associated with the database management system, and clarify the association relationship between tables according to the logical relationship.

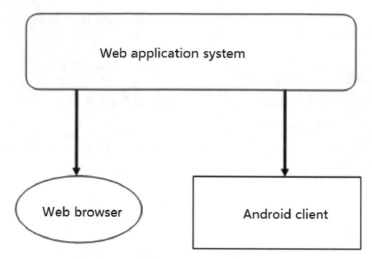

Fig. 2. Project application architecture diagram

The code of the control layer is designed according to the Convention of rails, so that it can closely cooperate with the classes of the model layer to realize the design function. The front-end view is supported by the CSS framework YUI compatible with mainstream browsers and based on the rails view template, so as to realize the front-end view page that can be well presented in mainstream browsers (including smartphone browsing) as far as possible.

The development of Android client is based on the premise that it can be compatible with the mainstream models of all grades in the current market to the greatest extent and can be put into use as soon as possible. Considering the tight time, heavy tasks and other factors, it is preliminarily decided that the development of Android client of the project strives to be simple. The web address of the web system is packaged to the client by using web control, and the client is processed concisely to meet the actual needs to the greatest extent.

4   Conclusion

Because the current requirements of the project are not clear, there are many factors that will affect the design and development of the system after it is actually put into use. Therefore, it is necessary to have more flexible characteristics in terms of technology to ensure that the response speed of system development can be improved and the robustness of Architecture and system can be ensured in the process of dealing with changes in the future. Ruby on rails has this series of features and has been tested by facts. As for English learning, the application of information technology has always focused on the application field of memorizing words. After trying this kind of application, it can be obviously found that mechanical memory is not conducive to improving English level, but also makes users feel pressure. After summarizing the successful experience, this design idea is found. Compared with the traditional memorization of words, this idea is an innovation of concept and an English learning project in line with human characteristics. Therefore, this inherent advantage will also provide enough innovation and development space for the development of this project.

References

1. Hall, C., Capin, P., Vaughn, S.: Understanding the development and instruction of reading for English learners with learning disabilities (2020)
2. Mccarron, S.P.: Post-secondary reading development and print exposure in L1 and L2 speakers of English (2020)
3. Hu, M., Xiao, L., University, Z.: The development and implementation of extended english courses. Curriculum Teach. Mater. Method (2020)
4. Amin, A.R.: Using repeated- reading and listening €"while- reading via text-to- speech APPs in developing fluency and comprehension. World J. English Lang. **12**, 211–220 (2022)
5. Lou, Y.E.: Teaching disciplinary literacy to adolescent english language learners: vocabulary development and reading within the disciplines. TESL Canada J. **37**(1), n1 (2020)
6. Rodriguez, S., Chaves, C., Quiroga, A.: Development of a swine health monitoring system based on bio-metric sensors. In: Cortes, D.F., Hoang, V., Trong, T. (eds.) AETA 2019. LNEE, vol. 685, pp. 244–251. Springer, Cham (2021). https://doi.org/10.1007/978-3-030-53021-1_25
7. Krepel, A., Bree, E., Mulder, E., et al.: The unique contribution of vocabulary in the reading development of English as a foreign language. J. Res. Reading **44**(3), 453–474 (2021)
8. Chang, W.C., Khelil, N., Hung, Y.C.: Exploring the differences between leaders' and designers' communication styles: insights from industrial design micro-enterprises in Taiwan. J. Enterprising Cult. **28**(3), 201–221 (2020)
9. Bond, M.H., Yan, L., Barrera, A.Z., et al.: Development and usability of a Spanish/English smoking cessation website: lessons learned (2020)
10. Fu, X., Lou, K., Zhu, Z., et al.: The development track and influencing factors of reading ability of chinese elementary english learners. Open Access Libr. J. **08**(4), 1–6 (2021)

Development Countermeasures of Artificial Intelligence in the Field of Architectural Cultural Heritage Protection and Utilization

Yuanyuan Shi[✉]

Changchun University of Architecture and Civil Engineering, Changchun 130607, Jilin, China
s307007005@163.com

Abstract. Artificial intelligence is a rapidly developing technology. Many people are interested in using it for cultural heritage protection and utilization, but they lack knowledge of the development of artificial intelligence and its application in cultural heritage protection. The application ability of this technology in the protection and utilization of architectural cultural heritage has become an important topic for the future development of architecture, which needs further research. In this article, we introduced the concept of "development countermeasures", which can help us better understand how artificial intelligence plays a role in the protection and utilization of architectural cultural heritage.

Keywords: Artificial intelligence · Development countermeasures · Protection and utilization of architectural cultural heritage · Digital save

1 Introduction

In the field of architecture, there is not only the early work of shape finding and optimization, but also the late work of architectural heritage protection.

In the field of architectural heritage protection, image recognition is relatively mature at present: through convolutional neural network deep learning, diseases or decorations can be identified and counted more quickly and accurately through surface image features.

For example, identify the through joints between bricks and stones in the masonry structure, the crispness, alkalinity, pulverization and missing of the bricks and stones, count the positions and percentages of various diseases, and simply determine the degree of diseases to assist the survey and design. When faced with large-scale and complex terrain masonry structures, how to model them in combination with UAV tilt photogrammetry? This working mode has been gradually applied to the protection of the Great Wall a few years ago.

The processing of building point cloud data obtained by three-dimensional laser scanning through artificial intelligence is also a major direction: for example, through training, computers can automatically distinguish components in the point cloud of the whole building (split the whole), or match the structural correspondence by identifying

J. C. Hung et al. (Eds.): IC 2023, LNEE 1045, pp. 164–169, 2023.
https://doi.org/10.1007/978-981-99-2287-1_24

the feature points of disassembled building components (virtual assembly), and even assist in the design of construction processes [1]. This kind of work can make protection workers more calm when facing a large number of similar components left over from history.

In addition to the assistance of artificial intelligence in survey, the monitoring and management of building heritage have an urgent need for artificial intelligence. For example, how to identify the distorted data in the sensor, how to analyze the massive monitoring data, how to judge and set the warning value in the monitoring data, how to scientifically design the emergency plan through historical data, how to calculate and set the maximum carrying capacity of visitors in a certain area or building, and so on need the help of research forces in the direction of artificial intelligence [2].

2 Related Work

2.1 Research Status of Digital Protection of Architectural Cultural Heritage

Cultural heritage is the carrier of human civilization, with high historical, scientific, cultural and artistic values. It is also a profound mark left by a nation's vigorous development in the long history. However, most of the ancient buildings in China are of wood structure system. Due to its materials, after a long period of time, they are much more fragile than brick and stone materials, and are more vulnerable to natural disasters or human factors. At present, the earliest surviving wood structure buildings in the world are also those of the Tang Dynasty, and the number is rare. It can be said that China is facing extremely severe challenges in the protection of architectural cultural heritage [3]. At present, the practice of digital protection projects of architectural cultural heritage at home and abroad can be roughly divided into two categories: the first category is digital reproduction of architectural cultural heritage with the help of artificial intelligence technology, which complies with the principle of "authenticity" and establishes a historical information database of architectural cultural heritage on this basis; The second type takes the information presentation of architectural cultural heritage as the main content, and uses various digital media as a platform to convey the historical information value behind architectural cultural heritage to the public in a more intuitive and comprehensive way.

China's research on virtual reality technology started relatively late, and there is still a certain gap compared with some developed countries and regions. However, with the development and maturity of artificial intelligence technology, the concept of digital protection of cultural heritage has gradually taken root. Virtual reality technology, as a widely used carrier of heritage information presentation technology in recent years, has received considerable attention in China, It has been widely used in the research of cultural heritage. At the same time, a large number of domestic experts and scholars have carried out interdisciplinary research and exploration in various application forms based on virtual reality technology, which provides rich reference materials for the project research of this subject. Since the first year of VR in 2016, civil level virtual reality technology has been widely popularized. Until now, virtual reality technology has become an extremely important technical means of digital information presentation and dissemination in the field of cultural heritage protection.

2.2 Artificial Intelligence Technology for Genetic Information Dissemination

The high-speed development and popularization of artificial intelligence technology have fundamentally changed the traditional concept and method of cultural heritage protection in the past. At the same time, the technical characteristics of artificial intelligence technology, such as immediacy, interactivity and globality, also enable the sharing of information resources of cultural heritage information in different countries and regions in a short time [4].

The wide application of artificial intelligence technology has also brought new opportunities and challenges to the dissemination of cultural heritage information. On the one hand, the display of cultural heritage is no longer confined to the closed protection space, but gradually turns into a more diversified and three-dimensional display platform; On the other hand, we should deeply consider how the cultural attributes of cultural heritage as an entity object should be displayed through artificial intelligence technology. But there is no doubt that the process of transforming architectural cultural heritage from physical resources to digital resources has broken through a series of limitations inherent in traditional architectural cultural protection methods. With the help of artificial intelligence technology, we can quickly complete the integration and management of historical information resources through the digital replication of architectural cultural heritage, realize the transformation from physical objects to virtual objects, and adapt to the diversified The demand for three-dimensional display has gradually changed the mode of cultural heritage information communication from the traditional one-way communication to the more personalized interactive communication, from the past point-to-point communication to the point-to-point communication, from the static communication based on visits to the dynamic communication based on interactive experience to break through the time-space barrier [5]. Therefore, the participation of artificial intelligence technology in the dissemination of heritage information not only conforms to the development trend of the digital era, but also is an important means to advance the dissemination of heritage information.

3 Development Countermeasures of Artificial Intelligence in the Field of Architectural Cultural Heritage Protection and Utilization

1. Use artificial intelligence technology to obtain historical information of architectural cultural heritage. At present, the artificial intelligence technologies mainly applied to the data acquisition of three-dimensional models include three-dimensional laser scanning technology, structured light scanning technology, tilt photogrammetry technology and so on. Using artificial intelligence technology to obtain the historical information of architectural cultural heritage has the following advantages: first, we can obtain the data and historical information on the surface of architectural cultural heritage in a non-contact way without destroying its surface, thus avoiding unnecessary damage to architectural cultural heritage in the actual operation process [6]; Secondly, artificial intelligence technology has a very high efficiency in obtaining historical information, which greatly reduces the time required for protection work

and saves a lot of human and material resources; Thirdly, the accuracy of the artificial intelligence technology in obtaining historical information is extremely high. Taking the structured light three-dimensional scanning technology as an example, the technical equipment can obtain a large amount of high-density point cloud data information projected onto the surface of the object to be measured in a short time by projecting grating stripes with phase information and acquiring the stripe image deformed on the surface of the object with a binocular camera, which greatly reduces the error in manually obtaining data [7]; Finally, with the help of historical information obtained by artificial intelligence technology, follow-up display and presentation can be carried out in a variety of ways, which is conducive to the smooth development of follow-up related work. Figure 1 below shows the genetic case of digital Yuanmingyuan architectural culture.

Fig. 1. Digital Yuanmingyuan architectural culture inheritance case

2. Use the digital information resources of architectural cultural heritage to build a database. Because the architectural cultural heritage itself covers a large area and has a large number of buildings, the information resources obtained in the digital protection work are also extremely large. It is necessary to establish a corresponding database system to manage and store the model data, text information, picture information, image information, etc. of the architectural cultural heritage in a classified and orderly manner, which is not only conducive to the subsequent viewing and application of relevant information resources [8], It is more convenient to share information resources worldwide and jointly promote the smooth implementation of digital protection of architectural cultural heritage.

3. Use artificial intelligence technology to obtain dynamic information related to architectural cultural heritage. With the help of artificial intelligence technology, we can present the historical information of architectural cultural heritage to the public through a more vivid and specific display method. In this process, three-dimensional

animation technology and motion capture technology in artificial intelligence technology play a vital role. On the one hand, with the help of three-dimensional animation technology, creators can dynamically display architectural cultural heritage in the virtual world, such as the construction process, development and change of architectural cultural heritage [9]; On the other hand, with the help of motion capture technology, creators can restore historical figures related to architectural cultural heritage, and substitute realistic acting skills of actors in the real world into the virtual world, so as to facilitate more vivid and interesting expression of historical information of architectural cultural heritage in the future.

4. Artificial intelligence technology is used to assist the restoration of traditional architectural cultural heritage. With the support of artificial intelligence technology, staff can build relevant three-dimensional models in advance with the help of acquired information resources, so as to observe and analyze the damaged parts from multiple angles and all directions, and select the most practical and effective repair means. At the same time, the participation of artificial intelligence technology can help staff to carry out virtual restoration of architectural cultural heritage in advance. While avoiding secondary damage, they can also predict the state after restoration and confirm the feasibility of restoration [10]. "It has greatly improved the protection technology of cultural relics and ancient sites and also improved the efficiency of protection."

4 Conclusion

Artificial intelligence technology has enriched the information presentation of architectural cultural heritage. Traditional cultural heritage exhibitions or open tours of historical buildings will be damaged more or less in the opening process due to the impression of sound, light and human factors. Compared with the physical reproduction of cultural heritage, digital reproduction is not only cheaper, but also can copy all the details of cultural heritage almost perfectly, ensuring the "authenticity" of architectural cultural heritage. At the same time, the highly restored three-dimensional model of architectural cultural heritage constructed by using information resources can also be viewed from multiple angles and all directions by the audience through virtual reality equipment. The unique interactive experience has narrowed the distance between the public and cultural heritage, and further aroused the public's awareness and enthusiasm for the protection of architectural cultural heritage. At the same time, because of the strong replicability of digital replicas, even large-scale historical building sites can be exhibited without time and space restrictions after one copy, reducing the cost of open exhibition and potential security risks.

Acknowledgement. This paper is the research result of the doctor and youth supporting project "value evaluation system of commercial historical blocks in Changchun" supported by social science foundation of Jilin Province, with the project number of 2021c103.

References

1. Wen, Y., Shuangshuang, L.I.: Research on protection and utilization of architectural cultural heritage in canal towns: a case study of daokou ancient town. J Landscape Res. **13**(5), 5 (2021)
2. Li, H., Chen, B., Li, J., et al.: Status, problems and countermeasures of artificial intelligence application in medical education. Chin. J. Evid. Based Med. **20**(9), 1092–1097 (2020)
3. Chen, S.: The impact of artificial intelligence and data fusion technology on the accounting industry and its countermeasures (2021)
4. Li, Y., Deng, K., Chen, X.: The application of artificial intelligence in psychological counseling based on "Treat Pre-Disease" (2020)
5. Shan-Bing, Y.I.: Application of artificial intelligence in the protection of intangible cultural heritage. Sci. Econ. Soc. (2020)
6. Fang, K.: The development dilemma and countermeasures of strong artificial intelligence in meeting human emotional needs. In: Stephanidis, C., Antona, M. (eds.) HCII 2020. CCIS, vol. 1224, pp. 631–640. Springer, Cham (2020). https://doi.org/10.1007/978-3-030-50726-8_82
7. Yu, X., Zhang, R., Zhang, B., et al.: Challenges of artificial intelligence to patent law and copyright law and countermeasures. Chapters (2021)
8. Sebastianelli, A., Mauro, F., Cosmo, G.D., et al.: AIRSENSE-TO-ACT: a concept paper for COVID-19 countermeasures based on artificial intelligence algorithms and multi-source data processing. ISPRS Int. J. Geo-Inf. **10**(1), 34 (2021)
9. Mathur, P., Pathak, A.K.: Intervention of artificial intelligence in agriculture: role, application and status. Indian J. Environ. Protect. (41–2) (2021)
10. Patricia, P.P., Helen, B.B.: Artificial intelligence and data protection: a comparative analysis of AI regulation through the lens of data protection in the EU and Brazil. GRUR Int. **71**(10), 924–932 (2022)

Development of Nursing Quality Evaluation System in Nursing Homes Based on Internet

Wenjing Wu[✉]

Wuhan Railway Vocational College of Technology, Wuhan 430064, China
maiwwj_0524@126.com

Abstract. As the number of elderly population in China increases year by year, the population dependency ratio is also rising rapidly. More and more elderly people will go out of their homes and enter nursing homes to spend their old age safely. At the same time, the state and local governments have also issued policies and guidelines to accelerate the development of the cause of aging, and the nursing home, as the main part of institutional elderly care, will develop rapidly. With the expansion of the scale of nursing homes and the improvement of facilities, the management of nursing homes has become a bottleneck restricting their development. Sanatoriums have been using the Internet as a tool to improve services. This is because it has become an important part of modern society and can be used in many ways. The Internet is useful not only for enterprises, but also for those who are looking for health care information or other topics related to this field. It makes it possible for people to learn more about different types of health care facilities and how they work. By doing so, patients will be able to choose the facilities they want to go to and the treatment they need.

Keywords: beadhouse · Nursing quality · Internet · Evaluation system

1 Introduction

Population aging is a major issue in the development of human society in the 21st century. For China, since 1999, the number of the elderly population has been increasing, and the degree of aging has continued to deepen. By the end of 2011, the number of elderly people aged 60 and over in China was 185MILLION, accounting for about 13.7% of the total population, larger than the total elderly population in Europe. It is predicted that the elderly population in China will continue to grow rapidly in the next 20 years [1].

As can be seen from Fig. 1, the development of population aging in China is characterized by a large population base and fast development speed. According to the prediction of the United Nations, in the first half of the 21st century, China will have the largest elderly population in the world, accounting for one fifth of the world's total [2]. By the middle of the 21st century, China will continue to become the world's second largest country with an elderly population of 300–400 million, second only to India. As for the development speed of aging, the Research Report on the prediction of the development trend of population aging in China points out that the population over 60 years old in

J. C. Hung et al. (Eds.): IC 2023, LNEE 1045, pp. 170–177, 2023.
https://doi.org/10.1007/978-981-99-2287-1_25

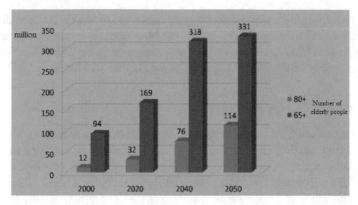

Fig. 1. Number of elderly people

China will increase at a rate of more than 3% per year, far higher than the natural growth rate of 6.6‰ of the total population. The recent report released by the center for strategic and International Studies (CSIS) of the United States also pointed out that China's aging will further accelerate and will become a country older than the United States by 2030 [3].

This paper is about the development of the nursing quality evaluation system of nursing homes based on Internet, exploring how to apply the "Internet +" technology to the field of home-based elderly care services in the future, promoting the intelligent, scientific and standardized development of home-based elderly care services in China, and making the cause of elderly care in China embark on the track of benign development [4].

2 Related Work

2.1 Population Aging

"Population aging" is translated from the phrase "aging of population", which refers to the dynamic process of the corresponding growth of the proportion of the elderly population in the total population due to the reduction of the number of young people and the increase of the number of older people caused by the reduction of fertility and the extension of life expectancy. In the book "population aging and its socio-economic consequences", the United Nations pointed out that when a country or region's population aged 60 and above accounted for 10% of the total population, or the proportion of the population aged 65 and above reached 7%, then it can be considered that the country or region has entered population aging [5]. Population aging is a form of population age structure change, which depends on two factors: first, the decline in fertility leads to the decrease in the number of children or the proportion of the total population, so that the proportion of the elderly population in the total population increases; Second, the decline in mortality and the extension of life expectancy have increased the absolute number of the elderly population and its proportion in the total population. Population aging is the inevitable result of economic and social development and progress [6]. Practical research shows that only when the economy develops to a certain extent and the people's living standards and medical standards rise, the birth rate and mortality rate will show a downward trend. China introduced the term "population aging" from the document of the world assembly on ageing held by the United Nations in Vienna in 1982.

2.2 Concept and Characteristics of Nursing Quality Evaluation Index System

The Joint Commission on Accreditation of health care organizations (JCAHO), as a professional organization implementing medical institution accreditation in the United States, has fully explained the evaluation indicators of nursing quality and believes that the essence of the indicators is a quantitative measurement tool. Therefore, it can be used for the monitoring and evaluation of health care services. Nursing quality evaluation indicators have become the most basic and important means to monitor nursing quality and ensure patient safety, and have been applied to the management and evaluation of nursing care activities [7]. The American Nurses Association (ANA) has set the basic conditions for screening nursing quality evaluation indicators, including high nursing specificity, indicator data that can be collected in the process of nursing practice, and widely considered to be closely related to nursing quality.

Chinese scholars also believe that the most basic and important means of quality management are indicators and indicator systems, which came into being with the development of management science. The index should include two parts: name and value, and the name of the nursing quality evaluation index must meet the two conditions of reflecting the characteristics of nursing work and being able to express with quantitative characteristics at the same time. Only by screening the nursing quality evaluation index can we evaluate the level of nursing quality and measure nursing work [8]. The index system is an orderly collection of indicators from many different sources and uses. A single index can only reflect a certain point or a certain aspect of the evaluation content. Only through the index system can we reflect the whole picture. In other words, only relying on the scientific nursing quality evaluation index system can we make an accurate and comprehensive evaluation of nursing quality and lead the development direction of nursing work [9].

Therefore, the characteristics of nursing quality evaluation indicators are summarized as follows: ① operability, which should be easy to observe and measure in the process of nursing practice; ② Sensitivity, which can reflect the actual quality of nursing work; ③ Specificity, which can reflect the important aspects of nursing work; ④ Objectivity, the screening and formulation of indicators are based on clinical practice; ⑤ Simplicity, simple and clear structure, and easy index system.

3   Nursing Quality Evaluation of Nursing Homes

3.1   Analysis on the Necessity of Constructing Nursing Quality Evaluation Index System

It has been more than ten years since China introduced Donabedian's result process result model as the theoretical framework to build a nursing quality evaluation system. Since the 21st century, everyone has paid increasing attention to the nursing quality evaluation index system. However, according to the recent survey of the current use, there is still a large gap in the understanding and application of nursing quality evaluation indicators among regions and hospitals.

Based on the platform of the National Alliance for the promotion of nursing quality, this study conducted a survey on the use of nursing quality evaluation indicators in 9 tertiary general hospitals in the index group of the alliance hospital. The statistical results showed that 7 hospitals had misstatements in the category of indicators, and some did not carry out structural indicator monitoring; The proportion of reporting process indicators is significant, and there are great differences in the items and quantities of monitoring process indicators among hospitals. It is urgent to build a standardized nursing quality evaluation index system to provide reference for nursing managers to evaluate the quality [10].

3.2 Basic Principles for Establishing Evaluation Index System

The establishment of evaluation index system is to comprehensively, accurately and effectively reflect the evaluation object, which is a complex system engineering. The evaluation index system of nursing process quality in geriatric hospitals is not just a simple combination of some indicators, but an index set established on the basis of certain principles. The principles it should follow are:

(1) Scientific principle

The construction of nursing process quality evaluation index system in geriatric hospitals must be guided by scientific theory, conform to the characteristics of disease spectrum and health service needs at this stage, and accurately reflect the existing problems and development trends of nursing process quality in geriatric hospitals. The establishment of each index should be scientific and reasonable. After sufficient demonstration and research, all links of the calculation process of the index should also be accurate and rigorous.

(2) Systematic principle

The index system should be reasonable in structure, clear in level, interrelated and coordinated, and can comprehensively reflect the integrity and essential characteristics of the evaluation object. The index system established in this study should reflect the main factors of nursing process quality in geriatric hospitals to ensure the comprehensiveness and credibility of the evaluation.

(3) Feasibility principle

The evaluation index system should be objective, reasonable and easy to operate. Therefore, when establishing the index system, this study makes statistics through scientific investigation methods, decomposes the index into three levels, and realizes the clarity, concretization and quantification of the index, so as to improve the operability, availability and practical value of the index system.

(4) Principle of consistency

The established evaluation system should be consistent with the evaluation objectives, which can comprehensively and fully reflect the intention of the evaluation activities, and should not contradict the objectives. The design of indicators at all levels of this study closely focuses on the two entry points of "geriatric hospital" and "nursing process quality".

(5) Principle of independence

The indicators at the same level of the evaluation index system should not have inclusion relationship, overlap and cause and effect relationship. Each indicator should be independent of each other and reflect the actual situation of the evaluation object from different aspects.

4 Internet Based Nursing Quality Evaluation System for Nursing Homes

4.1 Discussion of Indicators

Through consulting domestic and foreign literature and referring to relevant theories, after two rounds of expert consultation, this study believes that the quality of nursing process in geriatric hospitals consists of three aspects, namely, geriatric nursing practice, nursing safety management practice and nursing management practice. Elderly nursing practice and its subordinate indicators are the leading indicators of nursing quality management in geriatric hospitals, reflecting the degree to which nursing managers attach importance to quality management, the specific implementation of the responsibilities undertaken by nursing staff, and the nursing behavior implemented by nursing staff for elderly patients, as shown in Fig. 2. Nursing safety management and its subordinate indicators are the focus of nursing quality management in geriatric hospitals. Safety management implements the concept of giving priority to prevention, uses technology, facilities, education and management to fundamentally take effective measures to minimize the occurrence of errors and accidents, and create a safe and comfortable medical and nursing environment for patients. Nursing management practice mainly manages from the aspects of environment, personnel, first aid, nosocomial infection and nursing documents, which is the guarantee of nursing quality management. From the perspective of weight setting, the weight of elderly nursing practice and nursing safety management practice is 0.4, and the weight of nursing management practice is 0.2. Therefore, experts believe that in the quality management of nursing process in geriatric hospitals, we should pay more attention to elderly professional nursing and safety management to improve the quality of nursing service process.

Nursing evaluation is the first step in the nursing process and the first tool used in nursing patients. The quality of nursing evaluation is also an important indicator of

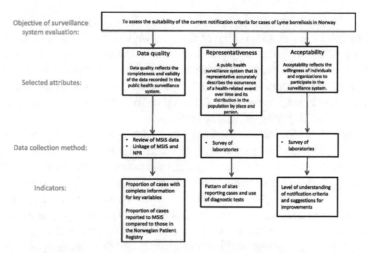

Fig. 2. Nursing quality evaluation system of nursing home

nursing quality evaluation. The evaluation of patients plays a very important role in the criteria of the joint committee international hospital evaluation. Good evaluation ability can help nurses identify patients' problems as soon as possible, so that the problems can be corrected when they are not very serious, so that prevention is better than treatment. The nursing evaluation carried out by nurses on patients should be carried out from the physiological, psychological, social and cultural aspects of patients. Only by comprehensively understanding and mastering the physical and mental status of patients can we provide the best nursing measures for patients. The aging of physical functions makes the physical, psychological and social needs of the elderly different from adults, and the focus and scope of assessing the elderly are not completely consistent with adults. The biggest impact on the elderly is not the disease itself, but the impact caused by functional and cognitive changes.

4.2 Application of Nursing Quality Evaluation System in Nursing Homes in Practice

According to the results of the above discussion, the observation method, interview method and questionnaire method were used to conduct practical tests through peer expert evaluation, patient evaluation and nursing staff self-evaluation, and a random sampling survey was conducted on the current nursing technical services of the corps to further verify that although the nursing industry in our district has made great progress, it is affected by personnel, management level and other aspects, There are still many problems in clinical nursing that do not conform to modern nursing concepts.

This study uses specific indicators of customer satisfaction in Internet technology to set up survey items. As a standardized nursing language, NOC has been widely concerned by the international nursing community because of its scientificity, progressiveness and practicality. Using the outcome of the Internet to formulate the satisfaction survey of the elderly and nurses, satisfaction does not represent the final result. It changes dynamically over time, and can reflect the state or behavior perceived or subjectively felt by patients or family members in a certain period of time, as well as the effect of nursing behavior in this period. The survey not only paid attention to the quality of basic nursing skills, but also paid attention to caring for patients and service concepts; We should not only pay attention to the protection of patients' rights, but also pay attention to nurses' own nursing according to law; At the same time, it puts forward certain requirements for the humanistic and social quality of nursing staff.

The survey adopts the standardized language with clear definition, effectiveness and credibility in the Internet to set up the survey items. The survey results are conducive to transforming into standardized, credible and easy to measure data, which can promote the development of nursing information system in nursing homes. The survey can evaluate the whole process of nursing service, directly reflect the effect of nursing measures and nursing plan implementation, and the change of score can be reported as the result of nursing measures, and can also be used for horizontal comparison of nursing service between different departments and nursing homes, which is conducive to continuous quality improvement and easy to be popularized and used in clinical practice.

5 Conclusion

The safety of the elderly is an eternal topic in nursing institutions and the main theme of nursing quality management. At different stages of development, nursing experts at different levels pay different attention to quality management. Since the 21st century, people have paid more and more attention to the nursing quality evaluation index system of nursing homes, but there is still a large gap in the understanding and application of nursing quality evaluation indexes of nursing homes among regions and hospitals. How to solve this objective problem? While the nursing center of the Hospital Management Institute of the national health and Family Planning Commission is committed to developing nursing sensitive quality index sets and establishing national nursing databases, for medical institutions, We should build a standardized nursing quality evaluation index system guided by patient safety and patient satisfaction based on the principle of focusing on key points, being operable and easy to promote, and comprehensively considering the three dimensions of structure, process and results.

References

1. Domhoff, D., Seibert, K., Stiefler, S., et al.: Associations between quality of care in informal provider networks and nursing home admissions in Germany: results of a retrospective cohort study using German health claims data. Appl. Netw. Sci. **7**(1), 1–24 (2022)
2. Role, J., Hong, C., Rosario, C., et al.: Inpatient staffing dashboard: a nursing–information technology collaborative project. Comput. Inf. Nurs.: CIN **39**(11), 772–779 (2021). Publish Ahead of Print
3. An, H.J., Choi, J.S., Min, R., et al.: The Korean version of the virtual patient learning system evaluation tool: assessment of reliability and validity. Nurse Educ. Today **106**(4), 105093 (2021)
4. Valls, A.C., Gallagher, C., Mont, L.L., et al.; Quality evaluation of patient educational resources for catheter ablation of atrial fibrillation. Eur. J. Cardiovasc. Nurs. (2021)

5. Archuleta, M., Mcgraw, C., D'Huyvetter, C., et al.: An educational outreach program: a trauma system's 5-year experience. J. Trauma Nurs. **29**(3), 152–157 (2022)
6. Ying, L., Ya, B., Ls, C., et al.: Construction of evaluation indexes of nursing students' quality and safety competencies: a Delphi study in China. J. Prof. Nurs. **37**(3), 501–509 (2021)
7. Sten, L.M., Ingelsson, P., Bckstrm, I., et al.: The development of a measurement instrument focusing on team collaboration in patient transfer processes. Int. J. Qual. Serv. Sci. (2021). ahead-of-print(ahead-of-print)
8. Singleterry, L.R., Caulfield, S.: Continuous quality improvement of writing assignments: a process for faculty development. Nurs. Educ. Perspect. **42**(2), 122–123 (2021)
9. Cheng, Q. Li, B., Zhou, Y.: Research on evaluation system of classroom teaching quality in colleges and universities based on 5G environment (2021)
10. Su, C.: The background significance and results combing of the research on the quality evaluation system of the party building work in colleges and universities in the new era. J. High. Educ. Res. **2**(4), 178–185 (2021)

Engineering Intelligent Construction Technology Based on BIM Technology

Xiaozhen Ni[1(✉)], Hongbo Wang[1], Guangjun Li[2], and Kezai Zheng[2]

[1] Department of Civil Engineering, Guangdong Engineering Polytechnic, Guangdong 510520, China
13902294795@139.com

[2] China Construction 4th Engineering Bureau 6th Corp. Limited, Guangdong 511497, China

Abstract. Intelligent building technology is a term used to describe the use of sensors and other technologies that allow remote monitoring and control of buildings. Intelligent building system is also known as smart home automation or smart home control. The goal of these systems is to provide automated solutions for controlling the lighting, heating, ventilation, safety and energy management of buildings. "Intelligent construction" is a new construction concept, which requires the development of the construction industry to take the road of low consumption, low pollution and sustainable development. At the same time, it can apply information technology to reform the traditional production mode of the construction industry. The core technology of "intelligent construction" concept is building information modeling (BIM). BIM Technology can change the phenomenon of low efficiency of information transmission and sharing among various departments and participants of engineering projects. It is the direct application of information means in the construction industry and the technical support for the construction industry to change the traditional construction concept.

Keywords: BIM Technology · Engineering · Intelligent construction technology

1 Introduction

The construction industry is still an important industry in China's national economy, but the characteristics of labor-intensive and on-site production make it difficult to improve the productivity of the construction industry; With the current requirements of economic and social development, the profits of the construction industry are getting thinner and thinner, and the employment management is becoming more and more difficult. The runaway investment, construction period, quality and management of engineering projects will be exacerbated.

It can be seen from this that the dilemma faced in the construction of engineering projects precisely needs to be solved intelligently [1]. The construction of smart city provides a very favorable opportunity for the intelligent construction of engineering projects. Whether it is application system, facility construction or information platform, it can provide a solid foundation for the intelligent construction; On the other hand,

J. C. Hung et al. (Eds.): IC 2023, LNEE 1045, pp. 178–184, 2023.
https://doi.org/10.1007/978-981-99-2287-1_26

the important economic status, high energy consumption and low productivity of the construction industry also means that smart cities that are not smart construction will have obvious shortcomings, that is, smart cities are inseparable from smart construction [2]. The "smart house" in the UK and the "smart building" in Amsterdam mentioned above need intelligent construction means. For example, the Internet of things technology is used to monitor the construction process, making the whole construction process efficient, low-carbon and intelligent.

Intellectualization is transformed from data and informatization. The intellectualization of construction mode, namely "intelligent construction", as an emerging construction concept, was proposed by Dr. Yang Baoming. It mainly expounds the two meanings of intelligent construction [3]. One is that the whole construction industry can take the road of sustainable development, ensure the efficient use of various resources in the whole construction process, realize the requirements of low-carbon and energy conservation, and save resources to the greatest extent Protect the environment and reduce pollution; The second is to use advanced information technology to realize the intellectualization of the whole construction process, so that all parties can work together, information and data can be effectively shared, and a win-win situation can be truly realized [4].

The realization of smart construction has the following significance and functions for the whole construction industry: first, the refined construction management requirements of smart construction can change the extensive production mode on the construction site, so as to save 5%–10% of the invested capital; Secondly, the requirements of intelligent construction on project quality can prolong the life cycle of the whole project, reduce unnecessary resource loss and realize the requirements of low-carbon construction [5]; Thirdly, the concept of smart construction requires all participants of the project to work together, so as to improve the information management level of enterprises, promote the management level of construction enterprises, and realize the economies of scale of the whole construction industry.

2 Related Work

2.1 Origin and Definition of BIM

(1) Origin of BIM

BIM (building information modeling) was developed by Georgia in 1975 Dr. Chuck Eastman of the Institute of Technology University, the "father of BIM", proposed that in the whole life cycle of an engineering project, the function of integrating all information, including the geometric characteristics of the building, functional requirements and the performance of building components, into the same building model can be called the building information model. The single building information model can also contain the progress, cost, resources and other information in the construction process [6].

In 2002, Autodesk officially put forward the concept of building information modeling. Since then, BIM has been widely spread. Major software companies such as Autodesk, graphisoft and Bently have successively put forward the definition of Bim and launched BIM design, analysis, simulation and construction software.

(2) Definition of BIM

The definition of Bim in the national building information modeling standards (nbims) is as follows: BIM is a digital model, which contains physical geometric information and functional characteristics. In the whole life cycle of engineering project, it can provide reliable basis from conceptual design to demolition stage. At the same time, each project participant can create, extract and update project information in the BIM model according to their own permissions. BIM is also a shared digital model based on specific standards, which can meet the collaborative work of all project participants [7].

According to the definitions of Bim in various versions at home and abroad, although they will have their own advantages in understanding, they all have the following common points:

① The information contained in BIM contains not only the functional attributes and physical characteristics of building components or equipment, but also all the information of all participants, serving the information management of the whole life cycle of engineering projects;

② BIM is a digital expression, which is parametric and computable;

③ Information sharing based on open standards.

2.2 Main Features of BIM

According to the previous description of the origin, definition and comparison with CAD of BIM, it is concluded that BIM has the following characteristics:

(1) Parametric representation of building components

Parameterization is an important idea in architectural design. It mainly includes two parts: parameterized element and parameterized modification engine. Parametric elements are expressed in the form of components. The differences between components are mainly reflected in the adjustment of parameters. Parameters save all the information as element components [8]. The parametric modification engine mainly provides the modification technology of modifying elements. The modification of any element by designers can automatically associate with other elements, and even the parameter changes caused by the deletion, movement or size change of components will change the parameters of relevant elements and associate them, without modifying the associated elements and views one by one.

(2) Dynamic display and adjustment of 2D, 3D and parametric models

The dynamic display and adjustment of two-dimensional, three-dimensional and parametric models is one of the important features of BIM. It can not only express elements in the traditional two-dimensional plane form, but also real-time building components in three-dimensional form, but also carry out analysis and calculation in a specific case.

(3) Diversification of information output forms

In addition to the model represented by BIM, the most important thing is the database that stores project information. BIM database can export information in corresponding format according to customer needs. Including planar 2D drawings, text, tables, 3D models, etc. The forms of information output can be roughly divided into two categories [9]. One is graphic data, such as plan, section, three-dimensional effect, elevation, etc.; The other is non graphic data, which will be output in the form of documents, such as Bill of quantities statistics, door and window class table, equipment information table, etc. BIM can also be dynamically modified during information output, that is, if the parametric information of a component element in the model is modified or changed, it can be dynamically reflected in the corresponding report in time. If the size of a beam changes, the size, length and volume of the beam will be automatically adjusted in the engineering quantity statistics [10]. The table does not need to be modified manually, which greatly improves the work efficiency.

The information sharing and communication mode of BIM is shown in Fig. 1 below:

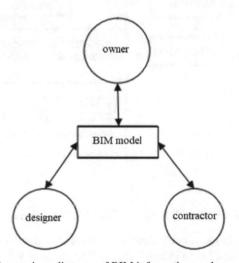

Fig. 1. Comparison diagram of BIM information exchange methods

3   Engineering Intelligent Construction Technology Based on BIM Technology

3.1   Intelligent Construction Architecture Based on BIM

According to the research significance of the framework system of intelligent construction, and combined with the horizontal relationship of the function realization of each subsystem of the intelligent construction system, build the intelligent construction system based on BIM as shown in Fig. 2. The construction system is based on the BIM system platform design, which can be divided into data layer, model layer and application layer. Through IFC data conversion processor, the construction party can obtain the 3D building information model provided by the designer and serve as the basis for the operation of the system platform. At the same time, the 4D building information model generated by linking the 3D building information model with the construction progress is used to realize the construction optimization control, dynamic construction management and dynamic construction simulation, and meet the progress objectives, cost objectives, quality objectives and safety objectives in the construction stage. At the same time, all participants of the project can also set different access rights through the network platform for project negotiation and coordination.

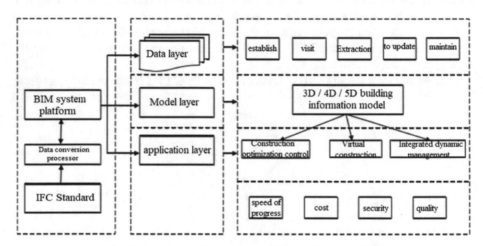

Fig. 2. Intelligent construction architecture based on BIM

3.2   Implementation Steps of Intelligent Construction Based on BIM

From the perspective of the globalization of construction industry, the development of BIM Technology has become a global hotspot. The value of BIM is not only reflected in collaborative design, but also plays an important role in the whole life cycle of engineering projects. In terms of housing industry research, BIM provides strong technical support for automation and large-scale housing industrialization. Its main role is reflected in the creation of three-dimensional information model, component assembly simulation and the whole process tracking and supervision of large-scale component procurement, manufacturing, transportation and installation. At the same time, BIM Technology can meet the green and low-carbon construction concept required by smart construction, mainly including sunshine calculation, energy consumption analysis, carbon emission index calculation and analysis of BIM Technology.

The wisdom based on BIM is of great application value, and some cities have taken positive action. For example, Shenzhen has done a lot of exploratory work to promote the application of BIM Technology. The Shenzhen Public Works Department has established a BIM work leading group and research group, established strategic cooperation with relevant BIM software providers, and actively studied the establishment of BIM standards of the public works department under the current conditions. In order to promote the promotion and application of BIM Technology in the industry, enhance the core competitiveness of enterprises and promote the healthy and sustainable development of the industry, Shenzhen housing and Construction Bureau organized the BIM Committee of Shenzhen engineering design industry to formulate the 2012 work plan for BIM promotion in Shenzhen design industry. The main work includes: organizing the preparation of the guidelines for BIM application research and development in Shenzhen engineering design industry; Carry out BIM application demonstration in the project; Hold BIM training lectures, etc.

Ningbo should actively learn from Shenzhen and other cities, take early action, make comprehensive planning, overall coordination, and formulate implementation steps and measures to promote smart construction in Ningbo. For example, set up Ningbo smart building leadership and promotion organization, formulate BIM application guidelines, issue incentive policies and incentives to promote BIM based smart building application, carry out BIM application demonstration in the project, carry out BIM application in the whole life cycle and whole process, and formulate the overall promotion and application plan and scheme of BIM based smart building.

4 Conclusion

"Smart city" is a new model of global urban development, and the construction of global smart city is inseparable from the comprehensive intellectualization of the construction field, in which smart construction is an indispensable and important part. "Smart construction" is a new construction concept. It requires the development of the construction industry to take the road of low consumption, low pollution and sustainable development. At the same time, it can apply information technology to reform the production mode of the construction industry. Combined with the concept of life cycle management and lean construction, this paper obtains the connotation of intelligent construction, improves resource utilization and realizes the requirements of low-carbon and energy saving; Take BIM as the technical core.

Acknowledgement. CIM Intelligent City Construction Innovation Platform for Industry-education Integration (Project No. 2021CJPT021).

References

1. Song, Y., Li, Z., Zhang, Z.: Research on the method for the construction of Space Situational Intelligent Cognitive ability based on knowledge engineering (2022)
2. Wang, X., Xu, X., Zhang, J., et al.: Research on intelligent construction algorithm of subject knowledge thesaurus based on literature resources. J. Phys.: Conf. Ser. **1955**(1), 012038 (2021)
3. Dong, S., Wang, L., Huang, W.: Research on intelligent construction intensive management based on building information modeling technology. IOP Conf. Ser.: Earth Environ. Sci. **783**(1), 012106 (2021)

4. Wen, Y.: Research on the intelligent construction of prefabricated building and personnel training based on BIM5D. J. Intell. Fuzzy Syst. **40**(4), 8033–8041 (2021)
5. Cunfa, L., Zhansheng, Z., et al.: Research on intelligent construction control method of prefabricated building based on LoRa technology (2020)
6. Hao, J., Yang, H., Zeng, C., Yang, D.: Research on construction of batch intelligent production line for micro/nano satellite. In: Wang, Y., Fu, M., Xu, L., Zou, J. (eds) Signal and Information Processing, Networking and Computers. Lecture Notes in Electrical Engineering, vol. 628, pp. 226–236. Springer, Singapore (2020). https://doi.org/10.1007/978-981-15-4163-6_27
7. Jiang, W., Zhang, N.: Research on characteristics of paper-plastic composite film based on intelligent optimization algorithm. Pers. Ubiquit. Comput. 1–13 (2021)
8. Maciej, S.: Intelligent prediction modeling of the post-heating mechanical performance of the brick powder modified cement paste based on the cracking patterns properties. Case Stud. Constr. Mater. **10**, e00668 (2021)
9. Wu, X., Wu, S.: Research on the location optimization of intelligent express self delivery cabinet on campus-take chongqing university of posts and telecommunications as an example. Int. Core J. Eng. **6**(5), 10–18 (2020)
10. Wu, Z., Wang, S., Yang, H., et al.: Construction of a supply chain financial logistics supervision system based on internet of things technology. Math. Prob. Eng. **2021**, 1–10 (2021)

English Online Learning System Based on Web

Jiamei Wu[✉] and Zhili Ni

Wuhan Railway Vocational College of Technology, Wuhan, Hubei, China
stefanie1251@163.com

Abstract. It is an inevitable trend of modern education to improve students' learning efficiency, promote the reform of students' learning methods and develop their autonomous learning ability. With the development of Internet and web technology and the deepening of its application in education, the learner centered online learning model provides a broad development space for students' autonomous learning, and has become an important way and development direction of modern education. This web-based English online learning system is a free and interactive English teaching tool, which provides learners with real-time opportunities to practice English. Learners can interact with native speakers of the target language through video and audio dialogues. The project, developed by the University of California, Berkeley, has been used by thousands of students around the world. Provide learners with a variety of real dialogues and activities aimed at teaching them how to use simple everyday expressions in dialogues. The software allows users to choose different types of courses: listening, reading or writing exercises.

Keywords: English · WEB · Online learning system

1 Introduction

With the development of science and technology, some new technologies have entered the field of education, which have played a significant role in promoting both teachers' teaching process and students' learning process. These technologies not only reduce the cost of practical application of pedagogical theory, but also generate new discoveries from the exploration process of other fields, and interact and integrate with the pedagogical field. With the continuous development of teaching theory, the application of the theories of "the zone of proximal development" and "the army of mastering learning theory" in teaching further requires that teaching should focus on the zone of proximal development of students, provide students with content that is appropriate to their difficulties, mobilize students' enthusiasm, give full play to their potential, and realize individualized teaching [1]. At the same time, with the continuous progress of computer artificial intelligence technology, various adaptive learning systems to realize the "zone of proximal development theory" have been designed. These systems can select teaching contents and teaching methods according to the actual situation of learners, promote the process of teaching and learning, establish an interactive platform for teaching and learning, and stimulate students' initiative and creativity.

J. C. Hung et al. (Eds.): IC 2023, LNEE 1045, pp. 185–191, 2023.
https://doi.org/10.1007/978-981-99-2287-1_27

Develop an online learning system specifically for English to promote students' deeper understanding of English and improve students' interest in English learning [2]. At the same time, it provides a platform for teachers and students to communicate. Teachers can provide students with test questions, materials and suggestions on English learning through this system. So as to better understand students' deficiencies in English learning, so as to better help students solve problems. Students can test their own ability in this system and fully understand their own shortcomings. You can also learn about the latest developments in English learning.

As an important component and development branch of online distance education, online learning has been booming in some developed countries abroad. People choose courses and take exams online. In particular, the popularization of Internet services and the construction of high-performance, low-cost computer network-based online learning have matured in terms of technical and economic conditions. However, in China, the application of the Internet in teaching is limited to online registration and online score query, and has not really formed the scale of online learning test [3]. The traditional way of learning can no longer meet the needs of modern learning. With the rise of online education in China, colleges and universities have developed their own online teaching platforms. As an important part of online courses, online learning systems have also emerged, such as the online teaching platform of Beijing Normal University and the online learning platform of Shanghai Jiaotong University.

2 Related Work

2.1 Online Learning and Its System

Some scholars believe that online learning refers to a kind of network-based learning that learners carry out around a certain knowledge content within a specified time in order to achieve a specific learning goal in the network learning environment composed of multimedia network learning resources, online learning community and network technology platform.

This study holds that online learning (also known as networked learning) is a new learning method that provides learners with a learning environment through the establishment of an education and teaching platform on the network, so as to enable learners to learn through the network. In the network-based learning environment, there are a large number of rich learning resources. Learners can download and use learning resources anytime and anywhere, and learners can also express their views and views through the network and share their resources with everyone [4].

Online learning refers to the learning environment and behavior, while online learning system refers to the digital technology application platform that online learning relies on. Through this platform, learners can learn online according to their own needs.

Promoting education through network and promoting the development of education through network have become a hot spot in the world. With the continuous development of network technology, computer and other multimedia technologies have also undergone great changes. It has gradually evolved into an important means of dealing with information, learning, work and entertainment from the initial use as a tool for people to deal with data [5]. Royal Ping DOM, an American Internet research organization, showed in its global survey report at the end of 2011 that as of November 2011, the total number of global Internet users had reached 2.1 billion (including 485million Chinese Internet users, ranking first in the world), and the total number of global Internet websites had reached 550million. Computers have become an important tool that affects people's lives. In this network age, with the rapid explosion of information, education is also facing a major and historic revolution. Today's education is facing the network, the future, the world and modernization. Education for all, quality education, lifelong education and personalized education are gradually coming to us.

2.2 System Theory

A system is an aggregate, which is usually a whole composed of many interrelated elements. A complete system usually includes four parts: Division Rules, structural framework, constituent elements and environment. The structural framework is mainly used to represent the basic structure of the system, and to visualize the various elements of the system and the relationship between the components; Constituent elements refer to the various elements that participate in the composition of the system; Environment refers to other things outside the system that can interact with the system [6]; The division rule is simply the division basis of the system, which represents the relationship between different attributes of the system. The basic structure of the system is shown in Fig. 1:

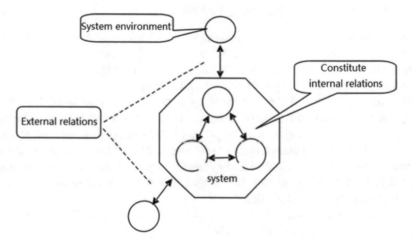

Fig. 1. Basic structure of English online learning system

The research on the system is usually carried out from the following three aspects:

(1) Properties of the system. The attributes of the system are mainly divided into two types: new attributes and inherited attributes. The new attributes of the system refer to the attributes in the original system, but they do not exist in its constituent elements. It can show the newly added characteristics of the system as a whole; Inherited attributes of a system mean that the system has the same attributes as its constituent elements.

(2) Hierarchy of the system. The level of the system mainly refers to the relationship between the constituent elements of the system and the system and each constituent element, which reflects the parallel relationship between local individuals or the total score relationship between the whole and the part. The elements of a system are the basis for the overall existence of a system. If the elements participating in the system itself are also a system, it is called a subsystem; Similarly, if the environment of the system is also a system, it is called the parent system of the system.

(3) Systematic view. The complete description of specific things should include two perspectives: function and composition. Both of these two perspectives include spatial relations, but spatial relations can also be used as an independent viewpoint of the observation system. The viewpoint of system can be divided into two types: the viewpoint of system composition and the viewpoint of system function. The viewpoint of system composition mainly inspects the attributes of the system from the perspective of each element of the system and the relationship between each element; The functional view of the system mainly examines the attributes of the system from the perspective of the relationship between the system and the environment [7].

3   English Online Learning System Based on Web

3.1   Components of Online Learning System

Online learning originated from distance education. With the rise of network technology integrating text, voice, video and other information, and the continuous popularization of Internet technology, distance education has been supported by powerful technologies and means and has a wide range of communication channels.

From the perspective of system theory, each system is composed of different elements, and the category of system is universal. There are various systems in the world, but no matter what kind of system is composed of elements, online learning, as an educational activity of human beings, is also a subsystem of human social system. It also has elements. According to the viewpoint of system theory, the selection of online learning elements should first adhere to the relationship between system and elements, and grasp the general characteristics of elements.

In essence, online learning is the same as traditional learning. They are organized and planned cognitive activities to enable learners to achieve specific teaching goals. However, from the perspective of teaching process, online learning is fundamentally different from traditional learning. Traditional learning pays more attention to the position and role of teachers in teaching [8]. The whole learning process of learners is under the supervision and control of teachers. In the whole teaching process, teachers occupy a major position and play a major role; In the network-based learning environment, learners' learning activities are mainly carried out through computer networks, and they communicate with the outside world through computers. In this case, students become the masters of the whole learning process, and teachers play a relatively small role, mainly as guides. The learning process of online learning can be represented by an atmosphere like structure, as shown in Fig. 2:

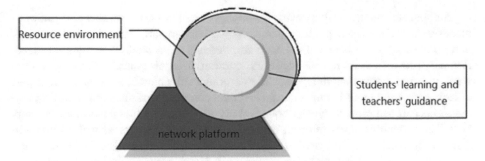

Fig. 2. Element model of online learning system

In the process of online learning, the central part is students' learning and teachers' guidance. It takes students' learning as the center of the whole activity, with teachers as the leading and students as the main body. What surrounds the outside is the resource environment, which is like the earth's atmosphere, which surrounds teachers' teaching and students' learning, and provides nutrients for the entire teaching activities. It is a necessary prerequisite for the existence of the online learning system. Without resources, the teaching activities between teachers and students will lack the possibility of existence. There is a space for communication between teachers and students within the resource environment, which is the essence of teaching implementation. The network teaching platform is the foundation to support other elements in the online learning system. It is the basic tool used by teachers and students in the process of network teaching and the basic means to display network resources [9]. The communication between teachers and students in the middle part, including teachers' guidance, students' cooperation, independent exploration and so on, is the intermediate link of teaching process elements. It shoulders the bridge connecting other elements and is also the core element to realize the demands of teaching value. The leading role of teachers and the main role of students are fully reflected here.

3.2 Basic Functions of Online Learning System

Compared with traditional education, online learning has remarkable characteristics. First of all, the network is a huge library of educational resources, through which teaching resources can be shared. In the networked teaching platform, there are abundant digital teaching resources composed of sound, text, pictures, various teaching software, expert guidance groups and other teaching information. Every learner in the online learning system can share this educational information resource, which cannot be realized by any other media and technical means; Secondly, online learners can get information resources of any subject, any field or even any country or region from the network. Online learning system can break through the limitations of time and space. In the network environment, learners' learning activities no longer have differences in teaching level, educational ability, teaching materials, etc. every online learner can get equal opportunities for education.

A complete online learning system usually includes five parts: learning part, communication part, management platform part, technical support part, evaluation and feedback part. The learning part is one of the main components of the student learning interface. It mainly provides learners with necessary learning materials (such as teaching courseware, video, teaching materials, etc.); The communication part is a supplementary part of learners' autonomous learning and an essential component of the online learning system. This part can provide synchronous and asynchronous communication for learners [10]. The communication can occur between learners and learners, as well as between learners and teachers. Mainly in the form of e-mail, forums, famous teachers' Q & A, etc. The management part is mainly for managers. It is the macro-control and micro grasp of the whole process of online learning, ranging from the control of the whole online learning system to the management of the learning progress of each chapter and unit; The technical part refers to the technical support service of the online learning system. If an online learning system wants to operate normally, the technical maintenance of the online learning system is an essential part; Evaluation and feedback: this part seems simple, but it plays an important role in online learning and plays a key role. Learners' academic performance is largely determined by evaluation and feedback. An appropriate and timely feedback can help learners grasp their own direction in the big environment of the network, so as not to get lost in the vast amount of learning materials. The evaluation and feedback part can be divided into two forms: self-evaluation and other evaluation. Self-evaluation is that learners understand their own learning through online testing, self reflection and other forms. The evaluator of other evaluation can be the learner's learning partner or teacher. The evaluation can evaluate the learner's learning in the form of communication or examination. Both self-evaluation and other evaluation can help learners better understand their mastery of knowledge, and enable teachers to timely monitor and help students' learning.

4   Conclusion

The purpose of this paper is to design and implement an online learning system for English grammar, mainly the design and implementation of the student model of the core system. The system establishes the corresponding knowledge base of English grammar teaching content according to a certain relationship model, and uses the student model to represent the students' learning characteristics and learning status, and automatically generates the teaching content; The system selects learning content according to students' different levels and learning situations According to the cognitive model, students' errors are diagnosed, analyzed, judged and remedied; It can automatically track the changes of students' learning status, and continuously improve teaching strategies to achieve students' recent development zone and individualized teaching according to the requirements of individual teaching strategies; The system can evaluate students' learning behavior.

References

1. Wang, Y.: Optimization of English online learning dictionary system based on multiagent architecture. Complexity **2021**(4), 1–10 (2021)

2. Peng, N.: Research on the effectiveness of English online learning based on neural network. Neural Comput. Appl. **29**, 1–12 (2021)
3. Sun, N., Zhao, H.: Construction of English multi-module online learning system based on intelligent analysis technology. In: Macintyre, J., Zhao, J., Ma, X. (eds.) The 2021 International Conference on Machine Learning and Big Data Analytics for IoT Security and Privacy: SPIoT-2021 Volume 1, pp. 692–698. Springer International Publishing, Cham (2022). https://doi.org/10.1007/978-3-030-89508-2_89
4. Wang, X., Zhang, D., Asthana, A., et al.: Design of English hierarchical online test system based on machine learning. J. Intell. Syst. **30**(1), 793–807 (2021)
5. Han, Y.: Evaluation of English online teaching based on remote supervision algorithms and deep learning. J. Intell. Fuzzy Syst. **5**, 1–12 (2020)
6. Atmojo, W.T., Adisaputera, A., et al. The quality of learning system by using online learning technology based on SIPDA during Covid-19 pandemic. J. Phys. Conf. Ser. **1811**(1), 012064 (7p.) (2021)
7. Chen, G., Chen, P., Huang, W., et al.: Continuance intention mechanism of middle school student users on online learning platform based on qualitative comparative analysis method. Math. Probl. Eng. **2022** (2022)
8. Cancino, M., Capredoni, R.: Assessing pre-service efl teachers' perceptions regarding an online student response system. Taiwan J. TESOL (2020)
9. Wang, Y.: Implications of blended teaching based on theory of semantic wave for teaching English writing in high school. J. Higher Educ. Res. **3**(2), 166–168 (2022)
10. Shen, G., Jie, S., Li, Z., et al.: Bypassing logits bias in online class-incremental learning with a generative framework (2022)

English Speech Recognition Based on Deep Machine Learning Algorithm

Aiyan Du[✉]

Qilu Institute of Technology, Jinan 250200, Shandong, China
550824859@qq.com

Abstract. With the deepening of artificial intelligence technology in people's life, English speech recognition, as an indispensable technical means, has become the focus of attention. In the past decades, the acoustic model combined with Gaussian mixed model hidden Markov model (gmm-hmm) has occupied a leading position in the field of English speech recognition technology. However, with the English speech data becoming more and more large and complex, the shortcomings of the traditional model are gradually revealed. English speech recognition based on deep machine learning algorithm is a branch of artificial intelligence, which mainly studies human English speech understanding. It uses pattern recognition, statistical modeling and computer vision to achieve this goal. Deep neural network (DNN) has been used as the basis of many other types of algorithms; English voice to text system, using DNN to convert spoken language into text; Automatic transcription system, using DNN to recognize English speech, and then convert it into text; Machine translation between different languages using DNN.

Keywords: Deep machine learning · English speech recognition · English

1 Introduction

With the continuous development of the times, computer technology is also developing by leaps and bounds. More and more information is presented to people's daily life and work in the form of multimedia. People always choose the most convenient and natural way to communicate. Natural language has undoubtedly become the first choice. English voice information cannot only inherit people's thoughts and express feelings, but also be simpler than words and pictures. It has become one of the most popular communication methods in today's society. Among them, the most simple, direct and effective way is to transmit information through English pronunciation [1]. Due to the development of artificial intelligence in recent years, there are more and more opportunities for people to communicate with machines. Whether in scientific research or daily life, computer has been integrated into all aspects of people's life and work, and the increasingly rich and powerful functions of computer are helping people complete all kinds of work [2]. If the computer can understand English voice and speak, it will save hardware such as keyboard, liberate hands and make it easier for people from different language regions to communicate. Therefore, human-computer interaction through simple English voice

J. C. Hung et al. (Eds.): IC 2023, LNEE 1045, pp. 192–198, 2023.
https://doi.org/10.1007/978-981-99-2287-1_28

instructions has been pursued for a long time, and English speech recognition has become the key technology to achieve this goal. Moreover, in recent years, the rapid development of deep learning has brought new ideas to English speech recognition [3]. Among them, the rise of neural network also makes English speech recognition technology attract more and more attention.

Because the human auditory system and pronunciation system are very complex, at present, countries have not developed an English speech recognition system as perfect as the human auditory system. On the other hand, the speaker's accent, feelings when speaking and the changes of the surrounding environment will have a significant impact on the performance of English speech recognition system. Therefore, in order to overcome these defects, optimize the performance of English speech recognition system and make it more practical and intelligent. At present, there are still many problems to be solved [4]. Therefore, on the basis of previous research, we need to further explore and study new technologies and theories.

2 Related Work

2.1 Deep Learning Background

In 2006, Hinton proposed a fast and greedy algorithm, which can only carry out deep belief network learning and training in one layer of network at a time. Compared with the best data discrimination algorithm, this algorithm can give better data classification. The proposal of this algorithm makes deep learning become a research hotspot.

Neural network of pattern recognition, which can be understood as "self-learning without supervision", and is called cognitive machine. Therefore, deep learning has attracted the attention of many researchers and experts. Back propagation is introduced into neural network and applied to identify numbers. However, due to the limited technology level at that time, the processing speed of the computer cannot meet the operation requirements of complex algorithms [5].

In the 1970s, many researchers tried to use back propagation algorithm to train supervised deep neural networks, but in the end, most of them failed. The problem of gradient disappearance is the main reason. Because of being challenged by various simple models, neural network has not been widely popularized and applied.

The problems mentioned above have been solved. Jurgen schmidhuber proposed a multi-layer network in 1992. The specific operation method is to first train each layer contained in the deep neural network through unsupervised learning algorithm, and then continuously fine tune the whole network according to back propagation (BP) 2 'algorithm. And Sepp Hochreiter et al. [6]. Pointed out that the simple recurrent neural network (RNN) and convolutional neural network (CNN) have been significantly better than the previous conditional random field (CRF) method with excellent performance. The long-short term memory (LSTM) neural network is adopted. The network includes input, output and forgetting gates. In the task of vocabulary annotation, It has better performance than simple neural network.

Hinton et al. Showed that all layers with deep characteristics are modeled by restricted Boltzmann machine (RBM). If the model with multi-layer network is trained, the multi-layer network structure model will become a generation model. In recent years, researchers usually rely on the powerful computing power of computers to solve the problems mentioned above, such as using graphics processing unit (GPU) to solve the problem of computing acceleration. In 2010, Dan ciresan and other technicians in the Swiss artificial intelligence laboratory verified that the gradient disappearance can be ignored by using GPU to run the anti BP algorithm. Foreign IBM, Google, domestic Alibaba, iFLYTEK, baidu Research Institute and other companies and research institutes are also studying the application of deep learning in English speech recognition system.

2.2 The Significance of Deep Learning in English Speech Recognition

The full realization of artificial intelligence is inseparable from a major problem: how to realize the communication between human and machine. At this time, English speech recognition technology is particularly important. However, due to the increasingly complex application scenarios of English speech recognition, the traditional English speech recognition technology can no longer meet the current needs, and many systems can not achieve the recognition effect that can be used at the application level. At this time, there is an urgent need for the support of new technical means and theoretical methods to break through the current bottleneck state; With its powerful ability of massive data modeling, deep learning model has undoubtedly become a hot research object [7].

If we want to seek a breakthrough in technical means, we undoubtedly need more novel theoretical methods. At the same time, the deep learning model has quickly entered the public's vision with its advantages in model structure. If we can combine English speech recognition technology with deep learning theory through research, whether it is deep learning itself or English speech recognition technology itself, it will undoubtedly push the two to another height at the same time. It can not only help us deeply understand the working principle of deep neural network itself, but also use the knowledge of deep learning to improve an important technical means [8].

From the aspect of practical application, the initial proposal of artificial neural network did not produce satisfactory results in application because of the lack of theoretical research and the lack of network training caused by the underdeveloped hardware facilities at that time. Until the restricted Boltzmann machine (RBM) and deep confidence network proposed by Hinton and his disciples after unremitting research, the deep neural network can be said to have reached the stage of practical application. Although the training time increases with each passing day due to the complexity of the network model, the introduction of GPU parallel operation can effectively alleviate this problem [9]. It is believed that large-scale parameter training tasks will become accustomed in the near future.

3 English Sound Recognition Based on Deep Machine Learning Algorithm

3.1 Model Composition and Training Process of Deep Learning

The nonlinear transformation in deep learning network is mainly realized by the nonlinear representation between the input and output of hidden layer units. In fact, the concept of network node we mentioned is the same as that of neuron in artificial neural network. According to the theoretical knowledge of neuron, the following model equation can be obtained. It can be seen that it represents the network output of the ith node of the hidden layer 1 (see Fig. 1 for the node model). Its input is the weighted sum of all node outputs from layer 1 to layer L-1, that is:

$$y_j = \sum_1^n w_{ij} - \theta_j \qquad (1)$$

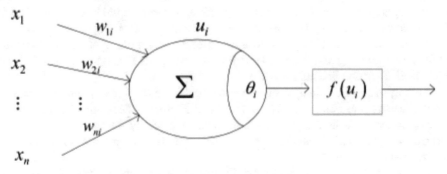

Fig. 1. Network node model

In the training process of deep learning, if all layers are trained at the same time, the complexity will be high, but if only one layer is trained at a time, the deviation will continue to increase [10]. Hinton proposed a training process in 2006, and the specific steps are as follows:

(1) Adopt bottom-up unsupervised learning. Firstly, single-layer neurons are constructed layer by layer, and then the wake sleep algorithm is used to adjust and optimize the parameters of each layer. Only one layer needs to be adjusted at a time, and it is adjusted layer by layer. The wake stage in wake sleep algorithm refers to the cognitive process, which mainly uses the external characteristics and upward weight to continuously generate the state of nodes, and continuously modifies the generation weight between layers by using the gradient descent method. At the same time, the process of continuously modifying the weights of sleep and wake at the bottom layer mainly refers to the process of continuously generating the weights of sleep and wake at the top layer.
(2) The top-down supervised learning method is adopted, that is, the data with labels are trained, the error is transmitted through the top-down method, and then the whole network is continuously fine tuned.

3.2 English Speech Feature Extraction

A good acoustic feature determines the quality of the recognition system. Only by working hard on the extraction of acoustic features can we get an ideal English speech recognition system. As for the English phonetic data itself, due to the differences of different people in gender, age, pronunciation habits, as well as physiological and psychological differences, it will also change with the passage of time, making people have more or less differences even when expressing the same content. If we can eliminate the specific characteristics of individuals in speaking style and habits and get the most direct expression of information, the English speech recognition system will be easier to read information, and the system performance will be improved.

The essence of acoustic feature extraction is waveform signal compression, or deconvolution, in order to achieve the most reasonable classification. Considering that the change speed of English speech signal is very slow in a short time (10 ~ 30 ms), and its features are relatively stable, the corresponding features should be extracted through short-time analysis in English speech signal processing. According to the current situation, the features selected by most English speech recognition systems are: cepstrum coefficient, Mel frequency cepstrum coefficient, perceptual linear prediction (PLP) and linear prediction coefficients (LPC).

Among them, the principle of cepstrum coefficient is homomorphic processing. Firstly, the discrete Fourier transform (DFT) of English speech signal is determined by calculation, then its reciprocal is calculated, and finally the result is obtained by inverse discrete Fourier transform (DFT). This method can provide relatively stable characteristic parameters.

Linear prediction analysis is based on human vocal mechanism. It uses the short tube cascade model between vocal channels. It is assumed that the signal data at time n can be described by the linear combination of previous signal data. When the mean square error between English speech sample value and linear prediction estimation value reaches the minimum value, the linear prediction coefficient is determined.

Mel cepstrum coefficient and perceptual linear prediction are obviously different from the above two methods. They refer to the mechanism of human ear perceptual audio signal to a certain extent and deconvolute in the frequency domain to obtain acoustic features. In the process of extracting MFCC features, firstly, the signal is mapped from time domain to frequency domain through FFT signal, and then its logarithmic energy spectrum is convoluted through a group of triangular filters evenly distributed on Mel frequency domain scale. Finally, the output is processed by discrete cosine transform method, and several coefficients in the front are retained. The operation process of PLP is: firstly, the LPC parameters are obtained by Durbin method, and then the log energy spectrum is discrete cosine transformed after calculating the autocorrelation coefficient.

After acquiring acoustic features, in order to improve the robustness of features, it is generally necessary to normalize them. The widely used methods include cepstral mean normalization (CMN), vocal trace length normalization (VTLn), Rasta filter 126, linear discriminant analysis (LDA) Maximum likelihood linear transformation (mllt) and heteroscedastic linear discriminant analysis (HLDA) are shown in Fig. 2.

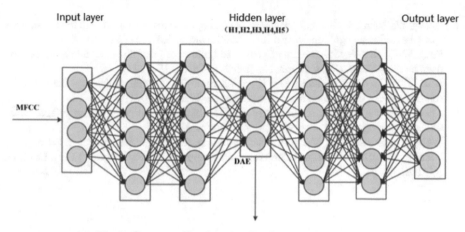

Fig. 2. Deep machine learning English sound recognition

4 Conclusion

As one of the hottest research fields at present and in the future, deep learning has achieved good results in the field of English speech recognition. The performance of English speech recognition system often directly affects the effect experience of most intelligent systems, so the future development direction must be to combine the two technologies to promote mutual progress. The acoustic feature extraction is taken as the research object, and the research work is carried out based on the depth automatic encoder model. Deep automatic encoder belongs to multi-layer network model, which is based on unsupervised training. It is widely used in the fields of data dimensionality reduction, feature extraction and so on. This paper focuses on the analysis and research of this deep learning model from the perspectives of feature data preprocessing, model structure and network training parameters.

Acknowledgements. Phrased achievements of the 2021 school-level educational reform project (kcsz202108) of Qilu Institute of Technology.

References

1. Song, Z.: English speech recognition based on deep learning with multiple features. Computing **102**(99), 1–20 (2020)
2. Wang, J.: Speech recognition of oral english teaching based on deep belief network. Int. J. Emerg. Technol. Learn. (iJET) **15**(10), 100 (2020)
3. Mustaqeem, K.S.: Speech emotion recognition based on deep networks: a review. In: Korean Information Processing Society (KIPS-2021) (2021)
4. Yang, Y., Yue, Y.: English speech sound improvement system based on deep learning from signal processing to semantic recognition. Int. J. Speech Technol. **2020**, 1–11 (2020)
5. Wang, Y., Wei, Z.: Research on speech enhancement based on deep neural network. J. Phys: Conf. Ser. **1650**, 032163 (2020)

6. Zheng, W., Zheng, W., Zong, Y.: Multi-scale discrepancy adversarial network for crosscorpus speech emotion recognition. Virtual Real. Intell. Hardw. **3**(1), 65–75 (2021)
7. Feng, W.K., Hua, H.X.: Research of image recognition of plant diseases and pests based on deep learning. Int. J. Cogn. Inform. Nat. Intell. (IJCINI), **15** (2021)
8. Kim, M.S.: Lipreading architecture based on multiple convolutional neural networks for sentence-level visual speech recognition. Sensors. **2021**, 22 (2021)
9. Chen, Z., Dong, R.: Research on fast recognition method of complex sorting images based on deep learning. Int. J. Pattern Recogn. Artif. Intell. (2020)
10. Wang, P.: Research and design of smart home speech recognition system based on deep learning. In: 2020 International Conference on Computer Vision, Image and Deep Learning (CVIDL) (2020)

Enterprise Human Resource Scheduling and Optimization Based on Big Data

Linjie Tong[✉]

China Railway First Survey and Design, Institute Group Co., Ltd., Xi'an, 710043, Shaanxi, China
linsey_ruc@163.com

Abstract. The scientific allocation of human resources to achieve the best match between people and things has always been an important goal pursued by human resources management. Big data is a mass of information that can be stored, analyzed and mined. It contains a lot of valuable information about business activities, and can provide in-depth understanding of user needs, user behavior trends and other relevant factors. The problem with this kind of data is that ordinary users cannot easily access it without special software or hardware. This makes it difficult to analyze for specific purposes, such as improving scheduling algorithms or optimizing human resource processes. However, some researchers have solved these problems, and they are developing solutions based on big data analysis technology to solve these problems.

Keywords: Resource scheduling · big data · human resources · optimization

1 Introduction

Human resource allocation has always been an important subject of human resource management. Whether the allocation process is reasonable and whether the allocation effect is good will affect whether employees can give full play to their own quality and whether the organization's work objectives can be achieved. In today's increasingly complex social contradictions and the increasing number of work tasks undertaken by employees, but the manpower has not yet increased significantly, it is undoubtedly a necessary measure to revitalize and optimize the existing human resources and let each employee play the greatest role in his own position [1]. From the current situation that the country vigorously promotes the big data strategy, enterprises successfully implement the data management mode, and big data has excellent performance in many fields, especially in the field of human resource management, using big data technology to optimize human resource allocation will be an innovative idea with both efficiency and science.

As far as the current research is concerned, the academic discussion on the allocation of human resources mainly focuses on the quantitative prediction of human resources, less on the structural and quality problems in the allocation of human resources, and the way of allocation and the effect investigation after allocation need to be improved

J. C. Hung et al. (Eds.): IC 2023, LNEE 1045, pp. 199–205, 2023.
https://doi.org/10.1007/978-981-99-2287-1_29

[2]. Moreover, in the research of enterprise big data, the main research direction is how to make big data technology serve external policing activities. There is no systematic research result on internal policing activities, especially the allocation of enterprise human resources.

The research focus of this paper is as follows: first, using empirical research method, comprehensively analyze the problems and causes of human resources allocation in Chinese enterprises; Second, through literature research and normative research, it discusses the necessity and feasibility of using big data to optimize human resources allocation, and clarifies the mechanism of big data promoting the optimal allocation of human resources [3]; Third, in view of the above problems and reasons, combined with the characteristics and functions of big data technology, explore the Countermeasures of using big data to optimize human resource allocation. It is hoped that the above research results will provide ideas for the enterprise to solve the problem of insufficient human resources, release the vitality of existing personnel and improve the combat effectiveness of the team.

2   Related Work

2.1   Human Resources

Broadly speaking, people with normal intelligence are human resources. In the narrow sense, human resources, also known as labor resources or labor force, refers to the total number of people who can promote the whole economic and social development and have the ability to work. The most basic aspects of human resources include physical strength and intelligence. From the perspective of practical application, it includes four aspects: physique, intelligence, knowledge and skills [4].

The status of human resources involves the quantity and quality of human resources. The quality aspect includes the age, education background, knowledge structure, ability, etc. for employees who master certain knowledge and can apply it to their work, management calls them knowledge-based employees. Entering the new century, we are entering the era of knowledge economy [5]. The social economy is characterized by relying on science and knowledge to create wealth. The more high-quality human resources an enterprise has, the stronger its ability to create wealth and survive and develop.

Human resources are renewable resources, also known as renewable resources. The regeneration of human resources is mainly based on the reproduction of population and the reproduction of labor force, which is realized through the continuous replacement of individuals in the population and the process of "labor consumption - labor production - labor re consumption - labor re production". Of course, the regeneration of human resources is different from that of general biological resources. In addition to abiding by general biological laws, it is also dominated by human consciousness and affected by human activities [6]. The renewable resources - human resources discussed in this paper are of the significance of labor reproduction.

(1)   Human character

Human character is an important aspect of human resources. People's character can generally be reflected through their speech and behavior. At the same time, there is neither absolutely good character nor absolutely bad character.

(2) Software development project HR type

The human resources involved in software development projects are classified from the perspective of skills. There are mainly the following types, which can be described by the pyramid structure. The project manager is at the top level, the system architect, system designer, development team leader and test team leader are at the second level, the senior system architecture engineer, senior system design engineer, senior software engineer and Senior Test Engineer are at the third level, and the software engineer is at the bottom level.

2.2 Dispatch

(1) Basic concept of scheduling

Scheduling is to allocate reasonable resources and time for various actions in the plan, solve various constraints in the real world, and coordinate system work after the plan has been formulated to ensure the correct completion of tasks. Scheduling can be regarded as an extension of planning tasks.

(2) Common scheduling methods

The research of scheduling originates from the needs of industrial production. Job shop scheduling is a typical scheduling problem. Scheduling problems often have many different representations, especially in the aspects of task size, resource consumption and scheduling constraints. Therefore, it is impossible to design a general scheduling algorithm, and special solutions must be developed for each type of scheduling problems [7]. Because the actual scheduling problem involves many constraints and has a large solution space, it is very necessary to introduce heuristic knowledge to reduce the search workload. Various artificial intelligence technologies can be applied to solve the scheduling problem. Several common scheduling methods are listed below:

① Production system. The heuristic knowledge is represented by condition action rules, so that the system can make various constraint satisfaction decisions in the generation process of scheduling schemes.

② Heuristic search. Design heuristic scheduling algorithm and search evaluation function, so that the system can better determine the search direction, select the next activity to be scheduled, and compare the advantages and disadvantages of multiple optional local scheduling schemes [8]. The algorithm and evaluation function often depend on the application field, which is not universal, but has high efficiency.

③ Opportunity reasoning. Using the blackboard knowledge source architecture, the CSP solving subtasks to be executed by the scheduling system are sorted by priority, and the corresponding knowledge source is always activated to solve the subtask with the highest priority.

④ Hierarchical scheduling. The complex scheduling problem is abstracted hierarchically, the problem is simplified, and the scheduling scheme is generated in a gradually refined way. The high level of hierarchical scheduling only considers the key constraints and ignores the details, so as to quickly generate rough scheduling schemes. These schemes become the guidance and basis for the low-level refined scheduling.

⑤ Constraint relaxation techniques. The realization of multiple goals may conflict with each other. When conflicts occur, some constraints must be relaxed to improve the efficiency of scheduling. In order to support the relaxation of constraints, constraints should be divided into hard and soft constraints, and the relaxation methods and limits should be specified for soft constraints.

3 Enterprise Human Resource Scheduling and Optimization Based on Big Data

3.1 The Role of Optimizing the Allocation of Human Resources in Enterprises

The optimal allocation of enterprise human resources directly affects the individual employees, and the employees' personal experience will reflect the allocation effect into the overall structure and operation activities of the public security organs. The efficient operation of the public security organ depends on the group resultant force generated by the mutual assistance between individual employees. The person post matching is to lay a foundation for the formation of the group resultant force by orderly arranging individuals and ensuring the scientificity and rationality of the organizational structure. Scholars have found that the poor effect of human resource allocation will lead to the increase of conflicts among members, which will lead to the decline of the quality of organizational decision-making, the decline of members' ability to implement decision-making and the emotional distortion of accepting decision-making, and ultimately lead to poor organizational performance [9]; The optimal allocation of human resources is conducive to improving the interpersonal relationship among organizational members, reducing personnel conflicts and enhancing organizational cohesion. It is a direct method to improve organizational performance. Reasonable human resource allocation enables employees to perform their respective duties and play their own role in a good organizational atmosphere, which not only optimizes the personnel structure of the public security organ, but also improves the work efficiency of the organ, and finally achieves the effect of successfully completing the task of maintaining national security and social stability.

The optimal allocation of enterprise human resources not only affects the individual employees, but also the public security organs as a whole. The public security organ is composed of individual employees. When the individual structure is reasonable, the overall function of the public security organ will be maximized; When the organ reaches the optimal state, it will have a series of positive effects on the individual. Individual employees and public security organs have a close relationship of mutual influence and survival, and the optimal allocation of enterprise human resources is a powerful link among them [10]. The function of optimizing the allocation of enterprise human resources is shown in Fig. 1.

3.2 Targeted Solutions of Big Data to Existing Problems in Enterprise Resource Allocation

To sum up, big data technology can play a role in the optimal allocation of enterprise resources through management thinking, management technology and management

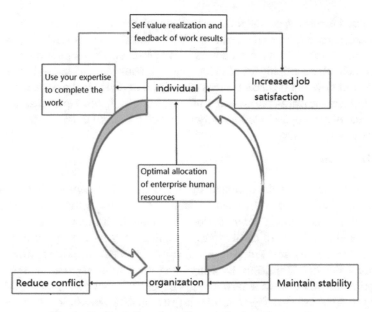

Fig. 1. Optimal allocation of enterprise human resources

mode. In view of the specific problems existing in the human resource allocation of Chinese enterprises, big data can be solved in the following ways: in terms of improving the insufficient allocation of the total amount, big data provides data basis and adjustment suggestions for the preparation competent departments to make preparation decisions by predicting the development trend of employees' work, changes in employees' responsibilities and task volume, as well as human supply and demand, On the whole, it can meet the demand of current enterprises for human resources. In terms of improving the uneven regional and hierarchical allocation, big data provides a reference for the public security organs to take measures to allocate human resources and reassign work tasks across regions and levels in a timely manner by predicting the supply and demand trend of human resources and dynamically monitoring the distribution of human resources and workload changes; By analyzing the data of enterprise resource allocation among different regions, we can find the regions with relatively reasonable allocation and set them as benchmarks to provide standards and experience for the scientific allocation of enterprise resources of the whole public security organ. In terms of improving the low degree of person post matching, big data has laid a solid foundation for "knowing the post" and "knowing the person" by collecting data, conducting all-round job analysis and personal quality analysis, providing scientific data support for the final matching link and improving the degree of person post matching. In terms of improving the imbalance between busy and idle work, big data monitors the change trend of human resource management workload, human resource distribution and staff working hours, so as to provide a basis for the public security organs to increase or decrease human resources, reassign posts and work tasks, and ensure that the workload of employees is in an appropriate range. In terms of the rigid configuration mode, big data encourages managers to timely adjust the existing personnel structure and job arrangement by actively collecting and reporting personnel changes, so as to achieve the purpose of enriching post strength and activating the configuration mode. As for the problem of poor configuration effect, big data timely alerts employees of errors through real-time collection and analysis of employee workflow and other data, urges employees or units to take corrective measures in a timely manner, learns the working methods of excellent employees, and uses the improvement of employees' work quality to ensure the configuration effect.

4   Conclusion

It is an inevitable choice to optimize the allocation of enterprise human resources and maximize the efficiency of existing human resources in the current situation of increasingly complex social contradictions and insufficient human resources. Nowadays, the rapid development of big data technology and its successful application in human resource management in various fields provide a new idea for optimizing enterprise human resource allocation. By combining the role of big data with the process of enterprise human resource allocation, this paper finds that on the one hand, big data technology can reverse the subjective thinking in human resource allocation and improve the objectivity and foresight of allocation decisions; On the other hand, by providing collection, analysis and visualization technology, it helps managers to do a good job in job analysis, personnel analysis, human resource prediction and work quality monitoring in configuration, and can timely warn employees of work deviation, so as to realize dynamic optimization of configuration effect. Moreover, from the perspective of the current social environment and the implementation of data policing, using big data to optimize the allocation of human resources has great feasibility.

References

1. Hung-Yi, T.: Research on the application of big data in enterprise human resource management. J. Phys. Conf. Ser. **1744**(3), 032241 (5p.) (2021)
2. Sun, C., Li, X.: Research on the relationship between human resource management activities and enterprise performance based on the supervised learning model. Disc. Dyn. Nat. Soc. **2021** (2021)
3. Li, X., Wang, Z.: Research on recruitment problems and countermeasures in enterprise human resource management—take Shenzhen Paijie company as an example. In: International Conference on Education, Economics and Information Management (ICEEIM 2019) (2020)
4. Bari, M.S., Al-Din, M.: Research on strategic human resource management innovation-oriented. Int. J. Market. Human Resourc. Manage. **1**(1), 14 (2020)
5. Kainzbauer, A., Rungruang, P., Hallinger, P.: How does research on sustainable human resource management contribute to corporate sustainability: a document co-citation analysis. 1982–2021. Sustainability. **2021**, 13 (2021)
6. Sun, G., Feng, Y., Li, X., et al.: Research on DHR talent training of human resource management major under big data era (2021)
7. Alshaikhmubarak, A., Camara, N.D., Baruch, Y.: The impact of high-performance human resource practices on the research performance and career success of academics in Saudi Arabia. Career Dev. Int. **25**(6), 671–690 (2020)

8. Wood, S.: Human resource management–performance research: is everyone really on the same page on employee involvement?. Int. J. Manage. Rev. (2020)
9. Zhao, Z.: Research on the construction of a performance management system for human resources. The Science Education Article Collects (2020)
10. Guo, J., Zhang, S.: Research on equilibrium scheduling of airborne network resource based on load Gini coefficient. Int. J. Inf. Commun. Technol. **16**(2), 162 (2020)

Evaluation Data of Poor College Students Based on Improved Apriori Algorithm

Xianqiang Hou[✉], Na Liu, and Jing tian

Shandong Institute of Commerce and Technology, Jinan 250103, Shandong, China
jqwhzzs@yeah.net

Abstract. In recent years, with the continuous expansion of college enrollment in China, more and more students have the opportunity to enter the university. However, due to the reform of the charging system in colleges and universities, the number of poor students in colleges and universities is increasing. In this worrying situation, our government and universities have taken a series of measures and made many achievements. In this study, we use the Apriori algorithm to extract evaluation data from poor college students' compositions. The purpose of this study is to examine how poor college students' writing performance is evaluated by taking their compositions as samples. We also investigated whether there were differences between male and female students in assessing their own performance. The results show that, compared with other college students, Apriori algorithm can effectively evaluate the paper quality of students whose academic ability is at least at the middle level. This is especially useful for papers with poor evaluation ability.

Keywords: Poor students · Apriori algorithm · Analysis of evaluation data

1 Introduction

With the implementation of the cost sharing mechanism of higher education and the acceleration of the popularization of higher education in China, the problem of poor college students has become a focus of attention from all walks of life. Generally speaking, poor students refer to those who are unable to pay for education or who are difficult to pay for education during their schooling due to their family's economic difficulties. According to the survey of the Ministry of Education, at present, students with financial difficulties account for 15% - 30% of the total number of students in colleges and universities, among which students with special financial difficulties account for 8% - 15% [1].

Different scholars have analyzed the causes of the poor college students in China's social transition period from different angles, and summarized the following main factors: 1. Natural environment factors: remote geographical location, poor remote areas and natural disaster areas where children from families are easy to become poor college students; 2. Family environment factors: the children of families with no fixed income, families with more children studying, families with laid-off main labor force and families with changes often cannot afford to go to college; 3. Social change factors: During the establishment and improvement of China's socialist market economy system, the imbalance of economic development between regions has gradually expanded, and the income gap between urban and rural residents has further widened [2]. At the same time, higher education implements the integration of enrollment. College students pay fees to go to school. The high tuition makes many families unable to afford it. Even if their children barely go to college, they also become poor students in colleges and universities; 4. School factor: At present, there are many problems in the funding system for poor college students in China, and a stable and effective working system and operating mechanism have not been established, so the problem cannot be fundamentally solved; 5. Other factors: for example, industrial family poverty, that is, college students from families with poor economic benefits become poor college students [3].

In recent years, with the development and wide application of information technology, most universities in China have developed or purchased their own student information management systems, including those for poor students. However, many systems only stay at the level of data information management, and do not meet the decision-making needs of management workers. In other words, data analysis technology and data mining technology are not used for in-depth analysis and processing of basic data. Therefore, the simple backup and query of a large amount of data is far from enough to meet the requirements of modern management, and we urgently need the poor college students management system to have the ability to assist decision-making [4]. After analysis and research, data mining technology is a feasible and effective method to solve this problem. At present, it is still blank to apply data warehouse and data mining technology to the field of managing and maintaining the basic information of poor students in China.

2 Related Work

2.1 Definition of the Concept of Poor Students

There is no uniform regulation on how to formulate a scientific concept of "poor students". Generally, it is divided into the following two defining standards:

(1) Qualitative criteria. Some views believe that whether students are poor students cannot be defined only by the identification materials issued by the township (or street) or civil affairs department where the student's family is located, but should be determined by the consumption of the students themselves. The determination method should be first determined by the class, and then reviewed by the identification working group of the department where the student lives, and the files of poor students should be dynamically managed and the personnel should be regularly adjusted [5]. Others believe that nowadays many colleges and universities implement "free lunch" for poor students, which attracts many non poor students to apply, and they do not hesitate to use fake materials to get through. The disadvantage of this method is that it cannot really meet the original intention of subsidizing poor students.

(2) Quantitative standards. According to the living consumption level line of students, poor students are identified. According to the usual method of studying relative poverty issues, the poverty line recognized by poor college students is generally calculated according to the following formula:

$PL = X/2$, where PL represents the poverty line and X represents the average consumption level of college students (the average value of college students' consumption related to college life and study). According to the PL value, the poverty rate (PR) can be calculated, that is, the proportion of students below the poverty line in the total number of students. The disadvantage of this algorithm is that there will be some deviation in actual operation, which may be caused by different perceptions of the rationality of poor students' consumption. First, the basic living standard of the college location should be fully considered. Poverty is a relative concept [6]. The criteria for identifying poor families will be different depending on the economic development of the student's family. The criteria for dividing poor students should not be simply limited to whether they can eat or not. They should also take into account the basic nutritional needs, basic communication needs, cultural and sports activities needs, etc. that students need for healthy growth. Some necessary expenses should be treated specifically. They should not simply be regarded as "high consumption" and "one size fits all", which should be strictly prohibited.

2.2 Approval Process of Financial Aid for Poor Students

Changzhou Engineering Vocational and Technical College adopts the method of combining the above mentioned quantitative standards with qualitative standards to define poor students.

Poor students are divided into three categories: 1. Special difficulties 2. Difficulties 3. General difficulties.

The procedure for the definition of poor students in the college is to first apply in writing by the students themselves. After receiving all the materials, the head teacher shall establish an assessment team in the class to investigate whether the contents filled in the form are true, and fill in the survey results, and then submit them to the department for confirmation. Finally, the list of poor students will be publicized after being approved by the leading group of the college's student aid work [7]. Finally, file the data of poor students in the Student Management Office of the College.

The archives of poor students are the objects to determine the national student loans, tuition reduction and relief, and difficult subsidies, in order to ensure the integrity, authenticity and accuracy of the archives. The identification process and results of poor college students are no longer static, but dynamic and adjustable. For the students who have established the files of poor students, they should regularly understand their situation, regularly inspect their study, life and family economic conditions, establish a tracking and efficiency mechanism, and realize dynamic management. According to the changes of the students' family economic situation, a moderate investigation can be conducted, and the decision can be made through democratic evaluation [8]. The advantage of this approach is that it cannot only update the database of poor students in a timely manner, but also delete the students who no longer meet the criteria for poor students from the database, and it can also help the students whose families encounter unexpected situations to join the database of poor students in a timely manner.

The problem of poor students is a hot issue that the country and society are generally concerned about at present, and the funding system for poor students is becoming increasingly perfect. As for the poor students themselves, some students try to solve their poverty problems through similar ways; However, a small part of the student union, under the influence of some psychology, is shy of explaining the real situation, which makes it difficult to obtain reasonable funding and brings some difficulties to the funding work.

3 Apriori Algorithm

At present, the most classical association rule algorithm is the Apriori algorithm proposed by Agrawal et al. in 1994. Apriori algorithm uses an iterative method called layer by layer search, which uses k item sets to explore and find $(k + 1)$ item sets. First, find the set of frequent 1 itemsets in the transaction database and record it as L. Then, use L to find the set Lp of frequent itemsets, and then use L to find L, where searching for each L requires scanning the database once, and then go on until you know that you can no longer find frequent k-itemsets, thus completing the mining of frequent itemsets [9]. It can be seen that the algorithm uses a priori knowledge of the properties of frequent itemsets, and the name of the algorithm is based on this fact.

In the classical Apriori algorithm, the database needs to be scanned many times. In this paper, we use a data mining algorithm based on matrix. Specifically, a mapping rule is used to map the transaction database to a corresponding matrix, so that the database scanning can be completed on this matrix, greatly reducing the I/O time cost. For the database we want to mine, establish such a mapping rule:

$$Sim_1\left(d_i, d_{1j}\right) = \frac{\sum_{k=1}^{M} W_{ik} \times W_{1ik}}{\sqrt{\sum_{k=1}^{M} W_{ik}^2} \cdot \sqrt{\sum_{k=1}^{M} W_{1jk}^2}} \tag{1}$$

This paper proposes a method based on dynamic memory allocation to store frequent itemsets, which has a small memory overhead. Every time a frequent itemset is generated, it can be directly saved to disk. After the frequent k-itemset is generated, the frequent (k-1) itemsets can be released to reduce the space occupation. In order to reduce the number of connections when generating frequent k-itemsets from frequent (k-1) itemsets, a sequential arrangement method is proposed to constrain the generation of candidate frequent itemsets to reduce the generated candidate frequent itemsets, as shown in Fig. 1. The experiment proves that the algorithm of association rule mining based on dynamic memory allocation proposed in this paper overcomes the shortcomings of generating a large number of candidate sets and needing to scan the database many times, and has good operability.

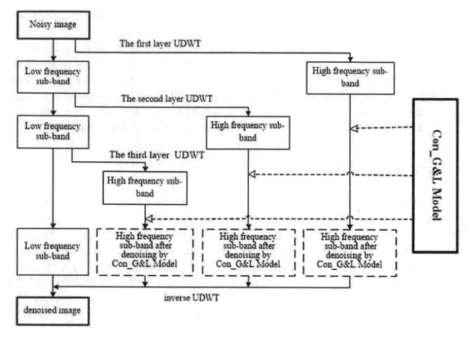

Fig. 1. Apriori algorithm flow

4 The Application of Improved Apriori Algorithm in Poor Students' Information System

This paper takes the engineering university students as an example to establish a relational database, which contains 8727 students' information. At present, Harbin Engineering University has established a campus all-in-one card system to replace student cards, examination cards and meal cards (including school canteen consumption, supermarket consumption and bath consumption). Therefore, the campus card database is mainly divided into two parts, one is the basic information database of students, and the other is the database of students' meal card consumption. Our purpose is to analyze the characteristics of poverty at a certain consumption level according to the consumption of poor students, so as to provide a reference for better funding work [10]. After the mining target is defined, after the data is cleared and collected, we finally get the six attributes we need for mining, which are poverty, consumption amount, only child, scholarship, loan and work study. The structure of the Apriori mining table can be designed as shown in Fig. 2.

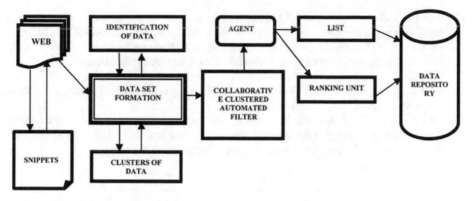

Fig. 2. Processing flow of poor college students' information system

5 Conclusion

There are a large number of basic information data of students' families in the database of the university student aid management system. Educational decision-makers can use the basic information to understand the family economic situation of a student or a group of students, so that they can carry out individual funding for specific problems of a student, but it is difficult to find the overall funding policy. Based on campus card consumption, this paper applies association rule discovery to the poor student aid system in colleges and universities, which is to mine the association rules in the basic situation data of the poor student group and discover the correlation and laws between poor student consumption. Therefore, the focus of association rules in the poor student aid system in colleges and universities is to discover the characteristics of a group of poor students, rather than a specific student. This chapter first sorts out and analyzes the data to be processed and clears up the required information. On this basis, the improved Apriori algorithm is applied to mine the knowledge we need and provide auxiliary information for efficient managers.

References

1. Xie, F., Geng, C., Jiang, Q.: Student Mental Health Evaluation System Based on Decision Tree Algorithm. Springer, Cham (2022)
2. Liu, J., Dong, X., He, Y.: Research on evaluation algorithm of the higher education system based on neural network. Trans. Comp. Educ. **3**, 1–4 (2021)
3. Yz, A., Hz, B.: Research on the quality evaluation of innovation and entrepreneurship education of college students based on extenics. Proc. Comput. Sci. **199**, 605–612 (2022)
4. Yan, W., Shabaz, M., Rakhra, M.: Research on nonlinear distorted image recognition based on artificial neural network algorithm. J. Interconnect. Netw. **22**(Supp06) (2022)
5. Ding, X., Xu, S.: Food safety pre-warning system based on robust principal component analysis and improved apriori algorithm (2021)
6. Miao, F., Wu, Y., Li, L., Liao, K., Xue, Y.: Triggering factors and threshold analysis of baishuihe landslide based on the data mining methods. Nat. Hazards **105**(3), 2677–2696 (2020). https://doi.org/10.1007/s11069-020-04419-5

7. Zhang, M., Chen, W., Yin, J., et al. Lithium battery health factor extraction based on improved douglas–peucker algorithm and SOH prediction based on XGboost. Energies. **15** (2022)
8. Guo, Y., Li, Q.: Ecological security evaluation and main effects analysis around dongting lake based on the improved pressure-state-response model. Res. Agric. Modern. **42**(01), 132–141 (2022)
9. Guo, J., He, X., Ma, X., et al.: Matching of small packages of traditional Chinese medicine based on improved RANSAC algorithm (2022)
10. Zhu, Y., Wang, Y., Liang, W.: Research on fog resource scheduling based on cloud-fog collaboration technology in the electric Internet of Things. Recent Adv. Elect. Electron. Eng. (Formerly Recent Patents Elect. Electron. Eng.) **14**(3), 347–359 (2021)

Evaluation Model of Students' Employability Based on Fuzzy Theory Algorithm

Nan Li[✉]

Department of Student Affairs, Jiangsu University of Science and Technology, Zhangjiagang Campus, Zhangjiagang 215600, Jiangsu, China
prince_rinoa@163.com

Abstract. The concept of "College Students' employability" has been put forward with the increasing attention of all sectors of society. In recent years, with the enrollment expansion of major colleges and universities year after year, the number of graduates has also increased year after year. Graduates can't find jobs after graduation, but enterprises can't recruit people under the banner of high salary. This is the so-called employment gap. Therefore, this paper studies the evaluation model of students' employability based on fuzzy theory algorithm. Fuzzy theory is a model used to evaluate the performance of students' employability. It was first proposed by m.s.guttman and r.l.schafer, who developed it as a tool to evaluate the effectiveness of training programs (Guttman and Schafer, 1969). The fuzzy evaluation model is based on two elements: "fuzzy set" and "fuzzy rule". Fuzzy sets are represented by a mathematical function called membership function, which can be applied to any data set under some conditions; Fuzzy rules are logical expressions that represent the judgment of membership.

Keywords: Fuzzy theory · Students' employability · evaluation model

1 Introduction

With the continuous development of China's economy and the increasing progress of science and technology, the country is also more strict in the selection and employment of talents, and the investigation is also more sufficient. Colleges and universities also began to pay attention to the development of students' comprehensive ability, in order to cultivate them into a talent who can adapt to the society and drive social development. At this stage, however, there is such a situation: major colleges and universities have expanded enrollment year after year, and the number of graduates has also increased year after year. With the development of socialist market economy, the competition for talents will become more and more fierce [1]. According to Chengdu Business Daily, by 2010, the number of college graduates will be nearly 10million every year, while the employment capacity will be only 8million, which means that 2million college graduates will face the problem of difficult employment. However, enterprises are always hungry for high-quality talents. Enterprises cannot recruit people, and college students cannot find jobs. This is called "structural contradiction". It is not difficult to analyze these

© The Author(s), under exclusive license to Springer Nature Singapore Pte Ltd. 2023
J. C. Hung et al. (Eds.): IC 2023, LNEE 1045, pp. 213–219, 2023.
https://doi.org/10.1007/978-981-99-2287-1_31

problems to find out the reasons [2]. Graduates who have experienced in the university stage do not seem to have the ability to meet the needs of the society. On the other hand, we also pay attention to what kind of talents enterprises need to achieve the prosperity of enterprises and social progress.

The concept of "College Students' employability" has been put forward with the increasing attention of all sectors of society. The current employment problem of college students is not only related to the realization of individual self-worth of students, but also related to the orderly and healthy development of higher education in China, as well as the harmony and stability of the whole society. Therefore, the cultivation and improvement of College Students' employability is no longer a unilateral concern of colleges and universities, organizations or college students. It has become a common concern of the three parties. At present, the main problem is that for the subject of College Students' employability, from the perspective of solving practical problems, effective measurement tools are first needed for more detailed and in-depth research [3]. However, the domestic research on the employability of college students is still in its infancy, the relevant measurement tools are rare, and the structure is not unified. Based on the above facts, this study attempts to summarize and analyze the relevant research on College Students' employability at home and abroad, conduct interviews from the perspective of employers and students' self cognition, compile college students' employability self-assessment scale, and explore the structure of College Students' employability in combination with modern statistics and measurement technology [4]. Then take the employment satisfaction as the benchmark, and try to predict the employment satisfaction of college students based on their employability. In addition, it also analyzes the current situation of College Students' employability.

2   Related Work

2.1   Employability

Up to now, there is no unified standard for the concept of "employability" in foreign countries. Overtool believes that "employability" cannot be described by specific work ability, but should be described as an ability applicable to all industries. Not only that, this ability is related to all occupations in each industry; The former Ministry of education and employment (DfEE) defined "employability" as the ability of individuals to make full use of the employment opportunities provided in the labor market to realize their potential from the two levels of obtaining and maintaining work. Therefore, employability includes the ability to obtain and maintain work. The international labor organization also defined the employability from the perspective of obtaining and maintaining work. At the same time, they also pointed out that employability is not only the ability of individuals to obtain and maintain work, but also the ability to develop in work and cope with various changes in life; Brown once interpreted employability as the ability of individuals to seek and obtain different employment opportunities, and then to maintain their existing employment [5]. Specifically, it refers to the ability and qualification that individuals acquire by strengthening the use of training and educational opportunities in order to obtain and maintain decent work, obtain promotion in existing positions, and be able to cope with changes in technology and the labor market at any time. This kind of ability and ability can be carried around.

In China, many scholars basically cite the concept of employability from foreign scholars. For example, Zhang Huili (2009) once quoted the concept of employability of British scholars Hiller and Pollard in her "Research on the composition of College Students' core employability", and defined employability from three aspects: obtaining employment, maintaining employment and re employment, and then carried out the next research. Wang Peijun also cited Hiller and Pollard in his research on employability [6]. At the same time, he also pointed out that employability is composed of knowledge, skills and attitudes, and this group of knowledge, skills and attitudes is attractive to employers. Zhao Dong once explained in his own research that employability is the ability and characteristics that can be competent for work tasks and meet organizational needs. These abilities and characteristics are the product of the interaction of social environment, political and economic environment and personal environment. Hu Zunli, Liu Shuo, Cheng Aixia have had the basis of research at home and abroad It is pointed out that the employability of college students refers to "the ability of college students to succeed in employment, which includes knowledge, skills, personality and other factors, and is a group of comprehensive abilities" [7].

2.2   Elements of Employability

Zheng Xiaoming, a domestic scholar, believes that college students' employability can be explained as a kind of ability to realize their own values and employment ideals. This ability can be improved through the accumulation of knowledge and the development of comprehensive potential during school. This view is endorsed by most domestic researchers. Their ideological ability, learning ability, application ability, practical ability and adaptability together constitute the employability of college students. Ideological ability marks the maturity of College Students' thoughts, which is embodied in their ability to distinguish politics and their insight into society; Learning ability is the cornerstone of employment, specifically refers to the ability to acquire knowledge; Practical ability is the external expression of the comprehensive application of other abilities, and adaptability is the ability of college students to adjust their physiology and psychology, which is the key to the smooth transformation of college students. Xiongshuyin and others pointed out that the employability of college students to apply for jobs and obtain employment opportunities and jobs is composed of solid professional knowledge and strong functional ability, honest, trustworthy and down-to-earth will quality, strong team awareness and the spirit of innovation, as well as certain practical experience [8].

The research of different scholars or institutions on the constituent elements of employability takes their own theoretical basis as the starting point, and the conclusions are different, independent and interrelated. For this reason, there is no final conclusion about the constituent elements of employability at home and abroad, and China is still in the preliminary stage in this regard, which is much weaker than foreign countries.

In order to further understand the dynamic relationship between the constituent elements of employability, foreign scholars have conducted a more in-depth study on employability by building a clear structural model. Among them, the latest and most perfect model is the careeredge model built by Lorraine Dacre pool and Peter Sewell

(2007), researchers of the employability center of the University of Central Lancashire in the UK. They proposed the model by summarizing existing theories and their own experience, as shown in Fig. 1.

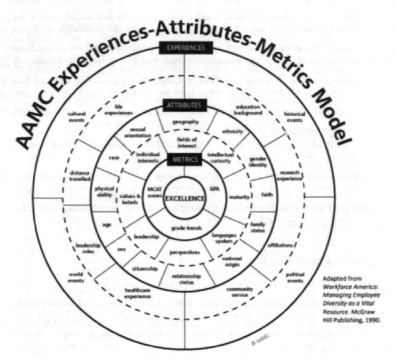

Fig. 1. College Students' employability model

3 Evaluation Model of Students' Employability Based on Fuzzy Theory Algorithm

3.1 Fuzzy Theory Algorithm

Fuzzy control is the most widely used and effective field of fuzzy theory. At present, it has helped solve many problems that cannot be solved or difficult to solve by traditional control theory. For example, in the automotive field, the comprehensive design of traditional automatic control controller needs to be based on the accurate mathematical model of the controlled object (i.e. state space model or transfer function model), but in practice, There are many factors that affect the system. For example, it is difficult to find an accurate mathematical model for the oil-gas mixing process and the combustion process in the cylinder. In this case, the birth of fuzzy control theory is of great significance, because fuzzy control is a mathematical model that does not need to establish a mathematical model or know the precise process in advance [9]. Obviously, to develop intelligent vehicles, we cannot do without fuzzy control technology, such as the temperature regulation control of automotive air conditioning. The same is true of cars and many things, which is also an important reason why fuzzy control research is so hot.

Fuzzy control simulates the thinking process of human brain through the basic principle of fuzzy mathematics, identifies the phenomenon with fuzzy characteristics and identifies the results, then gives an accurate result value and effectively controls the controlled object [10]. It is quite different from manual control and empirical control, as shown in Fig. 2:

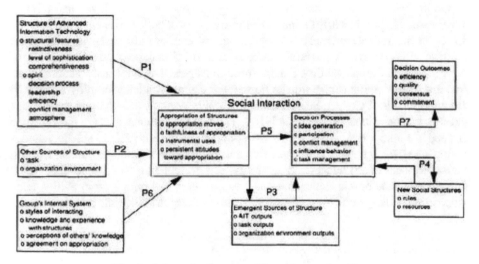

Fig. 2. Schematic diagram of fuzzy theory algorithm

Fuzzy comprehensive evaluation theory is a famous expert in fuzzy science L Professor azdah proposed it in 1965. Comprehensive evaluation is a reasonable overall evaluation of various factors that affect the attributes of things, while fuzzy comprehensive evaluation is proposed based on the nonlinearity of the evaluation process, while teaching process evaluation is a typical nonlinear evaluation. There are many evaluation objects, including leaders, peers and students, and the base number of students is relatively large. Everyone has everyone's thoughts and has different attributes, Then their evaluation descriptions are also different, and teaching is a process, which carries out the whole semester, and the evaluation at different times is also different. Fuzzy comprehensive evaluation is to evaluate things that cannot be described by a certain number by integrating various fuzzy indexes through a fuzzy means.

$$Sim_1(d_i, d_{1j}) = \frac{\sum_{k=1}^{M} W_{ik} \times W_{1ik}}{\sqrt{\sum_{k=1}^{M} W_{ik}^2} \cdot \sqrt{\sum_{k=1}^{M} W_{1jk}^2}} \tag{1}$$

3.2 Evaluation Model of Students' Employability

Fuzzy comprehensive evaluation method is based on fuzzy mathematics, using mathematical methods to study and deal with objective fuzzy phenomena, and make a general evaluation of things or phenomena affected by various factors with the help of fuzzy mathematics. Fuzzy comprehensive evaluation method was first proposed by Chinese scholar Wang peizhuang, which is deeply loved and widely used by researchers. Its method advantage is that the mathematical model is simple, easy to master, and can concretize multi factor and multi-level complex problems.

On the basis of summarizing and analyzing the concepts of employability at home and abroad, this paper clearly defines the employability of college students, further constructs the theoretical structure of College Students' employability through open-ended questionnaires, interviews and other methods, and compiles the self-assessment scale of College Students' employability. Finally, through empirical research and statistical analysis, this paper determines the structural model of College Students' employability. The structural model of College Students' employability consists of two main dimensions and 13 sub dimensions. As shown in Fig. 3. The two main dimensions are the ability to succeed in employment and the ability to maintain employment; The 13 sub dimensions are basic knowledge, practical knowledge and experience, planning awareness, information acquisition and understanding ability, self presentation ability, communication ability, extraversion, self-management ability, planning and organization ability, team cooperation ability, problem solving ability, sense of responsibility and self-confidence.

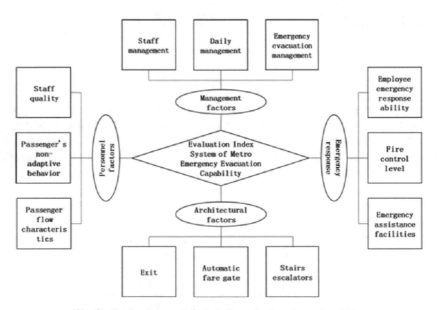

Fig. 3. Evaluation model of College Students' employability

4 Conclusion

From the perspective of the number of college graduates, the establishment of an employ-ability evaluation model suitable for current college graduates is helpful for college students to evaluate their abilities and simplify the recruitment process of employers. This paper uses ideological quality, professional knowledge and working ability The five evaluation indicators of professional quality and employment performance are used to build an application-oriented undergraduate employability evaluation model. The model is applied to practice, which proves that this model is feasible. At the same time, it also provides an important reference basis for colleges and universities and college students, and provides some reasonable suggestions. However, there are many factors that affect the employment of undergraduate students, and this paper is not comprehensive enough to determine the indicators, so the establishment of a relatively perfect evaluation model of College Students' employability needs more practical investigation, and this model needs to be improved, supplemented and improved.

References

1. Huang Y.: The evaluation of students' physical health based on the integration of family and school physical education. Revista Brasileira de Medicina do Esporte **27**(spe), 80–82 (2021)
2. Long, P.D., Nga, N.T.: Students' evaluation on field trips as a means to prepare graduate employability at a Vietnamese university. Human. Soc. Sci. Lett. **10** (2022)
3. Tan, L., Guan, Y.: Talent evaluation model of college students based on big data technology (2021)
4. Wang, S., He, Y., Lin, H.: Model of normal university students' education quality based on artificial neural networks. J. Phys.: Conf. Ser. **1828**(1), 012146 (2021). (8pp)
5. Glerum, D.R., Judge, T.A.: Advancing employability: applying training evaluation to employability development programs. Career Dev. Int. (2021)
6. Guo, H., Jia, Z.: Analysis on the frame model of poor student identification and funding management based on financial information sharing system. In: Hung, J.C., Yen, N.Y., Chang, JW. (eds.) Frontier Computing. FC 2021. Lecture Notes in Electrical Engineering, vol. 827. Springer, Singapore (2022). https://doi.org/10.1007/978-981-16-8052-6_57
7. Wu, F., Liu, X., Wang, Y., et al.: Research on evaluation model of hospital informatization level based on decision tree algorithm. Secur. Commun. Netw. (2022)
8. Ma, Y., Wang, R.: Human error evaluation model of civil aviation maintenance based on grey wolf optimizer. J. Phys.: Conf. Ser. **1848**(1), 012111 (2021). (8pp)
9. Siddique, S., Ahsan, A., Azizi, N., et al.: Students' workplace readiness: assessment and skill-building for graduate employability (2022)
10. Sokhanvar, Z., Salehi, K., Sokhanvar, F.: Advantages of authentic assessment for improving the learning experience and employability skills of higher education students: a systematic literature review. Stud. Educ. Eval. **70**(2), 101030 (2021)

Experimental Study on Triaxial Strength of Reinforced Soil in Loess Region Based on Sequence Reduction Algorithm

Cui Hao[✉] and Fan Yue

Xi'an Siyuan University Shaanxi, Xi'an 710038, China
cuihao0916@163.com

Abstract. As a new composite structure, reinforced soil is widely used in highway, water transportation, water conservancy, railway and other industries, and has achieved a wide range of social, environmental and economic benefits. Based on the sequence reduction algorithm, the triaxial strength of reinforced soil in loess area is tested, and the effects of reinforcement ratio, reinforcement type and dosage, water content and temperature on triaxial strength are discussed. The results show that when the reinforcement ratio is 1:3.0, the total reinforcement is 0.8 tons / square meter, the water content is 0%, the temperature is 20 °C, and when the reinforcement ratio is 1:4.5, the total reinforcement is equal to 2 tons, the maximum triaxial compressive strength is obtained.

Keywords: Reinforced soil · Sequence domestication · Loess region · Triaxial strength

1 Introduction

Loess is a kind of yellowish brown soft sediment accumulated and produced in the Quaternary. Its particle composition is mainly silt particles. Loess contains a lot of carbonate and has a macroporous structure. Loess is widely distributed in the world, covering an area of 13million km2. The distribution characteristics are mainly in arid areas, and some are also distributed in semi-arid areas. China's loess covers an area of 640000 km2, mainly in the central and western regions. Loess has loose soil, large pores and high strength, but some loess is prone to collapse and deformation when encountering water, and liquefaction is easy to occur when shaking, so the engineering property is relatively special. With the development of the "the Belt and Road" economic belt, the engineering construction and economic development of the loess areas along the economic belt have been restricted to a certain extent [1].

In the past 20 years, with the rapid development of science and technology and the needs of modernization, reinforcement technology has developed rapidly, and various forms of reinforcement projects have been built all over the world, such as the reinforcement of many slopes with soil nails, root piles and long fiber mixed sand; Many weak foundations have been reinforced with geotextiles, geogrids, reinforcing mesh and mesh

J. C. Hung et al. (Eds.): IC 2023, LNEE 1045, pp. 220–228, 2023.
https://doi.org/10.1007/978-981-99-2287-1_32

piles. The common feature of reinforcement engineering is that tie bars are embedded in the soil to improve the mechanical properties of soil structure. This technology is generally known as reinforcement technology. In a broad sense, reinforced soil technology is called geotechnical reinforcement technology, or geotechnical reinforcement technology. The common forms include fiber soil, reinforced soil, composite soil and modified soil.

The soil itself has a certain compressive and shear strength, but its tensile strength is very low. After adding or laying appropriate reinforcement materials in the soil, the strength and deformation characteristics of the soil can be significantly improved. If the reinforcement materials are buried in the soil, the stress of the soil can be diffused, the modulus of the soil can be increased, and the lateral displacement of the soil can be limited [2]. The rapid development of reinforced soil technology mainly lies in its novel structure and beautiful shape; Simple technology and convenient construction; Low structure and material saving; Short construction speed and construction period; It has the advantages of low cost, obvious benefit, strong adaptability and wide application.

2 Related Work

2.1 Research Status of Reinforced Soil

Reinforced soil is a composite composed of one or more horizontal reinforced material components laid alternately with filler in the soil, which closely combines the tensile strength of reinforcement with the compressive strength of soil, improves the overall compressive strength of soil, and enhances the stability of soil. At present, in the engineering field, it is generally composed of fill and reinforcement. It is also a composite material composed of multi-layer horizontal reinforcing materials and fillers laid alternately and compacted tightly. Reinforced materials increase the deformation modulus of soil, limit the lateral displacement of soil, and bear the lateral tension generated by soil, thus improving the stability of soil and related buildings [3].

The 1970s witnessed the rapid spread and development of reinforced soil technology. The corresponding research work and experiments are also carried out at the same time. At that time, the French bridge and road center and the University of California had the most research and achieved remarkable results. F.schlosser, n.c.long, i.juran of France and K.L. Lee and D P. McKittrick and other scholars have made outstanding contributions.

In the 1990s, the limit state method began to be applied in some countries. The ultimate limit state and service limit state were considered in the design, and the partial factor was used to replace the single safety factor. The limit equilibrium method was better improved, but it has not been widely used. Soil itself has complex mechanical properties, and there are many kinds of geosynthetics, with great differences in performance, creep and temperature effects. It is difficult to simulate the mechanical properties buried in soil in the test. In addition, it is also necessary to consider the role between soil and reinforcement. It is very difficult to rely solely on theoretical analysis [4]. At present, reinforced soil has not formed a set of reasonable and mature theory of application practice.

In the 21st century, scholars from all over the world have done a lot of experimental research on reinforced soil tests. The design and calculation theory and construction technology of reinforced soil have developed by leaps and bounds, and have broken through the norms of many countries in terms of filler restrictions and the scope of application of reinforcement materials. However, the development of reinforcement theory obviously lags behind the application of practice. In the 21st century, we should first solve the improvement of calculation theory and calculation method of reinforced soil The durability of reinforced materials and the application specifications to supplement and improve reinforced soil technology [5].

2.2 Domestic Research Results of Reinforced Soil

In our country, reinforced soil technology is widely used in engineering construction, which has played a great role in promoting the modernization of our country. Chinese scholars' research on reinforced soil is also increasing day by day, and have achieved fruitful research results.

Reinforcement leads to the change of soil stress field and displacement field. After reinforcement, the reinforced soil is no longer an isotropic structure, so the failure mode of soil has also changed fundamentally. Shen Zhujiang (1998) studied the influence of tendon friction on the change of stress state, and found that when the strength of the tendon is large enough and the fracture or pullout does not occur, circular arc sliding cannot occur, and its failure form can only be lateral extrusion failure accompanied by settlement [6].

Chinese scholars have also done a lot of research on the triaxial test of reinforced soil. Ding Wantao et al. (2007) studied the influence of moisture content on the strength of reinforced expansive soil, and found that the addition of reinforcement caused the disturbance of the structure and suction strength of expansive soil, weakening the influence of the structure of reinforced expansive soil on its strength. The strength of reinforced expansive soil decreases significantly with the increase of water content, and the decrease of friction angle is more obvious than the decrease of cohesion, indicating that the strength of reinforced expansive soil is sensitive to the change of water content. Cui TANGCAN et al. (2004) obtained from the triaxial test of reinforced sand that the stress-strain curve of unreinforced soil is softened, while the stress-strain curve after reinforcement is hardened, and reinforcement has a significant impact on the stress-strain curve of soil [7]. The reinforcement effect of reinforcement material on soil can be exerted only when it reaches a certain axial strain, and its reinforcement effect increases with the increase of reinforcement layers and decreases with the increase of confining pressure. Lei Shengyou reinforced loess studied the stress-strain and strength characteristics of reinforced loess by triaxial test method, and obtained the relationship between the strength index of reinforced soil and the strength index of plain soil. Through the study of the normalized behavior of strength, the effect of reinforcement under different confining pressures is analyzed; The reason for the increase of the strength of reinforced soil is explained by the viewpoint of composite material and stress circle. Fu Hua et al. (2008) studied the factors affecting the strength of reinforced soil through the indoor triaxial test method, and discussed the change law of sample strength under different reinforcement conditions [8]. The test results show that the effect of reinforcement is more obvious with the increase of axial strain; For the specimen with low strength, the strength of the reinforced specimen increases significantly; For the same sample, the higher the strength of the reinforced material itself, the more obvious the reinforcement effect; For the same kind of reinforcement material, the more the number of reinforcement layers is, the smaller the spacing of reinforcement layers is, the more obvious the reinforcement effect is; The higher the compactness of the sample itself, the more obvious the reinforcement effect.

From the above research results, it can be found that China has done a lot of research on reinforced soil in recent years, and its development started relatively late compared with the West. Few people consider the medium principal stress in the research and propose a reasonable design and calculation method for the medium principal stress. Chen qiunan et al. (2006) have proposed the design and calculation of reinforced soil retaining wall considering the role of the medium principal stress, but due to its complexity, it has not been applied to practice.

3 Basic Principle of Reinforced Soil Action

In the project, the loose sand can be piled up into a sand pile with a natural angle of repose, and the cohesive soil can be excavated into a slope with a certain height. If the horizontal reinforced materials are laid in layers in the sand, the sand can form a composite with the reinforced materials, which can keep the sand at a certain height without collapsing. Adding reinforced materials to the cohesive soil can better strengthen the soil strength. It shows that the reinforced composite formed after soil reinforcement has greatly improved the stability and mechanical properties of soil than unreinforced soil. In the practical application of retaining walls, building foundations, slopes and other projects, we can obtain certain practical significance and economic effects. The proposal of reinforced soil technology is put forward and developed from this idea [9].

In practice, the application of reinforced soil technology must obtain sufficient theoretical support. Vidal and others first carried out triaxial tests and field tests on reinforced soil, and put forward various theories to explain the action principle of reinforced soil. So far, the interaction principle between reinforcement and soil can be roughly summarized into two categories: one is friction reinforcement theory, and the other is quasi cohesion theory or cohesion principle.

3.1 Principle of Friction Reinforcement

In the reinforced structure, the wall panel will be subjected to the self weight of the soil and external forces to produce earth pressure. The earth pressure is transmitted to the tie bar through the wall panel. If the earth pressure is too large, it is possible to pull out the tie bar. If we want to ensure that the tie bar is pressed by the fill, there must be enough friction between the fill and the tie bar to prevent the tie bar from being pulled out. Based on this consideration, the principle of friction reinforcement is produced.

According to the basic structure between reinforcement and soil in reinforced soil, a micro segment is taken out of the reinforced soil to study. As shown in Fig. 1, let the length of the micro element taken out be DL, the stress on the left side of the brace section is T1, the stress on the right side is T2, and the normal stress on the brace is σ, The weight of reinforcement and soil mass of micro element are ignored. Assuming that the friction coefficient between reinforcement and soil particles is f, the width of reinforcement band is B, and the tensile force of soil in this micro segment is DT, dt = t1−t2, since there is friction on both sides of reinforcement, the total friction force between soil particles and reinforcement is:

$$dF = 2\sigma fbdl \tag{1}$$

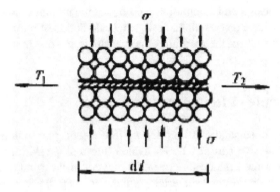

Fig. 1. Schematic diagram of friction reinforcement

3.2 Quasi Cohesion Theory

The quasi cohesion theory was introduced by Schlosser and long (broms.b.b et al.1992), which believed that the internal friction angle of reinforced soil was almost the same as that of unreinforced soil, but increased the cohesion of soil. Generally, the elastic modulus of tie bar in reinforced soil is much larger than that of fill, so reinforced soil structure can be regarded as anisotropic composite material. Due to the joint action of fill and tie bar, the external measured strength includes the shear resistance of fill, the friction between fill and tie bar, and the tensile resistance of tie bar. The strength of reinforced soil has been significantly improved. This can be verified in the triaxial tests of reinforced sand cylindrical soil samples and unreinforced sand cylindrical soil samples [10].

The strength of reinforced sand is higher than that of unreinforced sand, which can be explained according to Coulomb theory and Moore failure criterion. The triaxial comparative test of reinforced sand cylindrical soil sample and unreinforced sand cylindrical soil sample shows that if the unreinforced sand sample is σ 1 and σ 3. When the limit equilibrium is reached under the action, the reinforced sand sample is in the same size σ Under the action of 1, the limit equilibrium cannot be reached, and it is in the elastic equilibrium state, as shown in Fig. 2, which shows that the strength of reinforced sand samples has been improved.

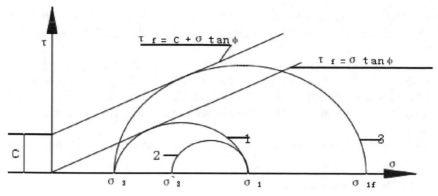

Fig. 2. Stress analysis of quasi cohesive theory

4 Triaxial Strength Test of Reinforced Soil in Loess Region Based on Sequence Reduction Algorithm

4.1 Test Plan

Taking the Loess of a brick factory in Wugong, Shaanxi Province as the research object, the triaxial tests of various stress paths under different dry density, moisture content, consolidation confining pressure and medium principal stress ratio were carried out with ts-526 triaxial test instrument. The test steps are as follows:

(1) Sample preparation: determine the number of reinforced layers of the sample according to the test plan, make reinforced soil samples, and use appropriate methods to make samples with a certain moisture content according to the test requirements.

(2) Check the instrument: turn on the power supply of the instrument and check whether the air compressor, pressurization system, measurement system and pressure pipeline work normally.

(3) Sample loading: after fixing the base and rubber membrane, place the boiled permeable stone, soaking filter paper, sample, soaking filter paper, permeable stone and sample top cap on the instrument in turn, and install the pressurizing device in the direction of small principal stress and medium principal stress, and finally install the axial pressurizing device.

(4) Install each pressure pipe line on the instrument, install the displacement sensor correctly, and check whether the displacement sensor works normally again.

(5) Sample consolidation: clear the measurement system, and then pressurize it by stages according to the consolidation confining pressure required by the test. The pressurization sequence is $\sigma 3 \backsim \sigma 2 \backsim \sigma 1$. Then start consolidation, and record the deformation in all directions after consolidation for 2 h.

(6) Shear: the test requires to control the value of the middle principal stress ratio B for shear. The test method is: fix the small principal stress unchanged in the test, manually adjust the size of the middle principal stress when the large principal stress increases to ensure that the value of B remains unchanged, and then record the data of the other four displacement sensors in sequence when the deformation in the direction of the large

principal stress reaches the required size. When the axial deformation of the test reaches 13mm or the deformation of the force ring reaches the maximum deformation, The test is over.

(7) Unload: unload by stages after cutting σ 2、 σ 3、 σ 1 direction pressure.

(8) Unloading: after unloading the pressure, remove the displacement sensors in the direction of medium principal stress and small principal stress, disconnect all pressurized pipelines, then unload the pressure plates in turn, and finally take out the sheared soil samples.

(9) End the test and turn off the air compressor, measurement system and power supply.

4.2   Test Results and Analysis

Principal stress difference between pure soil and reinforced soil in various cases during shear process (σ 1–σ 3) The relationship curve with axial strain is shown in Fig. 3. As shown in Fig. 3 (a) and (b), the peak principal stress difference (σ 1–σ 3) With the increase of confining pressure, the linearity and tangent slope of the initial section of the stress-strain curve increase greatly, showing obvious elastic properties. The comparison between unreinforced and reinforced soil samples shows that unreinforced soil has an obvious peak value under low confining pressure. After the peak value, the axial strain continues to develop and the principal stress difference changes greatly, showing the characteristics of brittle failure. With the increase of confining pressure, after reaching the peak value, the axial strain continues to develop and the principal stress difference changes little, transforming into plastic failure; Reinforced soil shows typical plastic failure characteristics under both low pressure and high pressure. The analysis of the test results shows that the reinforced soil can increase the axial strain when the soil is destroyed, and the axial strain when the unreinforced soil is destroyed is only 2.7%–9%, while the reinforced soil can still bear great stress after the peak value. At confining pressure σ When 3 is 200 kPa and 300 kPa, the principal stress difference curve of reinforced soil begins to approach that of pure soil. Because in the case of small strain and small confining pressure, the tensile performance of reinforcement does not play out. With the increase of confining pressure, the principal stress of reinforced soil is close to that of unreinforced soil, indicating that the performance of reinforced material has begun to play.

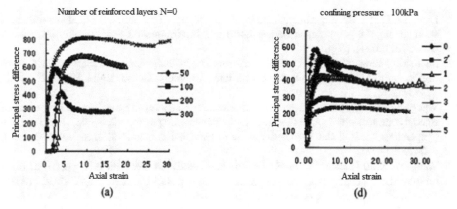

Fig. 3. Principal stress difference and axial strain curve of reinforced layer

5 Conclusion

Reinforced soil is a new type of geotechnical engineering material after reinforced concrete and prestressed concrete. Reinforced soil technology is being more and more applied to engineering, which requires that we should speed up and deepen our understanding of reinforced soil. In this paper, the deformation strength of reinforced soil under complex stress is studied by using ts-526 multi-functional triaxial apparatus. The effects of the number of reinforced layers, dry density, moisture content, consolidation confining pressure and the ratio of medium principal stress on the deformation strength characteristics of reinforced loess are mainly studied. Under the conditions of different reinforced layers, dry density, moisture content, consolidation confining pressure and medium principal stress ratio, the stress-strain curves of reinforced loess are hyperbolic. The reinforced loess measured in this test is a strong hardening stress-strain relationship curve under the three-dimensional stress state.

References

1. Saidi, T., Amalia, Z., Hasan, M., et al.: An experimental study on bond strength of abaca fiber as natural FRP material. In: IOP Conference Series: Materials Science and Engineering, vol. 1087, no. 1, p. 012020 (2021). (8pp)
2. Xiao, X., Gui, X., Hao, D.: Experimental study on bond strength failure of GFRP lap joints. J. Phys.: Conf. Ser. **2194**(1), 012039 (2022)
3. Li, S., Zhou, R., Yang, F., et al.: Effect of particle combination ratio of soil rock mixture on shear strength. IOP Publishing Ltd. (2022)
4. Kishore, G.: Experimental study on strength attainment of concrete containing silica fume and fly ash. In: IOP Conference Series: Materials Science and Engineering, vol. 1197, no. 1, p.012003 (2021)
5. Zhang, G., Chen, C., Li, K., et al.: Multi-objective optimisation design for GFRP tendon reinforced cemented soil. Constr. Build. Mater. **320**, 126297 (2022)
6. Raja, M., Shukla, S.K.: Experimental study on repeatedly loaded foundation soil strengthened by wraparound geosynthetic reinforcement technique. J. Rock Mechan. Geotechn. Eng (2) (2021)

7. Li, J., Li, L., Li, C.: Experimental study of triaxial test of unsaturated expansive soil shear strength. In: IOP Conference Series: Earth and Environmental Science, vol. 692, no. 4, p. 042010 (2021). (6pp)
8. Shen, Y.S., Tang, Y., Yin, J., et al.: An experimental investigation on strength characteristics of fiber-reinforced clayey soil treated with lime or cement. Constr. Build. Mater. **294**(1–2), 123537 (2021)
9. Wang, Y., Sun, J., Liang, Z., et al.: Experimental study on the mechanical properties of triaxial compression of white sandstone under the coupling action of chemical corrosion and temperature. In: IOP Conference Series Earth and Environmental Science, vol. 692, no. 4, p. 042009 (2021)
10. Wang, S., Xue, Q., Ma, W., et al.: Experimental study on mechanical properties of fiber-reinforced and geopolymer-stabilized clay soil. Constr. Build. Mater. **272**(4), 121914 (2021)

Exploration and Practice of Entrepreneurial Talent Training Mode Through Five Industries Based on BP Optimization

Hui Yuan[✉]

Shandong Institute of Commerce and Technology, Jinan 250103, Shandong, China
yuanhui5681@163.com

Abstract. Using the method of case analysis, this paper explores the training mode of entrepreneurial talents in five industries based on BP optimization. The purpose of this study is to explore the training mode of entrepreneurial talents through five industries based on BP optimization, and provide evidence for policy-making and education planning. In order to obtain data, researchers collected information from literature review, expert interviews and field observations. The results show that there are many factors affecting entrepreneurship, such as family background, education level, social environment, etc., but because different types of enterprise furniture have different characteristics, it is impossible to find any factors that equally affect all entrepreneurs.

Keywords: BP neural network · Talent training mode · case analysis

1 Introduction

When it comes to talent training, Sun Yat Sen once said: "governing the country and managing the country, talent is urgent". In any era, we should pay attention to talent training. The so-called talent training mode refers to the synthetic system of talent training objectives, training specifications and basic training methods. Through the optimal combination of constituent elements, we can build a variety of different training modes. The diversification of talent training modes is the inevitable requirement of China's social and economic development, In such an environment, how should we innovate the talent training mode?

First of all, the importance of talents is not only professional ability, but also quality-oriented education. There is a classic sentence in "University" that is, "the way of university is to be clear in virtue, new people and perfection." College education is to cultivate people. Don't train people into machines and do a lot of boring homework every day. We should cultivate students how to adapt to the society and cultivate students' quality, judgment ability and values. We should not be too lax in the University. We should have our own goals and don't do nothing all day. When leaving the University, we should not only have professional ability, but also be a qualified and harmonious person. Secondly, we should pay attention to practice, Pay attention to innovative education,

J. C. Hung et al. (Eds.): IC 2023, LNEE 1045, pp. 229–234, 2023.
https://doi.org/10.1007/978-981-99-2287-1_33

establish the idea of personalized education, and don't stifle students' creativity, which is the correct talent training.

When it comes to talent training, the significance of building a demonstration base of industrial big data platform is highlighted [1]. The significance lies in innovation and breakthrough, innovation and breakthrough of talent training mode, fission innovation and entrepreneurship, effectively promote industrial digitization, talent digitization and service digitization, give full play to the value of data, and carry out industrial services such as transformation of technological achievements, integration of industry and education, talent training, incubation of mass entrepreneurship and innovation, investment and intelligence attraction around integrated data, Build the integration operation mechanism of entity business and network platform; With industrial big data as the core, accelerate the resource integration among the government, enterprises and colleges, build a communication platform for collaborative operation, and promote the development and upgrading of regional industries. The established entrepreneurial talent training model based on BP model has played a pioneering and innovative role in improving the entrepreneurial talent training system in Colleges and universities.

2 Related Work

2.1 BP Neural Network Training Model

Neuron is the basic computing unit of neural network, also known as node or unit. It can accept input from other neurons or external data, and then calculate an output. Each input value has a weight, and the size of the weight depends on the importance of this input compared with other input values. Then execute a specific function f on the neuron, as shown in Fig. 1 below. This function will perform an operation on all input values and their weights of the neuron.

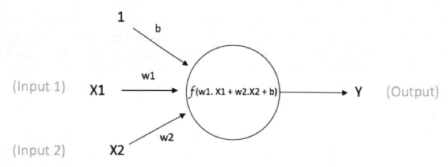

Fig. 1. Specific function f definitions

Under the guidance of the school running concept of "taking service as the purpose, employment as the guidance and taking the road of combination of production and learning", colleges and universities have determined the overall work goal of improving students' professional ability, improving graduates' employment competitiveness and entrepreneurial ability, and realizing full employment. It has established the task goal of cultivating high-tech applied BP optimization talents needed by the society for production, construction, management and service line. Under the guidance of such school running concept and BP optimized talent task goal, colleges and universities have carried out a lot of exploratory reform of BP optimized talent training mode [2]. However, whether it is the requirements of higher education theorists or social enterprises for professional BP optimized talents, the recognition of the training mode of higher education is not very high. Therefore, the training of BP optimized talents in higher education needs to be reformed. In the process of college education, our college has done a lot of exploratory work in optimizing the reform of talent training mode, and has made great achievements. However, the training of real professional BP optimization talents still needs to strengthen the reform [3]. After learning and summarizing, we realize that increasing the productive training in the school and increasing the training time of enterprises is the breakthrough of the reform of BP optimization talent training mode.

2.2 Talent Training Mode

Colleges and universities should deepen the reform of education and teaching, establish a diversified view of talents, and comprehensively strengthen the cultivation of College Students' ability and quality. "Improving quality is the core task of the development of higher education and the basic requirement of building a strong country in higher education. The structure of higher education is more reasonable [4]. A number of internationally renowned high-level colleges and universities with characteristics have been built. Several universities have reached or approached the level of world-class universities, and the international competitiveness of higher education has been significantly enhanced." It can be seen that China's colleges and universities must speed up the reform of education and teaching quality and take the road of connotative development, so as to improve the quality of talent training in higher education and better serve the social development of the new era.

Modern economic and social development needs talents. The connotative development path of higher education is to meet the needs of economic and social development and China's strategy of strengthening the country with talents. Higher education personnel training is the main way for higher education to serve the society. With the further deepening of China's modernization and reform and opening up, the economic structure has been continuously adjusted, the economic level has been continuously improved, and some high-tech industries have been emerging, which has promoted the continuous transformation of the mode of economic development, which has caused changes in the demand structure of human resources, and there is an urgent need for all kinds of talents of different types and abilities. The rapid development of modern science and technology leads to a comprehensive development trend in the development of various disciplines of higher education, which requires colleges and universities to take the way of connotative talent training by integrating the demand for talents in the development of economy and science and technology and the reality of their own reform. Education provides talent support for the development of modern economy and science and technology [5]. Therefore, colleges and universities should optimize and integrate all kinds of educational resources, timely adjust the discipline layout and specialty setting, and strengthen reform measures from the aspects of specialty setting, curriculum setting, talent training objectives, training program formulation and teaching methods, so as to meet the social and economic development and scientific and technological progress, Effectively promote the economic structure and scientific and technological level to high-end.

3 Exploration and Practice of Entrepreneurial Talent Training Mode Through Five Industries Based on BP Optimization

In the training process of "five industries through" entrepreneurial talents, the four aspects of industry, learning, research and innovation are twisted together through the training scheme of Higher Vocational entrepreneurial talents, running through the spiral line with DNA gene, reflecting the development process of both horizontal kink and vertical gradient between industry, learning, research and innovation. After five stages of study, occupation, industry, employment and entrepreneurship, the students have the employability and entrepreneurial ability through the cultivation of 3 national and 18 shared provincial training bases [6]. As shown in Fig. 2 below, the five industry penetration model is shown.

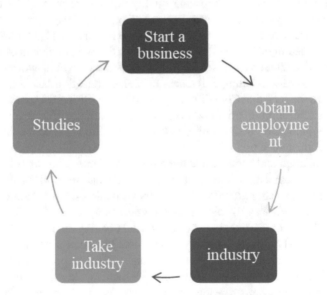

Fig. 2. Five industry linkage model

Build a scientific specialty setting and curriculum system. The training of BP optimized talents in Colleges and universities should formulate training strategies in combination with their own advantages and characteristics, and make clear provisions on the specifications and requirements of BP optimized talents from the aspects of knowledge, innovation ability and comprehensive quality in order to meet the needs of social and economic development. In the process of BP optimizing talent training, take effective ways to comprehensively improve the quality of BP optimizing talent training and meet the social development in the new era [7]. Adjust the professional structure and pay attention to the professional development system. Professional development directly affects the training quality of BP optimized talents. Colleges and universities should start from the demand of economy and Society for BP optimized talents, timely adjust the professional setting and enrollment scale according to the subjective and objective requirements, and better adapt to the needs of various fields based on their own school running positioning and training advantages. With the development of society, the demand for compound BP optimization talents is expanding. Therefore, we should adhere to the principle of scientificity 3 in specialty setting, not only limited to the professional knowledge, but also carry out the training of interdisciplinary and interdisciplinary BP optimization talents. In terms of curriculum, we should pay attention to basic courses [8]. Attaching importance to basic courses of disciplines and professional basic courses will broaden students' professional scope and lay a solid discipline foundation for students. In addition to basic courses, we should also focus on general education courses aimed at improving students' comprehensive quality. In the curriculum, we should reduce the class hours of theoretical courses and compulsory courses, and increase practical courses to pay attention to the cultivation of students' practical ability. Such as vocational skills, professional experiments, psychology and various elective courses, which can expand students' knowledge, enable students to actively participate in professional training, improve the practical application ability of knowledge and apply what they have learned. Bringing social practice into the teaching plan will strengthen extracurricular guidance, broaden the scope of majors, strengthen the cultivation of students' practical skills, and build a practice platform for students [9]. It will improve their scientific research innovation and subjective motivation, master both theoretical knowledge and practical knowledge, and can adapt quickly after entering the society.

The digitization of knowledge and the rapid popularization of modern network information technology have greatly expanded the access to knowledge. In this case, teachers should make innovative use of advanced science and technology, deepen classroom reform in the construction of curriculum system, and have a forward-looking vision for some new breakthroughs in artificial intelligence, communication and energy that can bring great changes to people's production and life, so that students can actively learn and explore scientific knowledge to meet the requirements of their own comprehensive ability development brought about by scientific and technological reform [10]. In order to adapt to the trend of scientific and technological progress, colleges and universities must deepen the reform of educational system, take the connotative BP and optimize the talent training mode.

4   Conclusion

This paper establishes the fitting model of BP neural network and five industries through entrepreneurship talent training, adopts a systematic point of view, uses scientific methods, establishes a reasonable model, carries out quantitative research, finds the restrictive factors, and adjusts the implementation links of secondary industry, occupation, industry, employment and entrepreneurship. There is a breakthrough in theory and can be operated in practice. Further optimize the curriculum system in the entrepreneurial talent training mode of "connecting five industries", and organically combine the professional curriculum system with the entrepreneurial curriculum system, which is in line with the law of human learning and growth and the formation of knowledge and skills. It can reflect the characteristics and attributes of higher vocational teaching and play a better leading role in the innovation of Higher Vocational entrepreneurship education system.

Acknowledgements. 1. C105, Research on the application of big data in student Service and management education of Higher Vocational Colleges,the open competition mechanism to select the best candidates key cultivation project of Shandong Institute of Commerce and Technology Double High School plan.

2. C201, High quality Party construction leads business high quality development path research, the open competition mechanism to select the best candidates key cultivation project of Shandong Institute of Commerce and Technology Double High School plan.

References

1. Liu, L., Dong, X.Y.: Discussion on the quality cultivation of entrepreneurial talent of art design major in engineering college. DEStech Trans. Soc. Sci. Educ. Hum. Sci. (eelss) (2020)
2. Zhang, Z., Liu, Y., Mao, G.; Research on the cultivation of innovative and entrepreneurial talents based on cooperative education in application-oriented universities. In: 2020 International Conference on Big Data and Informatization Education (ICBDIE) (2020)
3. Han, G.: Training mode and evaluation method of entrepreneurial talents in higher vocational education. In: Hu, Z., Petoukhov, S., He, M. (eds.) Advances in Artificial Systems for Medicine and Education V. AIMEE 2021. Lecture Notes on Data Engineering and Communications Technologies, vol. 107. Springer, Cham (2022). https://doi.org/10.1007/978-3-030-92537-6_38
4. Liu, F.: An Experimental study on cultivation model of interdisciplinary innovative and entrepreneurial talents in internet sports industry. J. Phys: Conf. Ser. **1744**(4), 042158 (2021)
5. Wang, Y.: Thoughts on the cultivation of innovative and entrepreneurial talents in colleges and universities based on the "Craftsman Spirit" based on big data analysis. J. Phys: Conf. Ser. **1648**, 032196 (2020)
6. Luo, D., Ye, Z.: Research on the application of VR technology in the cultivation of double innovation talents (2020)
7. Hu, Y., Li, N., Luo, C.: Quality evaluation of practical training of innovative and entrepreneurial talents in universities based on statistical learning theory after COVID-19 epidemic. J. Intell. Fuzzy Syst. **39**(1), 1–7 (2020)
8. Wei, Y., Lv, H., Chen, M., et al.: Predicting entrepreneurial intention of students: an extreme learning machine with gaussian barebone harris hawks optimizer. IEEE Access **23**, 76841–76855 (2020)
9. Khorram, M.R., Ghahderijani, M.: Optimal area under greenhouse cultivation for tomato production. Int. J. Vegetable Sci., 1–5 (2020)
10. Nithiya, E.M., Fenila, F., Vasumathi, K.K., et al.: Cultivation of scenedesmus sp. using optimized minimal nutrients and flocculants – a potential platform for mass cultivation. Environ. Technol. (2020)

Exploration and Practice of the Construction of Big Data Algorithm Excellent Course

Lu Dai[✉]

Wuhan Donghu University, Wuhan 430212, Hubei, China
minicat114@163.com

Abstract. The project of high-quality courses came into being under the background of the rapid increase in the number of students in Colleges and universities, the shortage of higher education resources and the need to improve the quality of higher education. Represented by national excellent courses, it is defined as a demonstration course with first-class teachers, first-class teaching contents, first-class teaching methods, first-class teaching materials and first-class teaching management. Excellent courses refer to excellent courses with characteristics and first-class teaching level. This course is the best course for you to learn big data algorithms. This course will help you understand and explore big data algorithms in detail. I think this is one of the best courses for those who want to learn big data algorithms. It teaches you how to build a good algorithm through a large number of examples and practical problems.

Keywords: Excellent courses · big data · Curriculum construction

1 Introduction

The Ministry of education and the Ministry of finance of China jointly implement the undergraduate teaching quality and teaching reform project. As one of the contents of this project, excellent courses have attracted much attention because of their large number of projects and huge investment, and many countries or organizations have already implemented similar open course projects. As one of the important contents of the education quality engineering project, the excellent course is an important means to guarantee the quality of ordinary higher education in China. The quality and operation effect of the project are concerned by all aspects. Therefore, it is very necessary to study the current situation of the construction of excellent courses in ordinary colleges and universities, find out the deficiencies and seek countermeasures [1].

In the face of the massive amount of information in the information society, the time for information users to screen, evaluate and apply information has obviously increased. At the same time, higher requirements have been put forward for the ability of users to screen, evaluate and apply information, and the demand for sharing the information that has been professionally screened and evaluated has correspondingly arisen [2]. At the same time, under the requirements of ensuring quality higher education and the requirements of social development and progress calling for the emergence of

© The Author(s), under exclusive license to Springer Nature Singapore Pte Ltd. 2023
J. C. Hung et al. (Eds.): IC 2023, LNEE 1045, pp. 235–242, 2023.
https://doi.org/10.1007/978-981-99-2287-1_34

new teaching channels and means, the gradual maturity of information technology has provided conditions for opening up new educational channels and carriers, and also made it possible to help people learn more efficiently through the popularization and application of information technology.

MIT has decided to make more than 2000 courses available to the world through the Internet, which has opened up a new vision for us [3]. The large-scale sharing of educational resources with complex contents is operable. The development of information technology makes it possible to introduce new ways and carriers for knowledge transmission. The achievements of a series of information technologies, such as network technology and storage technology, as well as the rapid popularization of the network, have provided technical support for knowledge dissemination, and it can be realized to acquire knowledge anytime and anywhere. It is the development of information technology that makes the sharing of high-quality curriculum resources into reality [4].

Since the implementation of the excellent course construction project in China, many achievements have been achieved. However, in the construction, some problems of one kind or another are inevitably exposed, which hinders the function of excellent courses. In order to make the construction of excellent courses play a better role, we must analyze and find the shortcomings in the construction of excellent courses in order to find countermeasures.

2 Related Work

2.1 Excellent Courses

In 2003, China officially launched the project of excellent courses. According to the notice of the Ministry of education on starting the construction of excellent courses of teaching quality and teaching reform project in Colleges and universities (JSH [2003] No. 1) issued by the Ministry of education, excellent courses can be defined as: excellent courses are exemplary courses with the characteristics of first-class teachers, first-class teaching contents, first-class teaching methods, first-class teaching materials and first-class teaching management in the selected level. It is an excellent course with characteristics and first-class teaching level. Its construction should be based on the talent training objectives, reflect the modern education ideas, conform to the scientific nature, progressiveness and the general laws of education and teaching, have distinctive characteristics, and can properly use modern education technology and methods, with remarkable teaching effects, and have the role of demonstration and radiation promotion [5]. This concept is also widely accepted and used by researchers.

After analysis, the requirements for the construction of excellent courses can be summarized as follows. First, leading. The quality of the teaching team, teaching content, teaching methods and means, teaching materials, experiments and mechanisms of the construction of excellent courses should be among the best in the country, and can become the benchmark of similar courses. The construction level in all aspects has a leading position in this field. The achievements and experience accumulated in the course of the construction of courses can be effectively and widely promoted to other similar courses. Second, pragmatism. The construction of excellent courses is not only completed by moving the courses to the Internet, but also by building the connotation of each content [6]. For example, in terms of teaching team construction, the courses should be in the charge of the main professors. There should be a echelon teaching team with reasonable structure, stable personnel, high teaching level, good teaching effect, and a certain proportion of tutors and experimental teachers. At the same time, excellent courses also require corresponding evaluation mechanisms and incentives to encourage teachers with different professional titles to continuously and orderly participate in the construction of excellent courses. Third, openness. This characteristic is reflected in two aspects. On the one hand, the network carrier is mainly used to publish the syllabus, teaching plan, exercises, experimental guidance, reference catalogue, etc. of the course, which is free and open to realize the sharing of high-quality teaching resources. These resources and teaching contents must be scientific and cutting-edge, and can timely reflect the latest scientific and technological achievements in the discipline. On the other hand, the results form and implementation process are open [7]. On the basis of publication, various achievements can also be formed in various forms. For example, in addition to the traditional paper platform, the teaching materials can also be built into three-dimensional teaching materials containing various media forms. The implementation process of the excellent course can break the previous experimental course form with fixed content, encourage undergraduates to participate in scientific research activities, and integrate comprehensive and innovative experiments into the course.

2.2 Research Status of Excellent Courses

The research on the construction of excellent courses in China began in 2002. According to the search results of China HowNet, the types of the construction of excellent course textbooks and the relevant compilation and publishing arrangements were mainly introduced from the perspective of the publishing house at that time. The introduction of the action plan for the revitalization of education in 2003–2007 formally puts forward the requirements for the construction of excellent courses, and affirms the necessity of excellent courses. At present, various circles of society have conducted a lot of research on the construction of excellent courses. Because there are many similarities between excellent courses and open courses implemented abroad, the research of excellent courses is often mixed with the research of open courses [8].

Some researchers seek the experience that can be used for reference in the construction of excellent courses by studying the concept, current situation, problems and development trend of open courses and excellent courses. Most of the open courses are introduced based on the open courses of the Massachusetts Institute of technology, and some other similar courses and their supporting organizations and departments are listed. The concept of excellent courses is unified with the expression of "five first-class" in the documents of the Ministry of education and the Ministry of finance. These studies generally analyze the advantages and disadvantages of open courses and their development trends, and also form a consensus in the following aspects of the construction of fine courses: mainly including the implementation of free sharing of educational resources in fine courses, the need to rely on information technology and use the network platform as the carrier, and the five elements are indispensable in the construction process. At the same time, these studies have noticed some shortcomings of the construction of excellent courses, such as lack of funds and technical support, lack of standardized evaluation mechanism and incentive mechanism, and no effective method for copyright protection [9]. These studies have proposed the need to improve the construction of excellent courses from the above aspects, but failed to put forward more specific measures to make up for these shortcomings.

Some studies have carried out targeted research on the problems exposed in the process of constructing excellent courses. For example, some studies draw on the evaluation strategies of similar foreign courses, and put forward the shortcomings in the evaluation process of excellent courses from the aspects of resource sharing, implantation and application of curriculum evaluation theory, and put forward improvement plans; Some study the cost of excellent courses - the inhibition and damage of excellent courses to the educational value, so as to establish the corresponding evaluation system. Another example is that some studies take the weakness of the technical force of the construction of excellent courses as a breakthrough. By analyzing how to apply information technology to the construction of excellent courses, we can make the smooth implementation and efficient operation of excellent courses, and thus solve the technical bottleneck of the construction of excellent courses. However, the measures proposed in these research results are still unclear.

3   Theoretical Basis of Excellent Course Construction

3.1   Management Basis of Excellent Course Construction

As the most direct foundation for the construction of excellent courses, management gives full play to the five functions of management: decision-making, planning, organization, leadership and control, so that the reasonable allocation of course resources and the improvement of the utilization rate of course resources can be realized.

Frederick Taylor's scientific management theory focuses on how to improve the production efficiency through improving management means. In order to complete the task of constructing excellent courses with high quality and efficiency, the management department uses its theory for reference and uses relevant contents to establish a relatively complete construction system according to the actual situation of excellent courses.

In the scientific management theory, the theory of establishing a special planning layer divides the management function and the executive function. So that the standards can be formulated in the construction of excellent courses to ensure the completion of the construction of excellent courses.

The principle of standardization summarizes the experience and skills, summarizes them into regular contents to guide the staff to work more effectively, and measures their work achievements, thus improving the work efficiency and quality. The application of scientific management theory in the construction of high-quality courses not only introduces scientization and standardization into its construction and management, but also hopes to improve the quality of higher education through the implementation of scientific management and the construction of high-quality high-quality courses [10]. While improving the quality of higher education and helping learners to obtain more high-quality education resources, excellent courses bring corresponding rewards to educators, stimulate personal potential and meet their personalized needs to a certain extent. Through the application of scientific management methods and means, especially the principle of standardization, in the construction and management of high-quality courses, the goal of promoting the quality of higher education in China and increasing the satisfaction of educators and educatees has been achieved.

3.2 Pedagogical Basis of Excellent Course Construction

On the basis of philosophical schools, pedagogy has formed different educational concepts and teaching methods. Based on the philosophical viewpoint of the school of analytical philosophy, Hurst proposed the implementation of liberal education, that is, several unique knowledge forms formed by human beings when they master the world and can be used as the basis for any learning. The emergence of excellent course construction provides the content basis and action possibility for the implementation of liberal arts education. Since the 1970s, the relevant theories of critical pedagogy in the educational circle have led the work and research of pedagogy. Although there are many schools of critical pedagogy with different viewpoints, the basic viewpoints of critical pedagogy are the same. First, teachers help learners to emancipate their consciousness by feeling the situation they are in; Second, educators should adopt the attitude and method of practical analysis to explain the situation in life in detail, so that each learner's thinking habit is more active and active, so that learners can use their subjective consciousness to analyze in the process of absorbing knowledge. The emergence of critical education provides a theoretical basis for the construction of excellent courses.

People's all-round development guides the construction of excellent courses. Human development is inseparable from knowledge. As a way to transfer knowledge, curriculum has become complex and diverse due to the complex and diverse relationship between human development and knowledge. The selection of course content, the understanding of the nature and the selection of teaching methods have become important links in human development. With the change of the nature of knowledge, the development of human put forward new requirements for curriculum, such as developing personality curriculum, strengthening humanities curriculum and opposing knowledge hegemony. These new requirements put forward expansion requirements for the teaching channels of the course. From the fundamental point of human development, excellent courses provide solutions for the dissemination of all kinds of knowledge and meet the new requirements of courses.

4 Exploration on the Construction of Excellent Courses Based on Big Data

4.1 Open up Information Communication Channels

Excellent courses need to establish a communication ring between managers, builders and users to achieve barrier free communication between any two of the three parties. Effective communication can more accurately understand whether the responsibility of each responsible party in the construction and maintenance of excellent courses has been completed, and make the evaluation more objective and fair so as to achieve effective supervision. At present, the supervision of excellent courses mainly relies on periodic results supervision and lacks process supervision. Moreover, it is not enough to realize effective process supervision only by users' feedback to the excellent course group through the excellent course webpage. Therefore, it is particularly important to supplement the feedback link of excellent courses. As shown in Fig. 1, in order to ensure the effectiveness of supervision, it is necessary to establish a model of "double feedback" mechanism for the construction and management of excellent courses. Double feedback refers to that there are two subjects receiving feedback in this feedback link, namely, the excellent course group responsible for the specific construction of excellent courses and the management departments at all levels; There are two user feedback channels, that is, users can feedback to the excellent course group or the excellent course management department. The solid arrow indicates the flow direction of the information flow of excellent courses.

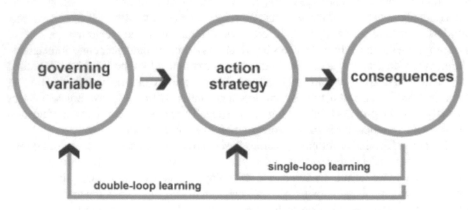

Fig. 1. Double feedback loop diagram of excellent course

4.2   Establish Diversified Incentive Mechanism

Only when the individual needs are consistent with the organizational goals can the incentive play a positive role. To realize the long-term construction mechanism of excellent courses, it is an effective way to establish a diversified incentive mechanism according to the needs of the construction team of excellent courses. The diversification of incentive mechanism refers to the diversification of incentive methods. The second is that the incentives of technical personnel and management personnel other than the teachers of excellent course construction cannot be ignored, otherwise it is difficult to form a stable construction team. According to the division of incentive objects, collective incentive mechanism and individual incentive mechanism can be established; According to the division of incentive forms, we can establish explicit incentive system, supervisory incentive system and implicit incentive system; According to the division of incentive content, we can establish material incentive system and spiritual incentive system; According to the time division, long-term incentive system, medium-term incentive system and short-term incentive system can be established. Figure 2 shows the conceptual model for determining the type of requirement. According to this model, the demand types are divided into 20 types. According to the motivation difference of different personnel, each person corresponds to one or several demand types. No matter who the incentive object is, the incentive for the construction of excellent courses needs to be carried out in a targeted manner, and often requires the comprehensive use of different types of incentive methods. Excellent course is a project of sustainable development. To ensure its normal operation, it is necessary to establish a long-term management mechanism. The diversified incentive mechanism can realize the harmony and unity of individual diversification needs and the goal of excellent courses. Therefore, the adoption of diversified incentive mechanism is conducive to the existence and implementation of the long-term construction mechanism of excellent courses.

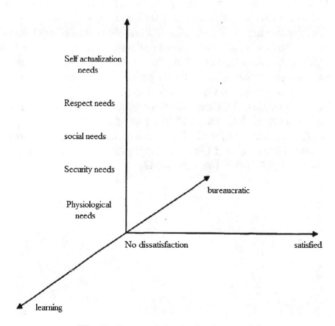

Fig. 2. Demand determination model

5   Conclusion

The construction of excellent courses is a long-term and arduous task. All members of the project team are confident to continue to do a good job in the follow-up work of the construction of excellent courses according to the construction plan of excellent courses, drive the continuous improvement of professional teaching quality through the construction of excellent courses, and make new and greater contributions to the scientific development strategy of "open school running, characteristic construction" and building a national demonstration high vocational college.

References

1. Zhang, W., Xu, S., Xu, Z., et al.: Exploration and practice of the construction of the first-class major applied chemistry of ECUST. Univ. Chem. **35**, 71–76 (2021)
2. Feng, X., Wang, P.: The exploration and practice of computer aided design in the construction of art education courses. J. Phys.: Conf. Ser. **1744**(4), 042036 (2021). (4pp)
3. Cui, D., Lv, X., Zhang, Q., et al.: Exploration and practice of reform of teaching mode of the electronic information engineering major based on new engineering. In: ICIMTECH 21: The Sixth International Conference on Information Management and Technology (2021)
4. Wang, X.: Construction of civil engineering teaching system based on data mining algorithm and big data technology. J. Phys.: Conf. Ser. **1852**(3), 032022 (2021). (6pp)
5. Zheng, J.: Construction and application of music audio database based on collaborative filtering algorithm. Discrete Dyn. Nat. Soc. **2022** (2022)
6. Deng, J., Qing, H.: Cross-border e-commerce course construction based on data mining algorithm. J. Phys.: Conf. Ser. **1852**(2), 022040 (2021). (7pp)
7. Song, J., Li, J.: Practice and exploration of collaboratively promoting curriculum ideological and political construction in higher vocational colleges. Open Access Libr. J. **9**(3), 6 (2022)
8. Heynemann, S., Philip, J., Mclachlan, S.A.: An exploration of the perceptions, experience and practice of cancer clinicians in caring for patients with cancer who are also parents of dependent-age children. Support. Care Cancer, 1–8 (2021)
9. Jw, A., Hz, A., Hua, L.B.: Research on the construction of stock portfolios based on multiobjective water cycle algorithm and KMV algorithm (2021)
10. Che, M., Zhang, X., Wang, Q., et al.: Exploration and practice of university curriculum evaluation from the perspective of big data—taking the course of database technology and application as an example. Educ. Study **3**(3), 342–347 (2021)

Financial Evaluation Model Based on Data Mining Algorithm

Nayi Zhang[✉]

School of Business and Management of Wuhan Business University, Wuhan 430056, Hubei, China
zhangnayii@163.com

Abstract. Enterprise financial evaluation refers to the evaluation and summary of the enterprise's financial status, operating results and cash flow, so as to provide relevant financial information for the enterprise's information users. The financial evaluation model based on data mining algorithm is a method to calculate the company value. The model is used to determine the value of the company and its shares. The valuation formula is based on the number of shares in circulation and their price per share. The model also considers other factors, such as dividends paid by the company, which will affect its overall value. In addition, it also considers the amount invested in the company through the purchase of shares or stock options given to employees and executives.

Keywords: Data mining · financial evaluation

1 Introduction

Following the emergence of the Internet of things and cloud computing, the emergence of big data has once again triggered a disruptive technological revolution in the field of information industry in the 21st century. Since 2010, big data has become the focus and main research object of all walks of life. In the era of big data, the environment changes with each passing day [1]. People realize that the original data mining methods are limited in many aspects such as data acquisition, storage, analysis and visualization, and can not be effectively applied. In order to adapt to the modern unpredictable market competition, enterprises must explore a new management mode based on big data analysis [2]. At present, some European and American high-tech enterprises and Chinese Internet companies such as Baidu and Tencent have also introduced and committed to analyzing and using big data, and achieved fruitful results. It can be seen that the world has ushered in the era of big data, and the future innovation and competition will rely on big data to a great extent.

At present, the research on big data mainly focuses on social computing, business intelligence, new media communication, information management architecture and other fields, while there are very few studies involving enterprise financial evaluation. By carrying out financial evaluation, enterprises can check the behavior results of managers and analyze the causes of their formation, so as to guide enterprises to operate scientifically

© The Author(s), under exclusive license to Springer Nature Singapore Pte Ltd. 2023
J. C. Hung et al. (Eds.): IC 2023, LNEE 1045, pp. 243–250, 2023.
https://doi.org/10.1007/978-981-99-2287-1_35

[3]. At the same time, all stakeholders can also understand the comprehensive situation of the enterprise and make correct decisions. However, the current financial evaluation has the shortcomings of single data source and lagging results. Today, with the rapid development of information technology, it is difficult to evaluate the financial situation of enterprises in a timely and comprehensive manner, and can not provide reference for the company to formulate correct financial strategy, which is not conducive to the development of enterprises. Therefore, it is urgent to study the financial evaluation of enterprises under the big data environment.

The defects existing in the current enterprise financial evaluation can not meet the new needs of enterprise managers. Big data has the advantages of extensive data sources and real-time comprehensive analysis, which can effectively make up for the short-comings of traditional financial evaluation. Therefore, it is very necessary to study the financial evaluation under the big data environment [4]. On the basis of revealing the development and application of big data and the current situation of financial evaluation, this paper demonstrates and analyzes the necessity and feasibility of enterprise financial evaluation based on big data, then puts forward the process and method of financial big data processing in the financial evaluation under the big data environment, constructs the factor cluster analysis model, and then makes an empirical research based on this model.

2 Related Work

2.1 Concept of Big Data

At present, the importance of big data has attracted widespread attention, but different experts and scholars have different understandings of the concept of big data, and there is no unified conclusion so far.

McKinsey's "big data acquisition and innovation" is defined as the first time that McKinsey's "big data acquisition and management software" will be used in the field of productivity and innovation in 2011. Baidu Encyclopedia believes that big data "needs to adopt a new processing mode to have stronger insight, decision-making ability and process optimization ability, so as to adapt to the characteristics of huge scale, high growth rate and diversity of information assets". Wikipedia defines big data as "the amount of data covered is so huge that it is impossible to use the commonly used software tools to obtain, process and integrate information within a certain time to help enterprises make business decisions" [5]. Merv Adrian believes that in the big data environment, the ability of traditional IT technology and software and hardware tools to collect, process and analyze data cannot meet the needs.

In its report, IDC defined big data as "creating a new technology and architecture system to discover its potential economic value through high-speed data collection, discovery and analysis". Through different interpretations of big data by many experts, I believe that the big of big data is not simply reflected in the size of its data, but also in the fact that it provides a new data analysis method to solve all problems [6]. That is, in such a huge and complex data, how to obtain, sort out and integrate data, so as to provide useful data for all walks of life, so as to promote the data to play its value, and finally bring huge profits to enterprises and faster development to society.

(1) Large data volume: the data volume of big data is usually more than 10TB. Because the Internet is fast and convenient, many devices and terminals can access it. These terminal devices will produce a large amount of data every day. For example, the call and Internet browsing records of mobile phones, computers, tablets and other terminals we usually use will form data and be obtained and used by some institutions. With the rapid development of data, the current data level has developed to ZB level [7]. IDC predicts that the global big data stock will grow rapidly at an annual rate of 40%. Based on this, it is estimated that the data volume in 2020 will be 45 times that in 2007.

(2) Speed of data generation: it refers to the high speed of data generation and processing, so it can form a continuous data flow, and managers can realize real-time query, so as to improve decision-making ability.

(3) Diversity: the conventional data type is relational database or data warehouse. The data type of big data also includes many new types, including many forms such as pictures, videos and so on.

(4) Low value density: it means that although the value quality of big data is very high, due to the huge amount of data, too much invalid data reduces the effectiveness of the data, making the value density very low.

With the continuous development of big data, when it reaches a certain stage, it will bring great value and qualitative leap to all aspects of society and life.

2.2 Data Mining

With the wide use of database technology and the rapid development of computing technology, computer performance and network, people are facing a difficult problem, that is, how to extract valuable information from massive data. The query function is far from meeting people's needs, and data mining came into being. Some people define data mining as a process of extracting implicit, previously unknown and potentially valuable information from data and database. However, some people believe that data mining, that is, knowledge discovery in database, is to quickly and efficiently find interesting rules from large data sets. Data mining is a new field of database research [8]. The mined knowledge can be used for information, management, query processing, decision support, process control and so on. In fact, for applications, how to define is not important. The important thing is that people's desire to extract valuable information from massive data can be realized through data mining [9].

As shown in Fig. 1, it shows the basic process of data mining. Data mining is a circular process, which usually involves the steps of data selection and transformation, model establishment, model evaluation, model interpretation, model application and model consolidation.

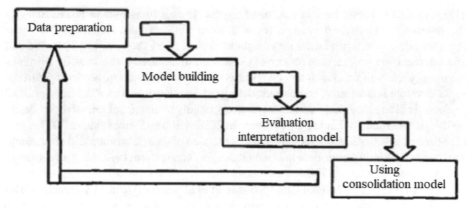

Fig. 1. Data mining process

3   Basic Theory of Enterprise Financial Evaluation

3.1   Overview and Methods of Financial Evaluation

Financial appraisal refers to the quantitative and qualitative analysis of the financial situation of an enterprise from the perspective of the enterprise, according to the current national accounting system, tax regulations and market price system, according to certain evaluation procedures and using corresponding evaluation methods, so as to judge and reveal the financial situation of the enterprise. Enterprise financial evaluation generally carries out quantitative and qualitative analysis from four aspects: profitability, solvency, operation ability and development ability. Enterprise financial evaluation helps enterprise managers take corresponding measures according to the evaluation results to promote the healthy and sustainable development of enterprises [10].

Common methods of enterprise financial evaluation include:

(1) Financial evaluation methods commonly used in western countries.

① DuPont analysis. DuPont analysis is a method to evaluate the financial status and economic benefits of enterprises based on the internal relationship between financial indicators. The core of DuPont analysis system is the return on net assets index. The return on net assets of an enterprise can be decomposed into multiple financial indicators, that is, the return on net assets = net profit / average net assets = net interest rate of total assets × Equity multiplier.

② Wall specific gravity scoring method. At the beginning of the 20th century, wall, an American scholar, first put forward the wall proportion scoring method. In his credit barometer research and financial statement ratio analysis, he selected seven financial indicators: current ratio, fixed asset ratio, property right ratio, accounts receivable turnover rate, inventory turnover rate, self owned capital turnover rate and fixed asset turnover rate, and connected them by linear relationship to determine the proportion of the seven indicators respectively, Then compare it with the standard ratio, calculate the score of each index and the overall cumulative score, and finally evaluate the credit level of the enterprise.

(2) Main modern financial evaluation methods.

① Balanced scorecard. The balanced score card is a set of financial evaluation system proposed by Professor Robert Kaplan of Harvard University and David Norton, executive director of noronton Institute. The Balanced Scorecard fully considers the internal and external factors affecting the enterprise and how to balance short-term and long-term interests from the perspective of enterprise finance, customers, internal processes, learning and growth, combined with financial and non-financial indicators, quantitative and qualitative indicators. It has the characteristics of comprehensive financial evaluation. After more than 20 years of development, the balanced scorecard has become an important tool for managing the group strategy and has a very important impact on the planning and implementation of the group strategy.
② Economic value added method. Economic value added is a well-known consulting company in the United States.

A set of financial evaluation standards registered and implemented by (Stern Stewart & Co.). It is an enterprise financial evaluation method based on the operating profit after tax and the total capital cost invested to obtain these profits. EVA = after tax net operating profit - cost of capital. Where, capital cost = total investment capital of the company × Cost of capital ratio. The basic idea of EVA is to require managers to consider the cost of capital whether it is debt operation or equity capital operation. Its idea is that if an enterprise wants to continue to operate, its operating income must at least exceed the capital cost invested by investors. Only when EVA is positive, it means that the enterprise has created value.

3.2 Specific Contents of Enterprise Financial Evaluation Index System

China's enterprise financial evaluation index system mainly includes profitability index, Solvency Index, operation ability index and development ability index.

(1) Basic contents of enterprise profitability index system.
Profitability refers to the ability of enterprises to obtain profits, which is specifically reflected in the amount of profits and the level of profits within a certain period. Whether the operators, owners, investors and creditors of enterprises attach great importance to profitability, because enterprises can only continue to operate if they make profits. Of course, the financial personnel of an enterprise also attach great importance to profitability. In order to achieve the goal of its financial management, an enterprise must improve its profitability.
(2) Basic contents of enterprise Solvency Index System.
The solvency of an enterprise is the ability of an enterprise to repay its due debts within the agreed period. By evaluating the solvency of an enterprise, it can reflect how much financial risk it faces. Creditors, investors and financial personnel of enterprises are very concerned about solvency. Solvency can be divided into short-term solvency and long-term solvency.

(3) Basic contents of enterprise development capability index system.

Development capability index refers to the evaluation of the future development trend and development capability of enterprises. Compared with the short-term income of enterprises, investors, especially long-term investors, usually pay more attention to the future development ability of enterprises.

4 Financial Evaluation Model Based on Data Mining Algorithm

4.1 Analysis of Enterprise Financial Big Data

The financial data involved in enterprise financial evaluation include: economic business data and comprehensive decision-making financial data. Under the big data environment, the economic and business data that enterprises need to obtain in financial evaluation is not single. It includes both financial data and non-financial data, including both structured data such as production and sales report data, as well as semi-structured and unstructured data such as contracts, letters, emails, archives, etc. Data sources are no longer only from accounting vouchers, but also from various business documents, such as contracts, sales lists, purchase orders, delivery/receipt documents, etc. there are also many types in form, including graphics, video, audio, web pages, etc. Moreover, in order to ensure the high quality of financial data, accounting elements should be strictly distinguished. For example, "inventory" can be strictly subdivided into "raw materials", "semi-finished products", "finished products", etc.

Financial data analysis in the big data environment is to explore new data analysis methods and analyze the financial big data stored in the data warehouse on the basis of improving the existing data analysis methods. As the core algorithm of data mining, the core idea of cluster analysis is to simplify the data processing through data modeling. Its characteristic is that it can process a large amount of data efficiently and quickly, so it can deal with the processing of financial big data well. Factor analysis is an analysis method for dimension reduction. For the multidimensional data in financial big data, factor analysis is used to reduce the dimension of sample data, which is conducive to cluster analysis. Therefore, the analysis of financial big data mainly adopts cluster analysis method, supplemented by factor analysis method. The specific analysis process is: financial data preprocessing - factor analysis - cluster analysis.

4.2 Cluster Analysis

Cluster analysis is a statistical method for multivariate variables or data, which is suitable for big data mining. The basic idea of cluster analysis is that the financial indicators or company samples studied are similar to varying degrees, so specific statistics can be selected according to multiple observation indicators of the samples to measure the similarity between company samples, and the samples are divided based on these statistics to aggregate the company samples with large similarity into one class, Aggregate less similar companies into different classes.

There are many methods of cluster analysis. The commonly used methods are: partition based method, hierarchy based method, density based method and grid based method.

(1) Method of division

The method of division refers to taking the distance as the judgment basis to determine the similarity between objects. After several iterative calculations, the data objects in the original data set are classified into k categories, in which each data object is classified into and belongs to only one cluster. The most commonly used methods are k-means algorithm and K-median algorithm.

(2) Hierarchical approach

According to the different forming directions of clustering hierarchy, hierarchy based method can be divided into aggregation method and splitting method. The core idea of the aggregation method is to first calculate the similarity between various data objects, and then merge the data objects with great similarity into a cluster, while the splitting method is just the opposite. Its basic idea is to first treat all data objects as a whole, and then gradually divide the data objects with great differences into different clusters according to the similarity between data. Hierarchical clustering method is no longer limited to division based on distance, but also clustering based on density.

5 Conclusion

As another major change in the field of information technology, big data has rapidly become the latest research hotspot of management. Based on the basic theoretical research and Application Research of data mining, this paper finds that big data has shown its great application value and development potential in various fields, and has become the focus of competition among enterprises in various industries. As one of the departments dealing with the most data, the emergence of data mining is not only a great opportunity but also a severe challenge for the enterprise financial department. In the era of big data, the ability of data collection and analysis will become the core ability of the enterprise financial department in the future. At present, there is no theoretical and systematic research on how the big data environment affects the enterprise financial evaluation and how the enterprise financial evaluation will change, but there is no doubt that the data mining algorithm will bring great changes to the enterprise financial evaluation. If enterprises want to grasp the opportunity and stand out in the fierce competition, they must be prepared, take active action and keep pace with the times.

References

1. Wang, A., Yu, H.: The construction and empirical analysis of the company's financial early warning model based on data mining algorithms. J. Math. **2022** (2022)
2. Tang, Y., Lan, Y.: Design of university financial decision-making platform based on data mining. J. Phys.: Conf. Ser. **1881**(4), 042063 (2021). (6pp)
3. Jiang, H.: Research on the development of an intelligent financial management system based on data mining technology. J. Phys.: Conf. Ser. **1578**(1), 012062 (2020). (6pp)
4. Geng, C., Xu, Y.,Metawa, N.: Intelligent financial decision support system based on data mining. J. Intell. Fuzzy Syst. (2), 1–10 (2021)
5. Gao, B.: The use of machine learning combined with data mining technology in financial risk prevention. Comput. Econ., 1–21 (2021)

6. Stertz, F., Mangler, J., Rinderle-Ma, S.: Temporal conformance checking at runtime based on time-infused process models (2020)
7. Mahdiraji, H.A., Tavana, M., Mahdiani, P., et al.: A multi-attribute data mining model for rule extraction and service operations benchmarking. Benchmark. Int. J. (2021). In Press
8. Khanchel, H.: Banking risk analysis in Tunisia: a case study of BTE Bank (2020)
9. Huang, P.: Big data application in exchange rate financial prediction platform based on FPGA and human-computer interaction - sciencedirect. Microprocess. Microsyst. **80** (2020)
10. Chunjiang, Y.U., Wenjun, L.I., Wang, D.I., et al.: A prediction method of defaulters of bank loans based on big data mining (2020). AU2020100708A4[P]

Hidden Dangers of Gynecological Nursing Based on Big Data Analysis

Ye Tian[⊠]

Yunnan Medical Health College, Kunming 650106, Yunnan, China
Ty15877977375@163.com

Abstract. The research on Gynecological Nursing hidden dangers based on big data analysis is a research based on scientific methods. It can be defined as a process in which two or more researchers work together to collect and analyze a large amount of information from different sources, such as literature, databases, surveys, etc. This research is based on big data analysis. Researchers used a large number of medical records and analyzed them to find out whether there were any hidden dangers in nursing practice. Then, statistical methods will be used to analyze this information to draw conclusions that will help improve health care services. This study is based on whether there are hidden dangers in nursing practice based on big data analysis, which will help nurses understand their mistakes when practicing gynecology.

Keywords: Big data analysis · Gynecological Nursing · Security screening

1 Introduction

A series of measures to ensure the physical and mental safety of patients in the treatment stage are called nursing safety. Obstetrics and gynecology is a high incidence area of medical accidents and disputes, which is related to the life safety of mother and baby generations, and has high requirements for obstetrics and gynecology nursing. With the prosperity and development of social economy, people's awareness of self-protection and legal concepts are increasing, and the requirements for the quality of medical care are also getting higher and higher. In the process of nursing, nurses not only need to master professional obstetrics and gynecology nursing knowledge and nursing skills, but also should have a strong sense of safety, so as to minimize errors in obstetrics and gynecology nursing, which is conducive to reducing medical disputes. This paper summarizes and analyzes the causes of the problems existing in obstetrics and gynecology nursing work, and puts forward countermeasures.

Obstetrics and gynecology nursing is a kind of occupational risk, and every clinical nurse in obstetrics and Gynecology should bear the responsibility and obligation. Because he runs through the whole clinical nursing process, through analyzing the possible nursing risks in the nursing process of Obstetrics and Gynecology, he formulates perfect nursing preventive measures to minimize the injury. Therefore, it has become one of the core issues for hospital managers to cultivate the safety awareness of innovative

J. C. Hung et al. (Eds.): IC 2023, LNEE 1045, pp. 251–257, 2023.
https://doi.org/10.1007/978-981-99-2287-1_36

obstetrics and gynecology nursing in clinical practice, formulate relatively standardized medical operation rules, and minimize the occurrence of medical and nursing errors [1].

This paper discusses the application of gynecological nursing safety hidden danger research based on big data analysis in gynecological clinical nursing, so as to enhance the harmony between nurses and patients and improve the quality of nursing. According to the needs of patients' condition and gynecological diseases, set up basic nursing service items and connotation, and plan the basic nursing service process, that is, collective basic nursing service process and single disease basic nursing service process. After its application, the patient satisfaction can be significantly improved, and the overall nursing quality of Gynecology can be improved.

2   Related Work

2.1   Big Data Analysis

Today is an era of continuous integration and development of information technology and mobile technology. The original Internet extends to the mobile Internet, and the computer as the client is gradually transformed into the mobile client with the mobile phone as the main medium. The convenience of mobile greatly accelerates the generation of data flow, and the data flow generated by "mobile" gradually dominates. Due to the innovation of data sensing technology, data transmission capacity There are fundamental changes in transmission speed and transmission accuracy. [2] The sharp increase in transmission capacity, the improvement of transmission speed and transmission accuracy have brought about the rapid growth of data flow. The growth of data flow is mainly reflected in two dimensions: one is the increase of data capacity; the other is the diversification of data types. The growth of data flow makes it difficult for current hardware facilities to effectively manage and utilize data.

$$\pi(\theta|x) = \frac{p(x|\theta)\pi(\theta)}{m(x)} = \frac{p(x|\theta)\pi(\theta)}{\int_{\emptyset} p(x|\theta)\pi(\theta)d_{\theta}} \tag{1}$$

The interests of patients are always the highest requirement of Gynecological Nursing.

Precise Gynecological Nursing has similarities with genomics and personalized Gynecological Nursing, and also shows unique aspects: first, gene analysis is the main technical means of precise Gynecological Nursing to treat diseases; Secondly, the accurate analysis and treatment of diseases will generate a large amount of data. The calculation, interpretation and utilization of big data are the key factors to achieve accurate disease diagnosis and treatment, involving bioinformatics, data science, translational medicine and other fields. At present, precise diagnosis mainly includes molecular diagnosis technologies such as gene sequencing, gene chip and fluorescence in situ hybridization [3]. Among them, gene sequencing technology has developed rapidly, the cost has been continuously reduced, and has gradually entered a broader stage of clinical application.

Big data analysis can promote the concept change of medical staff to a certain extent. In the past, medical staff mainly focused on treating patients, but at present, according to the analysis results of big data, they may pay more attention to the research on disease prevention methods in the future. Under this concept, the general public's view on health will also change. They will pay more attention to maintaining and protecting their bodies, and try to cope with some diseases that have not yet occurred with a healthier lifestyle [4].

But in the era of big data, these problems will be solved gradually. Big data analysis can make a more scientific and reasonable analysis of the incidence rate, complication rate and cause of a disease in a certain area. With the help of the results of the analysis, medical and health institutions can gradually change the previous simple treatment to both prevention and treatment, and suggest people to improve their lifestyle and eat more certain foods to prevent certain diseases, so as to reduce the incidence rate of major diseases from the root.

2.2 Gynecological Nursing Safety Hazards

The concept of legal awareness is weak, the awareness of self-protection is poor, the concept of nursing staff is obsolete, and they lack crisis awareness and emergency measures. In the process of nursing, they often ignore potential hidden dangers, ignore the psychological needs of patients and respect the rights and interests of patients, causing contradictions between nurses and patients from time to time. Nurses lack nursing work experience, are not familiar with the nursing system and process, and have insufficient sense of responsibility for nursing work.

The implementation of nursing rules and regulations and nursing operation process are not rigorous. In the process of nursing diagnosis and treatment, nurses lack due sense of responsibility: the requirements of Obstetrics and gynecology for nursing operation are very strict. Some nurses do not strictly implement the operation rules and regulations in the nursing process, and lack of sense of responsibility, resulting in careless postpartum observation of pregnant women, incomplete condition records, and failure to detect and diagnose abnormal conditions of mothers and infants in time; If obstetric drugs are not taken seriously, it is very easy to use the wrong drugs, and the consequences are quite serious.

Professional and technical quality and its influence because young nurses have limited understanding of nursing technology and lack of nursing experience; The observation and nursing of patients with severe pregnancy induced hypertension are not in place, which will lead to eclampsia in severe cases. Obstetrics and gynecology nursing requires strong professionalism [5]. Therefore, obstetrics and Gynecology nurses should have strong emergency response ability and high professional quality.

There is a lack of nurse patient communication. A good nurse patient relationship is based on effective communication. Some nurses have caused medical disputes due to the dissatisfaction of patients and their families due to their lax language expression and improper service attitude. In addition to the physical and pathological care of pregnant women, gynecological and obstetric care also includes psychological care.

3 Research on Gynecological Nursing Safety Hazards Based on Big Data Analysis

Nursing is a highly practical discipline, especially obstetrics and Gynecology, which is related to the safety of two people. The content of Obstetrics and gynecology nursing involves general nursing, surgical nursing, neonatal nursing, puerperal nursing and other aspects. There are many work items and heavy workload. Clinically, we should pay attention to improving the professional quality of nursing staff, strengthening the professional ethics education and nursing management of love and dedication, carrying out psychological adjustment, and relieving psychological pressure, Take the initiative to find and eliminate hidden dangers of nursing safety, so as to avoid patient nursing disputes. To improve their comprehensive competitiveness, hospitals must have a reasonable and safe management system [6].

Big data analysis has always been combined with the development of current mainstream information technology, such as the rise of cloud computing, which has promoted the development of information technology and the rapid growth of data. How to effectively find rules and find value from a large amount of data has become an urgent problem. At present, Hadoop technology is the mainstream technology in the data field. HDFS distributed file system, MapReduce computing framework, spark memory computing framework, HBase distributed database, hive data warehouse and other technologies are used to solve big data mining, data analysis and prediction analysis. Hadoop has high performance, high stability, big data application platform management, distributed system infrastructure, can make full use of cluster computing and high-speed storage nodes, simple and easy-to-use expansion, and it is more convenient to add distributed services.

$$S(i) = \frac{b(i) - a(i)}{\max\{a(i), b(i)\}} \tag{2}$$

$$h_{w,b} = f(W^T x) = f(\sum_{i=1}^{3} w_i x_i + b) \tag{3}$$

The data used in this study is divided into two parts: the basic data field based on the basic information of patients and the potential safety hazard data based on Gynecological Nursing. In order to achieve the purpose of standardized sorting, the two kinds of data are sorted out by formulating standards, classification, scope division and other methods. Basic data information category analysis [7]:

1) Forecast

Classification and prediction: classification refers to the processing of discrete data, and prediction refers to the processing of numerical data. Regression prediction: regression prediction can be divided into point prediction and interval prediction. Regression point prediction means that the given variable value x0 uses the regression value as the predicted value Y0 of variable y. However, there will be errors between the predicted value and the actual value, so it is necessary to obtain the possible deviation range to improve the reliability of prediction. This is the so-called interval prediction, that is, there is a certain probability to predict the change range near Y0.

2) Clustering

Assign a given set of data objects to different clusters without pre specified categories. K-means: hard clustering algorithm is a representative prototype based objective function clustering method. The algorithm accepts the input K, uses the criterion function and error square as the clustering criterion function, and takes the Euclidean distance as the measure, which is the best classification of the initial clustering center vector V, and the evaluation index J can reach the minimum value.

3) Abnormal value handling

In clustering algorithm, the anomaly is the background noise embedded in the cluster. In the anomaly detection algorithm, anomalies are points that have neither aggregation nor background noise [8].

The above three types of data analysis and processing models cover almost all big data. In the process of studying medical data processing, one or more algorithms are usually involved, which need to be combined with data and undergo a lot of data analysis and processing to meet the needs of model processing.

4   Simulation Analysis

In order to effectively solve the storage needs of multiple categories and large quantities of data, the platform adopts the branch deployment storage system for the unified management of data. At present, the mainstream storage methods all apply Hadoop framework technology. The following design ideas for data storage are from three aspects:

A. Structured data storage platform: this platform can realize the double writing of key data, ensure that the business platform will not cause application interruption due to the failure of any storage component or storage unit, and has high reliability. At the same time, the switching between storage devices is transparent to the upper application without human intervention.

B. Unstructured data storage platform: the unstructured data storage platform provides storage space for users' video, picture, medical record sharing storage applications of PACS, ERM and other systems [9]. The platform is based on distributed storage system. It is built with the industry's advanced cluster technology, multi copy technology, parallel read-write technology and scale out expansion technology. The underlying hardware adopts commercial standard units to provide users with a global single namespace. It has the characteristics of large capacity, high performance, convenient expansion, high reliability, easy management and low construction cost, and solves the storage problem of massive unstructured data in Gynecological Nursing industry. Figure 2 below shows the data analysis inline annotation test process.

Combined with the above storage scheme, Hadoop framework technology is used to store and process a large number of gynecological nursing data in a distributed manner. Hadoop implements a distributed file system (HDFS), which has the characteristics of high fault tolerance. Its core designers are HDFS and MapReduce. HDFS provides storage for massive data, [10] and MapReduce provides calculation for massive data. It is widely used in big data processing applications because of its natural advantages in data extraction, deformation and loading.

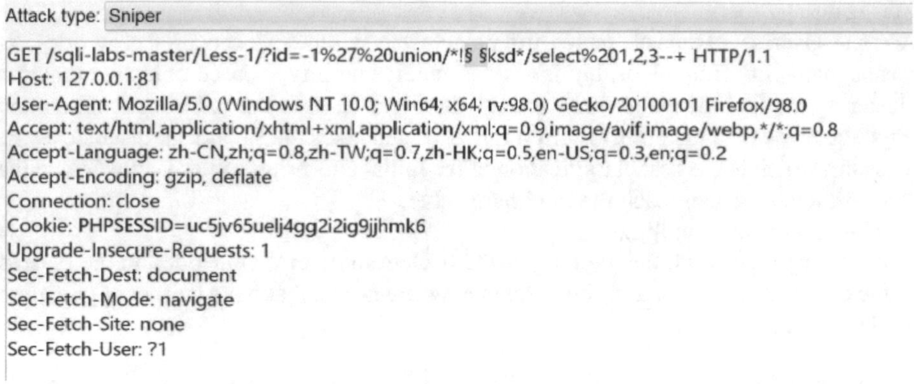

```
Attack type: Sniper
GET /sqli-labs-master/Less-1/?id=-1%27%20union/*!§ §ksd*/select%201,2,3--+ HTTP/1.1
Host: 127.0.0.1:81
User-Agent: Mozilla/5.0 (Windows NT 10.0; Win64; x64; rv:98.0) Gecko/20100101 Firefox/98.0
Accept: text/html,application/xhtml+xml,application/xml;q=0.9,image/avif,image/webp,*/*;q=0.8
Accept-Language: zh-CN,zh;q=0.8,zh-TW;q=0.7,zh-HK;q=0.5,en-US;q=0.3,en;q=0.2
Accept-Encoding: gzip, deflate
Connection: close
Cookie: PHPSESSID=uc5jv65uelj4gg2i2ig9jjhmk6
Upgrade-Insecure-Requests: 1
Sec-Fetch-Dest: document
Sec-Fetch-Mode: navigate
Sec-Fetch-Site: none
Sec-Fetch-User: ?1
```

Fig. 2. Data analysis inline annotation test process

5   Conclusion

The research on Gynecological Nursing hidden dangers based on big data analysis is a research that collects and analyzes a large amount of information from different sources. The purpose of this study is to provide the best solution to the problems related to Gynecological Nursing. The main benefit or purpose of this study is to provide the best solution to the problems related to gynecological care. This will help improve health care services and make them more effective and efficient.

References

1. Shuang, X.U., Gynecology, D.O.: Study on the current situation and countermeasures of gynecological nursing safety hidden danger. China Contin. Med. Educ. (2018)
2. Mao, R.J., Zheng, J.P., Jing, C.: Research and implementation of hidden trouble-shooting intelligent analysis system based on big data. Coal Sci. Technol. Mag. (2018)
3. Xia, F., Song, H., Tang, M., et al.: Research on hidden danger risk perception technology based on big data. In: CONF-CDS 2021: The 2nd International Conference on Computing and Data Science (2021)
4. Li, G., Wang, Y., Li, W., et al.: Influence of temperature characteristics based on big data on cable accessories under HV DC field. J. Phys.: Conf. Ser. **1852**(2), 022069 (2021)
5. Yang, P., Yang, M.: Research on the management model of university students academic early warning based on big data analysis. In: International Conference on Communications, Information System and Computer Engineering. North China University of Technology, Beijing University of Chemical Technology (2019)
6. Vo, A., Zr, A., Fy, B., et al.: Big data analysis methods based on machine learning to ensure information security (2021)
7. Jee, H., Jin-Hwan, Y.: Obesity and physical function-related parameters using exercise programs developed by ROC curve analysis for gynecological cancer patients: national health insurance sharing service big data analysis. In: International Sport Science Congress IN COMMEMORATION (2018)
8. Luo, Y., Li, J., et al.: MOOC course evaluation based on big data analysis (2018)

9. Yuan, K., Zhou, L., Li, H.: Statistic analysis of safety accidents in filling stations based on big data. In: DSIT 2020: 2020 3rd International Conference on Data Science and Information Technology (2020)
10. Zhang, C., Liu, R., Zhang, L., et al.: Investigation and research on food safety knowledge, attitude and behavior of university students in kunming based on big data analysis. J. Phys: Conf. Ser. **1648**, 022032 (2020)

Intelligent Evaluation of English Language Teaching Effect with Fuzzy Inference Algorithm

Tan Jinju[✉]

English Department, Jingchu University of Technology, Jingmen 448000, Hubei, China
t643689757@163.com

Abstract. The intelligent evaluation of English language teaching effect is the only way to improve the training quality of normal education in the era of intelligent education. The explosion of data and the updating iteration of machine learning algorithm make it possible to evaluate, classify and visualize the data of English teachers' teaching ability on a large scale. The intelligent evaluation of English teaching effect based on fuzzy inference algorithm is an intelligent system to evaluate the effectiveness of new methods or procedures. It will be used to measure the effectiveness of different methods and schemes to find out which one is more effective. The system can also be used to evaluate the effectiveness of individual teachers, courses, schools or institutions. Based on fuzzy logic technology, it can be applied to many fields such as education, medicine, engineering and so on.

Keywords: Fuzzy reasoning · English language · Teaching effect · Intelligent assessment

1 Introduction

Campus network can maximize the sharing of information resources. In the network teaching environment, modern information technology is used to provide convenience for school educational administrators in the actual implementation of teaching management. English language teaching effect evaluation is a value judgment of the teaching process, which is of great significance for teachers to improve teaching quality and strengthen teaching management. The school promotes construction and reform through evaluation [1]. On the one hand, the educational administration staff can monitor the teaching effect of teachers regularly, ensure the teaching quality of schools, improve the overall quality of teachers, and realize the standardization of daily teaching work; On the other hand, it also provides targeted information for teachers to improve teaching, and promotes teachers to improve teaching work. Through the evaluation of teachers' teaching quality, teachers should strengthen their sense of service, constantly improve their teaching level, and strictly and effectively control all links of the teaching process [2].

In the process of traditional teaching effect evaluation, school educational administrators usually use directional indicators, which often requires extensive collection of students' opinions, and manual investigation, statistics and analysis of teachers' teaching

J. C. Hung et al. (Eds.): IC 2023, LNEE 1045, pp. 258–263, 2023.
https://doi.org/10.1007/978-981-99-2287-1_37

quality, so as to facilitate quantitative analysis. This process is quite labor-intensive and material. In order to facilitate the school educational administration personnel to conduct scientific and efficient evaluation of teachers' teaching effect, reduce their workload, improve work efficiency, and better provide accurate and reliable data basis for teachers' actual teaching results and professional title evaluation, we need to find a better solution to the evaluation method of teachers' teaching effect formed by effectively implementing teaching activities and effectively collecting students' feedback under the network environment, So as to promote the quality of teaching more effectively [3].

The evaluation of English language teaching effect is an important way for school educational administrators to supervise the teaching quality of English teachers in school. Its purpose is to improve the overall teaching quality of the school. Therefore, teaching evaluation is an important part of school teaching management. With the development of computer network technology and the enhancement of enterprise informatization, the breadth and depth of the application of database technology have been greatly expanded [4]. Abandon the traditional investigation method of paper-based original data collection, essentially improve the work efficiency of academic staff, reduce the work cycle and difficulty, use scientific tools to collect information and process data, create a good operation mode and evaluation system for the evaluation of English language teaching effect, and establish an efficient English language teaching effect evaluation support system under the network teaching environment, Improve the efficiency of evaluation and the accuracy of data processing, and evaluate the scientific, comprehensive, objective and fair results.

2 Related Work

2.1 Research Status of Teaching Effect Evaluation

With the rapid development of IT technology, network teaching has become an important teaching means and teaching place. Compared with traditional teaching, the quality assurance system of network teaching is not perfect. To ensure the quality of network teaching and establish an effective evaluation model of network teaching has become an important topic of network teaching research. At present, the teaching evaluation module in the supporting platform of network teaching often only contains the test part, lacking the corresponding analysis and feedback [5]. At present, among the research and development achievements of education supporting systems in various countries around the world, many online teaching platform systems have gradually been applied from theory, such as WebCT developed by the computer department of Columbia University, pathware and virtual developed by Simon Fraser University in Canada. In these teaching systems, according to the research of actual teaching applications, there are corresponding online teaching evaluation functions. Under the influence of the general environment, network teaching and distance education have begun to develop generally. Since the president of Tsinghua University proposed to develop distance education teaching under modern technology in the 1990s, the Ministry of education has approved more than 60 key universities including Tsinghua University to carry out theoretical and Experimental Research on Distance Education under Network technology [6].

According to the current research and development of online teaching platform, it is not difficult to find that the practical work on online teaching is being carried out step by step. On this basis, the teacher evaluation system of online teaching platform has great research value and development space. The evaluation of teachers' teaching quality in Colleges and universities in China started in the mid-1980s. So far, it has developed into an evaluation system that uses the characteristics of teachers' effective teaching behavior as the standard to evaluate teachers' teaching quality. Practice has proved that it has distinct guidance and can effectively help teachers understand the teaching situation and promote the improvement of their teaching quality [7]. At the same time, according to the purpose of evaluation and the operational environment of evaluation, the sources of evaluation information are selected, and the scientific methods of education and psychometrics and statistics as well as modern educational technology are fully used. On the basis of in-depth research on the reliability, accuracy and effectiveness of various information sources and collected information, different evaluation questionnaires are prepared according to different information sources, which improves the accuracy and reliability of collected information. Foreign teaching quality evaluation started in the 1920s and developed earlier than China. By the 1980s, a fair, just and transparent teacher evaluation system had been established. The guiding ideology of its assessment indicators is to maintain a high degree of consistency with the positioning and development strategy of the school, and it is the direct embodiment of the implementation of the school's development goals to the individual teachers; The assessment index system should not only play a goal oriented role, which is conducive to the development of teachers themselves, but also ensure the quality of school teaching and the development of scientific research and service work, so as to promote the development of the school.

2.2 Fuzzy Algorithm

In 1965, American cybernetics expert Zadeh first put forward the concept of fuzzy set, which began a new era of fuzzy mathematics. Say that a is a fuzzy subset of u on a given universe. If for any u ∈ u, it corresponds to a unique real number a (U) on [0,1], the ordered pair {u, a (U) |ueu} can be regarded as the fuzzy set on u, and a (U) represents the membership of u to a, that is, a mapping is constructed: u → [0,1], u → a (U). The mapping is called the membership function of fuzzy set a: the closer a (U) is to 1, the higher u belongs to a; On the contrary, the closer a (U) is to 0, the lower the degree that u belongs to a. That is, the membership degree indicates the degree to which the element belongs to set a, which is different from the traditional set [8].

The determination of the membership function of a fuzzy set is a key step for us to use the fuzzy set theory to deal with practical problems, but this process depends more on our subjective understanding, so this paper uses a variety of parallel methods in the determination of the membership function, in order to get closer to real life.

(1) Membership function of fuzzy set
In order to describe the different tendencies of intermediate transition things towards the two, ZAD proposed the membership function. If there is a number a (x) on [0,1] corresponding to any element X in the universe u, a is called the fuzzy set on u, and a

(x) is called the membership of X to a. When x changes in U, a (x) is a function, which is called the membership function of A.

(2) Fuzzy relation

Universe x × The fuzzy relation in y = {(x, y) x ∈ x, y ∈ y is X × The value of the membership function of the fuzzy set of Y in the real axis closed interval [0,1] reflects the degree of correlation between X and y.

3 Intelligent Evaluation of English Language Teaching Effect Based on Fuzzy Reasoning Algorithm

3.1 Fuzzy Reasoning Model

Fuzzy system is composed of fuzzy rule base and fuzzy inference engine. The core fuzzy inference model in fuzzy inference engine is the inference form FMP: b* is derived from a → B and a*, where a, a* are fuzzy sets on domain x, and B, b* are fuzzy sets on domain y. Other reasoning models can be transformed into this simple reasoning model through appropriate mathematical processing [9]. In 1973, American cybernetics expert and founder of fuzzy mathematics ZAD proposed a compositional rule of inference (CRI) for solving FMP problems.

$$B^* = A^* \bullet R \tag{1}$$

B here is called CRI solution of FMP problem relative to operator R, or R-type CRI solution for short. Zadeh initially proposed to use implication operator R (a, b) = a'v (anb), while Mamdani, the founder of fuzzy systems, proposed to use implication operator R, (a, b) = anb. At present, operator R is still the most used in fuzzy systems. This also shows that the fuzzy reasoning system in the current fuzzy system is single.

$$E(t)\dot{x}_{d+1}(t) - E(t)\dot{x}_{k+1}(t) = E(t)\Delta \dot{x}_{k+1}(t) = f(t, x_d(t)) + B(t)u_d(t) -$$
$$f(t, x_k(t)) - B(t)u_k(t) = f(t, x_d(t)) - f(t, x_{k+1}(t)) + B(t)\Delta u_{k+1}(t) \tag{2}$$

The commonly used fuzzy systems based on CRI algorithm are understood as some interpolation method. At present, in the actual fuzzy control system, the CRI algorithm is usually used, and a certain fuzzy implication operator can be selected to construct the fuzzy controller according to the actual situation, and good application results have been achieved in different environments. However, from the perspective of general logic, CRI method is not perfect. In order to make up for the deficiency of CRI, the full implication triple I algorithm of fuzzy reasoning (hereinafter referred to as triple I algorithm) is proposed, and it is carefully analyzed from the perspective of logic [10]. It is considered that triple I algorithm is superior to CRI algorithm. We will compare and analyze it from the perspective of fuzzy control system application, as shown in Fig. 1.

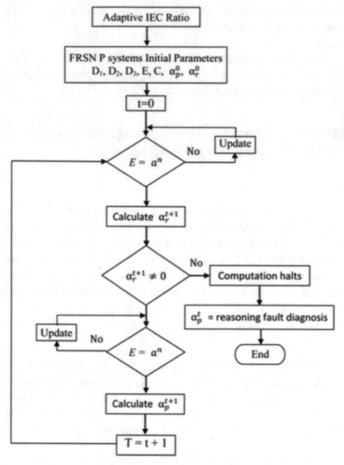

Fig. 1. Fuzzy reasoning algorithm flow

3.2 Intelligent Evaluation of English Language Teaching Effect

The application of fuzzy concept to data mining algorithm is to quantify the qualitative concept with fuzzy set, and then analyze and find the association rules with fuzzy attributes. The symbols or variables used in this algorithm are described as follows:

Minsupport: the minimum support given.

Minconfidence: the minimum confidence given.

50: The large item set of item (attribute) r, referred to as the large R item set, contains r items.

C_r: The candidate large item set of item (attribute) r, referred to as the candidate large R item set, contains r items.

The basic idea of fuzzy association rule mining algorithm is: first, convert each numerical value into the value of fuzzy variable expressed by membership function, that is, map the variation range of numerical value to the universe of the corresponding fuzzy set. Calculate the weights of fuzzy sets corresponding to each item (or attribute) in all transaction data. In the following mining process, only the fuzzy set with the largest weight is used for each item (or attribute), which ensures that the number of items (or attributes) is consistent with the original. The algorithm focuses on the most important fuzzy items (or fuzzy attributes), which reduces the time complexity. Fuzzy association rules can be found by applying fuzzy concepts to the mining process.

4   Conclusion

After the above 10 steps, the output is the association rules with fuzzy attributes, which can be used as potential meta knowledge in a given transaction database. This fuzzy reasoning rule is the most extensive and popular knowledge form of expert system. It is based on deductive reasoning, which is easy to understand and can ensure the correctness of reasoning. The fuzzy set theory is introduced into the research of association rules, and the fuzzy concept is used to summarize and abstract the data. By defining the concept of fuzzy association rules, the representation and application scope of traditional deterministic association rules are expanded. Using fuzzy association rules to express the relationship between numbers is more suitable for people's thinking habits and reasoning methods. Through this case study, schools and teachers can understand students' learning ability and interest from qualitative to quantitative analysis, so as to adjust or modify the teaching plan and teaching content, and then teach students according to their aptitude, so as to achieve better teaching results.

References

1. Xia, Z.:A convolutional network-based intelligent evaluation algorithm for the quality of spoken English pronunciation. J. Math. **2022** (2022)
2. Geng, L.: Evaluation model of college English multimedia teaching effect based on deep convolutional neural networks. Mob. Inf. Syst. (2021)
3. Wei, Y.: Study on the construction of English ICAI course based on BP neural network algorithm. J. Phys.: Conf. Ser. **1992**(2), 022185 (2021). (6pp)
4. Jjw, A., Hak, B., Shc, C., et al.: Effect of customization, core self-evaluation, and information richness on trust in online insurance service: intelligent agent as a moderating variable. Asia Pac. Manag. Rev. **27**(1), 18–27 (2022)
5. Lang, A.: Evaluation algorithm of English audiovisual teaching effect based on deep learning. Math. Probl. Eng. **2022** (2022)
6. Chen, J.: Analysis of intelligent translation systems and evaluation systems for business English. J. Math. **2022** (2022)
7. Zhang, L.: The effect evaluation of flipped classroom in college English translation teaching under the blended teaching mode (2021)
8. Chen, X., Wang, Y.: The existing problems of English teachers' classroom evaluation language and its countermeasures in junior middle school (2021)
9. Li, L.: Multi-evaluation system of English teaching guided by constructivism. In: CIPAE 2021: 2021 2nd International Conference on Computers, Information Processing and Advanced Education (2021)
10. Zhang, Y., Min, Y., Chen, X.: Teaching Chinese sign language with a smartphone. Virtual Reality Intell. Hardw. **3**(3), 248–260 (2021)

Interactive Media Design Method in Digital Exhibition of Art Museum Based on Big Data

Lingli Hu[✉]

Arts Academy, Hubei Polytechnic University, Huangshi 435000, Hubei, China
hulingli@hbpu.edu.cn

Abstract. With the development of science and technology, digital exhibition is widely used in art museums. Interactive media is favored by exhibition designers because of its unique nature. However, there are still some defects in the interactive media design of digital exhibition in Art Museum. There are also many researches on Interactive Media Design in the digital exhibition of art museum based on big data, but the actual results are poor. Starting from the characteristics of interactive media, this paper explores the digital exhibition mode in line with the art museum, studies its design method and process, and uses the collaborative filtering recommendation algorithm to recommend the exhibits of the digital exhibition of the art museum to visitors, so as to improve the experience of visitors participating in the exhibition of the art museum. The survey results show that the majority of visitors to the art museum are aged between 35 and 55, and there are more local visitors than foreign visitors. We hope to use big data technology to design a digital art museum exhibition with interactive media characteristics, so as to improve the public's sense of visiting experience.

Keywords: Big Data · Art Gallery · Digital Exhibition · Interactive Media

1 Introduction

Nowadays, the public has higher and higher requirements for the exhibition and display of art museums. It is necessary to have an artistic taste and an immersive feeling during the visit. This is also the development trend of art museums today. In response to this situation, it is necessary to optimize the structure of the art museum, enrich the exhibits of the art museum, and build a modern art museum system to play a positive role in promoting the artistic aesthetics of the exhibition. Therefore, the exhibition of art museums is of great significance to the construction and transformation of modern art museums.

There are many researches on interactive media design methods in digital exhibitions of art museums based on big data, and scholars have achieved good research results. For example, Lai Jie studied the application of multimedia technology in the exhibition design of art museums, and believed that the use of digital and networked high-tech means greatly reduced the labor intensity of art museum designers and helped better serve the public. The exhibition integrates cutting-edge technology and cultural creativity, which can increase the flexibility of exhibitions, break the time and space constraints of exhibitions, and enhance exhibition interaction to enhance the value of art museums [1]. Lee C F has established a lot of art exhibit resource databases, transformed exhibit information into digital form, and carried out protection work on it. He believes that using big data technology to assist the exhibition design of art museums to realize the digital virtual display of exhibits is art. The main direction of the development of museum culture [2]. Although the design methods of digital interactive media in art galleries based on big data at home and abroad have been studied many times, it is necessary to strengthen the research in this area and design better display methods so that the audience can obtain exhibit information.

This article expounds the relationship between interactive media and digital exhibition methods of art museums, defines the form and content of exhibitions based on the multidisciplinary knowledge of art museums, establishes a digital exhibition system of art museums, and presents exhibitions that combine technology and art. The exhibition space of the pavilion enhances the aesthetics and understands various arts and cultures, and carries out artistic edification.

2   Discussion on Interactive Media Design Method in Digital Exhibition of Art Museum Based on Big Data

2.1   Application of Display Methods in the Exhibition Design of Digital Art Museums

Digitization refers to the use of digital information processing technology in all links of something. The digital exhibition mode increases the amount of information displayed through video, network technology, mobile phone and other technologies. The audience experiences the exhibition effect through their own perception system, so as to improve the participation and interaction of visitors and increase the fun and entertainment of the exhibition process [3, 4].

(1) 360°Cinema
360°surround screen theaters usually use cylindrical projection screens to surround the audience in the center. General ring screen special effects include ring screen, dome screen, screen screen, fog screen projection, etc. The wide picture is fully presented in the audience's field of vision. The full range of stereo sound effects and the plot complement each other, so that the exhibit information perfectly fits the film and brings a wonderful and beautiful Immersive audiovisual enjoyment [5].
(2) Interactive map, sand table projection
Import the exhibit information introduction video into the multimedia screen. When visitors visit, the system will transmit the location signal of the visitor to the computer. The computer system responds to this signal and makes a corresponding video introduction of the exhibit through the driving influence. Visitors can click on the screen for an interactive demonstration. If no visitors nearby enter the system to capture the signal range, the system will switch to automatic demonstration mode [6].

(3) Mobile phone tour, voice guidance

Mobile phone navigation is realized through the combination of mobile phone remote control technology and large-screen projection technology. This new type of remote control technology is extremely interactive, allowing visitors to experience the theme and cultural connotation of the display while entertaining. Many art galleries now have special APP applications, which can be downloaded to the mobile phone and guided by the mobile phone to participate in art museum exhibitions more intuitively and freely. There are also some museums placing QR codes around the exhibits. It is particularly convenient to scan the QR code of the exhibits with a mobile phone to obtain related information. Visitors can choose to understand the basic information and cultural connotation of the exhibits according to their hobbies.

(4) Glasses-free 3D

3D is three-dimensional graphics. Naked-eye 3D uses the characteristics of the left and right eyes of humans to have parallax and near and far distances to create a three-dimensional effect. It does not require additional equipment (such as 3D glasses, helmets, etc.) to use 3D stereo space and depth to save.

2.2 Features of Exhibition Design of Interactive Media Art Museum

Interactive media, also known as interactive multimedia, usually refers to products or services based on computer systems that respond to user behavior through the presentation of content in the form of text, images, animations, and videos. It acquires physical information from the outside world through various sensors and transmits information through the interface, transforms the information into digital or analog signals that the computer can understand, and translates the computer program, combines the recorded signals with other materials, and presents them to the outside world through the physical interface. The whole process of information processing realizes human-computer interaction [7, 8]. In the information age, the media is regarded as a medium or tool for disseminating information. Extended to the art museum, the media refers to exhibition methods, exhibition methods, and display language, which are the exhibition methods that the art museum chooses to convey the exhibition information to the audience. It is a form of communication technology, and the content presented to the audience is called "exhibition" or "exhibition item". Different from traditional media information dissemination methods, interactive media emphasizes the interaction and communication between man and machine.

(1) Interaction between visitors and exhibits

Nowadays, people visiting art museums have changed from passive acceptance to active exploration. People have long disliked the traditional single display form, and the display form of ordinary cultural relics plus graphic explanation boards has become commonplace. Speaking of the most fundamental attribute of interactive media, we have to mention its interactivity. The combination of its fundamental attribute and the exhibition design of art museums can be said to be very harmonious. Visitors learn about the cultural information of the exhibits through various interactive methods such as touch, joystick, and manipulation, which improves the experience level of the viewers, thereby enhancing the cultural communication function of the art museum [9].

(2) Personalization of display effect

Interactive media is concerned with the information interaction between the computer and the external environment. In the way of interactive media exhibitions, the status of the audience has been greatly changed, and the audience gradually grasps the initiative of receiving information, viewing information according to their personal interests. In the way of interactive media exhibition, designers determine the organization of information according to the established purpose, and create a specific information dissemination environment for the audience. This kind of exhibition method has strong individual characteristics, which is convenient for the audience to understand the relevant knowledge in the art museum. However, it is precisely this kind of individuality that causes the audience to choose certain information according to their own interest and understanding ability when interacting with it. On this basis, reconstruct its meaning according to their own understanding, which is also prone to misunderstandings.

(3) Diversification of exhibition space

Digital interactive media technology has made the exhibition methods of art museums rich and diversified. The application of exhibition methods in digital media art museums has diversified the exhibition space. The reasonable use of digital media exhibition methods can well define and divide the exhibition space, so that visitors can observe the exhibition in a hierarchical and rhythmic manner, thereby obtaining more cultural information and knowledge.

(4) Diversity of interactive media

As an interactive media supported by digital technology, the support of science and technology is naturally indispensable. Multi-touch technology, sensor technology, virtual image system, etc., are all indispensable technical support in interactive media design. The interactive media in the digital exhibition use digital "sound, light, and electricity" to present the effect. Its principle of stimulating the senses is the same as that of traditional art, and there is no difference. What is changed is only the carrier of information, from the traditional physical modeling. Up to now, the development trend of electronic and digitalization. And with the continuous development and progress of technology, there are still more possibilities for interactive media [10, 11].

(5) The multi-sensory nature of exhibitors

Different from the singleness of sensory presentation in traditional media, interactive media has a multi-sensory nature. The concept of different aesthetics should not change with changes in means or methods. With the application of information technology in the field of interactive media, traditional aesthetic theories cannot be applied to new art forms. If the traditional media focuses on external physical form factors and transfers it to interaction, what needs to be realized is that the aesthetics in interactive media is no longer limited to visual aesthetic issues, but an experience that is closely related to all senses. Aesthetic experience is not only related to direct instinctive reactions, but also the interpretation of meaning and the emotions stimulated. It is a comprehensive experience.

2.3 Collaborative Filtering Recommendation Method Based on CoP Modeling

CoP (Community of Practice) refers to a group or community created to share knowledge, learn from each other, and connect with each other at work. CoP members obtain their interest characteristics within a period of time, and then make recommendations based on the weighted average of CoP and user interest functions. By calculating the similarity between different elements, you can accurately query the target user [12]. There are several ways to calculate similarity:

(1) Cosine similarity

The user score is regarded as a vector in the n-dimensional item space. If the user does not rate the item, the user score of the item is set to 0, and the cosine angle between the vectors is used to measure the similarity between users. The similarity calculation method between user u and user v is as follows:

$$sim(u, v) = \cos(u, v) \frac{u \times v}{|u| \times |v|} \tag{1}$$

In the formula, u and v represent the user's rating vector, and cos is used to calculate the cosine value.

(2) Related similarity

Assuming that the set of items rated by user u and user v is represented by I_{uv}, the similarity sim (u,v) between user u and user v is measured by Pearson correlation coefficient as shown in the formula:

$$sim(u, v) = \frac{\sum_{c \in I_{uv}} R_{u,c} \times R_{v,c}}{\sqrt{\sum_{c \in I_{uv}} R_{u,c}^2} \sqrt{\sum_{c \in I_{uv}} R_{v,c}^2}} \tag{2}$$

In the formula, $R_{u,c}$ represents the score of user u on item c, and $R_{v,c}$ represents the score of user v on item c.

3 Survey on the Number of Participants in the Digital Exhibition of Art Museums

3.1 Research Purpose

Through the application of interactive media in the digital exhibition of art museums in the data age, visitors can be immersed in the cultural scope of the exhibits when visiting the exhibition, and truly feel the charm of the exhibits and benefit from it.

3.2   Research Methods

This paper adopts the field research method to investigate the flow data of visitors to an art museum for a week, combined with literature analysis, sort out the data of interactive media in the exhibition design field in the data age, and draw up the interactive media design process in the digital exhibition of the art museum based on big data.

3.3   Data Acquisition

Through a field survey of an art museum, this article collected data on exhibitors in that place for a week, divided these visitors into 7 age groups, and distinguished between local visitors and non-local visitors.

4   Interactive Media Design Methods in Digital Exhibitions of Art Museums Based on Big Data

4.1   The Digital Exhibition of the Design Art Gallery Centered on Activities

The charm of interactive media exhibition methods comes from the interaction with the audience, and the core lies in the content of the exhibition. Art galleries need creative display content. This creativity is the interpretation and organization of existing content, which involves the understanding, organization and expression of exhibits and related cultural background knowledge and exhibition themes. The creativity of the exhibition does not lie in the form. Even if it is a successful exhibition method in other places, it does not mean that it will have the same effect in every art gallery. Therefore, the organization and expression of the content of the exhibition is particularly important. Designers need to determine the theme and content expression of the exhibition according to the needs of the audience, and strengthen their own overall understanding of the exhibition activities, with the goal of achieving the purpose of the event. Taking the activity as the center requires designers to comprehensively consider the relevant elements in the activity and their influence on the design, such as the subject, object, community and the division of labor, rules and tools of the activity. As shown in Fig. 1.

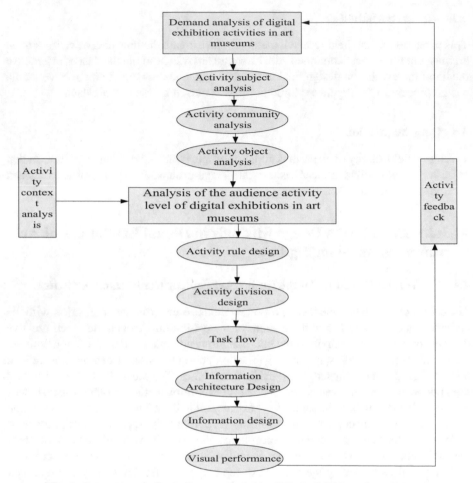

Fig. 1. Interactive media design process in the digital exhibition of art museums

4.2 The Audience Situation of the Audience in the Digital Exhibition of the Art Museum

It can be seen from Table 1 and Fig. 2 that within a week, there are not many visitors under the age of 15 to the art museum, including 6 foreign visitors, 17 local visitors, 28 foreign visitors between the ages of 15 and 25, 53 local visitors, 35 foreign visitors between the ages of 25 and 35, 64 local visitors, and a relatively large number of visitors between the ages of 35 and 45. There are 33 foreign visitors, 69 local visitors, 21 foreign visitors aged between 45 and 55, 48 local visitors, 14 foreign visitors aged between 55 and 65, 35 local visitors, 2 foreign visitors aged over 65 and 17 local visitors, and most of them are local visitors in each age group.

Table 1. The situation of exhibition audience

	Out-of-town visitors	Local visitors
Under 15 years old	6	17
15–25 years old	28	53
25–35 years old	35	64
35–45 years old	33	69
45–55 years old	21	48
55–65 years old	14	35
Over 65 years old	2	17

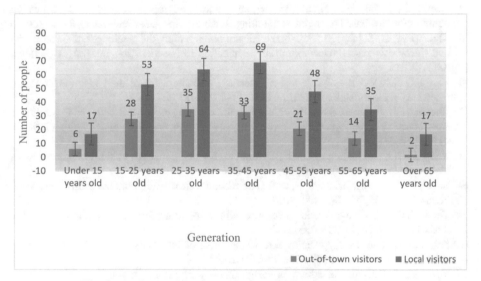

Fig. 2. Audience age group and corresponding number of visitors

5 Conclusions

With the advancement of big data technology, the exhibition space of art museums has been expanded. In the space environment of art museums, the display effect is a measure of the museum's level of cultural construction and art exhibition. The interactive experience facilities of art museums should make full use of digitalization to give full play to their characteristics, centrally transmit the information of art museum exhibits, improve the level of exhibition design, and achieve the effect of aesthetic taste and interactive fun that satisfy the public.

Acknowledgments. This work was supported by the youth project of philosophy and social science research project of Hubei Provincial Department of education, project name: "Research on the Application of AR Technology in the Inheritance of Yangxin Buiter of Huangshi Intangible Cultural Heritage", project number: 19Q197.

References

1. Jie, L.: Research on the application of digital media art in exhibition design. Design (024), 134–135 (2017)
2. Lee, C.F.: The changing nature of art museum education : how contemporary artworks and exhibition designs encourage visitors' interactive learning. J. Res. Art Educ. **19**(1), 79–96 (2018)
3. Zhe, H., Furong, L.: Research on digital library reference service model based on linked data. Univ. Libr. work **039**(005), 50–53 (2019)
4. Kim, M.J., Chung, H.J., et al.: The study of features and implications of coding education contents for kids based on digital storytelling. J. Korean Soc. Des. Cult. **23**(1), 21–31 (2017)
5. Marshalsey, L., Sclater, M.: Arts-based educational research: the challenges of social media and video-based research methods in communication design education. Int. J. Art Des. Educ. **38**(3) (2019)
6. Kwon, G.N.: Design language and architectural characteristics of the Chichu art museum designed by Tadao Ando. KIEAE J. **18**(5), 29–40 (2018)
7. Koebner, I.J., Fishman, S.M., Paterniti, D., et al.: Curating care: the design and feasibility of a partnership between an art museum and an academic pain center. Curator Museum J. **61**(3), 415–429 (2018)
8. Yang, J.: The application research of virtual reality art in modern exhibition space design. Heilongjiang Science **008**(024), 160–161 (2017)
9. Sugie, N., Kato, M., Nakatani, F., et al.: Architectural survey of Okura museum of art. AIJ J. Technol. Des. **25**(59), 487–490 (2019)
10. Yoon, Y., Oh, C.: Characteristics of design exhibition of domestic design museum since 2000. Arch. Des. Res. **31**(1), 225–235 (2018)
11. Fei, Y., Xueyong, L.: Application and development of information interaction in museum exhibition. Des. Art Res. **009**(001), 97–101 (2019)
12. Yuan, L.: A study of the exhibition design of the geological museum: exemplified by the geological museum of China. Acta Geosci. **038**(002), 304–312 (2017)

Local Culture Brand Building Method Based on Improved Apriori Algorithm

Qian Liu[1]([✉]), Yue Zhao[1], and Xing She[2]

[1] Shandong Women's College, Jinan 250000, Shandong, China
s717006070@163.com
[2] Anhui University of Technology, Maanshan 243000, Anhui, China

Abstract. This paper mainly discusses the idea of building local cultural brand, and then puts forward the realization method based on the improved Apriori algorithm. Through research, the improved Apriori algorithm has more advantages than the classical algorithm, and can better fit the local cultural brand building ideas to provide suggestions to manual workers, so as to ensure the quality of shaping results.

1 Introduction

The building of local cultural brand is related to the cultural image of the city and the overall cultural soft power of the country. It can bring a lot of economic benefits to the city and the country and expand the influence. These effects are particularly important in the context of international development, so the major local cities in China have actively carried out the building of local cultural brand. However, after a period of work, many cities find that there are problems in the cultural brand they shape, and the direction is not clear, the theme is fuzzy and other problems in the process of shaping, so people realize that the work is very complex and it is necessary to take mathematical methods to deal with it, which makes the Apriori algorithm highly respected. Although the classic Apriori algorithm can provide help in the building of cultural brands, the algorithm itself has defects and help efforts. Therefore, relevant fields believe that the Apriori algorithm can be improved first and then used to carry out work. Therefore, relevant research is needed.

2 Ideas for Building Local Cultural Brands

2.1 Basic Roadmap

The structure of local cultural brand itself can be divided into two levels: cultural theme and affiliated elements. The connotation of cultural theme is unchanged, but the form can change with the change of affiliated elements, which also indicates that the affiliated elements can change in many aspects such as quantity and form [1, 2]. On this basis, people need to establish various cultural brand images through the changes of affiliated elements in brand building, and ensure that the cultural brand image can play a positive

J. C. Hung et al. (Eds.): IC 2023, LNEE 1045, pp. 273–279, 2023.
https://doi.org/10.1007/978-981-99-2287-1_39

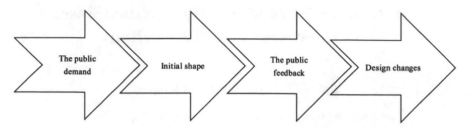

Fig. 1. Basic ideas of local cultural brand building

role in the development of urban economy or cultural soft power. From this perspective, the basic idea of local cultural brand building is shown in Fig. 1.

Combined with Fig. 1, first of all to understand the public demand in local cultural branding, then carries on the preliminary design, the second to the preliminary design result shows that the understanding of the public feedback, if the bad feedback, just adjust affiliated elements in the preliminary design result, continuously improve the quality of local cultural brand, also protect local culture brand and public demand match for a long time, only in this way can the local cultural brand always maintain a positive phenomenon, and the role can be fully played.

2.2   Ideas Implementation Problems

According to the basic idea, to realize this idea in the building of local cultural brand, we must solve two difficult problems: One is how to get a comprehensive understanding of the public demand, that is, the public demand is a macro elements, micro can be subdivided into many elements, generally are shown in Table 1, again there is a complicated relationship between each micro elements, and in a different perspective on the relationship between form also have differences, so to understand the public demand, and clarify the relationship is very difficult, at least artificial cannot rely on their own to do this, therefore, according to the thinking, if the public needs cannot be fully understood, the thinking cannot be realized, and the results of local cultural brand building are likely to be disconnected from the public needs, resulting in quality problems [3–5]. The second is how to correctly adjust the results, that is, assuming that the public needs clear, and the preliminary design result inconsistent with the public demand, then adjust the preliminary design result, adjustment method is to change existing accessory design element, and this will appear in the process of "how to identify changes to the elements and the public demand match", this is also a manual alone to deal with problems, the reason is that there are also many types of modified elements [6]. If the relationship between the modified elements and the public needs cannot be established, it is impossible to confirm whether the modified elements match the public needs. At this time, the modification cannot effectively solve the problem, which means that the result adjustment is not correct.

Table 1. Main micro elements of popular demand

The macro elements	The micro elements
The public demand	Aesthetic demand
	Cultural background
	Life pursuit
	Environmental quality

3 Improve the Apriori Algorithm and Implementation Method

In order to achieve the thinking goal of local cultural brand building, it is necessary to have a full understanding of the Apriori algorithm first, and make modifications according to its shortcomings, and then carry out work through the improved algorithm. The following analysis will be carried out according to these requirements.

3.1 Concept and Improvement of Apriori Algorithm

The basic concept and improvement method of Apriori algorithm will be discussed below.

3.1.1 Basic Concepts

Apriori algorithm is a data mining algorithm. Its main function is to mine association rules between data. The method is to decompose a huge data volume into several frequent itemsets and calculate each set [7]. The core of Apriori algorithm is a Boolean association rule. Under this rule, Apriori algorithm can obtain frequent itemsets through two steps: candidate set generation and node downward closed detection, and then mining the frequent itemsets. In the mining process, each frequent itemset is defined first, and then the relationship of each frequent itemset is calculated according to the definition. Therefore, Apriori algorithm can also be defined as a recursive method based on frequent itemsets. Apriori algorithm is a key part of the application of the data, division includes dimension rules, rules of level, two kinds of rules that can be either can not be used at the same time, the results will determine the quality of the frequent itemsets, namely if inappropriate, then the algorithm of frequent itemsets in support may not up to standard. Compared with other algorithms, Apriori algorithm is simple and can be directly searched and iterated layer by layer without tedious theoretical derivation, so it is easy to implement. Figure 2 shows the basic flow of Apriori algorithm.

Combined with Fig. 2, Apriori algorithm applications must first find all frequent itemsets, the method to confirm every ordinary itemsets in data in the whole frequency, the frequency compared with predefined minimum support, if the frequency of an item set and the minimum support are equal, or exceed the minimum support, so the ordinary itemsets are defined as frequent itemsets. Second use of frequent itemsets generated association rules, association rules has strong and weak points, the strong association rules refers to the correlation between itemsets is greater than the minimum support and

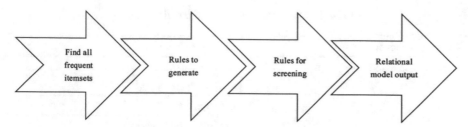

Fig. 2. Basic flow of Apriori algorithm

minimum confidence rules, if the association rules is the rules of the correlation is less than the minimum support and minimum confidence, combined with the standard, after the completion of find all standard rules, these rules include the rules of strength, when selected, it is converted to a desired rule, which contains only all frequent itemsets in the data set [8–10]. Thirdly, the rules are further screened to exclude all the rules that are less than the actual minimum credibility, and the remaining rules are retained. Finally, the relational model can be output according to the screening results of the rules, that is, all the retained association rules have the function of revealing the relationship form and correlation degree between the data and the target. According to the results, the target can know what it needs, what it does not need, and the degree of demand for different things. The expressions of minimum support and minimum confidence in Apriori algorithm are shown in Eqs. (1) and (2).

$$S(X \rightarrow Y) = \sigma(X \cup Y)/N \tag{1}$$

$$C(X \rightarrow Y) = \sigma(X \cup Y)/\sigma(X) \tag{2}$$

where, S and C are the minimum support and minimum confidence respectively, $\sigma(X \cup Y)$ is the support count or confidence of disjoint itemsets (X and Y are two disjoint itemsets), and N is the total number of transactions.

In addition, it can be seen from the above discussion that frequent item set is very important in the Apriori algorithm. Its main function is to reveal the variables in the data set and the occurrence frequency of each variable, which can provide necessary support for decision-making. He at the same time, with the Apriori algorithm also has many functions, such as association rule generation, correlation analysis, causal relationship between itemsets based parsing, sequence, design of partial periodicity, plot fragments decomposition, etc., and arbitrary function should be established on the basis of frequent itemsets, otherwise the function cannot be achieved, quality problems or function results. The existence of a variety of functions shows that the Apriori algorithm has a wide range of applications, currently it is mainly used in commercial consumption analysis, website intrusion detection and other fields.

3.1.2 Algorithm Improvement

Although the Apriori algorithm has a wide range of applications and good application value, it has some shortcomings that are difficult to overcome in some cases, as shown in Table 2.

Table 2. Main disadvantages and influences of the classical Apriori algorithm

Disadvantages	Impact
The database is scanned too many times. Procedure	This results in generalization
The intermediate term set is too large	It is difficult to define frequent itemsets accurately
Only unique support is supported	Be susceptible to interference
The adaptive ability of the algorithm is weak	Unable to adapt to dynamic environments

In view of the above defects, the classic Apriori algorithm will be affected by obstacles in the building of local cultural brand and cannot guarantee the quality of results. Therefore, it needs to be improved. The improvement methods are as follows: First, classic algorithms are much scanning database, because the algorithm itself no requirement to data level, the default for all data, and find a correlation between all the data, it needs to be improved in the data level limit, method to build data correlation standard, all lower than the standard data are not scan; Second, because the data magnitude is limited, the amount of intermediate item sets will also be reduced. Thirdly, cancel the unique attribute of support degree and replace it with multivariate function, so that multiple support degrees can be set up according to the needs. Thirdly, an adaptive mechanism is added to the classical algorithm. The method is to add a calculation termination condition, such as "when the operation of the algorithm is inconsistent with the actual environment, the calculation will be terminated, and the current results of the calculation will be retained, and the calculation will be recalculated at the inconsistent node".

3.2 Ideas and Implementation Methods

Combined with the improved Apriori algorithm and focusing on the two major problems in the realization of the basic idea, corresponding methods can be adopted to cope with and help shape local cultural brands. The specific methods are as follows.

3.2.1 Public Demand Analysis

First in order to provide improved Apriori algorithm environment support, need access to the public before culture brand shaping demand data, access to establish information collection channels: first, on the demand of the masses containing all of the micro elements, establish information acquisition path, and then set up an online platform, through the platform preliminary to collect all the information on the micro elements; Secondly, classifying information on each micro elements, methods for classification of "key words", which assumes that the information display A customer information contains the keywords "traditional culture", then the other users of information contained in the same keywords with A user category, and so on, thus complete the information collection.

Secondly, the user information in each category is converted into data, and the improved Apriori algorithm is used for calculation to know the specific needs of users. If the calculation results show that users A want to live in an environment rich in traditional cultural atmosphere, then local cultural brand building can take the needs of users A as the initial direction. As much as possible on this basis, in order to balancing the needs of all users, to rely on data in the algorithm middleweight limit method, user requirements to eliminate one is too small, or too little of the correlation between user requirements and local culture branding is not high, so it will be ruled out by algorithm, artificial can also supervise the process, to ensure the quality of results. Through the above steps, the public demand can be analyzed and the direction for local cultural brand building can be provided.

3.2.2   Matching Analysis of Elements and Requirements

Based on the improved Apriori algorithm, designers can display all design elements, and then assign labels according to the characteristics of each design element. For example, element A has labels such as "traditional culture" and "long history", and then use the algorithm to analyze the keywords of public demand. Result, user demand correlation between keywords and design element tag size, the assumption is contained in A user requirements "traditional culture" three key words, such as the three key words in A element tag all exist, is A element with A user demand correlation degree was 100%, representing both match exactly (the greater the correlation, the bigger the compatibility, conversely, the smaller). You can use the A element in your design, and you can do the same with later adjustments. On this basis, no matter the preliminary design or the subsequent adjustment design, the local cultural brand building can have a clear direction support with the help of the improved Apriori algorithm, and the final results are highly consistent with the public demand, so the quality can be guaranteed.

4   Conclusion

To sum up, local cultural brand building is beneficial to the development of urban economy and other aspects, so it must be actively carried out and the quality of results can be guaranteed. The improvement of Apriori algorithm can provide necessary help to make the direction of cultural brand building clear and keep up with the change of demand.

References

1. Zhang, Y., Wang, C.: Research on intelligent management of student achievement data based on improved Apriori algorithm. In: 2021 International Conference on Public Management and Intelligent Society (PMIS) (2021)
2. Pratama, A., Mulyawan, B., Sutrisno, T.: Implementation of improved Apriori algorithm for drug planograms in XYZ pharmacies and analysis of medicinal procurement planning using ABC critical index methods. In: IOP Conference Series: Materials Science and Engineering, vol. 1007, no. 1, p. 012186 (2020). (7pp)
3. Singh, P.K., Othman, E., Ahmed, R., et al.: Optimized recommendations by user profiling using apriori algorithm. Appl. Soft Comput. **106**, 107272 (2021)

4. Lakshmi, N., Krishnamurthy, M.: Frequent itemset generation using association rule mining based on hybrid neural network based billiard inspired optimization. J. Circ. Syst. Comput. **31**(08) (2022)
5. Krishna, B., Amarawat, G.: Data mining in frequent pattern matching using improved Apriori algorithm. In: Abraham, A., Dutta, P., Mandal, J., Bhattacharya, A., Dutta, S. (eds.) Emerging Technologies in Data Mining and Information Security. Advances in Intelligent Systems and Computing, vol. 813, pp. 699–709. Springer, Singapore. https://doi.org/10.1007/978-981-13-1498-8_61
6. Xiaoguang, Z.U., Chen, M., Liu, T., et al.: Present situation and prospect of comprehensive planting and culture in paddy fields in northeast China. Asian Agric. Res. **14**(2), 54–56 (2022)
7. Zhang, C., Teng, G.: Application of an improved association rule algorithm in rural development assessment in China. In: IOP Conference Series Earth and Environmental Science, vol. 772, no. 1, pp. 012089 (2021)
8. Lu, P.H., Keng, J.L., Tsai, F.M., et al.: An Apriori algorithm-based association rule analysis to identify Acupoint combinations for treating diabetic gastroparesis. Evid. Based Complement. Altern. Med. **2021**(17), 1–9 (2021)
9. Sokhangoee, Z.F., Rezapour, A.: A novel approach for spam detection based on association rule mining and genetic Algorithm. Comput. Electr. Eng. **97**, 107655 (2022)
10. Liu, Y., Hu, X., Luo, X., et al.: Identifying the most significant input parameters for predicting district heating load using an association rule Algorithm. J. Cleaner Prod. **275**(12), 122984 (2020)

Mixed Optimization Strategy of Resource Allocation in Higher Education System

Xin Li[(✉)]

School of Foreign Studies, Xi'an Medical University, Xi'an 710021, Shaanxi, China
lixin0507@hotmail.com

Abstract. Basic education is the foundation of education and the national quality education. Although China has achieved the grand goal of basically popularizing compulsory education, due to the unbalanced economic and social development, the gap between urban and rural areas, regions and schools in basic education still exists, and there is even an expanding trend in some places and in some aspects. The main reasons for this gap are uneven resource distribution and inefficient allocation. The mixed optimization strategy of resource allocation in higher education system is the combination of static optimization strategy and dynamic optimization strategy. It can be used to determine the best combination of resources (human and financial) in any given situation. The hybrid optimization strategy is used to find the optimal solution to maximize the total value and minimize the cost. The hybrid method consists of two main parts: the static part, which determines the optimal allocation of different types of resources; And a dynamic component that determines how these allocations change over time as new information becomes available.

Keywords: System resource allocation · Higher education · Hybrid optimization

1 Introduction

The most obvious characteristic of the era of popularization of higher education is its large scale and many styles. Specifically, it shows that the society has invested a lot in education, and the supply capacity of higher education is strong. It shows a trend of diversified development facing the market in many aspects, such as school running mode, education management methods, the goal of cultivating talents, quality evaluation standards and so on. As China's higher education has been carrying out elite education for a long time, talent training and scientific research and development activities are carried out in a relatively closed environment on a small and medium scale [1]. The academic community overemphasizes the self accumulation and development of knowledge and the level of the educated, and does not pay attention to the economic benefits of running a school, which artificially limits the investment and development scale of higher education, resulting in a waste of higher education resources Idle scientific research achievements and structural waste of talents. At present, China's higher education is

J. C. Hung et al. (Eds.): IC 2023, LNEE 1045, pp. 280–286, 2023.
https://doi.org/10.1007/978-981-99-2287-1_40

still in the primary stage of the Popularization Era. It is restricted by the elite education model in many aspects, such as education fund investment, resource allocation and market expansion [2]. If higher education wants to break through these restrictions and smoothly implement popularization, it must be based on regional open development. After all, no matter from the realization conditions or ways of education popularization, there is an inevitable connection between higher education and regional society. On the one hand, regional economic conditions, educational foundation and social and cultural traditions will affect the supply of higher education, including educational funds, good and sufficient students, on the other hand, regional society will affect the demand for higher education services. Therefore, it is necessary to carry out in-depth research on the optimal allocation of regional higher education resources, so that higher education can go out of the narrow space for running schools, take the initiative to face the regional market, study the regional social needs from the perspective of industrial management, and guide the allocation of higher education resources with the laws of market supply and demand, competition and value, so as to improve the economic and social benefits of resource allocation, Promote the sustainable development of higher education in the era of popularization [3].

In today's society, for a country or region, higher education is no longer just a public utility with high social benefits, but a strategic industry with high economic benefits. Although the "production" process of higher education has the characteristics of indirectness, internality, concealment and lag of benefit acquisition, its development will promote the fundamental transformation of production in other sectors of society, Promote the adjustment and upgrading of regional industrial structure [4]. Therefore, it is necessary to deeply study the optimal allocation of regional higher education resources, improve the use efficiency of regional higher education resources, improve the supply capacity of regional higher education for science and technology and talents, and realize the sustainable development of regional higher education and economy.

2 Related Work

2.1 Basic Concepts of the Theory of Optimal Allocation of Higher Education Resources

(1) Higher education

According to the higher education law of the people's Republic of China adopted at the fourth meeting of the Standing Committee of the Ninth National People's Congress on august29,1998, higher education refers to the completion of the foundation of higher secondary education_ The education implemented in the course of education includes academic education and non academic education. Corresponding to the revised international standard classification of Education (ISCED) issued by UNESCO in October, 1997, higher education defined in China's higher education law includes the fourth, fifth and sixth levels of education in the standard. The fourth level of education is non higher post-secondary education, equivalent to post-secondary education; the fifth level is the first stage of higher education, equivalent to college and undergraduate education; The sixth level is the second stage of higher education, which is equivalent to master's or doctoral education [5]. The fourth and fifth levels of education are divided into class A and class B courses. Class a courses are for preparing for further education, and class B courses are for preparing for directly entering the labor market for employment. The scope of higher education studied in this paper is relatively small, focusing on the fifth and sixth levels of higher education.

(2) Higher education resources

Higher education resources are the basis for higher education activities and the guarantee for the development of higher education. Higher education resources refer to all kinds of material and spiritual materials that can enable higher education to survive and develop, including tangible and intangible resources. Tangible resources such as: certain campus area (land), classrooms, office space, libraries, laboratories and information equipment, a certain number of teachers and students, school running funds, etc.; Intangible resources such as the reputation of the University, academic achievements, excellent management system, etc. Conventionally, in order to facilitate research, people divide higher education resources into broad sense and narrow sense [6]. The broad sense of higher education resources refers to the sum of all tangible and intangible resources, such as people, money, materials and information, which are closely related to higher education activities; In a narrow sense, educational resources only refer to tangible resources such as university assets and educational funds. The research of this paper is limited to the category of educational resources in a narrow sense.

(3) Resource allocation

Resource allocation is the core issue of economic research. The so-called resource allocation is the study of how economic organizations rationally allocate and use scarce resources among various economic uses, so as to solve the old and eternal problems of "what to produce, how to produce and for whom". The problem of resource allocation originates from the scarcity of resources. The purpose of studying this problem is to solve the contradiction between the scarcity of resources and the infinity of demand. Higher education resources are as scarce as economic resources, so there is also the problem of resource allocation.

2.2   The Connotation of Optimal Allocation of Higher Education Resources

Generally speaking, the optimal allocation of resources refers to the maximum proportional relationship between input and output or between cost and income. It embodies the concept of efficiency in economics. The output or income here refers not to any goods, but to useful goods that can provide people with satisfaction. From the perspective of economics, the final output is people's satisfaction, that is, utility. The input or cost, in a general sense, is the economic resources required to produce certain products under certain scientific and technological conditions, including labor resources and material resources [7]. The optimal allocation of resources is reflected in two aspects: the first aspect is "resource use efficiency", which means how a production unit, a region or a department organizes and uses scarce resources to make them play the greatest role and obtain the most valuable output; The second aspect is "resource allocation efficiency", which refers to how to allocate limited resources among different production units, different regions and different industries, that is, how to effectively allocate each resource to the most appropriate use direction. The efficiency of these two aspects is the relationship between micro allocation and macro allocation, and the two affect each other.

On the whole, improper allocation of resources will reduce the efficiency of resource utilization of micro entities. On the contrary, if the efficiency of micro entities is high, it will increase the total supply of economic resources and create conditions for further rational allocation of resources.

The optimal allocation of higher education resources also involves these two aspects: at the macro level, the state or region makes the scarce higher education resources reasonably distributed among regions and universities through certain systems and operating mechanisms to maximize the efficiency of higher education resource allocation [8]; At the micro level, each university redistributes the higher education resources obtained, trains qualified talents to the greatest extent, provides good scientific research results, and maximizes the use efficiency of higher education resources.

3 Mixed Optimization Strategy for Resource Allocation of Higher Education System

3.1 Allocation Mode of Basic Education Resources

The allocation of educational resources is the distribution of educational resources at all levels and regions. It is a multi-level and multi regional distribution choice of educational resources to find higher efficiency and benefits and balanced development of education. The main actors are the central government and governments at all levels. Educational resource allocation mode refers to the way or means to form the state of educational resource allocation [9]. As for the internal structure of resource allocation mode, it includes four parts: allocation subject, allocation object, allocation motive force and allocation decision-making mode (as shown in Fig. 1). The so-called different allocation methods refer to different combinations of allocation subject, motive force and decision-making methods under the condition that the allocation object remains unchanged.

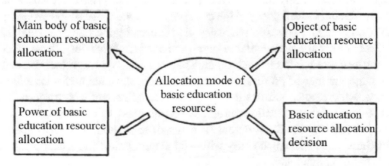

Fig. 1. Educational resource allocation system

The allocation mode of basic education resources determines the basic pattern of the allocation of basic education resources in a region, and then regulates the overall trend of the development of basic education in the region. Under the modern social and economic conditions, from the perspective of allocation mechanism, the allocation of educational resources can be divided into two basic resource allocation modes: planned allocation and market allocation. The practice of many countries shows that the effect of planned allocation or market allocation is not ideal. In the actual process of resource allocation, due to the failures and defects of the market and government plans, most of the resource allocation methods are mixed. The purpose is to achieve a reasonable state of resource allocation and achieve the goal of resource allocation. The government allocates educational resources through the educational budget mechanism, and guides social resources to invest in education, so as to promote educational development and achieve the social development goal of educational equity.

From an international perspective, in the stage of basic education, the allocation of educational resources is mainly controlled by the government power, and the market is a supplementary factor [10]. The practical way to combine the two is: the planning mechanism mainly formulates the long-term plan and strategy for the development of basic education from a macro perspective, determines the proportion and focus of investment, regulates the operation process of the school running body through specific administrative and legal means, and finally realizes the goal of basic education. The market mainly introduces the competition mechanism and efficiency mechanism from the micro resource allocation of basic education, so as to optimize the combination of various elements in basic education and make the most effective use of various resources, so as to realize the optimization of educational economic efficiency.

3.2 Optimize Configuration Structure

The research of industrial economics shows that economic organizations have the phenomenon of scope economy. Economies of scope refers to the savings caused by enterprises' reasonable production (operation) of several related products or services. Scope economy occurs because the production of several products within an enterprise can share common information, machinery, facilities, management, finance, marketing, etc. diversified operation greatly improves the utilization of enterprise resources, and is conducive to improving the efficiency of the enterprise. The theory of scope economy is also applicable to the higher education industry. As modern higher education shoulders the multiple tasks of teaching, scientific research and serving the society, these functions of higher education are related to each other. They are all knowledge-based services and require a large number of professional human resources, funds and scientific research facilities, which create conditions for the realization of economies of scope. It can be seen that regional higher education can achieve economies of scope and obtain structural benefits by optimizing the input-output structure of education. This is of great significance for China, a developing country with a large population and a lack of educational resources, to hold big education.

The input structure of regional higher education resource allocation mainly includes: ① resource variety structure, that is, the proportion of varieties within different types of higher education resources. Higher education resources can be generally divided into three types: human, financial and material resources. According to different functions, human resources can be subdivided into professors, associate professors, lecturers, teaching assistants, teaching assistants and administrative personnel. The proportional relationship between the varieties of resources within human resources directly affects the function and final output capacity of human resources in higher education. Similarly, financial resources can be refined according to the source and use direction of capital, and material resources can be classified according to different functions. The classification proportion of higher education resources in depth will directly affect the level, structure, form and total amount of education output. ② Resource combination structure, that is, the collocation relationship of different types of resources. A reasonable combination of human, financial and material resources can greatly save the total resource investment and improve the efficiency of resource allocation. ③ Resource distribution structure. That is, the distribution proportion of regional higher education resources in the region. Through the implementation of the balanced development strategy of regional higher education, actively and steadily develop "city universities" and take a rational and balanced app- roach to the possession, use and distribution of local sources of higher education, which can be divided into three states: starting point (condition) equity, opportunity equity and result equity.

4   Conclusion

With the establishment and improvement of the market economic system in China, the industrial economy under the planned economic system is gradually being replaced by the regional economy. The rise of the regional economy inevitably requires the acceler- ated development of regional higher education and the large-scale cultivation of high- quality regional characteristic talents who can promote the adjustment and upgrading of the regional industrial structure; In addition, with the promotion and deepening of the reform of the higher education management system, a new system has been basically formed in the field of higher education, with two levels of management, division of labor and responsibility between the central and provincial people's governments, focusing on the overall planning of the provincial people's governments and the organic combi- nation of sections and blocks. All these have created favorable conditions for promoting the development of regional higher education. However, China's higher education has implemented a centralized planning and management system for a long time. The gov- ernment has used administrative means to allocate higher education resources, artificially separating the relationship between colleges and universities and regional society. Col- leges and universities lack the power and channels to serve the local community, and the regional function of Higher Education cannot be properly reflected, which restricts the development of China's regional higher education theory and practice.

References

1. Sreethar, S., Nandhagopal, N., Karuppusamy, S.A., et al.: A group teaching optimization algorithm for priority-based resource allocation in wireless networks. Wireless Pers. Commun. **123**(3), 2449–2472 (2021)
2. Yao, J.: Research on optimization algorithm for resource allocation of heterogeneous car networking engineering cloud system based on big data. Math. Probl. Eng. **2022** (2022)
3. Hao, H., Xu, C., Yang, S., et al.: Multicast-aware optimization for resource allocation with edge computing and caching. J. Netw. Comput. Appl. **193**, 103195 (2021)

4. Dang, Q.L., Xu, W., Yuan, Y.F.: A dynamic resource allocation strategy with reinforcement learning for multimodal multi-objective optimization. Mach. Intell. Res. **19**(2), 138–152 (2022)
5. Bi, K., An, K., Li, X.: A resource optimization allocation strategy for China's shipbuilding industry green innovation system. Int. J. Innov. Technol. Manag. (2020)
6. Wang, H.N., Liu, J.Y., Zhang, J.M.: Research on joint optimization of edge computer offload strategy and wireless resource allocation. J. Phys: Conf. Ser. **1648**, 042079 (2020)
7. Maiti, M., Vukovic, D.B., Shams, R., et al.: Resource-based model for small innovative enterprises. Manag. Dec. (2020)
8. Hu, X., Xu, S., Wang, L., et al.: A Joint power and bandwidth allocation method based on deep reinforcement learning for V2V communications in 5G. (2021)
9. Mehrotra, S., Rahimian, H., Barah, M., et al.: A model of supply-chain decisions for resource sharing with an application to ventilator allocation to combat COVID-19. Naval Res. Log. (NRL) **67**(5) (2020)
10. Chidambaram, M., Shanmugam, R.: Optimization-based resource allocation for cloud computing environment (2021)

Mobile E-commerce Application Based on 5g Network

Zhijian Mai[✉]

College of Information Engineering, NanNing University, Nanning 530000, Guangxi Zhuang
Autonomous Region, China
maizhijian2005@163.com

Abstract. Mobile e-commerce application is the next generation of mobile commerce, which will be a major trend in the future. Mobile e-commerce applications can buy anything from anywhere at any time. It enables users to shop online anytime, anywhere. This is achieved by using smartphones as payment systems rather than credit cards or cash. The main advantage of this technology is that it can be implemented on all types of devices, such as laptops, tablets, smartphones, etc., which means that there are no restrictions on where people can use mobile phones to shop. The arrival of 5g era has introduced new vitality to the growth of China's mobile e-commerce, but at the same time, it has not improved the existing Internet security problems, especially the large-scale coverage of wireless WiFi, which leads to the leakage and illegal use of personal information. Its scientific and technological innovation and service diversification will make the majority of users face more serious hidden dangers of Internet and personal information security. Therefore, in the 5g network environment, the future opportunities and challenges of China's mobile commerce coexist, and the research on mobile commerce information security becomes particularly important.

Keywords: 5g network · Mobile electronic commerce

1 Introduction

With the rapid development of mobile communication technology, mobile terminals have been popularized in people's daily life. As a new business model, mobile e-commerce has also developed rapidly, becoming the primary element of economic model informatization and the trend of e-commerce research. Mobile e-commerce has laid the foundation for the rapid development of mobile e-commerce with its unique advantages of openness, interoperability, no time limit, convenience, rapid, personalized application, wide popularity and large-scale users [1]. At the same time, the popularity of 5g network, which has the characteristics of fast speed, strong timeliness and stable connection, promotes the rapid growth of mobile e-commerce, shows the convenience and introduces new impetus. With the popularization of 5g network technology and the growth of mobile e-commerce, intelligent terminals are more closely connected with our daily life, and are widely used in video connection, P voice, web browsing, search positioning, online

© The Author(s), under exclusive license to Springer Nature Singapore Pte Ltd. 2023
J. C. Hung et al. (Eds.): IC 2023, LNEE 1045, pp. 287–293, 2023.
https://doi.org/10.1007/978-981-99-2287-1_41

games, social platforms and mobile payment. However, there are security risks such as information leakage, illegal theft and loss in mobile intelligent terminals and internal networks, The information security of users has been greatly threatened [2]. As an important communication tool, the security of intelligent terminals has become a great challenge.

At present, the operation of mobile e-commerce on the network is far from enough, and the laws on the network are not perfect, which restricts the development of mobile e-commerce in the network. In such an environment, the introduction of safety standards and relevant laws will also become a trend. As a social public product, information security can only be provided by the government. Therefore, in the era of mobile Internet, the government's supervision of mobile information has also become an important part of ensuring social information security [3].

Therefore, it is very important and urgent to explore the current situation of mobile commerce information security at home and abroad from the perspective of 5g network promoting the development of mobile e-commerce, and to explore how to use 5g network technology to promote and lead the rapid development of mobile e-commerce and the countermeasures to solve security problems.

2   Related Work

2.1   Concept of Mobile Commerce

In a literal sense, mobile commerce is simply "mobile + business". It can be said to be business in mobile or mobile in business. Mobile is a means and business is a purpose.

From the perspective of users, individual consumers buy some entertainment information content, including pictures, ringtones, games, event results, etc. at present, mobile commerce is mainly business activities at the level of entertainment or mass messaging. People can use mobile phones and other mobile communication devices to surf the Internet anytime and anywhere, query information, buy products, and Book services, which is convenient, fast, and time-saving. Mobile commerce refers to the use of mobile terminals to achieve various activities, including operation, management, trading, entertainment, etc., by connecting public and private networks [4]. According to the type of end users, mobile commerce is divided into enterprise mobile commerce and personal mobile commerce. Mobile commerce is to give consumers more convenient business experience. For enterprise users, mobile commerce can provide them with fast and convenient information services, which are applied to internal office, external services, information release and targeted publicity [5].

From the perspective of technology, mobile commerce is not only a technological innovation, but also an innovation of enterprise management mode. Mobile communication devices such as mobile phones, pagers, personal digital assistants (PDAs) and laptops are connected with the background of enterprises, and online business activities are carried out through wireless communication technology, so that the mobile communication network and the Internet are organically combined, breaking through the limitations of the Internet, carrying out information interaction more efficiently and directly, expanding the field of e-commerce, saving human costs, and enabling enterprises to grasp market dynamics and trends in time. Mobile commerce makes full use of its mobility, eliminates the restrictions of time and region, provides convenience for e-commerce activities, and makes information transmission and commercial transactions possible anytime and anywhere. Therefore, mobile commerce is an innovative business model that uses all kinds of mobile devices and mobile communication technology to store, transmit and exchange all kinds of business information anytime and anywhere, and carry out business activities [6].

Mobile commerce essentially belongs to the category of e-commerce and information commerce. It is a new business model arising from technological development and market changes. Due to the relevance between mobile commerce and telecommunications services, it is different from wireless commerce in many aspects, such as business model, business revenue point and so on. With the continuous popularization and development of mobile communication, mobile commerce will become a new field of e-commerce growth and a new industry of wealth creation movement in China in the next five years.

2.2 Characteristics of Mobile Commerce

Mobile commerce is the extension and development of traditional Internet e-commerce in the mobile field. Although mobile commerce and e-commerce are very similar in many characteristics, it is one-sided to simply regard mobile commerce as an extension of e-commerce, because their service objects, service methods (including terminal equipment) and technical characteristics are very different. The most important one is the distinction between business characteristics, which is mainly reflected in the following four points:

(1) "Mobility" of service objects. People who need mobile commerce services are generally mobile. It is one-sided to only understand mobile commerce as mobile e-commerce, because mobile is not only mobile terminals, but also the movement of people and services.
(2) "Immediacy" of service requirements. Mobile commerce customers generally require to get the required information immediately.
(3) The "privacy" of the service terminal. Because mobile terminals are generally for personal use and will not be public, which brings unique advantages to mobile commerce, the development of business combined with private identity authentication is a promising direction.
(4) "Convenience" of service mode. Due to the limitations of mobile terminals, especially mobile phone buttons, mobile commerce services require simple operation and short response time.

3 5g Network and Mobile E-commerce

3.1 5g Network And Its Key Technologies

The development of mobile commerce is based on a complete mobile network. At present, China's mobile network is in the period of full popularization from 3G to 5g. The guarantee demand of 5g network has become an indispensable key to the information security of mobile commerce.

5g network integrates the advantages of 3G and WLAN, greatly improving the speed and quality of information, voice, video and image transmission. The access systems of various services in mobile commerce are different. The IP technology in the 5g core network system can enable users to seamlessly roam between 3G, 5g, WLAN and fixed networks[7]. The average download rate of 5g can reach more than 100Mbps, which can almost guarantee the network demand of all wireless business services. Obviously, 5g has incomparable advantages. 5g technology shows a breakthrough and rapid increase in application traffic due to its complete performance, widely used intelligent terminals, and customers' increased demand for the network.

There are four key points in 5g network:

(1) Orthogonal frequency division multiplexing technology first uses serial/parallel transformation to transform the high-speed serial data to be transmitted into low-speed data parallel on multiple subchannels, and then uses a large number of inter-working beams to coordinate and transmit them together [8]. At the receiving end, the coherent carrier is used for coherent reception, and then restored to the parallel / serial transformation of the original high-speed data.
(2) Applied radio technology is a technology that promotes the opening of various wireless communication system structures. It uses application loading to apply template and partitioned hardware functional components to ordinary hardware systems. Use software to define wireless functions as much as possible.
(3) Smart antenna is a kind of multi wave antenna with self-adaptive array or no switch-ing. Smart antenna is the key application of future mobile communication, which has the characteristics of suppressing signal disturbance, automatic tracking and digital beam modulation.
(4) Multiple input and multiple output technology enables the base station transmit-ter and mobile receiver to have multiple antennas, so that they can have multiple subchannels in the same frequency band to transmit signals and achieve diversity gain.

3.2 Security Issues of Mobile Commerce

The rapid growth of mobile technology network interconnection technology benefits from the popularity of mobile terminals. Mobile commerce application technology has changed from a single consumption mode to a business experience mode. However, there are still some problems in China's mobile commerce, such as potential safety hazards of mobile terminal equipment, information leakage, imperfect credit system, imperfect policies and regulations, and so on.

According to the current development form, mobile e-commerce security includes network security and mobile commerce security. Network security aims at these cur-rent problems that may appear in the computer network, such as the threat of Internet facilities, hackers' illegal access to personal information in cracking the system, the wide-ranging expansion of the virus, the inability to restore and back up information, and the unreasonable formulation of the authority of the management staff of the internal system of the network [9]. In order to ensure the security of network communication, these problems need to be solved. Various situations of mobile commerce information security are prone to appear in the process of online payment transactions on the Internet, such as the theft of personal information of customer banks, the counterfeiting of network evidence of mobile e-commerce, the possible malicious default in the payment process, and so on. In order to ensure the smooth progress of trade, we should strengthen the security of Internet credit as a condition to ensure the efficient operation of the security system.

The composition of mobile e-commerce security system is based on the composition of network security system. Because it has the characteristics of various security threats and the characteristics of the target system and scheme, the security model shown in Fig. 1 below shows and summarizes the overall management method of network information security:

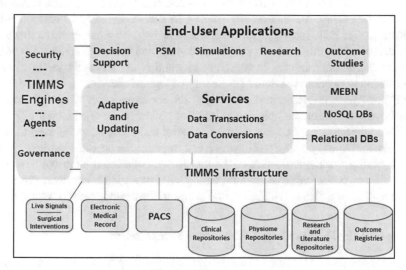

Fig. 1. Safety system

4 Application Research of E-commerce Logistics Management Based on 5g Network

4.1 Intelligent Application of Scheme Formulation

In the past, in the field of logistics management, in the process of formulating the logistics plan, a series of parameters that need to be entered are usually manually set and operated, which leads to the fact that there is no fixed parameter setting scheme in the process of formulating the logistics plan, and a series of human factors play a great guiding role. In this way, the formulation of the logistics plan will inevitably be due to the difference of the operator's personal knowledge level, ability Differences in work experience, operating habits, etc., lead to a series of problems such as the laxity and inconsistency of logistics scheme setting, which may even lead to the development of ten different logistics schemes for ten people [10]. Therefore, at present, there is a lack of global control in the formulation of logistics plans in China, which eventually leads to the emergence of problems such as time-consuming, labor-consuming, unreasonable order allocation, unreasonable choice of transportation mode, untimely information feedback, improper handling of unexpected situations, etc. in the formulation process of logistics plans, resulting in the inability to formulate a "optimal solution" logistics plan, but if 5g network technology is supported and applied, It can greatly reduce the occurrence and occurrence of the above situations, so as to achieve twice the result with half the effort.

4.2 Application of 5g Network in Improving Logistics Service Quality

Consumers are the main force of all business activities. Only by meeting the needs of consumers can we maximize the scale and economic benefits of enterprises. E-commerce logistics service is also one of the categories of the service industry. Compared with the traditional service industry, the logistics service industry pays more attention to the effect of service, constantly improves its own service level, and strives to achieve the competitive point that others cannot achieve, so as to meet the personalized shopping needs of consumers. 5g network in the field of e-commerce logistics services is reflected in using a more "intelligent, convenient and efficient" service experience to meet the different needs of consumers. For example, in the logistics picking behavior in the logistics link, ai artificial intelligence robots set with 5g network technology can be used to replace the traditional manual picking work, which can improve the accuracy and efficiency of logistics picking work, greatly shorten the logistics distribution time investment, reduce the use of labor costs, improve the automation level of logistics services, promote the improvement of logistics service level, and promote the development of enterprises.

4.3 Improve the De Stocking of Warehouse Management

As an important part of logistics management, warehousing management plays a great role in the development of enterprises. However, in the past, when carrying out logistics management, enterprises could not make accurate market forecasts, and the circulation links continued to increase, making the warehousing error rising, making it difficult to ensure the consistency of inventory and demand, which seriously affected the effective implementation of logistics management and caused great losses. According to the previous situation of logistics management, half of the cost of logistics management is used in warehousing management, so reducing inventory is an important way to improve the market competitiveness of enterprises.

5   Conclusion

Based on the above in-depth analysis and comprehensive research on the impact of 5g technology on logistics and supply chain management, we can understand more clearly that each reform of logistics and supply can make further breakthroughs in related technologies and further enhance the industrial level. At present, China's logistics industry is in an important period of transition between traditional mode and modern mode, especially in the environment of increasingly perfect supply chain. Through the effective application of 5g technology, logistics and supply chain have become the easiest and best application scenario of industrial Internet of things.

Acknowledgement. Research on mobile E-commerce based on 5G network. Project leader: Zhijian Mai. Supported by scientific foundation of Nanning University (Grant No. 2019JSGC02).

References

1. Zhu, Z., Bai, Y., Dai, W., et al.: Quality of e-commerce agricultural products and the safety of the ecological environment of the origin based on 5G Internet of Things technology. Environ. Technol. Innov. **22**(2), 101462 (2021)
2. Zheng, L., Liu, S.: Research on the strategy of mobile short video in product sales based on 5g network and embedded system. Microprocess. Microsyst. **82**, 103831 (2021)
3. Yao, W., Zhang, C., Deng, G., et al.: Research on urban electric vehicle public charging network based on 5G and big data. J. Phys. Conf. Ser. **2066**(1), 012045 (2021)-
4. Ge, Y.L.: Research on the influence of mobile e-commerce webcast on users' online shopping intentions. In: 2021 The 6th International Conference on E-business and Mobile Commerce (ICEMC 2021), 27–29 May 2021 (2021)
5. Zhao, J., Chu, S.: Research on flower image classification algorithm based on convolutional neural network. J. Phys: Conf. Ser. **1994**(1), 012034 (2021)
6. Yang, Y., Sha, C., Su, W., et al.: Research on online destination image of zhenjiang section of the grand canal based on network content analysis. Sustainability **14**(5), 27312022 (2022)
7. Huang, Y., Chen, L., Hong, J., et al.: Research on 5G communication slicing accessoptimization with constraints on communication delay of substation network protection. In: Society of Photo-Optical Instrumentation Engineers (SPIE) Conference Series. SPIE (2021)
8. Lv, P.: Research on the application of adaptive matching tracking algorithm fused with neural network in the development of E-Government. Math. Probl. Eng. **2022** (2022)
9. Hassan, A., Umair, M.: Recommendation system based on neural network models to improve efficiencies in interacting with E-Commerce PlatforMS.US 20210166104A1 (2021)
10. Wang, Y.: Research on E-commerce big data classification and mining algorithm based on bp neural network technology (2021)

Multi target Tracking Technology of Athlete's Physical Fitness Video in Football Match Based on KCF Algorithm

Yu Tianbo[✉], Man Xiaoni, Chen Xin, and Wang Yue

Sports Department, Shenyang Jianzhu University, Shenyang 110168, China
vivianjeccey@126.com

Abstract. The purpose of this research is to design and develop a multi - target tracking system for football players. The main goal of the project is to track the physical fitness level, movement and performance of football players in real time during the game. This research will develop three different parameters on the KCF algorithm: 1) number of targets; 2) Speed; 3) The size (distance) of the target. This will help coaches, coaches and athletes to evaluate their performance in the competition by analyzing their sports patterns. This method can be used as an effective tool to evaluate players' performance in real time by detecting and tracking different angles of motion. This method will be tested by using an experimental device consisting of two cameras placed in different positions in front of each player. The video clips captured by these cameras will be analyzed through a computer software application that can detect location, movement and physical condition.

Keywords: KCF algorithm · Multi target tracking technology · football match · Athlete fitness video

1 Introduction

With the development of the sports industry and the arrival of the new media era, the analysis technology based on artificial intelligence has gradually been deployed to sports data analysis to help the coach team carry out data analysis, technical and tactical development, auxiliary player training and injury assessment. AI technology is pushing major sports events to a new height in a more scientific and efficient way, and building a new platform for human beings to pursue outstanding sports performance. Therefore, the development and deployment of intelligent analysis and management solutions for sports video data has become an inevitable trend [1].

Multi target tracking is an important and widely used field in target tracking technology. Multi target tracking technology was first applied in the military field, which is a very complex problem. In multi-target tracking, a group of time varying observation data received from the target or other sources (clutter and false alarm) is provided to the filter to estimate the state of a group of targets. In a cluttered environment, tracking requires the use of observation models to identify the correct target from the measurement data, and track the target according to the state model [2]. It includes many aspects, such as track preprocessing, track initiation, data association, track maintenance, track elimination, performance evaluation, etc. Among them, data association algorithm is the core problem of multi-target tracking. Because the uncertainty of the target observation process and tracking environment increases the fuzziness of the corresponding relationship between the measurement information and the target source, the data association problem directly affects the performance of target tracking [3].

2 Related Work

2.1 Research Status of Football Video Analysis

Users who need intelligent analysis of football video can be divided into two categories: audience groups and professionals. Audience groups include TV users, mobile device users and network users. With the development of communication technology and the Internet, audience users on multiple media platforms have fast communication conditions. However, audience groups may not be able to spend a lot of time watching the direct broadcast of the game. In order to save time and quickly obtain event information, they prefer to have a targeted understanding of the game process [4]. Their needs include real-time game video summary and detection of specific events of interest; Professional users include athletes, coaches and commentators. How to extract the effective information of the team and athletes in order to make training plans efficiently and conveniently, evaluate the performance of players or analyze competition strategies is their concern. For these users, the content they are interested in is related applications of target detection and tracking, such as player trajectory extraction and team data analysis based on this.

Since 2000, football sports video analysis based on visual information has attracted much attention, and many researches have been carried out in the field. According to T D'Orazio concluded that the content of intelligent analysis of football video can be represented by the hierarchical structure as shown in Fig. 1. From different semantic levels, soccer video intelligent analysis tasks are classified into three categories: supplementary information, video summary and high-level analysis [5] (Fig. 1).

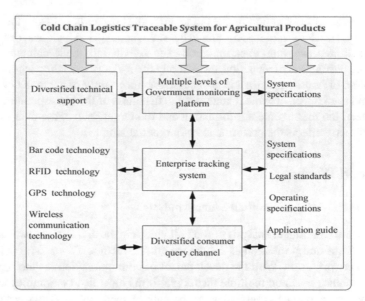

Fig. 1. Research content of intelligent analysis of football video

2.2   Research Status of Multi-target Tracking

The statistics of players' running data in football video belongs to the multi-target tracking task in dynamic video scenes. Its difficulties mainly include occlusion between individuals, complex tracking and matching logic, and high real-time requirements of the algorithm. Therefore, the multi-target tracking task in football video has high requirements on the robustness and operational efficiency of the algorithm [6].

Multiple Object Tracking is an important computer vision task, which has received much attention in both academic and commercial fields. Multi target tracking is different from single target tracking. In this tracking mode, the algorithm needs to match the existing target track according to the rectangular box detected by the target in each image frame. For new targets, new tracking sequence trajectories are generated; For a target that has left the field of view of the screen, it is necessary to stop tracking the track. In this process, the matching calculation process of target and detection can be regarded as object re identification. For example, when tracking multiple pedestrians, the pedestrian image set of the existing track is regarded as a gallery, and the newly entered target detection result is regarded as an image query entity [7]. The association and matching operation of target detection results and tracks can be regarded as the process of retrieving and querying images from the image library.

Due to the progress of target detection technology, the popular multi-target tracking framework is mainly based on the tracking by detection. In the detection based multi-target tracking problem, the algorithm needs to correlate the existing target tracks according to the detection results in each frame of image. It can be divided into online, near online 30 and offline offline methods. The online method can only use the current and previous detection results of the video during association matching; The offline method can use all the detection results of the video, while the near online method can use the detection information of the current frame and the following frames [8]. In order to meet the real-time requirements of practical applications, online multi-target tracking algorithm is of great concern. For example, Deep SORT, POI, CNNMTT LSST algorithms have achieved excellent results in MOT Challenge dataset.

Recently, the method of deep learning has made some progress in the field of target tracking. In the field of multi-target tracking, it is a trend to adopt generative network model and deep reinforcement learning, which can effectively enhance the adaptability of the tracking scene and improve the performance of the tracking algorithm. However, considering the interaction between objects and the complexity of tracking scenarios, many multi-target tracking problems are limited to the learning of matching metrics, and the practical application is far from sufficient research.

3 Analysis of Multi-target Tracking Algorithm

In the soccer video intelligent analysis system, this paper uses the multi-target tracking technology to solve the task of players' mobile data statistics. Multi target tracking (MOT) is an important research task in video analysis system, such as video pedestrian monitoring, athlete analysis, automatic driving, etc. Due to the progress made in the field of target detection, the recent multi-target tracking framework is mainly implemented based on tracking by detection. For the tracking by detection based multi-target tracking problem, the main problem of the algorithm is to match and correlate the existing target tracks according to the target detection results in the video image [9].

From data matching mode Seen from the above, multi-target tracking can be divided into online, near online and offline matching modes. The online method can only process current and previous information during association matching; The offline method can process all the information, while the approximately online method can process the information of the current frame and the following frames.

For multi-target tracking tasks in pedestrian monitoring, sports video analysis and other application scenarios, the real time performance of the tracking algorithm is an important indicator. Obviously, online multi-target tracking methods can better adapt to this scenario. The videos in these scenes have one common feature: the camera is fixed. We think that tracking tasks in fixed view video and mobile view video focus on different issues. The core challenge of MOT in mobile perspective video is to re recognize the target after long-term occlusion, and pay more attention to the apparent characteristics of the target. The core challenge of fixed perspective MOT is to focus on accurate matching among dense targets, and pay more attention to target motion characteristics. In this paper, we mainly study online multi-target tracking in fixed view video.

The classical online multi-target tracking algorithms SORTI and DeepSORT have their own advantages and disadvantages. DeepSORT has better performance but slower speed than SORT due to the use of apparent feature information; Although SORT has speed advantages, its performance is relatively poor. On this basis, this study proposes improvements. Inspired by the method mainly focusing on target motion features in SORT, the cosine distance measurement method of depth apparent feature in Deep-SORT is replaced by the square Mahalanobis distance measurement method of target rectangle [10]. At the same time, the initialization filtering mechanism is added and the parameters are fine tuned. After modification, the tracking algorithm achieves a balance between performance and efficiency. In the case of using high-precision SDP detection, the best MOTA tracking performance score among the three can be obtained in the MOT Challenge dataset. In addition, this method is more prominent in fixed shot video. An example trace is shown in Fig. 2.

Fig. 2. Research content of intelligent analysis of football video

4   Multi Target Tracking Technology of Athletes' Physical Fitness Video in Football Match Based on KCF Algorithm

The original intention of this paper is to provide intelligent analysis of football video and targeted data information for users (audience groups, professionals) with demand. According to the research status in the field of football video analysis, this paper selects two application scenarios: the statistics of football players' running distance and the query of football semantic event information. Therefore, the soccer video intelligent analysis system designed in this paper should include the above two applications. According to different functions, the business process of this system is shown in Fig. 3.

For the application of a player's running distance statistics system, due to the problem that users need to participate in the selection of key points in the calculation process of target tracking and running distance statistics, an interactive interface design is proposed. With the user's operation, the player's running track is gradually mapped to the field model through player detection, multi-target tracking and field registration, so as to analyze and calculate the running distance, and finally realize the data feedback function of the player's running distance and running track.

For the application of the second soccer semantic event information query system, the C/S architecture is proposed to realize the connection between the database and

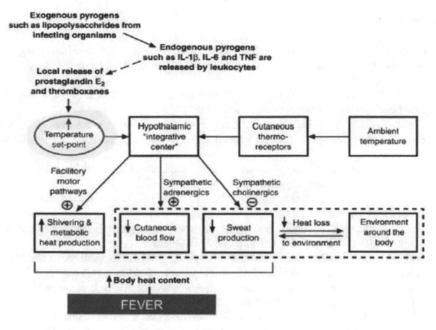

Fig. 3. Business flow chart of football video intelligent analysis system

the query interface. The query function is realized by the user inputting the statement information to be queried and feeding back the results and details of semantic event identification. The analysis results of semantic events are obtained from the analysis of the group behavior recognition algorithm designed in the fourth chapter of this paper and stored in the system database. On the basis of realizing the query function, the interface operation functions of browsing, adding and deleting database information are added.

5 Conclusion

With the increasing popularity of sports and the development of technology in the field of computer vision, it is possible to use intelligent analysis of football video to bring better experience to users. With the further maturity of the research results, it can be deployed in the future in the intelligent management system of video data, applied to the intelligent data analysis of team sports sports video data, and automatically generate wonderful clips of the game, providing technical support for real-time tactical analysis, post game technical and tactical guidance, intelligent play, etc. In the future, we can further combine natural language processing, recommendation system and other technologies to achieve automatic interpretation, sports video retrieval and personalized recommendation of sports videos. The research results of the project can also be applied to other types of video analysis tasks, such as video monitoring, video classification, etc., with a broad application prospect.

References

1. Qiwen, Z.: Research on athlete training behavior based on improved support vector algorithm and target image detection (Retraction of Vol 39, Pg 5725, 2020). J. Intell. Fuzzy Syst. Appl. Eng. Technol. **5**, 41 (2021)
2. Zhu, T., Zhang, C., Wu, T., et al.: Research on a real-time driver fatigue detection algorithm based on facial video sequences. Appl. Sci. **14**, 2224 (2022)
3. Zhu, L.Y., Deng, Y.J., Wei, D., et al.: Research on the uniqueness and algorithm of multi-parameter identification in coupling thermal acoustic and solid problem. Sci. Sin. Technol. **521** (2021)
4. Luo, C., Cao, S., Xu, C., et al.: Research on statistical algorithm of multi-region medicine bags based on refinement algorithm (2021)
5. Zhou, S.: Research on local topology tracking of power grid based on graph theory. Secur. Commun. Netw. **2021** (2021)
6. Wang, M., Zhang, Y., Zhang, D., et al.: Research on valley current prediction control algorithm based on buck converter. J. Phys.: Conf. Ser. **2354** (2022)
7. Wei, L.M., Li, K.K.: Research on the maximum power point tracking method of photovoltaic based on Newton interpolation-assisted particle swarm algorithm. Clean Energy **3**, 3 (2022)
8. Yuan, G., Lv, F., Xu, Z., et al.: Research on face tracking algorithm based on detection and supervision tracking. J. Phys.: Conf. Ser. **2209,** 012028 (2022)
9. El, A., Prl, A., Yong, P.A., et al.: Research on optimization operation technology of QT oil pipeline based on the Heuristic algorithm. Energy Rep. **8**, 10134–10143 (2022)
10. Wei, K.: Simulation research of nodes target tracking algorithm based on clustering in wireless sensor network. Int. J. Internet Protoc. Technol. **2**, 14 (2021)

Multi-objective Cost Optimization of Highway Engineering Based on Ant Colony Algorithm

Yanpeng Zhang[✉]

Linfen City Comprehensive Transportation Administration Had, Linfen 041000, Shanxi, China
Zyp7077@163.com

Abstract. In the process of traditional highway engineering project management, engineering cost, engineering progress and engineering quality are recognized as the three main control objectives of engineering construction, which is also the main research topic in the field of highway engineering. However, with the continuous improvement of engineering management system and requirements, the focus in this field is also expanding. Multi-objective optimization of highway engineering is a new research field. Ant colony optimization (ACO) algorithm has been widely used in the field of industrial production, but it has not been applied to road construction. This paper presents a multi-objective cost optimization method based on ant colony algorithm, and compares its performance with other methods. This paper also discusses the application scenario of ant colony algorithm in highway engineering.

Keywords: Highway engineering · Ant colony Multi objective cost optimization

1 Introduction

Since the beginning of the new century, our country has incorporated infrastructure construction into the important national economic strategic deployment and planning, especially the highway engineering, which has invested huge funds in related construction. The purpose is to strive to "connect villages and villages" and make people's lives more convenient. Although highway engineering construction is changing with each passing day and the general environment for highway construction is good, there are still many uncertainties from the perspective of construction enterprises [1]. Industry competition urges all construction enterprises to try their best to reduce the project cost to obtain the space for survival. However, blindly reducing the cost will inevitably affect the cost expenditure in all aspects of the project. Such an approach has a huge impact on a complete and reliable project [2]. Therefore, how to better carry out cost management so that enterprises can take into account the value of various aspects of the project when completing the project is an inevitable problem for the development of today's construction enterprises.

As far as the current development trend of Highway Engineering in our country is concerned, the main idea of enterprise cost management is still how to reduce costs and improve benefits. Other aspects of the project such as quality, progress, safety,

J. C. Hung et al. (Eds.): IC 2023, LNEE 1045, pp. 301–307, 2023.
https://doi.org/10.1007/978-981-99-2287-1_43

environment and so on are not considered enough. If the enterprise still gains greater benefits at the expense of project quality, ecological environment and safety protection measures, the development of the enterprise will inevitably be hindered [3]. How to use scientific means to optimize the quality, safety, progress, environment and other factors, so that they can be more coordinated and orderly, so that the cost of the enterprise can be reduced, and the benefits and project quality can be improved, which is a new way for the future development of the enterprise.

Most of the traditional cost management only makes overall improvement from the aspects of management ideas, technical methods and so on, without too much consideration of the influencing factors of other objectives. In this paper, the quality, progress, safety, environment and other objective factors faced by highway engineering are comprehensively considered in the overall cost control of highway engineering by reading the previous relevant literature [4]. It provides a basic method for multi-objective modeling of complex engineering construction in the future. In this paper, the ant colony algorithm, which is a relatively novel intelligent algorithm, is used to solve the multi-objective optimization problem, and the improved algorithm is tested to make the processing results more accurate. The application of intelligent algorithm in engineering is very meaningful for the intelligent development of engineering.

2 Related Work

2.1 Highway Engineering Cost Management

Today's highway engineering projects have strict requirements, complex engineering, and large construction machinery are emerging in endlessly. In order to enable enterprises to pursue interests while meeting other main control objectives, enterprises must have a strict process and characteristics for cost management:

(1) Sound organizational structure

An enterprise must have a complete organizational structure to control the project as a whole. In particular, highway engineering, a project with many factors, needs a sound and reasonable organization system of the project department and the enterprise headquarters [5]. In view of the unidirectional and non replicable characteristics of the construction project, the organization must be simplified and improved in the principle of cost saving and efficient office. Each post has a full-time person in charge, and there will be no "kicking the ball" in case of problems.

(2) Strict operation procedures

At the construction project site, strict operation procedures must be implemented for each goal of each process. Problems in each procedure should be specific to the corresponding responsible person. The formulation and operation of the procedures must be approved by the project responsible person at a meeting. For each sub goal, whether it is quality, progress, safety or environmental protection, the relevant person in charge should formulate a clear implementation process to ensure that each process of each goal can run stably and continuously [6].

(3)   Clear objectives and responsibilities

Since it is a sound system, the overall goal is directly connected with each sub goal. Each sub goal is specifically assigned to the project task to the individual workers. A meeting is held every day to determine today's work tasks and key points. Do not blindly carry out invalid work. Relevant training shall be conducted for each full-time person in charge of the task to ensure that the key points of the task are explained in place, so that every employee has a sense of responsibility. No carelessness is allowed in the project, ranging from property loss to life and health impact.

(4)   Standardized cost accounting and assessment

Regular cost accounting is an important link to control the target cost. The progress cost, quality cost, safety cost and environmental protection cost in this paper are the key points of regular accounting. Only through cost accounting can we find out whether the cost exceeds the standard, where it exceeds the standard, and then make rectification or change the construction scheme according to the problem.

2.2   Multi-objective Optimization Method

Multi-objective optimization originated from mathematical programming. After years of development, multi-objective optimization has been applied to many aspects, such as economy, management, military and so on. The application of multi-objective also promotes these aspects to get a good development. The application of multi-objective programming method in this paper is mainly to optimize the cost management of highway engineering. Up to now, multi-objective solution methods can be divided into two categories: one is the traditional solution method. In the final analysis, this solution method is still single objective solution. Its principle is to analyze the preference degree of multiple objectives, then attach different weights to each objective, and finally convert it into single objective solution to obtain the preference solution [7]. This method can not coordinate the relationship between various objectives well. The other is called production method. The application basis of this method is the cognitive experience that has no preference for multiple targets. The algorithm production method is used to solve multiple targets, and finally obtain multiple sets of solutions that meet the requirements.

There are many kinds of methods commonly used in the construction of evaluation function for multi-objective function, such as minimax, linear weighting, ideal point, square weighting, etc. [8]. This paper will not describe each method here, but simply explain one of the linear weighting methods:

The main idea of this method is to give an importance degree of the decision-maker's experience cognition for each function according to the importance degree of multiple objective function models, which is expressed by the weight coefficient. Each objective function is multiplied by its corresponding weight coefficient to redistribute the value of each objective function. Finally, each processed objective function is added to form a new single objective function, The optimal solution set is obtained by solving the single objective function. This method is also obvious for the subjective nature of decision makers.

$$\text{Of which}: d\left(x_{il}, x_{jl}\right)=\begin{cases} 0, x_{il} \neq x_{jl} \\ 1, x_{il} \neq x_{jl} \end{cases} \tag{1}$$

3 Ant Colony Algorithm

3.1 Overview of Ant Colony Algorithm

The ant colony searches for food by releasing pheromones on the ground it passes through. The route with short path accumulates more pheromones, and the ant will give priority to select, so as to find the optimal path.

Ant colony can not only find the optimal path, but also adapt to the external environment. If an obstacle suddenly appears on the optimal path, the path cannot continue to pass, and the ant will search for a new optimal path. At the beginning, ants randomly choose a route and leave a certain amount of pheromones to help later ants make choices. When later ants come to obstacles, they will choose the path with high pheromone concentration [9]. After a period of time, through the positive feedback mechanism of the ant colony, the ant colony has found a new optimal path.

According to this characteristic of ants, previous researchers have developed ant colony algorithm. The process of ants searching for food source is shown in Fig. 1:

Fig. 1. Principle of ant algorithm

3.2 Basic Principle of Ant Colony Algorithm

In order to facilitate understanding, the traveling salesman problem (TSP) is used to simply explain the algorithm model.

Suppose m ants randomly appear in n cities, the distance between city I and city J is D, and the amount of information on the path of city I and city J at time t is expressed by R (T). At the beginning, the amount of information on all paths is the same. Let R, (0) = c (C is a normal number), and ant K (K = 1,2... M) selects the direction to go and starts moving according to the amount of information on each path. The city set that ant K passes through is written in tabu (k = 1,2..., m), and the tabu table changes with the change of the algorithm. P (T) is the transition probability of ant K moving from city I to city J at time t:

$$
p_{ij}^k(t) = \begin{cases} \dfrac{\tau_{ij}^\alpha(t)\eta_{ij}^\beta}{\sum\limits_{j\in N_j^k} \tau_{ij}^\alpha(t)\eta_{ij}^\beta} \\[2ex] 0 \end{cases} \tag{2}
$$

4 Research on Multi-objective Cost Optimization of Highway Engineering Based on Ant Colony Algorithm

4.1 Cost Quality Relationship Analysis

In the cost input of highway engineering, quality should be considered first. If the input cost is too small, the construction party will certainly save the cost in other ways to ensure its own profits. For example, the cost will be reduced by cutting corners on work and materials. The result is that there are problems in the quality of highway engineering. Of course, this is only superficial. The most important thing is that it will cause other serious consequences. Conversely, the process with quality problems is bound to be repaired, In turn, it increases the expenditure of quality cost. If the input cost is too much, it will inevitably lead to excess quality. For enterprises, it is a loss of interest [10]. Therefore, only a more reasonable cost input can save the cost of enterprises on the premise of ensuring the project quality.

In this paper, the concept of quality cost is used for reference. Quality cost mainly refers to the cost consumed when the products meet the specified requirements. The cost of quality can be divided into two categories: the first category is called preventive identification cost. Including quality management training fee, quality improvement measures fee, quality level improvement reward fee, testing and identification management fee, etc. Category II loss cost. It includes the repair of nonconforming products, quality accident losses, quality accident handling fees, quality claims, quality litigation fees, etc.

4.2 Multi Objective Cost Dynamic Management System

The future trend of highway engineering construction projects must be to gain social recognition in a dynamic cost control. The dynamic control is mainly based on perfect rules and regulations. First, each target cost model is established respectively, and the cost optimization analysis is carried out to obtain the optimization cost results. The optimization cost results are taken as the planned values, and the progress cost, quality cost, safety cost and environmental cost are taken as the target planned values, Then use this target value to guide the construction process and prepare the construction plan. The construction project department will carry out construction in strict accordance with the construction plan, and check the actual cost of construction at regular intervals.

If there is a corresponding deviation, carry out specific deviation analysis, carry out targeted construction adjustment, and carry out model optimization again according to the specific situation to form the target value and cycle the review steps. If there is no deviation in the construction of this process, continue to control the cost of the next process. The specific procedure is shown in Fig. 2:

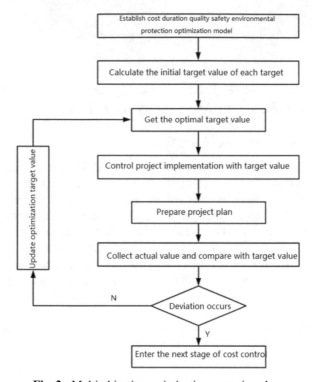

Fig. 2. Multi objective optimization execution chart

5   Conclusion

Based on the traditional duration cost model, this paper analyzes the actual highway project duration, and establishes a quadratic function duration cost model with reward and punishment coefficient; The quality cost of highway engineering is classified. Based on the assumption of Dr. Julan's quality cost model, a quality cost model about the process quality level coefficient is established. Based on the analysis of previous studies on quality duration model, the whole project duration is split and analyzed, and the relationship model between process duration and process quality level is obtained. According to the quality level coefficient of the process, a quality objective model between quality cost and construction period is established.

References

1. Zhang, X., Gong, J., He, J., et al.: Research on optimization of the container exchange mode of cross-border highway transportation between China and Vietnam. In: RSVT 2020: 2020 2nd International Conference on Robotics Systems and Vehicle Technology (2020)
2. Hosseini, R., Rashidi, M., Bulaji, B., et al.: Multi-objective optimization of three different SMA-LRBs for seismic protection of a benchmark highway bridge against real and synthetic ground motions. Appl. Sci. **10**(4076) (2020)
3. Ghanizadeh, A.R., Heidarabadizadeh, N., Mahmoodabadi, M.J.: Effect of objective function on the optimization of highway vertical alignment by means of metaheuristic algorithms. Civil Eng. Infrastruct. J. **53**, 115–136 (2020)
4. Tang, Y.W., Xianguo, C., et al.: Research on multi-objective optimization of concrete mix ratio based on convolutional neural network and genetic algorithm. Constr. Build. Mater. **253**(4), 119208 (2020)
5. Ye, M., Wang, L., Zhang, H.: Research on multi-objective optimization model of the planting structure based on TOPSIS. J. Phys: Conf. Ser. **1848**(1), 012109 (2021)
6. Wei, M., Yang, Y., Liu, Z., et al.: Research on multi-objective operation optimization of integrated energy system based on economic-carbon emission. IOP Conf. Ser. Earth Environ. Sci. **512**, 012014 (2020)
7. Song, Y., Zhang, Z., Mao, J., et al.: Research on fast intelligence multi-objective optimization method of nuclear reactor radiation shielding. Ann. Nucl. Energy **149**, 107771 (2020)
8. Zhang, Y., Atasoy, B., Negenborn, R.R.: Preference-based multi-objective optimization for synchromodal transport using adaptive large neighborhood search. Transp. Res. Rec. **2676**(3), 71–87 (2022)
9. Alsaadi, I., Wang, H., Chen, X., et al.: Multi-objective optimization of pavement preservation strategy considering agency cost and environmental impact. Int. J. Sustain. Transp. **15**(1) (2020)
10. Lin, B.: Optimization of express train service network: under the competition of highway transportation. IOP Conf. Ser. Earth Environ. Sci. **587** (2020)

Neural Network Technology for Electrical Fire Early Warning System

Lan Yu[✉]

School of Energy and Power, Changchun Institute of Technology, Changchun 130012, Jilin,
China
yulan_1025@163.com

Abstract. Electrification not only promotes the rapid progress of productivity
and human civilization, but also provides a greater possibility for the occurrence
of fire. Electrical fire is mainly caused by abnormal high temperature caused by
line overload, short circuit, poor contact, arc spark, electric leakage, lightning or
static electricity, which leads to ignition of surrounding combustibles or spontaneous combustion of cables. At present, the detection of electrical fire is mainly
through the detection of the surface temperature of various wires and cables and
the detection based on the electromagnetic principle, which has the problems of
high false alarm rate and difficult maintenance. Considering that most electrical
fires are caused by abnormal high temperature of the line due to electrical faults,
which makes the insulating layer self ignite or ignite the surrounding substances,
while the insulating layer made of plastic or rubber has the characteristics of high-
temperature thermal decomposition and will release specific gas when working
above the rated temperature, this paper proposes the technical research of neural
network on the electrical fire early warning system, The application research of
neural network technology in electrical fire early warning system is the application research of artificial intelligence and machine learning technology in fire
detection. Its main purpose is to use artificial intelligence to detect and predict the
occurrence of building fire, and then provide an early warning means to prevent
people from fire accidents.

Keywords: Neural network · Electrical fire · early warning system

1 Introduction

Since mankind invented and began to use electricity, our life and development have
been inextricably linked with it. With the gradual electrification of agriculture, industry,
national defense, service and other industries all over the world, social productivity has
made great progress. At the same time, human living standards have also been greatly
improved, and the degree of social civilization has also been greatly improved. However,
electrical appliances, power wires, electricians or power electronic equipment scattered
around our living and working environment will cause electrical fires due to various
abnormalities or faults, cause huge property losses to the people, and even pose a threat
to life.

© The Author(s), under exclusive license to Springer Nature Singapore Pte Ltd. 2023
J. C. Hung et al. (Eds.): IC 2023, LNEE 1045, pp. 308–315, 2023.
https://doi.org/10.1007/978-981-99-2287-1_44

Some data show that in China's urban power consumption, more than 80% have different degrees of potential safety hazards, and the electrical fires in the 1980s accounted for about 15% of the total fires in China at that time; After entering the 1990s, the occurrence of electrical fire has accounted for more than 20% of the total fire at that time [1]. In addition, according to the "national fire statistics from January to August 2009" released by the fire bureau of the Ministry of public security, the fires caused by electrical reasons are the most, resulting in the largest casualties and losses: the number of electrical fires accounts for 29.1% of the total number of fires, and the direct property loss is as high as 344.549 million yuan [2].

In addition to destroying buildings, causing property losses and casualties, electrical fires may also lead to large-area and long-term power supply interruption, and bring greater explicit and implicit losses to industrial production, scientific research and people's life. Moreover, due to the existence of electrical faults, there may be the risk of electric shock or electric shock. Therefore, it is of great significance to carry out the research on electrical fire, find out various electrical faults and physical phenomena causing electrical fire, alarm and even early warning of electrical fire, and provide a more safe guarantee for people's life, industrial production and the normal operation of various public institutions.

2   Related Work

2.1   Topological Structure of BP Neural Network

BP neural network is a multi-layer feedforward neural network. The main characteristics of the network are: positive propagation of input signal and back propagation of error signal. In the process of signal forward transmission, the input training signal is input from the input layer of the network, transformed and processed by neurons in each layer to the output layer. If there is a deviation between the training output of neural network and its expected output, it will enter the process of error back propagation [3]. During back propagation, the error signal is transmitted back from the original forward propagation path, and the weights and thresholds of neurons in each layer of the network are modified along the direction of the negative derivative of the error criterion function to the network parameters. After limited iterations, the error function tends to be the smallest, so that the training output of BP neural network is constantly approaching the expected output. In the process of signal transmission, the output of neurons in each layer only affects the state of neurons in the next layer.

The topology of a typical three-layer BP neural network is shown in Fig. 1:

2.2   Parameter Selection of BP Neural Network

Generally speaking, using a neural network to complete model identification requires three elements: appropriate network structure and parameters, appropriate training sample set and appropriate error criterion function. Specifically, it is necessary to determine the number of layers, nodes of each layer, excitation function and initial weight of neural network, input and output sample set, and consider the selection of error criterion function at the same time.

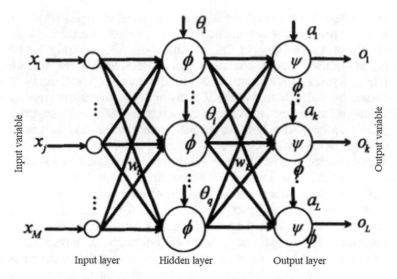

Fig. 1. Topological structure of BP neural network

(1) Select the model. The model is an approximate description of the actual system. Of course, the closer it is to the actual system, the better. However, if we blindly pursue the approximation accuracy, the structure of the model will become very complex, so that its original value will be lost in practical application. If the accuracy requirements are properly reduced and some secondary factors are ignored, the model can be simpler, the amount of calculation in the learning process will be reduced, and the solution will be relatively simple, so it will reflect its advantages [4]. Therefore, when establishing the model of the actual system, we should not only consider that the model can accurately approximate the input and output characteristics of the actual system, but also consider that the amount of calculation in the training process will not be greatly increased because the model structure is too complex. Reflected in the structure of the neural network, it is mainly the determination of the number of layers of the network and the number of neurons in each layer. In the selection of network layers, the general principle is: on the premise of effectively reflecting the input-output relationship of the actual system, less layers should be selected to make the network structure as simple as possible.

(2) Selection of neuron excitation function. In the multilayer feedforward neural network, in order to make the neural network have better generalization ability and convergence speed, the sigmoid function is generally selected as the excitation function in the hidden layer, and the linear function is generally selected as the excitation function in the output layer.

(3) Selection of initial weight. One of the main factors affecting whether the learning of the network can converge to the global minimum and whether it can converge quickly and smoothly is the size of the initialization weight of the neural network. If the initialization value of the weight is large, it is easy to make the weighted neuron input enter the saturation region of the s function, and the derivative of the saturation

region of the s function is very small, resulting in a small amount of adjustment of the network parameters each time, which makes the training process very slow and even stop [5]. From the perspective of global optimization, the weights should be initialized as close to the global minimum as possible. Therefore, in order to make the network converge as soon as possible and reduce the occurrence of "platform phenomenon", the initial weight is generally set to a small random non-zero value in BP algorithm.

3 Technical Research on Electrical Fire Early Warning System

3.1 Electrical Fire Monitoring System

According to the national standard issued by China in September 2005, the alarm and control system for electrical fire prevention is defined as a system that can send alarm signals, control signals and indicate the alarm position when some parameters of the line to be protected exceed the alarm set value. It is mainly composed of electrical fire monitoring equipment and electrical fire monitoring detector.

Electrical fire monitoring system belongs to early warning system, which is different from the traditional automatic fire alarm system. The traditional automatic fire alarm system mainly detects the temperature, fire light and smoke information after the fire through temperature, light and smoke detectors. When the detection value exceeds a certain threshold, it sends an alarm signal to remind people or staff of the occurrence of fire; Some fire alarm systems also add some auxiliary devices, such as automatic fire extinguishing devices, communication devices, etc., to play a certain degree of automatic fire extinguishing, sending fire information to the fire center and other auxiliary functions [6]. The traditional automatic fire alarm system is mainly to minimize the loss caused by fire. It is a passive fire protection mechanism.

The electrical fire monitoring system uses electrical fault detection and protection devices to detect the electrical phenomena that are easy to cause electrical fire or the physical phenomena shown by electrical faults, such as short circuit, electric leakage, fault arc or electric spark. When these phenomena appear or reach a certain degree, it will send out alarm or early warning information; At the same time, the area where the fault phenomenon occurs is displayed in order to eliminate the fault. The purpose of electrical fire monitoring system is to avoid the occurrence of electrical fire and its losses. It is an active fire prevention or protection method, which plays an early warning role in electrical fire to a certain extent.

3.2 Main Causes and Analysis of Electrical Fire

A large number of research data and electrical fire cases show that the main electrical faults causing low-voltage electrical fire include short circuit, poor contact, aging of power equipment, leakage, harmonic, overload and so on. The arc discharge, spark discharge and high temperature generated by these electrical faults at the fault point can easily cause the surrounding combustibles and insulating combustible materials of the line or equipment body to catch fire, which has the potential danger of causing electrical fire [7].

The latent period of electrical fire hazards is generally long and hidden, which is easy to be ignored. Generally, small faults or physical defects accumulate slowly. When encountering strong and sudden electrical faults, electric arcs, electric sparks and high temperatures will burst out suddenly, making the surrounding combustibles, electrical equipment and insulating materials of electrical lines meet the fire conditions and burn.

In recent years, in view of the great harm of electrical fire, the state has introduced the standard of electrical fire monitoring equipment and system. Modern low-voltage electrical lines and power lead ends of electrical equipment are generally equipped with low-voltage circuit breakers, leakage protectors, overcurrent relays and other protection devices, in order to provide power-off protection in case of serious short circuit, under-voltage, leakage, overload and other faults of electrical lines. However, the avoidance of electrical fires has not been much improved.

4   Technical Research on Electrical Fire Early Warning System Based on Neural Network

4.1   Similar Characteristics and Phenomena of Electrical Faults

(1)   Preconditions for electrical failure

According to From the above analysis of the formation mechanism of electrical fire, we can see that various electrical faults causing electrical fire are closely related, and under certain conditions, various causes can be transformed into each other. For example, the occurrence of short-circuit faults is mostly caused by the damage of line insulation by external force, or the scorching and rapid aging of line insulation materials caused by overload and high temperature, or the arc channel caused by poor contact contacts other line metal bodies [8]. The short-circuit further strengthens these faults and forms a vicious circle, which eventually leads to serious electrical faults, which has the potential to cause electrical fire. Another example is the occurrence of leakage fault. When the normal working current in the electrical line or electrical equipment is damaged due to the insulation of the conductor or equipment, part of the working current flows out of the line or equipment and flows through the external conductor to the earth to form a leakage circuit. The leakage current will produce high temperature, electric spark or arc at the fault point or grounding, resulting in electrical fire. Leakage is a short circuit to ground to some extent.

(2)   Electric arc and spark

Many research data show that most of the current electrical fires are caused by electric arc or electric spark caused by electrical fault. From the mechanism analysis, it can be seen that electric spark or arc discharge will occur at the physical fault point of the line, whether it is short circuit, open circuit, poor contact, arc grounding short circuit, leakage, overload, etc. Since arc and electric spark will not cause large current or voltage fluctuation in the line, some electrical protection devices are usually difficult to operate, so they have gradually become the main cause of low-voltage electrical fire in recent years [9]. Electric spark is a common discharge phenomenon in low-voltage electrical circuits. When the connection between wires and between wires and electrical protection devices is not firm and poor contact, it is often expressed in the form of "ignition"; When the electrical circuit has faults such as short circuit, open circuit and arc grounding short circuit, the larger current change rate will cause spark discharge due to the larger contact resistance of the physical fault point of the electrical circuit.

4.2 Fire Warning Signal Detection Parameters

Some modern electrical circuits or electrical equipment are generally equipped with protective devices, such as circuit breakers, leakage protectors, etc. These protective devices can act, cut off the power supply or send out alarm information when the electrical fault phenomena such as short circuit and leakage reach a certain degree.

However, according to the statistics of the Fire Department of the Ministry of public security, the number of electrical fires has not decreased in recent years, but has an increasing trend with the wide and deep use of electricity. This is because the cause of electrical fire is gradually becoming more hidden, such as arc short circuit or grounding fault. When the short circuit or grounding resistance is very high, the change of line current or voltage is not obvious, and the protection device cannot act; Another example is poor contact, the contact resistance is relatively small, and the change of line voltage and current is not obvious; But they have gradually become the main force causing electrical fires [10]. However, they will also show some external phenomena, such as arc and electric spark, temperature rise caused by arc or spark, charred smell, "Chi Chi" sound, etc. it is noted that when an electrical fault occurs in an electrical circuit or electrical equipment, the arc or electric spark has sound, heat, light, pressure, electromagnetic and other effects in addition to the harmonic components in its voltage and current waveform, Using these effects, various combined arc or spark detection devices can be developed.

Therefore, we can use ultrasonic detection and ultraviolet detection to detect the electric spark or arc as the criterion of whether the arc or electric spark occurs; Due to the need of the rapidity of the early warning system, the use of sound and light detection has better advantages than temperature rise and smell. In this paper, the current signal is selected for detection at the same time, and some electrical fault phenomena are basically judged by detecting the current harmonic change rate, which is used as an auxiliary judgment for the occurrence of arc or spark. For the detected ultrasonic signal, the schematic diagram of the designed signal detection circuit is shown in Fig. 2. Due to piezoelectric overload The output voltage signal of the acoustic sensor is very weak, generally mv level, so two-stage amplification is adopted in the circuit, and the amplifier uses a single power supply lm358 operational amplifier.

5   Conclusion

The traditional electrical fire monitoring or alarm system mainly judges the signs of electrical fire through the detection of leakage, temperature and other parameters. To a certain extent, it can predict the alarm and reduce or avoid the loss. However, the parameters it detects, such as temperature, are the result of the fault or abnormal development

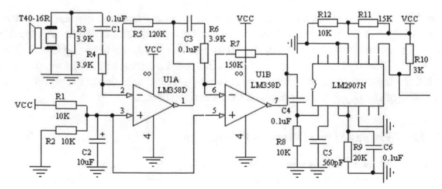

Fig. 2. Schematic diagram of ultrasonic signal detection circuit

to a certain extent. The early warning function has great limitations and is not effective in avoiding electrical fire. Therefore, this paper studies the low-voltage electrical fire early warning system, which not only meets the basic requirements of relevant national standards, but also tries to achieve timely and reliable early warning in order to avoid the occurrence of electrical fire.

Acknowledgements. 1. Project of Jilan Provincial Department of Education Science Research Fund (NO. JJKH20210677KJ)

2. Project of Building Control and Energy Saving Optimization Experiment Center Project (NO. BCES2019KF19)

3. Project of Housing and Urban-rural Development (NO. 2014-K8-064)

4. Project of Jilin Provincial Department of Science and Technology (NO. 20160204019SF)

References

1. Li, M., Wu, Q., Chen, D.: Research on air conditioning electrical energy saving control based on computer neural network. J. Phys. Conf. Ser. **1992**(2), 022022 (5pp) (2021)
2. Sun, Y., Fu, J., Ma, Q., et al.: Research on wear recognition of electric worker's helmet based on neural network. J. Phys, Conf. Ser. **1449**, 012057 (2020)
3. Zhou, E., Zhao, Y., Dai, Y., et al.: Research on temperature field prediction model of electric spindle based on improved bp neural network (2021)
4. Bicakci, S., Coramik, M., Gunes, H., et al.: A New artificial neural network-based failure determination system for electric motors. Arab. J. Sci. Eng. **47**, 835–847 (2021)
5. Hossain, M., Gopisetti, L.S.P., Miah, M.S.: Artificial neural network modelling to predict international roughness index of rigid pavements. Int. J. Pavement Res. Technol. **13**(3), 229–239 (2020)
6. Luo, J., Zhu, H., Jiang, Y., et al.: The 10.7-cm radio flux multistep forecasting based on empirical mode decomposition and back propagation neural network. IEEJ Trans. Electr. Electron. Eng. **15**(4) (2020)
7. Al-Gabalawy, M., Mostafa, M.A., Hamza, A.S.: Mitigation of electrical hazards on the metallic pipelines due to the HVOHTLs based on different intelligent controllers. IET Sci. Measure. Technol. 14, 1077–1087 (2021)

8. Lin, Y.Z., Li, X.M., Li, Y.Q.: Research on rapid detection technology and application of mortar compressive strength based on neural network. Mater. Sci. Forum **996**, 110–116 (2020)
9. Yao, Y., Wang, N.: Fault diagnosis model of adaptive miniature circuit breaker based on fractal theory and probabilistic neural network. Mech. Syst. Signal Process. **142,** 106772- (2020)
10. Zhang, A., He, J., Lin, Y., et al.: Recognition of partial discharge of cable accessories based on convolutional neural network with small data set. COMPEL – Int. J. Comput. Math. Electr. Electron. Eng. **39**(2), 431–446 (2020)

New Application of Improved Dynamic Programming Algorithm in Traffic Engineering System

Qun Zhou$^{(\boxtimes)}$ and Tao Wu

Cccc First Highway Consultants Co., Ltd, Zhuozhou 710065, Hebei, China
zzqq029@163.com

Abstract. This paper deeply studies some problems of highway traffic engineering system in China, including the composition and function of highway traffic engineering system, design method, construction and development direction, the setting principle of highway traffic engineering monitoring, toll collection and communication system, and the evaluation method of highway traffic engineering system. Dynamic programming algorithm is a method used to solve the minimum cost flow problem in the network. It has been applied to various problems in traffic engineering, such as determining the best path of vehicles and optimizing the road network. The dynamic programming algorithm solves the optimization problem by decomposing the optimization problem into smaller problems that are solved at the same time. This method reduces the computational complexity by allowing each subproblem to be solved independently, and then combining the results of multiple subproblems. In this way, it can find a solution quickly (as they originally need) compared with the computing resources required for all sub problems to be combined at the same time.

Keywords: Dynamic programming algorithm · Traffic engineering system

1 Introduction

Expressway traffic engineering system is set up to meet the fast, convenient, safe and comfortable traffic characteristics of Expressway and the needs of expressway management. It directly affects the operation effect of expressway, and has a direct and close relationship with the driving safety, traffic capacity, service level, traffic and operation management, investment benefits, etc. of expressway. It is an organic part of the main project of expressway. Practice shows that the expressway traffic engineering system plays a positive and important role in ensuring road safety, improving and maintaining road capacity and service level, improving expressway management means, giving full play to Expressway benefits and adapting to traffic development [1].

By the end of this year, China's highway mileage will exceed 19000 km, ranking second in the world, second only to the United States. The "five vertical and seven horizontal" 12 national trunk roads planned by the Ministry of communications will be completed by 2010. All provinces, autonomous regions and cities are also accelerating

© The Author(s), under exclusive license to Springer Nature Singapore Pte Ltd. 2023
J. C. Hung et al. (Eds.): IC 2023, LNEE 1045, pp. 316–323, 2023.
https://doi.org/10.1007/978-981-99-2287-1_45

the construction of their skeleton highway network. The national expressway network running north and south, connecting East and West, and the expressway network connecting the eastern provinces, autonomous regions and cities will be formed step by step [2]. After nearly 10 years of operation, the traffic volume of expressways built in the early stage of our country has increased rapidly, and they have faced the practical problem of expanding the traffic capacity of channels. The rapid development of Expressway in China has put forward a wide range of construction needs for traffic engineering system. The construction investment demand of traffic engineering system is large, accounting for about 1/10 of the cost of expressway. The construction of traffic engineering system itself also needs to consider its investment benefit [3].

However, despite more than ten years of research, exploration and construction practice, so far, there are still many problems for the design, construction and management of expressway traffic engineering system, including the relationship between expressway traffic engineering system and road main works, basic setting theory, reasonable scale of the system, future construction and development mode, etc. With the rapid development of China's national economy and highway construction, the above problems have been put in front of us realistically, which have attracted more and more attention and become the common concern of scientific research and design, investment decision-making and construction management departments and units.

2 Related Work

2.1 Development History of Highway Traffic Engineering System Construction

China's highway traffic engineering system is developed with the construction of expressway. The relatively complete system implementation began with the construction of Beijing Tianjin Tangshan expressway, including safety facilities, toll collection system, road monitoring system, communication system (emergency telephone), service areas along the line, etc. At the beginning of construction, the scale determination and technical standards of highway traffic engineering system in China mainly refer to Japanese standards and relevant technical specifications of the United States and Europe [4]. With the practice and exploration, our country has gradually formed its own standard system, and has successively compiled the technical specifications for the construction of highway safety facilities, the standards for road traffic signs and markings, the technical requirements for emergency telephone, the charging mode and charging system, the technical conditions of highway anti dazzle facilities, the land use indicators along the road, etc. in the technical standards for highway engineering issued by the Ministry, some provisions and requirements are also made for the implementation conditions of the road traffic engineering system. The formulation and promulgation of the above standards and specifications have played a positive role in promoting the construction of highway traffic engineering system in China [5].

At present, the construction of expressway traffic engineering system in China is in a period of rapid development. Compared with the beginning of construction, the implementation scale and technical system have changed greatly. For example, the setting standard of road safety barriers has been basically changed from the initial layout of some sections to the whole layout along the Expressway and the central separation belt; The charging system has changed from full manual charging to semi-automatic charging. The type of pass ticket has developed from paper ticket and magnetic ticket to two-dimensional bar code and IC card, and closed-circuit television monitoring and automatic license plate recognition system have been added; The monitoring system developed from nothing, and gradually changed from simple information collection to certain information release and traffic guidance control, and implemented the road section monitoring center system, tunnel and super major bridge monitoring system [6]; The communication system has transitioned from the initial PDM session mode to SDH optical fiber transmission system. Emergency phones are widely used, and 800 megabyte or 450 megabyte wireless trunking systems are also implemented in some sections.

2.2 Problems Faced by the Construction of Expressway Traffic Engineering System in China

Expressway traffic engineering system is a comprehensive system, involving road engineering, traffic flow, road capacity, computer system, control system, communication system, network engineering, information engineering, traffic management, etc. it is closely related to technological progress, as well as policy and management awareness. According to the construction practice and development of domestic expressway traffic engineering system, the following problems exist and face in the construction of expressway traffic engineering system in China at present:

(1)   Theoretical basis of system setting

The expressway traffic engineering system includes traffic safety facilities, toll collection system, monitoring system, communication system, management service maintenance facilities, etc. there is no systematic theoretical research and satisfactory answer to whether the above systems are set up together during road construction, the relationship with the main road works and road traffic composition, traffic flow characteristics, road capacity, service level, traffic volume growth, and its setting principles. Among them, Up to now, the site scale of the charging system, the layout spacing and scale of service facilities still apply Japanese standards; Whether the communication system must still be debated; There is no unified understanding of the scale and setting principles of the monitoring system [7]; All these have brought great confusion to the construction of highway traffic engineering system in China.

(2)   Networking construction of expressway traffic engineering system

With the acceleration of the construction process of expressways, expressways in many provinces have formed a network, such as Guangdong, Jiangsu, Shandong and other provinces. By the end of this year, expressways will exceed 1000 km, and expressways in Hunan, Hubei, Hebei, Beijing, Liaoning and other provinces and cities will also be connected into a network. Under the environment of Expressway gradually becoming a network, the networking construction of expressway traffic engineering system has become an important topic [8].

(3)   Information construction of traffic engineering system

Facing the 21st century, the international competition is mainly the competition of informatization. China has taken the informatization construction as the basic development strategy of the country in the 21st century. The highway traffic engineering system is an important part of the traffic informatization construction. For example, the communication network system of the highway will form the traffic information backbone network, and the monitoring system is a part of the traffic informatization management [9]. Therefore, how to promote the informatization construction in the construction of expressway traffic engineering system and the informatization construction of transportation industry through the expressway informatization construction will also become an important topic in the field of expressway traffic engineering system construction.

3   Function and Composition of Traffic Engineering System

3.1   Construction Objectives of Expressway Traffic Engineering System

Expressway is different from general highways, which is characterized by full closure, full interchange and strict control of access; Car only, speed limited traffic; Set up a central divider to drive separately; There are perfect transportation facilities and service facilities along the line. Moreover, due to the huge investment in expressways, the construction funds need to be raised through multiple channels, and China's expressways need to collect fees and repay loans.

Therefore, the construction goal of China's expressway traffic engineering system is: to set up scientifically and reasonably according to the principles and methods of traffic engineering, so as to ensure the maximum traffic capacity of expressway: the minimum traffic accidents; The fastest way to recover traffic after troubleshooting; It has the least impact on the ecological environment; Provide safe, fast, efficient and comfortable driving conditions for vehicles; Adapt to the modern management of expressway; And ensure the repayment of construction investment [10].

According to the basic concept of expressway traffic engineering system and the objectives to be achieved by the system, expressway traffic engineering system is a practical engineering concept. It is a variety of facilities set up according to the principles and methods of traffic engineering to give full play to the "high-speed, efficient, safe and comfortable" characteristics of expressway, mainly including expressway traffic safety facilities, toll collection system, monitoring system Communication system and management and maintenance facilities supporting these systems.

3.2   Composition and Function Analysis of Expressway Traffic Engineering System

Highway traffic safety facilities mainly include road traffic signs, highway guardrails, pavement markings, isolation facilities, anti dazzle facilities, sight guidance signs and construction safety facilities.

(1) Traffic signs. Traffic signs convey concise and clear information to vehicles and drivers through the changes of words, symbols, patterns and sign shapes, and ensure a certain visual distance, so that drivers can recognize and understand in a short time, so as to better use highways and improve highway transportation benefits. According to the function, the traffic signs of Expressway in China are divided into warning signs, prohibition signs, indication signs, direction signs and auxiliary signs.

(2) Traffic markings. Traffic markings are defined as lines, arrows, words, elevation marks, protruding road signs or other guiding devices, which are set on the road surface or other facilities to control and guide traffic. Traffic markings can ensure the traffic flow to separate lanes, guide the traffic direction, guide vehicles to enter the appropriate lane before convergence and diversion, improve the popular driving conditions of vehicles, and make rational use of the effective area of the road.

(3) Safety guardrail. The safety barrier is a facility set on the central divider or roadside to prevent vehicles from crossing the central divider and entering the opposite lane or crossing the way out. At the same time, the barrier also has the functions of restoring the vehicle to the normal driving direction, reducing the deceleration caused by the collision, reducing the damage to passengers and inducing the driver's line of sight. Safety barrier is a safety barrier provided by the highway when traffic accidents such as vehicle off the road caused by vehicle driving errors occur, which can avoid or reduce major fatal accidents caused by rollover or collision.

(4) Barrier. Barrier is a general term for artificial structures that isolate and close highways. Its purpose is to prevent irrelevant people and livestock from entering and crossing the expressway, eliminate traffic interference, so that the expressway can form a stable and fast traffic flow, and greatly improve the safety and efficiency of driving.

4 Traffic Engineering System Based on Improved Dynamic Programming Algorithm

4.1 Improved Dynamic Programming Algorithm

Combined with the idea of computer parallel processing, the one-way recursive algorithm is improved to a two-way recursive algorithm in order to improve the running speed without affecting the accuracy.

For a multi-stage decision-making problem with a certain starting point and end point and a moderate number of stage variables, the two one-way recursive methods in dynamic programming barely meet the requirements in terms of search speed, but if complex multi-stage decision-making problems are encountered, the search speed will slow down sharply. In order to ensure the search speed in complex multi-stage decision-making problems, it is very important to improve the dynamic programming model accordingly. The two algorithms are applied to a stage variable and method at the same time for recursive improvement.

The improvement idea is: the improved algorithm is to use the dynamic programming algorithm to recurse backward and forward stage by stage from the starting point and the end point at the same time. There must be a stage in the two-way recursion. If the recursion in the two directions coincides at this stage, the algorithm will be optimized at this time and the recursion process will end, which greatly reduces the operation time and improves the operation efficiency.

The basic judgment conditions of the improved algorithm are as follows:

$$\|\Delta u_{k+1}(t)\|e^{-\lambda t} \leq \rho\|\Delta u_k(t)\|e^{-\lambda t} + m_5 de^{-\lambda t} + m_4(\int_0^t e^{(pk_f+m_2+m_3)(t-\tau)}e^{\lambda(t-\tau)}e^{-\lambda t}$$

$$(m_1\|\Delta u_k(\tau)\| + pd)d\tau)$$

(1)

The flow of dynamic planning improved algorithm is shown in Fig. 1.

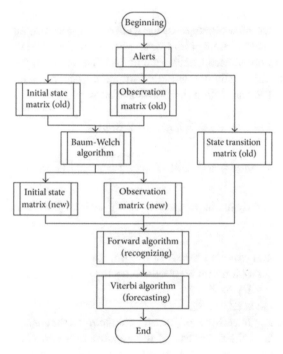

Fig. 1. Flow chart of dynamic programming algorithm

4.2 Application of Dynamic Programming Algorithm in Traffic Engineering System

In order to verify the superiority of the improved algorithm, the network diagram of the traffic engineering construction of a section of Shanghai Shentong Metro is intercepted, as shown in Fig. 2. The value between the two stations (nodes) represents the actual time required for the construction between the two nodes. The optimal construction period of the project is calculated through the improved two-way recursive algorithm, so as to reduce the cost.

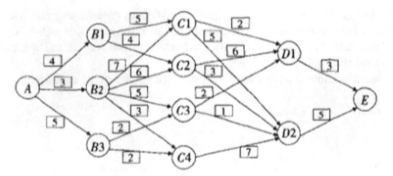

Fig. 2. Traffic engineering route network diagram

1) In order to reflect the advantages of the improved algorithm, the sequence solution method of classical dynamic programming is used to calculate the shortest construction period of the engineering network diagram. The solution process of the network diagram can be artificially divided into four stages according to the actual situation of the project. The calculation process is as follows:

When $k = 1$, according to the calculation equation:

$$\lim_{k \to \infty} \| \Delta e_{k+1}(t) \|_\lambda \leq \left(\frac{m_1 m_5}{b - \lambda} \frac{1}{1 - \hat{\rho}} + \frac{p}{b} \right) cd \tag{2}$$

(1) Use the improved dynamic programming algorithm to solve this decision-making problem:

According to the operation process of the improved algorithm, the optimal path of the decision-making problem can be obtained by the sequence tracking method again:
A $\to$ B$_3$ $\to$ C$_3$ $\to$ D$_1$ $\to$ E.
The optimal value is 12 (days).
Through the example verification, it can be concluded that using the improved algorithm can realize the optimal strategy of the construction period of rail transit project faster and more accurately, and find out the shortest construction path, that is, the shortest construction period, which has a certain practical application value for saving the cost of the construction project and ensuring that the project can be completed on time.

5 Conclusion

The design and development of daily safety management system of traffic engineering is related to the safety and stability of rail transit engineering construction. Using the improved dynamic programming algorithm in the leadership decision query module can properly solve the defect that the calculation speed is significantly reduced when the decision parameters reach the quantitative level. The daily safety management system of rail transit engineering is an integrated management information system independently developed. Its main function is to systematically analyze the data obtained from the safety inspection, so as to find the existing safety problems and hidden dangers as soon as possible, provide safety early warning for the safety of traffic construction engineering, guide decision-making, so as to put forward corresponding modification schemes, and control the safety problems in the bud. To further improve the system and apply it to specific practical projects, it needs further research and development.

References

1. He, M., Xiong, H., Kou, J., et al.: An improved dynamic programming method in the optimization of gas transmission pipeline operation. J. Phys. Conf. Ser. **1746**(1), 012088 (6pp) (2021)
2. Hendzel, Z., Kołodziej, M.: Neural dynamic programming with application to wheeled mobile robot. In: Szewczyk, R., Zieliński, C., Kaliczyńska, M. (eds.) AUTOMATION 2022. AISC, vol. 1427, pp. 213–222. Springer, Cham (2022). https://doi.org/10.1007/978-3-031-03502-9_22
3. Bakry, I., Lyubimov, V.V.: Application of the dynamic programming method to ensure of dual-channel attitude control of an asymmetric spacecraft in a rarefied atmosphere of Mars. Aerosp. Syst. **5**(2), 213–221 (2022)
4. Chen, L., Xiao, Y., Yang, T.: Application of the improved fast iterative shrinkage-thresholding algorithms in sound source localization. Appl. Acoust. **180**, 108101 (2021)
5. Liu, Y., Liang, J., Song, J., et al.: Research on energy management strategy of fuel cell vehicle based on multi-dimensional dynamic programming. Energies **15**, 5190 (2022)
6. Xu, J., Wu, S.: Analysis and application of dynamic programming. J. Phys. Conf. Ser. **1865**(4), 042023- (2021)
7. He, Z., Wang, C., Wang, Y., et al.: An efficient optimization method for long-term power generation scheduling of hydropower station: improved dynamic programming with a relaxation strategy. Water Resour. Manag. **36** (2021), Published for the European Water Resources Association (EWRA)
8. Nowak, M., Trzaskalik, T.: A trade-off multiobjective dynamic programming procedure and its application to project portfolio selection. Ann. Oper. Res. **4**, 1–27 (2021)
9. Liang, H., Lu, H., Feng, K., et al.: Application of the improved NOFRFs weighted contribution rate based on KL divergence to rotor rub-impact. Nonlinear Dyn. **104**(4), 3937–3954 (2021)
10. Yang, R.: Application of the improved cobweb model in China's new energy vehicle market. In: Li, X., Yuan, C., Kent, J. (eds.) Proceedings of the 5th International Conference on Economic Management and Green Development. Applied Economics and Policy Studies. Springer, Singapore (2022). https://doi.org/10.1007/978-981-19-0564-3_16

Online Course Construction of Higher Mathematics Based on Internet

Xiaomin Li[✉]

Shandong Xiehe University, Shandong 250109, China
crystal.xiaomi@126.com

Abstract. The construction of Higher Mathematics Network Course Based on Internet is a new teaching method. It was developed by educators and researchers in the past decade. They studied how people learn and create their own learning experience through computer technology. This approach supports students to actively participate in the learning process and allows them to work at home or wherever they want. The main advantage of this method is that it does not require students to use any special equipment, which allows everyone to use it without restrictions. The construction of higher mathematics online courses based on the Internet provides an interactive environment in which students can easily communicate with teachers and students. The main purpose of this research is to realize such a teaching design: the network teaching system can call the application on the Internet platform, and the data is still stored in the local server. According to the teaching requirements and teaching characteristics of the course of advanced mathematics, this paper constructs an online learning system. Based on Baihui Internet platform and API technology, Baihui writer, Baihui show and other Internet products are nested in the online learning system to improve the teaching quality.

Keywords: internet · Advanced mathematics · Online courses

1 Introduction

In the rapidly changing era of knowledge economy, Internet technology is developing rapidly towards broadband, high-speed and multimedia, and multimedia network education is also booming. Network education, which combines Internet technology and education, has gradually become a research hotspot in the field of education and has been widely concerned. It has also become an important field of education reform in the 21st century [1]. The Ministry of education clearly pointed out in the "action plan for revitalizing education in the 21st century": "relying on the modern distance education network, we should set up high-quality online courses, organize national first-class teachers to teach, and realize the sharing of educational resources across time and space." Multimedia network education has become an important topic in the reform and development of education all over the world. Therefore, China should actively keep up with the pace of the times and focus on the research and development of Multimedia Internet network

© The Author(s), under exclusive license to Springer Nature Singapore Pte Ltd. 2023
J. C. Hung et al. (Eds.): IC 2023, LNEE 1045, pp. 324–331, 2023.
https://doi.org/10.1007/978-981-99-2287-1_46

education, so as to improve the modernization and informatization level of education in our country [2]. At present, with the support of governments at all levels for educational policies, the increase of investment in education, the purchase of computers and the construction of networks, and the gradual maturity of hardware construction, the research on the construction of curriculum software related to online education, especially the relevant theoretical and practical issues related to the design and development of online courses, is not deep enough. This paper will deeply discuss and analyze the teaching design theory, the design and implementation of network courses, and the evaluation of network courses involved in the construction of network courses [3].

In the current education mode, mathematics can be said to be a very important discipline. It is not only the foundation of all science and engineering courses, but also shoulders the important mission of building learners' logical thinking. It can be said that the quality of mathematics education directly affects the development of the country and individuals. With the gradual deepening of learning, the difficulty of mathematical knowledge is also increasing. In the university stage, students mainly study advanced mathematics, linear algebra, probability statistics, etc. from the current learning situation of college students, these courses do have problems such as low pass rate and low classroom performance. In this case, most colleges and universities do not adopt more effective teaching methods, but mainly focus on classroom explanation and exercises. In addition, there is no scientific and effective statistical means applied to the right and wrong answers and the mastery of knowledge points, so it is difficult for teachers to clearly know what shortcomings each student has, and there is no sufficient basis to adjust the curriculum structure [4]. They can only rely on Teachers' subjective assumptions to explore, which is contrary to the concept of "student-centered". For this reason, this set of advanced mathematics assisted learning system came into being. It can help students use fragmented time to study mathematics, and achieve the purpose of improving mathematics performance through targeted topic exercises. In the process of answering questions, the concept and means of learning analysis technology are used to collect the learning data generated by learners and summarize it in combination with relevant algorithms, so as to present the feedback results with personalized guiding significance to learners, point out the shortcomings of learners in the process of higher mathematics learning, and realize the concept of personalized learning with the support of data analysis.

2 Related Work

2.1 Definition of Online Courses

(1) Definition of online courses

Courses are teaching contents and learning experiences organized according to certain educational objectives, Its definition is students (learner) "the whole of learning activities under the guidance of teachers, which includes a wide range of concepts such as educational objectives, contents, activities and evaluation methods. The so-called online courses, literally, are educational courses developed and used in the online environment [5]. Using online learners can obtain more learning opportunities, rich teaching resources, more convenient learning environment, and fully reflect the characteristics of the network Point. From a deep understanding, online courses refer to the sum of teaching activities designed for one or more learners under the guidance of advanced educational ideas and teaching theories, and carried out in a certain subject with the help of the network environment. Network education has changed the traditional mode of transmission of educational information, and thus led to changes in educational ideas, models, methods and so on.

(2) Differences between network courses and multimedia software, CAI courseware and network courseware

Online courses are different from multimedia software. Network course is a multi-factor project, which includes not only the knowledge system required by educational courses, but also the process of learning activities defined to achieve teaching goals. This kind of network course system usually includes course content, multimedia assistance, learning tools, simulation experiments, collaborative communication, evaluation and evaluation, etc. from the perspective of content, the network course system is generally composed of network courseware database, material case database and test question database. The general multimedia teaching software is relatively simple in function. It is mainly used to solve the key and difficult points in specific teaching or specific courses. The teaching content is closed and fixed, and it can not carry out convenient interaction, and the way of feedback is also lack of interactivity. Network courses are different from CAI courseware and network courseware [6]. Network courses can be composed of multiple courseware, database and database. When reflecting the teaching content, it can use text, video, audio and other multimedia methods. However, network courseware is often designed for knowledge points, and its strength is relatively small. It is generally self-contained and used independently. The so-called self-contained and independent use means that the network courseware should cover the selected field of teaching content. It should not only be comprehensive, but also the designed content should have the best way of presenting information, which can be used for teaching independently. Network courseware does not have the characteristics of combining multiple courseware, database and database to organize teaching content [7]. In terms of inclusiveness, online courseware can be said to be an element of online courses. It can be a part of online courses in the form of web pages, subsystems and vector animation courseware.

2.2 Characteristics of Online Courses

(1) Openness of teaching process

At present, the Internet has a very wide coverage, so as long as learners access the network, they can learn at any time and anywhere, so that learners can arrange learning content and progress according to their own time. Specifically, the openness of online courses is manifested in the openness of teaching resources and teaching process. The opening of teaching resources is manifested in two aspects: first, seek a wider range of information through links to relevant websites and browser queries, and jump learning can also be carried out between various knowledge points through hypertext and hypermedia links [8]; Second, teachers or system administrators can update or supplement teaching resources at any time according to the teaching progress.

(2) Sharing of learning resources

By using the Internet, excellent educational resources can be placed on the network, which overcomes the constraints of time and teaching resources in the traditional education mode. By extending the links of course related resources to the whole Internet and realizing the sharing of global network resources, every student can enjoy a high level of education.

(3) Interaction of network teaching

Interactivity means that in the process of online education and teaching, it can not only realize the human-computer interaction between the client and the server, but also enable students and teachers distributed in different places to exchange information directly through e-mail, forums, online chat rooms, video conference technology and virtual reality technology [9]. At the same time, learners in different places can also inspire and help each other, which helps learners develop their ideas, give play to their learning initiative and creativity, and thus master knowledge at a higher level, as shown in Fig. 1.

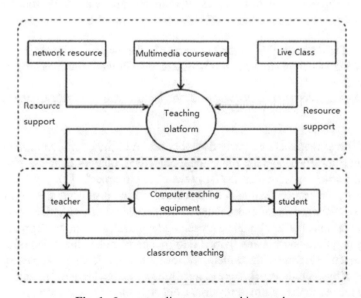

Fig. 1. Internet online course teaching mode

(4)   Diversity of presentation forms of teaching content

The main technical support of network teaching is shown in two aspects, namely, network technology and multimedia technology. Make full use of the rich information transmission advantages of multimedia, and use video, audio, text, charts and other forms to modify and package the learning content at multiple levels and angles, so that the originally boring knowledge becomes vivid and vivid. In this way, by stimulating learners' multiple senses, learners' interest in learning will be fully mobilized, their understanding and mastery of the knowledge learned will be deepened, and then the level and quality of teaching will be improved.

3   Problems in Online Courses

Through a large number of investigations on the online course system currently in use, it is found that there are still many problems in the current online course, mainly reflected in the following aspects:

3.1   The Presentation Form of Teaching Content is Single

At present, most of the online courses are presented in the form of simple web pages, which conforms to the characteristics of online education, but a large number of online courses in the form of static web pages are organized in a linear way, which is difficult to improve the update frequency. At present, there are three ways to present the teaching content:

(1)   Text and still image: this method is similar to the moving of books;
(2)   PowerPoint Lecture Notes: in this way, teachers' PowerPoint files are moved to the Internet;
(3)   The lecture videos and lecture notes of the lecturer are presented online at the same time.

The first two methods obviously do not conform to the cognitive laws of learners. They do not carefully design the educational objectives, objects, contents, and methods, nor plan the teaching activities in the network environment. The third way appears because many experts and scholars have recognized the disadvantages of the first two ways, and put forward that the teaching content should be diversified, and the teaching materials of online courses should be presented on the network in the form of streaming media. Many online courses are divided into two parts, one is the video of the lecturer reading the lecture in front of the camera, and the other is the PowerPoint corresponding to the content of the lecture. Psychological research shows that learners' attention retention is not only related to the nature of materials, but also related to the changes of materials. Therefore, when teachers' unchanging posture and teaching methods last for a long time, it is easy to cause learners' visual and psychological fatigue, and it is difficult to realize the charm of teachers' teaching, which reduces the authoritative position of the lecturer in the eyes of learners and affects the teaching effect of online courses, so this method is also not in line with the cognitive law [10].

3.2 The Navigation System is Not Powerful

The Internet provides learners with very rich hypermedia resources. If there is a lack of navigation system, learners will get lost in the learning process, produce a feeling of being at a loss, and reduce the efficiency of learning.

At present, the function of online course navigation system is not strong, mainly in the following aspects:

(1) There is no system for learners to understand the knowledge level, their own knowledge level, learning plan and learning method required for learning the course.
(2) The organizational form of the course is not hierarchical or reticular, but linear. It does not provide the function of query and retrieval between units, which is not convenient for jumping learning. The switching between learning units can only be achieved by moving forward, backward or starting from scratch.
(3) Lack of learning path tracking function, learners can only rely on memory to recall their learning path. Once learners interrupt the learning process, they can only start from scratch.
(4) Lacking the help system of course learning or the knowledge map of online courses, learners can only grope and operate according to their own experience.

4 Online Course Construction of Higher Mathematics Based on Internet

4.1 Investigation on the Demand of Network Teaching of Advanced Mathematics

It is the rapid development of information technology that mankind has entered the era of knowledge economy. Similarly, creativity, openness and diversity have also become the characteristics of education in the era of knowledge economy. Based on the rapid development of computer and communication technology and supported by multimedia and network technology, network education has shown its strong vitality.

As one of the compulsory courses for some engineering majors and mathematics majors, advanced mathematics is based on real number theory, and its main content is calculus. This paper aims to build an online teaching platform for advanced mathematics, and realize the services of online teaching, online communication, online homework, online evaluation and other functions, so as to enable learners to get rid of the limitations of time and space and realize autonomous learning.

In order to design a network teaching system that more meets the requirements of learners, a questionnaire survey is carried out for some students in Jilin University. Through the analysis of the survey data, learners are more inclined to autonomous learning, so creating a network learning environment of autonomous learning, collaborative learning and personalized learning is conducive to learners' autonomous learning.

4.2 The System Structure of Network Course of Advanced Mathematics

The production of the online course of advanced mathematics should be based on the theoretical and technical basis of the online course and the teaching design idea, abide by the design principles of the online course, rely on computer technology and network technology, and give full play to the imagination of the designer. This online course is applicable to the teaching and self-study of mathematics and other related majors that need the basis of mathematics. It belongs to the basic course of higher mathematics. It not only schematizes and visualizes some difficult abstract concepts, formulas and rules, but also provides teaching guidance and organizes teaching Q & A, so as to promote the understanding and application of the basic concepts and principles of higher mathematics and provide a free learning environment for learners, And with the help of computer technology and network technology, according to the principle of teaching design, a variety of learning activities are designed. The online course system of advanced mathematics includes five modules: learning content, guidance, detection and interaction between teachers and students. These five modules are organically formed into a unified whole, which is complementary, interrelated and inseparable. As shown in Fig. 2.

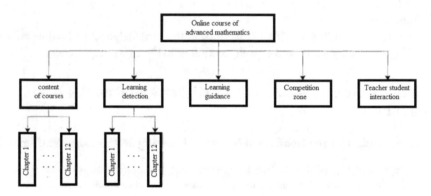

Fig. 2. Online course architecture based on Internet

5   Conclusion

Facing the rapid development of it and network technology, it has become the development trend of modern education to meet the needs of social development and comprehensively promote the construction of educational informatization. As the soul of network education informatization, online courses can spread and share educational resources across space and time, so that users can receive knowledge transmission anytime and anywhere. However, the quality of an online course directly affects the effect of talent training, which requires our online course designers to have the teaching experience of front-line teachers and advanced network technology knowledge at the same time. In order to meet the requirements of higher mathematics curriculum reform and modern teaching, and give full play to the initiative of learners, we have studied and analyzed several popular learning and teaching theories, and designed the network course of higher mathematics. Using other theories as a supplement, fully apply it to the design of online courses.

References

1. Xu, J.: Research of online courses construction based on the theory of interaction. J. Higher Educ Res. **3**(1), 26–28 (2022)
2. Jin, H.: Research on the path of mathematics discipline construction based on demand orientation: A case study of Tianjin Agricultural University. Asian Agric. Res. **12** (2021)
3. Yu, J., Zhang, J., Chen, Y., et al.: A course construction mode of "one body-two wings" for software engineering major under background of new engineering. J. Phys. Conf. Ser. **1883**(1), 012135- (2021)
4. Qiao, Y.: Research on the course construction of "japanese extensive reading" based on the multimedia under the background of micro-class. J. Phys. Conf. Ser. **1744**(4), 042096 (4pp) (2021)
5. Lee, J.E., Recker, M.: The effects of instructors' use of online discussions strategies on student participation and performance in university online introductory mathematics courses. Comput. Educ. **162** (2021)
6. Bai, J., Che, L.: Construction and application of database micro-course knowledge graph based on Neo4j. (2021)
7. Alarfaj, M., Sangwin, C.: Investigating a potential format effect with two-column proofs. Teach. Math. Appl. Int. J. IMA **41**, 198–217 (2021)
8. Zhang, Z.: Construction and empirical study of learner portrait in online general education course. Discrete Dyn. Nat. Soc. **2022** (2022)
9. Limone, P., Toto, G.A., Cafarelli, B.: The decision-making process and the construction of online sociality through the digital storytelling methodology. Electron. **10**(20), 2465-(2021)
10. Deng, J., Qing, H.: Cross-border E-commerce course construction based on data mining algorithm. J. Phys. Conf. Ser. **1852**(2), 022040 (7pp) (2021)

Optimization of Logistics Industry Organization Management System in Digital Intelligence Era

Ying Guo[✉]

School of Wuhan, Business University, Wuhan 430056, China
LYgg1011@163.com

Abstract. With the rapid development of the world economy and the progress of modern science and technology, the logistics industry, as a new service industry in the national economy, is developing rapidly all over the world. Internationally, the logistics industry is considered as the artery and basic industry of national economic development. Its development level has become one of the important symbols of measuring a country's social and economic development level and comprehensive national strength, and is known as the "accelerator" to promote economic development. The optimization research of logistics industry organization management system in the digital intelligence era is a research field. Its main focus is to develop new technologies and find out how to implement new technologies. Research in this field requires an understanding of all aspects, including technical, social, economic, political and other related issues. It also requires an in-depth understanding of the current trend of technology development and its application to solve problems.

Keywords: Logistics industry · The age of digital intelligence · Organizational management · System optimization

1 Introduction

As an economic activity, logistics is formed with the development of commodity economy. In today's increasingly globalized economy, modern logistics, as the third profit source and an important part of the tertiary industry, is receiving more and more attention and is facing unprecedented development opportunities. As far as its original meaning is concerned, logistics refers to the physical transfer or space-time transfer of materials. With the development of social productive forces and the refinement of social division of labor, the circulation industry has gradually separated from other industries and become a bridge between production and consumption [1]. With the development of this industry, logistics is also moving towards modernization. At the same time, the demand of modern enterprises for modern logistics is getting higher and higher.

With China's entry into WTO, the pattern of market competition has changed greatly, and the competition faced by modern enterprises has become increasingly fierce. The demand of enterprises for logistics is not only to solve the differences in commodity production and consumption location and time, but also to reflect the demand for

J. C. Hung et al. (Eds.): IC 2023, LNEE 1045, pp. 332–338, 2023.
https://doi.org/10.1007/978-981-99-2287-1_47

raw materials, intermediate inventory The planning, implementation and control of the effective flow and storage of the final products and related information from the starting place to the consuming place, and the core of the logistics demand gradually turns to the substantive flow, substantive storage, information flow and management coordination [2].

Since modern logistics almost covers the whole process of commodity space movement, under the background that the supply of most commodities exceeds demand or the balance of supply and demand, there is no hope of raising prices and increasing efficiency, the constraint of resource scarcity is becoming stronger, the wage is rigid, it is difficult to reduce material consumption, and it is difficult to build new scientific and technological innovation projects, the adoption of modern comprehensive logistics by modern enterprises will be a way to reduce material consumption and improve dynamic productivity [3].

2 Related Work

2.1 Industry Background of Logistics Industry

At present, China's logistics industry is still at the primary level, and there is still a lot of room for improvement in the future. Internationally, the ratio of total social logistics costs to GDP is usually taken as an important indicator to measure a country's logistics operation efficiency and modernization. After comparing the operation efficiency of China's logistics industry with that of foreign countries, it is believed that the index of China's logistics industry is on the high side, and there is a big gap between China and developed countries [4].

The logistics industry has been growing well in recent years. From 2004 to 2009, the total amount of social logistics has been growing steadily. According to the prediction of experts, the total amount of social logistics in China will exceed 1 trillion yuan in 2010, and it will continue to grow at a steady rate in the next few years [5].

E-commerce can be divided into B2B, B2C and C2C models according to different trading objects. The latter two are collectively referred to as online shopping and are typical applications of e-commerce. In this part, iResearch consulting will mainly take online shopping as the starting point to compare and analyze the development of e-commerce at home and abroad. The so-called online shopping mainly refers to the process of transferring goods or services from merchants (sellers) to individual users (consumers) through the network The popularization and application of Internet is a necessary condition for the development of online shopping [6]. According to iResearch's research, there is a high positive correlation between the popularity and application of the Internet and the development maturity of the online shopping market in the region.

2.2　Demand Analysis of Logistics Industry

Based on the research of many local logistics companies and Taobao buyers and sellers, we believe that there are mainly the following demands in the industry:

(1)　Demand for fund payment and settlement

With the gradual rise of cash on delivery, collection and payment on behalf and other businesses, the demand of logistics companies for payment and settlement is growing, and users are also inclined to use convenient settlement methods such as credit card settlement and mobile payment settlement. According to the statistics of logistics companies, at present, about 10–15% of businesses need to carry mobile POS machines to complete payment, and this ratio is still expanding.

(2)　Requirements of logistics process management

The logistics service process is composed of many links, which involves a series of relationships between entrustment and entrustment, agent and agent. Both the logistics company and the user are very concerned about whether the node information in the logistics service process is fed back in time, such as where the goods are stored and where the goods are being shipped. Both the user and the logistics company are concerned about the information of each link in the service process. The personnel mobility of logistics companies is large. With the help of information system, the management cost will be effectively saved and the management level will be improved. The combined use of mobile data terminals will effectively provide good Sichuan experience and generate brand effect [7].

(3)　Demand for logistics order management

E-commerce is a rapidly developing emerging industry. E-commerce has also brought a large number of logistics orders. How to better realize the access to the e-commerce platform, how to allocate these orders in a low-cost manner, and how to better perform the performance control have gradually become the concerns of logistics companies.

(4)　Demand for integrated intelligent terminals

In recent years, the informatization of large logistics companies has developed rapidly, and intelligent equipment such as logistics PDA can be widely used to assist business development. With the development of payment demand in the logistics process, more and more logistics companies begin to equip POS to reduce the cash management cost of the collection business, improve the capital turnover rate and reduce the risk of counterfeit currency. However, with the normalization of payment demand, customers gradually find it inconvenient and costly to carry two types of equipment. Therefore, a large number of logistics companies hope to have integrated intelligent terminal solutions to reduce the workload of operators and reduce the cost of logistics companies.

The logistics service platform is based on the needs of the above industries. It allows small and medium-sized logistics companies to access the platform in an open platform, and realizes the logistics standardization service by combining the platform software and yitihua intelligent terminal.

3   Problems Existing in the Logistics Industry at the Present Stage

Logistics service is a series of relationships between entrustment and entrustment, agent and agent, which is completely based on the credit system. Logistics services often involve various complex rate coordination mechanisms and various settlement entities, so the fund settlement is complex. In addition, the transfer and control of property rights also need a risk control mechanism. Based on the previous research on logistics companies, this paper believes that there are several major problems in the logistics industry:

(1)   Data flow is not integrated, which makes management more difficult

The logistics service process is complex and often involves the coordination of multiple information flows. As shown in Fig. 1, it is a traditional settlement method (traditional cash settlement and traditional credit card settlement). Both of these two methods have the problem that the capital flow and the business flow are disconnected. For example, if you buy a set of cups, the order value is 88 yuan and the freight is 8 yuan. The descendants of the express company received 96 yuan for home delivery. However, in the traditional settlement method, the express company does not know which order 96 yuan corresponds to, and often needs to be reviewed manually. However, the logistics company has a huge order volume, which brings great difficulties to financial statistics [8]. The root cause is that the data flow is not integrated, and the capital flow and business flow are not matched. The disjunction of various information flows has brought great difficulties to the refined management of logistics companies, and increased the efficiency of management.

(2)   The degree of informatization is not high, affecting business expansion

In addition to several large-scale express companies, many small and medium-sized logistics companies have a low degree of informatization and a large number of manual links. For example, if the method of cash collection is adopted, the problem of change collection is often encountered during the settlement, which affects the work efficiency, and there is also the risk of carrying too much cash and receiving counterfeit money. In addition, the collection of cash collection funds is entirely manual, which increases the cost of fund management and reduces the capital turnover rate. Many risks and management costs have caused many small and medium-sized logistics companies to dare not easily intervene in the business of collecting payment for goods [9]. Relying on the manual logistics process, it is difficult to achieve transparent monitoring and poor user experience, which also affects the business expansion of the company. Therefore,

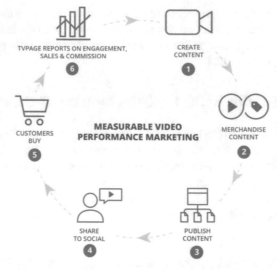

Fig. 1. Traditional payment and settlement methods

the low level of informatization makes it difficult for small and medium-sized logistics companies to expand their business, and the user service ability and work efficiency are also low.

4 Optimization of Logistics Industry Organization Management System in Digital Intelligence Era

The intelligent logistics system with the Internet of things as the core has built a three-tier SOA system, including the service management layer, the application layer and the infrastructure layer, as shown in Fig. 2.

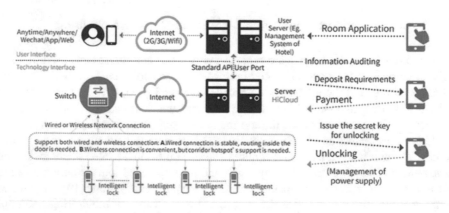

Fig. 2. IOT intelligent logistics system

This paper makes a detailed analysis from three aspects: the overall structure analysis, the infrastructure development mode and the application layer development mode.

(1) Establishing logistics service oriented information architecture

SOA system refers to a service system that closely connects the application software of various functional services through network services, logistics service interfaces and contracts. SOA itself is a programming language, hardware platform and general operating system independent of the actual service. It defines the connection interface through a third-party neutral method. Its role is to reorganize the logistics data and the logistics services of enterprises to customers, help the logistics industry establish its own business model, and thus improve the quality of logistics services. However, different logistics companies have different modes of operation and development, and the organizational forms of their systems are also different [10]. Therefore, when building the architecture of the intelligent logistics system, specific analysis should also be carried out according to the actual situation.

(2) Development mode of infrastructure layer

The infrastructure layer development model is an information platform composed of software and hardware based on the physical development architecture. Among them, through the combination of software and hardware equipment, three different functions can be realized: first, information sharing. Enterprises can provide logistics enterprises with information sharing services including software, hardware and system maintenance through third-party operators. Logistics enterprises do not need to build their own network information platform to maintain relevant information during the development process. Second, it can directly share information. It is a basic information service platform established in the form of big data. An independent firewall is established through the internal network and the external network to ensure security. However, this model also has its shortcomings, that is, a large amount of funds must be invested to carry out the corresponding construction, so there are not many enterprises currently. The third is the mixed information model, which is a mode combining sharing and private information. It not only takes the advantages of the former, but also avoids the disadvantages of the latter. Therefore, more and more people pay attention to it, and it will be the main way of its development.

At present, the construction of the intelligent logistics system is still at the initial stage, and many problems have been encountered in the construction. It is necessary to strengthen the exchange of information. To ensure the efficient operation of logistics, we must establish a unique, safe and stable network system of logistics industry; Secondly, it is necessary to build a more complete online information platform that conforms to the characteristics of the logistics industry to ensure the smooth development of relevant work. Finally, on this basis, the relevant work of the industry is further standardized to better meet the needs of users, so as to achieve the purification of the entire logistics industry environment.

5 Conclusion

With the passage of time, the development and reform of social economy, logistics has had a profound impact on human life, and the development of logistics industry has become increasingly diversified. The emergence of various means of transportation such as sea, air and land transport has made the transportation of goods more convenient. In daily life, people's demand for logistics and transportation is also increasing. For example, some users hope that the goods they buy will not be found during transportation, while others put forward new requirements for transportation speed. When using the Internet of things to build a smart logistics system, enterprises should also fully understand the needs of users, and establish the smart logistics platform that users love according to the actual needs of users to meet the needs of users.

Acknowledgements. Study on optimization of logistics Industry organization management System in the age of digital Intelligence, No: 2022CSLKT3-142.

References

1. Li, J., Wan, M.: Sensor-based mountain landslide sensitivity and logistics supply chain management optimization. Arab. J. Geosci. **14**(16) (2021)
2. Yang, A., Liu, J., Wang, Y.: Research on optimization strategy of cold-chain logistics car scheduling. J. Phys: Conf. Ser. **1910**(1), 012033 (2021)
3. Wenzhi, S., Weipeng, Z., Zhang, H., et al.: Hierarchical energy optimization management of active distribution network with multi-microgrid system. J. Ind. Prod. Eng. **39**(3), 210–229 (2022)
4. Negro-Calduch, E., Azzopardi-Muscat, N., Krishnamurthy, R.S., et al.: Technological progress in electronic health record system optimization: Systematic review of systematic literature reviews. Int. J. Med. Informat. **152**, 104507- (2021)
5. Vukobratovi, M., Mari, P., Horvat, G., et al.: A survey on computational intelligence applications in distribution network optimization. Electronics **10**(11) (2021)
6. Zhao, X.: Research on the construction of emergency logistics system in Dalian. Am. J. Ind. Bus. Manag. **11**(12), 11 (2021)
7. LiRuicheng, C.: Research on automation control of university logistics management system based on wireless communication network. Wirel. Commun. Mobile Comput. **2022,** 1–8 (2022)
8. Pan, C., Liu, M.: Optimization of intelligent logistics supply chain management system based on wireless sensor network and RFID technology. J. Sensor **2021** (2021)
9. Li, S.: Empirical study of the optimization of the lean production system in China. Int. J. Sci. Bus. **8** (2022)
10. Zhang, H., Li, X., Kan, Z., et al.: Research on optimization of assembly line based on product scheduling and just-in-time feeding of parts. Assem. Autom. **41**(5), 577–588 (2021)

Paperless Circulation of Business Management Documents Based on Electronic Signature Technology

Xuekai Zhang[1](✉), Cheng Zhang[1], and Yiqun Wang[2]

[1] State Grid Shandong Electric Power Company, Jinan 250001, Shandong, China
84190464@qq.com
[2] Economic and Technology Research Institute, State Grid Shandong Electric Power Company, Jinan 250000, Shandong, China

Abstract. The research on Paperless circulation of enterprise management documents based on electronic signature technology aims to study the current trend and problems of paperless circulation of enterprise management documents, including the importance of electronic signature technology in Paperless circulation. It also studies various methods for the electronic circulation of commercial documents, such as electronic signatures, digital signatures and other related technologies. The research also focuses on how to apply these technologies to the paperless circulation of business management documents. Paperless circulation is a step towards the goal of completely eliminating paper from business management documents. Using electronic signature technology for document authentication and verification is a way for organizations to achieve this goal. Electronic signatures have not been widely accepted, but as a substitute for handwritten signatures, they are becoming more and more popular. Handwritten signatures are easy to be forged or changed.

Keywords: Electronic signature technology · Paperless · file management

1 Introduction

In the information and networking era, the online office mode is more and more widely used. Most of the electronic documents popular in online transactions and online government affairs are presented in the form of dynamic web pages. How to ensure the security of electronic documents has become an urgent problem for all kinds of office automation systems. As an implementation of electronic signature, digital signature can effectively solve the security problem of electronic documents, but the digital signature itself is not visible to the signer, and Chinese people have a deep-rooted habit of using seals. Therefore, it is of great significance to study the web page oriented electronic signature technology for realizing safe and reliable paperless office.

Paperless business management has transformed the traditional business management into electronic business management, which is a subversive change for the business management market and has a great impact on the business management registration and holding mode. Compared with the theoretical level, this impact is more reflected in the practical activities of the business management market: from the perspective of investors, paperless business management means the change of the traditional physical business management form and the way of rights protection [1]; From the perspective of business management registration and settlement, paperless business management means that the legal basis of traditional business management needs to be changed accordingly; From the perspective of regulators, paperless business management means that traditional business management law, guarantee law and company law are facing new challenges. This profound change has a great impact on the legal effect, legal relationship and responsibility system of business management registration. It is necessary to study and discuss the special legal issues in the business management system under the paperless background, and improve the relevant system design in China, so as to solve the problems in the business management practice under the paperless background.

However, the current situation is that China has not made corresponding changes in the system design or business management legislation, which leads to the vague and uncertain rules in the confirmation, transfer, mortgage and other aspects of business management ownership. Under the background that paperless business management is in urgent need of its corresponding legal system to regulate, sorting out the views of the academic community, discussing the rights and interests of paperless business management, and comprehensively understanding the opinions and suggestions on reforming and improving China's paperless business management system will surely better promote the development of China's business management system and meet the requirements of the development of the paperless business management market [2]. Therefore, it is of great practical significance to discuss the business management rights and interests under the current background.

2   Related Work

2.1   Digital Signature

Cryptographic technology started early. After decades of development, various encryption technologies are becoming mature. From the perspective of application, they are mainly divided into symmetric key system and asymmetric key system. Unlike the symmetric cryptosystem, the key pair used in the asymmetric cryptosystem is composed of an encryption key and a decryption key, and the other cannot be calculated from one of the keys. Asymmetric cryptosystem meets the requirements of open network, does not need to transfer personal decryption secret key, simplifies the management of secret key, and can easily complete the digital signature and identity verification of both communication parties. The user can disclose the encryption key to the outside, and the private key for decryption needs to be kept strictly. Anyone can obtain the public key of the user and send the data encrypted by the public key to the user. After receiving the data, the user can restore the message using the private key. Only the owner of the private key can interpret this information. Therefore, theoretically, no unauthorized person can decrypt this message.

Digital signature is developed from asymmetric cryptosystem. It provides an identity authentication method and plays an important role in the fields with high security requirements such as banking, online trade and online government affairs. The basic principle is that the information sender uses the hash algorithm to generate a fixed length one-way hash value for the message content, that is, the digital digest. Then the sender encrypts the digital digest with the private key to obtain the digital signature, and then the sender sends the signature together with the data to be sent to the data receiver [3]. The receiver recalculates the digest of the received original data, and then decrypts the digital signature attached to the data information through the public key to obtain the decrypted information digest. If the digital signature can be correctly decrypted with the public key of the sender, This can ensure that the sender cannot deny having sent these data. Compare the calculated digest with the decrypted digest. If they are the same, the received data can be proved to be complete.

2.2 USBKEY Technology

In the asymmetric key system, the private key should be kept strictly confidential and must be kept properly from generation to cancellation. When the user obtains the key and digital certificate through CA's certificate issuing system, e-mail, website download, etc., it must be kept properly. There are many methods for storing and protecting keys. Among them, floppy disk is a key storage medium used by Ca in the early stage. The floppy disk completes the data reading and writing operation through the floppy disk drive. Although it is convenient to carry, its access speed is slow and it is easy to be damaged during use [4]. The appearance of hard disk solves the shortcomings of slow access speed and easy damage of floppy disk, but its volume is large and its mobility is poor. There is also a fatal disadvantage in using the above two media to store keys: the digital certificates stored therein are very easy to be stolen and destroyed by hackers or illegal elements. In practical applications, the best way to save data such as keys is to store them in a simple, portable and secure space. Therefore, using USBKEY key key disk to store keys is an ideal solution.

USBKEY is a USB interface hardware device with built-in microcontroller or smart card chip [125], which comprehensively adopts smart card technology, PKI technology and USB technology, and provides a secondary development interface conforming to PKI standard, which can achieve the purpose of storage, encryption and decryption and signature [126]. The internal storage space can be used to store the user's private key, digital certificate and user related information in the electronic signature system. USBKEY saves its internal data information in the form of files, and can assign different access states to files according to needs. In order to ensure the security of files, a state machine is designed inside. When you want to obtain or modify the contents of a file, the status opportunity obtains the access status of the file through the state transition. Only after obtaining the access status can you continue the operation [5]. The state transition of the state machine can only be triggered when the pin code verification or identity authentication of the user is successful, thus effectively ensuring the security of the internal data. In order to prevent hackers from brutally cracking the pin code, the USBKEY manufacturer controls the number of attempts by setting an internal calculator, thus effectively improving the reliability of the device.

3   Research on Paperless Circulation of Business Management Documents Based on Electronic Signature Technology

If we want to realize paperless office, we must modify the existing PDF documents and fill in specific data. If we want to analyze and modify the specific content of PDF, only the above basic understanding is not enough [6]. Even if we need to analyze and modify with the help of a third-party development package, it is a huge and complex project. How can we most conveniently and quickly modify PDF documents?

Here we can take advantage of the characteristics of PDF itself, because it is a structured format composed of many objects, so we do not need to analyze the existing data content, as long as we add a new data object to it. As for the layout description information, it is also a ready-made supported overlay mode. Simply put, we can achieve our goal by adding an object, telling it the layout position, and overwriting the original content. Of course, we still need to use open-source third-party development packages to realize it, which can be more convenient. Now the more successful open-source free development packages include iText and pdfbox, and the basic functions can be realized. Here we choose iText to use and explain [7].

LText is available in Java and C # versions. The description in this document uses the Java version. The following functions have been implemented in the SDK of the R & D center, including Java and C #.

According to the above ideas, we will encounter and need to solve the following problems:

1. The area found shall be covered with white background color, as shown in Fig. 1 below.

```
PdfCanvas canvas = new PdfCanvas(page.newContentStreamAfter(),
                                 page.getResources(),
                                 pdfDoc);
canvas.saveState();
canvas.setFillColor(Color.WHITE);
Rectangle area = new Rectangle(x, y, w, h);
canvas.rectangle(area);
canvas.fill();
```

Fig. 1.   Project code:

In order to prevent our modification from being displayed together with the original content, causing mixed and unsightly effects, we first cover the modified area with a white background color before overwriting the new content to ensure that the filled content can be displayed normally and clearly.

How to easily and visually determine the area to be filled and covered. The simplest way is to open the existing PDF file and draw the area we want to modify and fill with the cursor. This involves the conversion between the screen view coordinates and the PDF's own coordinate system [8]. The position coordinates of the PDF page are generally the lower left corner of the PDF page, the horizontal is the X axis, and the height is the Y axis. Therefore, the coordinates obtained from the selected area on the screen and the coordinates used for PDF modification need to be converted accordingly. Of course, before this, we must first determine the page number to be modified.

2. Fill in handwritten signature picture or text content

Create a new paragraph, set handwritten signature pictures or text contents in the paragraph, and then set the corresponding border attribute (border), margin attribute (margin) and spacing attribute (padding) to adjust the display effect. However, if it is a picture, it does not need special treatment. If it is a text content, it also needs to involve the effect of line feed display.

Because the typesetting requirements of PDF will be more beautiful, when controlling the text typesetting, if simple and common processing is adopted, it will inevitably affect the beautiful typesetting and fail to meet the requirements in the face of various situations. Therefore, many times we need to distinguish between single line text, multi line text, left and right up and down alignment format, and when the text exceeds the display range, use reduced font and line feed display to achieve better results [9].

The processing method is to cycle the font from large to small, and judge whether so many words can be displayed in this area. There are two kinds of judgment logic. The first is to use the function provided by text to judge, as shown in Fig. 2:

```
IRenderer renderer = p.createRendererSubTree().setParent(cnv.getRenderer());
LayoutArea layoutArea = new LayoutArea(1, area);
LayoutResult result = renderer.layout(new LayoutContext(layoutArea));
if(result.getStatus() == LayoutResult.FULL){
    break;
}
```

Fig. 2. Function judgment provided by text

The values of layoutresult are full, partial, and notifying. This is a method brought by iText that can be used to determine the filling status of the area. It is convenient to use. However, the use method may change due to the different versions of iText. Then, if there is any special format requirement, it may need to be processed separately according to the situation.

The seal uses pictures. In order to better fit the actual situation, it is better to make the background color of the seal picture transparent before use. If more than one seal is to be sealed, it is necessary to note that the name of the seal cannot be duplicated, otherwise the signature will fail. In the case of multiple seals, the first seal can verify whether the contents of the document have been modified, and can also know that the entire PDF document has been modified, because the following signature object is added.

If the digital certificate used here is a formal external document, it must be a formal digital certificate issued by a third-party Ca, which needs to be purchased at a cost. Each certificate has its validity period [10]. If it is internal data, you can use an internally recognized self signed certificate.

4 Conclusion

The combination of the above parts can basically realize the modification of paperless electronic document (PDF) format, as well as the anti-counterfeiting of the document, and can also record who authorized the document. This has the same effect as the actual paper document. With the improvement of informatization and automation technology, I believe this will be the only way to upgrade the office environment.

References

1. Tang, Q., Yuan, J., Qunsheng, M.A., et al.: Implementation and application of paperless filing system for medical records based on electronic signature. China Med. Dev. (2018)
2. Xiaoying, N., Li, M., Kang, M., et al.: Hospital paperless construction based on electronic signature. Cyberspace Secur. (2018)
3. Liu, S.: Research on the circulation of maritime documents based on blockchain technology. IOP Conf. Ser. Earth Environ. Sci. **831**(1), 012066 (2021)
4. Pysarenko, V., Dorohan-Pysarenko, L., Kantsedal, N.: The method of identity verification when signing electronic documents based on biometric means of identification. IOP Conf. Ser. Mater. Sci. Eng. **568**, 012103 (2019)
5. Cui, X., Lei, M., Yang, T., et al.: Design of experiment management system for stability control system based on blockchain technology. In: Intelligent Computing and Block Chain. FICC 2020. Communications in Computer and Information Science, vol. 1385. Springer, Singapore (2020). https://doi.org/10.1007/978-981-16-1160-5_24
6. Hu, C., Wang, X., Wang, Y.: Research on equipment management system based on robot laboratory. J. Comput. Commun. **8**(007), 23–31 (2020)
7. Xiaoping, D.U., Yong, Z., Tao, L.I., et al.: Application of electronization management system of document settlement based on OCR and electronic signature technique. J. State Grid Technol. College **5** (2018)
8. Jin, P., Wei, J., Meng, Y.: Implementation of paperless medical record filing in perioperative period based on digital authentication technology. Chin. Health Qual. Manag. (2019)
9. Zhao, X.Y., Cao, L.Q., Liu, M.T., et al.: Research on closed-loop management of hospitalization prepayment gold line based on blockchain technology. Health Econ. Res. (2019)
10. Kademaunga, C.K., Phiri, J.: Factors affecting successful implementation of electronic procurement in government institutions based on the technology acceptance model. Open J. Bus. Manag. **07**(4), 1705–1714 (2019)

Personalized Course Recommendation Model of University Environmental Design Based on Collaborative Filtering Algorithm

Yiyang Li[✉]

The Tourism College of Changchun University, Changchun 130607, Jilin, China
475579514@qq.com

Abstract. With the development of information technology, the teaching management system of colleges and universities has undergone corresponding changes. In the practice of teaching management in Colleges and universities, a large number of elective courses are provided for students, and there are structural deficiencies and shortcomings in course categories and majors. For example, under the organization and management mode of the course resource center and most of the current course selection methods, it is difficult for students to choose suitable courses that meet personal professional development and personality needs. In view of this, we apply the personalized recommendation technology to the course selection system to provide students with reasonable, scientific and personalized course selection recommendations according to their learning needs and interest preferences. This is a research project aimed at developing a personalized course recommendation model for college environment design. The goal is to provide students with the best learning experience and an environment conducive to their academic success. So as to avoid the blindness of students' course selection and improve the utilization rate of curriculum resources and the quality of course selection.

Keywords: Collaborative filtering algorithm · Environmental design · individualization · Course recommendation model

1 Introduction

In the information age, network is an important way for people to obtain information. With the rapid development of the Internet, network resources are growing exponentially. In the ocean of information, how can users find the information they need in a short time? How can information producers make the information they produce stand out in a short time and get the attention of users? And the times require the cultivation of diversified and personalized talents and the construction of an innovative country. In view of this, personalized recommendation system came into being [1]. The task of personalized recommendation system is to connect users and information well. On the one hand, it provides users with personalized, accurate and new information needs in a short time according to their different interests and hobbies; On the other hand, we should make

J. C. Hung et al. (Eds.): IC 2023, LNEE 1045, pp. 345–352, 2023.
https://doi.org/10.1007/978-981-99-2287-1_49

full use of information and show it to users who are interested in it, so as to achieve a win-win situation between users and information producers [2].

In the information age, colleges and universities began to implement the course selection system and credit system. Elective courses are offered to coincide with the talent training methods in the information age, that is, to cultivate diversified and personalized innovative talents. At the same time, the report of the 18th CPC National Congress clearly pointed out that "we should comprehensively implement quality education, deepen comprehensive reform in the field of education, strive to improve the quality of education, and cultivate students' innovative spirit". At present, there are a large number of elective courses in the course selection system of colleges and universities[3]. Facing this situation, how can students find courses suitable for their interests in a short time? For this thorny problem, some scholars apply personalized recommendation technology to college course selection system, and design a course selection system with personalized recommendation function to provide personalized services for students. Personalized information service refers to the service that enables users to obtain information that meets their specific interests and preferences. It has been widely used in e-commerce and some web search engines [4]. However, there are still data sparsity and cold start problems in the practice of personalized recommendation technology, which affect the accuracy of recommendation.

Based on the above analysis, this paper proposes an architecture of personalized course selection recommendation system based on course attributes and attribute value preference matrix, which provides a solid foundation for the next development of a course selection system with the function of personalized recommendation of different elective courses for students according to their interest preferences and learning needs.

2 Related Work

2.1 Personalized Information Service

Personalized information service is a service that provides users with product information that meets their unique personality needs according to their habits, usage behaviors, hobbies and characteristics. It was first proposed at the artificial intelligence association.

Personalized information service, first of all, is a service that analyzes the needs of users and actively organizes information sources to meet their individual needs; Secondly, it is also a service to cultivate personality and guide needs, which helps individuals find personality and promotes the diversification and diversified development of the information society. It provides services for users in the following three ways: first, users customize the information resources and services they need according to their own needs and hobbies, and create a personalized information environment [5]; Second, according to the user's characteristics and personality, the information provider takes the initiative to provide him with the most important information resources and services, and dynamically updates the information provided to him as the user's interests and needs change; Third, readers can easily and quickly log in all digital library series that are most similar to their own needs and interests according to their needs, which is the basis of personalized information services.

At present, many websites provide personalized services for users, such as movie finder, Amazon, Taobao, etc. With the development of personalized information services, libraries also continue to introduce personalized services. Foreign relatively perfect systems include mygateway and MyLibrary. At the same time, domestic personalized libraries are not willing to lag behind, For example, "my digital library is based on personalized integration and customization portal" established by the National Science Digital Library of the Chinese Academy of Sciences and "digital library personalized information service system (kingbasedl)" of the library of Renmin University of China. Providing students with personalized information services in all aspects is conducive to realizing personalized education and cultivating personalized, diversified and innovative talents.

2.2 Composition of Personalized Recommendation System

The recommendation engine part, data input part and result output part are the three major components of the personalized recommendation system, and the core part of the system is the recommendation engine, which determines whether the data recommended to users is suitable for users' interests and needs.

The recommendation engine part is based on the data input part, using personalized recommendation technology and related technologies to analyze it and predict the recommendation results. Personalized recommendation technology and its corresponding recommendation algorithm will be introduced in detail later in the article.

The data input part is the interface to obtain information such as user interests and needs. In terms of acquisition methods, information is divided into: display information and implicit information. Display information refers to the basic information of individuals, the scoring information of some items, and the questionnaires filled in the system or website, etc. Implicit information refers to users' browsing methods, access paths, favorites and other behaviors in the website, which can be automatically recorded in the log file by the system, and users' interests and needs can be inferred from it by using data mining technology [6].

The result output part mainly displays the results generated by the calculation of the system recommendation section to the user. Its display modes are:

(1) It is recommended to recommend top-N products for users according to their browsing history and interests, such as the "you may also like" recommendation column on Taobao and Zhuoyue.
(2) The prediction scores the prediction of the recommended project according to the user's evaluation of other projects. For example, the column "customers who buy this product buy it at the same time" on zhuoyue.com.
(3) Individual scores output the scores and average scores of other users on the target items. For example, on Taobao, customers rate products.
(4) Comments output other users' comments on the descriptive text of the target project.

3   Collaborative Filtering Algorithm

3.1   Overview of Collaborative Filtering Algorithm

Content based filtering technology comes from the field of information acquisition. The algorithm calculates the similarity between the user model and the information model in the system, and recommends resources with large similarity to users. At present, some recommendation systems use this technology to realize personalized recommendation, such as personal web watcher, newseeder, infofinder and other systems.

Content based recommendation technology produces recommendations for users from two aspects: on the one hand, it recommends items with high matching degree to users through the comparison of user model and item feature model; On the other hand, it is to classify the items, judge the category of users' scores or purchased items, and then recommend similar items to users.

The extraction of information features, the description of user personality characteristics and the calculation of similarity between them determine the accuracy of the recommendation algorithm [7]. The algorithm has high requirements for the description of user interest file and the feature extraction of resources. If the user interest file fails to accurately describe the user's interest, the recommendation results calculated by using it may not be consistent with the user's real interest, and the feature extraction of resources is the same. The construction standard of user's personality description and project's feature description should be consistent, otherwise it will affect the calculation of similarity between them.

The similarity between user model and information model in content-based recommendation algorithm is the core of the algorithm. A function that calculates this similarity is as follows:

$$P(i,j) = \overline{R}_i \frac{\sum_{n \in Ni} sim(i, N) \bullet (R_{n,j} - \overline{R}_N)}{\sum_{n \in Ni} sim(i, N)} \tag{1}$$

The features of content-based recommendation algorithm are as follows:

(1)   The recommendation results are easy to understand and do not need domain knowledge;

(2)   User interest description files, project feature extraction and their similarities can be carried out offline;

(3)   Only the displayed information of resources and users can be analyzed, and it is difficult to accurately obtain implicit information such as the style of users and resources;

(4)   Because the recommendation of this algorithm is based on the user's previous interest description file, it is difficult to find the user's new interested resources due to the lack of user feedback.

(5)   The recommended resources are severely limited by the feature extraction ability of the recommended object, which is limited to document resources and cannot effectively extract the features of multimedia resources.

(6)   The cold start problem of new users, because the system is difficult to obtain accurate information of users.

3.2 Recommendation Technology Based on Collaborative Filtering

Collaborative filtering algorithm is a widely used algorithm in personalized recommendation technology, and has become a research hotspot in academia. Collaborative filtering is also called social filtering. The research of this technology is mainly divided into two categories: Global based collaborative filtering and model-based collaborative filtering.

(1) User based collaborative filtering technology

The emergence of this algorithm marks the birth of recommendation system, because it is the oldest algorithm in recommendation system. It was proposed in 1992 and then applied to mail filtering system and news filtering system[8]. The user based collaborative filtering algorithm is to recommend the resources that the target user may like by calculating the similarity of the user's access behavior, as shown in Fig. 1. The user based collaborative filtering algorithm is mainly divided into two steps:

I.   find a set of users with similar interests to the target users;
II. find items in this collection that the user likes and the target user has not heard of, and recommend them to the target user.

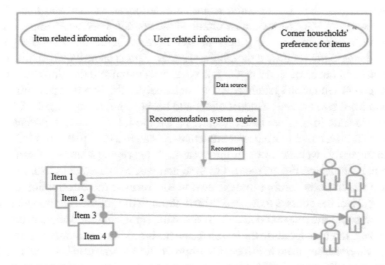

Fig. 1. Principle of course recommendation based on collaborative filtering algorithm

(2) Project based collaborative filtering technology

Item based collaborative filtering was proposed by Professor sarwr in 2001. The algorithm idea is to calculate the similarity between user behaviors from the original user based collaborative filtering algorithm to calculate the similarity between items to find the nearest neighbor of the recommended object. It is currently used in personalized recommendation systems, such as Netflix, excellence, youtube and so on [9]. The steps of Project-based Collaborative filtering algorithm are as follows:

a. Calculate the similarity between items;

    I. build a project recommendation list for users according to the similarity of the project and the historical behavior of users. Using this algorithm, the

4 Personalized Course Recommendation Model of University Environmental Design Based on Collaborative Filtering Algorithm

4.1 Functional Requirements of the System

This system is a personalized course resource recommendation system combined with the interests of users. Its users are mainly student users and system administrators. In order to determine the needs of users, we have made a detailed survey of learners before the development of the system, such as: the ways and methods for students to know the course information, and whether they are satisfied with the obtained course information. After that, the research results are summarized, and the main focus of learners is whether the system can recommend the courses they want, and whether the course information can meet their personalized needs. Therefore, when designing this system, we should not only meet the common needs of most users, but also fully consider the personalized needs of learners. On this basis, we design and develop the curriculum resource recommendation system. Next, we will analyze the functional requirements of the system from two aspects: Student client and background management [10].

The main functions of the client include registration, login, course retrieval, course recommendation, course scoring, etc. Student users can use the browser on PC or mobile phone to enter the website to open the system. After registration, they can log in to the system by entering the correct user name and password. Student users can search the course resources according to their own needs, browse the information of relevant courses, or select the courses they need to learn through the courses recommended by the system, and score the relevant courses. At the same time, student users can also upload some course related materials in the learning materials module, or choose to download the learning materials they need to effectively improve their learning efficiency. In addition, students can modify some of their own information in the "user center" module, such as gender, college and other information.

4.2 Model of Personalized Course Selection Recommendation

With the continuous popularization of the credit system, colleges and universities have generally established a course selection system, which includes many courses, including public elective courses and compulsory courses. Facing so many courses, how can students choose the courses they like? This is a difficulty faced by the current course selection system. In view of this, according to the personalized needs of learners' learning style, learning needs and learning ability, this paper adds a personalized recommendation function to the course selection system to effectively solve this problem, so as to avoid the blindness of students' course selection. Its model architecture is shown in Fig. 2.

Student model: the student model in the personalized course selection recommendation model includes the basic personal information and personality characteristics of learners. Get the basic information of learners from the registration information, analyze the learning style measurement scale filled in by learners, get the academic information of learners in the process of interaction with the system, and store these three aspects of information in the student model to provide the basis for the recommendation engine.

Course model: it refers to describing courses from the relevant attributes of courses, so that learners can comprehensively understand each course as a whole, choose courses according to their own interests and needs, and provide data basis for the recommendation engine.

Recommendation engine: according to the information in the course model and student model, and the calculation of the similarity between the two models, we can get the similar neighbors of the course and students or the similarity between the course and students. Collaborative filtering recommendation technology is used to provide new, popular and personalized courses for learners, which not only meets the needs of students' personalized learning, but also expands students' vision.

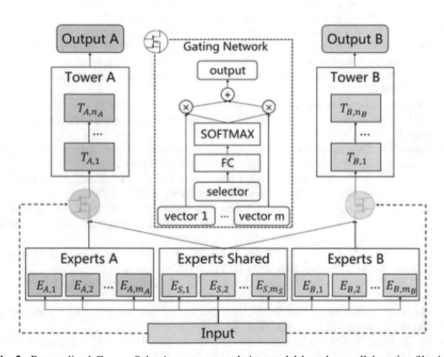

Fig. 2. Personalized Course Selection recommendation model based on collaborative filtering algorithm

5   Conclusion

The personalized college course selection recommendation system based on collaborative filtering algorithm calculates the similarity according to students' interests, degrees of learning and majors, and then recommends the corresponding courses according to the similarity. The use of personalized course selection recommendation system in Colleges and universities can effectively improve students' learning interest and school teaching quality, help students choose appropriate courses scientifically and reasonably, and provide effective learning methods for students' personalized development.

Acknowledgement. 1. The second batch of industry university cooperation collaborative education project "Research on the construction of curriculum system of environmental design specialty in private colleges and Universities under the background of new liberal arts construction" of the Department of higher education of the Ministry of education in 2021.

2. In 2021, the "golden course" construction project of the school of tourism of Changchun University, the course name is "indoor hand-painted performance techniques".

References

1. Cheng, H., Gan, B., Zhang, C.: Research on personalized recommendation method based on social impact theory. J. Phys. Conf. Ser. **1848**(1), 012128 (7pp) (2021)
2. Wu, H., Meng, F.: Research on the application of personalized course recommendation of learn to rank based on knowledge graph. In: Li, W., Tang, D. (eds.) Mobile Wireless Middleware, Operating Systems and Applications. MOBILWARE 2020. LNICS, Social Informatics and Telecommunications Engineering, vol. 331. Springer, Cham (2020). https://doi.org/10.1007/978-3-030-62205-3_2
3. Zhang, S.: Research on recommendation algorithm based on collaborative filtering. In: ICAIIS 2021: 2021 2nd International Conference on Artificial Intelligence and Information Systems (2021)
4. Song, H.Y., Zhang, H., Xing, Z.H.: Research on personalized recommendation system based on association rules. J. Phys. Conf. Ser. **1961**(1), 012027 (10pp) (2021)
5. Ni, Z., Ni, F.: Research on knowledge graph model of diversified online resources and personalized recommendation. J. Phys. Conf. Ser. **1693**(1), 012178 (7pp) (2020)
6. Li, J., Qi, S., Chen, L., et al.: Research on personalized recommendation based on big data technology. In: 2020 5th International Conference on Mechanical, Control and Computer Engineering (ICMCCE) (2020)
7. Liu J , Choi W H , Liu J , et al. Personalized movie recommendation method based on deep learning. Math. Probl. Eng. **2021,** 1–12 (2021)
8. Rui, D.: Research on the practical teaching mode of environmental design based on the "ntegrated" Educational Whole Value Chain. **2021**(2) (2021)
9. Zhang, T.: Research on environmental landscape design based on virtual reality technology and deep learning. Microprocess. Microsyst. **81**(4), 103796 (2020)
10. Hu, X., Wang, Y., Chen, Q.B., et al.: Research on personalized learning based on collaborative filtering method. J. Phys: Conf. Ser. **1757**(1), 012050 (2021)

Personalized Recommendation of Literature Resources in University Library Based on Abstract Content Filtering Algorithm

Yumei Dang[✉]

The Library of Guangxi Normal University for Nationalities, Chongzuo 532200, Guangxi, China
dymglg@163.com

Abstract. The purpose of this study is to recommend the literature resources of University Libraries Based on content filtering algorithm. This study uses a method that combines the concept of abstract content filtering with personalized recommendation. To achieve this goal, we use a database containing information about books and articles in various fields such as the humanities, social sciences and natural sciences. The main purpose of this study is to understand whether the use of abstract content filters can improve the recommendation quality of different types of users, especially those who are interested in certain topics or have specific needs. In order to achieve these goals, we used two methods: one is based on comparing the results obtained from the two.

Keywords: Content filtering · library · Personalized recommendation · literature

1 Introduction

With the arrival of the big data era, the explosive growth of information has become more and more intense. The data is rich but the key information is poor, and the information is submerged in the general data. In the face of massive Internet database resources, users are more and more concerned about how to quickly and effectively obtain useful information and the latest information, which has also become a hot area for researchers. In this case, the traditional service mode of the library as the information center of university literature resources can no longer meet the requirements of university teachers and students [1]. Especially under the impact of big data technology, university libraries are facing various severe challenges. As a characteristic service of university libraries, it is very necessary to improve the personalized recommendation of literature resources and innovate the original service mode, To meet the needs of different users.

In general, users solve the problem of information overload through the information retrieval function, that is, inputting keywords. However, this traditional information retrieval mode does not take into account the personalized needs of users. In the face of more and more diversified information needs of readers, the traditional information retrieval mode appears to be more and more powerless. In addition, in the face of the explosive growth of new resources, the demand of readers is also changing greatly with the growth of information. The traditional service model can no longer meet the needs of readers for personalized information. Due to the lack of timely and accurate access to the required resources, a large number of document resources are wasted because they can not be used, which has a great impact and challenge on the existing service mode of the library. Information recommendation technology has brought a breakthrough to the solution of this problem.

The information recommendation technology establishes the user interest model by describing the user interest characteristics. The description basis is the user behavior, needs and other information contained in the user's past information retrieval records. Finally, the appropriate recommendation algorithm is used to match the object to be recommended with the user interest model, and recommend the literature resources of interest to the user [2]. And dynamically track the update of user needs in a timely manner and recommend useful information to users in a real-time manner. Personalized information recommendation has become a hot topic in information service research. It helps users to obtain literature information timely and accurately, and provides personalized information recommendation services for different users. It is more and more popular.

2   Related Work

2.1   Collaborative Filtering Recommendation

Collaborative filtering recommendation needs to build a user item score matrix, calculate the target user or the neighbor of the target item, predict the target user's score on the resource according to the neighbor's score on the resource, and finally form a top-N recommendation list to recommend 36 for the user. Its advantage is that it does not depend on the specific content of resources, only needs user rating data, and can handle video, audio and other resources that are difficult to extract features, such as Facebook, Amazon, shrimp music and Douban. In the collaborative filtering recommendation algorithm, it is divided into user based collaborative filtering and Project-based Collaborative filtering according to the different objects to be considered. The collaborative filtering based on users calculates the user similarity according to the user item score matrix, searches for the nearest neighbor for the target user, predicts the score of the target user on the resource according to the evaluation of the neighbor set on the resource, and takes it as the basis for recommendation to the target user [3]. Collaborative filtering recommendation based on items needs to calculate the similarity between resources first, and predict the user's score on the target resources according to the user's score on similar resources. The difference in application occasions is the main difference between user based collaborative filtering and Project-based Collaborative filtering. The former can reflect the degree of preference of projects in small groups including users, while the Project-based Collaborative filtering is more personalized and stable, and can reflect the inheritance of users' interests, so as to facilitate the exploration of users' potential interests.

2.2 Text Word Segmentation

Word segmentation refers to the process of dividing words with independent meanings from natural language. The divided words are the smallest units of computer language processing. The Chinese text is not like the English text, which naturally uses spaces to separate words. In Chinese, the boundary between "word" and "phrase" is fuzzy. Therefore, in the process of Chinese text processing, it is first necessary to segment the Chinese text into one word or phrase. This technology becomes the Chinese word segmentation technology [4]. There are three kinds of existing word segmentation algorithms: word segmentation based on statistics, word segmentation based on understanding and word segmentation based on string matching. Each of the three word segmentation algorithms has its own characteristics. Different word segmentation algorithms should be selected according to the characteristics of the text. Sometimes, it is necessary to integrate multiple word segmentation algorithms to process the text content, so as to give play to the advantages of each algorithm, learn from each other's strong points to meet the needs of different users.

Word segmentation is realized by word segmentation. The accuracy of keyword extraction by word segmentation has a great impact on improving the accuracy of recommendation. Currently, the popular Chinese word segmenters include text abstracts and keywords, which are developed based on different program languages. At present, the most popular Chinese word segmentation device is Jieba word segmentation, which is encoded in Python language [5]. In order to meet the needs of different users, the word splitter provides a simplified mode, a full mode and a search mode for users to choose. In addition, the word splitter also supports a user-defined dictionary with user personalized characteristics, and also supports traditional Chinese words for users to choose.

2.3 Keyword Feature Extraction

TF-IDF (term frequency inverse document frequency) algorithm is an important text feature extraction technology, which is widely used in the fields of document classification and search, and is used to evaluate the importance of words to the text in the document set or corpus. TF stands for word frequency, indicating the frequency of a word in its document; IDF represents the inverse document frequency, which is used to evaluate the distribution of words in the whole document library (i.e., the universality of words to the document library); The TF-IDF value is obtained by multiplying the TF value by the IDF value. In a document, the higher the TF value of a word, the higher the frequency of occurrence of the word in the document. The larger the IDF value, the more concentrated the distribution of the word in the whole document library. In short, the higher the TF-IDF value, the stronger the ability of the word to distinguish the content attributes of the document, and the greater the weight. The calculation formula is as follows:

$$\arg \min_{SC} \sum_{i=1}^{k} \sum_{x \in C_i} |X - \mu_i|^2 \tag{1}$$

The TF-IDF value of the words is calculated by the above formula. This calculation method can make the words representing the characteristics of the document get a higher weight, and ordinary words (words that cannot reflect the characteristics of the document) get a lower weight [6]. According to this feature of TF-IDF function, words with large weight in the document can be extracted as feature words that can represent the characteristics of the document.

3 Research on Personalized Recommendation of University Library Literature Resources Based on Abstract Content Filtering Algorithm

Vector space model is widely used in the field of text vectorization, which mainly involves the extraction of feature words and the calculation of their weights (the main algorithm is TF-IDF). However, for a large resource library such as university library, if electronic documents are expressed in the form of vectors, the dimension will be very large, which is easy to cause a dimension disaster and make the recommendation system inefficient. In this paper, the method of combining the cosine value R and the matching degree value SIM is proposed to improve the original method of recommendation based on vector space model. Where R is obtained by extracting the word vector of the feature word with large weight from the user interest information and the document resource information, and then calculating the cosine value of any two word vectors by the cosine formula. Combining the corresponding feature word weight, the matching degree value SIM of the user interest and the document resource is further calculated, and the documents with high SIM value are recommended to the target user in a list. Compared with the original recommendation based on vector space model, on the one hand, the feature words with large weight are screened out, which can not only represent the characteristics of literature resources, avoid the problem of dimension disaster caused by text vectorization, but also reduce the computational complexity and improve the operational efficiency of the recommendation model [7]; On the other hand, for users, extracting feature words with large weight from the user interest information to express their features can more accurately express the user's interests, and avoid that the recommendation range is too wide due to the generalization of user interests (the number of feature words is large) and weaken the significance of personalized recommendation. The following Fig. 1 shows the filtering process of the personalized recommendation system in the recommended literature resources.

Description of main contents of the model:

1. Data preprocessing. Data cleaning and text type conversion shall be carried out for the collected user reading record information and documents to make them meet the data requirements of the model, so as to facilitate word segmentation, feature extraction and word vector training.
2. Participle. Text word segmentation is a very important part of document processing. The accuracy of word segmentation has a great impact on improving the accuracy of recommendation. At present, the most popular Chinese word segmentation device is Jieba word segmentation, which is encoded in Python language. In order to meet the needs of different users, the word splitter provides a simplified mode, a full mode and a search mode for users to choose [8]. In addition, the word splitter also supports

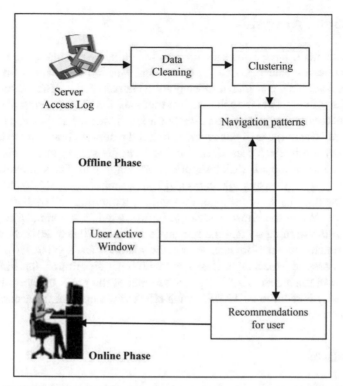

Fig. 1. The filtering process of personalized recommendation system in recommended literature resources

a user-defined dictionary with user personalized characteristics, and also supports traditional Chinese words for users to choose.

3. TF-IDF。TF-IDF algorithm is used to extract the features of user interests and documents. This algorithm was first proposed by Salton. Its main idea is that the more times a word appears in the document, and the smaller the scope of occurrence in the whole document set, the stronger the ability of this word to distinguish the content attributes of the document. The weight calculation formula of feature words is w = TF * IDF.

Word2vec。Word2vec is a word vector training tool, which can train word vectors for documents after word segmentation. The tool provides cbow and skip gram training models, and combines hierarchy softmax and negative sampling optimization technology to efficiently express words into vectors [9].

⑤ Literature recommendation basis calculation module. This module is the core content of the improved model. Then, the SIM is normalized to make it between 0 and 100 for observation and processing. Finally, the first n literatures with large SIM values are recommended to the target users.

4   Simulation Analysis

For this experiment, three schemes with certain similarity calculation are considered, namely Euclidean distance, cosine similarity and Pearson correlation. Some specific methods have not been considered to weight the characteristics. This factor is set to 1. The formation of neighbors is applied to two methods, that is, the correlation weighting threshold and the maximum neighbor. Finally, three schemes are designed in combination with the score. The purpose of this article is to build a digital library and help users find the books they are interested in. Users can log in to the system after registration, and relevant evaluation standards shall be applied after logging in. The simulation system is used to form a set of 200 users, 200 items and 750 evaluations, of which 600 evaluations are used as the training set of the algorithm and the remaining 150 evaluations are used as the test set. When the above evaluation is processed in the simulation, it will be distributed to the training set and the test set at a ratio of 80% to 20%. In the analysis of the experiment, two performance evaluation methods are applied [10]. The first is accuracy, the second is coverage. Compared with other algorithms, the MAE index of the algorithm in this paper is 9.8% higher than that of the previous algorithm, and the time complexity is reduced by 19.7%, which effectively improves the accuracy of book recommendation.

5   Conclusion

The purpose of this study is to study the effectiveness of personalized recommendation system in recommending literature resources. Literature resources can be any type of information that supports academic research, such as journal articles, books, and other types of publications. The main purpose of this study is to evaluate the personalized effectiveness of content-based filtering algorithms. In addition, it aims to provide suggestions for improving the quality and efficiency of library use. To achieve these goals, we developed an automated method for extracting abstracts from web pages using machine learning algorithm (I-Net) to select relevant literature resources.

References

1. Liang, Y.: Study on Library Big Data Literature Service Scheme of Big Data Colony Algorithm. Springer, Cham (2022)
2. Wang, X., Xu, X., Zhang, J., et al.: Research on intelligent construction algorithm of subject knowledge thesaurus based on literature resources. J. Phys. Conf. Ser. **1955**(1), 012038 (10pp) (2021)
3. Cong, H.: Personalized recommendation of film and television culture based on an intelligent classification algorithm. Pers. Ubiquit. Comput. **24**(2), 165–176 (2020)
4. Chaabi, Y., Ndiyae, N.M., Lekdioui, K.: Personalized recommendation of educational resources in a MOOC using a combination of collaborative filtering and semantic content analysis. Int. J. Sci. Technol. Res. **9**(2), 3243–3248 (2020)
5. Li, S.: Research on the propagation characteristics of negative news information based on personalized recommendation algorithm. Math. Probl. Eng. **2022** (2022)

6. Chen, S., Huang, L., Lei, Z., et al.: Research on personalized recommendation hybrid algorithm for interactive experience equipment. Comput. Intell. **36**(3), 1348–1373 (2020)
7. Li, L.: Learning recommendation algorithm based on improved bp neural network in music marketing strategy. Comput. Intell. Neurosci. **2021**(2021)
8. Wang, Y., Qiu, T., Nikodem, M., et al.: A systematic literature review of revealed preferences of decision-makers for recommendations of cancer drugs in health technology assessment. Int. J. Technol. Assess. Health Care **38**(1), e36- (2022)
9. Chen, J., Wang, Z., Zhu, T., et al.: Recommendation algorithm in double-layer network based on vector dynamic evolution clustering and attention mechanism. Complexity **2020**(3), 1–19 (2020)
10. Fu, L., Ma, X.M.: An improved recommendation method based on content filtering and collaborative filtering. Complexity **2021** (2021)

Pharmacology Database and Analysis Based on Cloud Computing Technology

Renhui Feng[✉] and Tao Feng

Zaozhuang Vocational College of Science and Technology, Zaozhuang 277599, Shandong, China
guitarist86@163.com

Abstract. As a hot topic in the field of research and application in recent years, cloud computing is considered by most IT enterprises and industry insiders as the core architecture of the next generation computer network application technology. Pharmacology database and analysis (PDAC) is a web-based application that allows users to access the latest pharmacology information of more than 6000 drugs. The database contains drug profiles, including chemical structure, molecular weight, molecular formula, CAS registration number and other relevant data. Users can also search for drugs by name or drug action category. In addition, users can perform various analyses by themselves or in collaboration with others using the powerful tools of PDAC.

Keywords: Cloud computing technology · database pharmacology

1 Introduction

In the 1990s, the network, as a novel and convenient information medium, was gradually recognized by people. People are aware of its huge scale of computing resources, fascinated by its huge application prospects, and set about to study how to use these resources efficiently and conveniently. At present, the new knowledge economy based on Internet has become an important goal pursued by developed countries. Obviously, the development and expansion of the Internet economy has become one of the most important indicators to measure the level and quality of a country's modernization development [1]. During the 12th Five Year Plan period, China will comprehensively improve the informatization level, promote the deep integration of informatization and industrialization, realize the integration of three networks, and build a next-generation national information infrastructure with broadband integration and security. This is the proposal of the Central Committee of the Communist Party of China on formulating the 12th Five Year Plan, which was released a few days ago. It was adopted at the fifth plenary meeting of the 17th Central Committee of the Communist Party of China on october18,2010 [2].

In the field of biomedicine, the digitalization of various instrument platforms and numerous digital sensors are producing a large amount of data all the time. In the biological information industry, with the development of sequencing technology and the enhancement of computer computing power, the sequencing price of the whole genome has dropped from hundreds of millions of dollars a decade ago to thousands of dollars

© The Author(s), under exclusive license to Springer Nature Singapore Pte Ltd. 2023
J. C. Hung et al. (Eds.): IC 2023, LNEE 1045, pp. 360–366, 2023.
https://doi.org/10.1007/978-981-99-2287-1_51

today, which makes it possible for more people and species to obtain DNA information [3]. Jason Bobby, head of the personal genome project at Harvard Medical School in the United States, believes that 50million people will have personal genetic maps by 2015. With the rapid development of mobile devices and mobile Internet, portable physiological devices are becoming popular. If individual health information can be connected to the Internet, the resulting amount of data is immeasurable. The European Bioinformatics Research Center (EBL), located in the UK, is part of the European Molecular Biology Laboratory and one of the largest bioinformatics data centers in the world. At present, it has 20pb of data, including genome information, protein information, small sub data, etc. There are about 2PB of genome data in EBI, and the amount of genome data is increasing by 2PB every year. BGI is one of the largest genomic data producers in the world, producing 6tb of genomic data every day. Since 1982, the data in GenBank has doubled almost every 18 months [4]. The amount of raw data generated by the "thousand human genome project" will be close to the level of Pb.

At present, with the popularization and application of cloud computing technology, people have the ability to use large-scale distributed computing resources in the network. As a hot topic in the field of research and application in recent years, cloud computing is considered by most IT enterprises and industry insiders as the core architecture of the next generation computer network application technology. In the cloud computing environment, users can no longer spend high hardware and software costs to have powerful computing resources and huge storage capacity, all of which can be completed by cloud computing service providers. It not only saves the cost, but also does not need to spend a lot of energy [5].

2   Related Work

2.1   Concept and Scope of Traditional Medicine

In recent years, traditional medicine has received renewed attention all over the world. According to the survey, 80% of the population in developing countries depend on traditional drugs provided by plants or animals. In addition, people The failure rate of research and development of synthetic drugs and toxic and side effects show a worrying growth trend. Therefore, people's attention also tends to traditional and natural drugs. However, traditional drugs have complex mechanisms and components, which have been the key obstacles to the development of traditional drugs [6]. Network pharmacology provides new ideas and methods for analyzing the scientific connotation of the complex system of traditional medicine, contributes to the new drug design and research and development of traditional drugs, and points out the direction for the modern research and development of traditional drugs. Network pharmacology will certainly become a useful help to promote the inheritance and innovative development of traditional medicine.

Traditional medicine refers to the drugs from the natural world used by a certain race or nation through generations and forming traditional experience It has the following characteristics: nationality, which is closely related to the culture, religion, customs and habits of the nationality; Regionality is closely related to the region, flora and fauna and natural resources where the nationality lives; Tradition is closely related to the historical and cultural conditions of the nation [7]. Therefore, traditional medicine is more from the perspective of the experience accumulated by different ethnic groups in using natural drugs. Their substance bases and letans are natural products. Once these substances are proved to have medical value, they are natural drugs. For example, commodity production, they are crude drugs. The discipline of studying natural drugs or medicinal materials (crude drugs) is commonly known as "pharmacognosy".

2.2 Study on the Material Basis and Action Mechanism of Traditional Drugs

The research on the material basis and its action mechanism is the basis for the formulation of the quality standards of traditional drugs, the discovery of innovative drugs and the research in various fields of the modernization of traditional drugs. At the same time, it also plays an important role in the development of the international market for traditional drugs. Although the research on the material basis of traditional drugs has a history of nearly 100 years, the material basis of most traditional drugs has not been clarified yet [8]. The main reasons are as follows: first, the composition of traditional drugs is complex and the research is difficult; Second, the degree of attention is insufficient, and there are few organized and systematic studies; Thirdly, chemistry and pharmacology can not be well coordinated; Fourth, due to financial reasons, although many traditional drugs have been preliminarily studied, there is no systematic and in-depth research.

Although the research on the action mechanism of the material basis is the weakest field in the research of traditional drugs at present, it is the most important key point that can best explain the action characteristics of traditional drugs and let the world know about traditional drugs. The rapid development of modern biotechnology, especially the continuous elucidation of disease pathogenesis and drug action targets, as well as the development of genomics, proteomics and metabonomics, provides an effective method for the study of traditional drug action mechanism. The research on the action mechanism of the material basis of drugs can be carried out from two aspects [9]. First, under the guidance of its traditional medical theory and in combination with the action characteristics of natural drugs, the whole animal model is used to study the action mechanism in vivo (for example, the compatibility of drugs is studied by using the traditional Chinese medicine theory of "monarchs, ministers and envoys"; Secondly, the mechanism of action in vitro was studied at the cell level and enzyme level to clarify the target.

The chemical components of traditional drugs are complex. Even a single traditional drug has dozens or even hundreds of components, while the components of traditional compound drugs are more complex. At present, people in the industry basically agree that traditional drugs have the characteristics of multi-component and multi-target action, but there is no effective method to study the characteristics, material basis and mechanism of action of multi-component and multi-target in the complex system of traditional drugs.

3 Pharmacology Database and Analysis Based on Cloud Computing Technology

3.1 Cloud Computing Technology

Cloud computing can provide users with the computing power, storage capacity and application capacity allocated on demand. The final purpose is to facilitate users and greatly reduce users' software and hardware procurement costs. Cloud computing is the comprehensive development of distributed processing, parallel processing and grid computing. It is also the result of virtualization, SaaS (software services), has (hardware services), PAAS (platform services) and other comprehensive applications.

Or the commercial realization of these computer science concepts. Many multinational information technology companies such as IBM, Yahoo and Google are using the concept of cloud computing to sell their products and services. As long as we have a mobile phone or a pen and notebook, we can get the services we want through the browser client, even including services such as supercomputing. In fact, users are the owners of cloud computing in this respect. Simply put, cloud computing is to take advantage of the ability to process data on the Internet and large-scale data computing center software to separate complex computing from stand-alone computers and run on the Internet.

(1) Storage data intensive cloud platform

Storage data intensive cloud computing platform is a cloud computing platform that mainly provides data storage and search services, and wins customers by providing customers with safe and convenient cloud storage services, as shown in Fig. 1. Cloud storage uses the powerful storage capacity of the server cluster in cloud computing to save data for customers. Users do not need to know whether their files are stored on one server node or among multiple nodes, nor do they need to know whether the nodes are trusted. These will be handled by the cloud server. There are no technical barriers to the implementation of cloud storage. It requires cloud devices, cloud software, cloud services and so on to be organically integrated to provide users with barrier free cloud services [1].

Existing cloud computing providers provide basic cloud storage services, which are based on their own distributed file storage systems. Google has the largest information base and knowledge base today, and has its own uniqueness in mass storage. The proposed GFS file storage system can realize real-time monitoring, fault-tolerant detection, automatic recovery and other functions of the file system. It is a relatively good file system built on the storage conditions of untrusted nodes. It is efficient and highly optimized for the management of large files, but it does not provide an effective optimization scheme for the storage of small files. So it can not fully adapt to the massive small file storage in the cloud computing environment. Fastdfs is an open source file system. It also does well in mass storage and load balancing, but it is still not reasonably optimized in small file storage.

(2) Computing computing intensive cloud computing platform

Computing computing intensive cloud computing platform is a cloud computing platform mainly focusing on data computing and processing services, providing users with a corresponding level of high-performance computing environment. Users can also select the corresponding computing power according to their own needs. Through the high-performance computing power of the cloud computing platform, users and enterprises can obtain computing power comparable to the existing mainframe, and carry out large-scale data processing and calculation, which is convenient for enterprises and individual users.

(3) Integrated cloud computing platform

The integrated cloud computing platform is to effectively integrate the powerful storage of cloud computing and super capacity computing. While reasonably utilizing the storage space of the storage nodes of the cloud cluster, it does not waste the computing power of each node, and realizes the integration of cluster storage and computing power through corresponding strategies to process and calculate data.

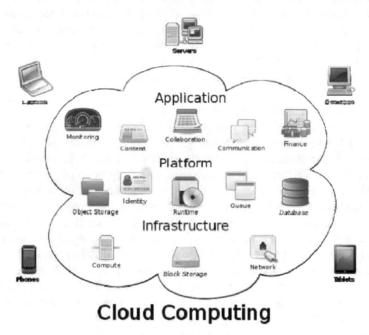

Fig. 1. Conceptual structure of data intensive cloud platform

3.2 Pharmacology Database

1) Data storage form

As the name suggests, a document oriented database is composed of a series of self-contained documents. This means that all relevant data is stored in the document, not in the relational tables of the relational database. In fact, there are no tables, rows, columns or relationships in document oriented databases, which means that they are schema independent and do not need to define strict schemas before actually using the database. If a document needs to add a new field, it only needs to include the field, so it does not affect other documents in the database. Therefore, the document does not have to store empty data values for fields that do not have values.

Suppose that a large number of business cards need to be stored in the database. The information on each business card includes the name of the person, the name of the company, the position and contact information (which may include landline number, mobile phone number, fax), etc. In a relational database, you need to use more than four tables to store these data: a "person" table, a "company" table, a "contact details" table, and a table for storing the business card itself. These tables have strictly defined columns and keys, and use a series of joins to assemble the data.

2) Unique identifier

Another difference between the two databases is the storage of unique identifiers. You can usually use primary keys in relational databases, which are generated by an auto increment feature or sequence generator. Of course, these identifiers are unique only to the table or database used, and can be used by other tables or databases. If two databases on different networks are updated at the same time, the two databases will not get the next unique identifier accurately at the same time. CouchDB has no automatic increment or sequence feature. Instead, it assigns a universal unique identifier (UUID) to each document, which prevents other databases from accidentally selecting the same unique identifier.

CouchDB can adapt to a wide range of application scenarios. In some applications that are occasionally connected to the network, we can use CouchDB to temporarily store data and then synchronize. It can also be used as a large-scale distributed data storage in the cloud environment. Next, we will analyze CouchDB in detail from its architecture 1. Its architecture is shown in Fig. 2:

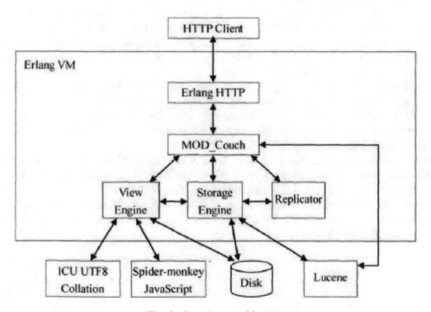

Fig. 2. Database architecture

4 Conclusion

The core problem of cloud computing is data management. However, the demand for cloud computing is subversive to the traditional data management technology, because the fundamental assumption of the traditional data management technology is no longer tenable. In the cloud computing environment, according to the characteristics and requirements of cloud computing applications, it has become an important research topic to study, design and develop new data management technology for cloud computing environment by using the experience of traditional data management technology for reference.

References

1. Li, Y., Wei, G., Zhuang, Z., et al.: A Network Pharmacology-based study on active ingredients of corydalis rhizoma on coronary heart disease (2020)
2. Liu, P., Yang, P., Zhang, L.: Mode of action of shan-zhu-yu (cornus officinalis Sieb. et Zucc.) in the treatment of depression based on network pharmacology. Evid.-based Complement. Altern. Med. **2020**(1), 1–10 (2020)
3. Zhu, Y., Yu, J., Zhang, K., et al.: Network pharmacology analysis to explore the pharmacological mechanism of effective chinese medicines in treating metastatic colorectal cancer using meta-analysis approach. Am. J. Chin. Med. **49**(08), 1839–1870 (2021)
4. Wu, S.S., Hao, L.J., Shi, Y.Y., et al:. Network pharmacology-based analysis on the effects and mechanism of the wang-bi capsule for rheumatoid arthritis and osteoarthritis (2022)
5. Liu, Y., Wei, P., Qiu, Z., et al.: Study on mechanism of Shufeng Jiedu granules in treating novel coronavirus pneumonia based on network pharmacology. In: AIP Conference Proceedings. AIP Publishing LLC AIP Publishing (2020)
6. Jingshuo, A.N., Sun, W., Wang, Y., et al.: Analysis of the Mechanism of Sanqi (Panax Notoginseng) on hepatocellular carcinoma based on network pharmacology (2020)
7. Li, D.H., Su, Y.F., Sun, C.X., et al.: A Network pharmacology-based identification study on the mechanism of xiao-xu-ming decoction for cerebral ischemic stroke. Evid.-Based Complement. Altern. Med. **2020**(4), 1–8 (2020)
8. Kropiwnicki, E., Evangelista, J.E., Stein, D.J., et al.: Drugmonizome and Drugmonizome-ML: integration and abstraction of small molecule attributes for drug enrichment analysis and machine learning. Database J. Biol. Databases Curation 2021 (2021)
9. Chugh, N., Sharma, D.K., Singhal, R., et al.: Blockchain-based decentralized application (DApp) design, implementation, and analysis with healthcare 4.0 trends. Basic Clin. Pharmacol. Toxicol.y **126**(S4):139–140 (2020)

Power Engineering Cost Prediction Based on Clara Algorithm to Optimize SVM Parameters

Yanqin Wang[1]([⊠]), Zhen Dong[2], Na Li[1], Yong Wang[2], Ning Xu[1], and Hongshan Zhang[1]

[1] State Grid Hebei Electric Power Company Economic Research Institute, Shijiazhuang 050000, Hebei, China
wangyanqinsjz@126.com
[2] State Grid Hebei Electric Power Company, Shijiazhuang 050000, Hebei, China

Abstract. In the long-term engineering practice, various engineering related units have accumulated a large number of project cost example data, because there is no appropriate application technology, the data can not be used directly and effectively. The project cost has always been calculated according to the quota calculated by the national statistics, and there are often very big differences in the budget. Therefore, the main objective of this paper is to apply the optimal control method to power optimization by using an extended version of the classical model. The method is divided into two steps: first, we propose a new method to calculate power consumption; Second, we take it as the input of adjusting control parameters. To this end, we use a new model, which includes electrical and mechanical components. In addition, it enables us to estimate the electrical and mechanical losses due to its different mechanisms with high accuracy.

Keywords: Clara algorithm · SVM parameters · Electricity · Project cost · forecast

1 Introduction

After half a century of development, computer and information technology have brought great changes and impacts to human society. Among the three elements (energy materials and information) that dominate human society, information increasingly shows its importance and dominance. It pushes human society from the era of industrialization to the era of informatization, and makes all major institutions in modern society involved in the wave of data and its processing (data collection, storage, retrieval, transmission, analysis and representation). With the expansion of the scope of human activities, the acceleration of the pace, and the progress of technology, people can obtain and store data in a faster, easier and cheaper way, which makes the amount of data and information grow exponentially [1]. However, human activities are based on human wisdom and knowledge, that is, observation and understanding of the external world, correct judgment and decision-making, and taking correct actions. Data is only the raw material

© The Author(s), under exclusive license to Springer Nature Singapore Pte Ltd. 2023
J. C. Hung et al. (Eds.): IC 2023, LNEE 1045, pp. 367–373, 2023.
https://doi.org/10.1007/978-981-99-2287-1_52

obtained by people observing the external world with various tools and means, and it has no meaning in itself. From data to wisdom, we need to go through the process of analysis, processing and refining. Data is raw material, it just describes what happened What matters, it does not provide a reliable basis for judgment or explanation, and action [2]. People analyze the data, find out the relationship, and give the data a certain meaning and relevance, which forms the so-called information. Although information gives some meaningful things in the data, it is often not related to the task at hand and cannot be used as the basis for judgment, decision-making and action. Only through reprocessing and deep insight can we obtain more useful information, that is, knowledge The so-called knowledge can be defined as "a group of logical connections in information blocks, whose relationships are found through the closeness of context or process." Understand its mode from information, that is, form knowledge [3]. On the basis of a large amount of knowledge accumulation, summed up into principles and rules, we form the so-called wisdom How to acquire knowledge from data and how to use these data and knowledge to provide productivity pose new challenges to the field of computer and information technology.

Engineering cost estimation is an important topic in the engineering industry. In the long-term engineering practice, our country has accumulated a large number of historical case data. Because there are many influencing factors involved in the project cost, and the influence of these factors on the overall project cost is relatively complex, it is difficult to form accurate and effective knowledge to guide the decision-making of the project. The relevant departments of the state use statistics to produce various quotas to guide the project cost budget and estimate to support decision-making [4]. Due to various reasons, this pricing method is not very ideal. Under the influence of backward pricing technology, the large design estimation error has become a common problem, and the reliability of estimation is very low.

This paper focuses on the historical examples of engineering cost, and some key problems of knowledge utilization of engineering data. It is committed to establishing a reasonable case engineering cost knowledge representation model. Intelligent technology and fuzzy mathematics are applied to solve the key problems of Clara in the field of project cost estimation. Design and implement a project cost estimation system based on Clara.

2 Related Work

2.1 Engineering Cost Theory

The essence of project cost is the total cost of the project from planned construction to completion. Through the comparative analysis of the relevant expenses incurred in the budget and the actual cost consumed in the construction of specific projects, it is confirmed that the comprehensive and effective utilization of human, material, equipment, funds and other resources can be achieved in the construction activities of electric power projects. During the specific implementation of the project, determine the total budget, specify the standard quantity, effectively control the total cost, and estimate the items, key facilities and staff required for the construction. Promote the management of project principal and the improvement and utilization of cost [5]. In the process of effectively managing the cost of power projects, we must grasp every step of construction activities from the details, and implement cost management activities from the design of the project to the commencement of the project and then to the completion of the project. Let them play their due role, control the construction progress, minimize errors, and improve economic benefits. In this regard, the cost of power engineering has crucial research value: strictly follow the requirements of the standard to carry out the cost management activities of engineering projects, which will promote the rapid development of power engineering construction.

Under the social background of the continuous improvement of the development level of China's socialist market economy, the demand for power resources has also increased significantly. The construction of power engineering projects is developing towards the trend of large-scale, high-quality and high-level. According to reliable data, during the "13th five year plan" period, the total investment price of China's power grid was twice that of the "12th Five Year Plan" period, and it was still increasing [6]. Therefore, doing a good job of cost control of power project activities can not only improve the quality of project construction, but also improve the progress, and can achieve healthy construction, but also reduce costs and improve the economic effectiveness of the project. In today's society, economy is becoming more and more important in life, so the demand for electricity is also growing. Therefore, in the construction of power engineering projects, it is a problem that every project cost designer should consider to effectively control the project cost and improve the cost efficiency of the project. Some special power projects, such as HVT system, have the characteristics of complexity, difficulty and high consumption, which leads to the different characteristics of power engineering projects from general engineering projects:

(1) Complexity: the project cost is not a matter of one unit or department, but involves multiple organizations, such as the power company, construction unit, supervision unit, material company and design unit in this paper. A complete power engineering cost process requires the cooperation and cooperation of the above units.

(2) Integrity: the important contents of project cost management are estimation, budget, settlement, etc. They not only have a clear division of labor, but also are inseparable. Only when they are treated uniformly in project cost management, can we make project construction activities and resources reasonable, efficient and practical, and make the continuous improvement of project economic development.

(3) Dynamic: the market prices of materials, equipment, energy and other materials involved in power engineering project activities are also changing. In this case, in order to ensure the accuracy of project cost management activities, changes need to be made according to the relevant contents of different construction stages [7]. Therefore, from another perspective, the project has a strong dynamic, and the cost management that pays attention to project activities also has a high dynamic.

(4) Phased: according to the relevant national regulations, the project activities of any project have certain steps and stages, so that the project is divided into multiple closely related but different stages and links. The significance and value of each stage are different. Among them, investment estimation can make the

smooth development of power engineering projects, and has important guidance and demonstration effect on the development of power engineering projects.

2.2 Basic Principles of Support Vector Machine (SVM)

We can describe the definition of support vector machine as follows: find a hyperplane, which can separate the data in the training set, minimize the training error of samples, and have the shortest weight vector. Those training samples close to the interface are the support vectors to determine the optimal hyperplane.

In other words, the implementation of support vector machine is to map the input vector to a high-dimensional feature space by implementing selective nonlinear mapping, and construct an optimal classification hyperplane in this feature space. Usually, in the linear case, the dual problem is a convex quadratic programming problem, and its solution only involves the inner product operation of vectors. Therefore, in order to construct a hyperplane in the feature space, it is not necessary to consider the feature space in the form of display, but only to know the inner product operation in this space [8]. The output is a linear combination of intermediate nodes, and each intermediate node corresponds to a support vector, as shown in Fig. 1 below.

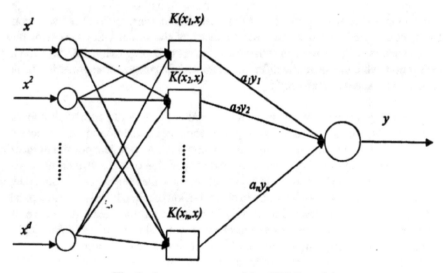

Fig. 1. Support vector machine (SVM) model

If all vectors in the training set can be correctly divided by a hyperplane, and the distance between heterogeneous vectors closest to the hyperplane is the largest (i.e. maximum marginalization), then the hyperplane is the optimal hyperplane. The heterogeneous vector nearest to the hyperplane is called support vector. A set of support vectors can uniquely determine a hyperplane.

3 Power Engineering Cost Prediction Based on Clara Algorithm to Optimize SVM Parameters

3.1 Functional Requirements Analysis

Through the investigation of the engineering projects of the power supply company, it is found that it mainly includes the construction of substation and transmission line, and the voltage level ranges from 220 V to 500 kV. For the convenience of unified management, the construction of substations and transmission lines are collectively referred to as transmission and transformation projects in this paper. The construction of power transmission and transformation projects requires a lot of money. Generally, the construction is carried out in a phased mode, which is mainly divided into investment decision-making stage, bidding stage, design stage, construction stage and completion settlement stage. Therefore, cost management must be scientifically managed according to the business of each stage. The author mainly participated in the research and development of engineering management, budget estimation/budget management, material management, system management and other modules [9].

The system management module includes four management functions: unit information, department information, user information and user group information. The system management function can only be accessed and operated by the system administrator. It is mainly responsible for maintaining the basic data of the system, and provides the operations of adding, modifying, deleting and querying the basic data information. The demand of budget management is to reasonably control the budget data generated in different stages of the project. Similar to the budget management, the budget management is still the budget report data and related engineering information submitted by the project centralized department, and the budget management department reviews the submitted budget data. If it is approved, it will be distributed to relevant departments for implementation. If the approval fails, the submitted data will be returned with approval comments, and the reporting department needs to continue to modify the budget data. The budget management module of this paper mainly provides two functions: reporting budget information and querying budget information.

The non functional requirements of the system are different from functions. Functions affect whether the system can complete business operations, while non functions generally affect the interaction with system users. For example, whether it is convenient for users to query information, whether it is fast to publish information, and whether there are hidden dangers of information leakage. The non functional requirements of this system mainly include the following points:

(1) The scalability, maintainability and reliability of the system are extremely important. When the system is unstable or suddenly disconnected, whether the system has corresponding emergency measures, whether the business logic changes during the use of the system, whether the system can quickly change to meet the requirements, and whether the system can recover data when some data is lost due to system failure or improper user operation are all issues that the system needs to consider.
(2) The usability of the system is also an issue to be considered in this project. Ease of use involves interactive design, user interface, user behavior and other aspects.

There are three principles of ease of use, namely, easy to learn, easy to see and easy to use. Easy learning requires users to help with systematic learning through online documents; It is easy to see that the function operation of the system can be reflected in the interface, and it is easy for users to operate correctly; And easy to use is the operating system that users can quickly become familiar with.

(3) The security of the system is the feature that each system should ensure. Through the installation of firewall and other technologies, it can prevent the loss of data and viruses in the system, resulting in unnecessary losses.

3.2   Design of Power Engineering Cost Management System

Software and data structure are the two modules of the transformation requirements of system design. The purpose of software structure design is to divide the complex system into modeling, confirmation interface and activation interface according to functions. The structure design describes the data characteristics and designs the database. This chapter analyzes and studies the system design in detail, including the overall architecture, outline architecture, detailed functions and database model of the system [10]. This paper selects the three-tier architecture technology to develop the system. The overall architecture of the system is shown in Fig. 2, which mainly includes the presentation layer, business layer, data layer and external system interface.

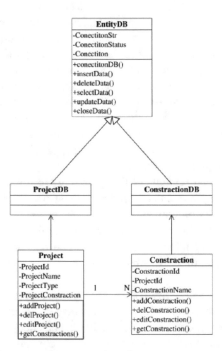

Fig. 2. Structure of power engineering cost management system

The system network structure mainly refers to the physical topology of the network connection between the nodes in the system, which is generally described from the perspective of the underlying physical transmission. The network structure of this system refers to the network topology connecting the company's servers and workstations. The important nodes in the topology include servers, switches, clients, firewall workstations. The web server and database of this system are built in the central machine room of the power supply company, and the workstations in the company directly access this system through the LAN. For project management parties outside the company, such as design units, construction units, material units, etc., access to the company's router through a dedicated line, and then enter the company's intranet through a protective wall filter, and finally achieve access to this system. The design of this network structure not only allows all units of project management to access the system, but also ensures the security of intranet data.

4 Conclusion

Information technology is penetrating every corner of society. Using information technology to improve the social productivity of traditional industries is a topic of concern in society. It is believed that many traditional fields will present a new situation in the application of information technology. The research of this paper is to explore the application of intelligent computing technology to improve the traditional engineering cost estimation and mine the use value of engineering cost case data. This paper creatively applies CBR and other intelligent computing technologies to the estimation and reasoning of project cost, and achieves relatively satisfactory results.

References

1. Xu, X., Peng, L., Ji, Z., et al.: Research on substation project cost prediction based on sparrow search algorithm optimized BP Neural network. Sustainability **13**, 13746 (2021)
2. Ma, L., Wang, Q., Liao, S., et al.: Research on cost prediction of power transmission and transformation project based on combination prediction model. In: IOP Conference Series: Earth and Environmental Science, vol. 769, no. 4, p. 042019 (7/pp)
3. Wang, H., Liang, Y., Ding, W., et al.: The improved least square support vector machine based on wolf pack algorithm and data inconsistency rate for cost prediction of substation projects. In: Mathematical Problems in Engineering, vol. 2020 (2020)
4. Tan, P., Duc, T., Tran, C.M., et al.: Continuous QoE prediction based on WaveNet. In: ICCAE 2020: 2020 12th International Conference on Computer and Automation Engineering (2020)
5. Zhang, Y., Pan, G., Chen, B., et al.: Short-term wind speed prediction model based on GA-ANN improved by VMD. Renew. Energy **156**, 1373–1388 (2020)
6. Chen, W., Qi, X., Fang, J., et al.: Research on single tower cost prediction technology of transmission line engineering based on 3D Design. J. Phys: Conf. Ser. **1601**, 022035 (2020)
7. Chen, Y., Su, Y., Du, X.: Research on equipment life cycle cost prediction based on GA-LSSVM. J. Phys: Conf. Ser. **1605**, 012089 (2020)
8. Wang, X., Liu, S., Zhang, L.: Highway cost prediction based on LSSVM optimized by intial parameters. Comput. Syst. Sci. Eng. **36**(1), 259–269 (2021)
9. A mathematical method for operation and maintenance cost prediction based on transfer learning under non-stationary power data. Adv. Appl. Math.**10**(1), 98–108 (2021)
10. Maitanova, N., Telle, J.S., Hanke, B., et al.: A machine learning approach to low-cost photovoltaic power prediction based on publicly available weather reports. Energies **13**, 375 (2020)

Power Prediction of Solar Photovoltaic Power Generation Based on Matrix Algorithm

Wenbo Yang[✉]

Lanzhou Resources and Environment Voc-Tech University, Lanzhou 730010, Gansu, China
mayn555@163.com

Abstract. Photovoltaic power generation is affected by complex and changeable weather conditions, which has the characteristics of volatility and intermittence, resulting in difficulties in photovoltaic grid connection. In order to ensure the stable operation of power system after large-scale photovoltaic grid connection, it is necessary to accurately predict the photovoltaic power in advance. Different weather conditions will lead to great differences in the irradiance actually received by photovoltaic arrays, so a single prediction model is not enough to cope with the prediction of photovoltaic power generation under various complex weather changes. The precision of photovoltaic data clustering is not the key to improve the precision of photovoltaic data classification. To solve the above problems, this paper proposes a matrix algorithm based on solar photovoltaic power prediction.

Keywords: Matrix algorithm · Solar energy · Photovoltaic power generation · Power prediction

1 Introduction

The problems of environmental pollution and limited storage of traditional energy force mankind to vigorously develop new energy technologies. At present, most electricity is generated from conventional power resources, such as fossil fuels. The traditional energy is fossil fuel, which is a kind of hydrocarbon or its derivatives, including oil, natural gas, coal and other natural resources, which has made a great contribution to the development of the national economy. However, in the process of its development and application, it has brought a series of environmental problems, such as acid rain, soot, greenhouse effect, atmospheric soot and destruction of the ozone layer. Moreover, the storage of traditional fossil energy is limited, which is non renewable energy. For the sustainable development of mankind, mankind must vigorously develop new energy. As a revolution in energy development, the development and application of green energy is being carried out all over the world [1]. The so-called green energy refers to the electric energy converted from renewable energy such as solar energy, water energy, hydrogen energy and wind energy through specific power generation devices. Its biggest advantage is that it does not discharge or rarely discharges pollutants such as waste gas and wastewater harmful to the environment in the production process. The most successful use of solar energy is the solar cell manufactured by applying the photoelectric conversion principle, that is, photovoltaic power generation [2].

Solar photovoltaic power generation system is a power generation system that converts the energy radiated by the sun into electric energy through solar cells. The amount of photovoltaic power generation mainly depends on the total radiation received on the solar panel, but this radiation is not uniform over time. The direct current output from the photovoltaic power station shall be converted into alternating current through the inverter and incorporated into the power system. The variability and intermittence of solar energy make it difficult for photovoltaic power to be connected to the power grid. The variability of solar energy resources and the uncertainty related to prediction are the root causes of most problems that must be solved to maintain the stability of power grid. Part of the wave is determined and explained by the rotational and translational motion of the earth relative to the sun, which is accurately described by physical equations. However, there are also unexpected changes in the solar irradiance reaching the earth's surface, mainly due to the existence of clouds, which randomly block the sunlight, making the prediction of photovoltaic power generation uncertain. The instability of solar power generation may lead to unnecessary increase of rotating reserve and operation cost [3]. This makes people need to accurately predict the solar photovoltaic power generation at different time intervals, so as to ensure the stability of the power grid by balancing supply and demand while maintaining low cost. It is very important to accurately predict the output power of photovoltaic system, which has been considered as the key challenge of large-scale photovoltaic integration.

2 Related Work

2.1 Deep Learning Method

Deep learning algorithm has achieved great success in the fields of computer vision, natural language processing, speech recognition and so on. Deep learning, a simple generalization, is to enhance its feature learning ability by constructing a deep neural network. The main deep learning algorithms include RNN, LSTM, Gru, CNN, etc. The LSTM model is constructed to predict the short-term photovoltaic output, which only takes the weather description, temperature and humidity as the model input, and explores the influence of LSTM layer capacity change, FC layer capacity change, dropout mechanism and different activation functions on the prediction accuracy of the model [4]. Compared with multilayer perceptron, cyclic neural network and bidirectional cyclic neural network, the prediction accuracy of LSTM photovoltaic power generation model is second only to bidirectional cyclic neural network, and the training time is shorter than bidirectional cyclic neural network. Although deep learning has achieved great success in image processing and text mining, it still has many shortcomings. First, there are many super parameters in deep learning, and the learning performance depends heavily on fine parameter adjustment; Second, the training of deep learning model needs a large amount of data. Therefore, deep learning is difficult to apply to the learning task of small-scale training data, and sometimes even fails because of medium-scale training data; Third, neural network is a black box model, its decision-making process is difficult to explain, and its learning behavior is difficult to analyze with theory. Moreover, before training the deep learning model, we must determine the neural network architecture and predict the complexity of the model. Therefore, the deep learning model is usually more complex than the actual model. Deep neural network increases the complexity of the model by increasing the depth of the network [5]. The larger the variance of the model, the better, but it will not lead to excessive complexity in the model learning.

2.2 Combined Model Method

Develop different prediction models for different types of photovoltaic data. Firstly, the k-means algorithm is used to cluster the weather data, and a variety of clustering performance metrics (including calinski harabasz index, silhouette coefficient and Davies bouldin index) are used to determine the optimal number of clusters K. After determining the division of photovoltaic data, multiple neural networks are trained for each type of photovoltaic data. The number of hidden layers of each neural network is different, and the photovoltaic power prediction value is equal to the median of all neural network prediction values. This method can predict the photovoltaic power output of the photovoltaic power station every half an hour the next day according to the previous photovoltaic power generation and weather forecast data. Compared with the previous method, it can predict the output of all power of the day at the same time, rather than making iterative prediction according to the previous prediction value[6]. The prediction duration is 7.00am to 4.00pm per day. The evaluation index is measured by mean relative error (MRE). The experimental results show that the prediction accuracy of integrated neural network is 3.6% higher than that of single neural network.

In order to integrate the advantages of multiple algorithm models, mind evolutionary computation (MEC) is used to optimize BP neural network, particle swarm optimization (PSO) is used to optimize support vector machine, and the limit learning machine model is established. Then the variance covariance weight dynamic allocation method is used to combine the three models to predict the 31 day photovoltaic data as a whole. In unknown prediction, multiple algorithms are organically assembled to take advantages and avoid disadvantages, complement each other, and ensure reliability and adaptability. The experimental results show that the prediction accuracy of the combined method is closest to that of the best single model, but it is not better than that of the best single model.

The photovoltaic training set data are divided into three categories by K-means clustering. In the training stage, each kind of data set is trained by random forest algorithm; Each random forest model is used to predict the whole training set, and the three prediction value vectors are input into ridge regression as feature vectors for training, so as to determine the combined weight of the three random forests. In the test stage, three random forest models are combined with weights for combined prediction. The advantage of this method is that it does not need to classify and predict the photovoltaic data of the test set[7].

To sum up, there are two strategies to improve the prediction accuracy of photo-voltaic power generation: first, strengthen the cluster analysis of photovoltaic data and realize fine classification; second, optimize the training of prediction model and select complementary algorithm model for combined prediction.

3 Solar Photovoltaic Power Prediction Based on Matrix Algorithm

3.1 Matrix Performance Measurement

The purpose of the matrix is to make the data in the same category as similar as possible, while the data in different categories should be separated as far as possible, that is, "birds of a feather flock together". For the matrix results, to know the quality of the matrix effect, we must measure the performance. Matrix performance measures can be roughly divided into two categories: one is to compare the matrix results with a "reference model" to become an external indicator; the other is to directly evaluate and investigate the matrix results without using any reference model to become an "internal indicator". Internal verification indicators are usually based on the following two criteria: first, cohesion. It mainly measures the close relationship between objects in the same group[8]. A series of measures evaluate cluster compactness based on variance, and a lower variance indicates a better compactness. In addition, there are many methods to estimate the compactness of the matrix according to the distance, such as the maximum or average pairwise distance and the maximum or center based average distance. Second, the degree of separation. It measures how different or separate a cluster is from other clusters. For example, the pairwise distance between cluster centers or the pairwise minimum distance between objects in different clusters is widely used as a measure of separation, as shown in Fig. 1. In addition, density based measures are used in some indicators.

$$S(i) = \frac{b(i) - a(i)}{\max\{a(i), b(i)\}} \tag{1}$$

3.2 Random Forest

In ensemble learning, according to the different learning methods of multiple weak learners, it can be divided into bagging algorithm and boosting algorithm. Through the repeated extraction of the training set samples, each sample sampling constitutes a sampling set, and each sampling set obtains a weak learner through training. If we sample t times, we will generate T sampling sets, train and generate T weak learners, and finally combine t weak learners into a powerful learning. Bagging's combination strategy is relatively simple. For the classification problem, the simple voting method is generally used to obtain the category with the most votes as the final output; For the regression problem, the simple average method is generally used. The regression results predicted by T weak learners are arithmetically averaged to obtain the final model output.

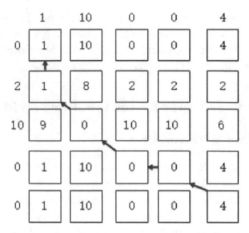

Fig. 1. DTW distance matrix and optimal path

Random forest E3 (RF) algorithm is an integrated learning algorithm based on bagging. Its weak learner is composed of a series of decision trees. And the random forest improves the generation of each decision tree. Assuming that the total number of sample features is n, in the generation process of ordinary decision tree, the best feature among n sample features is selected as the division basis of left and right subtrees when the node is split, while each decision tree of random forest randomly selects K features from n features (k is less than n), and then selects the best feature among K features to divide the left and right subtrees. The random forest algorithm carries out repeated sampling of the training set for many times. Each sampling trains a decision tree to generate a model [9]. Finally, multiple decision trees are used for classification or regression prediction. The model structure is shown in Fig. 2. In the training model stage, the k-means algorithm based on Euclidean distance is used to cluster the actual photovoltaic power of the historical data set, and the photovoltaic historical data set is divided into k classes. Each class establishes a photovoltaic power prediction model based on random forest [10]. The prediction vector of K random forest models on the whole photovoltaic historical data set is input into ridge regression to determine the combined weight of K random forests. In the prediction stage, the weather data of the prediction day is input, K prediction vectors are obtained by K random forest models, and the weighted combination prediction is carried out by using the weight determined in the training stage.

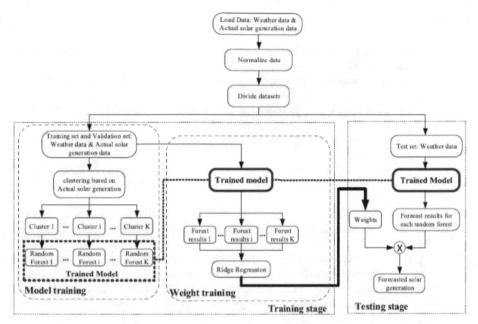

Fig. 2. Photovoltaic power generation prediction model based on matrix algorithm and combined random forest

4 Conclusion

Solar photovoltaic power generation is affected by the alternation of day and night, meteorological conditions and other factors, and has the characteristics of fluctuation and randomness. This characteristic will lead to large-scale photovoltaic power generation and grid connection, which will do harm to the power system. If the photovoltaic power can be accurately predicted in advance, it is of great significance to the power system dispatching. Therefore, this paper proposes a matrix algorithm based solar photovoltaic power prediction. According to the visual comparison between the predicted photovoltaic value and the actual photovoltaic power, it can be found that the prediction method proposed in this paper can predict the photovoltaic power in stable weather quite accurately, but there are deficiencies in the prediction of photovoltaic power in severe weather fluctuations, and it has not been able to accurately predict the photovoltaic power in severe sudden weather. The main reason here is that the actual irradiance changes randomly in the case of severe weather changes. Due to the lack of cloud information in the data set, it is very difficult to accurately predict the actual irradiance.

Acknowledgements. 2021 Innovation fund project of Gansu Provincial Department of Education (Item No: 2021B-429) ; 2021 General project of Gansu Academy of Educational Sciences(Item No: GS[2021]GHB1768) ; 2020 drought Meteorological Science Research Fund Project(Item No: IAM202009).

References

1. Sarin, C.R., Mani, G.: Demand Response of a Solar Photovoltaic Dominated Microgrid with Fluctuating Power Generation (2021)
2. Bae, S.: Solar photovoltaic power prediction using big data tools. Sustainability **13**, 13685 (2021)
3. Anqi, A.E.: Small-scale solar photovoltaic power prediction for residential load in Saudi Arabia using machine learning. Energies **14**, 6759 (2021)
4. Zazoum, B.: Solar photovoltaic power prediction using different machine learning methods - ScienceDirect (2022)
5. Carrera, B., Min, K.S., Jung, J.Y.: PVHybNet: a hybrid framework for predicting photovoltaic power generation using both weather forecast and observation data. IET Renew. Power Gener. **14**, 2192–2201 2020
6. Lin, G.Q., Li, L.L., Tseng, M.L., et al.: An improved moth-flame optimization algorithm for support vector machine prediction of photovoltaic power generation. J. Clean. Prod. **253**, 119966 (2020)
7. Zhang, S., Dai, H., Yang, A., Shi, Z.: Environmental parameters analysis and power prediction for photovoltaic power generation based on ensembles of decision trees. In: Shi, Z., Vadera, S., Chang, E. (eds) Intelligent Information Processing X. IIP 2020. IFIP Advances in Information and Communication Technology, vol. 581, pp 78–85. Springer, Cham (2020). https://doi.org/10.1007/978-3-030-46931-3_8
8. Azka, R., Soefian, W., Aryani, D.R., et al.: Modelling of photovoltaic system power prediction based on environmental conditions using neural network single and multiple hidden layers. In: IOP Conference Series Earth and Environmental Science, vol. 599, p. 012032 (2020)
9. Liu, Y.B., Ying-Li, W., Zhang, W.: Design of small solar power generation system based on GA-BP prediction algorithm. Sensor World **40**, 304–321 (2020)
10. Pereira, S., Abreu, E., Iakunin, M., et al.: Prediction of solar resource and photovoltaic energy production through the generation of a typical meteorological year and Meso-NH simulations: application to the south of Portugal (2020)

Prediction and Analysis of Network Literature Value Based on Ant Colony Algorithm

Na Zhao[1,2]($\boxtimes$)

[1] Northwest Minzu University, Lanzhou 730030, Gansu, China
106363500@qq.com
[2] Lanzhou University of Arts and Science, Lanzhou 730030, Gansu, China

Abstract. The value prediction and analysis of network documents based on ant colony algorithm is a technology to determine the value of a book, including its marketability. This technology uses an algorithm based on ant colony theory. It is used to predict the value of books according to their marketability. The main idea behind this method is to use ants to simulate human beings and their behavior towards objects or other things they are interested in. Ants are famous for their frugality in choosing objects that are more suitable for them than other objects, so we can say that they are good at making them to solve many problems, such as predicting the value of network literature, finding the best path from one node to another, or predicting which node to choose for a task. The main concept behind the algorithm is that it uses ants to find out how and where they can get food and water. Ants can do this because they have wisdom that other animals and creatures do not have. This means that ants can quickly learn methods that are useful to them and use their knowledge accordingly.

Keywords: Ant colony Network literature · Forecast analysis

1 Introduction

Nowadays, the threshold of online writing is low and the efficiency is high. Tens of thousands of words a day are not a problem for GaN di. What he pursues is nothing more than being cool, crazy, proud and killing. However, the content of tens of thousands of words is often very empty, and the tweeting is serious, as is the daily diary of primary school students. Moreover, the plot can't stand scrutiny, the logic collapses, and the protagonist often has serious double labels. What's more, he is a Versailles patient as soon as he comes on stage. I liked such works 100% when I was a student, but after being severely beaten by the society, I found that some gods were embarrassed out of the universe. Because this kind of work only satisfies the fantasy of some readers and even the author (the wealth is invincible, and the beauty is like clouds), it is naturally shallower in depth. The plot is in a straight line, and people with a certain experience want more than pleasure.

Besides, why don't you want to write without depth? It's not true. Some great gods write small white texts, with a clear positioning[1]. First of all, to write works with a certain depth, the author must study the plot, and it is certain to constantly overturn and rewrite, which takes a lot of time. Calvin is even more casual, and his efficiency will become low, which is not in line with commercial value. Secondly, the writing style is not only to tell stories smoothly, but also to have amazing psychological descriptions, scenery descriptions, fighting descriptions and even overall view descriptions, etc. the words are beautiful and accurate (think Luo Shen's Fu is written in vernacular, do you still think it's beautiful?), Without a lot of reading, it is impossible. In other words, the efficiency will become low again, which is not in line with commercial value. Therefore, xiaobaiwen is the best choice. From the perspective of readers, they are impatient and unwilling to move their heads when reading. Whether it is foreshadowing or foreshadowing, it is a serious waste of time in the eyes of readers. It is best to directly destroy the world after enjoying it. Therefore, xiaobaiwen is still the best choice. Based on this, this paper studies the prediction and analysis of network literature value based on ant colony algorithm.

2   Related Work

2.1   Ant Colony Algorithm

Basic idea of ant colony algorithm description ant colony algorithm is a new bionic algorithm derived from the biological world of nature. It is a positive feedback process of learning information. It is a swarm intelligence algorithm designed according to the ant foraging principle, that is, the search mechanism when ants find the shortest path from ant nest to food. When understanding ant colony algorithm, we can take the traveling salesman problem as an example, which is referred to as TSP here. TSP gives different city locations and the distance between cities, in order to find the nearest route that can pass through all cities, and each city only goes once. The path analysis of ant colony algorithm is shown in Fig. 1 below.

The difference between traveling salesman problem and Hamiltonian path:

Hamiltonian path is a cycle (circle), which is now also called hamilton's puzzle. It is to find out whether there is a ring that can walk once at each place, but the purpose of TSP is to find out whether there is a shortest ring in apricot.

$$\pi(\theta|x) = \frac{p(x|\theta)\pi(\theta)}{m(x)} = \frac{p(x|\theta)\pi(\theta)}{\int_\emptyset p(x|\theta)\pi(\theta)d_\theta} \tag{1}$$

Pheromones, also known as pheromones, refer to substances secreted by an individual into the body and detected by other individuals of the same species through olfactory organs (such as accessory olfactory bulb and vomeronasal organ), so that the latter shows certain changes in behavior, emotion, psychology or physiological mechanism [2]. It has communication function. Almost all animals have proved the existence of pheromones.

Ants will release a substance called "pheromone" in the process of moving, and other ants will have the ability to perceive "pheromone". They will walk along the path with high concentration of "pheromone", and every passing ant will leave ° pheromone on the road", which forms a mechanism similar to positive feedback. After a period of time, the whole ant colony will reach the food source along the shortest path.

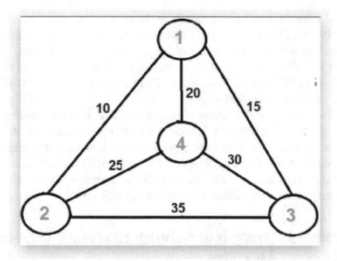

Fig. 1. Path analysis of ant colony algorithm

The pheromone value is modified at runtime and represents the cumulative experience of the ant colony, while the heuristic value is the problem related value. In the case of TSP, the heuristic value is set to the reciprocal of the edge length.

2.2 Value of Network Literature

Popular works are usually not recognized by classical literature, which is the case in history, but it does not mean that popular novels cannot reach the level of classical literature. It can only be said that they have different pursuits, resulting in mutual dislike.

Classical literature pursues more expression, the author's inner nature and pure artistry. Its writing purpose itself is not for everyone, so the readers it expects are readers who can resonate with themselves and have certain aesthetic appreciation ability. Commercial success is only a by-product of this kind of literature.

Popular novels, represented by online novels, are reader oriented works. Students who read online novels should understand how strong this guidance is. In those days, after the misty journey became popular [3], Xiuzhen novels were everywhere. After infinite terror became popular, there was infinite flow everywhere, and nine out of ten rushed. This is the market-oriented situation. If it is popular, more than n people will imitate it to share a cup of pods. From the perspective of classical literature, this is a standard evil Road, and it can't be killed on the table. It's not surprising that the two ideas are similar.

In fact, popular novels are not only online novels, but also detectives and suspense.

However, there is a way from popular novels to classic literature in this world. I mention a name, Raymond Chandler. Chandler is a master of tough guy detective mystery novels. He is the only one who writes in classic literature on the road of detective novels. He has a long farewell, which is highly praised by Eliot, Haruki Murakami, Oden, Qian

Zhongshu and others (I don't know that their classmates only need to know that these people are very good, (W'). He is the Daniel who connects the popular with the classic.

Detective story is also a kind of popular novel. In the concept of classic literature, it is also not on the table. However, since someone can walk through this road, it shows that this road is not blocked. As long as it is not blocked, there will always be so many successes for millions of people [4].

Judging from the current situation of online novels, it was a barbaric growth in the early years, but now it can only be regarded as a small success, and it is very early from maturity. Some of the books I've been paying attention to recently, such as the story of the shepherd, 40000 years of cultivation, and the Lord of mysteries, are basically of a type that didn't exist before, indicating that the unknown space is still large. In the future, there may be a better brother who will really make this road through.

3   Prediction and Analysis of Network Literature Value Based on Ant Colony Algorithm

Network literature originally refers to the literary form corresponding to oral literature and written literature. It is a new literary form that takes the network as the carrier and takes the network as the media and tool in the whole cycle from creation to dissemination. Logically speaking, this should be a qualitative change in the form of literature. The network has the potential to promote the vigorous development of literature. Now, the reason why we equate network literature with Xianxia literature, fantasy literature, crossing literature and so on can only explain that the form of "network literature" is the most mature and developed from creation to market. The achievements of this mature and developed field of "network literature" - whether in mass communication, cultural influence or commercial achievements, are obvious to all.

The basic idea of ant colony algorithm is to place several ants randomly at the position of data points, each ant starts to find the optimal path, selects the next node according to the amount of pheromone on the path in the process of finding, and continuously modifies the pheromone with the length of the path, so as to realize the function of communicating with each other by pheromone, that is, to realize the feedback of information [5]. Obviously, the more times a path is traversed, the more pheromones remain on the path, which means that the better the path is, the greater the probability of being selected, so as to achieve the purpose of optimization.

First, K ants are randomly placed on K data points, and each ant selects the candidate node with the highest probability as the next selection standard. When calculating the transition probability between nodes, it not only calculates the distance between two points on the basis of different attribute discreteness of each dimension, but also comprehensively analyzes the differences between data by comprehensively considering the distribution of data.

$$mbest(t+1) = \frac{1}{M} \sum_{i=1}^{M} P_i(t) = \left(\frac{1}{M} \sum_{i=1}^{M} P_{i1}(t), \frac{1}{M} \sum_{i=1}^{M} P_{i2}(t), ..., \frac{1}{M} \sum_{i=1}^{M} P_{iD}(t) \right)$$

$$(2)$$

When all ants have completed all nodes, they choose the best path to store, and only update the pheromone on the optimal path [6]. This positive feedback mechanism of information ensures the accuracy of the results and improves the speed. When the number of cycles reaches the threshold set by the user, the cycle is ended, and the best ones in general are selected as the basic elements of the image from the saved optimal path of each cycle. Among them, there may be more than one edge in the optimal path. The number of times it exists should be saved as the weight. When cutting each time, the weight will be subtracted by 1. Only when the weight of the edge is, the edge can be regarded as deleted, and the initial transfer probability stored for each edge should be saved for subsequent work.

The main cost of the algorithm is to calculate the distance between points and the path finding process of all ants during the cycle [7]. Because NC cycles are set and K ants are placed in each cycle, the time cost of the algorithm is much higher than that of other general composition algorithms, but the accuracy is also greatly improved.

4 Simulation Analysis

The core idea of improved ant colony algorithm and K-means algorithm is combined to realize the function of clustering. With the help of the unique pheromone and transition probability of ant colony algorithm, there is no need to define the number of clusters generated after clustering, which eliminates the error caused by manually setting the K value of K-means algorithm, and can achieve better clustering results. When evaluating the pheromone, it is set according to the length between the point and the center of the cluster, which increases the convergence speed of the algorithm; Under a certain number of cycles, the maximum ratio of inter class distance and intra class distance is taken as the best clustering result output [8]. Combining the higher accuracy of ant colony algorithm without setting the initial number of clusters with the higher time efficiency of K-means algorithm, we get rid of their shortcomings, so that we can cluster the data quickly and effectively. When mining outliers, the outlier coefficients are calculated by comparing the temporal changes of each point, so as to find the outliers determined by the similarity of neighbors.

In order to improve the accuracy of detection and reduce the time required, the algorithm is mainly divided into two stages. First, the time attribute of the data is temporarily ignored, and the distance between the data is calculated by the weighted Euclidean distance. Ant colony and K-means algorithm are combined for clustering, which improves the accuracy of clustering and reduces the time consumption of the traditional ant colony algorithm [9]. At the same time, the clustering quality is comprehensively measured from the intra class distance and inter class distance to find the optimal clustering result. In the second stage, the distribution of time points of each variable in the cluster after clustering is considered, that is, to determine the timing vector of each variable [10]. Taking the trend consistency between variables as the principle, the outlier vector is determined from the perspective of different changes from neighbors, and then the outlier coefficient of variables and the outlier coefficient of variables at a certain time point are calculated. According to the order of outlier coefficient from high to low, temporal outliers can be mined.

5 Conclusion

Ant colony algorithm is a technology that uses ants as the main agent to find the best path in the graph. Its working principle is to find the shortest path between two nodes by considering all possible paths and then calculating its cost. The algorithm can be used to solve problems such as finding the shortest path, calculating distance and routing algorithm. If there is a problem with your network, it is very important to understand the problem. You can use this tool to predict and analyze the value of the network based on ant colony algorithm. This tool will help you understand the value of the network.

References

1. Chen, Y.: Application of intelligent algorithm based on genetic algorithm and extreme learning machine to deformation prediction of foundation pit. Tunnel Constr. (2018)
2. Wang, P., Xu, X., Liu, C.: An Improved adaptive genetic algorithm and its application in intelligent course scheduling system. In: 2019 6th International Conference on Information Science and Control Engineering (ICISCE) (2019)
3. So, G.B.: Design of an intelligent NPID controller based on genetic algorithm for disturbance rejection in single integrating process with time delay (2020)
4. Zhang, Y., Jie, Z., Pei, W.: Research and implementation of intelligent test paper composition based on genetic algorithm. In: 2018 9th International Conference on Information Technology in Medicine and Education (ITME). IEEE Computer Society (2018)
5. Yang, A., Liu, H., Chen, Y., et al.: Digital video intrusion intelligent detection method based on narrowband Internet of Things and its application. Image Vis. Comput. **97**, 103914 (2020)
6. Li, H., Li, G., Huang, Z., et al:. Application of BP neural network based on genetic algorithm optimization. In: ICIIP 2019: 2019 4th International Conference on Intelligent Information Processing (2019)
7. Yongbin, Y.U., Yang, C., Deng, Q.等.: Memristive network-based genetic algorithm and its application to image edge detection. J. Syst. Eng. Electr. **32**(5), 9 (2021)
8. Liu, P., Wang, Z.: A fault diagnosis intelligent algorithm based on improved BP neural network. Int. J. Pattern Recognit. Artif. Intell. (4) (2018)
9. Wanli, S.: Online test paper composition based on genetic algorithm (2018)
10. Li, Z., Guo, L., Gao, H., et al.: Instance-based ensemble deep transfer learning network: a new intelligent degradation recognition method and its application on ball screw. Mech. Syst. Sig. Process. **140**(18–9456), 106681 (2020)

Prediction of General Aviation Industry Development Prospect of Hainan Free Trade Port Based on BP Neural Network Optimization Model

Xiaoshuo Zhao[✉]

Civil Aviation Transportation School, Sanya Aviation and Tourism College, Sanya 572000, Hainan, China
258365548@qq.com

Abstract. The development of general aviation industry is the need of economic and social development, and helps to develop the whole high-tech industry and enhance national defense strength. The purpose of this study is to predict the development prospect of general aviation industry in Hainan free trade port based on BP neural network optimization model. The data used in this study are from the following sources: (1) the general aviation industry development prospect survey conducted by the China Bureau of statistics; (2) A database containing information on assets owned by general aviation and its foreign investors in China; (3) An online database containing airport information in China. The prediction method used is BP neural network optimization model. After applying this method, we get a score value according to the specific characteristics of each airport (such as location, size).

Keywords: BP neural network · Hainan free trade port · General aviation industry · Forecast

1 Introduction

It is generally believed that civil aviation is divided into public air transport and general air transport. The history of the development of general aviation in the world shows that the higher the scientific and technological level of any country, the more developed the national economy of any place, and the more prosperous the general aviation industry of any country or place. As an advanced production tool and technical means, general aviation will play an increasingly important role in economic and social development [1]. General aviation is flexible, fast and efficient. In industrial and agricultural production and other fields, it has incomparable advantages and irreplaceable role of ground machinery and manual operation. The development of general aviation also plays an important role in promoting the growth of investment and consumption, which can promote economic growth and drive the development of related industries [2]. According to relevant estimates, investing 1 yuan in general aviation can drive economic growth by

J. C. Hung et al. (Eds.): IC 2023, LNEE 1045, pp. 387–392, 2023.
https://doi.org/10.1007/978-981-99-2287-1_55

7 yuan, while investing 1 yuan in automobile industry can drive economic growth by 4 yuan. In addition, the development of general aviation industry is the need for the development of the entire civil aviation industry, which is also conducive to the development of the entire high-tech industry and the strengthening of national defense strength [3].

Under the background that the central government takes Hainan free trade port as a pilot to further promote reform and opening up, it focuses on the innovative development of service trade in Hainan free trade port, and empirically explores the key factors affecting the development of service trade in Hainan with relevant data. On the basis of relevant research by domestic and foreign scholars, through multi angle analysis, it explores how Hainan can better use the innovative development of service trade to drive the regional economy in the national free trade port strategy, Verify the impact mechanism of the development of service trade in Hainan free trade zone through actual data, provide sustainable momentum for the sustainable growth of service trade in Hainan, help Hainan free trade port improve the competitiveness of service trade, and promote the transformation and upgrading of service trade in Hainan free trade port, which is the practical significance of this study [4].

2 Related Work

2.1 Free Trade Port

It has been more than 400 years since the first free trade port appeared in Europe. According to Wikipedia, a free trade port is a special economic zone set up near a port or island. Its main purpose is to develop trade. Goods can be unloaded, transported, produced and re exported in this area without jurisdiction. Encyclopedia Britannica also explains the free trade port, which refers to the area where foreign goods can be loaded, unloaded, transported, re exported, processed and stored for a long time without customs inspection and taxation.

Relevant domestic researchers also elaborated their views. Li Jianping (2013) believes that a free trade port refers to an area with a port as the center, where goods, funds and population can flow freely without Customs interference and without tariffs. Li Jiuling (2014) believes that a free trade port refers to a port that is outside the customs territory of a country, where all foreign goods can flow in and out freely. The "port" here not only refers to the port near the sea, but also covers inland ports, inland airports, etc. In essence, the free trade port is an area with more standardized and convenient advantages in trade and investment than other regions [5]. Generally speaking, we can think that the free trade port is the most open special economic zone in the world, "inside and outside the customs" and "freedom of movement" are its most prominent characteristics.

Before exploring the establishment of Hainan free trade port, the pilot free trade zone has been implemented in other regions of China for nearly five years. In 2013, China's first free trade zone was established in Shanghai. Since then, Guangdong, Fujian and Tianjin free trade zones have emerged. Then, in March 2017, a total of seven pilot free trade zones were established in six inland provinces except Zhejiang. Although China has set up many free trade zones, Hainan free trade port is still an indispensable part. The biggest difference between Hainan's "double free trade" construction and other domestic free trade zones is that Hainan is a free trade zone and a free trade port [6]. Hainan's "double free trade" construction covers the whole island, including both urban and rural areas.

There are countless models of free trade ports. At present, there are mainly the following kinds in the world: the first is industrial and commercial. The second type is entrepot trade. The third type is tourism and shopping The fourth type is comprehensive. The comprehensive free trade port has diversified functions, which can not only carry out entrepot trade, but also extend its territory to tourism, finance and other modules. Its scope of economic activities and development direction have obvious international characteristics, of which Hong Kong, China and Singapore are the most exemplary The construction trend of contemporary free trade port is approaching to the comprehensive type, which is mainly reflected in the continuous integration of various service functions, including not only trade services but also productive services [7].

2.2 General Aviation Industry

General aviation industry is the sum of various economic departments and economic activities engaged in general aviation flight activities. According to the National Bureau of statistics, the general aviation industry includes general aviation services, air traffic management, flight schools, etc. In accordance with the provisions on the administration of general aviation business license, the business items of general aviation enterprises can be divided into the following three categories: (1) class a onshore oil services, offshore oil services, helicopter off load flights, artificial precipitation, medical aid, aviation exploration, air tourism, official flights, private or commercial pilot license training, helicopter pilotage operations, aircraft custody business, rental flights General aviation charter flights: (2) class B aerial photography, aerial advertising, ocean monitoring, fishery flights, meteorological detection, scientific experiments, urban fire control, air patrol; (3) Class C aircraft seeding, aerial fertilization, aerial spraying of plant growth regulators, aerial weeding, prevention and control of agricultural and forestry pests and diseases, grassland deratization, prevention and control of health pests, aerial forest protection, aerial photography [8]. The categories of business items not included in the above three categories shall be determined by the CAAC. Rescue and disaster relief are not restricted by the classification of the above three categories of projects, and shall be implemented in accordance with the relevant provisions of the Civil Aviation Administration of China.

In 1951, the Civil Aviation Administration established the first general aviation flight team in New China and carried out field pest control and forest fire prevention operations. At that time, the concept of "general aviation" was not used, and the aviation for industrial and agricultural services was collectively referred to as professional aviation. In 1956, the Civil Aviation Administration established a professional aviation office, which was responsible for the unified management of general aviation affairs throughout the country. In 1980, CAAC established a professional aviation bureau, which raised the management of general aviation to a higher level, indicating the importance and support of the state. In 1986, the State Council officially used the term general aviation to replace professional aviation, marking the integration of China's general aviation industry with the world[9]. After the civil aviation system reform in 1987, general aviation was introduced to the market as an independent department or independent business entity of major airlines, and the development of China's general aviation industry has made some achievements.

3 Prediction of General Aviation Industry Development Prospect of Hainan Free Trade Port Based on BP Neural Network Optimization Model

3.1 BP Neural Network Model

In the mid-1980s, David rnmelhart, Geoffrey Hinton, Ronald Williams, David Parker and yannn Le Cun independently discovered BP algorithm. BP network is composed of input layer, hidden layer and output layer. The number of hidden layers can be one or more, and each layer is composed of multiple neurons. The number of neurons in the input layer is the dimension of the input signal, and the number of neurons in the output layer is the dimension of the output signal. Its characteristics are: neurons in each layer are only connected with neurons in adjacent layers; There is no connection between neurons in each layer; There is no feedback connection between neurons in each layer. The input signal propagates forward to the hidden node. After the excitation function, the output of the hidden node propagates forward as the input of the output node [10]. The output layer processes the received input and gives the output result. Generally, the S-type function or hyperbolic tangent is used as the activation function in the hidden layer, while the linear activation function is used in the output layer.

It has been proved theoretically that a three-layer BP network model can realize any continuous image. The training structure is shown in Fig. 1:

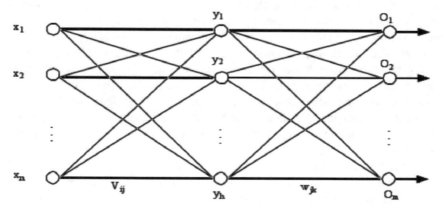

Fig. 1. BP neural network model

3.2 Prediction Model of General Aviation Industry Development Based on BP Neural Network

To predict the aviation industry, we must first admit that there is a certain repeatability in the stock, that is, there are laws in the aviation industry, which are hidden in historical data. From a mathematical point of view, they can be described as a functional relationship, and prediction is to find out these laws. Kolmogorov theorem proves that BP network has the ability to approach nonlinear mapping with arbitrary accuracy, and the learning ability of neural network and the ability to store data rules through learning are very suitable for aviation industry prediction. Compared with other methods, it has certain advantages. The stock price contains a large number of internal laws and characteristics that determine the change of stock price, which is highly nonlinear. The artificial neural network can autonomously find the laws and characteristics between parameters from the complex data by learning the historical transaction data. Therefore, the application of neural network to the prediction of stock price trend can achieve good results.

The work needed to establish neural network model for stock prediction includes:

(1) Determine the input and output variables of neural network

Hainan free trade port has a large number of historical data, including the opening price, closing price, highest price, lowest price, trading volume, etc. According to principal component analysis combined with Dow Theory, the closing price is the most important of all prices; The closing price represents the final outcome reached by the long and short sides after a day of competition, which is the balance point of both sides; Other prices, such as the highest price and the lowest price, represent the price for a short time. Therefore, the daily closing price of stock index is selected as the input and output variable here;

(2) Data preprocessing

Too large input layer data value will cause paralysis of neural network, so it is necessary to preprocess the data. Preprocessing must maintain the characteristics of the original data, generally using linear normalization method. Normalization processing algorithm: calculate the maximum value ymax and minimum value Ymin of the data sample, and then process the original data {y} according to formula (1) to calculate the processed input variables {u,}. Using the processed data to train the artificial neural network can find the internal relationship between variables more easily and quickly;

$$E(w, b) = \sum_{j=0}^{n-1} (d_j - y_j)^2 \tag{1}$$

(3) Select the number of hidden layer nodes

The determination of the number of hidden layer nodes is a very important step in the process of determining the topology of BP neural network model. Reasonable selection of the number of hidden layer nodes can effectively avoid the phenomenon of "over fitting" in training, so that the network has good generalization ability. However, there is still a lack of universal and scientific methods to determine the hidden nodes. Generally, the selection is based on empirical functions, and the structure as compact as possible is selected on the premise of achieving the required accuracy.

$$\sup_{0 \leq t \leq T} \|I - L(t)C(t)P^{-1}(t)B(t)\| \leq \rho_1 < 1 \tag{2}$$

4 Conclusion

China's general aviation industry has developed from scratch, especially since the reform and opening up. But on the whole, it is still in its infancy, the internal and external market environment is not perfect, and the huge economic potential it contains is far from being explored. The legal environment of the general aviation industry has deficiencies in the legal system, macro-control, market operation, security services and other aspects to varying degrees. We should learn from foreign advanced experience, reform the disadvantages, introduce more specific general aviation industry development plans, formulate national industrial policies, strive to broaden investment channels, relax restrictions on the entry of private capital, and encourage private capital and foreign investment in the construction of general aviation airports General aviation operation and maintenance enterprises should reform the financing methods of general aviation industry, further strengthen the position and role of the government in general aviation industry investment, and create a good investment environment.

References

1. An, X., Zhao, F.: Prediction of soil moisture based on BP neural network optimized search algorithm. In: IOP Conference Series: Earth and Environmental Science, vol. 714, no. 2, p. 022046 (9pp) (2021)
2. Wu, L., Cao, Y., Zhang, Z.: Analysis and prediction of China's future pension industry based on fitting algorithm and BP neural network. J. Phys: Conf. Ser. **1952**(4), 042141 (2021)
3. Zha, A., Tu, J.: Research on the prediction of port economic synergy development trend based on deep neural networks. J. Math. **2022** (2022)
4. Wan, B., Shen, Y.: Stock trend prediction of communication industry based on lstm neural network algorithm and research of industry development strategy: -a case study of zhongxing telecommunication equipment corporation (2021)
5. Zhang, M.: Prediction of rockburst hazard based on particle swarm algorithm and neural network. Neural Comput. Appl. **34**, 2649–2659 (2022). https://doi.org/10.1007/s00521-021-06057-9
6. Zhang, L.: Research on collaborative innovation network mechanism of general aviation enterprises based on complex network. In: IOP Conference Series: Earth and Environmental Science, vol. 632, no. 5 p. 052099 (4pp) (2021)
7. Liu, T., Liu, K.: research on the development and trend prediction of china's sports industry based on organizational structure. Acad. J. Humanit. Soc. Sci. **3**(11), 140–144 (2020)
8. Lz, A., Hh, B.: Share price prediction of aerospace relevant companies with recurrent neural networks based on PCA. Exp. Syst. Appl. **183**, 115384 (2021)
9. Liao, H., Wang, C., Liu, X., et al.: Economic development forecast of china's general aviation industry. Complexity **2020**, 1–8 (2020)
10. Wang, J., Zhang, T., Luo, C.: Health prediction of airborne system based on grey prediction. J. Phys. Conf. Ser. **1616**, 012091 (2020)

Quality Regression Coefficient of UAV Structure Based on Fuzzy Clustering Algorithm

Yuyuan Guo[✉], Lu Dai, and Ziyi Zang

Beijing Satellite Manufacturing Co. Ltd., Beijing 100094, China
`fengzhengnuaa@126.com`

Abstract. Uav structure high quality leads to unmanned aerial vehicle (uav) flight flexibility and controllability is reduced, the structure quality is too low will lead to uav cannot against external force such as high altitude wind, also affect the actual operation, so in order to find the uav structure mass balance, this paper will be based on the fuzzy clustering algorithm, regression coefficient of the quality of the structure were studied. This paper introduces the basic concept of fuzzy clustering algorithm, and then discusses the analysis scheme of UAV structure quality regression coefficient. Finally, a UAV structure is taken as an example, and the analysis is carried out by fuzzy clustering algorithm according to the scheme. The results show that the UAV structure quality is good, and the analysis result is accurate. It shows that fuzzy clustering algorithm can be used as a tool to analyze the quality regression coefficient of UAV structure and has good application value.

Keywords: Fuzzy clustering algorithm · Uav structural quality · Regression coefficient analysis

1 Introduction

Regression coefficient analysis is a widely used quality analysis method, which is often used in the structural quality analysis of mechanical equipment, and UAV belongs to mechanical equipment. Therefore, in order to ensure the structural quality of UAV, people will analyze it through regression coefficient. However, there are many specific forms of regression coefficient analysis, which mainly depend on different algorithms. The difference of algorithms determines the function and application conditions of corresponding regression coefficient analysis. Therefore, when using regression coefficient analysis, it is necessary to make a reasonable choice based on the actual problem type. This condition, the uav structure quality which involves complex factors, so from a mathematical perspective, the fuzziness of its structure has an obvious quality problem is a typical fuzzy mathematical problem, therefore, carries on the regression analysis, in order to better choose to deal with the fuzzy mathematics method, and fuzzy clustering algorithm is a kind of representative method of fuzzy mathematics. Therefore, in order to understand the specific application method of fuzzy clustering algorithm in the regression coefficient analysis of UAV structural quality, it is necessary to carry out relevant research.

© The Author(s), under exclusive license to Springer Nature Singapore Pte Ltd. 2023
J. C. Hung et al. (Eds.): IC 2023, LNEE 1045, pp. 393–400, 2023.
https://doi.org/10.1007/978-981-99-2287-1_56

2   Basic Concept of Fuzzy Clustering Algorithm

Fuzzy clustering algorithm is a kind of algorithm improved from the classical clustering algorithm, that is, the classical clustering algorithm as the basis, which is integrated into the fuzzy mathematical method to get the fuzzy clustering algorithm [1–3]. There is no difference in basic functions between fuzzy clustering algorithm and classical clustering algorithm. Both of them can process huge data, then extract corresponding features, and finally cluster according to the distance between data features and classification items to achieve accurate classification (see Fig. 1 for the clustering principle) [4]. However, there are differences in application conditions between the two algorithms: Classical clustering algorithm can only to itself has clear characteristics of the digital data processing, if data digitization characteristics are not clear, can't use the algorithm to realize the clustering, opposite fuzzy clustering algorithm to make up for the defects, can not clear in the data of fuzzy characteristics were extracted, then classify, visible both applicable condition is different, Among them, fuzzy clustering algorithm is specially used to deal with fuzzy mathematical problems.

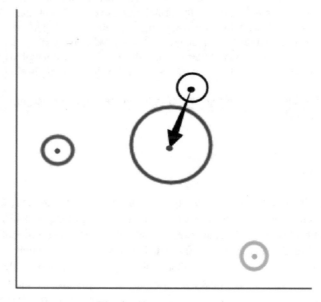

Fig. 1. Clustering principle

Combined with Fig. 1, red circle is a classified project, the rest three colors circle respectively represent the different data, confirm the classification of project and the characteristic value between different data, the characteristic value as the center of the data, then any characteristic value and classification of a data item as the characteristic value of subtraction, if the results show that the eigenvalues of the data and classification project membership to a minimum, In the figure, the distance between the center point of the black circle and the center point of the red circle is the smallest, indicating that the black circle belongs to the red circle classification, so the black circle is close to the red circle. This process has the feature of "clustering" in form, so it is called "clustering" [5–7].

It is worth mentioning that the fuzzy clustering algorithm itself and can be divided into two types, respectively is the system clustering analysis method, the system clustering method, the difference between the two need to get the fuzzy relationship model, and then clustering analysis, Fig. 1 is a typical system the principle of fuzzy clustering algorithm, which is a rough first distinguish between data samples, are classified in accordance with the principle of most again, After each classification, rationality test will be carried out. If the result is not reasonable enough, classification will be carried out on this basis. Through continuous iteration of this process, the best model will be finally found [8–10]. The two types of fuzzy clustering algorithms have their own characteristics, but the system clustering analysis method is more common and can deal with most problems, including the regression coefficient analysis of UAV architecture quality. Therefore, this paper will also choose the system fuzzy clustering algorithm for analysis. The basic flow of the system fuzzy clustering algorithm is shown in Fig. 2, and formula (1) is the expression of the algorithm.

$$\sum_{i=1}^{c} u_{ij} = 1, \nabla j = 1, ..., n \tag{1}$$

In the formula, c is the coefficient, u is the membership degree of eigenvalues, ij is the center point of classification items and data respectively, so u_{ij} refers to the size of the membership degree u between the center point of classification center i and the center point of data j. The larger u is, the lower the membership degree is, and the higher the vice versa. n is the set of j, representing the whole calculation range.

3 Analysis Scheme of UAV Structural Quality Regression Coefficient

3.1 UAV Structural Quality Modeling

To use fuzzy clustering algorithm to analyze the regression coefficient of UAV structural quality, it is necessary to get the mathematical model of UAV structural quality first, so as to clarify the fuzzy mathematical relationship between various indicators of structural quality and drive the operation of the algorithm. The UAV structural quality modeling is generally divided into three steps, namely, the establishment of structural quality indicators, the data of indicators, and the establishment of fuzzy relations. The detailed contents of each step are as follows.

3.1.1 Establishment of Structural Quality Index System

Uav structure quality is a comprehensive index, and several secondary indexes can be obtained after refinement. See Table 1 for details.

There is an interaction between the three second-level indicators in Table 1, that is, in order to improve the stability or load capacity of the UAV, it is necessary to improve the

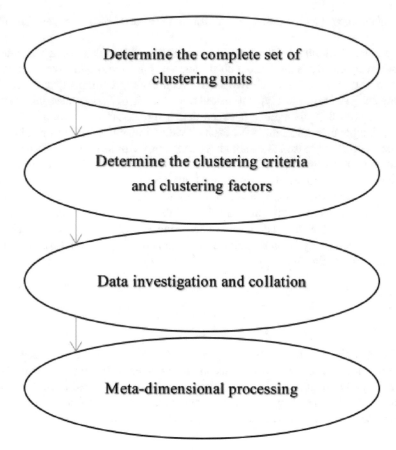

Fig. 2. Basic flow of systematic fuzzy clustering algorithm

Table 1. Secondary index of UAV structure quality

The name of the	Explain
The weight of the	The weight of the overall structure
The stability of	Tolerable range of external forces
Load capacity	The maximum weight that can be carried at the limit value of the influence of external forces

density and thickness of the structure, etc. These operations will lead to weight increase, on the contrary, the stability or load capacity will be weakened, which will lead to weight decrease. Under the influence of the three factors, it can be preliminarily seen that the UAV structure quality problem is relatively fuzzy, and it is difficult to find the balance point among the three factors by ordinary methods, but the fuzzy clustering algorithm can solve this problem.

3.1.2 Index Datalization

The purpose of index datalization is to convert each index into a numerical value so that fuzzy relationship modeling can be carried out from a mathematical perspective. The specific implementation method of index datalization is data collection, but it is worth mentioning that in data collection, except for weight index, which can be directly collected by measurement method, stability and load-bearing capacity are all non-structured data, that is, they will have different performances under different circumstances. Aiming at this characteristic, need to collect by means of simulation, the specific method is: first, in combination with actual conditions to establish the simulation environment model, load model, which can simulate the real environment of various external force on the structure of the unmanned aerial vehicle (uav) influence strength, direction and so on, which is mainly based on the actual weight of the object for unmanned aerial vehicle (uav) structure to vertical downward pull; Secondly, according to the measurement results of weight index and the current structure framework of UAV, the UAV simulation model is established. Third, the unmanned aerial vehicle (uav) simulation model and simulation environment model, load model respectively, to observe first unmanned aerial vehicle (uav) simulation model is affected by the other two models respectively, which assumes that the unmanned aerial vehicle (uav) simulation model in the simulation environment model can't remain stable, is a sign of unmanned aerial vehicle (uav) current structure stability is not enough, or unable to carry vertical tensile load model, It means that the current structural load capacity of UAV is low. According to this result, the parameters of the simulation environment model and the load model are adjusted, and then the cyclic test is conducted to know that the current UAV simulation model can maintain stability in the simulation environment and has a good load capacity performance position in the load model. Fourthly, combine the three model together, simulation structure of unmanned aerial vehicle (uav) under the condition of load is influenced by the external force of the scene, if the structure of unmanned aerial vehicle (uav) cannot influence under the condition of load under external force, appear unstable phenomenon, will begin to adjust stability, or to reduce the tension of the load model, until the uav structure can influence under the condition of load under external force, remain stable for a long time. Combining with the results of the fourth step, the data of stability and bearing capacity are extracted and combined with the weight index data to complete the index datalization.

3.1.3 Establishment of Fuzzy Relationship

Let the value of UAV structure mass after data index be 100, 200 and 200 respectively. According to the mutual influence of the three indexes, the decrease/rise of weight will lead to the decrease/rise of stability and the decrease/rise of load capacity, and the value of the decrease or rise will remain unchanged. According to this condition, quality of unmanned aerial vehicle (uav) structure of fuzzy relation is: the weight of index is set to the classified project, taking the value as the center, stability and load capacity of data, take the same value as the center, the three influence each other, the fuzziness is to don't know weight down/up concrete leads to stability, load capacity down/up the numerical change much.

3.2 Regression Analysis

According to the established fuzzy relationship, the fuzzy clustering algorithm can be used for regression analysis. The methods are as follows: first, the weight of UAV structure is adjusted down/up with the value of 1 for each adjustment, and the changes of stability and load capacity are observed; Secondly, combined with the real standard requirements, to ensure stability and load capacity is not lower than the standard, giving weight index down/up regulating rules, if cut operation, stability and load capacity is lower than standard weight, cut operation is not feasible, to raise, instead just cut operation not result in stability and load capacity is lower than standard, It means you can keep going down without going up. In this way can be found based on the standard minimum weight, the whole process in the model, the fuzzy clustering algorithm is stability index or load capacity to weight index, the principle of process, the assumption by weight index 2, resulting in a decline in the numerical stability of the structure of unmanned aerial vehicle (uav) decrease 3 load capacity, stability and the distance between the weight of shorter, Therefore, the membership degree of stability in weight is higher, which means that the current UAV structure quality focuses on stability and weight, and the load capacity will be relatively insufficient, but it is already the most extreme load capacity under current conditions. In this way, the quality of UAV structure can be evaluated from the functional perspective.

4 Empirical Testing

4.1 Test Scheme

Chose two completely different weight, stability, load power of unmanned aerial vehicle (uav) as test object, through the standard data confirmed that both weight, stability and load capacity of concrete numerical value, and then adopt fuzzy clustering algorithm, simulation model was calculated with both, understand the characteristics of the unmanned aerial vehicle (uav) quality of the original structure, again to two unmanned aerial vehicle (uav) is the same of the original structure model of transformation, The above process is repeated to judge whether the structural change direction of the two UAVs after transformation meets the expectation and whether the specific degree meets the requirements. If it meets the expectation and meets the requirements, it means that the fuzzy clustering algorithm successfully gives the UAV structural quality assessment result, which provides a strong basis for the UAV structural transformation.

4.2 Test Results

Through pre-test, it is confirmed that the weight, stability and load capacity of UAV A are 133, 312 and 309 respectively, so the stability of UAV A is the most prominent, while the weight, stability and load capacity of UAV B are 124, 266 and 273 respectively, so the load capacity of UAV B is the most prominent. Now it is necessary to find the balance between the weight, stability and load capacity of A and B UAVs under external force environment conditions, so the transformation is carried out. As requested, the fuzzy clustering algorithm is used for regression analysis, the results show A unmanned aerial vehicle (uav) based on the original structure, when the weight decrease/(10, its stability will drop down / 5, load capacity will decrease/(3, B whenever weight/decline in 10, its stability will drop down / 7, load capacity will decrease / 4. According to this result, reducing the weight of UAV A by 80 and reducing the weight of UAV B by 20 requires that the stability and load capacity of the two should not be lower than the standard value. The test results are shown in Table 2.

Table 2. Test results

Name of UAV	Standard values	The results of
A	Stability: 280	Stability: 282
	Load capacity: 280	Load capacity: 285
B	Stability: 280	Stability: 280
	Load capacity: 280	Load capacity: 281

5 Conclusion

To sum up, the relationship between various indicators of UAV structure quality is relatively fuzzy, and it is difficult to adopt other mathematical methods other than fuzzy mathematical method for regression coefficient analysis. Therefore, fuzzy clustering algorithm can be selected to solve the problem. Fuzzy clustering algorithm can well deal with the fuzzy relationship between indicators, through the clustering principle to confirm the respective impact of indicators, so as to confirm the index balance point, make the UAV structure quality more comprehensive.

References

1. Cai, W., Xu, S., Zhang, L.J., et al.: Pairwise constraints cross entropy fuzzy clustering algorithm based on manifold learning and feature selection. J. Phys. Conf. Ser. **1948**(1), 012033(7pp) (2021)
2. Tang, Q., Zhao, Y., Wei, Y., et al.: Research on the mental health of college students based on fuzzy clustering algorithm. Secur. Commun. Netw. **2021**(3), 1–8 (2021)
3. Zhao, F., Zeng, Z., Liu, H., et al.: Semisupervised approach to surrogate-assisted multiobjective kernel intuitionistic fuzzy clustering algorithm for color image segmentation. IEEE Trans. Fuzzy Syst. **28**(6), 1023–1034 (2020)
4. Lata, S., Mehfuz, S., Urooj, S., et al.: Fuzzy clustering algorithm for enhancing reliability and network lifetime of wireless sensor networks. IEEE Access **8**, 66013–66024 (2020)
5. Zhang, H., Liu, J., Chen, L., et al.: Fuzzy clustering algorithm with non-neighborhood spatial information for surface roughness measurement based on the reflected aliasing images. Sensors **19**(15), 3285 (2019)
6. Sun, J., Dai, Y., Zhao, K.: DBFCM: a density-based fuzzy c-means with self-regulated fuzzy clustering parameters. In: 2020 39th Chinese Control Conference (CCC) (2020)

7. Hu, J., Xie, C.: Research and implementation of e-commerce intelligent recommendation system based on fuzzy clustering algorithm. J. Intell. Fuzzy Syst. **3**, 1–10 (2021)
8. Qu, Y., Wang, Y.: Segmentation of corpus callosum based on tensor fuzzy clustering algorithm. J. X-Ray Sci. Technol. **29**(5), 1–14 (2021)
9. Gosain, A., Dahiya, S.: An effective fuzzy clustering algorithm with outlier identification feature. J. Intell. Fuzzy Syst. **3**, 1–12 (2021)
10. Li, S.B.: Analysis of human-land coupled bearing capacity of qiangtang meadow in northern Tibet based on fuzzy clustering algorithm. Math. Prob. Eng. **2020**(2), 1–9 (2020)

Recommendation Method and System for Fitness of Children and Adolescents Based on Ant Colony Algorithm

Haibo Dou[✉]

Xianyang Normal University, Xianyang 712000, Shaanxi, China
douhaibo1111@126.com

Abstract. With the national attention to health and the public attention to their own health, fitness exercise has become one of the most concerned issues. As the future and hope of the country, teenagers' physical health is not optimistic at present. How to improve teenagers' interest in fitness exercise and provide complete and effective fitness exercise theory for teenagers' fitness exercise should become the focus of our attention. The health recommendation method and system for children and adolescents based on ant colony algorithm is a new method to improve the health level of the body. In this case, it is a computer program that uses ants as a model. The main idea behind this method is that there are many methods to find the best solution to any problem, but in order to know which method is the most effective, we need some standard or rule (in our example: ant colony algorithm).

Keywords: Children and adolescents · Ant colony Bodybuilding · Recommended method

1 Introduction

Since the reform and opening up, the physique of young students in China has shown a serious downward trend, which has aroused widespread concern of the whole society. In December2006, the national school physical education work conference held in China put forward several measures and policies to improve students' physical health, including the nationwide development of sunshine sports [1]. The purpose of this activity is to attract young students to go to the playground, into the nature and under the sun, actively participate in physical exercise, and mobilize the enthusiasm of young students to actively participate in physical exercise through a variety of extracurricular sports activities. In the past three years, with the continuous development of sports, the physical condition of students has indeed been improved, but there are also many problems in the specific implementation process [2]. For example, there is a lack of scientific fitness program guidance in the exercise process, as well as an objective evaluation of the exercise effect.

The purpose of fitness is to strengthen the physique, improve the condition of body function, and improve the functional ability of various organ systems. To verify the good

J. C. Hung et al. (Eds.): IC 2023, LNEE 1045, pp. 401–406, 2023.
https://doi.org/10.1007/978-981-99-2287-1_57

effect of exercise on body function, it is necessary to objectively evaluate the effect of exercise. The effect of exercise refers to the adaptive changes and good responses of various organ systems in morphology, structure and function under the influence of physical exercise [3]. The measurement and evaluation of exercise effect is a very important issue. Through the measurement and evaluation, we can see the effect of exercise, which can better stimulate the enthusiasm of exercisers, and provide the necessary scientific basis for future exercise. At present, people mostly use the method of physical fitness test to evaluate the fitness effect, but only from the fitness results. Physical fitness test is a very effective method to evaluate the national physical condition, but it has certain defects and inapplicability to evaluate the fitness effect. In China, the evaluation of sports test results is mainly based on the comparison of the scoring standard table, scoring each test result respectively, obtaining the grade and score of the corresponding evaluation index, and then giving a total score and grade [4]. The traditional evaluation system can not get obvious evaluation on the individual differences before and after exercise, but two sets of health evaluation standards, pre exercise evaluation and post exercise evaluation, have been established through this experiment. In terms of evaluation results, China mainly focuses on the overall evaluation of a large group of teenagers and children or national physique, rather than individual evaluation. In the test of students' reaching the standard, the performance of sports quality is equal to the level of physical fitness. In fact, good sports performance does not necessarily mean good physique [5]. This is not conducive to individuals' reasonable and correct understanding of their health status. Therefore, there is a lack of evaluation of individual test results in the study of students' physical health in China. According to the tasks and requirements of teenagers and children, following the laws of physical and mental growth and development of teenagers and children, combined with the principles of exercise and fitness, this study attempts to explore a more scientific method to evaluate the fitness effect through the comparative study of several different fitness effect evaluation methods. To establish an individual evaluation system for the effect of fitness exercise, through the individual comparison of the exercise effect of young students before and after the experiment, in order to serve the establishment of an evaluation system for the effect of fitness exercise among young people and children.

2   Related Work

2.1   Fitness Information Service

After entering the 21st century, Japan has launched the third nationwide fitness activity, and online fitness plays an indispensable role. Many high-tech manufacturers have invested human and material resources in online fitness and developed a large number of interesting and diverse online fitness equipment. In the United States, the monitoring of physical activities is often in the form of scales rather than remote monitoring. Among them, the fitness market for obesity and people who lose weight is particularly popular. Therefore, there are many studies on the weight loss model and literature based on the Internet. In the past 10 years, the American online fitness industry has flourished, and fitness websites cover a wide range and provide comprehensive services. Nike + training and ABS work out are widely used in foreign countries, but the types are relatively single. Most people use them because of the brand Nike [6]. According to the talking data 2014 mobile Internet data report, the number of mobile health management users on IOS and Android platforms is HTC, and the growth rate is increasing. The number of APP users such as Gudong and ledi running has exceeded 10 million, which shows that it has a huge user base, has been widely used among the crowd, and is increasing at a high speed. Fitness app assisted exercise has changed the traditional way of physical fitness, requiring a fixed period of time to exercise for different sports or different equipment. It is more suitable for today's stressful life and heavy work, and is welcomed by more people [7].

In China, the current number of sports fitness apps is not particularly large. And in the form of several dominant companies. The platform is less innovative. For example, keep is currently commercialized seriously. Feel highlights the monitoring of physical health indicators. It is also similar to yuepao, which only highlights its social advantages. As the largest online fitness platform in China, muscle net has the most kinds of fitness training programs and has made a good display in news, diet and other aspects. However, there is no research on item exchange, and it is not possible to recommend the next step for users according to their status and previous exercise history. In other words, there is no good user system interaction. For example, a user who used to be a marathon runner may need to run more to meet his own sports needs.

2.2 Recommended Technology

The earliest data recommendation originated from a collaborative filtering algorithm, which was proposed by Bob et al. In 1992. The collaborative filtering algorithm was originally designed to filter spam, and was later applied to data recommendation. The core of collaborative filtering algorithm is to find group members who are similar in some aspect among groups with different characteristics, so as to recommend each other for these group members. In 2006, for the first time, there was an application-oriented film recommendation competition, and the greatly improved recommendation algorithm mainly served the situation of laboratory research. In the film recommendation competition hosted by Netflix, the winner is a recommendation algorithm based on matrix decomposition. In fact, in 1990, some people have proposed various recommended algorithms based on decomposition dimension reduction, such as singular value decomposition. It was not until 2006 that singular value decomposition method was used in matrix decomposition. Although the recommendation based on decomposition has good effect, it has high complexity in time and space [8]. In order to solve the problem of high time and space complexity, the decomposition recommendation of least square method is proposed. This algorithm has improved in storage space and computing time, but it still faces the problem of data sparsity. In order to solve this problem, clustering recommendation algorithm began to appear. The earliest clustering recommendation algorithm is k-nearest neighbor algorithm. Its principle is to classify information according to user behavior and item data to complete the expression of clustering. The clustering based recommendation method solves the problem of data sparsity well, and its representative k-nearest neighbor recommendation algorithm is also practiced in its recommendation system by Gongde Guo.

The recommendation performance of a single collaborative filtering algorithm can not be compared with clustering methods, but the combination of multiple collaborative filtering algorithms can also play a better role. A single algorithm has obvious shortcomings. The combination of multiple algorithms can be applied flexibly according to the actual data of the system. For example, some algorithms are good at solving the cold start problem, and some algorithms are good at solving the data sparsity problem [9]. At present, the application system that only uses a single recommendation algorithm basically does not exist. Different recommendation algorithms can be adopted according to the needs to better solve the data recommendation problem, which is the so-called hybrid recommendation technology. The collaborative filtering recommendation technology, weighted combination recommendation technology, cross harmonic recommendation technology proposed by Thomas are relatively mature hybrid recommendation algorithms.

3 Fitness Recommendation Method for Children and Adolescents Based on Ant Colony Algorithm

3.1 Ant Colony

Ant system is an artificial system that imitates the behavior of ant colony. In nature, ants are almost blind, but they can find the shortest distance between food and ant nest. Ecologists' research shows that ants achieve this by means of a special secretion - pheromone. Pheromone is a chemical substance secreted by ants. Ants secrete this substance in the process of looking for food. Ants use this substance to communicate and cooperate with other ants to find a shorter path. The more ants passing through a certain path, the greater the strength of this pheromone on the path. When ants choose the path, they tend to choose the direction with high pheromone intensity. This is why ants can find a shorter path back to their nest or food. On the one hand, each ant leaves a certain amount of information material on its path, and the information material left on the path gradually decays with time. On the other hand, later ants can perceive this pheromone and guide their behavior with the strength of the residual pheromone on the path[10]. The greater the strength of the pheromone, the greater the probability of being selected. Obviously, the process of ant colony searching for food source is a process of positive feedback of pheromone (positive feedback is based on the release of pheromone and the tendency of ants to take the route with rich pheromone, so as to quickly find the optimal solution). The flow of ant colony algorithm is shown in Fig. 1.

3.2 An Ant Colony Algorithm Based Fitness Recommendation Method for Children and Adolescents

By judging whether a recommendation system can provide users with sports prescription items that really meet their needs, we can reflect the success of the recommendation system. Therefore, when designing the sports prescription recommendation system, this paper deeply analyzes the user's information, because after cluster analysis, it plays a vital role in the recommendation of sports prescriptions. Through a large number of repeated

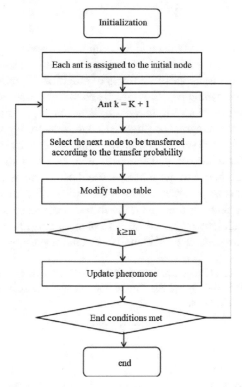

Fig. 1. Recommendation process based on ant colony algorithm

experiments, we can get a better clustering effect. The results of recommendation should be personalized. It is not only necessary to do a lot of research on recommendation methods, but also essential to understand the field of sports prescription. It is necessary to understand not only the formulation principles of exercise prescription, but also the specific contents of exercise prescription, so that the recommended exercise prescription results will be more ideal. The characteristic information of the target user reflects the interest preference. When designing the recommendation system, this paper adds the user's interest preference information to the recommendation process, so that the quality of the recommendation results is higher and can better meet the needs of users.

In this paper, the user preference sports prescription recommendation system based on K-means clustering is designed to meet the personalized needs, mainly in the following two aspects.

(1) The first is to use clustering algorithm to classify the initial data, so that similar users can be divided into the same class. In this way, we can calculate the similarity with the users in the class and get an initial recommendation result.
(2) On the basis of the initial recommendation results, the user's interest preference information is analyzed and extracted. Because each person's interest preference may be different, the requirements for exercise prescription are different, so this paper integrates the user's interest preference information.

4 Conclusion

With the progress of society and the rapid development of economy, great changes have taken place in living habits. Although the living standard is constantly improving, the national health problem has become the focus of attention. People have high incomes, but their bodies are overwhelmed. They are faced with a good economic income but no good health. Therefore, more and more people began to pay attention to the field of sports health, hoping to change their physical conditions through sports intervention, rather than through drug intervention. At present, there are many exercise prescriptions to refer to, such as fat reduction and rehabilitation, but they still can not meet people's needs. The public hopes to get more personalized exercise prescriptions that are more in line with their own conditions. Through exercise, they hope to obtain very ideal exercise intervention effects. In this paper, the fitness recommendation method and system for children and adolescents based on ant colony algorithm, which integrates user preferences and recommends more personalized exercise prescriptions.

Acknowledgements. Special Project of Scientific Research Plan from Department of Education in Shaanxi Province (19JK0915); Special Funds for Scientific Research Project of Xianyang Normal University (XSYK17047).

References

1. Salim, D., Perdana, N.J., Mulyawan, B.: Application of the case based reasoning & sorensen-dice coefficient method for fitness exercise program. In: IOP Conference Series Materials Science and Engineering. IOP Publishing Ltd, p. 012188 (2020)
2. Al-Motairy, B., Al-Ghamdi, M., Al-Qahtani, N., et al.: Building a personalized fitness recommendation application based on sequential information. Int. J. Adv. Comput. Sci. Appl. **12**(1) (2021)
3. Pavelski, L.M., Delgado, M., Kessaci, M., et al.: Stochastic local search and parameters recommendation: a case study on flowshop problems. Int. Trans. Oper. Res. (2020)
4. Sahu, S., Kumar, R., Mohdshafi, P., et al.: A hybrid recommendation system of upcoming movies using sentiment analysis of youtube trailer reviews. Mathematics **10**(9), 1568 (2022)
5. Mulay, A., Sutar, S., Patel, J., et al.: Job recommendation system using hybrid filtering (2022)
6. Button G.J., Michael, A., Sueann, R., et al.: Utilizing a "crawl, walk, run" training model to enhance field sanitation capabilities for peacekeeping forces: a recommendation for the department of defense global health engagement enterprise. Military Med. (2022)
7. Oliveira, S., Silva, G., Gorgonio, A.C., et al.: Team Recommendation for the pokémon go game using optimization approaches. In: Brazilian Symposium on Computer Games and Digital Entertainment - SBGames 2020 (2020)
8. Gandin, C., Matone, A., Ghirini, S., et al.: Alcohol, youth and sport: recommendation and good practice examples from the FYFA European project. J. Sports Med. Phys. Fitness **61**(5), 732–742 (2021)
9. Pandey, G., Narayana, V., Ramanna, R.: Method and system for dynamic recommendation of experts for resolving queries:, US10586188B2 (2020)
10. Zimovnov, A.V., Sokolov, Y.A.: Method of and system for determining user-specific proportions of content for recommendation:, US20200089724A1 (2020)

Removal of Cationic Malachite Green Dyes in Waster Water by PGS-AMPS-AM Hydrogel

Ting Yuan[1(✉)], Xianmao Zhang[2], Dongmei Zhang[3], and Mingye Wang[4]

[1] Chemical Products Inspection Institute, Gansu Institute of Product Quality Supervision and Inspection, Lanzhou 730050, Gansu, China
15719330708@163.com
[2] Gansu Institute of Product Quality Supervision and Inspection, Lanzhou 730050, Gansu, China
[3] Dingxi Anding District Agricultural Products Quality and Safety Testing Center, Dingxi 743000, Gansu, China
[4] Quality Inspection Department, Qingyang Energy and Chemical Group Ward Petroleum Technology Co. LTD., Qingyang 745000, Gansu, China

Abstract. A PGS-AMPS-AM (Attapulgite clay, 2-acrylamide-2-methyl propane sulfonic acid and acrylamide) hydrogel was produced by free radical crosslinking copolymerization. The cationic dye Malachite Green adsorption capacities of the hydrogel in waste water was investigated. The result shows that PGS-AMPS-AM hydrogel can be used as a superior absorbent for waste water treatment and the maxmium adsorption capacity could reach as high as 3632 mg/g.

1 Introduction

With the development of industry, water pollution has become more and more complicated. Various chemical reagents are poured into rivers leading to the result that a wealth of new disease kept intruding meanwhile causing human body's recession. One of them contaminated dyes is malachite green (MG) [1] Which many studies have revealed its high toxicity, high residue and carcinogen.

Hydrogel [2] has a large number of applications, for example, drug delivery [3] bioengineering [4] and absorption [5–7]. In terms of the adsorption, hydrogel was applied in waster water to remove metal ions, dusts, impurities, even dyes. In this manuscript, a hydrogel adsorbent(PGS-AMPS-AM) was prepared using the method of graft copolymerization to exhibit a high adsorption quantity. Many chemcial groups such as $-NH_2$, SO_3H, $-CONH_2$ was induced into the hydrogel, which could interact with the $-N^+$ of the MG at the same time the typical structrue of the hydrogel 3D porous network could provide a large interaction space between PGS-AMPS-AM hydrogel itself and MG. The equilibrium isotherm was analyzed and the parameters was also calculated. The results shows that the PGS-AMPS-AM hydrogel is successfully produced and the maxmium adsorption capacity is as high as 3632 mg/g. Maybe this satisfactory result also provide a new thinking of adsorbent structure design.

2 Methods

2.1 Materials and Instruments

Attapulgite(ATTP) from Jiangsu tire; 2-acrylamide-2-methylpropyl sulfonic acid (AMPS, analytical pure), Sinopharm Chemical Reagents Co.LTD; Malachite green (MG, analytically pure), Sinopharm Chemical Reagent Co.LTD., N, N-methylene bisacrylamide (BIS, analytical pure), Shanghai Zhongqin Chemical Reagent Co.LTD; Ammonium persulfate (APS, analytical pure); Sodium bisulfite (SBS, analytical pure), Yantai shuangshuang Chemical Co.LTD;

WE-3 Water Bath Constant temperature oscillator (Tianjin Onuo Instrument Co, LTD.); Shanghai pHS-3C ph meter; UV757CRT Spectrophotometer (Shanghai Keheng Industrial Development Co.LTD.).

2.2 Pretreatment of PGS

2 g of PGS and 100 mL HCl (3 mol/L) were add into the 250 mL flask with magnetic stirrer at room temperature for 3 h. Then, the suspension was centrifuged at 4000 rpm, washed with double distilled water and dired at 60 °C in vacuum drying oven to a constant weight.

2.3 Synthesis of PGS-AMPS-AM

The graft product was synthesized by free radical crosslinking polymerization. Accuray weigh PGS(1.0 g), AMPS(0.0025 mol), AM(0.025 mol), were dissolved into deionized water (10 mL) with magnetic stirrer for 10 min. Then oxygen-free nitrogen was purged through the solution for 30 min. Then, APS(0.05921 g), SBS(0.02960 g), BIS(0, 08882 g), were taking into the reaction mixture followed by further purging with nitrogenand heating at 40°Cfor 6h to obtain the target super absorbent. The resulting product was washed by double distilled water for serval times a week (change the water two times a day to remove the substance that didn't participate the polymerization). After that, the product was put under a nitrogen atmosphere at 60 °C for the desired time. The polymer were smashed and limited in the range of 200 mesh. The polymers were pured into the soxhlet extractor with a mixture of methanol and acetic acid (9:1,v/v) to extract the impurities for 24h. Finally, the polmer were washed with methanol and double water and dried in vacuum dried. The chemcial reaction mechanism was shown as Fig. 1.

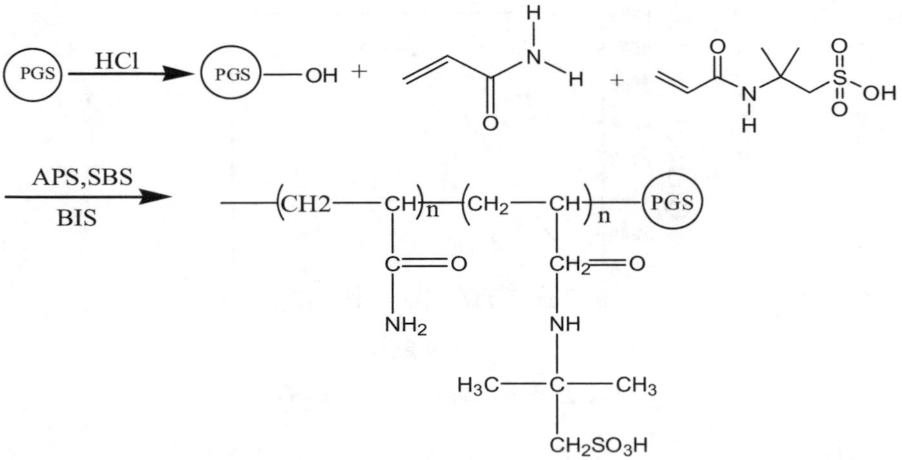

Fig. 1. Diagram of PGS-AMPS-AM polymerization procedures

2.4 Property Investigation

Here we studied the ability of the superabsorbent to the cationic MG dyes of the simulation wastewater. Accuracry pre-weighed power sample 5 mg were put into a 50 mL erlenmeyer flask that contain 30 mL malachite green 100–700 mg/g in each experimen, then the container were immersed into a WE-3 water-bathing constant temperature vibrator for 24 h. The absorption spectrum of the dye was recorded on UV/VIS spectrophotometer to detection. The colour was measured as absorbance at the maximum absorbance wavelength. The absorb ability(Q) is calculated by using the following expressions

$$\frac{(C_0 - C_e)}{m} V \tag{1}$$

where C_0 is the intial concentration of MG (mg/g), C_e is the final or equilibrium concentration of MG (mg/g), V is the volume of MG solution (mL), m is the weight of hydrogel (mg) and Q is the amount of MG solution adsorbed onto the unit amount of the hydrogel (mg.g^{-1})

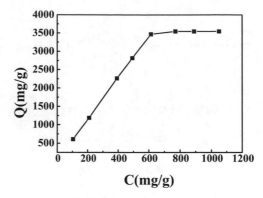

Fig. 2. Adsorption isotherm of MG

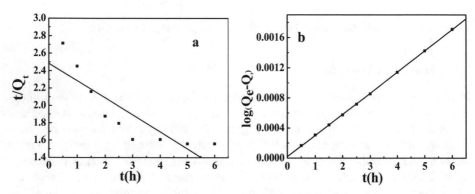

Fig. 3. Pseudo-first-order kinetic for the adsorption(a); Pseudo-second-order kinetic for the adsorption(b)

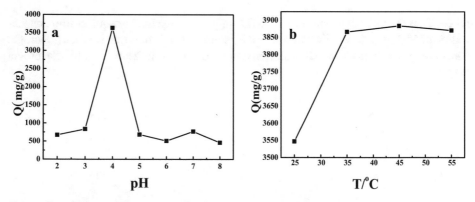

Fig. 4. the effect of pH on adsorption capacity(a), the effect of temperature on adsorption capacity(b) °C

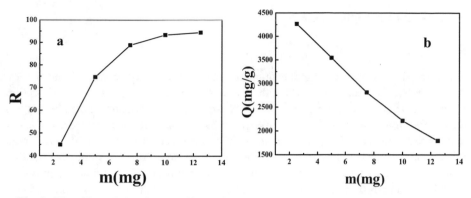

Fig. 5. The effect of absorbent quality to the removal rate(a) and the adsorption capacity(b)

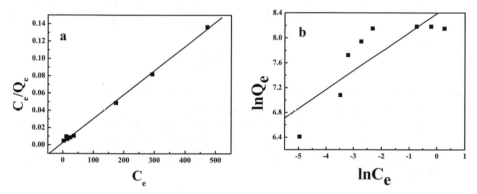

Fig. 6. The fittings of adsorption isotherm data for malachite green to the linearized Langmuir (a) and Freundlich (b)adsorption models.

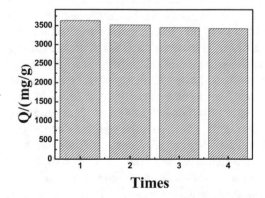

Fig. 7. The adsorption results of regeneration

3 Results and Discussion

3.1 Dynamic Adsorption

The adsorption capacity is a significant index to evaluate the performance of a adsorbent. As can be seening the Fig. 2, when the dyes concentration increased, the adsorption amount also increased heavily until the concentration of the dyes reached to 700 mg/L. Out of the limit concentration, the adsorption to the cationic malachite green dyes is maintained constant. The maximum static adsorption capacity of the malachite green is 3547 mg/g, which indicated the polymer possesses an efficient absorbent capability. Figure 2 also displays the influence of intial concentrations of malachite green on the equilibrium adsorption capacities. A well-known model for this adsorption process was presented as follows:

$$\log(Q_e - Q_t) = logQ_e - \frac{K_1}{2.303}t \tag{2}$$

$$\frac{t}{Q_t} = \frac{1}{Q_e^2 K_2} + \frac{t}{Q_e} \tag{3}$$

where Q_e is the amount of malachite green dyes of e adsorbed quilibrium; Q_t is the amount of a constant time; K_1 is a constant of the pseudo-first order kinetic mode; t is the time of vibration; K_2 is the pseudo-second order kinetic mode [8] constant. Via the Fig. 3 the process of the absorption coincidented with the pseudo-second order kinetic mode, the constant of this model beyond 0.99. At the moment, the process of this adsorption is a chemical behavior.

3.2 Effect of pH value

Based on what we has mastered that absorption is a significant phnomenon which is influenced by the value of pH. Thus, we has discussed this impact factor to further analyze the process of adsorption. The charge properties of absorbent is become one of most significant elements to investigate form 2 to 8 are shown in Fig. 4(a). It is noted that the capacity of adsorption increased as pH value decreased, and high efficiency was observed at pH = 4.0. It shows a tendency of decline after the maximum capacity of adsorption. The lower adsorption capacity is abserved in highly alkline aqueous solutions, while the higher adsorption capacity was absorved in a relative value of pH. The most probable reason for this is that the positive charges of adsorbent rised can absorb abundant MG as the value of pH increased. There are minority positive when the pH is less than 4.0 which was due to the $-H$ has an significant negative effect on the $-SO_3H$ of adsorbent to dissociate. With the value of pH going up, the degree of dissociation has a certain extent enhanced until alkaline surrounding. At the same time, the solution tends to be alkaline, resulting in a large amount of malachite green molecules precipitation, resulting in the solution color becomes pale, adsorption capacity presents a trend of decreasing.

3.3 Effect of Temperatures

In this paper, the relationship between temperature and adsorption capacity in the temperature range of 25 °C ~ 55 °C was investigated. The initial concentration was 800 mg/L and the amount of adsorbent was 5 mg. The relationship between temperature and adsorption capacity was shown in Fig. 4(b). It is suggested that the rise of temperature lead to a subsequent increase of adsorption ability. The absorblity is slightly descend and stable above 45 °C. In other words, the maximum ability is at 45 °C. The phenomenon might be due to the increased activity of active sites. An excellent reason is that the adsorbent can dissociate better in a high temperaturethan a low condition to produce a large number of $-SO_3{}^{2-}$ to interact with even more dye molecules.

3.4 Effect of the Amount Absorbent

The amount of absorbent is an element to effect the removal rate, this experiment has elect 2.5, 5.0, 7.5, 10.0, 12.5, 15.0 mg as points to investigate the relation. From the Fig. 5(a), we can observe that the removal rate has increase in some degree with the amount of absorbent. The relative adsorption capacity shows an increase phenomenon in Fig. 5(b), in terms of the quality of absorbent. The removal rate is calculated by the equation as follow:

$$R = \frac{C_0 - C_t}{C_0} \tag{4}$$

3.5 Adsorption Isotherms

The Langmuir model assumes that the adsorption is a monolayer adsorption as shows by the following equation:

$$\frac{C_e}{Q_e} = \frac{1}{bQ_{max}} + \frac{C_e}{Q_{max}} \tag{5}$$

where C_e(mg/L) is the equilibrium of concentration of MG in solution, Q_e(mg/g) is the amount adsorbed, Q_{max}(mg/g) is the adsorption capacity and b(L.mg-1) is the langmuir constant.

Another model is that the Freundlich isotherm which is a adsorption occurs on all kinds of layers and asymmetry areas. The equation is given as follows:

$$\ln Q_e = \ln K_F + \frac{1}{n}\ln C_e \tag{6}$$

where C_e denotes the equilibrium concentration(mg/L) of MG, Q_e is the mass of adsorbent(mg/g), K_F(L/mg) is a constant of this model, n is a index of absorption ablility.

According to the data of experiment, the absorption process is well fit with the Langmuir iostherm model [9], the correlation contents all beyond 0.99. While the Freundlich model wich n is between1 and 10, is not satisfied the experiment data. Meanwhile, the

Table 1. The parameter of Langmuir and Freundlich

isotherm	parameter	MG
Langmuir	Qm(mg/g)	3595
	R2	0.9996
	RL	0.107
Freundlich	Kf((mg/h(L/mg)))1/n	4398
	n-1	0.30506
	R2	0.7103

data even suggests that this process is a monolayers absorption. Through the Langmuir isotherm calculate the Q_{max} are similar to the actual absorption ability.

The dismensionless parameter RL of Langmuir isotherm equliibrium is defined:

$$R_L = \frac{1}{1 + bc_0} \qquad (7)$$

Where C_0 is te initial concentration of MG(mg/L).The value of R_L is regard as an index to evaluate the process of absorption whether is benificial or not.It is an inferior phenomenon to absorb when the R_L is more than one. While there is an perfect absorbption process when the value is between zero and one. Otherwise it is an irreversible process as the R_L is one.

The Gibbs free energy is calculated by the following formula:

$$\Delta G = -RT \ln R_L \qquad (8)$$

The thermodynamic parameters are documented in Table 1. The absorption of MG is endothermic in nature because of the value of enthalpy change is positive.

Otherwise, a possible adsorption process is a chemical adsorption.

3.6 Regeneration and Cyclic Adsorption Capacity

Transfering a certain concentration 30 mL to a 100 mL conical flask in a superior condition by the above experiment at pH=4,.the quality of absorbent is 5 mg , 25 °C in terms of the actually surrounding. The stability and regeneration of PGS-AMPS-AM was poured into in 20 mL 1 mol/L hydrochloric acid for one day to obtain a repeat product. The adsorption results of regeneration are shown in Fig. 7. The absorbent can be reused some times after extraction by hydrochloric acid in the adsorption capacity for MG. The experiments suggest that PGS-AMPS-AM is expected to be used as a recycling absorbent in the treatment of waste water.

4   Conclusions

In this study, a novel PGS-AMPS-AM hydrogel was synthesized by free radical crosslinking copolymerization. The absorbent has high adsorption capacity to MG. It responds faster to MG than some physical methods and is much cheaper than chromato-graphic methods used for conventional purposes. The adsorbent can provide a reliable and effective solution for the recovery and utilization of MG in industrial wastewater.

References

1. Shah, H., Ahmad, K., Naseem, H.A.: Water stable graphene oxide metal-organic frameworks composite (ZIF-67@GO) for efficient removal of malachite green from water. Food Chem. Toxicol. (4) (2021)
2. Sarkar, N., Sahoo, G.: Reduced graphene oxide decorated superporous polyacrylamide based interpenetrating network hydrogel as dye adsorbent. Mater. Chem. Phys. **250** (2020)
3. Liu, Y., Fan, Q.: Construction of nanocellulose-based composite hydrogel with a double packing structure as an intelligent drug carrier. Cellulose (2021)
4. Smith,E.E., Yelick, P.C.: Bioengineering tooth bud constructs using GelMA Hydrogel. Methods Mol. Biol. (Clifton, N.J.) (2019)
5. Hijab, M., Parthasarathy,P.: Minimizing adsorbent requirements using multi-stage batch adsorption for malachite green removal using microwave date-stone activated carbons. Chem. Eng. Process., 108318 (2021)
6. Hosseinzadeh, H., Ramin, S.: Fabrication of starch-graft-poly(acrylamide)/graphene oxide/hydroxyapatite nanocomposite hydrogel adsorbent for removal of malachite green dye from aqueous solution. Int. J. Biol. Macromol. (2018). S014181301730524X
7. Baloch, F.E., Afzali, D., Fathirad, F.: Design of acrylic acid/nanoclay grafted polysaccharide hydrogels as superabsorbent for controlled release of chlorpyrifos. Appl. Clay Sci. **211**(2), 106194 (2021)
8. George,G., Saravanakumar, M.P.: Correction to: facile synthesis of carbon-coated layered double hydroxide and its comparative characterisation with Zn–Al LDH: application on crystal violet and malachite green dye adsorption—isotherm, kinetics and Box-Behnken design. Environ. Sci. Pollut. Res. **25**(30) (2018)
9. Le, H.Q.: Modeling of CO2-activated adsorption on chitosan hydrogel for dye removal in aqueous solution. J. Chem. Eng. Jpn **52**(8), 671–679 (2019)

Research and Analysis on the Influence of Emotion on Sleep Based on Physiological Signals

Xianfeng Zeng(✉)

Hainan Vocational College of Political Science and Law, Haikou 571100, Hainan, China
wxbzzbj@126.com

Abstract. The study is based on physiological signals of sleep. The main purpose of this study is to determine the impact of emotion on sleep by analyzing physiological signals during sleep. For example, you can determine whether there are any differences in heart rate and breathing patterns between people who have good night rest and those who do not. It can also determine whether the emotions that occur before bedtime affect our sleep conditions, or whether they cause us to wake up more frequently at night. This study involves using EEG to collect data from participants, analyze the impact of emotional state on sleep quality, and investigate whether anyone's sleep quality can be predicted by his or her own emotional state. In addition, we will try to find out how different emotions affect a person and what their impact on sleep is.

Keywords: Physiological signals · Emotion · Sleep effects

1 Introduction

Emotions can affect human behavior and play an important role in daily life. Many mental disorders are related to emotions, such as autism and depression. Therefore, emotion is often used as a reference to evaluate patients' mental disorders. More and more researchers focus on EEG analysis of different emotions caused by specific stimulation patterns. The research mainly focuses on designing experiments using multimedia materials (including images, sounds, texts, etc.) to stimulate the brain and expose its cognitive activities for emotional classification. Most of the existing EEG emotion models only use the time domain, frequency domain, spatial domain information of EEG signals alone or the combination of the above two characteristics [1]. These models ignore the complementarity between the time-frequency and space features of EEG signals, which will limit the performance of EEG classification models to a certain extent. How to make use of the complementarity between time-frequency-space features in EEG is a challenge. The acquisition of emotional information depends on the design of experimental paradigm and the improvement of advanced hardware acquisition equipment technology. For example, on the one hand, the quality of expression images and videos depends on the lens of shooting equipment, the capacity of storage equipment, the version of post

J. C. Hung et al. (Eds.): IC 2023, LNEE 1045, pp. 416–421, 2023.
https://doi.org/10.1007/978-981-99-2287-1_59

software processing and the degree of processing; On the other hand, it also depends on the illumination degree of the face and the individual psychological state of the subjects at that time; The understanding of emotion involves the reasons for people's emotion and the feedback of intention, which is the research scope of psychology and cognition; Emotional interaction is embodied in the communication between human and computer and the expression of empathy. These aspects are not only related to psychology and cognition, but also inseparable from environmental sociology. As an important branch of emotion computing, emotion recognition focuses on signal processing, artificial intelligence, brain like computing, pattern recognition and so on [2]. This paper focuses on the research and analysis of the influence of physiological signals on sleep.

2 Related Work

2.1 Physiological Feature Extraction

The acquisition and processing of multimodal physiological signals will involve three aspects: time, frequency and space. To obtain comprehensive physiological characteristics, we need to consider four important attributes: time/frequency characteristics, sample window, channel distribution and interaction between attributes.

In terms of time-domain features, the most common is the extraction of statistical features. Isik first performed wavelet transform on physiological signals, and then performed EOG, EMG, GSR, resp and plethysmography on volume pulse wave respectively Temperature signal extracts features such as mean, variance and standard deviation. Samarth extracts its statistical features (maximum, minimum, mean and variance, amplitude range) for wavelet coefficients, and then uses depth neural network and convolution neural network for emotion recognition respectively [3]. Wu extracts 24 features including mean, median, amplitude range, standard deviation, maximum and minimum values, maximum and minimum value ratio, and minimum value ratio from the original signal of skin electricity and its first-order difference samples and second-order difference samples, and then analyzes the impact of immune mechanism on emotion recognition performance.

For feature extraction in frequency domain, it is necessary to map physiological signals to different frequency bands, and then extract power spectral density (PSD), energy and some hybrid derived features within each frequency band. For example, calculate their energy ratio between different frequency bands to expand the sample feature dimension. The calculation of power spectrum mainly includes linear estimation (autocorrelation estimation, autocovariance estimation, Welch method, periodic graph method, etc.) and nonlinear estimation (maximum entropy method and maximum likelihood method).

2.2 Influence of Emotion on Sleep

At the psychological level, the stimulation and presentation of emotion is not a sufficient and necessary condition for the process of emotional arousal. It proves that after receiving subjective or objective stimulation, emotion will be naturally, quickly and may be automatically awakened unconsciously. The psychological definition of emotion is a psychological state stimulated by the comparison between external requirements and their own objective or independent needs in a given environment [4]. This state can reflect the relationship between individuals and their environment through certain evaluation criteria.

Modern people are under great pressure in work and life, and their work and rest are increasingly irregular, and the number of people with sleep disorders is gradually increasing. Among the many factors that affect sleep, psychological problems are an important aspect. Psychological phenomena such as anxiety, irritability, uneasiness and so on can lead to insomnia.

In daily life, we can also clearly understand the impact of psychological state on sleep quality. Nervous, bored, angry, feeling pressure often toss and turn, unable to sleep at night, even if you fall asleep, you often have more dreams and sleep uneasily. Therefore, improving sleep quality should start with improving psychology.

At the same time, sleep also has a significant impact on mental health. Sleep disorders are easy to cause irritability, anxiety, irritability, depression, and even psychological disorders and mental diseases. According to a study, sleep is an important factor among the seven factors that affect human life span, which shows the importance of sleep and its great impact on health.

Although it is not surprising that people have difficulty falling asleep when they are emotional, the relationship between emotional disorders and high-quality sleep is complex and bidirectional. Just as negative emotional states make it impossible to sleep well at night, frequent interruptions or lack of sleep can also lead to depression or anxiety. Whichever comes first, the final result is that depression and bad sleep complement each other [5].

Everyone has experienced occasional terrible dreams, but frequent nightmares are related to depression and anxiety, as well as poor sleep quality and low quality of life. This is a cycle that is difficult to break: disturbing or negative dreams will wake you up from sleep, making it difficult for you to return to sleep; Then, being unable to sleep at night will make you feel bad the next day and affect your sleep the next night.

3   Research and Analysis of the Influence of Emotion on Sleep Based on Physiological Signals

EEG signals are activated by time-frequency-space features in different emotional states, and there are some discriminant local features. We designed a space-frequency time attention mechanism (sst-/attention)) to dynamically capture these valuable local features. SST attention is composed of two sub components: spatial attention mechanism and frequency band/time attention mechanism.

Figure 1 below shows the physiological signal feature model.

Specifically, in this study, it is defined as the data of EEG electrodes with a length of time points. Where, is the data of all electrodes at the time point. Using the spatial position information of each electrode, we will transform it into a 2D time plan and the height and width of the 2D plan respectively. By stacking the two-dimensional plans formed at all time points, we can get the 3D spatiotemporal representation of the signal [6].

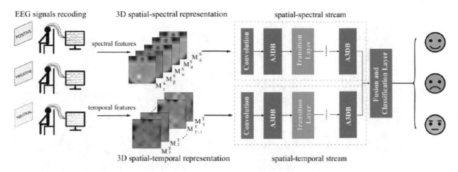

Fig. 1. Physiological signal characteristic model

We define it as the frequency domain characteristics of EEG signals containing frequency bands. Where, is the data of all electrodes in the frequency band. Similar to the time domain, it will be transformed into a d-frequency domain plan. By stacking the 2D plans formed under all frequency bands, we can get the 3D frequency space representation of the signal. Here, we use the EEG frequency domain characteristics of five frequency bands, namely.

Emotion recognition is to use computer signal processing and analysis methods to extract and classify the characteristics of psychological, physiological or physical behavioral parameters in various emotional states, so as to confirm the emotional state of individuals. At present, there are mainly two ways of emotion recognition: ① external behavior measurement: recognition through external behavior characteristics such as facial expression, voice or posture; ② Physiological signal measurement method: measure physiological signals such as respiration, heart rhythm, EEG or body temperature for identification [7]; Although the acquisition of physiological signals is not as simple as the former, it is spontaneous, not controlled by human factors, and can more objectively and truly reflect people's emotional state.

4 Simulation

Welch method is used to extract the baseline between the test frequency and 3 ~ 47Hz, and the window is 256 samples. Then, the baseline power was subtracted from the test power to obtain the power change relative to the pre stimulation period. These power changes are averaged over the frequency bands (3–7 Hz), (8–13 Hz), (14–29 Hz) and (30–47 Hz). For correlation statistics, we calculated the Spearman correlation coefficient between power change and subjective score, and calculated the p value of left tailed (positive) and right tailed (negative) correlation tests. We did this work for each participant separately, and then the 32 P values of each relevant direction (positive/negative), frequency band and electrode will be merged into one p value through the fisher's method platform. As shown in Fig. 2 below, the wakefulness and overall score are compared with theta (4–7 Hz).

The titer showed the strongest correlation with EEG signals, and correlation was found in all analyzed bands. In low frequency theta and alpha, the increase of potency leads to the increase of power [8]. This is consistent with the results of the pilot study.

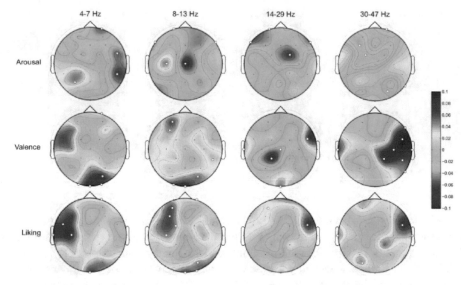

Fig. 2. Arousal and overall score vs. theta (4–7 Hz)

These effects are in the occipital region, so the position on the visual cortex may indicate that these effects are relative inactivation or top-down inhibition, which is caused by participants' focus on pleasant sounds. About β In frequency band, we found that the central descent and the power increase of occipital bone and right temporal bone were also observed in pilots. For the enhancement of the right temporal part β Ability is related to positive emotional self induction and external stimulation. Similarly, a positive correlation between valence states and high-frequency power has been reported, including those from the anterior temporal brain β and γ Belt. Accordingly, we observed a significant increase in left, especially right temporal gamma horsepower. However, it should be mentioned that EMG (muscle) activity is also prominent at high frequencies, especially on the anterior and temporal electrodes.

Similar correlations were found in all analyzed bands. About θ and α Refractive power, we observed an increase in the left central frontal cortex. Liking may be related to motivation. However, higher preferences were observed leading to left α [9] The increase in strength conflicts with the discovery of activation of the left frontal lobe, resulting in α Decrease, which is usually reported as emotion related to entry motivation. When considering the contradiction caused by some disliked clips that are likely to cause feelings of anger (because you have to listen to them, or just because of the content of the lyrics), this contradiction may be solved, which is also related to the motivation to enter, so it may lead to the reduction of alpha to the left. Stay β and γ The correct temporal enlargement found in the band is similar to the observed potency, and caution should be taken. Generally, the distribution of potency and preference correlation shown in Fig. 2 looks very similar, which may be the result of high correlation between the above scales [10].

5   Conclusion

The research and analysis of the influence of emotion on sleep based on physiological signals is a research aimed at understanding how emotion affects sleep quality. It is important for people to know what affects their sleep quality, so that they can take measures to improve their sleep quality. Sleep disorders are very common in humans, affecting about 30% of adults at some stage of life (Sleep Disorders Association). There are many factors that can lead to poor sleep quality, such as stress, anxiety, depression and hormonal changes (Sleep Disorders Association). However, there are also many factors that help maintain good sleep quality, such as physical activity and healthy eating habits.

Acknowledgements. Project supported by the Education Department of Hainan Province"Research on the emotional Labor strategy of the staff of Hainan Government Affairs Center facing the first-class Business Environment", project number: Hnky2022ZD-24.

References

1. Wang, Y., Feng, Q., Zeng, Y., et al.: Analysis on the influence of emotional management for negative emotion and sleep disorder of patients with cerebral infarction. World J. Sleep Med. (2019)
2. Zhou, W., Zhang, F., Wang, X., et al.: Analysis of the influence of politics on work emotion in workplace. Aggress. Violent. Beh. **2**, 101563 (2021)
3. Zhao, S., Ding, G., Han, J., et al.: Personality-aware personalized emotion recognition from physiological signals. In: Twenty-Seventh International Joint Conference on Artificial Intelligence {IJCAI-18} (2018)
4. Tago, K., Jin, Q.: Influence analysis of emotional behaviors and user relationships based on Twitter data. Tsinghua Sci. Technol. **23**(1), 10 (2018)
5. Lin, S., Xie, J., Yang, M., et al.: A review of emotion recognition using physiological signals. Sensors **18**(7), 2074 (2018)
6. Han, Y., Xu, Y.: The research of emotion recognition based on multi-source physiological signals with data fusion (2022)
7. Chen, M.F., Lin, S.Y., Chen, L., et al.: Influence analysis of stepwise psychological guidance on negative emotion of puerpera. J. Clin. Nursing (2018)
8. Wang, W.R.: The influence analysis of emotional fluctuations and social support degree on the post-traumatic growth in breast cancer patients. Nurs. Pract. Res. (2018)
9. Lu, H.M., Wu, Y.F., Art, S.O.: The influence of happy emotion on visual communication design. J. Jiamusi Vocat. Inst. (2018)
10. Zhou, D.K., Zheng, S.J., Song, L.M., et al.: The influence of the Insight-HXMT/LE time response on timing analysis. Res. Astron. Astrophys. **21**(1), 005 (2021)

Research and Application of Chinese Literature Automatic Abstract Extraction Based on Textrank Algorithm in Characteristic Resources Construction

Yumei Dang[✉]

The Library of Guangxi Normal University for Nationalities, Chongzuo 532200, Guangxi, China
dymglg@163.com

Abstract. The purpose of this study is to study the application of automatic abstract extraction of Chinese literature based on text ranking algorithm in feature resource construction. The objective of this project is to provide a system for automatically extracting general features from Chinese literary works as an effective method for constructing feature resources. A large amount of data has been collected from the Internet, including novels, short stories, essays and other texts. Based on these data, we developed a new method to automatically extract general features using text ranking algorithm and machine learning technology. Research paper on Automatic Abstract extraction based on text ranking algorithm in feature resource construction. It is an important part of Chinese Literature Automatic Abstract extraction system, which can be used for automatic extraction and classification of Chinese Literature Feature resources. This paper mainly introduces the characteristics of the new method and presents some application examples.

Keywords: Textrank algorithm · Construction of characteristic resources · Abstract extraction

1 Introduction

The process of using computers to process a large number of texts and produce concise and refined content is text summary. People can grasp the main content of the text by reading the summary, which not only greatly saves time, but also improves reading efficiency. However, manual summarization is time-consuming and labor-consuming, and can not meet the growing demand for information. Therefore, automatic summarization with the help of computer for text processing came into being [1]. In recent years, the research of automatic summarization, information retrieval, information filtering, machine recognition and so on has become a hot topic.

There are two main methods of automatic summarization: extraction and abstraction.

Among them, extraction is an extraction automatic summarization method, which forms a summary by extracting the existing keywords in the document; Abstract is a generative automatic summarization method. Abstract is formed by establishing abstract

J. C. Hung et al. (Eds.): IC 2023, LNEE 1045, pp. 422–428, 2023.
https://doi.org/10.1007/978-981-99-2287-1_60

semantic representation and using natural language generation technology. Because the generative automatic summarization method needs complex natural language understanding and generation technology support, its application field is limited. Therefore, I have also learned the extraction automatic summarization method.

In this paper, the average accuracy rate P, the average recall rate R and the average F value of the automatic summary evaluation indexes are used to compare the automatic summary effects of the improved sentence weight algorithm and textrank algorithm. The results show that the generation quality has been improved to a certain extent; Using the snowball method to determine the parameter 6 of the final sentence weight of the improved algorithm [2]; The coverage of the improved algorithm and textrank algorithm in the artificial summary sentence set is tested at different extraction numbers. The test results show that the effect of the improved automatic summary method is improved to a certain extent.

2 Related Work

2.1 Textrank Algorithm Flow

The key to understanding textrank algorithm is to understand PageRank algorithm. PageRank is an iterative algorithm used by Google to calculate and sort the weight of web pages included in the search engine.

The core idea of PageRank is that the value of a web page is determined by the value of each web page linked to this web page and the corresponding weight. In the PageRank algorithm, each web page is represented by vertices in the graph, and the link weight between web pages is represented by edges[3]. The iterative algorithm is expressed as:

$$AVF(x_i) = \frac{1}{m} \sum_{f=1}^{m} f(x_{ij}) \tag{1}$$

$$\arg\min_{SC} \sum_{i=1}^{k} \sum_{x \in C_i} |X - \mu_i|^2 \tag{2}$$

After understanding PageRank, it is easier to look at textrank. Textrank regards text data as nodes of a graph, and uses the relationship between text and text to establish an adjacency matrix. According to different objectives, textrank can realize keyword extraction.

(keyword extraction) and sentence extraction. Generally speaking, the process of textrank algorithm is divided into the following steps:

1. Specify the task objective, and add the text unit corresponding to the task as the vertex of the graph;
2. Add the relationship between text cells as the edge of connecting nodes in the graph. It can be a directed edge or an undirected edge, but it can be a weighted edge or an edge without weight. At this time, the establishment of the adjacency matrix is completed;
3. Iterate the textrank algorithm until it converges, and calculate the score of each node;

4. Sort the nodes according to the final scores, and extract Top-k as keywords or key sentences according to the sorting results.

Generally speaking, for the keyword extraction task, the text unit is each word after sentence segmentation, and the adjacency matrix is the number of times that words and words appear in adjacent positions (normalization processing); For the key sentence extraction task, the adjacency matrix considers the text similarity between sentences. Figure 1 below shows the flow of textrank algorithm.

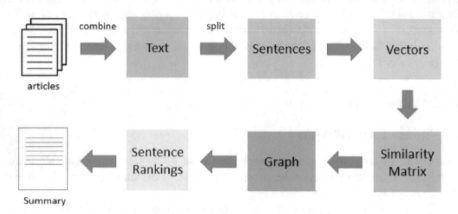

Fig. 1. Textrank algorithm flow

2.2 Abstract Extraction

Abstract sentence extraction aims to automatically extract several important sentences from a given text that can represent the main content of the whole text. Abstract sentence extraction method based on textrank algorithm takes sentences as nodes and the similarity between sentences as edges to establish a graph model; The established graph model is substituted into textrank formula for iterative calculation until convergence, and the weight score of each sentence is obtained; Finally, a certain number of sentences are extracted from the sentences with the highest scores to form a summary. In the calculation of sentence similarity, textrank algorithm is based on the overlap of two sentences. In addition, the author also puts forward some methods such as string kernel function, cosine similarity and maximum common subsequence [4]. Other methods for calculating similarity include editing distance based and semantic dictionary based.

The main steps of abstract sentence extraction are:

(1) Pretreatment. Including sentence segmentation, word segmentation, part of speech tagging and part of speech filtering. The preprocessing process is shown in Fig. 2 below.
(2) Extract abstract sentences. Rank the sentence weight scores calculated in step 3, and select the most important (highest score) t sentences as summary sentences.

(3) Composition summary. According to the requirements of sentence length, number of abstract sentences and compression ratio, sentences are extracted from the candidate abstract sentences to form a summary [5].

At the same time, it is found that the probability distribution of the abstract keywords at the beginning and end of the paragraph is uneven, only about 43% of the probability. Therefore, it is not possible to conclude that the sentence at the beginning of the paragraph is a summary sentence. Therefore, when considering the sentence position weight, the influence of the sentence position on the sentence weight will be appropriately reduced.

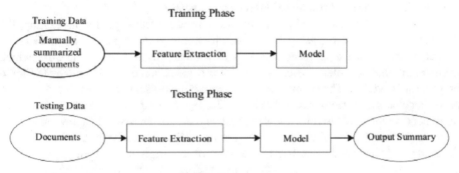

Fig. 2. Pretreatment

The traditional textrank algorithm only considers the relationship between sentences, and the calculation of sentence similarity is based on the method of content overlap rate between sentences. It does not consider the text structure, the relationship between sentences and titles, and the position [6]. For texts with text structure, such as papers and periodicals, other features of sentences should be added. Based on the traditional textrank algorithm, this paper integrates Title Similarity, abstract keywords and sentence position into the calculation process of sentence weight, and comprehensively considers the final sentence weight to generate the summary.

3   Research and Application of Automatic Abstract Extraction of Chinese Literature Based on Textrank Algorithm in Characteristic Resource Construction

The characteristic resource database is to build an information resource database with Chinese characteristics, local characteristics, higher education characteristics or resource characteristics, and provide services on the Internet, promote the development of teaching and scientific research in Colleges and universities, and support the construction of local national economy. If one or more words can express the meaning of some words when abstracting the abstract through TF-IDF algorithm, these words are named as keywords. The more keywords there are, the more important the sentences are. The important sentences will be selected as the text summary. This algorithm selects keywords through word frequency and inverse document frequency, and finally selects the sentences. Only the word frequency is considered, and other factors such as the position information of the sentences are not considered, So it is not accurate in abstract extraction.

The construction of the characteristic resource database is an important task of the digital resource construction of the library [7]. After many years of exploration, the characteristic database of the library has changed from the bibliographic database to the content database, from the collection of literature resources to the network resource link, from the isolated construction of each library to the joint development of multiple libraries, and from the self-sufficient resource service mode to the sharing service of Online Thematic Resources. When you click on the library website, you will often see columns such as "Digital Library", "special collection", "special library" and "self built database", and you can browse various special databases.

In the training process, skip gram model is used for training. Before training, the data set can be divided into two parts, one of which is used as the pre training work of the model, which is called external document, and the other is used as the testing work, which is called internal document. After the data set is divided, the data set documents should be preprocessed first. The specific steps are: removing noise, removing pictures, tables and other charts that are not easy to be recognized by the computer. After the removal, the document is segmented to form a sentence set, and the sentence set is segmented and the stop words are removed. During the skip gram model training, the word sequence after the sentence is arranged according to the order of the original document. The purpose of this is to more accurately obtain the context semantics and grammar information of the text. The vector dimension can be set manually [8]. The vector dimension set in this algorithm is 200. Put the preprocessed external documents into the skip gram model for pre training. After the training is completed, input the preprocessed test documents into the trained skip gram model to obtain the sentence vector set representing the syntax semantics and its context information.

4   Simulation Analysis

Textrank algorithm is widely used in automatic text summarization technology. The improved text summarization algorithm based on TF-IDF and texttank is based on the traditional textrank algorithm. For news text, TF-IDF is introduced to vectorize sentences and calculate sentence similarity, build a graph model, add sentence position weight to the final scoring formula, and finally obtain the sentence score to generate a summary [9]. The evaluation results of average accuracy, average recall and F value show that the improved text summarization algorithm based on TF-IDF and textrank is better than the classical textrank and TF-IDF algorithms. The project code of textrank algorithm is shown in Fig. 3 below.

In addition, the text summary generation algorithm based on the joint scoring of word2vec and textrank is also improved on the basis of the textrank algorithm. The idea of the algorithm is to integrate the word frequency and the semantic relationship of sentences. The algorithm trains external documents through word2vec to obtain the trained word2vec model, and applies it to test documents. After vectorizing the sentences, the sentence similarity matrix is obtained, At the same time, the test document is vectorized again by TF-IDF and the sentence similarity matrix is calculated. Finally, the two similarity matrices are fused to obtain a new sentence similarity matrix, and the new sentence

```
import numpy as np

def pagerank(M, num_iters=100, d=0.85):
    N = M.shape[0]
    R = np.random.rand(N, 1)
    R = R / sum(R)
    for i in range(num_iters):
        R = d * M @ R + (1 - d) / N
    return R

M = np.array([[0, 0, 0, 0, 1],
              [0.5, 0, 0, 0, 0],
              [0.5, 0, 0, 0, 0],
              [0, 1, 0.5, 0, 0],
              [0, 0, 0.5, 1, 0]])
R = pagerank(M, 100, 0.85)
print(R)
```

Fig. 3. Textrank algorithm project code

similarity matrix is brought into the scoring formula to generate a summary [10]. The evaluation results of average accuracy rate, average recall rate and F value show that the joint scoring text summary generation algorithm based on word2vec and textrank is better than the classical textrank and TF-IDF algorithms.

5 Conclusion

The purpose of this study is to construct automatic abstract extraction based on text ranking algorithm in feature resource construction. Feature resources are the most important part of the literature. They contain information about a large number of works, so we can use them to extract some data automatically. This paper presents an automatic method of extracting Chinese literature abstracts using text ranking algorithm, and applies it to constructing feature resources.

References

1. Qiu W , Shu Y , Xu Y . Research on Chinese multi-documents automatic summarizations method based on improved TextRank algorithm and seq2seq[C]// BIC 2021: 2021 International Conference on Bioinformatics and Intelligent Computing. 2021
2. Xintao X U , Chai X , Xie B , et al. Extraction of Chinese Text Summarization Based on Improved TextRank Algorithm[J]. Computer Engineering, 2019
3. Song, Z.: Chinese Language and Literature Intelligent Teaching System Based on Data Mining Algorithm[J]. Springer, Cham (2022)

4. Lyu L , Han T . A Comparative Study of Chinese Patent Literature Automatic Classification Based on Deep Learning[C]// 2019 ACM/IEEE Joint Conference on Digital Libraries (JCDL). ACM, 2019
5. Tang Y . Chinese Language and Literature Relational Database Mining Based on Association Rules Algorithm[C]// The International Conference on Cyber Security Intelligence and Analytics. Springer, Cham, 2022
6. Guo Q , Xiong A . Chinese News Keyword Extraction Algorithm Based on TextRank and Word-Sentence Collaboration[C]// 2020
7. Dawes A , Doughty M , Saines S , et al. Development and application of criteria to evaluate written CBT self-help interventions adopted by Improving Access to Psychological Therapies services[J]. The Cognitive Behaviour Therapist, 2022, 15:e28-
8. Chen W , Shao P , Zhang Y , et al. The application framework of big data technology during the COVID-19 pandemic in China[J]. Epidemiology and Infection, 2022, 150:e71-
9. Xiong A , Guo Q . Chinese News Keyword Extraction Algorithm Based on TextRank and Topic Model[J]. 2019
10. Li L , Wang Y , Xu Y , et al. Meta-learning based industrial intelligence of feature nearest algorithm selection framework for classification problems[J]. Journal of Manufacturing Systems, 2022(62-)

Ring Oscillator Optimization Design Model Summary

Xuhao Ye[✉], Zixuan Gao, Rongkai Cheng, Shuaiteng Liu, and Kaiwen Zheng

Shijiazhuangtiedao University, Shijiazhuang 050000, Hebei, China
yxh2962261206@163.com

Abstract. Ring oscillator is an important structure of digital clock chip, the speed, area, power consumption of ring oscillator determines the performance of digital clock chip. This article focuses on the optimization design of the ring oscillator in a digital clock chip.

First of all, in this paper, the oscillation frequency model of the toroidal oscillator delay is obtained, and the relationship between the output frequency and the capacitor charge and discharge time is obtained, and the magnitude of the output frequency of the inverter under different parameters is determined. Then, a one-target optimization model based on adaptive learning step search is established, and the optimal gate length and gate width and the optimal number of inverters are discussed with the lowest possible power consumption.

Keywords: adaptive learning step search · Single-objective optimization model · power consumption optimization model

1 Introduction

Chip is a collective term for semiconductor component products, which is a kind of silicon wafer containing integrated circuits. However, the manufacturing process of chips is very complex, going through thousands of processes and manufacturing through complex processes. Digital chips, in particular, are designed to be optimized as the process size continues to shrink.

The ring oscillator is a ring-shaped machine that consists of three non-gate or more odd non-gate outputs and input terminals connected end-to-end, which is an important structure in the digital clock chip. There are three important indicators of speed, area and power consumption in its design that need to be considered, and the faster the toroidal oscillator, the better the performance; The smaller the area, the lower the cost; The lower the power consumption, the longer the service life of the device.

2 Ring Oscillator Delay Oscillation Frequency Model

2.1 Model Preparation

The inverter is divided into two stages when working: the linear region and the saturation region. When the ring oscillator is operating, each inverter passes through a linear and

J. C. Hung et al. (Eds.): IC 2023, LNEE 1045, pp. 429–435, 2023.
https://doi.org/10.1007/978-981-99-2287-1_61

saturation region.

$$I_d = \begin{cases} K\frac{W}{L}[(V_{gs} - V_{th})V_{ds} - \frac{1}{2}V_{ds}^2], & V_{ds} < V_{gs} - V_{th} \\ \frac{1}{2}K\frac{W}{L}(V_{gs} - V_{th})^2, & V_{ds} > V_{gs} - V_{th} \end{cases} \tag{1}$$

wherein, the voltage between the gate sources is represented, the voltage between the leakage sources is represented, and the threshold voltage is represented; $V_{gg}V_{ds}V_{th}W$ indicates the gate width and L represents the gate gate length; K is the scale factor in different types of COMS tubes.

The load capacitance of the inverter is proportional to the gate area of the inverter of the next stage, and we can list the following equations

$$\begin{cases} C_i = kS_{i+1}, & i = 1, 2 \cdots N - 1 \\ C_N = kS_1 \end{cases} \tag{2}$$

where C_i is the load capacitance of the ith inverter and S_{i+1} is the gate area of the i + 1st inverter.

2.2 Time Delay Model Establishment

The inverter is made up of a PMOS tube and an NMOS tube in series. The NMOS tube pulls the output down when the input is high, and the time from high to low is the falling edge delay. The PMOS transistor is low at the input, which raises the output voltage, and the conversion time from low to high is the rising edge delay of the inverter. That is, we can think that the PMOS tube has no falling edge delay, and the NMOS tube has no rising edge delay. Different types of MOS tubes are solved by establishing time delay models.

Step1: Establish a PMOS rising edge delay model

In the CMOS inverter, the entire process of pulling up the PMOS tube to charge capacitor C is in a linear and saturated state, and we can get the operating current according to Eq. (1). I_d, can be approximated as constant current, so the voltage is

$$V_{out} = \int_0^{V_{mid}} dV + \int_{V_{mid}}^{V_{out}} dV = \frac{1}{C}\left(\int_0^{T_{mid}} I_d dt + \int_{T_{mid}}^{T_m} I_d dt\right) \tag{3}$$

Among them, V_{out} it is the output voltage of the inverter, V_{mid} the clamping voltage between the linear region of the inverter and the saturation region, representing the load capacitance, and T_{mid} is the critical time at the breakpoint between the linear region and the saturation region.

Combined with the formula (1) (2) (3), the finishing can be obtained

$$\begin{cases} T_{mid} = \dfrac{LCV_{mid}}{K_pW_p\left[(V_{gs}-V_{th})V_{ds} - \frac{1}{2}V_{ds}^2\right]} \\ T_r = \dfrac{2CL(V_{out} - V_{mid})}{K_pW_p(V_{gs}-V_{th})^2} + T_{mid} \end{cases} \tag{4}$$

where K_p is the scale coefficient of the PMOS tube; L is PMOS gate length, W_p is PMOS gate width, C It is the size of the load capacitance and the V_{mid} clipping voltage between the linear and saturation regions of the inverter.

That is, when the final output voltage reaches saturation, it is V_{dd}, that is, the above formula can be sorted.

$$\begin{cases} T_{mid} = \dfrac{LCV_{mid}}{K_p W_p\left[(V_{dd}-V_{th})V_{ds}-\frac{1}{2}V_{ds}^2\right]} \\ T_r = \dfrac{2CL(V_{dd}-V_{mid})}{K_p W_p(V_{dd}-V_{th})^2} + T_{mid} \end{cases} \tag{5}$$

NMOS descending edge delay model as above.

Step2: Oscillation frequency calculation

The delay time of a single-stage inverter can be expressed as shown in the following equation using rising and falling edge times.

$$t_{pd} = \frac{T_r + T_f}{2} \tag{6}$$

T_r where the PMOS T_f rising edge time is the N MOS falling edge time.

　Available

$$f = \frac{1}{(N \times (T_r + T_f))}. \tag{7}$$

2.3　Solving the Model

Based on the above model, we calculate the output frequency of the ring oscillator designed in Table 1 for different scenarios.

Table 1. Output power result table

The number of inverters	PMOS aspect ratio	NMOS aspect ratio	Supply voltage/V	Output frequency/MHz
11	400n/100n	200n/100n	1.2	27.6091
11	800n/200n	400n/200n	1.2	22.8489
11	1.6in/0.4in	0.8u/0.4u	1.2	16.9902
31	200n/100n	400n/100n	1.2	8.5052
31	400n/200n	800n/200n	1.2	7.0388
31	0.8u/0.4in	1.6u/0.4u	1.2	5.2340
51	500n/100n	500n/100n	1.2	10.3775
51	1000n/200n	1000n/200n	1.2	8.5883
51	1.8u/0.3in	1.8u/0.3u	1.2	8.7903
99	2in/0.5in	1u/0.5u	1.2	1.6733

3   Establish a Power Consumption Optimization Model Based on Single-Objective Programming

3.1   CMOS Dynamic Power Model

Step1: Dynamic power dissipation caused by charge-discharge capacitors.

When C is charged through the PMOS, V_C rises from 0V to V_{dd} and the circuit is from the power supply A certain amount of energy is drawn on V dd, part of which is used for consumption on the PMOS and the other part is stored by capacitor C. When V_{out} is flipped from high to low, capacitor C is discharged through NMOS, and the energy previously stored in capacitor C is NMOS consumption.

Using the micrometry, the current is *set to* i_{vdd}, the output voltage is V_{out}, which changes with time, which can be regarded as a time function, and the $0 \sim dt$ micro element that extracts energy from the power supply in the time range is

$$dE_{Vdd} = i_{Vdd}(t) \times V_{dd}dt \tag{8}$$

where E_{vdd} is the $0 \sim dt$ micron that draws energy from the power supply over time, V_{dd} is the supply voltage, and $i_{vdd}(t)$ is the current function.

The time it takes for load capacitor C to charge to V_{dd} is infinite, so the energy extracted from the power supply during load capacitor C charging is

$$E_{V_{dd}} = \int_0^\infty i_{Vdd}(t) \times V_{dd}dt = CV_{dd} \int_0^{V_{dd}} dV = CV_{dd}^2 \tag{9}$$

where C is the load capacitance and V_{dd} is the supply voltage.

The energy microelement stored by capacitor C over the time frame is $0 \sim dt$

$$dE_C = i_{Vdd}(t) \times V_{out}(t)dt = \int_0^\infty i_{Vdd}(t) \times V_{out}(t)dt \tag{10}$$

E_C stores energy over a certain time frame, *and* $V_{out}(t)$ is the output voltage that changes over time and can be seen as a function of time.

After the above formula is sorted out and simplified, it is obtained

$$E_C = C \int_0^{V_{dd}} V_{out} dV_{out} = \frac{1}{2}CV_{dd}^2 \tag{11}$$

As obtained from the above formula, only half of the energy extracted from the power supply during capacitance C charging is stored to capacitor C and is consumed by NMOS when the capacitor is discharged, and the other half is consumed by PMOS. The energy consumed by the CMOS inverter after each switching cycle is f times per second, and the CV_{dd}^2 power consumption is f times per second

$$P_{ext} = \frac{E_{V_{dd}}}{T} = CV_{dd}^2 f \tag{12}$$

where T is the switching cycle and f is the switching frequency.

Step2: Power Dissipation Due to Simultaneous on of NMOS and PMOS

In fact, the rise and fall time of V in cannot be zero. When the CMOS inverter is in a state where the NMOS and PMOS are on at the same time, the power supply has a direct flow path to ground. The DC pulse can be approximated as a triangle and the input level rises and falls in equal time, and the energy consumed by each switching cycle T is set to E_{dp}, i.e.

$$E_{dp} = E_{dpr} + E_{dpf} \tag{13}$$

wherein, E_{dpr} the direct flow path consumption during the rise of *the V* VE_{dpf} in level is *the* conversion energy consumption of V in during the high to low level.

It can be found that $0 \sim \frac{t_{sc}}{2}$ the instantaneous current during this time is

$$i(t) = \frac{2 \times I_{peak}}{t_{sc}} t \tag{14}$$

where, I_{peak} determined by the saturation current, t_{sc} is the time for PMOS and NMOS to turn on.

During $\frac{t_{sc}}{2} \sim t_{sc}$ this time the instantaneous current is

$$i(t) = -\frac{2 \times I_{peak}}{t_{sc}} t \tag{15}$$

Then V_{in} energy consumption during low to high level conversion is

$$E_{dpr} = \int_0^{t_{ce}} i(t) V_{dd} dt = 2 \int_0^{\frac{t_\varphi}{2}} i(t) V_{dd} dt \tag{16}$$

Tidy up and simplify, get

$$E_{dpr} = \frac{1}{2} V_{dd} I_{peak} t_{sc} \tag{17}$$

Similarly, V_{in} energy consumption during high-to-low transition is

$$E_{dpf} = \frac{1}{2} V_{dd} I_{peak} t_{sc} \tag{18}$$

rule

$$E_{dp} = V_{dd} I_{peak} t_{sc} \tag{19}$$

So the average energy consumption in a cycle T is

$$P_{dp} = \frac{E_{dp}}{T} = V_{dd} I_{peak} t_{sc} f = C_{sc} V_{dd}^2 f \tag{20}$$

where C_{sc} is the equivalent capacitance during this period.

t_{sc} represents two types of CMOS tubes that are on at the same time, and for a straight line input slope, its approximate value is obtained

$$t_{sc} = \frac{V_{dd} - 2 \times V_{th}}{V_{dd}} t_s \tag{21}$$

where V_{th} is the threshold voltage and t_s is the ascending edge of the inverter.

3.2 CMOS Static Power Model

Ideally, the CMOS inverter will not turn on pmOS and NMOS at the same time when it is stationary, which means that there is no path between the power supply and ground at steady state, no path current is formed, and the quiescent power consumption is zero.

In practice, there will be some leakage current I_{start} flowing through the reverse bias diode, where the quiescent power dissipation is

$$P_{start} = V_{dd}I_{start} \tag{22}$$

3.3 Establish a One-Objective Optimization Model

Set the total inverter power consumption to P_{sum}, i.e.

$$\min P_{sum} = P_{ext} + P_{dp} + P_{start} \tag{23}$$

You can determine which of the current constraint is

$$s.t. \begin{cases} P_{ext} = CV_{dd}^2 f \\ P_{dp} = V_{dd} I_{peak} t_{sc}f \\ P_{start} = V_{dd}I_{start} \\ N = 1, 3, 5 \cdots 99 \\ 0.06 < L < 100 \\ 0.12 < W_p < 100 \\ 0.12 < W_n < 100 \\ f = 5\,\text{MHz} \end{cases} \tag{24}$$

Step1: Preliminary Search

For the number of inverters N, set its step size to 2, we take 0.01um as the step, search for L in [0.06,100] and W_P and W_n in the [0.12,100] interval, respectively. Save each search value and determine whether the constraints are met. Using each search value, the minimum power consumption is saved, if the next search is worth the minimum power consumption is less than the current power consumption, that is, update, otherwise not updated. The resulting optimal fence length and width are preserved. Then do a precise search. The optimal PMOS gate length obtained by preliminary search, the optimal NMOS gate length obtained, the optimal gate width obtained is the optimal number of inverters obtained. $W_p = 0.15\,\mu\text{m}$ $W_n = 0.12\,\mu\text{m}$ $L = 0.37\,\mu\text{m}$ $N = 5$.

Step2: Precise Search

In steps of 0.0001um, search within W_p the [0.14,0.16] interval around the resulting optimal solution, and search within W_n the [0.12,0.13] interval, L respectively [0.36,0.38] Search within intervals, repeating the preliminary search steps.

According to the adaptive step algorithm, the one-objective optimization model is searched, and when the required area of the ring oscillator will reach the minimum, the optimal gate length of PMOS is finally obtained $W_p = 0.1403\,\mu\text{m}$, and the optimal NMOS gate length is $W_n = 1.202\,\mu\text{m}$ obtained The optimal gate width is $L = 0.3610\,\mu\text{m}$, the optimal number of inverters obtained is $N = 5$, and the minimum power dissipation is 23.4314MJ/s.

4 Evaluation, Improvement and Promotion of Models

In this paper, the adaptive learning step search is used, and the preliminary search and accurate search are carried out to improve the solution speed, and the repeatable utilization rate of the algorithm is high. Using the single-objective optimization model and the multi-objective programming model, there is a clear idea in the model solution. Layer by layer, the model is more generalizable. The model is robust and suitable for general models.

In the process of model establishment, combined with the reality, the area occupied by the connection between the inverters is added to the established model. Other different heuristics such as simulated annealing algorithm, particle swarm algorithm, etc. are used to maximize the ability to jump out of the local optimal solution.

References

1. Clara, D.: When transferring a CMOS inverter. https://blog.csdn.net/clarad/article/details/120799530.2022.08.06
2. Wang, D., Yi, S.: An improved CMOS ring oscillator with frequency stability, pp. 370–373, October 1999
3. Rabaey, J.M.: Circuits, Systems and Design of Digital Integrated Circuits (Second Edition). Publishing House of ElectronicsIndustry (2010)
4. Digital integrated circuits: Power consumption of CMOS inverters. https://zhuanlan.zhihu.com/p/411386762.%202022.08.06
5. yzn, k.: Mathematical Modeling - Pallet Optimal Placement Solution. https://blog.csdn.net/yzn_keshe/article/details/74762382.%202022.08.06
6. Xiang, Q., Lu, Y.: Characteristic analysis and simulation of CMOS inverter based onCadence, pp. 59-60. 2021.02.014
7. Ke, Z., Xuefeng, D., Haitao, Z.: Design of a low-power frequency stable CMOS ringoscillator, pp. 23–26. 2017.05.007

Risk Assessment Model of Accounting Resource Sharing Management Based on Genetic Algorithm

Yanhua Huang[✉] and Ting Shen

School of Business and Management of Wuhan Business University, Wuhan 430056, Hubei, China
hyh651018@163.com

Abstract. The risk assessment model of accounting resource sharing management based on genetic algorithm is a method to identify the risk factors related to processes and products. It is an analytical tool that helps to determine the level of risk factors for a product or service. This method also helps to determine the key points and measures for product or service improvement. The main purpose of this method is to reduce the risks associated with the production process by using the right resources at the right time. In order to achieve this goal, it involves various steps, such as data collection, determination of key performance indicators (KPIs) and calculation of probability distribution function. Although risk management has been a part of strategic management accounting, expanding risk management into a strategic management function of management accounting may better reflect the characteristics of management accounting in this era, so that enterprises can not only adapt to the changing economic environment, but also carry out scientific and accurate prediction, decision-making, planning and control from a strategic perspective, So as to lay the foundation for the long-term development of enterprises and the promotion of core competitiveness.

Keywords: Sharing management · Accounting resources · Genetic algorithm · risk assessment

1 Introduction

With the acceleration and deepening of the process of world economic integration, globalization and internationalization, especially the outbreak of the financial crisis, both business and academic circles are clearly aware of a global problem. If enterprises want to survive and develop and maintain permanent competitive advantage, risk management must be put on the agenda. In the information age and the modernization of management science, the internal and external environment faced by enterprises is becoming more and more complex, changeable environment and fierce competition [1]. Although it provides unprecedented development opportunities for enterprises, it increases the business risks of enterprises. For the increasing and increasingly complex risk factors, enterprises are difficult to form a sufficient risk response psychology, and lack a complete risk

© The Author(s), under exclusive license to Springer Nature Singapore Pte Ltd. 2023
J. C. Hung et al. (Eds.): IC 2023, LNEE 1045, pp. 436–443, 2023.
https://doi.org/10.1007/978-981-99-2287-1_62

management system to control the risk. How to adapt to the current complex external environment and improve the ability of enterprises to defend against risks has become a realistic and severe problem faced by enterprise managers [2]. The increasing demand for risk management in practice makes the academic community gradually explore and explore how enterprises can really do a good job in risk management from the perspective of theory and technology, and what theory, conditions and technical support are needed to realize risk management.

The development stage of accounting has changed from strategic oriented and management oriented to value oriented and management oriented. The proposal of strategic management accounting means that management accounting is no longer an information system that only deals with financial information, but a management information system that takes financial data as the main content and combines non-financial information to provide relevant information and technical support for enterprises to form and enhance their core competitiveness. Its essence is to serve the strategic management of enterprises. Strategic management accounting gives full play to the prediction decision-making, planning control and assessment evaluation functions of management accounting from the strategic level, so as to provide support for the long-term survival and development of enterprises and the realization of core competitiveness [3]. Management accounting has developed to the stage of strategic management accounting, and its strategic management function has become increasingly apparent. Humanistic management, life cycle management, total quality management and social effect management have become branches of strategic management function. However, if enterprises want to realize these functions, they want to really maintain permanent competitive advantage and expand risk management as a strategic management function of management accounting, It is also a necessity to comply with social and economic development [4]. The realization of the strategic management function of risk management not only needs the risk management theory as the basis for the realization of this function, but also needs the support of the basic functions and technical conditions of management accounting. Therefore, this paper takes risk management as the supporting point, around the center of risk management, extends risk management to a strategic management function of management accounting, and improves the methods and technologies of management accounting to support this function, so that enterprises can better manage the risks of enterprises based on management accounting and using the theories and methods of management accounting.

2 Related Work

2.1 Risk Management Concept

The risks and opportunities involved in risk management affect the creation and maintenance of enterprise value. It is implemented by the board of directors, management and other personnel of an enterprise. It is used to formulate and implement the enterprise's strategy in a process of the whole enterprise, and a series of beliefs and attitudes towards risks in this process. It aims to determine the potential matters that may affect the enterprise, manage them, and provide a reasonable guarantee for the realization of the enterprise's objectives [5].

Overall, enterprise risk management is an ongoing process that runs through the whole enterprise. It is a process that organically integrates the business activities, various management departments and the highest decision-making level of the whole enterprise; Enterprise risk management is influenced by its personnel at all levels. It is not the task of a person, nor the responsibility of a level and unit, but the common task and responsibility of all personnel in the whole enterprise; Risk management aims to identify potential events that may have an adverse impact on the enterprise. The best enterprise can always grasp the future, because the enterprise needs to always look forward and pay attention to risks, so as to be in the best state when dealing with risks [6]. A good risk management system is to look for potential risk events in the future and deal with uncertainty. Risk events may or may not have an impact on the enterprise, The key is to identify these "potential" and "possible" matters; Risk management provides a reasonable guarantee for enterprises to achieve their goals, but it cannot provide an absolute guarantee. Risk management is to control the bad risk, excavate the favorable factors and make use of them. However, it can not guarantee that anything in the enterprise can run normally at any time. It can only ensure that most business activities of the enterprise can be carried out normally in most cases. Sometimes, some events that seem not very important may occur, but have a significant adverse impact on the enterprise. Therefore, risk management can only provide a reasonable guarantee for the enterprise, not an absolute guarantee.

With the development of society and the continuous change of business environment, many new strategic concepts suitable for modern economic development have been deduced based on the above basic concepts of risk management [7]. The proposal of comprehensive risk management and strategic risk management and the construction of the framework enable managers to manage various risks faced by enterprises from a strategic perspective, not only pay attention to the risks faced by enterprises now, but also pay more attention to the risks faced by enterprises in the future, not only pay more attention to the local risks of enterprises, but also pay more attention to the overall risks of enterprises, so as to make risk management cover all levels of enterprises and take into account the short-term interests and long-term interests of enterprises, An organic circulation system that can maintain and create value for enterprises.

2.2   Elements of Risk Management

Determining whether an enterprise's risk management contributes to the realization of enterprise objectives is based on a subjective judgment based on the correct evaluation of the design and implementation of risk management elements. In order to be effective and adapt to the strategic development of the enterprise, the design of enterprise risk management must include all elements, and these elements must affect and contact each other and be effectively implemented. Enterprise risk management includes six interrelated elements: (1) internal environment. The internal environment is the basis of all elements of enterprise risk management and can have an impact on all aspects of other elements; (2) Goal setting. The report believes that the management of the enterprise must establish objectives before it can meaningfully assess risks, analyze corresponding risks according to different objectives, and have a set of goal setting process that can closely link the enterprise objectives with the enterprise mission and consistent with the enterprise's risk tolerance and risk preference; (3) Event identification. According to the report, events can be divided into three types: positive impact, negative impact or both; (4) Risk assessment. COSO defines risk assessment as the identification and analysis of important risks in the process of achieving objectives. It is the basis for deciding how to manage risks. Once the risks are identified, they should be analyzed and evaluated; (5) Risk countermeasures. Risk countermeasures refer to various strategies and measures taken by the management to prevent, control, transfer and compensate risks after evaluating relevant risks; (6) Control activities. COSO defines control activities as policies and procedures that help ensure that management instructions are implemented 7.

3   Genetic Algorithm

3.1   Basic Idea of Genetic Algorithm

Genetic algorithm (GA) is a computer simulation study of biological systems in nature by American scholar pr. John Holland and his students. It is a kind of optimization algorithm that simulates the process of biological evolution based on Darwin's theory of biological evolution and Mendel's theory of genetic variation. It is a global optimization search algorithm with adaptive heuristic.

Genetic algorithm starts with a population representing the possible potential solution set of the problem, and a population is composed of a certain number of individuals encoded by genes. Each individual is actually an entity with characteristic chromosomes [8]. Chromosome is the main carrier of genetic material, that is, the collection of multiple genes. Its internal expression genotype is the arrangement and combination of certain genes, which determines the external expression of individual morphology. For example, the feature of black hair is determined by the combination of certain genomes in the chromosome that control this feature. Therefore, it is necessary to map and encode from phenotype to genotype at the beginning. Because the work of imitating gene coding is very complex, it is generally simplified in the form of binary coding [9]. After the emergence of the first generation population, according to the principle of survival of the fittest and survival of the fittest, generation by generation evolution produces better and better solutions. In each generation, individuals are selected according to the fitness of individuals in the problem domain, and combined crossover and mutation are carried out with the help of genetic operators of natural genetics to produce a population representing a new solution set. This process will lead to the offspring population produced by the population like natural evolution, which is more adaptable to the environment than the original population. After decoding, the optimal individual in the last population can be used as the approximate optimal solution of the problem.

The genetic algorithm adopts the natural evolution model and introduces its basic terms, such as selection, crossover, mutation, etc. the following is a brief introduction to its basic terms.

3.2   Components of Genetic Algorithm

Traditional genetic algorithm is generally composed of the following four parts: genetic operator, fitness function, control parameters and coding mechanism.

(1) Genetic operator

Genetic operator, also known as genetic operation, imitates three basic processes in nature, namely reproduction, mating and mutation (referring to gene mutation). These three basic processes are very common in nature. It is these three basic genetic processes that ensure the evolutionary stability of organisms and the diversity of species. Therefore, in genetic algorithm, there are three basic genetic operators: replication operator, crossover operator and mutation operator. These three basic operators will be described in detail in the next section.

(2) Fitness function

Survival of the fittest is the basic law of natural evolution, so how to judge the advantages and disadvantages of individuals? This requires a measurement standard. The same is true in genetic algorithm. It is necessary to set a standard to evaluate the quality of individuals, which we call fitness function. It is an objective function. The purpose is to take this as the standard to evaluate which individuals in the population are good, then retain or carry out the next operation, and the individuals evaluated as unsuitable will be eliminated [10]. This process is also in line with the principle of "survival of the fittest" in nature.

(3) Control parameters

The main purpose of control parameters is to improve the optimization effect. It includes string length, population capacity, crossover rate, mutation rate and other parameters. These parameters are briefly introduced below:

The string length is a fixed length, which is recorded as l; the group capacity is recorded as size; The crossover rate is recorded as P.; the variation rate is recorded as P. There are two situations in their relationship:

1. The population capacity is small: for example, if size $= 30$, P. $= 0.6$, P $= 0.01$ will be taken
2. When the population capacity is large: for example, if size $= 100$, then p $= 0.3$, P. $= 0.001$

(4) Encoding mechanism

Genetic algorithm does not directly discuss the research object, but uniformly endows the object to the string arranged by specific characters in a certain order through some coding mechanism. Just as the study of biological genetics in nature starts with chromosomes, which are composed of genes arranged in strings. In the system global area (SGA), the character set is composed of 0 and 1, and the code is also a binary string. For the general genetic algorithm, of course, it can not be subject to this restriction. Strings are gathered together to form a whole, and the individual is a string. In the optimization problem, each string corresponds to a feasible solution; In the classification problem, the string can be interpreted as a rule. The first half is the antecedent or input, and the second half is the consequent, conclusion or output. The crossover operator is shown in Fig. 1.

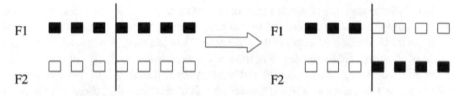

Fig. 1. Schematic diagram of single point intersection

4   Risk Assessment of Accounting Resource Sharing Management Based on Genetic Algorithm

Risk management is a comprehensive management project. Its basic idea is to improve the profitability and enterprise value of the enterprise from a strategic perspective by uniformly analyzing all risks inside and outside the enterprise, considering how to deal with, coordinate and arrange the internal factors of things to deal with the impact of external uncertainty on the operation of the enterprise, formulating management strategies to analyze various uncertainties and manage and control various risks. In order to give full play to the function of risk management and formulate effective risk management plans, management accounting cannot do without the coordination of various departments and levels, and of course, it cannot do without the assistance of the basic function of management accounting. The realization of the basic function of management accounting needs the support of the function of risk management. It can be seen that the functional system of management accounting is a complete circular system, Only when the system elements cooperate, contact and support each other can they give full play to the function of management accounting. Therefore, integrating the risk management function of management accounting with its basic function is the premise for management accounting to realize its function, as shown in Fig. 2.

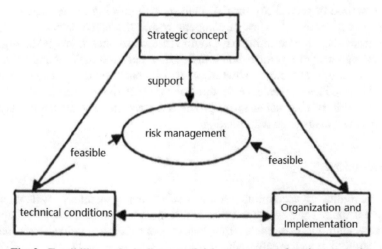

Fig. 2. Feasibility analysis diagram of risk management function expansion

First of all, management accounting is an information system. Its function is reflected in the transmission, analysis, processing and processing of information. On this basis, it carries out the business prediction of the enterprise. This function runs through the whole process of all economic management of the enterprise. Obviously, it also provides information and processing scheme for the risk management function. The system analysis function of risk management will analyze all aspects of risks faced by the enterprise, find the advantages and disadvantages of the enterprise, and adapt to the changes of the environment. The acquisition of these risk information can help the management make better and more accurate prediction and lay the foundation for subsequent decision-making and control.

Secondly, the advantage of management accounting decision support system, its decision support function is further reflected in assisting operators to make the best and most effective allocation and utilization of resources. Because the risks of operation and finance are everywhere in the process of resource allocation, management accounting must consider the risk factors and do the best risk management when playing its decision-making function, so as to truly realize the best allocation of resources. Therefore, the risk management function is the cornerstone for enterprises to make correct decisions. The decision support function of risk management is the key to effective risk management. It can make effective decisions to control risk according to the existing risk information and predict the future risk information. Therefore, the decision support function and decision function of risk management function are complementary. Although they are different, they are inextricably related.

Finally and most importantly, playing a role in the uncertainty of enterprise operation is the strength of risk management. The certainty in enterprise operation comes from all aspects of the enterprise. According to the Balanced Scorecard theory, these aspects can be summarized into four dimensions: customer, finance, internal process and learning and growth. Enterprises should also focus on these four aspects to analyze and control risks, and then evaluate the effectiveness of analysis and control. In the process of its development, management accounting has long begun to pay attention to risk factors, adopt the method of probability for risk analysis, and develop to the stage of strategic management accounting. In view of competitors and the fierce external market environment, management accounting pays more attention to uncertainty when organizing and processing enterprise information transmission, decision-making and control, performance evaluation and other activities, so that its basic functions can consider risk factors, This provides convenience for the realization of risk management function. In turn, the function of risk management provides a guarantee for the better play of the basic function of management accounting.

5   Conclusion

Because the traditional accounting resource sharing management risk evaluation model has the problems of long evaluation time and low evaluation accuracy, this paper constructs the accounting resource sharing management risk evaluation model based on genetic algorithm, which greatly reduces the evaluation time, and the effect of accounting resource sharing management risk evaluation is good, which can provide a more effective guarantee for the safety of accounting resource sharing management. Due to the difference between the simulation environment and the actual situation, there are some deviations in the experimental data of the model. In order to obtain more accurate experimental conclusions, it is necessary to further optimize the model.

References

1. Zhou, M., Feng, X., Liu, K., et al.: An alternative risk assessment model of urban waterlogging: a case study of Ningbo City. Sustainability **13**, 826 (2021)
2. Oksana, P., Nataliya, K., Valentyna, A., Serhii, O.: Audit risk assessment model in automated accounting systems of enterprises in Ukraine. In: Alareeni, B., Hamdan, A., Elgedawy, I. (eds.) ICBT 2020. LNNS, vol. 194, pp. 1192–1204. Springer, Cham (2021). https://doi.org/10.1007/978-3-030-69221-6_90
3. Li, Y., Yang, R., Qu, X., et al.: Study on the risk assessment and forewarning model of groundwater pollution. Arab. J. Geosci. **13** (2020). Article number: 407. https://doi.org/10.1007/s12517-020-05395-7
4. Filusch, T.: Risk assessment for financial accounting: modeling probability of default. J. Risk Finance **22**(11), 1–15 (2021)
5. Seitshiro, M.B., Mashele, H.P.: Quantification of model risk that is caused by model misspecification. J. Appl. Stat. **49**(5), 1065–1085 (2022)
6. Delavar, M., Eini, M.R., Kuchak, V.S., et al.: Model-based water accounting for integrated assessment of water resources systems at the basin scale. Sci. Total Environ. **830**, 154810 (2022)
7. Mujalli, A., Khan, T., Almgrashi, A., et al.: University accounting students and faculty members using the blackboard platform during COVID-19; proposed modification of the UTAUT model and an empirical study. Sustainability **14**(4), 2360 (2022)
8. Yoon, A.: Discussion of "audit firm assessments of cyber-security risk: evidence from audit fees and SEC comment letters." Int. J. Account. **55**(03), 1950018 (2020)
9. Liu, L.: Oil spill risk assessment of submarine oil pipeline based on fuzzy comprehensive evaluation and accounting. Arab. J. Geosci. **14**(18) (2021)
10. Teo, P.C., Lim, K., Ho, T., et al.: Cosmetic surgery industry in Brazil: an assessment using cause and effect model and risk assessment matrix. Int. J. Acad. Res. Account. Finance Manag. Sci. **10**(1), 157–164 (2020)

Steel Cutting and Blanking Problem in Steel Manufacturing Industry

Jiaying Lei[✉]

Yunnan University of Finance and Economics, Kunming 650221, Yunnan, China
lemontea0807@163.com

Abstract. With the wide application of steel structure products, it is urgent for steel structure enterprises to improve the utilization rate of raw materials. In this paper, the application of rectangular steel blanking and layout is the background, focusing on the formulation of steel cutting scheme under different demands. By establishing the model and solving the model, the optimal plan of cutting steel is obtained, so as to save raw materials, reduce product costs and improve the economic benefits of enterprises.

Keywords: Steel Cutting · Constraint Equation · NP-hard Problem · PSO Algorithm

1 Introduction

With the continuous improvement of economic development and social productivity, the scale of steel production continues to expand, steel enterprises demand more and more raw materials. Secondly, with the continuous development of steel industry and modern technology, steel manufacturers are paying more and more attention to the maximization of corporate profits. To maximize corporate profits, the cost of raw materials in the production process must be solved first. In the steel cutting process, it often causes a large amount of waste of raw materials, making the production cost increase. In addition, the number of knife changes caused by the poor layout of steel cutting is too much, which increases the labor cost. Therefore, how to improve the utilization rate of raw materials under the premise of meeting the production requirements, and then save the product processing cost, has very important practical value. In this paper, according to different sizes of raw materials and different order needs, the traditional cutting technology to improve, not only save the production cost, but also effectively improve the utilization rate of raw materials and cutting efficiency. Based on the actual production and life of cutting steel scene, the model is established and solved, and the optimal cutting scheme is calculated.

2 A Description of the Problem Studied

Disc shears use rotating disc blades to continuously cut longitudinal motion of raw materials. Before cutting the disc, it is necessary to arrange the knife according to the

© The Author(s), under exclusive license to Springer Nature Singapore Pte Ltd. 2023
J. C. Hung et al. (Eds.): IC 2023, LNEE 1045, pp. 444–450, 2023.
https://doi.org/10.1007/978-981-99-2287-1_63

order cutting scheme. It is assumed that the rows of knives can be arranged at arbitrary spacing on the rack, but the number of knives is limited, and the upper limit of a row of knives cannot exceed 5 knives (that is, a raw material can be cut into 2–6 lines of orders after a disc cutter, as shown in Fig. 1).Orders cut with the same row knife scheme are called a group of orders, cutting different groups of workers need to rearrange the knife, known as a tool change. Cutting each roll of raw materials need to change the knife. The cutting head is "one knife cut", that is, the whole steel plate is completely cut horizontally (as shown in red dotted line in Fig. 2).

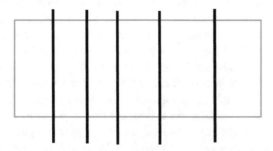

Fig. 1. Knife Arrangement Diagram of Disk Scissors   **Fig. 2.** Transverse Cutting

3   Model Assumptions and Conventions

Considering steel raw materials and orders as rectangles. In the cutting process, the thickness of steel material and slit width are ignored. The material distribution of this batch of steel raw materials is uniform, which will not affect the effect of cutting products. Regardless of the error caused by residual material loss caused by technology when cutting.

4   Development of Cutting Scheme Under Scenario 1 Constraint

Scenario 1 is designed to validate the cutting plan: the minimum number of raw materials required to achieve a high yield in order to meet the requirements of the cutting plan. It can be concluded that this is a multi-objective optimization problem, and the first step of analyzing the problem needs to clarify the objective function and its constraints.

Let the number of sheets of the raw material $i(i = 1,2, \ldots, n)$ be x_i, then the objective function of solving this problem is:

$$\min \sum_{i=1}^{c} x_i$$

Suppose that the number of each order $j(j = 1,\ldots,d)$ on raw material i is $y_{i,j}$, and the maximum number of sheets of raw material i is X_i, let the demand of processing material j be p_j, L_i and W_i represent the length and width of the i th raw material i respectively, l_j, w_j represent the length and width of the material j respectively [1].

According to the order demand exactly meets this condition, the constraint condition can be obtained (1):

$$\sum_{i=1}^{c} x_i y_{i,j} = p_j \tag{1}$$

The quantity of each raw material must not exceed its stock, the constraint condition can be obtained (2):

$$x_i \leq X_i \tag{2}$$

In the analysis of this problem, it is easy to calculate the yield if any cutting scheme is given, but to find the cutting scheme with the minimum number of raw materials and the high yield, it is necessary to check and compare all the schemes. If the permutation and combination is carried out directly, the situation involved will be very large, the amount of data is very large and difficult to calculate, which belongs to NP-hard problem and cannot be solved directly. Therefore, it is considered to classify and filter the data first to reduce the amount of data and simplify the calculation. This paper introduces a new definition of "template" to classify and filter data. Because this paper is to consider the cutting problem, the first consideration is to analyze the large rectangular block formed by the cutting, as a template, respectively, to consider the form of the order in the template. The black line in Fig. 3 represents the "one-size-fits-all approach" position, and the blue brackets represent a template.

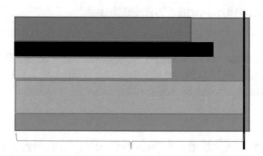

Fig. 3. Schematic Diagram of Template

After analysis, it is found that a template can accommodate 2–6 lines of orders, so consider the case that the template contains 2–6 lines of orders, arrange and combine the orders respectively, find out the arrangement form of orders in the template, get the form of one template, and find out the corresponding area utilization rate. By setting the standard of area utilization rate, the template is screened and the amount of data is reduced in this way.

In other words, let the various types in the template add a material category as m_i, then the length of the template is determined by the processing material with the longest length, that is, the template length $l = \max\{l_{m_i}\}$, the width of the template is the width of the corresponding raw material, so that the area utilization rate can be calculated as:

$$\sigma = \frac{\sum S_{m_i}}{l * W_i}$$

When the area utilization is set above a certain value (0.7 in this case), the template is saved to greatly reduce the amount of data. Calculate the corresponding template in the raw material yield situation. In order to minimize the number of raw materials as much as possible, the models after screening are combined to keep the rate of production and minimize the number of raw materials. Assuming that the number of order j contained in the t template on i is $q_{i,t,j}$, constraint conditions (3) and (4) are obtained:

The order shall be arranged on the raw materials to meet the width requirements:

$$\sum_{t,j} q_{i,t,j} {}^* w_j \leq W_i \tag{3}$$

The order shall be arranged on the raw materials to meet the width requirements:

$$\sum_{t,j} q_{i,t,j} {}^* l_j \leq L_i \tag{4}$$

In conclusion, the multi-objective optimization problem is as follows:

$$\min \sum_{i=1}^{c} x_i$$

$$\text{s.t.} \begin{cases} \sum_{i=1}^{c} x_i y_{i,j} = p_j \\ \sum_{t,j} q_{i,t,j} * w_j \leq W_i \\ \sum_{t,j} q_{i,t,j} * l_j \leq L_i \\ x_i \leq X \end{cases}$$

PSO algorithm is a global optimization algorithm based on the idea of swarm intelligence. It uses the collective behavior of fish and birds to imitate and uses the cooperation and competition between particles to calculate the optimal solution. PSO algorithm is used to screen out the optimal cutting scheme.

5 Development of Cutting Scheme Under Scenario 2 Constraint

Scenario 2 is designed to validate the cutting plan: for all raw materials, in order to meet all the requirements of the order, as far as possible to improve the yield and minimize the number of knife changing and cutting on small machines, to reduce labor costs. The minimum number of knife changes, that is, the cutting scheme used in cutting, typesetting order width as much as possible the same. The minimum number of small machine cuts is required, that is, the order length of the cuts should be as identical as possible. Combined with the two, the optimal cutting scheme is that a piece of raw material is uniformly cut into one order type, as shown in Fig. 4. As shown in Fig. 4, the white rectangle is the order of the same model, L is the total width of the same order model that a piece of raw material can be cut, l is the width of raw material. The remaining blue part is the leftover material or scrap after cutting.

Scenario 2 requires minimal use of raw materials, order fulfillment and yield. First of all, the number of sheets using raw materials should be minimized, and then the total yield should be considered in the case of the least raw materials. In order to meet the minimum number of raw materials, the required raw materials should be sorted according

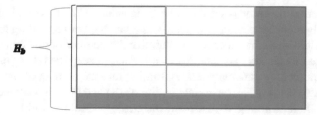

Fig. 4. Cutting Schematic Diagram

to the area, with the large area preferentially used. Under the condition of maximum raw material area, only cutting all raw materials with the same type of order can maximize the utilization rate of materials, as shown in Fig. 4. A piece of raw material is all cut into the same order model. There are two calculation methods. The first one is based on width (as shown in Fig. 5), and the second one is based on length (as shown in Fig. 6).

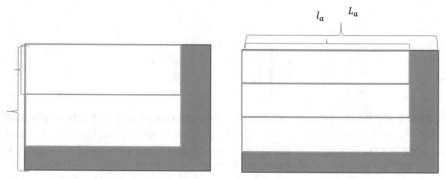

Fig. 5. Cutting Schematic Diagram with Width as Standard

Fig. 6. Cutting Diagram with Length as Standard

Take the width of the order as the standard, calculate the maximum proportion of the width of all orders to the width of the selected raw material, that is, $\frac{l}{L}$ is the largest. Then length as the standard, calculate the length of all other orders in the length of the raw material proportion is the largest, and compare the yield of both. To compare the size of the yield, under the same raw material area, that is, to compare the size of the waste area. In this paper, the remaining area of all orders (i.e. the blue part in the figure) under the condition of using the same raw material is calculated first, and then the area is sorted and the size of the yield is compared. Take the order with high yield as the standard, cut all raw materials into the model of this order. The same as above for other orders. The calculation method based on length is basically the same as above, but the relevant parameters are changed, so it will not be described here. Although the principle of the same two calculation methods is similar, but considering that the use of different standards may affect the cutting method, and then affect the calculation results, the two methods are adopted in this paper, and the calculation results are compared to obtain the optimal cutting scheme.

5.1 Typesetting with Width as Standard

Taking the width as the standard, suppose that there is a batch of rectangular order sample P_1, with the same length and width, and the set of its area is $\{S_i\}$. In addition, there is a batch of sample Q_1, with the same length and width as the raw material steel plate, whose area set is $\{W_j\}$.The arrangement of steel is actually to place this batch of samples P_1 properly on board Q_1. The so-called steel layout model is to find the most reasonable placement way to obtain the maximum utilization of steel materials. So that the objective function is at its maximum:

$$\max \frac{\sum_{i=0}^{n} S_j}{\sum_{j=0}^{m} W_j} \tag{5}$$

Assuming that the area of the remaining part after shearing is $\sum_{i=0}^{m} A_{i1}$, the formula (5) can be transformed into the minimum value of the following objective function:

$$\min \frac{\sum_{i=0}^{n} A_{i1}}{\sum_{j=0}^{m} W_j} \tag{6}$$

As shown in Fig. 4, the minimum value of the objective function is the blue part of the rectangular box. Because the sorting of raw material area is the ergodic standard calculation, that is, $\sum_{j=0}^{m} W_j$ is the fixed value, so when calculating the formula (6), we only need to calculate the minimum value of $\sum_{i=0}^{m} A_{i1}$.

Assume that the length of each order is l_a, and the width is h_b. The length of the material is L_a, and the width is H_b. The area of the blue part, namely $\sum_{i=0}^{m} A_{i1}$, can be calculated by:

$$\sum_{i=0}^{m} A_{i1} = (H_b - 2h_b) * l_a + (L_a - l_a) * H_b$$

Sort by the value of $\sum_{i=0}^{m} A_{i1}$ in descending order, and then cut in order. In the process of cutting one of these orders, there may be a problem that the order needs have been met, but the last few blocks do not use up the whole sheet of raw material. To solve this problem, calculate the maximum number of raw material sheets (a) that can all be cut into the same order, and save the remaining orders for last calculation. Divide the number of orders (x) that can be cut from a piece of raw material and the total number of the same order (t) required, and take the remainder of the two. The integer part is the number of raw materials required for the current cut order, and the remainder is the remaining order to be cut at the end.

5.2 Typesetting with Length as Standard

Taking length as the standard, suppose that there is a batch of rectangular order sample P_1, with the same length and width, and the set of its area is $\{S_i\}$. In addition, there is a batch of sample Q_1 with the same length and width as the raw material steel plate, whose area set is $\{W_j\}$. The layout of steel is actually to place this batch of samples P_1 on the raw material Q_1 properly. The steel layout model is to find the most reasonable placement way to obtain the maximum utilization of steel materials. So that the objective function is at its maximum:

$$\max \frac{\sum_{i=0}^{n} S_j}{\sum_{j=0}^{m} W_j} \tag{7}$$

Assuming that the area of the remaining part after shearing is $\sum_{i=0}^{m} A_{i2}$, the formula (7) can be transformed into the minimum value of the following objective function:

$$\min \frac{\sum_{i=0}^{n} A_{i2}}{\sum_{j=0}^{m} W_j} \tag{8}$$

As shown in Fig. 4, the minimum value of the objective function is the blue part of the rectangular box. As raw material area ranking is the standard for calculation, that is, $\sum_{j=0}^{m} W_j$ is the fixed value, so the calculation formula (8) only needs to calculate the minimum value of $\sum_{i=0}^{m} A_{i2}$.

The area of the blue part of the picture, $\sum_{i=o}^{m} A_{i2} = (H_b - 2h_b) * l_a + (L_a - l_a) * H_b$

(This is because the length and width of the order are different when taking width as the standard and length as the standard, that is, l_a and h_b are different. So even though $\sum_{i=o}^{m} A_{i1}$ and $\sum_{i=0}^{m} A_{i2}$ have the same formula, they have different results).

References

1. Wu, Y., Zhang, Y.: School of Computer Science and Engineering. Anhui University of Science and Technology. https://doi.org/10.19392/J.Cnki.1671-7341.202017089
2. Zhu, Y., Wu, H., Liu, Q.: Research on material cutting design based on linear programming. Technol. Market (11), 95 (2018)
3. Zhou, K.: Mathematical modeling of optimal cutting scheme of rectangular plate. Neijiang Sci. Technol. **41**(06) (2020)
4. Yuan, Z., Wang, Y., Zhao, H.: Optimization layout and cutting path planning of rectangular slates. Manuf. Autom.
5. Zhang, G.: Linear Programming. Wuhan University Press, Wuhan 200 Foundation: Suqian College Education Reform Project (SQC2018JG06)
6. Yu, X., Ding, Y., Jin, L., Liu, L., Zhang, Z.: Research on two-dimensional cutting scheme of stepwise optimization based on integer programming. J. Nantong Vocat. Univ. https://doi.org/10.3969/j.issn.1008-5327.2020.04.017

Teaching Effect Evaluation System of Practical Design Course Based on Computer Assistance

Ni Cai[✉]

Kunming University of Science and Technology Oxbridge College, Kunming 650106, China
caini313@163.com

Abstract. With the end of the 12th Five Year Plan and the beginning of the 13th five year plan, China's economic development has entered a new normal, and the quality of higher education has faced new challenges. Teaching is still the core task of higher education. The key to improving the teaching quality of colleges and universities is to promote the reform of sustainable development. Nowadays, in order to improve the teaching quality in Colleges and universities, especially the newly-built undergraduate colleges and universities seek the dislocation development with the established public colleges and universities, and actively seek transformation development and reform. The research on the teaching effect evaluation system of computer-aided practical design course is a new way to evaluate the quality of teachers. It can be used to evaluate the effectiveness and efficiency of instructors in training courses using computer-assisted instruction (CAI) technology. The research purposes are as follows: 1) to explore the role of computer-assisted instruction in improving students' academic performance; 2) Determine whether Cai improves students' learning motivation, learning attitude, achievement and education satisfaction; 3) Analyze how Cai affects students' academic performance, including knowledge acquisition ability, problem-solving ability and critical thinking ability.

Keywords: computer-aided · Practical design · Course teaching · impact assessment

1 Introduction

As China enters a new stage of socialist economic construction and development, higher education, which affects the national economy and development, has faced new challenges. Minister Yuan Guiren, Secretary of the party leading group and Minister of the Ministry of education, pointed out at the 2014 national education system working conference that we should always grasp one theme and one task, one of which refers to comprehensively deepening comprehensive reform; The first task is to promote education equity and improve quality, pay more attention to overall planning, gather strength, overcome difficulties, and strengthen system construction. Minister yuan stressed that the further modernization of education has entered a critical period. One of the key points mentioned is "to adjust the educational structure, improve the quality of education, and

J. C. Hung et al. (Eds.): IC 2023, LNEE 1045, pp. 451–457, 2023.
https://doi.org/10.1007/978-981-99-2287-1_64

provide strong intellectual and talent support for economic transformation and upgrading. We should accelerate the development of modern vocational education, promote the transformation and development of local undergraduate colleges and universities, accelerate the construction of first-class universities and first-class disciplines, and accelerate the development of continuing education."

Deepen teaching reform, improve teaching quality, It is directly related to the quality of talent training". Teaching quality is the lifeline of the survival and development of higher education. The core mission of higher education is talent training, and the main task of talent training is teaching. The key to the reform and development of colleges and universities is to improve teaching quality. The evaluation system of teachers' classroom teaching quality is a key link to ensure the quality of classroom teaching. An important measure to continuously improve the teaching management and teaching quality of colleges and universities is to create a scientific teaching quality of colleges and universities Quantity evaluation system. At present, colleges and universities still take classroom teaching as the main body, so improving the quality of classroom teaching has become the primary task, because its quality reflects and determines the quality of education in Colleges and universities. Teaching quality is the foundation of survival and development that a university must adhere to. The improvement of teaching quality in Colleges and universities must be achieved by strengthening the management of teaching units. Specific to each major, it is to pay attention to the quality of curriculum construction, and the most basic key point is that teachers should teach each course they are responsible for. Therefore, it is the key work for colleges and universities to carry out self-evaluation and establish a teaching quality evaluation system. How to make full use of all kinds of data in the teaching process, make the existing data into knowledge, and improve the management level and education quality are the core issues that should be considered first in establishing the evaluation system. In order to improve the quality of education, colleges and universities, especially newly-built undergraduate colleges and universities, are seeking dislocation development with established public colleges and universities, and actively seeking transformation, development and change. The success of teaching reform depends on its effect and whether it promotes the improvement of education quality.

2   Related Work

2.1   The Significance of Curriculum Teaching Evaluation

(1) Improve the quality of course teaching

At present, network teaching and network universities are springing up. As a new thing in the period of China's educational reform, whether network teaching can achieve greater development will ultimately depend on its teaching quality and teaching level. Course teaching is the resource and basis of network teaching, and its quality determines the quality of network teaching. Carrying out curriculum teaching evaluation and establishing curriculum teaching evaluation system are conducive to standardizing curriculum teaching, guiding the design and development of curriculum teaching, and improving the quality of curriculum teaching. At the same time, the evaluation of course teaching is conducive to the sharing and interoperability of course teaching among different systems. In short, curriculum teaching evaluation can standardize curriculum teaching, improve the quality of curriculum teaching, and finally promote the benign development of network teaching.

(2) Accelerate the promotion of course teaching

The significance of evaluation is also reflected in the promotion of curriculum teaching. Fair, objective, scientific and quantitative evaluation results are the direct basis for education departments to decide to accept and support online teaching methods. The evaluation of network teaching software enhances the credibility of network teaching and improves the assurance of the success of network teaching. Only the great development of network teaching can lead to the great development of network education software. Nowadays, the teaching of courses on the Internet is complicated. How teachers and students choose their own and efficient courses depends largely on the implementation of evaluation work. The scientific results of the evaluation work and the regular or irregular authoritative evaluation reports will be an important basis for people to choose online teaching software. This work also standardizes to a certain extent, purifies the online teaching market, and promotes the improvement of the overall level of online education software and course teaching.

(3) It is conducive to the correct decision-making of teaching management

To do a good job in teaching management, we must first make correct decisions. How to make all kinds of schools at all levels effectively implement quality education and improve teaching quality, we must use evaluation means. Curriculum evaluation provides effective information for network teaching decision-making, so as to make correct decisions.

(4) It is conducive to testing the level of network teaching

Course teaching is an important part of network teaching. Course teaching evaluation is the starting point and destination of the whole network teaching evaluation. Only by solving the problem of course teaching evaluation can we do a good job of network teaching evaluation. So as to finally judge whether our network teaching is successful and where the shortcomings of network teaching are, so as to be corrected.

2.2 The Role of Curriculum Teaching Evaluation

The role of curriculum teaching evaluation is multifaceted, and the main roles are as follows:

(1) Feedback regulation

Evaluation can provide feedback on teaching activities, so that teachers and students can adjust teaching and learning activities, so that teaching can always be carried out effectively. Teachers can adjust teachers' teaching work through information feedback, so as to improve students' learning effect.

(2) Diagnostic guidance

Evaluation is the process of analyzing the teaching effect and its causes, by which we can understand the situation of all aspects of teaching, and judge its effectiveness, defects, contradictions and problems. It finds out the reasons for the problems existing in the evaluation objects such as network teaching, learning resources, teaching activities and curriculum quality, and then provides improvement ways and measures for these reasons.

(3) Strengthen the incentive function

The results of evaluation, on the one hand, provide information for decision makers, on the other hand, it also gives feedback to the evaluated person or the evaluated unit. Its purpose is to improve work and improve the quality of education. If the work is done well, it will give people the motivation to carry forward their achievements and some spiritual satisfaction, which can better promote people's initiative in work or study, and encourage people to devote all their energy to work or study; If the work is not done well, it can urge people to improve their deficiencies by proposing where it is bad and how to improve it.

(4) Enhance communication

In the process of evaluation and evaluation conclusion feedback, due to the mutual contact, exchange and information communication between evaluators and evaluators, evaluators and evaluators, evaluators and decision makers, as well as between evaluators and evaluators, evaluators and decision makers, we can see the strengths of others, but also notice our own shortcomings, which is conducive to learning from each other, learning from each other, and moving forward together.

3   Computer Aided Practical Design Course Teaching Effect Evaluation System

3.1 Evaluation Method and Content

Nowadays, department evaluation, supervision evaluation and peer evaluation are the main evaluation methods of college teaching at present. At the same time, they are equipped with traditional quantitative evaluation methods - questionnaire method, online evaluation method, e-mail and so on. On the basis of many methods, products in the information age are now used to participate in teaching evaluation, such as SMS platform, wechat platform and so on. But no matter what kind of evaluation method, the content of evaluation is closely centered on teaching factors. Teaching evaluation includes all educational and teaching activities.

Nowadays, quality education is advocated. From the perspective of teaching needs, teaching evaluation is usually evaluated from two subjects, namely teachers and students, and form, namely teaching process. Through the above classification, teaching evaluation is realized from these three aspects. The specific contents are as follows.

(1) Students' evaluation. Students are the main body of teaching and the object of training. Although it is found in practice that there are still some extremely unfair factors in students' evaluation of teaching, it should be affirmed, On the whole,

students' evaluation is effective and credible [34. Students' evaluation should be comprehensive in many aspects, not limited to a transcript to evaluate their learning effect, but also include ability evaluation, professional quality evaluation, innovation ability evaluation, as well as moral and psychological quality evaluation. For many years, most of them only pay attention to the evaluation of grades, and ignore the evaluation of other aspects, so as to ignore the real purpose of education, the real value of training, and evaluation The limitations of have a direct impact on the scientificity and effectiveness of the results.

To transform external evaluation into the motivation of students' autonomous learning, students must first recognize the results of evaluation and recognize the objectivity, scientificity and effectiveness of evaluation, so that students can use the results of evaluation to adjust their student methods, modify learning strategies and achieve the best results. In the process of teaching, students have individual differences and diversified behaviors. Teachers can obtain a lot of information by observing and evaluating students' performance in the learning process, classify and summarize them, and make use of them to improve teaching quality and teaching effect.

(2) Teachers' evaluation. In traditional teaching, teachers play a leading role. With the continuous development of teaching, the main body in the teaching process is also changing. The main body is no longer the only, students and teachers are the main body in teaching activities. In the process of teaching, teachers should pay attention to the changes of their own roles and constantly guide, promote and help students.

To evaluate teachers, we should first evaluate the teaching level and teaching attitude, mainly including the management ability of engaging in classroom teaching and the ability to carry out teaching activities independently, such as formulating teaching plans, determining teaching contents according to teaching objectives, arranging and organizing teaching materials, language expression, mining the meaning of teaching materials, blackboard writing skills, information feedback, modern teaching methods such as the use of multimedia courseware, and even homework assignment Extracurricular guidance and teacher-student relations are also included in the evaluation. Work achievements include classroom teaching effect, teaching reform results, teaching experience summary, teaching papers, student assessment results, etc. It is an important aspect of evaluating teachers' professional ability and quality.

(3) Evaluation of teaching results. The most common thing in teaching evaluation is to evaluate the teaching results. Summative evaluation is usually used to evaluate the teaching effect, because the evaluation of teaching results needs to evaluate the progress, depth and other aspects of students' learning

3.2 Classroom Effect Evaluation

In the process of teaching, in order to achieve the established teaching goals, we should follow the principles of step-by-step, teacher-student communication and interaction. When students begin to learn a new course, it is generally not easy for them to suddenly face unfamiliar knowledge fields or professional terms. At this time, if teachers can guard against arrogance and impatience, they should have clear ideas, explain in simple terms, gradually introduce local style to explain the meaning of relevant professional terms to students, and pay attention to the role of guidance, so that students can change from passive acceptance of knowledge to active learning, as shown in Fig. 1. To teach professional knowledge, teachers should actively seek communication and interaction with students, give professional terms and professional knowledge in combination with specific cases, try to realize the interaction between teachers and students and students, and cultivate students' habit of independent thinking and active expression. If you have the conditions, you can try to divide students into study groups to strengthen the training of students' teamwork ability and communication ability, so as to achieve better results.

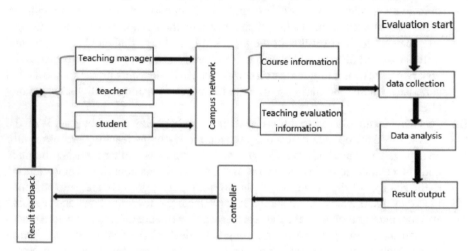

Fig. 1. Evaluation system of teaching effect of practical design course based on computer assistance

When carrying out classroom effect evaluation of college courses, we should strengthen the interaction between students and teachers, that is, master relevant professional knowledge through daily course learning, and fully activate the classroom atmosphere and actively mobilize students' enthusiasm and initiative through the flexible use of various teaching methods and means such as guided, heuristic and case study. Encourage students to dare to express their ideas and views. Expressing professional knowledge and answering questions is only a means and way of teaching. Learning and mastering professional knowledge is the fundamental purpose of teaching. Similarly, we should not abandon the basics. We should put an end to the extreme tendency of students who only prosper on the surface of the classroom, and we need the school to monitor the quality of the course throughout the whole process.

4   Conclusion

As a very important part of educational resources, curriculum teaching is the supporting condition for the development of network education and the main means to realize network education. It is also the basic unit of network teaching. High quality course teaching is the basic guarantee to realize high-quality network education. Since the construction of curriculum teaching in China is still in its infancy, there are deficiencies in the design, development, evaluation and implementation of curriculum teaching, so the requirements of curriculum teaching evaluation system are put forward. Aiming at this need, this paper studies the methods and procedures developed according to the standards, adopts the index system method, and references the relevant literature to establish a curriculum teaching evaluation index system including five aspects: curriculum content, teaching design, usability, artistry and technology.

References

1. Li, Y.: Research on the construction of practical teaching evaluation system of mechanical and electrical specialty in colleges and universities based on computer multimedia technology. J. Phys: Conf. Ser. **1648**, 032151 (2020)
2. Luo, L.: Research on practical teaching of virtual reality integration of Engineering Management major from the perspective of New Engineering. In: E3S Web of Conferences. EDP Sciences (2021)
3. Yin, B., Tang, M.: Research on the innovation and entrepreneurship of college students under the background of practical teaching system reform. DEStech Trans. Soc. Sci. Educ. Hum. Sci. (2020)
4. Jiajia, F.U., Zhang, Z., Guo, Y., et al.: Research on sharing mechanism of practical teaching resources from value-added perspective. Exp. Technol. Manag. (2020)
5. Zou, S., Xie, H., Xiao, C., et al.: Research on the construction and practice of practical teaching system in applied-oriented undergraduate universities. J. Shaoyang Univ. (Nat. Sci. Ed.) (2020)
6. Xie, C., Lu, H., Shi, D.: Research on undergraduate professional training mode of safety engineering under multidisciplinary crossing. Int. J. Soc. Sci. Educ. Res. **2**(12), 13–18 (2020)
7. Zhu, L.: Computer vision-driven evaluation system for assisted decision-making in sports training. Wirel. Commun. Mob. Comput. **2021**(7), 1–7 (2021)
8. Tian, Y.: Teaching effect evaluation system of ideological and political teaching based on supervised learning. J. Interconnection Netw. **22**, 2147015 (2022)
9. Wang, J., Li, Q., Chen, M.: Mixed method research on technological teaching system based on human resource management practices (HRMP): study pertaining to majors in nursing undergraduate programme. Technol. Invest. **13**(2), 13 (2022)
10. Zheng, Y.P., Zhang, R.N.: Research on the practical teaching system reform of public administration specialties in the context of new liberal arts. In: International Conference on Mental Health and Humanities Education (ICMHHE 2020) (2020)

Teaching Quality Evaluation in Universities Based on Apriori Algorithm

ZhaoWen Chen[✉] and RongXin Zuo

Hunan Technical College of Railway High-Speed, Hengyang 421002, Hunan, China
chenzhaowen2021@163.com

Abstract. The purpose of this paper is to establish a framework for teaching evaluation in colleges and universities by using decision trees and related rules, to study the relationship among educational factors, educational chain factors and educational effects, and to provide a decision framework for teaching evaluation. The study will take into full account many aspects that affect the quality of teaching, which is the combination of subjective factors (such as status, academic experience, age, gender, teaching methods and resources) and objective factors (such as type of course, course series, number of students, total hours per week), and make a comprehensive analysis of the relevant factors.

Keywords: decision tree algorithm · colleges and universities · evaluation system of teaching quality

In recent years, with the rapid development of database technology and the wide application of management systems, more and more data has been collected, making it very important to find useful information in database. Therefore, there is increasing interest in data digging technology as an effective solution to this problem, and we have got great achievements. Today, it is widely used in business, finance, industry and commerce. With the continuous development and application of data digging technology, the application of data mining technology in higher education will play an important role in improving the teaching management.

Teaching evaluation is a systematic study and evaluation of school teaching according to educational goals and standards to evaluate the value of activities and learning results in the learning process. It is an important part of school education management and an important tool to measure educational results. In each semester, schools conduct educational assessment surveys and collect a large amount of data. However, the current education evaluation is mainly based on the numerical calculation for the promotion of teachers and the evaluation of staff, which does not fully tap into the value of reflection and data. In this article, we analyze various data digging methods, apply the decision tree algorithm to the evaluation of teaching in colleges and universities, and conduct data digging to study the factors that affect the educational effect and their relationship, so as to better understand the relationship between teachers and staff.

J. C. Hung et al. (Eds.): IC 2023, LNEE 1045, pp. 458–463, 2023.
https://doi.org/10.1007/978-981-99-2287-1_65

1 Decision Tree Concepts and Mining Algorithms

1.1 Decision Tree Concepts

The most prestigious ID3 algorithm was introduced by Quinlan in 1986. The ID3 algorithm is a decision tree algorithm that uses Entropy information to select cases based on a set of eigenvalues. The C4.5 algorithm is based on the ID3 algorithm and was developed by Quinlan in 1993. To meet the requirements of processing large amounts of data, researchers have proposed several advanced algorithms, the most famous of which are SLIQ and SPRINT algorithms.

Generally speaking, it can be divided into two phases in the use of decision tree to classify. In the first stage, decision trees should be constructed by recursively generating decision tree forming sequences. In the second stage, input data should be classified using decision tree models. The process of building decision trees is very important. In essence, the generation of decision tree is a greedy algorithm. Each unclassified node is tested to find a set of example attributes (test attributes) starting with the top node. Based on test results, the training instance is divided into several subsets, each of which forms a new node and is repeated until a closed condition of the new node is reached. An important part of building a decision tree is the selection of test features and the distribution of sample sets. Different decision tree algorithms use different methods for this purpose. Several decision tree algorithms have been developed gradually, such as CLSJID3, CHAID, CART, FACT, C4.5, GINI, SEES, SLIQ, SPRINT, and others. The most famous algorithms are the ID3 and C4.5 algorithms proposed by Quinlan.

1.2 ID3 Algorithm

The ID3 algorithm is an optimal selection method of descriptive attributes based on the entropy subtraction theory. The attribute to be tested is the one with the highest information value in the current sample set. The sample is divided into as many subsets as possible because the values of the attributes to be tested are different and new nodes corresponding to the sample are added to the decision tree. This approach reduces the number of tests required to classify projects and ensures that only one simple (not necessarily the simplest) tree can be used.

$$I(s_1, s_2) = -\sum_{i=1}^{m} pi \log(pi)$$

In the figure above, pi is an arbitrarily sample belonging to the probability of C; using s;/s to estimate. Note: The logarithmic function is based on 2 because the information is encoded in binary.

$$E(A) = \sum_{j=1}^{m} \frac{s_{1j} + s_{2j} + \ldots + s_{mj}}{s} I(s_{1j} + s_{2j} + \ldots + s_{mj})$$

In the above formula, $I(s_{1j} + s_{2j} + \ldots s_{mj}) = -\sum_{i=1}^{m} p_{ij} \log 2(p_{ij}) = s_{ij}/|s_j|$ is served as the j-th $(s_{ij} + s_{2j} + \ldots s_{mj})/s$, which is the sample in s_j belongs to the probability of C_i.

In this way, the information gain obtained by dividing the corresponding sample set of the current branch nodes by using attribute A is:

$$Gain(A) = I(s_{1j} + S_{2j} + ...s_{mj}) - E(A) .$$

2   Application of Decision Tree in Teaching Evaluation

Considering the current technical level, this paper focuses on the introduction of ID3 algorithm, which has a wide range of applications and is easy to use, and is often used as a decision tree in the teaching evaluation system. Based on the existing evaluation results and the prior knowledge of teachers, we aim to develop a model of excellent teachers.

The stages of classified data analysis usually include data collection, data preprocessing, data classification, classification rule analysis and knowledge application. In this paper, this stage is used to describe the process of developing a decision tree model for teaching evaluation.

College teaching pays more and more attention to teaching quality, focuses on academic management, and increasingly use data mining technology to manage education. By applying membership standards to educational policies, college teaching leaders have a solid foundation to make the right decisions. As a part of the continuous improvement of teaching information management in colleges and universities, more and more evaluation data are collected into the online quality evaluation system. The evaluation database is connecting with other databases and a wide range of information about teaching in universities. The understanding of these sources not only provides the evaluation results for the teaching management department of colleges and universities, but also provides other important information.

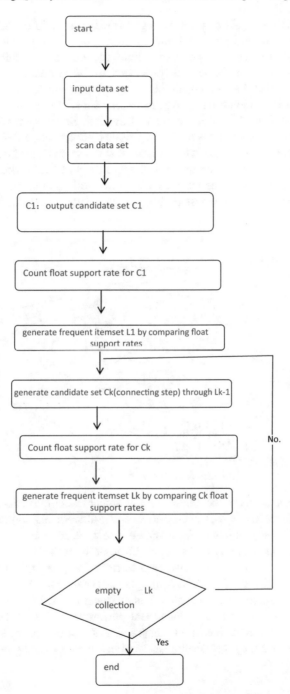

We have used association rules to test the prediction model of decision tree, and the results are basically consistent, which partly reflected the high reliability and applicability of the prediction model.

The correlation analysis of teachers' age, grade, academic qualifications, gender and evaluation results shows that they are consistent with reality, that is, the older the teachers' age and the higher the grade, the richer the teachers' teaching experience is, and the gender of teachers is not always related to the evaluation results. However, the analysis shows that the relationship between teachers' academic qualifications and evaluation results is asymmetric for the following reasons. Firstly, there are many retired teachers and most of them have academic qualifications in colleges and universities, so the evaluation results are good. Secondly, only three teachers have a PhD (statistics show that they are studying for a PhD), which means that the sample size is too small and the results are distorted. Third, in recent years, most of the teachers working in schools are new graduates. And most of the teachers who have worked in schools in recent years are newly graduated students, whose evaluation results are distorted due to their lack of teaching experience.

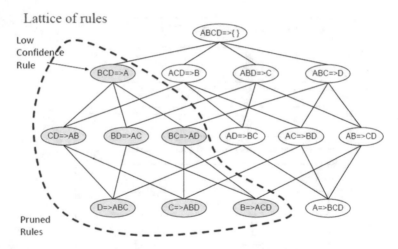

Secondly, the analysis of the relationship in actual teaching (including department, course type, number of students, teaching effect and evaluation results) shows that classes with less than 30 students has better teaching effect, elective courses are better than required courses, practical courses are more difficult to teach than theoretical courses, and the effect of teaching evaluation is also terrible. Teaching quality is the external expression of educational effect. Teaching quality evaluation aims to guide, support, motivate and standardize teaching quality. It is an important part of school management and an important tool to evaluate educational performance. Using data mining technology in quality assessment has important practical significance, because it provides a quantitative basis for colleges to develop and support classroom assessment and improve teaching quality.

3   Conclusion

In order to better describe the conditions required for a good teacher, the model needs to supplement certain indicators, such as workload, research skills, and so on. In addition, other data mining algorithms, such as related rules, can also be used to extract teaching information, which should be further analyzed. Data mining technology has been applied in many fields, and its early application shows that it has great potential in teaching and can help solve the problems of the current education system. Based on the general trend of education, data mining technology has a very broad prospects for data analysis in the field of education.

Acknowledgement. This work is sponsored by Hunan Province Educational Science "Thirteenth Five-Year Plan" Project "Research on the Construction of Multi-evaluation System for Teaching Quality of Mobile Smart Higher Vocational Teachers" (Project No: XJK18BZY064);

Hengyang science and technology research projects, "Development of Information System for Multi-evaluation of Teachers' Teaching Quality" (Project No: 202 002 072 253).

References

1. Chen, Q.: Application of decision tree algorithm in teaching quality evaluation system of colleges and universities. Teaching in Southwest Communications Universities (2010)
2. Yuan, Y.: Application of decision tree algorithm in teaching evaluation system of universities. J. Zhejiang Ocean Univ. (Nat. Sci. Ed.) 440–444 (2006)
3. Juanjuan: The evaluation of teaching quality and the deepening of teaching reform in colleges and universities. J. Teach. Natl. Univ. Cent. South China 286–387 (2003)
4. Liu, T., Su, B., Liu, J.: Current status and exploration of teaching quality evaluation of teachers in medical colleges and universities. China High. Med. Educ. 3–4 (2002)
5. Wanlan: Application research of data mining technology in teaching evaluation. Master's thesis of Teaching in Liaoning Engineering and Technology University (2003)
6. Linyang: Potential value of data mining in educational informationization. Mod. Educ. Technol. 65–67 (2002)
7. Zeng, J., Peng, A.: Diversification of evaluation index system of teaching quality. High. Educ. Dev. Eval. 32–35 (2006)
8. Li, X., Wei, G., Wei, J.: Research and exploration on the construction of evaluation index system of experimental teaching quality. Lab. Sci. 5–23 (2006)
9. Sun, Y., Sheng, H.: Study on KDD-based teaching quality evaluation system. J. Teach. Zhejiang Normal Univ. 110–113 (2005)
10. Meng, Y., Huang, Z.: Application of data warehouse technology in university education management. Xuzhou Normal Univ. Teach. J. (2003)
11. Huang, J., Pan, P., Wan, Y.: Research on system framework of data mining. Comput. Appl. Res. (2003)
12. Chen, Y.: Data Warehouse and Data Mining Technology. Electronic Industry Press (2002)
13. Liu, B., Zhang, Y., Fang, J.: Teaching evaluation method based on data mining. Comput. Modernization 87–89 (2005)
14. Qin, W.: Decision tree algorithmic analysis in classification technology. J. Shenzhen Vocat. Tech. Coll. Inf. 54–58 (2004)
15. Shen, C.: Study on classification algorithm of decision tree. J. Yancheng Inst. Technol. Nat. Sci. Ed. 22–24 (2005)

Teaching Reform of New Media Data Analysis Course Based on OBE Concept

Xianmei Liu[✉] and Dongmei Ning

Nanchang Institute of Technology, Nanchang, China
`xianm_liu@sina.com`

Abstract. The research on teaching reform of new media data analysis course based on OBE concept is a research project. It is conducted by the computer science teaching and learning research center of the school of information technology, Nanyang Technological University (NTU), Singapore. This study aims to study the effectiveness of using real-time online collaborative learning environment (OBL) as a teaching tool to teach students how to analyze data collected from social networking sites such as Facebook and twitter. The main purpose is to evaluate the impact of this method on students' performance, because they can learn how to conduct effective analysis. Researchers will be able to understand the teaching methods and practices that effectively improve students' learning results by analyzing their experience of using OBE based teaching methods. This study will also provide a theoretical framework for future research related to OBE based teaching methods. The main purpose of this study is to study how teachers use OBE (observation, brainstorming, evaluation) as a teaching tool.

Keywords: OBE concept · reform in education · New media data analysis

1 Introduction

As we all know, in the era of new media, content is king. Data analysis is one of the means to determine whether the content can be done well. What is data analysis? What is the purpose of data analysis? How to do data analysis? Is there any recommended tool for data analysis? What is data analysis? In short, it refers to the monitoring and statistics of a series of behaviors of users on the published content after we publish the content. For example, you published a dry goods article on official account. Simple data analysis refers to: how much exposure, reading, retention, conversion, etc. [1]. There are some disadvantages in this course, such as strong theoretical nature and students' lack of perceptual knowledge of production practice related work. How to cultivate students' interest in learning,

Moreover, it is an urgent problem to apply the learned knowledge to practice, so as to cultivate students' craftsmanship spirit, innovation consciousness and other excellent qualities, and provide guarantee for the integration of industry and education, the construction of new engineering and other aspects. As an advanced educational concept, OBE has been accepted by teachers engaged in higher education and applied to teaching.

© The Author(s), under exclusive license to Springer Nature Singapore Pte Ltd. 2023
J. C. Hung et al. (Eds.): IC 2023, LNEE 1045, pp. 464–469, 2023.
https://doi.org/10.1007/978-981-99-2287-1_66

It is a new educational concept with students as the main body and teaching results as the guidance. Through the analysis of the current situation of the new media data analysis teaching, the positioning of the course is clarified, and the teaching objectives are established in combination with the OBE education concept. Through the practice of teaching reform, this paper expounds the use of hierarchical and differentiated teaching system in teaching, and analyzes the data obtained, and interprets the integration effect of OBE education concept in the teaching of new media data analysis course [2]. It is pointed out that curriculum reform is a long-term process, and only by continuous improvement can we achieve good results.

2   Related Work

2.1   New Media Data Analysis and Operation

New media operation mainly involves the use of various network channels to promote enterprises or products, monitor the feedback of network users in real time, and analyze the feedback to guide the operation strategy. Most of the data covered here are user access and interaction data, which can help operators understand users' needs in depth and adjust operation contents and strategies in real time according to their needs. To analyze these data, we can abstract these data into various KPIs in combination with specific businesses, and observe the changes of these KPIs from various dimensions, such as the current week value, the comparison between week and week, the prediction of next week, the comparison between the current week value and planned value, and the change trend of the current month [3]. We can also introduce some competitive products for reference. There are many specific KPI dashboards on the network. The following is just a list for reference. Figure 1 below shows the analysis of website traffic:

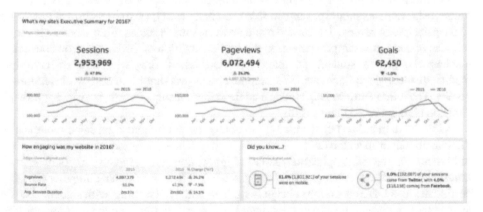

Fig. 1. Website traffic analysis

As long as the data of these dashboards is complete, they can be easily completed by using any BL tool. Of course, excel is no exception. In addition to excel, other Bi tools include tableau and powerbl, which are relatively more intelligent and have free versions. The dashboards can also be exported into PPT and PDF.

2.2 Current Teaching Situation of New Media Data Analysis Course

The course "new media data analysis" is highly theoretical, and needs the knowledge of advanced mathematics, complex variable functions and other disciplines as the basis. A large number of formula derivation causes students to be overwhelmed and difficult to master; The traditional teaching mode mostly presents the characteristics of "cramming", that is, teachers teach and students receive passively, which is often easy to ignore the role of students as the main body, and lack of necessary interaction and discussion; The connection between the chapters is loose, and students do not know how to apply the knowledge learned to the analysis and design of the actual system; At present, the final examination is still the main form of assessment [4]. In a limited time, it is impossible for the examination content to cover all the key knowledge. In addition, the assessment and evaluation requirements for practical ability and innovative spirit are low. Therefore, the teacher team found that in the actual teaching process, negative phenomena such as "students are not clear about the key points of learning", "the knowledge learned is boring" and "theory is divorced from practice" often occur, which seriously hit the students' enthusiasm for learning.

2.3 OBE Teaching Concept

OBE (outcome based education) is abbreviated to OBE, also known as ability oriented education, goal oriented education or demand education. It was proposed by Spady et al. In 1981, and soon received recognition and attention from everyone, and became the mainstream thought of educational reform in Europe and the United States. It emphasizes the students as the main body, the teaching results as the guide, and the learning results obtained by students after teaching in this way.

OBE education concept takes students as the main body; Guided by teaching achievements, achievements should pay attention to their practicality; Design teaching according to the principle of reverse design, and evaluate the results in stages [5]; It has the characteristics of continuous improvement. In the implementation of OBE education concept, teachers should help students complete four core issues: first, what kind of learning results students should achieve. The second is why we should achieve such learning results. The third is how to help students achieve these results. Fourth, how to know that students have achieved these results.

With the demand of talent training in society, the traditional rigid injection teaching can no longer meet the needs of today's society. The society requires graduates not only to master practical application skills and innovation ability, but also to obtain some ability test certificates, such as CET-4 and CET-6 certificates, national computer CET-2 certificates, etc. OBE education concept is student-centered teaching, with student needs as the goal orientation. It follows the principle of reverse design, and the training objectives are determined by needs, which ensures the consistency of educational objectives and results to a great extent.

Therefore, under the guidance of OBE education concept, through the study of new media data analysis during the University, help students develop the teaching objectives of new media data analysis course on the basis of mastering basic skills [6]. Take competition as learning, competition as teaching, take the content of computer grade examination as the teaching focus, and realize the teaching objectives at different levels according to the training needs of different majors.

3 Research on Teaching Reform of New Media Data Analysis Course Based on OBE Concept

When designing the new media data analysis course, we should first consider the four core issues that must conform to the OBE education concept. First, design teaching objectives. Design the ability level that students ultimately need to achieve by learning this course. Second, consider teaching needs. That is why we should learn these contents. Starting from the reality of students, we should analyze how to design the curriculum structure in the teaching of new media data analysis. Because students' majors are different, their emphasis on content is also different, so there are differences in teaching. Therefore, we should have a clear aim to truly enable students to find a platform for self realization in professional learning, so as to realize their own value [7]. Third, ensure the implementation of teaching. The new media data analysis course is a course combining theory and practice, which has an important impact on students' future study, employment and work. Therefore, after having the teaching objectives, we should ensure the implementation of the objectives. Fourth, the final teaching evaluation. The final learning achievements of students should be assessed and evaluated. On the one hand, it is the assessment of students' learning, on the other hand, it is also the assessment of teachers' teaching design. Through the test of both sides, we should constantly improve the teaching process, correct the imperfect content, and ensure to finally achieve the expected goal.

After clarifying the learning results, it is necessary to build a curriculum system suitable for the results, because the learning results ultimately need to be achieved by curriculum teaching. The construction of curriculum system adopts the principle of "reverse design", which starts from the knowledge (ability) structure. The curriculum system structure should correspond to the knowledge (ability) structure required to achieve the achievement goal one by one. Each knowledge (ability) should be supported by a clear curriculum, so that students can finally achieve the peak achievement. In short, each course of the curriculum system is essential to achieve the goal of learning outcomes.

Results oriented teaching attaches great importance to "student-centered", emphasizing personalized teaching rather than flooding teaching. Therefore, teachers should adopt more interactive and research-based teaching, make good use of strategies such as demonstration, diagnosis, evaluation, feedback and constructive intervention, accurately grasp students' learning characteristics [8], learning foundation and learning process, formulate appropriate teaching plans according to the relevant requirements of the curriculum, and encourage students to "self-study" in the teaching process, through "autonomous learning", "cooperative learning" and "Inquiry Learning", Gradually guide and assist students to achieve the expected results.

4 Build an Evaluation System

Result oriented education especially emphasizes the evaluation principle of "learning by teaching". Teaching evaluation mainly focuses on learning results, rather than teaching content and other aspects. At the same time, OBE tends to achieve evaluation rather than comparative evaluation, that is, it emphasizes students' self comparison rather than comparison between students. Therefore, teachers should give different grades from unskilled to excellent according to the degree to which each student can meet the educational requirements, and carry out targeted evaluation [9]. The commonly used course teaching evaluation methods include classroom questions, mid-term tests, questionnaires, etc. The evaluation results of project design, operation review, research report, etc. are often expressed by "conforming/nonconforming", "achieved/not achieved", "passed/failed", etc. As shown in Fig. 2 below, OBE teaching results are oriented.

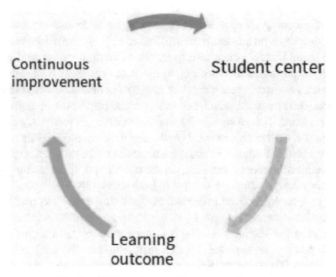

Fig. 2. OBE teaching achievement orientation

In OBE, after completing the whole course learning process, students can achieve the final results by achieving the learning achievement goals at all stages, from primary to advanced. Because the final results are completed step by step, a perfect continuous improvement mechanism needs to be established in the whole course teaching process, including the continuous improvement of training objectives, graduation requirements and teaching activities. It is necessary to ensure that students with different learning abilities will use different time, through different ways and means to achieve the same final goal.

The characteristic of OBE result oriented education is "reverse design and positive implementation". Starting from the final results, we reverse the curriculum system and teaching design, stratified by objectives, teach students in accordance with their aptitude, gradually deepen, and achieve the final teaching goals, from teacher centered to student-centered [10]. Teaching is no longer simply to instill knowledge, but to teach students to "enjoy learning", "learn", "learn", Pay more attention to cultivating students' autonomous learning ability.

5   Conclusion

The research on the teaching reform of new media data analysis course based on the concept of OBE is an experiment to evaluate the usefulness of OBE in improving students' ability to understand and analyze data. In this study, we tried to evaluate the effect of using OBE when teaching new media data analysis courses in universities. The purpose is to determine whether there is a difference in the scores of students who have participated in the courses taught with OBE and students who have not participated in the OBE courses in the tests to measure their knowledge of statistics and statistical methods. We also studied whether these results were consistent in different learning styles (vision/language/movement).

Acknowledgements. Nanchang Institute of Technology natural science research topics, Item No: NLZK-21-10.

Research on teaching reform of new media data analysis course based on OBE concept.

References

1. Liu, Y.F., De-Zhang, X.U., Liang, L.D., et al.: Exploration and practice of teaching reform of "calculation method" based on OBE concept in science and engineering major. In: Education Teaching Forum (2018)
2. Huang, S., Cao, H., Yan, L., et al.: Teaching reform of obstetrics and gynecology nursing course in higher vocational colleges based on OBE education concept from the perspective of big data. Biomed. J. Sci. Tech. Res. **37**, 29266–29276 (2021)
3. Liu, X.: An analysis on the teaching reform of chemistry course in university based on OBE concept. Yunnan Chem. Technol. (2019)
4. Yan, J.Y., Humanities, S.O.: Teaching reform of television program production course based on OBE concept. J. Guizhou Univ. Eng. Sci. (2019)
5. Wang, X.M., Huang, G., Jia, Y.T., et al : Teaching reform of textile material course based on OBE concept. Prog. Text. Sci. Technol. (2019)
6. Zhou, Y.Q., Wang, X.Z., Huang, G.P.: Discussion on teaching reform of bridge engineering course based on OBE concept. Sci. Technol. Vis. (2019)
7. Dong, J., Wang, Q., Liu, Y., et al.: Exploration on the teaching reform of food analysis based on the OBE concept. Theory Pract. Innov. Entrepreneurship (2019)
8. Li, D., Dong, G., Liu, Y., et al.: Teaching reform of optical fiber communication course based on OBE concept. Sci. Educ. Art. Collects (2018)
9. Wei, J., Teng, Z., Liu, S., et al.: Teaching reform of construction engineering cost based on OBE concept. Shanxi Archit. (2019)
10. Lou, G.: Exploration on the reform of advanced training teaching of interior design course based on "OBE" concept—taking the environmental design major of Northwest A&F university as an example. Guide Sci. Educ. (2018)

The Applicability of Fama-French Multifactoral Model in the Stock Investment of China's New Energy Industry

Shanshen Li, Qimeng Hao, Yaqian Liu[✉], and Jiaqi Meng

School of Economics and Management, Xi'an Shiyou University, Xi'an 710065, Shaanxi, China
chary_2006@163.com

Abstract. Based on the stock trading data of new energy industry in China's Shanghai and Shenzhen A-share markets from 1999 to 2021, this paper compares and analyzes the application of Fama-French five factor model and three factor model in the prediction of excess return. The research finds that: (1) Fama-French five factor model is effective in China's new energy industry market, and the five factor model is better than the three factor model in predicting the average rate of return of China's new energy industry; (2) The average abnormal return rate of Chinese new energy industry stocks can be explained by market factor, market value factor, profitability factor and investment level factor. The book to market ratio factor is a "redundant factor"; (3) In contrast, market effect and scale effect are the leading factors affecting the stock return of China's new energy industry, and the effects of profitability and investment level are relatively limited.

Keywords: Fama-French multifactoral model · China's new energy industry · Applicability of stock investment

1 Introduction

The development of new energy is related to the transformation and upgrading of energy and the implementation of future energy strategy. At the same time, accelerating the development of new energy industry is also of great significance to promoting the upgrading of industrial structure and promoting sustainable economic development. Emerging market countries attach great importance to the development of the new energy industry and strive to seize the opportunity under the development trend of the global low-carbon economy. As a large country with rapid economic development and large energy consumption, China has long realized the importance of the new energy industry in promoting green development and industrial upgrading. The government has provided many policy facilities and preferential treatment for the development of the industry. Many private capital has also actively entered the new energy industry, and the value chain of the new energy industry is becoming increasingly perfect. At present, a number of new energy listed companies with initial scale have been formed in China's capital market. They have made remarkable achievements in technology research and development, corporate governance and business promotion. The stocks of these companies are increasingly sought after by investors and even received close attention from overseas investment funds. Then, can the abnormal return rate of new energy industry stocks be predicted? What factors affect the rate of return? These issues are the focus of investors' global asset allocation. This paper enriches the relevant research in the field of asset pricing in China, combines the frontier of stock asset pricing theory with the actual situation of stock pricing in China's new energy industry, and systematically analyzes the applicability of Fama French multi factor model to the prediction of stock return in China's new energy industry, which complements the lack of research in this field.

2   Literature Review

The applicability of Fama and French (1993, 1996, 2015) [1] five factor models has been tested in major developed countries. It is of great practical significance to study the application of the five factor model in China's securities market. However, at present, the application of Fama French model in the securities market is still dominated by three factor model. Liao Li and Shen Hongbo (2008) [2] used Fama French (1993, 1996) [3, 4] three factor model to study the market effect of China's non tradable share reform and its influencing factors. Gou Dongning and Wang Weijia (2016) [5] conducted an empirical test on 16 listed banks in Shanghai and Shenzhen based on Fama French three factor model. Zhang Bing and Chen Xiaoying (2017) [6] tested the low price stock effect under the framework of Fama French three factor model. Yin Liya (2018) [7] added emotion factor, premium factor and growth factor to the classic Fama French three factor model for innovative research. There is not much literature on the study of China's securities market according to the construction method of profitability and investment style factors in Fama and French (2015). The research conclusions of only a few relevant articles are inconsistent. Zhao Shengmin et al. (2016) [8] tested the validity of the five factor model proposed by Fama French by using the data of China's stock market and found that the empirical results are opposite to those of the United States, and the three factor model is more suitable for China's capital market. Li Zhibing and Yang Guangyi (2017) [9] studied the application of the five factor model in different periods of China's stock market, and found that the five factor model has strong explanatory ability, and its performance is better than CAPM, three factor and Carhart four factor models. Ma Runping and Shen Jie (2021) [10] used the Fama French five factor model analysis method to test the fitting degree of Fama French five factor model to the GEM market by referring to the improvement ideas of domestic scholars.

Therefore, the discussion on the applicability of Fama-French multi factor model in the Chinese market needs further empirical evidence, especially for the emerging market, which is characterized by many uncertain factors and high volatility of stock price. Whether the multi factor model can still play its pricing advantage and forecasting ability is worth further exploration. In view of this, this paper takes the new energy industry in China's A-share market as the research object to verify the applicability of Fama French multifactoral model to the prediction of investment return in this emerging industry, and further enrich the relevant research on asset pricing methods in specific industries.

3 Sample Selection and Theoretical Model

This paper selects the monthly stock data from 1999 to 2021 of the seven concept sectors of nuclear power, photovoltaic, hydrogen energy, combustible ice, biomass energy, geothermal energy and wind energy in the wind database as the research sample. The data is from the wind financial terminal. Among them, the selection of the return rate of the stock market is consistent with the research of most scholars. The monthly return rate of the stock weighted by the market value of the stock circulation is taken as the return rate of the market, and the one-year lump sum deposit and withdrawal interest rate of RMB is converted into the monthly interest rate as the short-term risk-free interest rate. In order to make the information reflected in the data sample true and effective, this paper excludes all st and st* stocks, and excludes new stocks that have been listed for less than three months. The negative price to book ratio usually shows that the company's book assets can no longer offset its debts, so the stocks with negative price to book ratio are excluded. Based on the above selection criteria, by the end of 2021, a total of 159 new energy industry stocks were selected as the research samples.

In order to better evaluate the effectiveness of Fama French five factor model in the new energy industry of the A-share market, this paper takes the performance of Fama French three factor model as a reference. Fama French three factor model (1993) is as follows:

$$R_t - RF_t = \alpha + \beta(RM_t - RF_t) + sSMB_t + hHML_t + \varepsilon_t \tag{1}$$

where, R_t represents the expected rate of return of the investment portfolio in period t, RM_t represents the rate of return of the stock market in period t, RF_t represents the risk-free rate of return of the market in period t, the excess rate of return $RM_t - RF_t$ represents the market factor of the stock return, SMB_t (the average rate of return of small stocks minus the average rate of return of large stocks) represents the market value factor, HML_t (the difference between the average return of the stock portfolio with a high book to market ratio and the average return of the stock portfolio with a low book to market ratio) represents the book to market ratio factor. β, s and h are the sensitivity coefficients of the three factors respectively, X is the intercept term and y is the random error term.

Fama-French five factor model (2015) is expressed as follows:

$$R_t - RF_t = \alpha + \beta(RM_t - RF_t) + sSMB_t + hHML_t + rRMW_t + cCMA_t + \varepsilon_t \tag{2}$$

Among them, RMW_t (the difference between the return rate of the stock portfolio with strong profitability and that of the portfolio with weak profitability, which can also be called the difference between the return rate of the stock of the high-quality company and the low-quality company) represents the profitability factor. CMA_t (the difference between the average returns of high investment stocks and low investment stocks, also known as the difference between the aggressive and conservative corporate returns); α is the intercept term and ε_t is the random error term in t period; β, s, h, r and c are the sensitivity coefficients of the five factors respectively.

In order to evaluate the forecasting ability of Fama French five factors on the stock return of China's new energy industry, this paper uses the time series regression method of Griffin(2002) [11]. According to the ideas of Fama and French (2015), this paper replaces the book to market ratio factor *HML* with a new variable *HMLO* (orthogonal *HML*). Where, *HMLO* is the sum of the intercept term and residual term of the book to market ratio factor *HML* on the regression of $RM - RF$, *SMB*, *RMW* and *CMA* factors. Therefore, the new five factor model using *HMLO* as the factor regression of the book to market ratio is as follows:

$$R_t - RF_t = \alpha + \beta(RM_t - RF_t) + sSMB_t + hHMLO_t + rRMW_t + cCMA_t + \varepsilon_t \quad (3)$$

4 Empirical Analysis

Table 1 shows the regression coefficients of each factor of the five factor model. Considering that there is a short selling mechanism in the mature securities market, we pay more attention to the absolute value of the regression coefficient rather than the positive and negative of the coefficient. The results show that: (1) the market factor $RM - RF$ regression coefficient, whether in the three factor model or the five factor model, the six portfolio classification regression converges to 1, which has a very significant positive effect on the rate of return and has strong statistical significance. (2) As for the role of the scale factor *SMB*, there is asymmetry: if it is a low market capitalization stock portfolio, no matter whether the book to market ratio belongs to the high, medium or low group, it has a positive effect on the stock return. The maximum regression coefficient is reflected in the SL (low market capitalization - low book to market ratio) portfolio in the five factor model, reaching 0.868; However, for the stock portfolio with high market capitalization, in the subdivided groups with high, medium and low book value ratio, the scale factor regression coefficient is negative in almost half of the cases, and the absolute value of the coefficient is generally small, indicating that the scale factor *SMB* has a greater positive effect on the return rate of low market capitalization new energy stock portfolio, and there is a "small company" effect. (3) For the book to market ratio factor *HMLO* after orthogonalization, the study found that the regression coefficient was relatively stable after the transformation from the three factor model to the five factor model. When the book to market ratio factor and other factors were used to explain the excess return of new energy industry stocks, there was no significant change. (4) By observing the *RMW* regression coefficient of the profit level factor, it is found that the profit factor of the stock portfolio with high book to market ratio has a positive explanation for the stock return of the new energy industry, while the stock portfolio with low book to market ratio has a negative explanation. (5) The investment level factor *CMA* has the same effect as the profit level factor. Generally speaking, from the comparison of the absolute value of the regression coefficient, the market factor $RM - RF$ coefficient is the largest, and the scale factor *SMB* coefficient is the second. The impact of these two factors on the abnormal average return of new energy industry stocks is far greater than that of the other three factors.

In addition, from the perspective of model goodness of fit, the regression adjustment R^2 of three factor and five factor models are concentrated at 0.853–0.929, and the goodness of fit of five factor model is significantly improved compared with three factor model, which indicates that the prediction ability of the monthly average return of stocks in China's new energy industry has been improved.

Table 1. Regression results of Fama French three factor model and five factor model from 1999 to 2021

Category	Factor	Three factor mode			Five factor model		
		Regression coefficient	t Statistic	Adjusted R^2	Regression coefficient	t Statistic	Adjusted R^2
SH	α	−0.384	−1.937[*]	0.876	−0.400	−2.007[**]	0.878
	$RM - RF$	0.968	43.629[***]		0.965	43.379[***]	
	SMB	0.367	5.343[***]		0.468	4.916[***]	
	$HMLO$	0.276	2.920[***]		0.283	2.962[***]	
	RMW				0.103	1.076	
	CMA				0.020	0.189	
SN	α	0.092	0.581	0.912	0.089	0.555	0.929
	$RM - RF$	1.037	58.272[***]		1.042	58.363[***]	
	SMB	0.876	15.921[***]		0.543	7.119[***]	
	$HMLO$	−0.458	−6.037[***]		−0.460	−6.008[***]	
	RMW				0.085	1.110	
	CMA				−0.010	−0.117	
SL	α	0.728	3.542[**]	0.890	0.723	3.532[***]	0.892
	$RM - RF$	1.051	45.692[***]		1.041	45.516[***]	
	SMB	0.181	2.543[**]		0.868	8.869[***]	
	$HMLO$	0.753	7.680[***]		0.760	7.753[***]	
	RMW				−0.229	−2.340[**]	
	CMA				−0.128	−1.202[*]	
BH	α	−0.484	−2.294[**]	0.853	−0.465	−2.225[**]	0.858
	$RM - RF$	0.957	40.481[***]		0.965	41.379[***]	
	SMB	0.256	3.508[***]		−0.231	−2.315[**]	
	$HMLO$	−0.294	−2.917[***]		−0.302	−3.019[***]	
	RMW				0.224	2.245[**]	
	CMA				0.281	2.582[***]	
BN	α	0.097	0.584	0.913	0.106	0.640	0.915

(*continued*)

Table 1. (*continued*)

Category	Factor	Three factor mode			Five factor model		
		Regression coefficient	t Statistic	Adjusted R^2	Regression coefficient	t Statistic	Adjusted R^2
	$RM - RF$	1.019	54.985***		1.014	54.740**	
	SMB	−0.322	−5.614***		−0.057	−0.723	
	$HMLO$	0.598	7.572***		0.599	7.538***	
	RMW				0.024	0.300	
	CMA				0.162	1.867*	
BL	α	0.808	3.915**	0.887	0.767	3.814***	0.893
	$RM - RF$	1.074	46.481***		1.069	47.633***	
	SMB	0.006	0.091		0.274	2.856***	
	$HMLO$	−0.183	−1.860*		−0.170	−1.762*	
	RMW				−0.186	−1.935*	
	CMA				−0.452	−4.306***	

Note: *, ** and *** respectively represent significant at 10%, 5% and 1% significance level.

5 Conclusion

This paper uses the 275 month trading data of China's new energy industry stocks from January 1999 to December 2021 to test the effectiveness of Fama French five factor model in the A-share market. The main conclusions of the study are as follows: (1) by comparing the intercept terms of the three factor and five factor models under the $Size - B/M$ portfolio, this paper finds that the intercept term of the five factor model converges more to 0, and the goodness of fit of the five factor model is relatively high, which indicates that the prediction ability of the five factor model for the average return of China's new energy industry is significantly improved compared with the three factor model, It is suitable for effective prediction of excess return rate of China's new energy industry. (2) The factor decomposition test of the five factor model found that the average excess return of Chinese new energy industry stocks can be explained by market factor $RM - RF$, market value factor SMB, profitability factor RMW and investment level factor CMA. The book to market ratio factor HML is a "redundant factor", which has little impact on the prediction ability of the five factor model for the stock return of China's new energy industry. (3) On the whole, market effect and scale effect play a dominant role in the stock market of China's new energy industry, and the effects of profitability and investment level are limited.

Acknowledgements. This work was supported by the Social Science Foundation of Shaanxi Province of China under Grant 2019D009 and the Shaanxi Province of China, Department of Education Project of Philosophy and Social Sciences Key Research Base under Grant 19JZ052.

References

1. Fama, E.F., French, K.R.: A five-factor asset pricing model. J. Financ. Econ. **116**(1), 1–22 (2015)
2. Liao, L., Shen, H.: Fama French three factor model and research on the effect of non tradable share reform. Res. Quant. Econ. Tech. Econ. (09), 117–125 (2008)
3. Fama, E.F., French, K.R.: Common risk factors in the returns on stocks and bonds. J. Econ. **33**(1), 3–56 (1993)
4. Fama, E.F., French, K.R.: Multifactor explanations of asset pricing anomalies. J. Finance **51**(1), 55–83 (1996)
5. Gou, D., Wang, W.: Empirical test of China's listed bank stocks based on Fama French three factor model. Stat. Decis. Making (21), 158–161 (2016)
6. Zhang, B., Chen, X.: Study on the effect of low price stocks in Chinese stock market – based on Fama & Test of French three factor model. Financ. Forum **22**(10), 7–20 (2017)
7. Yin, L.: An empirical study on the influence of investor sentiment on stock returns – based on Fama French three factor model. Friends Account. (06), 51–56 (2018)
8. Zhao, S., Yan, H., Zhang, K.: Is Fama French five factor model better than three factor model – empirical evidence from China's a-share market. Nankai Econ. Res. (02), 41–59 (2016)
9. Li, Z., Yang, G., Feng, Y., Jing, L.: Empirical test of Fama French five factor model in Chinese stock market. Financ. Res. (06), 191–206 (2017)
10. Ma, R., Shen, J.: Research on the effectiveness of China's growth enterprise market – analysis based on Fama French five factor model. Econ. Issues (09), 46–52 (2021)
11. Griffin, J.M.: Are the Fama and factors global or country specific? Rev. Financ. Stud. **15**(3), 783–803 (2002)

The Application Level of Educational School Informatization Based on FCM Clustering Algorithm

Qian Chen[✉]

Big Data School of Fuzhou University of International Studies and Trade, Fuzhou 350202, Fujian, China
353192067@qq.com

Abstract. The rapid development and universal application of information technology have achieved the improvement of the quality of special education teaching. The application of information technology also provides a new learning method for special students, and provides a relatively unique auxiliary function for special education schools to carry out personalized auxiliary teaching. The research on the application level of educational school informatization based on FCM clustering algorithm is the application research of educational school informatization based on FCM clustering algorithm. This study focuses on applying this method to different types of applications. The main purpose of this study is to study how to apply it to different fields and how to use it according to the needs of different users. Therefore, this research will help those who are interested in using it for their own purposes, or who want to know more about its advantages and disadvantages, so that they can choose the most suitable one for themselves.

Keywords: FCM clustering · Educational schools · promotion of information technology

1 Introduction

Information technology also plays an important role in the field of special education. The application of information technology has promoted the transformation of special education methods, provided new teaching and evaluation strategies, and further improved the learning methods of special students. In recent years, relevant departments at all levels and all sectors of society have continuously increased their attention to special education. In 2010, the outline of the plan proposed to accelerate the development of special education and closely integrate the development of special education with the local economy, which fully illustrates the importance of the development of special education [1]. In 2017, the second phase of the promotion plan pointed out that special education teachers should continue to strengthen the application ability of information technology in special education. It can be seen that in the process of special education teaching, special education teachers, as the leaders and promoters in the informatization development of special education schools, need to constantly strengthen their own information

© The Author(s), under exclusive license to Springer Nature Singapore Pte Ltd. 2023
J. C. Hung et al. (Eds.): IC 2023, LNEE 1045, pp. 477–485, 2023.
https://doi.org/10.1007/978-981-99-2287-1_68

technology application ability and use information technology for better personalized teaching [2].

With the promulgation of the two special education promotion plans and the "China education modernization 2035" and other documents, the promotion effect of the application of information technology on special education in China is booming. The information-based auxiliary technology can improve the life of special people and play a positive role in the learning and communication mode of special people [3]. For example, Xu Hongmei and others have successively used modern education technology, touch map technology and other technologies to optimize visual defect compensation, so as to achieve the purpose of improving the education level of visually impaired children. The rapid development and universal application of information technology have brought new opportunities for the improvement of the teaching quality of special education [4].

In recent years, with the vigorous promotion and development of special education teaching in China, the number of special education schools in China has also increased year by year. According to the education statistics of the Ministry of education from 2014 to 2019, the scale of special education teaching in China has been expanding, and the number of special education students has also been increasing (as shown in Fig. 1). By the end of 2019, China has 2192 special education schools, including 27 schools for the blind, 396 schools for the deaf, 539 schools for the mentally retarded, 1230 other special education schools, and 794,612 students in special education [5]. It can be seen that the development of special education is still very important.

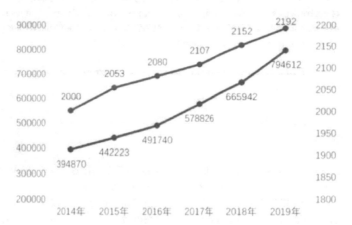

Fig. 1. Number of special education schools and students in school in recent years

In the development of information application in special education schools, special education schools should continue to strengthen the application of information technology in the teaching process, and use information technology to meet the personalized learning needs of disabled students. With the continuous introduction of relevant policies and regulations on the informatization of special education schools in China, the informatization development of special education schools has been vigorously promoted, but the informatization application level is still insufficient. Special education teachers have not yet met the integration needs of information technology and disciplines when teaching disciplines [6]. In addition, the informatization of special education in China also has some problems, such as the serious shortage of the application of informatization rehabilitation instruments and equipment, the lack of complete supporting digital education resources for special education courses, and the shallow application level of informatization assisted classroom teaching and rehabilitation training. It can be seen that accurate evaluation of the current situation of informatization application level in special education schools in China plays a vital role in improving the quality of informatization application level in special education level by level.

2 Related Work

2.1 Application of Informatization in Domestic Special Education

(1) Application of educational informatization

Fu Rong pointed out that the application of educational informatization is actually the innovation of education by using information technology. The application of educational informatization can improve the unbalanced development of high-quality educational resources in urban and rural areas, and provide a new form for the sharing of educational resources [7]. Nanguonong mentioned that the development of educational informatization not only depends on infrastructure and resources, but also constantly improves the level of informatization application. Teachers can improve the teaching quality by using information technology in the teaching process, and promote students to realize the all-round development of autonomous learning and personalized learning. Zhu zhiting believes that the application of educational informatization can effectively promote the further improvement of campus digital resource construction, improve teachers' information technology application ability and improve teaching quality.

The development of educational informatization includes the development of school infrastructure, teachers' information technology application ability, digital resource construction and other aspects [8]. The application of educational informatization can enable teachers and students to jointly improve information literacy in the information-based teaching environment and use information technology to improve learning efficiency.

(2) Application of informatization in special education

Informatization application has become an essential key link in the development of school education informatization. Teachers and school managers can use information technology to achieve high-efficiency and high-quality education, teaching and management activities in the process of education, teaching and management. The development of educational informatization in China gradually attaches importance to the integration and innovation of classroom teaching by using information technology. In 2019, relevant documents of the Ministry of education pointed out that the full application of informatization aims to promote the balanced development of high-quality education. Teachers should improve the application ability of information technology. The wide application of informatization can help disadvantaged groups such as students with mental retardation, students with visual impairment and students with hearing impairment receive education normally and improve their own educational level. Relevant documents of the State Council clearly point out that we should run special education well, promote the full coverage of education for school-age disabled children and adolescents, comprehensively promote integrated education, and promote the integration of medicine and education. It can be seen that the state attaches great importance to the application of information technology in the field of special education, meets the learning needs of special students, and gradually improves the inconvenience of special students' learning and life through information application means, which is urgent.

2.2 Foreign Special Education Informatization Application

Information technology is considered as an auxiliary factor of special education. It has been widely used in special education in developed and developing countries, and has been paid attention to in the teaching process. The application of information technology plays a vital role in meeting the normal learning needs of hearing impaired students in the Thai special school environment. Lersilp found through a questionnaire survey that in deaf schools, the application of information technology can help hearing impaired students participate in various activities, including talking with relatives and friends on social networks, and the application of information technology can help hearing impaired students have normal learning and life communication. In the application of information-based assistive technology, theeratorn investigated the situation of disabled students using information-based assistive technology to obtain learning information. The research shows that students with visual and hearing disabilities can better obtain the required learning information by using different types of information-based assistive technology.

In the research on the informatization policy planning of special education, Japan has issued policies on the informatization application of special education, requiring systematic and professional training on the application of information technology for different types of special students. Through the use of information technology tools, we can communicate with local medical institutions and students' parents and share the health status of special students, so as to help disabled students adapt to social life faster. In the relevant policy planning, Japan also proposed that the information application of special education schools should be closely combined with the specific learning situation and personalized development needs of special students; The use of various ICT equipment should help special students to integrate into social work and life normally in the future. It can be seen that foreign countries not only attach importance to the application and development of information technology in special education, but also hope that special students can improve their own shortcomings through personalized auxiliary technology, and independently participate in social work [9]. The application of information technology can not only improve the special teaching methods, but also make up for the physiological and psychological defects of special students through personalized auxiliary technology, and help special students realize rehabilitation training.

3 Introduction to Clustering Algorithm

Clustering is a model of statistical data analysis. Because of its simplicity, efficiency, easy to understand, easy to implement and other characteristics, it has received more and more attention. Because it has a huge data set for different needs,

3.1 K-Means Clustering Algorithm

(1) Clustering

K-means clustering (K-means) is a clustering algorithm based on prototype. It selects the clustering center according to the objective function, and the convergence speed is fast, so it is also the most widely used. Ma Jiajun used a simple and effective K-means clustering algorithm for grade evaluation when analyzing the ability of students in a certain dimension, in order to let students know the position of their ability in a certain aspect in the whole student group.

(2) Clustering algorithm

K-means clustering algorithm mainly comes from a vector quantization method in signal processing, and now it is more popular in the field of data mining as a clustering analysis method. The standard of K-means clustering is to divide the data into different index data into limited categories. K-means algorithm is a clustering algorithm based on the division. It uses the mean value of the values contained in the cluster to calculate the center of each cluster. Users need to specify the classification value of the division [10].

(3) Algorithm idea

The core of K-means algorithm is to group data according to some data sets given by users and some characteristics of data. In the k-means algorithm, the number of clusters K should be determined first, and then the cluster centers should be randomly selected. Calculate the distance from each data to the center of each cluster, which is usually expressed by a simple Euclidean function. The distance formula is used to compare a single data point with all cluster centers. Move the data points to the cluster with the shortest distance. Then re estimate the centroid. Similarly, each data point is compared with all centroids, and this process continues until the center convergence no longer changes. Each cluster must have a data object, and each data object can only belong to the nearest cluster center. Data objects in the same cluster have the greatest similarity, while data objects in different clusters have less similarity.

The distance formula is:

$$\mu_{rs}(x, z) = \{\max\{\min[\mu_R(x, y), \mu_s(y, z)]\}\} \tag{1}$$

3.2 FCM Clustering Algorithm

(1) Fuzzy clustering

Fuzzy clustering is mainly a method of processing data by giving the membership value of each data point in the data set. The value range of membership degree of fuzzy set is 0–1. Fuzzy clustering is basically a multivalued logic, which allows intermediate values, that is, the members of a fuzzy set can also be members of other fuzzy sets, and there is no sudden change. Bezdek combines Fuzziness with the concept of mathematics and puts forward fuzzy theory. Fuzzy theory is mainly used to explain some fuzzy information concepts in language. Based on this, fuzzy set theory has been widely developed, and Ruspini proposed fuzzy partition. Nowadays, FCM clustering algorithm is also applied to various fields.

(2) Clustering algorithm

In FCM clustering algorithm, the classification is obtained by membership matrix, and the final clustering effect is obtained after continuous iteration.

(3) Algorithm idea

FCM is the process of dividing the elements in the set into multiple self similar classes, and the data objects contained in each class have great similarity. Assuming that the finite sample data set x = {x1, X2, X3,…, xn}, each data sample and the Y features contained in the data sample xj = {xj1, XJ2, Xj3,…, xjn}.

$$\begin{cases} E(t)\dot{x}_k(t) = f(t, x_k(t)) + B(t)u_k(t) + d_k(t) \\ \qquad\qquad y_k(t) = C(t)x_k(t) \end{cases} \tag{2}$$

4 Application of Educational School Informatization Based on FCM Clustering Algorithm

4.1 Data Preprocessing of Informatization Application Level in Special Education Schools

The online questionnaire survey method based on the development of informatization application level in special education schools exports the questionnaire system data and collects and sorts out the original data of the development of informatization application level in special education schools. However, when manually filling in the questionnaire data, there are individual mistakes, which may cause the input data to be inconsistent with common sense and data errors. Therefore, the original data obtained may be incomplete, inconsistent or even partially invalid. In order to improve the quality of the obtained data, data preprocessing technology can be adopted.

(1) Data cleaning

After exporting the required original data through the questionnaire system, this data can not be directly used as the specific data for data analysis, Some invalid or missing data in the original questionnaire data need to be processed to ensure that the data type or format in the questionnaire data is consistent [80. In addition, it is also necessary to de duplicate the questionnaire data according to the actual situation. For example, this study is based on the current situation of information application level in special education schools. Therefore, in the questionnaire survey, special education schools are taken as the main body to understand some specific application index data about information application in their special education schools. Each special education school only needs the leader of the school or the chief teacher of the information office to fill in a questionnaire To ensure the authenticity and validity of the questionnaire data of information application in special education schools.

(2) Data standardization processing

In many informatization application index systems, due to the different nature of each informatization application index, there are different dimensions and orders of magnitude of informatization application index data. The standardized processing of information application index data of special education schools can quantify the specific application index data. Before cluster analysis and processing of data, it is a common processing method to standardize the original data. Common data standardization processing methods include min - Max standardization, log function conversion, etc.

4.2 FCM Cluster Analysis on the Informatization Application Level of Special Education Schools

FCM clustering algorithm is a kind of soft clustering based on fuzzy theory. FCM can provide more flexible clustering results than K-means clustering. Under the index data of special personalized teaching application possessed by special education schools, the objects in the data set cannot be directly divided into distinct clusters. FCM defines the concept of membership degree, which is used to measure the membership degree of an object to each cluster.

FCM fuzzy clustering algorithm process:

(1) Standardized data matrix;

(2) Establish fuzzy similarity matrix and initialize membership matrix;

(3) The algorithm starts to iterate until the objective function converges to the minimum value;

(4) According to the iteration results, the class of the data is determined by the final membership matrix, and the final clustering results are displayed.

The algorithm flow chart is shown in Fig. 2.

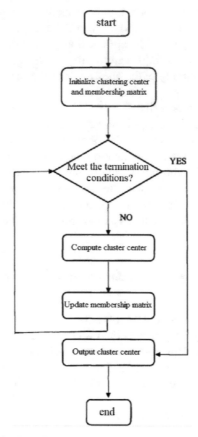

Fig. 2. FCM clustering algorithm flow chart

5   Conclusion

In order to better describe the characteristics of the development of informatization application level in special education schools, then fuzzy c-means clustering (FCM) algorithm is used to cluster and analyze the clustering results, and variance analysis is used to test the significant difference. Finally, FCM clustering algorithm is more suitable for the difference research of informatization application level in special education schools. Based on the analysis results of FCM clustering algorithm, the development of informatization application level of special education schools in different categories is obtained. The informatization application level of special education schools is divided into three different categories and grades. By comparing the development level differences of three different types of schools in three application dimensions, such as informatization infrastructure application, informatization teaching application and management informatization application, Three different levels of special education schools are named as "good", "moderate" and "insufficient" according to the development characteristics of special education informatization application level.

References

1. Gao, C., Jin, S.S., Science, S.O.: Research on the application status of sensor in intelligent road. Sens. World (2020)
2. Zhou, X., Li, C., Wang, C., et al.: Research and practice on training mode reform of applied talents in environmental engineering technology under industry-teaching integration training base. In: IOP Conference Series: Earth and Environmental Science, vol. 692, no. 3, p. 032036 (10 pp) (2021)
3. Raudenbush, S.W., Schwartz, D.: Randomized experiments in education, with implications for multilevel causal inference. Ann. Rev. Stat. Appl. 7(1), 177–208 (2020)
4. Choktanaprasit, N., Chantarasombat, C., Agsonsua, P.: The development of teacher development innovation for enhancing students' learning achievement in Rajaprajanugroh 50 school under the office of special education administration. World J. Educ. 10(2), 203–213 (2020)
5. Lu, L., Zhou, J.: Research on mining of applied mathematics educational resources based on edge computing and data stream classification. Mob. Inf. Syst. 2021(7), 1–8 (2021)
6. Sadiq, S., Anasse, K., Slimani, N.: The impact of mobile phones on high school students: connecting the research dots. Technium Soc. Sci. J. 30, 252 (2022)
7. Zhu, L.Y., Deng, Y.J., Wei, D., et al.: Research on the uniqueness and algorithm of multi-parameter identification in coupling thermal acoustic and solid problem. Sci. Sin. Technol. (2021)
8. Fernández-Cruz, F.J., Rodríguez-Mantilla, J.M., Díaz, M.J.F.: Impact of the application of ISO 9001 standards on the climate and satisfaction of the members of a school. Int. J. Educ. Manag. 34(7), 1185–1202 (2020)
9. Lai, L.: On the characteristics of Taizhou school's educational thought. Int. J. Soc. Sci. Educ. Res. 3(5), 163–167 (2020)
10. Ma, Y.: The application of schema theory in the teaching of English reading in senior high schools[J]. Reg. Educ. Res. Rev. 3(3), 17–20 (2021)

The Application of Computer Technology in Art Creation

Ying Wang[✉]

College of Arts and Information Engineering, Dalian Polytechnic University, Dalian 116499, Liaoning, China
1135629050@qq.com

Abstract. The application of computer technology in art creation is a research field that pays attention to the application of computer in art. It is used to produce works of art such as music, visual arts and movies. Like all research fields, it has its own history and development. In 1956, John McCarthy coined the term "artificial intelligence" at Stanford University; However, it was not until 1959 that alanturing proposed the famous Turing test of machine intelligence (Turing 1960). The computer provides a new technical means for the development of traditional art, expands the field of art design, facilitates art design methods, improves work efficiency, and enriches art creation.

Keywords: Computer technology · Artistic creation

1 Introduction

With the development of science and technology, computer technology is getting closer and closer to people's life. In the 1960s, Ivane Sutherland of Lincoln Laboratory of Massachusetts Institute of technology published a doctoral thesis entitled "Sketchpad", in which the word "computergraphics" was used for the first time, which laid the theoretical foundation of computer graphics, and thus determined the independent status of computer graphics as a new discipline. In order to explore the surface of the moon to help Apollo successfully land on the moon, NASA has perfectly combined computer digital technology with photos and images, opening up the way for computer art. In 1961, the American magazine "computer and automation" organized the world's first computer painting competition. Since 1965, computer has gradually become the performance medium and creative tool of artists [1]. In 1966, the hqwaro wis Gallery in New York held the first worldwide exhibition of computer painting. In 1968, the United Kingdom held the "cybernetic treasure" computer painting exhibition, and formally accepted computer painting works in the Venice Biennale in 1970, opening up the perfect combination of computer art into the palace of art, and opening up the way for computer art.

At the end of the 1970s, computer art design was introduced into China. First, it developed and expanded rapidly in Colleges and universities. While computer art design developed, it also brought great impact to the art industry. The habits formed over

J. C. Hung et al. (Eds.): IC 2023, LNEE 1045, pp. 486–492, 2023.
https://doi.org/10.1007/978-981-99-2287-1_69

thousands of years were almost destroyed overnight. People who are used to painting brushes have to pick up a mouse completely different in shape and mode of use to create art [2]. It is really difficult for the habits formed over the years. In the late 1980s, three-dimensional animation software was introduced into China, and began to use it to make TV program titles and advertisements. In the 1990s, 386 or 486 microcomputers were used for plane or three-dimensional visual art design. In recent years, the development and application of various practical design software have presented a splendid new world in front of all designers: Windows2000, coreldraw8.0, Photoshop 3DMAX, PageMaker, founder Feiteng, wending font library, language input, etc. all kinds of software are constantly upgraded to bring forth the new [3]. With the rapid development of computer software technology, the combination of computer technology and art will become closer and closer, which will have a profound impact on the future development of art.

2 Related Work

2.1 The Convenience of Computer Technology and Art Design

From the perspective of social function, fine arts can be divided into practical fine arts and appreciative fine arts. With the rapid development of computer technology, it was first applied to practical art in the art field. Computer has become an effective tool and working partner for practical art designers, which is obviously reflected in its ability to greatly improve work efficiency and design colorful pictures. Many scholars have discussed the computer as a high-tech product to improve the development of art design [4]. Xubangyue emphasized in his article thinking about computers that the functions of computers provide artists or designers with a variety of production ways and creative spaces. Designers or artists can easily change the color and style of the design by simply moving the mouse. Computers can quickly capture the instantaneous ideas of designers; Computer modification is fast, WYSIWYG, easy to delete, convenient to restore, accurate size and other functions, so that designers can modify repeatedly, expand and strengthen their own design creativity, and greatly shorten the time from creative design to the completion of works [5]. Pan Li further showed the convenience of computer in art design activities in his essay computer art essays. The paper points out that in the art design activities, repetition is a high proportion of content, such as the massive reuse of signs and standard fonts in the enterprise integrated design system (CIsystem). It can be easily copied by using the computer graphics design function, and the commonly used graphics, hues and other data can be stored in the computer to establish a personal graphics database for repeated use in the future. In addition, in the traditional design process, after the design scheme is passed, the production drawings must be drawn, which is heavy and boring. After the computer graphic design is adopted, once the scheme is determined, the whole set of high-quality design drawings can be output by the printer or plotter, thus the design cycle can be greatly shortened [6]. Here, the traditional drawing tools, materials and performance techniques have retired. Only designers rely on their own aesthetic thinking to design creativity through computers to create ideal creativity and the best visual effects. "Compared with the freehand drawing system, the computer processing of images and words is simple and accurate, and it can also timely convert the screen effect into printed matter. The computer provides a visual means to express and reflect the designer's creativity. When the designer skillfully master and operate the computer, using various graphic software and tool stunts, some unexpected graphic changes will occur on the screen, and So that designers can get inspiration from it and create ideal works." In his article art in the era of science and technology, leqihong said that in a broad sense, computer technology has enabled computer art workers to skip technical training and directly enter creative work. So as to give play to greater creativity and greatly improve the efficiency of work, which is unprecedented. No matter what step the staff are creating, even the most complex composition, such as the change of lines and the choice of structure, can be modified and adjusted in the modeling software. Due to the accuracy of the computer, in addition to directly observing the display with the naked eye, the color judgment in the creation process can also judge the description with specific numbers in a specific color mode, so as to achieve an accurate and standardized effect.

2.2 The Impact of Computer on Traditional Painting Language

The preservation methods of various artistic works are also different, and they all have relatively strict and scientific preservation methods, and they are also different in the display of works. It is limited by the type, size, display space, light and environment of works, and the preservation of works has also been damaged due to the change of time. When the computer is used in art activities, the situation is different. The painting tools are replaced by the host, keyboard, mouse, display and application software. The works are accurate digital graphics, and can be copied in a large number according to needs [7]. Because they exist in digital form, they are much more convenient to store, and there is no worry of deterioration and damage. The display can convey the communication between the media and people through a wide variety of visual display sizes. Works can be stored in different memories and can be carried around with a large amount of storage, such as floppy disk, removable hard disk, CD, etc. His works can be accessed into the square inch rooms of thousands of families through the Internet, which provides great convenience for people to understand art. In terms of techniques, traditional painting techniques are the experience summarized by painters after a lot of practice and understanding, and the maturity of a technique is often the result of years of painstaking efforts by the author. In computer art, the understanding of a technique is only the understanding of the results that can be produced by a certain command program. For example, in the perspective method of traditional painting, in order to draw an accurate perspective with various contents, the author must master a lot of perspective knowledge and practical experience, and spend a lot of time and energy [8]. However, in the computer, as long as you know the program command of making perspective, it is a matter of flicking your fingers. The change of color in traditional painting also needs to master a lot of color theoretical knowledge and practical experience. Of course, in the computer, as long as you select the desired color to make the result command, no matter how complex the color is, you can achieve the desired effect through multiple rendering commands. It has high accuracy in operation and does not need a lot of time-consuming and laborious training. The development of computer has found a new way out for those difficult problems in traditional painting, and simplified some complex techniques. Through the above comparison, the article points out that computer art is very different from traditional painting in terms of tools and techniques. Each has its own creative design. It should combine artistry and technology, and use computers to express the creativity of works more perfectly.

3 Art Creation Based on Computer Technology

3.1 Computer Art Creation Medium and Its Characteristics

The simplification of creation tools and materials is one of the important differences between computer art and traditional art. The computer art medium is mainly computer software and hardware, and the mouse, keyboard, display, scanner, digital camera, printer and plotter are indispensable tools. The works produced are accurate digital graphics, which can be copied in large quantities according to needs. It is very convenient to store, and there is no worry of deterioration and damage, as shown in Fig. 1. Works can be stored in a variety of memories and carried around. They can also be displayed and communicated with viewers through a wide variety of high-tech media systems. From this point of view, the creation tools of computer artists have been greatly simplified. The main convenience lies in the fact that they are all used for "traditional Chinese painting", "oil painting", graphic design and 3D modeling [9]. A "drawing board" is a computer screen, and a pen is a mouse or a stylus. Computer software provides various strokes, lines, pigments and materials, and the printer outputs the final work. Throughout the creative process, artists use "electronic" tools and materials provided by computer software. There is no need to spread paper, pens and pigments on a table and pile them up in a room full of spots and colors everywhere. The simplification of tools and materials enables artists, especially designers, to concentrate on conception and creativity and free themselves from the tiring production work.

Although the creation tools and materials of computer art have been simplified, its means of expression are more diverse and more powerful, which is beyond the reach of traditional art creation. Art is visual art. Computer technology has many advantages in visual expression for art. For example, by inputting various materials into the computer and combining them with graphics and fonts, the tactile texture can be introduced into the visual communication, associating with the specific experience of materials, producing a feeling of soft and hard, thick and thin contrast, or the visual features of shiny and dark, transparent and opaque, making the works more appealing. People have long simulated the three-dimensional effect in the two-dimensional space, trying to produce an illusory sense of space or visual depth. Graphic design software has given these methods to express the sense of space through overlap, size change, tilt change, curve change, texture thickness change, projection and penetration change, and face-to-face connection, so as to make the picture have richer levels and quickly and effectively increase the visual depth. Modern painters can not only draw new images by themselves, but also directly use ready-made images for "grafting" or adaptation. They can also deform, arrange and intersperse images to create novel works with visual dislocation.

3.2 Application of Computer Technology in Art Creation

As a complex information processing system, computer and application environment constitute a system, as well as developers and users. Hierarchical relation is a basic

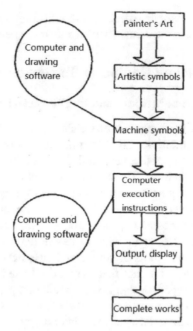

Fig. 1. Computer technology art creation process

concept for computer systems. Its establishment helps users to make clear which level of system resources it is dealing with, so as to explore how to enter this level, how to use the resources at this level, and how to communicate with other levels through this level. It is precisely because of the principle of dividing the system into different levels of hierarchy that the system design method can be implemented and unified in the way of thinking and working. Designers working in such a system environment can not only give full play to the advantages of the system in solving problems, but also improve their own work efficiency. In the traditional division of labor, from concept conception to implementation modeling is divided into multiple levels, and designers should participate in the work of each stage. The introduction of computer-aided design system makes the work that can be described in a certain language and reflected in a certain rule be replaced by the computer [10]. In another sense, it highlights the designer's creative ability, and the creation, evaluation and organization of design become the main work of the designer. Design bid farewell to the hardware era and entered the human era after the software era. The needs of users have become the main factor of product performance. Enterprises have injected more life factors into products, making technology more intelligent and flexible for users with different cultural backgrounds. For this reason, many companies employ sociologists to work with their own product designers. Instead of engaging in professional research, they employ them to make use of their own qualities to contribute to product design, because they must understand people before making any product. In Japan's high-tech industry, people without technical background will not participate in technical research. The reason why they are allowed to participate in the work is that the company's executives hope that they can solve problems with an open mind, rather than from the aspects of technology or computer science. For design, more and more generalists are needed. This development direction shows that the traditional design training is not suitable for the design work in the information society. The penetration of computer into design makes the designer's knowledge structure, professional skills, work procedures, design management and other aspects have to be adjusted.

4 Conclusion

In the 21st century, information has become a powerful productivity, and the torrent of information is changing rapidly. If a designer does not have this information in time, he will be far behind in this era. The greatest advantage of contemporary designers is that they enjoy the same information resources as the world's top designers at the first time. All kinds of media, magazines, the Internet, forums, lectures and exhibitions, we should pay attention to and participate in them in time, which can not only bring infinite skills and creative nutrition to a designer, but also ensure that you will always be the most advanced. Computer art is the product of the combination of computer technology and art, and its production is highly technical. The future art works will continue to develop and innovate under this high-tech technical means, showing the characteristics of high technology, popularization, diversification of artistic styles, and mass production and collectivization of works.

Acknowledgements. Research on Fine Arts Practice and Innovative Talent Training in Universities, JFYB2557.

References

1. Feng, L.: Research on the application of computer technology in software technology talents training system in higher vocational colleges. In: Journal of Physics Conference Series, vol. 1915, no. 3, p. 032035 (2021)
2. Chen, W., He, Y., Pei, Q.: Research on the design of electrical automation control system based on the application of computer technology. In: Journal of Physics: Conference Series, vol. 1992, no. 3, p. 032139 (2021)
3. Shen, H., Wang, J.: Research on application of computer technology in the optimization of department information system. In: Journal of Physics: Conference Series, vol. 1982, no. 1, p. 012139 (6 pp) (2021)
4. Li, H.: Research on the application of computer statistics technology in the educational information management system of colleges and universities (2021)
5. Zhang, B.: Research on the application of speech recognition in computer network technology in the era of big data. Int. J. Speech Technol. **26**, 259 (2023). https://doi.org/10.1007/s10772-021-09936-7
6. Li, M., Li, Y., Guo, H.: Research and application of situated teaching design for NC machining course based on virtual simulation technology. Comput. Appl. Eng. Educ. **28**(3), 658–674 (2020)
7. Sinclair, R.K.: The application of computer technology to the logistics process. In: 20th Annual Canadian Transportation Research Forum, Toronto, Ontario, 28–31 May 1985. Canadian Transportation Research Forum (CTRF) (2020)

8. Hao, T., Chen, X., Song, Y.: A topic-based bibliometric analysis of two decades of research on the application of technology in classroom dialogue. J. Educ. Comput. Res. **58**(7), 1311–1341 (2020)
9. Shi, X., Tang, K., Lu, H.: Smart library book sorting application with intelligence computer vision technology. Libr. Hi Tech **39**(1), 220–232 (2021)
10. Jia, Y.: Research on the application of computer aided technology in graphic design visual aesthetics. In: Journal of Physics Conference Series. JPhCS (2021)

The Application of EDI in International Trade and the Security of Its Integration with Internet

XunShuang Sun(✉)

College of International Trade, Anhui Institute of International Business, Hefei 231100, Anhui, China
happybetty6923@126.com

Abstract. The main purpose of this study is to understand the role of EDI in international trade and its integration with the Internet. It will also study security issues related to these two technologies and their impact on economic development. It also has many branches, including chambers of Commerce, trade associations and professional organizations. To promote trade between Member States and the rest of the world through electronic means such as electronic data exchange or Internet-based trading systems. This paper summarizes the existing literature on this subject and discusses some key issues related to it, such as: (1) how to use EDI to protect data flow; (2) How to design security network; (3) What are the risks involved in using these technologies? This study aims to contribute to a better understanding of these important issues by proposing an analysis based on theoretical findings.

Keywords: EDI · international trade · INTERNET · Safety study

1 Introduction

EDI system can realize the automatic and safe exchange of business documents between enterprises without manual participation, reduce errors, realize high-speed and safe supply chain data exchange, increase the work efficiency of both sides of trade and establish trade trust.

Common application documents include: order, inventory query, invoice, delivery notice, receipt confirmation, sales report, material demand, etc.

It is an international standard introduced and used by the international organization for Standardization (ISO). It refers to a method of electronic transmission from computer to computer, which is a business or administrative transaction processing and forms a structured transaction processing or message format according to a recognized standard. It is also a business language that can be recognized by computers [1]. It is widely used in the commercial trade of developed countries. For example, the exchange of purchase order, packing list, bill of lading and other data in international trade.

At present, due to the difference of information systems between domestic enterprises, EDI has not been fully applied to all supply chain systems. In foreign countries, EDI technology and application have been very extensive. Each company can exchange

J. C. Hung et al. (Eds.): IC 2023, LNEE 1045, pp. 493–498, 2023.
https://doi.org/10.1007/978-981-99-2287-1_70

documents with all upstream and downstream trading partners through EDI system, greatly saving labor and time costs, simplifying supply chain inventory and improving revenue. We Zhixing is a software company specializing in the development and implementation of EDI system. The rssbus EDI system developed has independent intellectual property rights, and has realized the communication protocols and specifications required in all EDI industries. It has realized the direct supply chain connection with trading partners for more than 100 enterprises in China [2], including international trade, logistics, retail, chemical industry, etc. Based on this, this paper studies the application of EDI in international trade and the security of its integration with Internet.

2   Related Work

2.1   EDI

The full name of EDI is electronic data interchange, and its Chinese name is electronic data interchange, also known as "paperless trade".

EDI system is a software system that supports EDI international standard transmission protocol and can transmit international standard messages. The international standard EDI system usually includes three functions: EDI message transmission, EDI message translation/mapping and integrated business system. It realizes external connection with trading partners, internal file translation and system integration, and external to internal automatic data transmission.

EDI message transmission – transmits business documents with trading partners, supports any international standard transmission protocols, such as as2, oftp (2.0), and can transmit any format file. EDI message conversion/mapping - realizes the conversion between different format files, supports the conversion between EDI standard messages and user-defined format files, and integrates business systems – EDI systems are integrated with business systems, reducing manual operations, simple configuration, and can quickly integrate sap and other business systems. Figure 1 below shows EDI international trade.

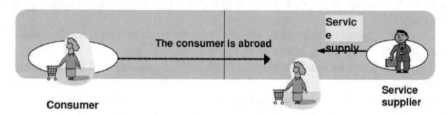

Fig. 1. EDI international trade

EDI complies with an international standard, which enables business data to be transmitted electronically from one business system to another through the network in a structured or standard message format, and enables computers to automatically transmit orders, invoices, inventory reports and other business documents in electronic form in a standard format. For example, a company can send orders, invoices and other business documents to another company through the edli system [3]. Similarly, it can also receive different types of business documents. The whole data exchange process is automatic, which reduces a lot of manual repeated operations and effectively improves the business processing efficiency.

EDI has been widely used in international trade, logistics, retail, medicine, electronics, chemical industry and other industries. The rapid development of EDI has affected the world. From individual retailers to international trade OEMs and international logistics hubs, EDI can be used for data transmission. Most domestic customers first contact or implement EDI in response to the requirements of foreign trading partners. With the deepening of trade cooperation at home and abroad and everyone's understanding of EDI technology, it has gradually changed from passive implementation to active promotion [4]. More and more domestic enterprises take the initiative to build EDI systems, connect with upstream and downstream trading partners, improve the automation level of enterprise data transmission, and improve the market competitiveness of enterprises.

2.2 Internet Convergence Security

The network security status of industrial control system (ICS) and its key element priorities, concerns and challenges were investigated. The background of the survey participants can be roughly divided into operational technology (OT) and information technology (it) professionals. Countries with developed network technology, relative to countries with backward network technology, occupy a dominant position in this field. Even a small event may cause a chain reaction at the national level.

In this context, international governance in cyberspace has experienced many twists and turns and achieved some results, but there are still some problems [5]. The multilateral complementary pattern of democracy, transparency and common development will be the general trend of international governance in cyberspace. In the evaluation report issued by the secure mobile application evaluation system, each evaluation item is described in detail, and the evaluation process is more detailed to the specific operation process, so that the reader can reproduce according to the operation process, eliminate the targeting, and make the evaluation system transparent, open and reasonable.

Although there are many important data in the report, it is undoubtedly more important to pay attention to the relationship between OT and it. Nearly 80% of the surveyed companies believe that the growing interconnection between OT and it is a challenge, mainly due to the digitalization of OT (especially industrial network), which may expose industrial systems and equipment that may not be adequately protected against network threats. It and ot teams usually have different security priorities and different system maintenance and improvement objectives. In addition, cultural differences and lack of communication between departments can also exacerbate this problem.

Credibility is a feature that has been used for many years to define the characteristics of it and ot systems. For it, credibility mainly involves security, reliability, privacy and resilience, while security is a lower priority. On the other hand, the credibility of OT mainly involves safety, reliability and elasticity [6]. Security is only handled in a small amount, and privacy is not within the scope of any ot. Solving the key system features missing in it and ot systems and focusing on the five key features of llot trusted paradigm will solve many it/ot integration problems, especially in terms of security, security and privacy.

3   Research on the Application of EDI in International Trade and Its Integration with Internet Security

To a certain extent, most of the main engine factories in international trade and their related enterprises in the supply chain have realized the use of automatic production equipment to replace a large number of manual operations on the assembly line, which has improved the industry competitiveness of enterprises. However, this is far from enough [7]. To improve the competitiveness of enterprises, it is not only necessary to improve the level of production automation, but also need to combine advanced communication technology, software, system services, etc. to achieve real-time information interaction, ensure the safe transmission of data, and improve the transmission efficiency of business information between enterprises.

At present, the business of major international trade OEMs is still expanding worldwide. In addition to traditional production line optimization, they also efficiently transmit data with upstream and downstream enterprises, which is also very helpful for enterprises to achieve cost reduction and efficiency increase.

The trade manufacturer BMW and the supplier use oftp2.0 transmission protocol to establish EDI connection channel, and use EDI system to transmit un/edifact standard messages. The main message types involved include: delfor delivery forecast, orders purchase order, desadv delivery notice, invoice, etc.

As shown in the overall solution in Fig. 2 below, BMW suppliers have SAP system [8]. In order to realize the automatic integration of EDI and SAP system, it is necessary to convert standard EDI messages into IDoc files through EDI system, and establish a connection with SAP system through simple interface configuration to improve the implementation efficiency.

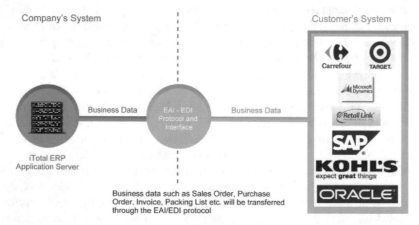

Fig. 2. EDI international trade solutions

After long-term practice and technical precipitation of major foreign trade manufacturers, EDI technology is recognized by the industry as an effective means to effectively help enterprises realize automatic information transmission with trading partners. EDI is the abbreviation of electronic data Interchange, which means electronic data exchange in Chinese. Enterprises exchange computer-to-computer business information in standard electronic file format, also known as "paperless trade". Next, let's take a look at how major international trade OEMs apply EDI!

At first, when the supplier receives the Tesla EDI demand email from the trading manufacturer, it is required to connect to EDI through van [9]. For the sake of data security, if you want to build an EDI platform on your own server without going through a third-party EDI service provider, you can communicate with Tesla by email and choose to connect directly to EDI. Tesla, a trading manufacturer, will reply by email that it can use as2 transport protocol to realize EDI direct connection.

Tesla, a trading manufacturer, establishes EDI system connection with suppliers, mainly involving five message types: 830 material demand forecast, 862 delivery plan, 856 delivery notice, 810 invoice and 824 application notice. If a supplier has an internal business system, it is the best choice to integrate the EDI system with the business system.

EDI connects with the trading manufacturer Berlin factory. When the trading manufacturer sends 810, the supplier that has established EDI connection with the trading manufacturer will continue to maintain the original connection channel, such as direct connection to as2 or van. For suppliers that have not established EDI connection, van is required.

Before EDI access, suppliers can generate shipping labels through the trading manufacturer Tesla portal [10]. Once the EDI system is online, the trading manufacturer Tesla will stop the label generation function of the portal. Therefore, suppliers need to find a label system that can generate shipping labels. The bridge of knowledge and practice has the ability to generate labels. The Zhixing label system can create content labels and create, add, delete, modify, and query 6J, 5J, and 1jlabels.

4 Conclusion

EDI studied in this paper is a system that allows the exchange of information between trading partners. The Internet is also used as a platform for electronic commerce and electronic data exchange. Electronic data interchange (EDI) is an electronic method to exchange business documents between companies or organizations through computer and telecommunication networks (such as the Internet). EDI allows enterprises to send standardized messages electronically, which can be processed automatically. These standards allow data to be automatically transferred from one trading partner to another by electronic means. In addition, there are standard formats for documents such as invoices and purchase orders, which can be automatically used with other companies in the network; This reduces the cost and time spent on paper.

Acknowledgements. This work was supported by Research on the competitiveness of service outsourcing industry in Hefei (SK2018A0952).

References

1. Seo, W.S.: A study on securing stability following the proposal and application of integration procedure following the diversification process of information security policies. J. Korea Inst. Electron. Commun. Sci. **13**(2), 405–410 (2018)
2. The application of business English in international trade (2021)
3. Toapanta, M., Xavier, G., Mafla, E.: Security model for the integration of the ministry of telecommunications and the information society with a public organization of Ecuador. In: 2020 the 4th International Conference on Information System and Data Mining (ICISDM 2020) (2020)
4. Liakopoulos, D.: Application and integration of principles and uses of international trade in regulatory systems. Revista CES Derecho **11**(1), 55–88 2020
5. Chen, R., Gao, X., Yu, H., et al.: The application of the integration of theory and practice in the professional course of automobile maintenance. Autom. Appl. Technol. (2018)
6. Murcia, A., Vargas, H., León, C.: International trade networks and the integration of Colombia into global trade. BIS papers chapters 100 (2018)
7. Laszuk, M., Šramková, D.: Challenges of Customs Law during the Paradigm of "Facility and Security" in International Trade. Białostockie Studia Prawnicze **26**(5), 9–21 (2021)
8. He, Z.: Discussion on the popularization and application of the standardization of international trade documents (2018)
9. Cerioli, A., Barabesi, L., Cerasa, A., et al.: Newcomb-Benford law and the detection of frauds in international trade. Proc. Natl. Acad. Sci. **116**(1), 106–115 (2018)
10. Bhat, A., Nor, R.M., Amiruzzaman, M., et al.: Methodology and analysis of smart contracts in blockchain-based international trade application (2021)

The Application of Ink Animation Based on Artificial Neural Network Style Transfer

He Xi[✉] and Hou Zhi

Xi'an Academy of Fine Arts, Xi'an 710065, Shaanxi, China
hexiyouxiang2021@163.com

Abstract. From the practical perspective of ink animation art, this paper aims to study the specific ways of artificial intelligence technology to intervene in the field of ink animation creation. Through the function introduction, model optimization and parameter adjustment of the image style migration algorithm, the creation idea of the animation image style migration for the light dormitory ink effect was explored, and we conducted detailed experiments. The verification results show that the method proposed in this paper can effectively meet the creation needs of ink animation, and provide technical support and thinking innovation for the ink animation creation in the era of artificial intelligence.

Keyword: ink animation creation of artificial intelligence image style migration of night ink

Ink animation is the most representative type of "Chinese school" animation, its vivid ink image, hazy meaningful such as poetic artistic conception, interpret the rich and profound aesthetic connotation of the Chinese nation, in the style of the world animation stage harvest high praise, become an important milestone of Chinese animation "nationalization road". Ink animation was already created in Shanghai Art Film Studio in 1960, but due to its complicated technology and high cost, only a few classic works were produced, such as "Tadpole Looking for a Mother", "Shepherd Flute" and "Landscape", and then experienced ten years of stagnation. With the rapid development of computer imagery, non-realistic rendering has brought opportunities for the revival of ink animation. Especially in recent years, artificial intelligence technology has triggered a revolutionary breakthrough in the field of non-realistic rendering, and expanded the new ideas for dealing with computer vision problems. Therefore, the targeted guidance of artificial intelligence technology to intervene has become a new opportunity to solve the current predicament of ink painting animation creation.

1 Background Overview

(1) Principle and development of the image style migration algorithm

Leon A, 2015 Leon A. Gatys and other scholars [1] creatively proposed using convolutional neural networks for image-style migration learning, which broke through the

J. C. Hung et al. (Eds.): IC 2023, LNEE 1045, pp. 499–504, 2023.
https://doi.org/10.1007/978-981-99-2287-1_71

previous nonparametric image migration methods limited to physical models. The so-called style migration refers to the process of the computer by extracting the source image texture as the style expression, and then using the style to reconstruct the target image. Gatys et al. use the pre-trained VGG model to separate the image features, quantify the features through the Graham matrix, and materialized the abstract artistic concept of "style", which is equivalent to algorithmically coding the artist's personal style, so as to simulate the artist's individual creation behavior.

Based on the ideas of Gatys, in 2017, Justin Johnson and other scholars [2] proposed the style migration method based on the iterative optimization generation model, which effectively improved the computational efficiency of image style migration, known as fast style migration. In the process of rapid style migration, an exponential improvement in real-time generation efficiency is achieved because the main computational tasks have been completed in the model learning stage. Although the rapid style migration is still limited by weak controllability and high style expansion cost in the application, the creative needs of ink painting animation can be met.

We try to apply artificial neural network in the field of ink animation creation, using fast style migration to realize the freehand ink painting effect of image expression, experimental results show that the method can automatically generate meet the requirements of specific ink art style animation video paragraphs, significantly improve the efficiency of ink painting animation creation, broaden the implementation of ink animation creation.

2 Thoughts of Ink Animation Creation Based on Image Style Migration

In the creation process of traditional animation, the mid-term drawing part is the difficulty of the traditional ink painting animation production process, and it is also the part with the longest cycle and the most costly labor cost. This paper tries to integrate the mid-term and some late processes of ink animation based on image style migration, so as to replace the tedious and repetitive labor in traditional production.

(1) The style setting and analysis of ink painting

In the ink language, "night ink" refers to the old ink standing still for a long time. As a unique style and technique in traditional Chinese painting, it has an indissoluble relationship with ink animation. In 1989, the ink painting animation Wu Shanming "Landscape" produced by Shanghai Art Film Studio invited the Chinese painter to design it, presenting a unique style of light sleeping ink, which complemented the humanistic spirit of ink animation. This paper chooses the style migration experiment of ink animation in the style of light dormitory ink, which also includes the following two reasons:

First of all, from the perspective of artistic expression, light night ink and ink animation creation are more suitable. Light night ink emphasizes the simple line modeling, reducing the dependence on color, clumps and other modeling techniques. These characteristics are consistent with the craft tradition of hand-drawn animation refining lines for the original painting creation, which is conducive to the integration and translation of the style of the animation art.

Secondly, from the perspective of style extraction, the good artistic recognition of this style is conducive to the feature identification and effect test of computer style migration. Its visual features include: first, the color characteristics with low contrast and high transparency; second, the light lodging ink is mostly refined with rich and clever short term, with written structural features; third, the light lodging ink forms a faint trace and obvious radioactive water stains, with ethereal and subtle hazy beauty.

(2) Style migration network model and optimization

The fast style migration model adopted in this paper is composed of two neural networks: image generation network and loss function network. According to the characteristics of ink animation creation technology, the traditional rapid migration model is optimized below:

First, the learning rate, as an important hyperparameter in the model training, determines whether the target function can converge to the local minimum value and its speed, and setting the appropriate learning rate can effectively improve the model training efficiency. In this experiment, the exponential learning rate was used instead of the fixed learning rate to avoid excessive shock when the loss function was convergent late in training. The initial learning rate was set to 0.001 and the base number to 0.9, and it was set to be one epoch per 50 steps: $Lr = 0.001 \times 0.9^{epoch}$

Second, introducing PSNR and SSMI values in the validation module to quantify the migration effect of the generated images. The PSNR value of this experiment was calculated from generated and content maps and used to quantify the content difference between generated and content maps; the SSMI value of this experiment was calculated from generated and style maps, which were used to quantify the style difference between generated and style maps.

Third, in order to meet the needs of animation production, the OpenCV cv2 library adds the video verification module to the model, and input the picture frame sequence or video during the validation. The model can generate the video files directly after the style migration, and customize the frame rate of the output video.

3   Experimental Process and Results Analysis of Ink and Wash Image Style Migration

(1) Experimental environment

This experiment was carried out under the 64-bit windows10 Professional Edition operating system, building network models using Pycharm software, Python3.6 language and PyTorch deep learning framework, training and validation using NVIDIA Quadro RTX 3000 GPU, and visually presenting the learning rate and loss function during training using tensorboard. The dataset selected the COCO dataset released by Microsoft (2014), with the training size of 400 * 400 pixels, and more than 600 animated drafts and pictures suitable for ink animation were selected as the verification set.

(2) Parameter debugging

1. Determine the content loss and style loss feature layer

   Through the conv2_2 layer of VGG19 model, the feature map that can better retain the image structure and semantic information, so that the edge contour of the object is richer and meets the elegant and clever line characteristics of light dormitory ink. Therefore, the conv2_2 layer is used to calculate the content loss function;

   As for the style feature layer, style features output from different convolution layers. The first three feature layers used in this experiment in constructing the style loss function were consistent with the original fast style migration model, while the fourth layer compared the results of the conv4_1 and conv4_3 outputs. The generating network was trained for 1,000 iterations with the 1:0.2 content and style weight ratio. Compared with the effect of the relevant feature layer, when using the conv4_1 layer, the excessive noise details at the edge are effectively omitted, and the integrity of the picture is stronger.

   Therefore, this experiment uses the conv1_1, conv2_2, conv3_3, conv4_1 layers in the VGG19 network.

2. Determine the weight ratio of content and style

   The generating network was trained in 1,000 steps at 1:0.1,1:0.2, and 1:0.4.The test found that when the content and style weight ratio is 1:0.1 and 1:0.4, the style characteristics and object structure are not obvious, and when the content and style weight ratio is 1:0.2, the generation map can better retain the content information of the content map, which also reflects the dim and elegant style, so the final selected content and style weight ratio is 1:0.2.

3. Determine the weight ratio of each style characteristic layer

   The style characteristics of different convolutional layers in VGG net-work are different, so the different weights of different feature layers and the proportion of different feature layers help to better restore the ink style.

   Four o n v c o 1 _ 2, conv2_2, conv3_3 and conv4_1 in VGG network to construct the style loss function, set the content and style weight ratio to 1:0.2, and conduct 1000 step training on the generating network. The test found that in the conv3_3 layer, the picture of the migration map is mainly composed of color blocks with light ink effect. Although the lack of structural contour bars, the effect of this layer is closest to the dark ink style. Using the validation set of the generative network trained using four feature layers, comparing the mean PSNR and SSMI, from conv1_2 to conv3_3 layers, as the network layer deepens, PSNR value decreases, and SSMI value gradually increases, indicating the drawing style gradually close to the style drawing, consistent with our direct observation; conv4_1 layer to indicate the content structure with more obvious contour, reducing the difference from the content map, but also a higher style.

   Through subjective judgment and objective scoring, it can be seen that with the deepening of the network, the overall ink painting effect of the generated graph has changed from wet to dry, from block surface to hook line. The overall style also goes from closer to the content map to closer to the style chart.

According to the experimental results, after multiple ratio tests, the weight of each characteristic layer was finally set to 0.2:1.9:1.5:0.4, respectively.

(3) Results analysis

After determining the V G G feature layer structure, content, style weight ratio and ratio of 19 network, the model was trained in 2000 steps and the generated network was validated with the test set. According to the two main creation contents of characters and scene design in ink animation, the generation effect of hand-drawn line draft and actual shot scenery is selected for comparison, as shown in Fig. 1:

内容图　　　　　　　　生成图　　　　　　　　生成细节

Fig. 1. Generates the results

It can be seen that in the processing of hand-drawn draft, the trained migration model can increase the change of the contour lines, make the original uniform and smooth lines show the sense of calligraphy stroke, and show the hollow lines and seepage ink brush characteristics; it can properly spread the light ink in the picture, adding a rich sense of hierarchy to the monotonous black and white draft. For landscape real photos, model can accurately identify the object contour, to short-term structure, the original complex realistic color into the ink style of light ink effect, weaken the strong light contrast in the original photos, and the picture for a certain degree of abstract and blank, generated with the combination of artistic texture.

Epilogue

This paper optimizes the fast style migration model on the production requirements of ink animation, and demonstrates the influence of different network feature layer structure, content and style weight ratio, and weight ratio of each style characteristic layer on the migration results through specific experimental training. The model trained after parameter debugging was used to migrate the hand-drawn line draft and landscape real photos of the characters in the validation set, and to explore the unique techniques of animation style migration for the effect of light dormitory ink. From the perspective of the specific practice of ink painting animation art creation, this paper conducts in-depth exploration and artistic extension of the image style transfer algorithm based on the field of artificial intelligence, which is conducive to enrich the aesthetic connotation of ink painting animation art and inject fresh scientific and technological vitality into the comprehensive revival of the "Chinese school" animation.

Acknowledgements. Practical research on 'immersion experience' of animation from the perspective of somaesthetics (NO: 20JK0258).

References

1. Gatys, L.A., Ecker, A.S., Bethge, M.: A neural algorithm of artistic style. J. Vis. (2015). 1508.06576
2. Johnson, J., Alahi, A., Fei-Fei, L.: Perceptual losses for real-time style transfer and super-resolution. In: Leibe, B., Matas, J., Sebe, N., Welling, M. (eds.) ECCV 2016. LNCS, vol. 9906, pp. 694–711. Springer, Cham (2016). https://doi.org/10.1007/978-3-319-46475-6_43

The Construction of Computer Education Resource Platform Based on Personalized Recommendation Algorithm

Qian Chen[✉]

Big Data School of Fuzhou University of International Studies and Trade, Fuzhou 350202, Fujian, China
353192067@qq.com

Abstract. Computer education resource platform based on personalized recommendation algorithm, which is a network-based platform, allows users to create, share and access educational resources, such as videos, podcasts, books and so on. It is designed to improve the learning effect by providing personalized suggestions according to users' interests. Educational resources can be recommended according to users' needs and preferences. It uses machine learning algorithms to analyze user behavior and preferences, and also provides various tools to help teachers and students create their own content and share it with others. Then recommend relevant educational resources for them. The algorithm also allows users to rate the recommended content to improve its quality.

Keywords: Personalized recommendation · computer · Educational resource platform

1 Introduction

In the current wave of information society development driven by information technology, the development of educational informatization, which aims to cultivate talents to meet the requirements of the information society, is in full swing. Theoretically, educational informatization can be said to be a comprehensive reflection of the "information society" theory and the "modernization" theory in the field of school education. Inkles once pointed out that "human modernization is an indispensable factor in national modernization…. Without the modernization of education, there will be no human modernization" [1].

Under the guidance of modern educational thought, the informatization of teaching management in Colleges and universities is to use information management theory and information management methods, take modern information technology as the core technology, fully consider external variables and information, organize and allocate teaching information resources, and carry out informatization teaching management activities, so as to achieve the established teaching goals efficiently [2]. In other words, college education informatization mainly uses information technology to improve the efficiency of

© The Author(s), under exclusive license to Springer Nature Singapore Pte Ltd. 2023
J. C. Hung et al. (Eds.): IC 2023, LNEE 1045, pp. 505–511, 2023.
https://doi.org/10.1007/978-981-99-2287-1_72

college management process and change the organization mode of management, accelerate the transmission and feedback process of management information, and finally improve the operation efficiency of college management.

College education informatization includes: teaching management informatization, scientific research management informatization, personnel management informatization, student management informatization, experimental equipment management informatization, administrative office and public service management informatization, etc. Using network technology, computer technology and communication technology to manage all kinds of school information resources in a comprehensive, scientific and standardized way, and to integrate and integrate these information resources is the focus of the construction of university education informatization.

The common fields of recommendation system include traditional retail industry, Internet industry such as e-commerce Amazon, Dangdang, social network service (SNS) Facebook [3]. However, with the increasing demand for learning and information, the public's dependence on online education resources is also increasing, and online information education has become a new trend. It can be predicted that the application of recommendation system in teaching assistant system will be more and more in the future. Educational informatization uses new technical means to comprehensively strengthen the management function, leading to the reasonable reorganization of management business processes and the transformation of management institutions, so as to further integrate various interrelated management functions, actively promote the scientific and democratic management, and further promote the construction of university management team [4].

2   Related Work

2.1   Research Status Abroad

After practice, the recommendation system has been very mature in some fields, and the feedback is good. According to the application fields of the recommendation system, it can be divided into: e-commerce recommendation, film and video recommendation, music recommendation, social recommendation, reading news recommendation, email recommendation, educational resources recommendation. The development of foreign recommendation systems is as follows:

(1) E-commerce recommendation: Amazon is an active user and user in the field of recommendation system, and has been rated as the "king of recommendation system" by RWW. Its recommendation application is mainly personalized product list and recommendation list of related products. It also collaborates with Facebook to make friend based recommendations [5]. Among them, the related recommendation purchase also purchase, browse also browse, and package sales mainly use collaborative filtering algorithm and content-based recommendation algorithm.

(2) Movie and video recommendation: netflic movie recommendation includes the title poster of the recommended movie, giving the recommendation reason and user feedback module. After using the item based recommendation algorithm, youtube,

the largest video website in the United States, found that the click through rate of personalized recommendations is twice that of popular videos.

(3) Music recommendation: the internationally famous music websites are Pandora and last fm。 Pandora's recommendation algorithm is mainly based on tag recommendation. Experts mark the melody, rhythm, arrangement and lyrics of the song, and calculate the similarity of the song based on these music genes [6]. Last. FM uses item based collaborative filtering combined with social networks to work together to recommend songs.

(4) Social recommendation: the most successful are Facebook and twitter. It can not only recommend friends, use edgerank for conversation recommendation of information flow, but also launch API for article recommendation.

(5) Read news recommendations: Google reader allows users to follow people they are interested in and see articles shared by users. Zite collects users' preferences and interests for articles, and uses collaborative filtering recommendation based on items. Digg news reading website uses user based collaborative filtering algorithm.

2.2 Domestic Research Status

Taobao, jd.com, etc. are well-known in the field of e-commerce recommendation in China, including popular recommendations, relevant recommendations, buy and buy, read and see. In the field of film and video recommendation, Youku and iqiyi are mainly popular video recommendations and tag based related recommendations. In the field of music recommendation, for example, cool dog music is classified according to the melody, composition and lyrics of music, and the similarity of songs in content is calculated. In the field of reading news recommendation, Douban's book reading recommendation, Baidu's popular news and related news recommendation are better [7].

The development of recommendation system for educational resources in China: first, it started late, and second, it has poor user satisfaction. In the past decade, imitating the intelligent recommendation system of foreign education, some educational resource websites have also appeared in China, such as Li Junwei's research on personalized recommendation of campus educational resource network based on collaborative filtering in 2009. In 2015, the design and implementation of jinzhifu's personalized recommendation system of educational resources based on big data of Chinese Academy of Sciences. Hu Xiaonan 2010 research on personalized recommendation of learning content based on knowledge points. Design and implementation of Chaihua 2015 Personalized Education Resource Recommendation System of Beijing University of technology [8]. However, most of them are immature and cannot provide good recommendation services to learners, and even confuse the learning direction of learners due to the messy classification of educational resources. Many educational resource websites and distance education websites have not realized personalized recommendation. In recent years, experimental results of adaptive teaching and adaptive recommendation of educational resources have also begun to appear in experimental centers of some universities. However, its recommendation efficiency is not high enough, and the research content is not landing enough, resulting in poor social promotion.

In general, the development of domestic recommendation system is also strong in the field of e-commerce and weak in the field of education. However, this also shows that the promotion of domestic educational resources needs to be developed, and there is also a large space for development. And should focus on the research of recommended technology.

3   Construction of Computer Education Resource Platform Based on Personalized Recommendation Algorithm

3.1   Personalized Recommendation System Algorithm

There are many kinds of recommended algorithms, but each algorithm has its own advantages and disadvantages, and also has its own limitations. Therefore, we should measure it before making decisions. It is necessary to understand the concept and working principle of these algorithms and have an intuitive impression of them. In practice, we may need to test a variety of algorithms to find the most suitable one for users. The principles and techniques of several typical recommended algorithms will be introduced below.

(1)   Memory based collaborative filtering recommendation algorithm

Collaborative filtering recommendation algorithm is to recommend content that conforms to users' interests according to the content of items or the relevance of users. The input content only depends on the data used, such as evaluation, purchase, download, user preferences, etc. [9]. Its advantage is that it does not need to have a deep understanding of domain knowledge and the characteristics of users and items. In most cases, the results are satisfactory enough. The disadvantages are: cold start problems, products that need standardization, high demand for user to item ratio of 1:100, will be affected by popularity, and the long tail performs poorly, so it is difficult to provide explanations. The collaborative filtering recommendation algorithm is subdivided into two categories, and its structure is shown in Fig. 1:

The memory based recommendation method takes each item or user as a vector and calculates the similarity of all other items or users with it by performing the nearest neighbor search. Find the relationship between items, that is, item based collaborative filtering between items or the relationship between users, that is, user based collaborative filtering between users [10]. With the pairwise similarity between Item or user, prediction and recommendation can be carried out. Model based recommendation mainly adopts statistics, machine learning, data mining and other methods, and adopts different algorithm models according to users' historical records to generate recommendations for users. The most common method is matrix factorization.

$$P(i,j) = \overline{R}_i \frac{\sum_{n \in Ni} sim(i, N) \cdot (R_{n,j} - \overline{R}_N)}{\sum_{n \in Ni} sim(i, N)} \tag{1}$$

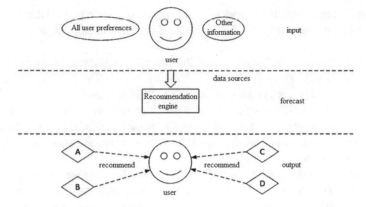

Fig. 1. Principle and structure of personalized recommendation system

(2) Content based recommendation algorithm

The recommendation idea of Itemcf is to recommend items similar to their favorite items to target users. The similarity of items is based on the fact that items appear in user behavior records at the same time.

The iltemcf algorithm consists of two steps:

Step 1: calculate the similarity between items.
Step 2: generate an item recommendation list with high similarity to the items that the target user likes.

The recommendation idea of content-based recommendations (CB) is to recommend items similar to the items that users liked in the past (collectively referred to as items in this article).

The input content only depends on the content or description of the item and the user, and does not include usage data. Advantages of CB: user independence, good transparency, new item can be recommended immediately, and there is no popularity preference. The disadvantages of CB are that it is unable to mine users' potential interests (over specialization) and generate new user problems for new users.

3.2 Computer Education Resource Platform

The topic recommendation system of computer education resources is a system developed based on the campus cloud management platform system. The campus cloud platform management system is a comprehensive campus management platform for school student performance, student daily management, teacher file management, and parent online communication management. As a part of this system, our educational resource topic recommendation system is nested in this system. The main functions of the system are as follows:

(1) Initiate examination: it is edited and filled by the teacher who initiated the examination. Enter the test name, fill in the start and end time of the test, and click "initiate test".

(2) Question type knowledge points and difficulty Management: teachers can add and delete knowledge points, and provide search function. The difficulty of the topic is counted and managed, and the search function is provided.

(3) Personal achievement check: it mainly provides students with the ability to check the monthly examination results of various subjects, and check the test papers and standard answers. Students can check their total score, the score of each question and their answers.

(4) Student performance analysis and report: it mainly provides students to view the number of knowledge points lost in a certain course and an exam, the corresponding loss of knowledge points, and to view the arrangement of knowledge points and the scores occupied by knowledge points in the whole exam paper. The score proportion of knowledge points in the analysis of the whole test paper is displayed in the form of a histogram.

(5) Topic recommendation: recommend topics with similar content and the same difficulty according to the students' examination and practice mistakes.

Thus, the overall architecture of the education resource topic recommendation system is summarized, as shown in Fig. 2:

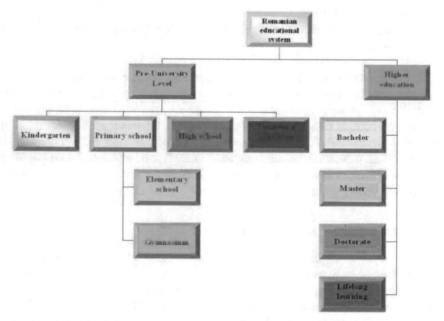

Fig. 2. Structure of computer education resource platform

4 Conclusion

Based on the analysis of computer education, this paper studies an education platform with reasonable computer education resource management strategy, education mechanism and education concept by using personalized recommendation technology. Combined with the actual situation of the investigation and analysis and the impact of the complex environment on education in the information age, on the basis of giving full play to school education, provide a reasonable, scientific and healthy online education platform, expand the space and time of family and student education, enrich the form and content of Education, and maximize the development and utilization of educational resources from all parties. Improve the modern education system to promote the comprehensive and healthy development of students' body and mind in the new era, guide the orderly development of computer education, strive to build a harmonious and healthy social environment for the growth of students, and escort the healthy growth of the educated.

References

1. Sun, M., Zhang, J.: Research on the application of block chain big data platform in the construction of new smart city for low carbon emission and green environment. Comput. Commun. **149**, 332–342 2020
2. Wang, C., Sha, Z., Jia, L.: Research on digital resource system construction of smart library based on computer network and artificial intelligence. J. Phys. Conf. Ser. **1952**(4), 042018 (2021)
3. Liu, L.: Research on the cultivation of applied talents and the construction of ideological and political education based on computer. J. Phys. Conf. Ser. **1744**(3), 032163, 5 p. (2021)
4. Wang, Z., Wei, J., Feng, J., et al.: Research on the construction of the general control platform for yunnan rare and precious metal materials genetic engineering. J. Micromech. Mol. Phys. (2020)
5. Peng, J., Wang, M., Peng, C., et al.: Research on extremely short construction period of engineering project based on labor balance under resource tolerance. PLoS ONE **17** (2022)
6. Xiang, X.: The construction of research network management platform based on BS structure (12) (2022)
7. Xu, J.: Research of online courses construction based on the theory of interaction. J. High. Educ. Res. **3**(1), 26–28 (2022)
8. Liu, R., Shao-You, L.I.: Construction and research of numerical control technology training room based on skill competition platform. Sci-tech Innov. Prod. (2020)
9. He, Z.: Research on the construction of teaching information platform in local normal colleges with the mode of "internet + education" (2020)
10. Feng, M.: Research on the construction of student ability evaluation system based on computer application. J. Phys. Conf. Ser. **1915**(2), 022037, 5 p. (2021)

The Construction of Shaanxi Intangible Cultural Heritage Translation Platform Based on Internet

Lu Zhang[✉] and Cong Wang

School of English Language and Literature, Xi'an Fanyi University, Xi'an 710105, Shaanxi, China
liu396273633@163.com

Abstract. The research on the construction of Shaanxi intangible cultural heritage translation platform based on the Internet is a research project combining basic research and applied research. It is to establish a database of intangible cultural heritage resources and related information, so that users can search for information in the database. The project team will also provide services, such as searching information about intangible cultural heritage resources, consulting experts and researchers, and providing suggestions on how to protect these resources. This study aims to explore new ways of translating Shaanxi intangible cultural heritage, and provide information support for local governments, scholars, researchers and other parties concerned about the protection of Shaanxi intangible cultural heritage. This study aims to provide an effective method for Internet technology research, analysis, evaluation and promotion of intangible cultural heritage protection in Shaanxi. It is also an important part of the intangible cultural heritage protection mechanism.

Keywords: Shaanxi intangible cultural heritage · Translation · Platform construction · internet

1 Introduction

Intangible cultural heritage is the full name of intangible cultural heritage. Intangible cultural heritage refers to various forms of traditional cultural expressions (such as folk activities, performing arts, traditional knowledge and skills, as well as related utensils, objects, handicrafts, etc.) and cultural spaces that are inherited from generation to generation by people of all ethnic groups and are closely related to people's lives. Intangible cultural heritage refers to the folk cultural and artistic heritage with national historical accumulation and wide representation, mainly by oral transmission [1]. Culture is an important foundation and basis for a nation to stand in the forest of nations in the world, and intangible cultural heritage (hereinafter referred to as intangible cultural heritage) is a more concentrated and specific expression of national culture.

From a historical point of view, the inheritance and protection of intangible cultural heritage in Shaanxi is the only way to continue the 5000 year old excellent culture of the Chinese nation. Shaanxi intangible cultural heritage inheritors are known as "living treasures of mankind". Due to their diversity, Shaanxi intangible cultural heritage

J. C. Hung et al. (Eds.): IC 2023, LNEE 1045, pp. 512–517, 2023.
https://doi.org/10.1007/978-981-99-2287-1_73

in China has different requirements for inheritors. Some Shaanxi intangible cultural heritage inheritors need to have basic qualities, but also have special requirements in appearance characteristics, body shape, sound characteristics, etc., and there are very strict standards for the evaluation of inheritors. The inheritors of Shaanxi intangible cultural heritage need to have enough talent, have the spirit of hard work and hard study, and be good at learning and understanding. The high requirements of comprehensive quality make it more difficult to obtain Shaanxi intangible cultural heritage inheritors, which are already very scarce. The possibility of complete inheritance of Shaanxi intangible cultural heritage is reduced, and the risk of no successor is increased. Shaanxi intangible cultural heritage inheritance is facing the crisis of fault, which ultimately reduces the richness of Shaanxi intangible cultural heritage in China and makes it difficult to continue the inheritance of traditional culture [2].

Chinese culture has a long history, and Shaanxi intangible cultural heritage is diverse and rich. The protection and inheritance of Shaanxi intangible cultural heritage in China is the key to retain and carry forward the root and soul of the Chinese nation. In the context of "Internet+", the research on the construction of Shaanxi intangible cultural heritage translation platform, the realization of the new development of Shaanxi intangible cultural heritage protection and inheritance, and the giving of new era characteristics to Shaanxi intangible cultural heritage is of great significance to improve the international influence and national cohesion of our culture.

2 Related Work

2.1 Intangible Cultural Heritage Protection Under the Internet

In the past, the dissemination and protection of intangible cultural heritage are faced with a presentation problem, that is, how to express or reproduce, how to make people feel. In the past, the main presentation of traditional culture was transmitted to the public through oral, printed paper, stage performance, video, audio and video, but in the digital era, due to the development and development of new media and various network applications, the presentation has also been constantly innovated. The new presentation of cultural heritage in the digital media environment has two characteristics: in terms of carriers, one is the multi-mode presentation of digital media. In the era of digital media, in addition to the coexistence of oral communication, printing paper, stage, film, music and other traditional media, intangible cultural heritage can also be spread through mobile phones, networks, outdoor advertising screens, digital magazines, digital radio, digital television, etc., of which the Internet is a particularly important digital media. Through these digital media, traditional culture appears in the form of text, pictures, audio, video and their combinations (such as virtual reality VR, augmented reality AR, source reality MR). Intangible cultural heritage can be reorganized and arranged [3]. The public does not have to receive information step by step, but can jump between cultural information according to their own needs. Because this information organization method is more in line with the characteristics of the human brain, it will be more conducive to the presentation of intangible cultural heritage information. Digital media has the experience ability to reproduce the intangible cultural heritage completely. Now ordinary multimedia technology has been able to combine vision and hearing well. The three-dimensional sensing devices used in virtual reality technology have also been able to track the changes of actions, and digital reproduction and display have promoted cultural communication.

Through the Internet three-dimensional scanning technology, digital video technology, digital imaging technology and other new digital technologies, the information carrying intangible cultural heritage can be obtained and preserved with high precision. The purpose is to use the prepared digital material network platform for information sharing, protection, repair and in-depth research.

2.2 Intangible Cultural Heritage Translation Platform Technology

The uploading and downloading of translated documents and corpus of Shaanxi intangible cultural heritage translation platform will involve the storage of large files. In order to reduce the pressure of web server and the disadvantage of limited concurrency of web server, the traditional storage method adopts to separate the file server and place the files on the FTP server. All the uploading and downloading of files adopt the FTP service provided by nginx, which really reduces the pressure of web server, To a certain extent, it provides the fault tolerance of the platform to save corpus, but after testing, this architecture is not high enough for the fault tolerance and parsing efficiency of large files, so the platform adopts the hadoop+spark scheme. Hadoop provides the ability of cloud storage [4]. The storage of large files will be divided into blocks by Hadoop. Machines in different Hadoop clusters back up data blocks with each other. Even if the data block in a machine is lost, it will not affect the access of translators to the data. Due to the addition of Hadoop cluster monitoring, problems in the machine can be found in time through the monitoring software. Spark uses streaming computing, It is very convenient to analyze the corpus files on Hadoop, and then put them into es search engine in batches. For translators.

There are many places in Shaanxi intangible cultural heritage translation platform that use caching. In terms of storage, the mybatis framework is used for database access. Compared with the traditional writing of SQL statements to access data, the data queried by SQL statements cannot be cached. Mybatis framework provides the mapping between Java objects and relational data, and operating Java objects is equivalent to operating relational databases. Mybatis provides a level-1 cache and a level-2 cache. The level-1 cache is a sqlsession level cache. As long as the session is not closed and DML statements are not used, the same data will be obtained from the level-1 cache when the program accesses it again. If the first level cache does not get data, there is also the second level cache. The second level cache is called the mapper level cache [5]. The scope of the cache is larger than that of the first level cache, and it is cross session level. Even if the session is closed, the program can access the data in the second level cache. The first level cache in mybatis is on by default, and the second level cache is not on by default, so it needs to be manually set in the mybatis configuration file. Key statements <setting name = "cacheenabled" value = "true"/>, and set the cache policy in the namespace.

3 Research on the Construction of Shaanxi Intangible Cultural Heritage Translation Platform Based on Internet

In this paper, Shaanxi intangible cultural heritage translation platform can be divided into client, server, database server, elasticsearch search engine server and load balancing server. Figure 1 is the overall topology of the system:

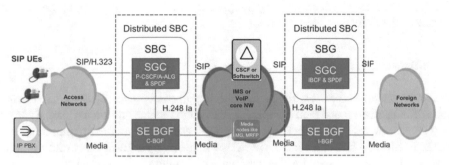

Fig. 1. Overall system topology

The form, content and value of intangible cultural heritage will mutate in the process of spreading. The migration of all nations in the world, large or small, will bring cultural variation. The dissemination of intangible cultural heritage will also get a new interpretation of its significance after different inheritors communicate with different audiences. There are a thousand Hamlets for a thousand spectators. People's cognition of the same thing often produces deviation. Different audiences will modify the prototype to meet their own needs. These changes have become the source of intangible cultural heritage [6]. In the process of translation, the deviation of translators and readers will not have a great impact on intangible cultural heritage.

"Intangible cultural heritage translation" Shaanxi intangible cultural heritage translation platform is by no means a simple extension of the cultural heritage exhibition hall, nor a mechanical Porter of intangible cultural heritage projects. It is a cultural website based on digital virtual museum technology to disseminate and protect Shaanxi intangible cultural heritage projects. Based on the intangible cultural heritage information collected by professional digital image and video technology and centered on the digital network platform, this website has become the transmission carrier of Shaanxi endangered intangible cultural heritage in the new environment. Compared with traditional cultural websites in various regions of Shaanxi Province, the design concept of "intangible cultural heritage translation" website is more advanced, which is more conducive to the inheritance of endangered intangible cultural heritage in Shaanxi Province, and breaks the barriers in time and space when protecting intangible cultural heritage in various regions [7]. The "intangible cultural heritage translation" website takes the digital way as the technical means, relies on the information dissemination theory, and protects the promotional language of intangible cultural heritage projects and successors through the Internet, eliminating the distance between users and various intangible cultural heritage projects in time and space. Users can know relevant information about Shaanxi intangible cultural heritage projects and successors anytime and anywhere through the Internet. According to the survey, there are no cultural communication websites focusing on the information of intangible cultural heritage inheritors in Shaanxi at present [8]. The emergence of "intangible cultural heritage translation" Shaanxi intangible cultural heritage inheritor website makes up for this gap. Intangible cultural heritage inheritors are a very important part of the dissemination and protection of intangible cultural heritage projects. Shaanxi intangible cultural heritage database only provides simple information for inheritors, which cannot play a higher-level role in publicity and protection for inheritors themselves.

4   Simulation Analysis

Shaanxi intangible cultural heritage translation platform is specially designed for small and medium-sized translation companies, translation teams and translators, aiming to solve the problems of low efficiency and difficult coordination of traditional translation methods. Shaanxi intangible cultural heritage translation platform is not only an online auxiliary translation system for Shaanxi intangible cultural heritage translation, but also a good partner for translation practitioners.

This system uses springmvc to build Java Web applications [9]. The main idea is that the springmvc framework provides a dispatcherservlet as the front-end controller to dispatch requests, while providing flexible configuration handler mapping, view parsing, locale and topic parsing, and supporting file uploading. The following focuses on the implementation process of springmvc separation controller.

Configure controller based on controller interface

Dispatcherservlet acts as a front-end controller in springmvc. Its core function is to distribute requests, which will be distributed to the corresponding Java classes, which is called handle in springmvc. The code is shown in Fig. 2 below:

```
@RestController
@RequestMapping("/corpus")
@Api(description = "语料导入、导出接口")
public class CorpusController {
    @Autowired
    private ICorpusService CorpusService;
    @ApiOperation(value = "导入记忆或术语", produces = "application/json")
    @ApiResponses(value = {@ApiResponse(code = 400, message = "验证错误") })
    @RequestMapping(value = "/parse", method = RequestMethod.POST)
    public SimpleResult parseCorpus(int corPusId,  String filePath, String
labelList,String sourceLanguageKey, String targetLanguageKey, int corpusType,
String lineid, int operatorId) {
     if (CheckParam.checkIsBlank(filePath, labelList, sourceLanguageKey,
     targetLanguageKey)) {
       return   SimpleResult.error ("参数有误"));
     }
```

Fig. 2. Translation code

Execute SQL script in MySQL database to complete the operation of creating database and table. In this system, the database is operated in an object-oriented manner, and the database is directly added, deleted, and modified through Po (persistent object) [10]. At the same time, mybatis is responsible for converting this operation into the operation of formulating database tables.

5   Conclusion

The construction of Shaanxi intangible cultural heritage translation platform based on the Internet is to establish a digital interactive network for the research and development of Shaanxi intangible cultural heritage, provide a convenient channel for information exchange and sharing among researchers from different countries, promote international cooperation in this field, and further strengthen the protection and promotion of intangible cultural heritage. In order to achieve these goals, we will develop an integrated database system to integrate historical documents, photos, video materials and other related resources into a unified organization through online search services; Develop an online resource center with integrated search capabilities.

References

1. Zhang, L., University S N: Research on the protection of national intangible cultural heritage based on cultural community. Guizhou Ethn. Stud. (2019)
2. Tian, Y.Y., Humanities, S.O., University S: On the C-E translation of culture-specific items in publicity texts of Shaanxi intangible cultural heritage. J. Hubei Univ. Educ. (2018)
3. Qi, X.: Research on the construction of unstructured data center based on cloud storage in universities——taking Shaanxi normal university as a case (2018)
4. Zhang, J., Liu, Y.: Research on translation strategy of intangible cultural heritage publicity based on internet information technology. J. Phys. Conf. Ser. **1648**, 022115 (2020)
5. Huang, W.B., Zeng, R.M., Xiang, H.D., et al.: Visual communication design of intangible cultural heritage based on Internet+ **9**(5), 5 (2019)
6. Jia, X.: Exploration of the redevelopment strategy of traditional handicraft industry under the background of "intangible cultural heritage" protection (2018)
7. Zhang, F.: Research on the construction path of the first-class major of journalism in Shaanxi universities under the concept of educational certification (2019)
8. Zhang, H., Art, D.O., University S: The exploration of Shaanxi intangible cultural heritage as curriculum resources in preschool art education. J. Shaanxi Xueqian Norm. Univ. (2018)
9. Yiting, R., Mengmeng, L.I., Qiong, Y.U., et al.: Construction of Shaanxi agricultural special product platform based on "internet plus big data". Farm Prod. Process. (2019)
10. Yin, F.C., Liu, X.Q.: Protection and inheritance of intangible cultural heritage in Yangling District of Shaanxi province. J. Yangling Vocat. Tech. Coll. (2018)

The Design and Application of College Students' Mental Health Automatic Evaluation Model Based on Multimodal Data Fusion

Zheng Yan[(⊠)]

Chongqing College of Architecture and Technology, Chongqing 401331, China
zhengyan77777777@126.com

Abstract. With the continuous increase of people's work pressure, the require-ments for mental health recognition are also higher and higher. There are various forms of mental health, and the traditional mental health evaluation mode rely-ing on experts has certain subjective deviation. Therefore, this paper proposes a recognition model based on facial, speech and gait behavior characteristics to realize intelligent recognition of mental health status, so as to reduce the impact of subjective factors on the recognition results. The design of automatic evalu-ation model of College Students' mental health. The purpose of this study is to design an automatic evaluation system, which can evaluate the mental health of college students in a simple and easy way through their writing performance. In order to achieve this goal, we adopted two methods: (1) according to the concept that each writer has his own style and personal characteristics, we compiled a self-assessment questionnaire; (2) This study uses the data collected in previous studies on writers' styles and personality characteristics.

Keywords: College Students' mental health · Multimodal data fusion · Automatic evaluation model design

1 Introduction

This paper focuses on people's mental health factors, aiming to effectively reduce the probability of accidents in key posts through systematic, in-depth and scientific research on mental health models. The research data show that the occurrence of a large number of power operation accidents is closely related to the psychological state of the power operators at that time. Therefore, it is very necessary to establish a real-time and effective mental health identification system to correctly judge the psychological problems of power workers and take timely psychological intervention treatment. It is related to the physical and mental health development of power grid workers, and more importantly, it can avoid the occurrence of dangerous power accidents [1].

In the past, most of the discrimination methods of mental health status were in the form of self-evaluation or other evaluation, such as using SCL90 symptom checklist for mental health assessment. Although this questionnaire type collection and analysis method is effective, if it is repeatedly used for a long time, the examinees may have the

J. C. Hung et al. (Eds.): IC 2023, LNEE 1045, pp. 518–523, 2023.
https://doi.org/10.1007/978-981-99-2287-1_74

behavior of recording questions, or they may not answer the questions according to their own actual situation, but answer the questions according to the results, thus leading to the distortion of the evaluation results. In addition, the results of his evaluation method depend on the professional ability of psychological experts, and the psychological factors are complex and difficult to quantify, so the evaluation results of his evaluation method contain strong human subjective factors. Therefore, this paper proposes a mental health automatic recognition model based on human behavior patterns [2]. This model combines human face, voice and gait information to comprehensively judge the current mental health status of the tested person. Compared with the use of evaluation scale or expert evaluation, this method can achieve the output of mental health status in a short time. Due to the small human interference, the reliability of mental health status recognition results is improved to a certain extent, and the model proposed in this paper has strong reusability and popularization.

2 Related Work

2.1 Research Status of College Students' Mental Health

Many researches on the mental health of college students show that in recent years, the overall mental health level of college students has shown a gradual downward trend, The number of students with various psychological problems or psychological barriers is increasing. The research results show that 31.13% of college students have psychological problems with different symptoms and different degrees of severity [3]. Existing studies have shown that 8.7% of college students have serious mental health problems, and the proportion of college students with moderate or more mental disorders is even as high as 20%. Many domestic research conclusions show that although the overall mental health level of college students is higher than that of the general population, the psychological problems in some symptoms such as compulsion, interpersonal relations, anxiety and hostility are more prominent. The research shows that the psychological problems are moderate and serious The proportion of college students with obsessive-compulsive symptoms is as high as 9.77%, the proportion of college students with moderate severity of interpersonal sensitivity is 7.33%, and the proportion of college students with moderate severity of anxiety and hostility symptoms is 3.76%. Moreover, many research results show that the mental health problems of college students mainly focus on interpersonal relations and emotions. Luo Xiaolu and others pointed out that college students have serious psychological confusion in interpersonal sensitivity, compulsion and bigotry [4].

The related research on the mental health of college students in our country has gradually reflected the characteristics of strong specialization, and the research heat is also rising year by year. The mental health of college students has received greater attention. Liu Xin and others have found that most of the research on the evaluation of College Students' mental health is based on a series of mental health scales, and a variety of mental health scales are applied to practice [5]. The research on how to take timely and effective intervention to college students' psychological problems has also shown an upward trend year by year. People began to pay more attention to the factors that affect mental health, and investigated the factors that affect mental health through the implementation of the scale, and a large number of investigation reports on mental health have appeared. Most of the scales used are representative ones that we usually contact with, such as symptom checklist 90 (SCL-90), Eysenck Personality Questionnaire (EPQ), Minnesota Multiphasic Personality Inventory (MMPI), etc. these scales are widely used to study the symptom performance or behavior problems of mental health.

2.2 Theoretical Basis for Automatic Evaluation Model Construction of Multimodal Data

Multimodal data refers to data containing two or more different forms or different sources. Words and images are the external manifestations of the results of human mind and psychology, which can reflect the psychological state of individuals. In the modern society with highly developed Internet, college students tend to publish text, images, expressions and other modal data on social platforms to express their personal ideas and emotions. Different modal data have complementary effects and can provide more explanatory information. By integrating and understanding the multimodal data, a more comprehensive and systematic analysis and evaluation of students' mental health can be realized [6].

This study builds an automatic assessment model based on Ecological transient assessment (EMA) and deep learning to realize real-time assessment of students' psychology in natural environment. Ecological instantaneous assessment is an ecological method proposed by psychologist Shiffman to sample and measure the relevant behaviors and experiences of the subjects in the natural environment in real time. 2 it has high authenticity and dynamics, and can more accurately reflect the psychological characteristics of the subjects. Deep learning is a hot and important research topic in the field of artificial intelligence [7]. It extracts features by constructing multi-layer neural networks, and then combines low-level features to form more abstract high-level features, so as to learn the internal laws and expression levels of various types of data, and acquire their hidden deep-level semantic knowledge.

3 Research on the Design and Application of College Students' Mental Health Automatic Assessment Model Based on Multimodal Data Fusion

3.1 General Method

In order to solve the problem of insufficient information in single-mode mental recognition mentioned above, multi-modal fusion can be used to synthesize the feature information of face, voice and gait for mental recognition. Multimodal fusion can form information complementarity among the three modes, thus improving the final psychological recognition effect. The most widely used multimodal fusion method is the fusion method based on direct cascade. Cascade fusion firstly extracts the features of each single data mode, then combines the features of different modes into a whole feature through feature fusion, and finally sends the whole feature to the classifier for recognition [8]. As shown in Fig. 1, firstly, the corresponding features are extracted for the three single modes of face, voice and gait, and then the features of the three modes are cascaded and fused together to obtain a fused feature. Finally, the classifier classifies the mental health state according to the fused feature, and outputs the mental health recognition result based on multi-modal fusion.

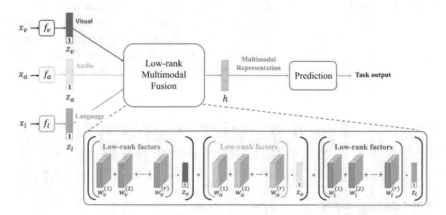

Fig. 1. Multimodal fusion mental health recognition based on direct cascade

At present, the most commonly used fusion method is direct cascade fusion. Cascade fusion makes full use of all modal information, and gives full play to the characteristics that modes can complement each other. Therefore, the recognition rate of mental health obtained by multi-modal method is higher than that of single-modal method, and the model performance will be improved to a certain extent.

Although this multi-modal fusion method makes full use of all modal information, the direct cascade method defaults that the mental health characteristics of all modes have the same impact on the final mental health recognition results. Obviously, when people's internal mental state is externalized, the proportion of each expression form is different, and the impact of the three modes of face, voice and gait on the final mental health recognition results is also different [9]. The judgment basis of each mental health problem is different. For example, facial information is more important to judge whether people suffer from depression than voice and gait information, and voice information is more important to judge whether people have hostile emotions than face and gait information; Gait information is more important than facial and voice information in judging whether people have OCD. Therefore, it is necessary to assign a reasonable weight to the information of each mode, so that those modes that have a great impact on the identification results can get a relatively large weight value. On the contrary, the weight value of the modes that have a small impact on the identification results should be set relatively small. How to reasonably set the weight value of each mode is also a problem worth considering.

3.2 Model Construction Ideas

In order to solve the problem that the direct cascaded multi-modal fusion method can not reflect the impact of each modality on the final mental health recognition results, this paper proposes a multi-modal fusion method that introduces attention mechanism to construct the mental health model. The attention mechanism is introduced in the feature fusion, so that the three modes of face, voice and gait can obtain reasonable weight values respectively, and the modes that have great influence on the recognition results are assigned greater weight values, so that the model can reasonably use the information of the three modes in the mental health recognition, instead of simply combining the features of multiple modes [10].

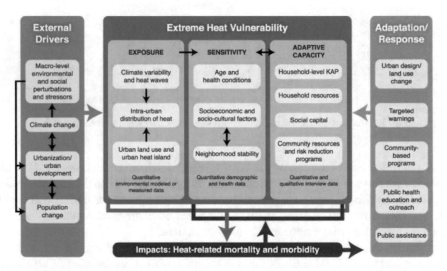

Fig. 2. Multimodal fusion mental health recognition with attention mechanism

The feature fusion process after introducing the attention mechanism is shown in Fig. 2. The feature extraction process of each modality is consistent with the cascade fusion method. The corresponding features are extracted from the face, speech and gait data, and the feature values of the three modalities are obtained. After obtaining the features of each modality, a weight matrix [W1, W2, W3] is initialized to represent the weight value of each modality feature, and WL is assigned to the facial modality, W2 to the speech modality, and W3 to the gait modality. The weight matrix will be optimized with the training of the model, and then the final weight value will be multiplied by the mental health feature matrix of the corresponding modality before fusion, and the obtained fusion features will be sent to the classifier for recognition. Finally, the multi-modal fusion mental health recognition result based on the attention mechanism will be obtained. Since the weight matrix is continuously adjusted and optimized with the training of the model, and the weight matrix corresponding to the data distribution can be learned on different data sets, the generalization ability of the fusion model based on the attention mechanism will also be stronger.

The introduction of attention mechanism can enhance the influence of relatively important features on mental health recognition results and reduce the proportion of unimportant or irrelevant features in the fusion process. Therefore, the distribution of facial, speech and gait features in the model after the introduction of attention mechanism will be more reasonable, thus improving the recognition accuracy and performance of the multimodal fusion mental health model.

4 Conclusion

According to the characteristics of students' social network platform data, this paper constructs a multi-modal fusion computing model for college students' mental health automatic assessment. The experimental results of this model on ja-ipad data set show that the accuracy of the model with multi-modal data is significantly higher than that of single-modal data. This shows that in addition to text information, image information is also an important basis for college students' mental health assessment. The average accuracy of the fusion model reached 84.85%, indicating that the model can accurately grasp the students' mental health level and effectively reveal the continuous change trend of students' psychological characteristics. At the same time, this is also in line with the development direction of "artificial intelligence + education", and brings new opportunities for the intelligent development of mental health education in Colleges and universities.

References

1. Wang, Q., Pan, F.: Research on the design of computer scoring system for chinese college students' English translation. J. Phys. Conf. Ser. **1992**(3), 032085 (2021)
2. Zhang, P.: Research on the application of user interest model and apriori algorithm in college students' education recommendation. In: Jan, M.A., Khan, F. (eds.) BigIoT-EDU 2021. LNICST, vol. 391, pp. 186–190. Springer, Cham (2021). https://doi.org/10.1007/978-3-030-87900-6_23
3. Zhou, L.: Applications of deep learning in the evaluation and analysis of college students' mental health. Discrete Dyn. Nat. Soc. **2022** (2022)
4. Li, X., Shi, X.: Design and application of mental health intelligent analysis system for college students majoring in physical education. J. Phys. Conf. Ser. **1852**(3), 032049, 7 p. (2021)
5. Hou, S.: Research on the application of data mining technology in the analysis of college students' sports psychology. Hindawi Limited (2021)
6. Qin, R., Zhao, L., Li, D., et al.: Research on design and application of power dispatch based on blockchain (2021)
7. Wang, Z., Wu, Q.: Research on automatic evaluation method of Mandarin Chinese pronunciation based on 5G network and FPGA. Microprocess. Microsyst. **80**(3), 103534 (2021)
8. Shan, S., Liu, Y.: Blended teaching design of college students' mental health education course based on artificial intelligence flipped class. Hindawi Limited (2021)
9. Wu, X.: Research on the reform of ideological and political teaching evaluation method of college English course based on "online and offline" teaching. J. High. Educ. Res. **3**(1), 87–90 (2022)
10. Yuan, F., Hu, D.: Research on the application of big data technology in college students' innovation and entrepreneurship guidance service. J. Phys. Conf. Ser. **1883**(1), 012171 (2021)

The Design of College Chinese Module Teaching System Under Multimedia Environment

Xiaocong Sui[1(✉)], Xiaohui Sui[2], and Xiangping Shen[1]

[1] Weifang Engineering Vocational College, Qingzhou 262500, Shandong, China
`suixiaocong123@163.com`
[2] Changle County West Lake Kindergarten, Changle 262400, Shandong, China

Abstract. This study aims to design a new teaching system for College Chinese. The main goal is to develop a module based teaching system, which can be used in distance education with multimedia environment. Methods and tools are mainly based on cognitive science, human-computer interaction and communication theory. Use multimedia teaching materials to provide effective learning experience, facilitate students to master the basic knowledge of Chinese, enhance students' enthusiasm through interactive activities, improve students' oral level through audio-visual demonstration, and use advanced technology to improve students' fluency. The purpose of this study is to design a college Chinese module teaching system based on Multimedia environment. The main objective is to design a modular teaching system of College Chinese that can be used in a multimedia environment, and to investigate and evaluate the effectiveness of different learning strategies such as self-study, cooperative learning, computer tutoring, etc. Investigate and evaluate the effectiveness of different types of learning materials (such as audio-visual materials, reading materials, writing materials, etc.). Analyze students' grades by using various evaluation methods.

Keywords: Multi-Media · College Chinese · teaching system

1 Introduction

College Chinese education plays an irreplaceable role in cultural inheritance, students' healthy development and humanistic quality improvement. However, corresponding to it, College Chinese education has long been in an embarrassing situation of neglect and stagnation: the courses are opened at will, the class hours are too short, and the course status is low; The textbooks are confused in edition and repetitive in content, which are not attractive to students; Teachers' good and bad are intermingled, their status is not high, and their motivation for work and research is insufficient; The teaching means are single, the methods are obsolete, and there is no scientific teaching effect evaluation [1]. The existence of these problems requires us to pay attention to college Chinese education with a more developmental multimedia vision.

The integration of multimedia environment and educational concept has given birth to the educational multimedia environment, which provides a feasible way to break

J. C. Hung et al. (Eds.): IC 2023, LNEE 1045, pp. 524–529, 2023.
https://doi.org/10.1007/978-981-99-2287-1_75

through the dilemma of College Chinese education. The study of multimedia in College Chinese education has important theoretical and practical value, whether it is to enrich the educational theoretical system of College Chinese, achieve educational goals, or improve the current situation of College Chinese education and improve the effectiveness of the curriculum.

The crisis of College Chinese education is unbalanced in the perspective of multimedia environment, and its unbalanced performance is different in different multimedia positions. The exploration of the unbalanced performance and reasons lays the foundation for the macro multimedia research and micro curriculum construction of College Chinese. This paper regards college Chinese education as an educational multimedia system, and discusses the expression and application of "two basic ideas" - the idea of multimedia system and dynamic balance, "three basic theories" - the theory of multimedia circle, the theory of comprehensive and harmonious development, the theory of sustainable development, "three basic ideas" - the concept of integrity, the concept of harmony and the concept of systematic development in College Chinese education [2]. There are many factors in educational multimedia system. This paper believes that the benign multimedia factors that have a great impact on College Chinese education are mainly "two subjects" and "three environments": educators and educatees, natural social environment, school family environment, individual internal environment; "Two relations" and "Three Laws": the relationship between people, the relationship between people and the environment, natural laws, social laws, and educational laws.

2 Related Work

2.1 College Chinese

Looking back on history, it is the collective College Chinese with hundreds of schools of thought contending, and it is the infinite enthusiasm for Chinese education that promotes the maturing and perfection of the concept of Chinese education. But at the same time, we also find that there is a serious disconnect between Chinese education theory and teaching practice. If you quarrel with yourself, I'll bury my head in teaching. There is a sharp contrast between the enthusiasm of theoretical discussion and the indifference of teaching practice. Therefore, in order to achieve the prosperity and development of Chinese education, more educational researchers and practitioners need to work together.

Taking history as a mirror, from the hundred year development history of Chinese language education concept, we can more clearly realize that Chinese education in the new era should not only conform to the background of the political and historical era, but also maintain its own discipline independence; We should not only learn from the experience and lessons of foreign countries, but also consider the national conditions of our country, inherit the crystallization of our ancestors' College Chinese and integrate them; We should not only reach a certain consensus in theory and practice, but also allow a hundred flowers to bloom and move forward in contention [3].

What we are talking about above is the century old concept change of Chinese education. Whether it is basic education or higher education, there is a development history of this concept. How can we make college Chinese education more subject independent, more in line with the pulse of the times, more suitable for China's national conditions, pay more attention to the all-round development of people, and pay more attention to the benign interaction between theory and practice? These problems are not only the starting point and foothold of our thinking about College Chinese education, but also the internal driving force and pursuit of this study. Only in this way, the concept of College Chinese education will not be more lost in the width of the world and the thickness of history [4].

2.2 Multimedia Teaching System Technology

The combination of multimedia teaching system and computer network can enable students to put forward the teaching content on the server at any time through the network with the help of network terminals for learning. At the same time, they can also interact with computers and remote classmates and teachers through the network. In addition, students are more proactive in place, time and learning progress, and can perfectly integrate work and learning. Hadoop is a large-scale distributed data processing platform, which implements a distributed file system HDFS (Hadoop distributed file system) and a distributed computing model MapReduce. As a file system, HDFS is mainly deployed on low-cost hardware devices, and has high fault tolerance and high reliability [5]. The high transfer rate it provides is very suitable for those applications with large data sets, and provides the form of stream access to file system data. MapReduce, a distributed computing model, is a programming model for processing and generating large-scale data sets. It was originally designed and implemented by engineers of Google, and later it slowly derived a better version. The main principle of MapReduce is to decompose the tasks of the application into small pieces. In addition, in order to enhance reliability, HDFS replicates multiple copies of each data block as backup, and places them in different computing nodes of the server cluster, so that MapReduce can process these data on their nodes. The brief workflow diagram of MapReduce is shown in Fig. 1. It can be seen from the figure that the data realizes the parallel task decomposition through the function of map, and then is allocated to the corresponding computing nodes through the mapping function for parallel operation and processing, and then the calculation results are counted and summarized through the function of reduce to get the final result.

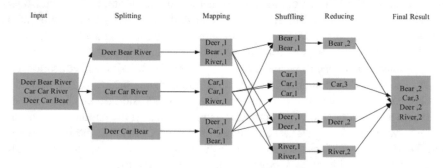

Fig. 1. Workflow of MapReduce

With the joint action of the three core components of HDFS, MapReduce and yarn, Hadoop ecosystem realizes complex workflow such as data access, resource scheduling, data fragmentation and fault tolerance mechanism, and has the advantages of strong scalability, low cost investment, good flexibility, efficient and rapid processing, and strong fault tolerance [6].

3 Research on the Design of College Chinese Module Teaching System in Multimedia Environment

The combination of multimedia teaching system and computer network can effectively break the boundary between work and learning, and allow people to learn through the network at any place and at any time. Every learner can consult the first-class teachers through the network, and also search for materials in the world's most famous and abundant libraries. Therefore, through the combination of multimedia teaching system and computer network technology, we can realize the sharing of various teaching resources and achieve a larger area of interactive teaching. Therefore, the networking of multimedia teaching system is a very important development mode.

When designing the multimedia teaching system of College Chinese, we should choose the appropriate multimedia materials according to the teaching objectives and teaching contents, and express the teaching contents according to the characteristics of the multimedia materials [7]. Multimedia materials mainly include text, graphics and images, sound, animation, video, etc. Different multimedia materials have their unique characteristics, which are suitable for the expression of different information. Generally speaking, multimedia materials can be divided into observable abstract and empirical symbols. The following Fig. 2 shows the multimedia data accelerated access process.

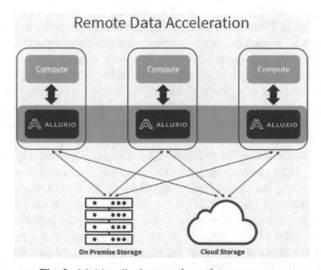

Fig. 2. Multimedia data accelerated access process

The design of College Chinese multimedia teaching system should be guided by modern educational thoughts and theories. The obtained system should conform to the cognitive characteristics of students, meet the teaching rules, and serve the teaching objectives of College Chinese. In order to cooperate with the multimedia teaching system of College Chinese teaching, it should be designed and made according to the requirements of College Chinese teaching content system and syllabus [8]. Teachers can use the College Chinese multimedia teaching system to organize teaching activities, so that students can use the designed multimedia teaching system to learn the basic knowledge, theory and skills of College Chinese courses, so as to master the difficulties and key contents of College Chinese courses, and achieve the purpose of College Chinese course teaching.

4   Database Design

The multimedia based College Chinese teaching system proposed in this paper uses the relevant theories of cloud computing and SOA design mode, combines it with a variety of network technologies, analyzes the business needs and solutions of educational informatization, and constructs a college Chinese teaching system with progressiveness, reliability, security and scalability [9]. The system can realize the unified management and deployment of massive educational resources and services, solve the problems of uneven distribution of educational resources and difficult sharing of resources, and meet the actual needs of educational institutions, students, teachers, educational managers and other aspects.

In the multimedia teaching system of "one point animation analysis of College Chinese", the relevant data of the test questions are stored in the access database system. At present, there are many database management systems on the market. The main reason for using access database is that access database is small in size, low in cost, and the amount of data in the multimedia teaching system of "one point animation analysis of College Chinese" is small, and the structure is relatively simple. Therefore, access, a lightweight database management system, is mainly used to manage the data in the system. According to the specific needs of the multimedia teaching system of "one point animation analysis of College Chinese", its "self-improvement" module mainly includes multiple-choice questions, blank filling questions, question answering questions and other question types. According to different question types, different data tables are used to store information.

Teaching materials are the basic conditions in the process of teaching activities. The core of the College Chinese teaching system lies in the application of the College Chinese teaching platform, and the core of the College Chinese teaching platform is the database [10]. With the advent of the era of big data, the database can store a large number of teaching resources, and these teaching resources are important materials stored in the database after a series of screening and sorting, It is through these materials that the College Chinese teaching system forms teaching materials in the application process, so the application of the College Chinese teaching system can realize the convergence of teaching materials.

5 Conclusion

The most effective way to learn Chinese is through multimedia resources, such as audio-visual materials and computer software programs, which can be used to enhance students' understanding of Chinese culture and language. However, these resources are not widely available in China or other Putonghua speaking Asian countries. Therefore, they are not very popular among learners of this language. Therefore, it will be beneficial to develop an interactive system with multimedia function. The purpose of this study is to study the design of Chinese module teaching system in multimedia environment. Its basic idea is to use multimedia learning materials, such as CD-ROM, videotape and audiotape, in order to improve students' understanding and memory ability by using different media content. This project will be divided into two parts: (1) study the design of Chinese module teaching system; (2) Research on interactive computer games to improve students' language ability.

References

1. Li, W., Fan, X.: Construction of network multimedia teaching platform system of college sports. Math. Probl. Eng. (2021)
2. Xia, Z., An, Y., Yu, X.: Practice and application of collaborative innovation design of management experiment teaching system under new media environment. In: 2020 International Conference on Modern Education and Information Management (ICMEIM) (2020)
3. Wang, M.: Design of college physical education teaching system based on artificial intelligence technology. JPhCS (2021)
4. Wang, Y., Tong, W., Lv, Q., et al.: Flipped classroom teaching system design under the background of subject reform based on information technology. J. Phys. Conf. Ser. **1852**(2), 022041 (2021)
5. Li, J., Li, S.: Design of multimedia assisted teaching system for basketball theory course. J. Phys. Conf. Ser. **1992**(3), 032081 (2021)
6. Wu, X.: Theory and practice of multimedia courseware design for ideological and political theory courses in colleges and universities (2020)
7. Li, J.: Based on multimedia and network social environment–a study of pragmatic acquisition in college English teaching **5**(4), 4 (2021)
8. Zhang, X.: Research on the design of college English lessons based on hybrid-styled teaching **4**(1), 5 (2020)
9. Gao, N.: Research on remote control method of English multimedia online teaching system in big data environment. J. Phys. Conf. Ser. **1486**(5), 052010, 8 p. (2020)
10. Wang, J.: Research on college English teaching system based on computer big data. J. Phys. Conf. Ser. **1865**(4), 042141, 10 p. (2021)

The Development of Human Resource Management Based on BP Neural Network Algorithm

Tingdan Zheng$^{(\boxtimes)}$

Business School, Beijing Institute of Technology Zhuhai Campus, Zhuhai 519085, Guangdong, China
zheng_td@126.com

Abstract. The development of human resource management based on BP neural network algorithm is a new type of artificial intelligence. This technology has the characteristics of "learning" and "adaptive". Learning ability can be learned from real-time data, and adaptability can adapt to different situations. In addition, it is easy to use and maintain, which makes it suitable for large-scale applications. The main purpose of this study is to find out how to use effective and efficient BP neural network algorithm in the field of HRM. Many studies have been carried out to understand what types of problems BP neural network can solve and how it can help improve the performance and efficiency of different aspects related to HRM.

Keywords: BP neural network · human resources · Development income

1 Introduction

The research on the development of human resource management has been carried out by many researchers for many years. In order to understand this research, it is necessary to understand the history of human resource management and its evolution from time to time. History of human resource management: the first use of the term "human resource" can be traced back to 1938, when Douglas McGregor published a book called the human side of enterprise, in which he defined human resource as "the sum of all abilities, skills, knowledge, attitudes and values possessed by employees of an organization". The research on the development of human resource management based on BP neural network algorithm is a research to improve human resource management based on the implementation of artificial intelligence (AI). It also includes developing new methods of human resource management and improving existing methods. The main goal of this research is to create an artificial intelligence system that can be used by companies as a substitute for human resource managers. This will enable them to focus more time and energy on their core business rather than waste on managing employee performance.

In 1984, Wernerfelt completely expressed the "resource - based view" (RBV) for the first time [2]. In 1991, Barney further developed and elaborated this view. This

J. C. Hung et al. (Eds.): IC 2023, LNEE 1045, pp. 530–535, 2023.
https://doi.org/10.1007/978-981-99-2287-1_76

view holds that the resources and capabilities of enterprises are heterogeneous, which determines the performance of different enterprises in an industry. According to RBV, enterprises can only create value in a rare and difficult way for competitors to imitate, so as to develop competitive advantage. Human resource is such a resource. When it penetrates deeply into the operation system of the organization, it can not only create value and increase the strength of the enterprise, but also this competitive advantage is difficult to imitate [3].

Although more and more enterprises realize the importance of human resource management, how to measure the performance of human resource management department has always been a big problem perplexing enterprises. The research on the development of human resource management has been carried out by many researchers for many years. The focus of this study is the improvement and development of human resources in the organization, but it also includes performance evaluation, salary system and other contents. The purpose of this study is to find out how to develop people and what improvements should be made to improve them. Because of the importance of this topic in today's society, many studies have been carried out on this topic.

2 Related Work

2.1 Human Resource Management Demand Analysis

As the old saying goes, "if you want to create things, create people first". Excellent talents of enterprises are the precious wealth of enterprises. Human resources are mobile resources, and the talent market is an open market. Today's increasingly fierce brain drain storm gradually makes enterprise managers realize that when the resignation of excellent employees is a normal that enterprise human resources must face, what enterprises need to do is no longer just manage individual cases, but must rise to the height of strategy [4].

Since we take employee turnover as the research object, we first need to define the concept of turnover. Generally speaking, the termination of the employment relationship between the employee and the employer and the termination of the labor relationship between the employee and the original unit are called resignation. In nature, resignation can be divided into voluntary resignation and involuntary resignation. Voluntary resignation includes resignation and retirement of employees; Involuntary resignation includes dismissal of employees and collective layoffs [4]. In the above classification, retirement is a kind of welfare treatment for employees who meet the statutory retirement conditions. Under normal circumstances, its number and proportion are predictable, and its occurrence has positive value for the age structure of enterprise renewal personnel; Collective layoffs only occur when the enterprise has serious difficulties in operation and can only reduce costs through layoffs. It is an accidental behavior, which is generally not considered in the resignation analysis.

2.2 Problems in Human Resource Management

The research on the development of human resource management is a new field that has developed rapidly in recent years. In this field, researchers focus on the following aspects: (1) how to develop human resources; (2) What measures should managers take; (3) How to manage people effectively. The basic concepts are as follows: people are not only material entities, but also social creatures. They can adapt to different environments and conditions through psychological characteristics such as ability, will, experience and wisdom.The turnover rate of the company has remained high in recent years, and this phenomenon is taken as the research object. Through investigation, it is found that there are many reasons for this phenomenon, but the main reasons are as follows:

Salary level of employees. According to statistics, it is found that salary is one of the important reasons for the high turnover rate of employees. The higher the salary, the lower the turnover rate of employees. During the investigation, it is found that the employees of the company are generally dissatisfied with the salary and think that the labor pay is inconsistent with the labor income.

Employees are not optimistic about their future. A promotion system is not very perfect, causing some employees to feel confused about their development future. Maslow's hierarchy of needs theory also shows that the needs of employees are multi-level. Once a good development opportunity is found, some employees will leave without hesitation. This is more obvious.

3   Significance of Human Resource Management Benefit Evaluation

It is of great significance to evaluate the benefits of human resource management from the perspective of organization, human resource management department or human resource manager:

The full authorization, trust and support of the CEO have a great relationship with the success of the operation of the human resources department. It is increasingly common for senior managers to demand results from human resource development departments. At the same time, the human resource management department should also prove its contribution to the front-line business department to the organization. If the human resource management activities cannot be measured and evaluated quantitatively, the work effect of the department cannot be displayed and the sincere recognition of each business department and superior management department cannot be obtained [5]. As long as the work of human resource management department is recognized and praised by everyone, the superior leaders are willing to work in human and material resources With financial support, human resource management can embark on the track of a virtuous circle.

Measuring the benefits of human resource management can provide human resource management departments with a lot of information conducive to improving their own management level. Each measurement system has clear and clear objectives. Once goals are set in terms of cost, time, quality, quantity and customer satisfaction, employees can understand what the organization expects them to do, know the priorities of different tasks, and know where to focus their energy. Experience has shown that as long as the measurement system is appropriate, employees will compete with each other in their work to meet or even exceed their goals. At the same time, the evaluation is also conducive to the comparison of peers by the human resources management department. When the human resources management department sees that another human resources management department has made progress in this field, it will feel obliged to make the same achievements. This pressure is an integral part of any professional organization [6].

All employees of the enterprise hope that the impact of their work on the company can be quantified, because it is directly related to their performance appraisal, as well as the sense of satisfaction and achievement brought by their career. From the perspective of hierarchy of needs theory, after the basic needs of employees are met, more senior needs need to be met. Employees hope to understand their performance and know how they are doing through effective ways, rather than relying on their own speculation; Employees hope that their work performance can be recognized and respected by others, and want to know how others evaluate themselves; Employees also need to know what needs to be improved to improve their abilities and skills. This trend of psychological needs for evaluation also applies to human resource managers.

4 Simulation Analysis

At present, artificial neural network theory has been widely used in the fields of function nonlinear optimization, computational pattern recognition, intelligent control and information processing. Artificial neural networks is a mathematical model established by simulating the behavioral characteristics of human brain neural networks. The model is composed of many neurons according to a certain network structure, which can carry out distributed parallel information processing. It has strong self-adaptive ability, and can adjust the interconnection relationship between nodes in the network according to the complex situation of different systems, so as to realize the purpose of information analysis and processing. In addition, it also has strong autonomous learning ability [7]. It can analyze and learn the laws of their mutual existence through the preset input and output data, and then form a stable function and network structure according to these laws. This learning and analysis process is the "training" of the network.

Similar to the neural network of human brain, the network model is also composed of three parts: transfer function, neurons and connection weights between neurons. The transfer input formula between neuron I and other neurons J is:

$$I_i = \sum_{j=1}^{n} w_{ji} x_j - \theta_i \tag{1}$$

BP (back propagation) artificial neural network, also known as forward feedback artificial neural network, is an information analysis and processing system formed by simulating the neural function and structure of human brain, but its specific algorithm adopts error reverse transmission. Its structure is relatively simple, which is composed of input layer, hidden layer and output layer.

The first thing to be solved in the evaluation of human resource management benefits is the evaluation index or index system of human resource management benefits. The index system is generally composed of multiple indexes in different influence aspects of several levels [8]. According to the relationship between the indexes of each level, it can form either a tree structure, as shown in Fig. 1, or a network structure.

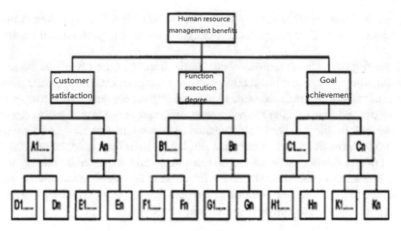

Fig. 1. Tree structure of index system

Artificial neural network is a complex network composed of a large number of simple information units called neurons, which is used to simulate the structure and behavior of human brain neural network. Neural network is a large-scale continuous time adaptive information processing system with high nonlinearity [9]. Theoretically, a three-layer feedforward network can approach any nonlinear function, and the multi-layer feedforward neural network is a powerful learning system with simple structure and easy programming. This research will be done by using intelligent algorithms and artificial intelligence, which are used to predict the future behavior of people in organizations [10]. The main purpose of the project is to understand how much employee performance can be predicted by using artificial intelligence (AI) and its accuracy as an effective tool for managers. The results of this study will help managers make more effective decisions about employees, thereby improving productivity.

5   Conclusion

BP network has strong mapping ability and self-learning and adaptive characteristics, while genetic algorithm has good global optimization ability, which can solve the defect that BP network is easy to fall into local minima. Through experimental simulation, the results show that the weight search of the system is feasible. Therefore, as long as there are enough training samples, the use of the system can better evaluate the benefits of human resource management. Because the neural network determines the network based on the learning of samples, if the error of samples is large, it may affect the evaluation effect of this model. Therefore, how to solve the impact caused by sample error needs further research.

References

1. Zhu, S., Liu, Y.: Analysis of human resource allocation model for tourism industry based on improved BP neural network. J. Math. **2022** (2022)

2. Li, Y., Zhao, D., Yan, S.: Research on travel agency human resource crisis early warning model based on BP neural network and computer software. J. Phys. Conf. Ser. **1744**(4), 042061 (2021)
3. Zhao, B., Xu, Y., Cheng, J.: Evaluation and image analysis of enterprise human resource management based on the simulated annealing-optimized BP neural network. Comput. Intell. Neurosci. (2021)
4. Zhang, P., Wang, C., Jiang, C., et al.: Resource management and security scheme of ICPSs and IoT based on VNE algorithm (2022)
5. Wan, S.: Agricultural water resources utilization and management under agricultural safety aim based on fuzzy neural network algorithm. Asian Agric. Res. **13** (2021)
6. Liu, Z., Ma, Y., Zheng, H., et al.: Human resource recommendation algorithm based on improved frequent itemset mining. Future Gener. Comput. Syst. **126**(1) (2021)
7. Zhao, N., Tsai, S.B.: Research on prediction model of hotels' development scale based on BP artificial neural network algorithm. Math. Probl. Eng. (2021)
8. Yan, J., Pan, Z., Tan, J., et al.: Assessment of water quality by firefly algorithm based on BP neural network model (2020)
9. Wang, C., Zhang, Z., Xi, Z.: A human body based on sift-neural network algorithm attitude recognition method. J. Med. Imaging Health Inform. **10**(1), 129–133 (2020)
10. Jia, J., Zhao, X., Zhao, R.: How human resource management strength affects employees' proactive behavior: based on self-determination theory. Hum. Resour. Dev. China (2020)

The Development Strategy of Commercial Banks Based on Big Data

Haixia Zhao[✉], Yangyang Guo, and Huisui Xing

School of Mathematical Sciences, The University of Jinan, Jinan, Shandong, China
935748819@qq.com

Abstract. This paper first analyzes the problems and challenges faced by commercial banks under the background of big data technology, then makes a model test on the role of big data in promoting the development of commercial banks, and finally puts forward several suggestions for promoting the development of big data technology applications in commercial banks.

Keywords: big data · Commercial banks · Econometric model

With the rapid development and popularization of computer technology and the breakthrough of data storage and application technology, the era of big data has brought great opportunities and challenges to commercial banks. Internet technologies such as big data and cloud computing are constantly updating. As the core of the Internet, big data always affects the development of commercial banks. Big data technology continuously injects innovation vitality and stimulates application potential into commercial banks, which helps commercial banks better analyze and summarize various personalized needs of customers and improve the efficiency of financial services.

1 Problems and Challenges Faced by Commercial Banks in the Context of Big Data

Under the background of big data technology, commercial banks mainly face the following problems.

(1) With the rapid popularization of big data technology, the professional talent reserve of commercial banks is seriously insufficient, and most commercial banks have too few professionals in big data to quickly complete the rapid transformation in the era of digital economy.

(2) Problem with data sharing. To make full use of big data, we must need massive amounts of high-value data. Because of trade secrets, competitiveness and other reasons, profitable commercial banks and enterprises will not easily disclose their own commercial customer data, big data statistical analysis process and modeling methods. The customer data of commercial banks can only be used for themselves, not for the entire commercial banking industry or society.

J. C. Hung et al. (Eds.): IC 2023, LNEE 1045, pp. 536–543, 2023.
https://doi.org/10.1007/978-981-99-2287-1_77

(3) The protection of massive data is also a problem that cannot be ignored. At present, the development of big data technology in China is not fully mature enough, and some small and medium-sized commercial banks have also developed applications related to big data. However, due to insufficient reserves of professionals and irregular applications of technology and other problems, there may be a risk of leaking customer data and some criminals can steal these customer information for financial fraud, laying hidden dangers for the future.

2 Empirical Research on the Impact of Big Data on the Development Strategy of Commercial Banks

2.1 Establishment of Econometrics Model

The development of big data has enhanced the profitability of commercial banks, the innovation ability of financial business and the ability to control risks. In order to verify the impact of big data technology on the development ability of commercial banks, this paper establishes a big data technology econometrics model to analyze. When establishing the econometric model, the differential control variables are introduced, and the efficiency evaluation method is used to express the development capacity TFP of commercial banks. Because TFP has certain viscosity effect, lagging TFP is adopted. TFP of commercial banks' development capacity is shown in Formula 1.

$$\text{TFP}_{it} = \alpha + \beta \text{Fintech}_t + \gamma X_{it} + \delta \text{TFP}_{it-1} + \mu_{it} \tag{1}$$

In formula (1), TFP_{it} indicates the TFP change of commercial bank I in year t. When TFP_{it} is greater than 1, it indicates that the TFP of the commercial bank in year t is higher than that in year t-1. When TFP_{it} is greater than 1, it indicates that the TFP of the commercial bank decreased in year t compared with that in year t-1. When TFP_{it} is equal to 1, it indicates that the TFP of the commercial bank is the same as that of the previous year. Fintech$_t$ is the China Fintech Index in year t. X_{it} refers to the control variables, including macroeconomic indicators, the maturity of the financial market, monetary policy indicators (i.e. indicators of the strength and weakness of the development of monetary policy capital markets) and some differential indicators at the level of commercial banks, such as the scale of commercial banks, risk management and control ability, profit earning ability, resource allocation ability, innovation ability, whether they are listed banks, etc. μ_{it} is the bias term. Due to the introduction of the lag period of TFP_{it-1}, a dynamic data panel model is formed. Dynamic data panel model estimation will inevitably encounter endogenous problems and in order to deal with this problem, this paper adopts Differential Generalized Method of Moments(DIF-GMM).

2.2 Econometrics Model Variable Descriptions

(1) Development capacity of commercial banks
      This paper uses TFP as the expression variable of the development ability of commercial banks. When calculating the TFP of commercial banks, we must consider the choice of input indicators and output indicators of commercial banks [1].

Quantifying the input and output of commercial banks is a challenge, because the characteristic of commercial banks is uniqueness [2]. For example, unlike the processing plant companies that produce physical goods, the products of banks include both intangible intermediate services and composite services composed of a series of products. Previous studies have proposed many solutions for the quantification of bank efficiency output, such as the number of deposit and loan accounts and the income of per user account [3]. Because this paper mainly studies the development strategy of commercial banks, rather than the changes of deposits and loans, it uses the services provided to customers as the output of commercial banks.

There are two main methods to judge service products: the production method and the intermediary method [4]. Production method refers to comparing banks to production workshops, which can produce many different types of deposit and loan accounts. The output of this workshop, that is, the output of banking business, is determined by the number of transaction vouchers and the type of transaction vouchers [5]. However, in general, it is difficult to obtain these data. Therefore, in daily production and life, we usually only use the number of deposit accounts and loan accounts to measure the output level of banks. In the intermediary theory, the bank, as a financial intermediary, has the function of transferring funds between depositors and lenders. In this theory, the output level of a bank is expressed by the number of bank loans and investments, and the input of a bank is expressed by the size of its employees and bank deposits. Both of these measurement standards are aimed at one aspect of the characteristics of the bank, and can be used as a method to judge the ability of different aspects of the bank. The operating cost of banks is reflected through the production method. This measurement standard is very effective for studying the cost efficiency of commercial banks, while the intermediary method takes into account the total cost and interest cost of banks, which is very effective for analyzing the economic differences between banks, evaluating bank efficiency and conducting boundary analysis.

(2) China fintech index

China's fintech index can be evaluated across different dimensions, including payment and settlement, resource allocation, risk management and network channels. With the help of Baidu Index database, the China fintech index is calculated by using data mining method. The keywords of these dimensions are shown in Table 1.

Search the above keywords in Baidu Index database, and calculate their annual average Baidu search index according to their monthly Baidu search index between 2011 and 2020. Finally, the financial technology index of China is calculated by factor comprehensive analysis method.

(3) Scale of commercial banks

For the quantification of the scale of commercial banks, this paper calculates the scale of commercial banks according to the value of their total assets. Research shows that large commercial banks use economies of scale and scope to reduce transaction costs and thus improve operational efficiency. However, some studies show that the expansion of commercial banks makes management more complex and instead reduced efficiency. Big data technology affects all aspects of commercial

Table 1. Key words of four dimensions of China's fintech index

dimension	explain				
Payment and settlement	Third party payment	Online payment	Online payment	mobile payment	two-dimensional barcode payment
Resource allocation	Online lending	Online loan	Online lending	Network investment	P2P lending
risk management	Internet insurance	Internet Finance	Network financial management	Online financial management	Network insurance
Network channel	Mobile banking	Internet Banking	Online Banking Service	Electronic banking	Online banking

banking business. The larger the asset scale, the stronger the ability of commercial banks to introduce big data technology and reform and innovate traditional business models, and the greater the potential of development capacity.

(4) Capital adequacy ratio of commercial banks

The research shows that the capital adequacy ratio of a commercial bank is related to the risk resistance of the commercial bank. Therefore, the higher the capital adequacy ratio of a commercial bank, the stronger the bank's ability to deal with the financial crisis, so as to maintain the robustness of the bank itself. The capital adequacy ratio of a commercial bank can measure the development ability of the commercial bank.

(5) Profit margin of commercial banks

Commercial banks mainly earn profits by retaining capital and through the interest rate difference between deposit and loan business. Therefore, the more capital commercial banks have, the stronger their ability to expand capital and risk management. Therefore, the profit margin of a bank has a positive impact on its development ability.

(6) Loan deposit ratio of commercial banks

The loan-to-deposit ratio of a commercial bank can be calculated by dividing the total loans by its total deposits in the same period. The efficiency of resource allocation of commercial banks can be measured by their total deposits and loans. The loan-to-deposit ratio of commercial banks represents the financing capacity of commercial banks, which can scientifically reflect whether the capital liquidity of commercial banks is active, and can also show its financing effect in making up for loan losses and customer withdrawals.

(7) Other business income

With the development of science and technology, commercial banks have derived many other businesses, such as consulting business and equity sale, in addition to the income obtained through traditional businesses, such as deposit and loan business. Other business income can reflect the innovative development strategy of commercial banks.

(8) Other macro control variables

   In order to make the results more objective, this paper also increased the GDP growth rate to reflect the impact of macroeconomic changes on the total factor productivity of commercial banks.

2.3 Econometrics Model Results

This paper selects 113 domestic commercial banks from 2011 to 2020 as the research object, including six large domestic commercial banks, twelve joint-stock banks, some urban commercial banks and some rural commercial banks. Through the calculation and analysis of the model, the TFP value is greater than 1, and obtained the data description is shown in Table 2 and Table 3.

Table 2. Data description of main variables of commercial banks' development capacity

variable	Observed value	mean value	standard deviation	minimum value	Maximum value
Development capacity	1244	1002	0.123	0	2.567
China fintech index	1130	0.595	0.329	0	1
scale	1130	7.598	1.625	4.407	12.531
capital adequacy ratio	1130	13.011	2.303	3.25	40.4
profit margin	1130	5.192	23.106	−59	208
Loan-to-deposit ratio	1130	69.020	11.53	13.34	110.02
Other business income	1130	12.194	8.677	0	48.2
GDP growth rate	1130	7.97	1.546	6.7	10.5

Table 3. Correlation coefficient matrix of main variables

	Development capacity	China fintech index	scale	capital adequacy ratio	profit margin	Loan-to-deposit ratio	Other business income	GDP GDP growth rate
Development capacity	1.000							
China fintech index	0.044	1.000						
scale	0.033	0.330	1.000					
capital adequacy ratio	0.053	−0.057	−0.203	1.000				
profit margin	0.024	−0.051	−0.183	0.078	1.000			
Loan-to-deposit ratio	0.026	0.005	−0.006	0.011	−0.003	1.000		
Other business income	0.067	0.167	0.015	−0.001	−0.018	−0.020	1.000	
GDP growth rate	−0.006	−0.616	−0.303	0.093	0.048	−0.012	−0.162	1.000

2.4 Econometrics Model Test

The hypothesis that fintech (big data technology) index improves the development ability of commercial bank is experimentally verified by DIF-GMM regression method. The experimental results are shown in Table 4.

It can be seen from Table 4 that the financial technology (big data technology) index of commercial banks is significantly positive at 1%, indicating that the development of big data technology has promoted the competitiveness of China's commercial banks, and then verified the basic hypothesis.

The variable of capital adequacy ratio of commercial banks is significantly positive, indicating that the higher the capital adequacy ratio of commercial banks, the stronger the risk control ability of commercial banks.

The significantly positive estimated coefficient of the loan-to-deposit ratio indicates that banks with stronger resource allocation ability can convert more deposits into loans, expand the source of bank income, and then improve their operational ability.

The estimated coefficient of GDP growth rate is significantly negative. This may be because people are more willing to invest in bank deposits and loans under the condition of rapid growth of gross national product, while commercial banks only rely on stable deposit and loan interest margin income, which has reduced their sense of hardship and innovation to a certain extent, leading to a decline in development capacity.

Table 4. DIF-GMM regression results

Fintech (big data technology) index	DIF-GMM			Significance level
	1.788	1.587	1.687	1%
standard deviation	0.330	0.380	0.392	–
Development capacity in the previous period	−0.421	−0.420	−0.413	1%
standard deviation	0.050	0.049	0.051	–
scale	–	−0.000	0.002	–
standard deviation	–	0.014	0.015	–
capital adequacy ratio	–	0.007	0.007	5%
standard deviation	–	0.003	0.003	–
profit margin	–	0.041	0.037	–
standard deviation	–	0.025	0.025	–
Loan-to- deposit ratio	–	0.002	0.002	1%
standard deviation	–	0.001	0.001	–
Other business income	–	0.000	0.000	–
standard deviation	–	0.001	0.001	–
GDP growth rate	−1.202	−1.137	−1.203	1%
standard deviation	0.207	0.216	0.218	–
Observed value	678	678	678	–

3 Development Strategies for Promoting the Application of Big Data Technology in Commercial Banks

The application of big data technology by commercial banks should continue to be strengthened. Through the above analysis, the following suggestions are put forward in promoting the application of big data technology in commercial banks.

(1) The development of big data financial technology in China is not mature enough, and all aspects may not be standardized and perfect, and the development differences between regions are large. Therefore, the state can strengthen the establishment of digital economy infrastructure, accelerate the digital reform of commercial banks and promote the popularization and application of Internet technology. In addition, more attention should be paid to fintech talents, and a scientific and effective training system should be established to provide more specialized talents for the society.

(2) Emerging Internet technologies represented by big data are growing rapidly, the corresponding regulatory system has not been improved, and there is a lack of relevant industry constraints and norms. Therefore, the bank regulators should improve relevant legislation, update industry norms and norms related to the application

of Internet technology as soon as possible, improve the regulatory system, and accelerate the combination of regulation and advanced technology.

(3) Commercial banks should optimize the concept of development and reserve more technical professionals. Strengthen technical cooperation with leading Internet enterprises and learn from them advanced application experience, such as data processing and protection. Based on the needs of emerging technologies, strengthen the reform and supervision of relevant businesses.

There are both opportunities and challenges for commercial banks to apply big data technology, and commercial banks should fully seize this development opportunity. Through the scientific and effective use of big data technology, aiming at some problems existing in the traditional business model of commercial banks, we should make improvements and optimization, so as to improve their investment and financing efficiency in the information age, enhance market share, make commercial banks innovate in the development of the digital economy era, and contribute to economic development.

References

1. Brei, M., Gambacorta, L.: Are bank capital ratios pro-cyclical? New evidence and perspectives. Econ. Policy **31**(86), 357–403 (2016)
2. O'Mahony, M., Timmer, M.P.: Output, input and productivity measures at the industry level: the EU KLEMS database. Econ. J. **119**(538), 374–403 (2009)
3. Bolton, P., Santos, T., Scheinkman, J.A.: Cream-skimming in financial markets. J. Finance **71**(2), 709–736 (2016)
4. Chamley, C., Kotlikoff, L.J., Polemarchakis, H.: Limited-purpose banking-moving from "trust me" to "show me" banking. Am. Econ. Rev. **102**(3), 113–119 (2012)
5. Favara, G.: An empirical reassessment of the relationship between finance and growth. In: IMF Working Paper, pp. 1–47 (2003)

The Inevitability of the Application of Computer Technology in Vocal Music Teaching

Lizhong Zhang[✉]

School of Design and Art, Nanchang JiaoTong Institute, Nanchang 330100, Jiangxi, China
565188013@qq.com

Abstract. With the continuous development and progress of society, computer technology is more and more widely used in basic education. Computer technology provides good conditions for teaching reform. It mobilizes the enthusiasm of teachers and students, organically combines the knowledge and interest of classroom teaching, and is conducive to the cultivation of students' ability and the realization of autonomous learning. The inevitability of the application of computer technology in vocal music teaching is a problem raised by many people. The answer is very simple and clear. Whether professional singers or amateur singers, this will be an indispensable part of music education. The reason why this question is raised is that there are several factors that make the use of computers in vocal music teaching inevitable; In fact, singing is not a natural skill; It requires training and practice to develop good technical and performance skills. In order to achieve these goals, one must first obtain high-quality teaching materials.

Keywords: Computer technology · vocal teaching

1 Introduction

Whether computer technology can play its real effect in basic education depends mainly on the changes in educational ideas, models and methods caused by it. Although computer technology seems to be powerful, it cannot give full play to its advantages unless it is under the guidance of scientific thought and careful design. Only according to the specific teaching situation, can computer technology play its powerful function in basic education. The decision of the CPC Central Committee and the State Council on deepening educational reform and comprehensively promoting quality education clearly requires "We will vigorously improve the modernization of educational technology and the informatization of education [1]. The State supports the construction of a modern distance education network based on the China Education and research network and satellite video system, strengthen the construction of economic and practical terminal platform systems and campus networks or local area networks, make full use of existing resources and various audio-visual means, and continue to do a good job in diversified audio-visual education and computer-assisted instruction."

The main goal of this paper is to study the application of vocal music Mu class in real life, and through investigation and analysis, study the traditional vocal music teaching,

J. C. Hung et al. (Eds.): IC 2023, LNEE 1045, pp. 544–550, 2023.
https://doi.org/10.1007/978-981-99-2287-1_78

summarize the advantages and disadvantages, and compare it with vocal music Mu class. Make some inferences about the future development direction of vocal music Mu class, and make a good prospect for the future of vocal music Mu class [2]. The significance of this paper is to let more people know the vocal music Mu class, know the "forest of Bel Canto" vocal music Mu class and understand the vocal music Mu class. In the future vocal music learning process, there will be more ways to learn, which can better improve the vocal music level of learners.

This article mainly studies the network vocal music teaching, compares the vocal music teaching with the traditional vocal music teaching mode, summarizes some objective problems existing in the current vocal music teaching in Colleges and universities, as well as the advantages and disadvantages of the network vocal music teaching, and comes to the conclusion that the vocal music network teaching should become the revolutionary progress of vocal music teaching [3]. Taking Zhang Meilin's vocal music Mu class "the forest of Bel Canto" as an example, this paper introduces the development, teaching mode and the impact on vocal music learners of the "the forest of Bel Canto" vocal music Mu class, analyzes the learning and practice in the Mu class, so that more people can understand the online vocal music Mu class teaching, and through the research of vocal music Mu class, it can play a certain role in improving their own vocal music generation methods and methods. And the author hopes that through this paper, more people can understand the "forest of Bel Canto" vocal music Mu class, have a certain understanding of the scientific nature of the Mu class, and play a certain role in promoting the development of the "forest of Bel Canto" Mu class [4].

2 Related Work

2.1 Overview of Vocal Art Teaching

Vocal music is a long-standing and ancient art discipline, and its final form of expression is to form a beautiful singing. It came into being with the emergence of human civilization. Up to now, there are three typical performing methods of vocal performance art in China: national singing, Bel Canto and popular singing. If you want to learn vocal music well and even teach it well, you must have a basic understanding of these three methods.

Bel Canto: mixed sound resonance, unified sound area, wide range, beautiful timbre, pleasant voice, grand volume and magnificent momentum. The training method of bel canto is very strict. It has a systematic training method. It scientifically mobilizes and uses various vocal organs in the human body to make sound. It attaches great importance to cavity resonance and has strong skills. Chinese bel canto was introduced to the West [5]. Bel canto was born in Italy in Europe and is called "Belcanto" in Italian. Its translation means "wonderful singing". After its introduction into China, it is called "Bel Canto". In the west, bel canto is the main form of vocal art. Bel canto is not only a method of vocalization or singing, but also a style and School of singing. Bel canto is very different from our national singing in performance. It does not attach great importance to expression and some body language as national singing. National singing pays great attention to the need to publicize and beautiful body language in performance, and singing should be full of vitality. On the contrary, how to portray the image of characters and express the main feelings of works in Bel Canto are shaped by the voice of performers, What highlights is the beauty of the essence of sound. Bel canto's vocalization is of Tut's scientific nature, and its voice is very soft and beautiful. Moreover, the repertoire and operas of Bel Canto all over the world are very rich, which also shows that bel canto has a considerable influence in the world. Most national scholars are committed to studying bel canto, including the vocal practice of national singing in China, which is also based on the vocal method of bel canto.

Therefore, as a student studying vocal music, he must have scientific vocal skills and methods, so that he can go further and further on the road of learning vocal music. In the process of learning vocal music, we can't ignore -- some theoretical studies, such as music theory, Solfeggio and ear training, also need us to spend a lot of time learning. These theoretical courses are the necessary prerequisites for us to learn vocal music, practice songs and express songs. When we learn the theoretical knowledge of music, we must work hard and understand the content flexibly on the basis of recitation [6]. Only by laying a solid foundation can we better learn vocal music and sing every song we learn. Solfeggio, as a compulsory course for students studying vocal music, has very strict conditions, that is, we must have a good theoretical foundation to learn Solfeggio and ear training well. Only when the musical theory knowledge reserve reaches a certain amount, can we learn Solfeggio and ear training well. Solfeggio and ear training is very important for learning vocal music performance. We must have a very good level of Solfeggio and ear training to grasp the scale of vocal music performance, so as to better interpret each vocal music work. Solfeggio teaching can better understand students' qualifications, discover and explore their potential in music, so as to better improve their comprehensive quality. Through vocal music teaching, students' creativity and musical mastery can be well exercised, their aesthetic ability and musical expressiveness can be enhanced, and their musical literacy and comprehensive ability in music creation can be improved [7]. Therefore, Solfeggio teaching is an indispensable and important link in the process of vocal music education. Basic knowledge of music theory and solfeggio practice can be said to be the necessary conditions for vocal music learning, just like the foundation of high-rise buildings. If there is no solid foundation of music theory knowledge, it can be imagined that high-rise buildings without foundation may collapse at any time.

2.2   Current Situation of Vocal Music Teaching

Today, the vocal music teaching mode in Colleges and universities still adopts the one-to-one oral teaching mode, which is commonly known as "master with apprentice", especially in some colleges and universities specializing in music and vocal music, vocal music teaching pays more attention to a "school" or "school", which is even more limited. It requires the professional vocal music teaching in Colleges and universities to adopt the "one-to-one" teaching mode, and from the perspective of the development history of Western professional vocal music, Vocal music teaching is mainly passed down to this day through oral teaching and imitation. In the process of vocal music teaching and learning, teachers' explanation and demonstration occupy a very important position and leading role. Therefore, on the road of vocal music learning, as long as conditions permit, most learners will still visit famous masters and teachers in order to get better guidance and teaching. In China, the holy land of vocal music learning is divided into North and south, represented by Shanghai Conservatory of music and Central Conservatory of music, because these two professional colleges have strong teachers, such as Guo Shuzhen, Jin Tielin, Zhou Xiaoyan and so on. Most of the students of these famous teachers can now be active on the vocal music stage in China and even the world, which also makes it a great honor for vocal music learners to accept the guidance of famous teachers on the road of vocal music learning. However, the traditional vocal music teaching method has been spread so far, and has formed an inherent teaching mode, which limits the number of students brought by vocal music teachers[8]. Although in recent years, many schools have adopted the collective teaching mode in vocal music teaching, they still need to adopt the small class system or even the one-to-one teaching mode if they want to really study vocal music pertinently. This also makes many vocal music learners and vocal music lovers have a headache in finding teachers, because there are few professional vocal music teachers, and they have to adopt targeted teaching mode, which makes time and space have great limitations, raises the threshold of vocal music learning, and makes most people have a passion for vocal music learning, but they can only flinch in the face of realistic conditions.

3 Vocal Music Teaching Based on Computer Internet Technology

3.1 Overview of Computer Internet Teaching

The growth process of computer teaching is not achieved overnight. It has experienced many years of development and improvement from the first simple teaching to the current powerful internet teaching. From the beginning, the only single page practice has grown into today's networked teaching stage. As early as the 1960s and 1970s, computer-aided language appeared. "At that time, the teaching content was limited to providing guidance on grammar and vocabulary. Like tutors, computers provided a large number of sentence pattern exercises and language test exercises, and made instantaneous judgments on students' answers, so that students could memorize and master the grammatical structure of the second language in repeated exercises. "more than ten years later, the second generation of computer-assisted instruction appeared [9]. Multimedia technology was added to the original foundation, and developers spent most of their energy on analysis and reasoning, rather than just using computers as a tool to store data. After this reform, students can use computers to solve problems in the process of learning, and test their assumptions, based on the original knowledge Based on it, we can explore new things and become new knowledge points. By this time, it is no longer the computer that dominates the human brain, but the opposite person begins to dominate the computer. After changes again and again, between the 1970s and 1990s, computers at this time can create a simulated learning environment, dominated by learners, and use students' own creativity to solve various difficult problems. In the 1990s, computer-assisted instruction gradually matured, but also has a higher goal, and strive to achieve the goal of using computers to simulate a near real language environment, so that people can communicate with each other in virtual space-time [10]. The above development of computer teaching can be regarded as a very scientific theoretical support for today's Internet teaching, and internet teaching has been improved step by step through practice. Because computer,

Internet and the technology combined with teaching are gradually maturing, the timeliness, interactivity, multimedia and unlimited sharing characteristics brought by internet teaching provide more advanced tools for language teaching.

3.2 Computer Internet Vocal Music Teaching

Traditional vocal music teaching and learning is an excellent form of education, but there are also many limitations in practical application: (1) vocal music learning is a process in which students slowly experience vocal music skills through vocal training and continue to make progress. In this process, the teacher is the most important guide and crutch. It is the responsibility of every vocal music teacher to guide students to find a way to scientifically train their voices by teaching students the necessary vocal skills. In fact, teachers cannot appear in every training of students, so students' autonomous learning is particularly important. At the beginning, learners lack sensitive self listening ability, and cannot accurately judge and grasp their own vocal state. They often cannot distinguish whether their vocal method is correct or not. Many times, they will not distinguish the quality of their voice, and will practice with what they think is the so-called "correct method". Therefore, it is easy to get into mistakes when practicing themselves and take many detours on the road of learning. (2) Unable to freely choose the place and learning time. Traditional vocal music teaching is face-to-face teaching, and teachers should teach students in accordance with their aptitude according to the different conditions and sound characteristics of each student. However, it is precisely because vocal music teaching has such strict requirements for the actual teaching environment that many students who are not qualified to learn face-to-face with teachers have lost the opportunity to learn. (3) In traditional vocal music teaching, teachers should not only give necessary guidance in vocal methods, but also in the analysis of works and the grasp of singing style. In the original teaching mode, vocal music teachers' elaboration and analysis of works has become the most important way for students to understand works, but to a certain extent, this restricts students' ability to understand works and learn, and also causes students' dependence and laziness on teachers.

The emergence of the Internet has not only played a revolutionary role in the teaching of other traditional disciplines, but also played a very good complementary and expanding role in vocal music teaching. To some extent, under the influence of the Internet, the means and teaching methods of vocal music teaching have undergone major changes. The Internet will play a positive and beneficial supplement to the traditional vocal music teaching. In view of the limitations of the traditional vocal music teaching mentioned above, we can make an effective remedy and a useful new attempt under the Internet-based teaching mode. First, internet teaching has a strong practical significance to make up for the timeliness and spatiality of traditional vocal music teaching. In the past, students who didn't have the opportunity and conditions to study face-to-face can conduct online vocal music teaching in the vocal music teaching forum on the Internet. This kind of vocal music teaching requires certain technical equipment. Through the real-time transmission of video and audio, it can break the limitations of space. Today's Internet technology can fully support the implementation of this educational means. Second, we can use the convenience brought by the Internet to collect the work background, literature, video and audio, which also forms a sharp contrast with the difficulty and tediousness of finding information in traditional vocal music learning. We don't need to spend a lot of time looking for relevant materials, and then screen them one by one. In the process of learning, we only need to input the data keywords we need in various search engines on the Internet, and we can get all kinds of information related to the materials we want to find. This makes students greatly improve the spatial nature of autonomous learning in the process of learning, and if we can make good use of Internet resources, We can also get authoritative and first-hand relevant literature and audio and video, as shown in Fig. 1.

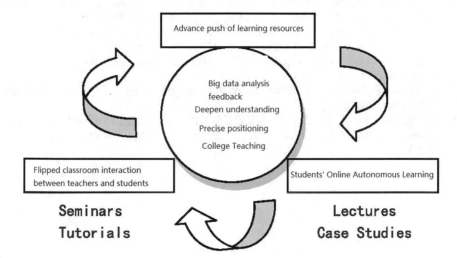

Fig. 1. Computer Internet vocal music teaching mode

4 Conclusion

With the development of science and technology, the application and popularization of computer technology, and the continuous innovation of teaching concepts and ideas; The development and status of Higher Vocational Education in China are improving year by year. With the launch of new technology assisted teaching activities, the voice of developing a comprehensive university is increasing year by year. The application of computer technology in vocal music teaching in higher vocational colleges will become a trend and meet the practical needs. In recent years, relevant classroom research has also increased year by year, and has also achieved certain results. Under the pressure of the general trend, higher vocational colleges have to carry out teaching reform. The application of computer technology to the classroom has not only improved the teaching quality, but also reduced the teaching pressure of teachers, and significantly improved the learning motivation and interest of students.

References

1. Shahnia, F.: Microgrids and their control. Hybrid Renew. Energy Syst. Microgrids (2021)

2. Zhu, S., Zeng, Y., Raj, V.: Explore the improvement of the management of China's international film festivals based on artificial intelligence. In: AICSconf 2020: 2020 Artificial Intelligence and Complex Systems Conference (2020)
3. Mahmud, M.M., Freeman, B., Bakar, M.: Technology in education: efficacies and outcomes of different delivery methods. Interact. Technol. Smart Educ. **19**, 20–38 (2021)
4. Fosch-Villaronga, E., Poulsen, A., Sraa, R.A., et al.: Gendering algorithms in social media. ACM SIGKDD Explor. Newsl. **23**(1), 24–31 (2021)
5. Koushick, V., Divya, C., Lakshmi, G.: L/C/X triple band compact dipole array antenna for RADAR application. J. Phys. Conf. Ser. **1432**, 012082 (2020)
6. Zhang, J., Luo, H., Xu, J.: Towards fully BIM-enabled building automation and robotics: A perspective of lifecycle information flow. Comput. Ind. **135**, 103570 (2022)
7. Zhang, Y.L.: Research on application of blended teaching pattern co-operated with E-learning in higher vocational education. J. Guangdong AIB Polytech. Coll.
8. Xiao, S.: The application and prospect of the state monitoring technology in steel plant. Instrum. Stand. Metrol. (2007)
9. Zhang, S.: Probe into the application of acousto-optic technology in memorial hall. Heilongjiang Sci. (2018)
10. Liu, Y., Miao, F., Su, W., et al.: Design of NB-IoT smoke sensing terminal based on photovoltaic and supercapacitor power supply. J. Phys. Conf. Ser. **1885**(5), 052008 (11pp) (2021)

The Law of International Trade Promotion Based on Growth Curve Algorithm

Zhentang Sun[✉]

College of Economics, Sichuan Agricultural University, Chengdu 611134, Sichuan, China
SunchuanTing710@163.com

Abstract. The law of international trade promotion based on growth curve algorithm is a new concept of the law of international trade promotion. The main purpose of this study is to find out how much the international trade promotion law has affected the promotion of natural resource poverty and developing countries, and whether it can be used as an effective tool for economic development. The study was conducted using data from different sources such as the World Bank Database and UNCTAD database. The research is based on the growth curve algorithm, which is a mathematical model used to analyze economic development. The main purpose of this algorithm is to predict the future development of economy. It can be used for forecasting and analysis in different fields, such as economics, finance, engineering and many other fields. It has been successfully applied in various applications, including stock market forecasting and financial planning. This method can also be used to predict future trends in the economy or business sector without collecting any data.

Keywords: Growth curve algorithm · International trade · Economic mathematical model

1 Introduction

International Economic and trade exchanges are not only the need for countries to complement each other's strengths and exchange needed goods, but also an important engine to promote world economic growth and human development and progress. Their resilience and vitality are a strong force to overcome various difficulties and challenges. Looking at the current situation of China's development, although it has achieved initial results in terms of economic level, it still lacks corresponding systems and measures. Compared with other developed countries in the world, there is still a certain gap, and the improvement of economic level needs to be further strengthened [1]. Therefore, the development of international economy and trade can effectively strengthen the external trade, cooperation and exchanges between China and other countries. China can learn from each other's strengths in the process of foreign trade, learn from foreign excellent economic systems, and constantly formulate forward-looking strategies in combination with the development form of the times, optimize China's economic system, improve China's economic model, and promote the diversified development of

J. C. Hung et al. (Eds.): IC 2023, LNEE 1045, pp. 551–556, 2023.
https://doi.org/10.1007/978-981-99-2287-1_79

China's economic model, So as to realize the double promotion of China's economic level and comprehensive national strength.

Close international trade relations are the primary driving force to promote benign competition among enterprises. The cooperation between overseas enterprises and local enterprises can effectively strengthen the exchange and learning of both sides, help employees' work enthusiasm and innovation spirit to effectively improve, stimulate the competitive potential of enterprises, promote enterprises to deeply explore the development direction and objectives, and help enterprises' sustainable development.

Secondly, the improvement of the economic benefits of enterprises is closely related to the improvement of the national economic level. The international trade under the background of globalization has increased the number of competitive objects of enterprises, which has brought some fresh blood and put some pressure on enterprises. In order to gain a foothold in the highly competitive market, enterprises should constantly examine themselves, follow the law of market development and the needs of the audience, and improve the service quality and production and operation quality, In order to improve their comprehensive competitiveness and occupy a certain advantage in the fierce market competition [2].

Economic globalization has promoted the efficient development of international trade, which has given it a new development direction and development goal. China should clearly understand the current international trade development situation, face many challenges in the process of foreign trade with an objective attitude, deeply grasp the market development law, look for breakthroughs and development opportunities, and then optimize the form of China's foreign trade with corresponding policies to improve the quality and ability of China's international trade, Achieve sustainable economic development. Based on this, this paper studies the law of international trade promotion based on the growth curve algorithm.

2 Related Work

2.1 Growth Curve Algorithm

Growth curve in transcriptome analysis, principal component analysis (PCA) is often a very important means to reflect the results. Principal component analysis is a commonly used dimensionality reduction algorithm in data mining. It was proposed by Pearson in 1901 and later developed by hotellingr in 1933. Its main purpose is "dimensionality reduction". By extracting the largest individual differences shown by principal components, it can also be used to reduce the number of variables in regression analysis and cluster analysis, which is similar to factor analysis. (to tell the truth, I didn't quite understand it).

Take a small example: for example, if you want to do an analysis on the factors of obesity, then you design 50 indicators that you think are very important. However, these 50 indicators are too complicated for your analysis [3]. At this time, you can use the method of principal component analysis to reduce the dimension. There will be one kind of connection and influence among the 50 indicators. After the principal component analysis, three or five principal component indicators are obtained. At this time, these principal component indicators cover most of your 50 indicators, which simplifies your analysis (from 50 dimensions to 3 and 5 dimensions).

If it is applied to the analysis of biological information, it can be understood as follows: we have obtained an expression profile data, which contains a lot of differentially expressed gene information. (generally, the number of these genes is very large, thousands). So it's definitely not possible to analyze so many genes. So we need to find the most representative to distinguish.

There are SVD Analysis Methods (singular value analysis method: the main application of SVD is to compress the data, and only the most important data is retained.), NIPALS analysis method (partial least squares PS regression), probabilistic PCA ppca: (that is, it is considered that the observed high-dimensional variables are actually generated by low-dimensional latent variables through a generalized linear model (this low-dimensional to high-dimensional mapping can be analogous to the curve equation in three-dimensional space, and the one-dimensional independent variable is mapped to the three-dimensional function value YY) [4]. Our purpose is to infer the underlying latent variables (low-dimensional) through the observed values (high-dimensional), In this way, the effect of data compression is realized).

This paper takes the volume of foreign trade goods in international logistics as the primary index to measure China's international logistics, and takes the growth curve function as the calculation model. Growth curve function model is also called logistic function model, which is widely used in modern commerce, production industry, biological science and so on.

2.2 Promotion Effect of International Trade

Thinking from a new angle of modern economic theory. According to zhouzhenhua, a domestic scholar. Opening up on the basis of structural benefits can improve the domestic industrial structure through international industrial linkages. This structural improvement is manifested in two aspects: first, to eliminate structural bottlenecks and promote structural rationalization; Second, strengthen the ability of structural transformation and promote the upgrading of structure. This requires that there must be selectivity in opening to the outside world, and the degree of trade openness should be determined according to the requirements of its structural optimization. At the same time, the state should make necessary intervention in import and export trade, and implement limited protection and support for the development of emerging industries on the premise of conforming to international norms. This is an open strategy for developing countries to actively participate in the international division of labor and make use of international industrial linkages to achieve leapfrog development. 2. The form of international trade promoting the growth of domestic industrial structure international trade promotes the growth of domestic industrial structure through the mechanism of comparative interests, which varies according to the national conditions of each country and the different development stages of each country's economy [5]. It can be roughly divided into two types: - it starts with the import of a certain product and uses this imported product to open up the domestic market and trigger the development of the industry in China. When the industry has developed to a certain extent, economies of scale have been fully utilized, and production costs have decreased significantly, the product can be exported by taking advantage of the comparative advantages of some domestic factors of production, and the development of the industry can be further promoted through the development of the international market. In the practice of foreign trade in China, the "import and export" should belong to this type. Import and export refers to the use of foreign raw materials and technologies to process and produce finished products for re export [6]. There is potential in domestic production and good sales abroad. However, carrying out the business of "import and export" under the condition of shortage of raw materials and parts can give full play to the advantage of rich labor force, tap the potential of equipment and technology, expand the production of export commodities, and form a labor-intensive factor combination.

3   Research on the Law of International Trade Promotion Based on Growth Curve Algorithm

This paper expounds the impact of international trade on the growth of a country's industrial structure, focusing on analyzing the dynamic effect of other countries' industrial structure changes through international industrial linkages to change their own industrial structure, so as to promote economic growth. Different from the general analysis of the relationship between economic growth and trade changes, this analysis must go through the intermediary link of perfecting the domestic industrial structure [7]. The theoretical assumption implied here is that international trade may bring short-term economic growth without improving the domestic industrial structure or even leading to the deterioration of the domestic industrial structure. This growth is at the cost of a large number of inefficient use of domestic resources, and will eventually hinder economic growth. Its growth curve model:

$$O_{v,j} = h\left(\sum_{i=1}^{f_{k-1}} \sum_{u \in N[v]} w_{i,j,u,v} x_{u,i}\right), (j = 1, ..., f_k) \tag{1}$$

The impact of China's international logistics on China's trade shows a typical growth curve: (1) in the early stage of China's international logistics development, the total logistics volume and foreign trade cargo transportation volume (x) are small, and the role in promoting China's trade is not obvious; (2) With the continuous accumulation, the overall scale of China's international logistics continues to expand Its strength has been strengthened, and its role in promoting China's international trade has become more and more obvious. At this time, it shows a rapid rising curve; (3) When the overall scale of China's international logistics reaches a certain level, its contribution to China's international trade reaches a saturation value, which is difficult to rise again and stays within a relatively stable range near the maximum value [8]. It can be seen that the growth curve function has a high similarity with the influence curve of China's cargo transportation volume on China's international trade amount. Therefore, it is feasible to use the growth curve function model to analyze the role of China's international logistics in promoting China's trade. Figure 1 below shows the trend of growth curve driven by international trade.

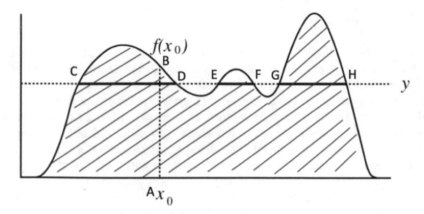

Fig. 1. Growth curve driven by international trade

The basic idea of one-dimensional slice sampling of the impact of international logistics on China's trade is: first, two-dimensional sampling (Z, g) is carried out from the shadow area between the probability curve f (a) and y = o, and then the point is projected onto the x-axis (that is, y is directly removed). The correctness of this method is actually very easy to understand, but the problem is, how to sample from this area?

Because it is two-dimensional sampling, a natural idea is to use Gibbs sampling to extract P (azlg) and P (g/z) respectively. How can these two probability distributions be determined for the impact of international logistics on China's trade? In fact, there are many ways to select P (Z g) and P (gaz) that can be guaranteed to be sampled from the shaded area. However, the most intuitive way is to define P (XLY) as the uniform distribution on the horizontal line and P (YX) as the uniform distribution on the vertical line. Given the initial point O, the specific steps are as follows [9]:

1. determine the vertical line (0, f (AO)) of O, and uniformly sample from this line segment (AB in the figure).
2. Find the horizontal line contained in the shaded part, that is, the line segments CD, EF and GH, and conduct uniform sampling from these three line segments to obtain the next sample 1.
3. repeat the above two steps.

It seems that the whole algorithm of international logistics to China's trade has been completed. The vertical line is easy to find because it is internally continuous; Horizontal lines are composed of many different line segments. How can we find them! The difficulty of this algorithm is also here. In fact, there is a feeling similar to the "random walk behavior" in the MH algorithm mentioned earlier, that is, how to make the algorithm explore the distant landscape (line segment). However, slice sampling can be regarded as a special case of MH algorithm [10]. Like HMC, it is just a bit of a miracle of transition kernel.

4   Conclusion

The research on the law of international trade promotion based on the growth curve algorithm is a research aimed at predicting the future development trend of international trade promotion. The main purpose of this study is to develop and use a new forecasting method, which can be used as an effective tool to promote foreign trade. The research was conducted by experts from many fields, such as economics, statistics, finance, law and sociology. It also includes some research related to this topic, such as forecasting market demand, forecasting commodity price changes, etc. Determine whether the export volume increases or decreases by comparing with the import volume. It also determines whether exports increase or decrease compared to imports. The analysis can be completed by using statistical data obtained from different sources such as customs statistics, foreign exchange rate data, etc. This type of research has been used by economists and statisticians around the world for many years because it provides very accurate results on how much income will be generated by exporting goods and services. The purpose of this study is to analyze the current situation and trend of promoting international trade.

References

1. Song, H.X.: Ensuring and implementing the cultural rights of citizens: based on the law of culture promotion. J. Henan Univ. (Soc. Sci.) (2018)
2. Daniel, C.: Compatibility in the law of treaties and stability in international law. Br. Yearbook Int. Law (2022)
3. Yousif, A.Y.: Legal protection for the safety of the traveler in commercial space flights: a study in uae law comparing international treaties and American law (88) (2021)
4. Nwoke, U.: Imposition of trade tariffs by the USA on China: implications for the WTO and international trade law. J. Int. Trade Law Policy **19**, 62–84 (2020)
5. Li, Q., Li, M., Guo, A., et al.: Research progress on growth curve fitting analysis of goose (2022)
6. Lv, Y., Wang, L.: A study on overall promotion mechanism of rule of law and regulations-based governance of the party—based on dynamic analysis of socialist rule of law with Chinese characteristics. Henan Soc. Sci. (2018)
7. Zhu, T., Chen, X., Gao, Q., et al.: Law of load growth in different industries based on logsitic regression model and adaptive density clustering algorithm. Power Syst. Clean Energy (2019)
8. Barral Martínez, M.: When public international law meets EU private international law: an insight on the ECJ case-law dealing with immunity Vis-À-Vis the application of the Brussels Regime. In: In: Sooksripaisarnkit, P., Prasad, D. (eds.) Blurry Boundaries of Public and Private International Law, pp. 217–238. Springer, Singapore (2022). https://doi.org/10.1007/978-981-16-8480-7_12
9. Wang, Y., Fang, Y., Xiaohui, W.U., et al.: Predicting method of the developing indexes based on the growth curve for extra-high watercut period of Daqing Placanticline oilfields. Petrol. Geol. Oilfield Dev. Daqing (2019)
10. Guo, J., Tang, C., Wang, M., et al.: Study on growth curve fitting of Holstein cows. J. Domest. Anim. Ecol. (2018)

The Training Path of Management Accounting Talents Under the Background of Great Wisdom Moving to the Cloud

Yanling Zhang[1](✉) and Guanshao Wu[2]

[1] Chengdu College of University of Electronic Science and Technology of China, Chengdu 611731, Sichuan, China
zhangyanling@163.com

[2] Chengdu Junchuang Tiancheng Fiscal and Tax Management Co., Ltd., Chengdu 610213, Sichuan, China

Abstract. The main idea of this paper is to explore the training path of management accounting talents under the background of great wisdom moving to the cloud. The author thinks that: 1) management accounting is an important profession in modern society. It is not easy for people without accounting ability to become managers; 2) Great wisdom is one of the main requirements for becoming a manager, but it is also difficult for anyone who is not good at thinking; Therefore, we can only rely on cloud computing technology and big data analysis technology to improve our thinking ability. The training of management accounting talents is a key factor for the development of enterprises. The purpose of this study is to understand whether the talent training path has been improved and whether it can be used as an effective method to improve the quality of management accountants. In this article, we will first introduce some background information of management accounting, and then introduce our methods and results.

Keywords: Cloud background · Management accounting · Talent training path

1 Introduction

Management accounting talents are becoming more and more important to enterprises and other aspects of society. Therefore, the cultivation of management accounting talents needs a good social environment and enterprise environment. Therefore, starting from the main responsibility, all social units and departments should change the previous financial accounting training ideas, pay attention to the training of management accounting talents, create a favorable environment for the training of management accounting talents from practical actions, and create more practical opportunities for the training of management accounting talents.

In view of the requirements for talents from the perspective of big data, enterprises must quickly recognize the characteristics of big data, change their thinking, form a new concept for the training of compound accounting talents, speed up the establishment of

J. C. Hung et al. (Eds.): IC 2023, LNEE 1045, pp. 557–563, 2023.
https://doi.org/10.1007/978-981-99-2287-1_80

new objectives for the training of accounting talents, adjust the curriculum, integrate curriculum resources, study new elements of the training of compound accounting talents, and study the training of accounting talents from a new perspective [1].

The teaching staff of enterprise accounting major should take the initiative to understand the talent impact of big data on enterprises, refine the new trends and requirements of accounting talent development in combination with the guidance of government policies, formulate the development plan of accounting professionals in line with the enterprise, formulate the curriculum system, and organize internal and external training and learning for relevant excellent accounting workers, so as to meet the needs of professional development as soon as possible. Actively explore the cultivation of talents in Colleges and universities.

As an important force in the training of accounting talents, colleges and universities have made some positive attempts in the training of management accounting talents in recent years. For example, the Central University of Finance and economics has specially established the Department of management accounting at the undergraduate stage, enrolling more than 50 students every year [2]. By means of professional training and organizing students to participate in practical teaching at the internship base, it has trained more than 200 undergraduates majoring in management accounting. A number of universities such as Shanghai University of Finance and economics have strengthened the development of management accounting courses, took management accounting as an important part of accounting teaching, and set up a research direction of management accounting in the graduate stage, teaching students management accounting knowledge, and providing enterprises with management accounting talent reserves to a certain extent. Based on this, this paper studies the training path of management accounting talents under the background of the great wisdom moving to the cloud.

2   Related Work

2.1   Background of Smart Mobile Cloud

"Great intelligence of cloud and things" is the core focus of future scientific and technological development, "cloud" is cloud computing, "things" is the Internet of things, "big" is big data, "intelligence" is artificial intelligence.

With big data as the core, it has built a comprehensive information system, realizing the perfect penetration and integration of people and things, things and machines, and machines and information.

So under the information system, what role does "cloud intelligence" play? Today, I'm going to give you a simple and intuitive analysis of the differences and connections between "cloud intelligence".

"Cloud" generally has a considerable scale. Some well-known cloud providers such as Google cloud computing, Amazon, IBM, Microsoft, Alibaba, etc. also have millions of servers. The "cloud" built on these distributed servers can provide users with unprecedented computing power. Figure 1 below shows the background of Dazhi mobile cloud.

"Great intelligence moving to the cloud" refers to big data, intelligence, mobile Internet, cloud computing and Internet of things technologies. The wide application of

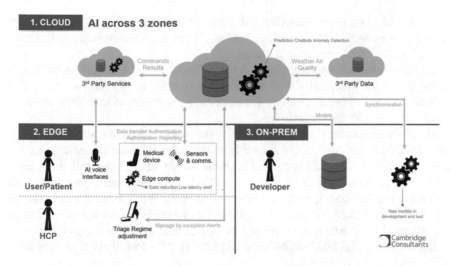

Fig. 1. Cloud moving background of Dazhi

"great wisdom moving to cloud" technology in the fields of production and social life has had a profound impact on the competitive environment, business model and enterprise management mode, and promoted the process of enterprise transformation [3]. As an important supporting means of enterprise transformation, financial management is an effective tool to realize the optimal allocation of resources under Organizational Transformation and business transformation. 1. The traditional financial work adopts the decentralized, closed and manual mode. The complex transaction behavior is continuously compressed in the accounting subjects. Each compression is the loss of information value until the information is compressed to the minimum data set. In this information processing process, the financial department discards the process data reflecting the business operation status of the enterprise, and only records the operation results.

Needless to say, big data provides high-performance data analysis and calculation through the management and processing of massive data. The processing of large amounts of data requires the support of hardware equipment, and cloud computing is the "supplier" of infrastructure. As a carrier connecting everything in the world through sensors, the Internet of things provides an available "data source" for the analysis of big data.

2.2 Management Accounting Personnel Training

This article mainly talks about the development of cost accounting, general ledger accounting and other accounting, assistant office clerk of the Department, etc. to management accounting. In fact, the real management accounting is the department head and the employees of each position, because each position in the enterprise is a responsibility center and cost center [4].

Every post needs to spend money and create value, but some posts create value that can be seen directly, and some posts create value by turning a few corners (indirectly).

The latter is often more important but also the most difficult to control. No matter what position, as long as you spend money, it is the cost center. Cost management is a very important part of management accounting.

It is inevitable for domestic financial personnel to transform to management accounting, but it may not be that all data are collected and sorted out by the financial personnel. Management accounting mainly uses and analyzes data to discover the essence of problems and predict the future.

All the data are collected and sorted out by the business department according to the actual situation in their daily work. However, management accounting needs to unify the calculation caliber of financial data and business data in advance (the definition should be clear). At present, this work is usually operated by the financial department, so that the Finance and business have a data interface to avoid duplication of work. The data generated by the business department will not be disorderly and independent, but will be continuously and orderly concentrated to the financial department every day.

After unifying the interface between financial data and business data, it is necessary to cultivate the cost awareness of business department heads and employees, and decompose all performance, costs and expenses into departments and posts. It should be noted that when training the cost awareness of employees, we should try our best to avoid using these data to assess employees [5]. Once the assessment is conducted, the timeliness, accuracy and integrity of the data will be problematic, and unexpected obstacles will be encountered in the process of data collection. When the historical data is relatively timely, accurate and complete, and has reference value, it will be considered to be included in the assessment.

3   Research on the Training Path of Management Accounting Talents Under the Background of Great Wisdom Moving to the Cloud

Establish an evaluation mechanism for management accounting talents, actively build a framework for the training of management accounting talents and an evaluation system for management accounting talents. On the one hand, the evaluation system can promote the training of management accounting talents in the process of evaluation, timely feed back the imperfections in the training of management accounting talents reflected in the evaluation results, and timely update and improve the training plan and objectives of management accounting talents, Realize the long-term talent development plan of management accounting, and keep the management accounting work continuously and effectively serving the social development; On the other hand, it also provides an effective way for enterprises and institutions to select excellent management accounting talents.

In addition, we should encourage and promote the construction of management accounting qualification examination and certification system, learn from the current financial accounting assessment system, which is divided into primary management accounting, intermediate management accounting and senior management accounting, organically combine the examination and evaluation system, further select typical management accounting cases to be open to the public, actively carry out case seminars, and constantly improve the management accounting talent training mechanism [6]. The following Fig. 2 shows the training path of management accounting talents.

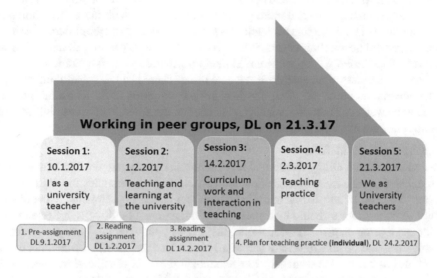

Fig. 2. Training path of management accounting talents

However, it is not possible to simply rely on Finance (Management Accounting) to make business department heads and employees have cost awareness. What management accounting can do is to analyze historical data, re evaluate whether the policies being implemented by the enterprise (including performance policies) match the development direction of the enterprise, and whether it can guide the whole enterprise to pay attention to cost management according to the actual internal and external conditions [7]. It would be best if a systematic solution could be put forward to the problems existing in the enterprise policy. If the solution could be approved, the business department heads and employees could be cost conscious.

In a word, management accounting is actually every employee in the enterprise. The management accountant in the financial department is more like a project manager who is responsible for the management accounting business of the whole company. The heads of departments and employees are like members of the project team. If the financial department wants to be valued, it must start the management accounting business. As mentioned earlier, the management accounting thinking is the thinking of management, business and boss [8]. Management accounting report (internal management report system) is what the boss most wants to see but can not describe the needs in detail.

4   Management Accounting Requirements

Familiar with data analysis and information system: the essence of management accounting is the quantitative management of enterprises. The ability of data analysis is very important for any management accountant. Management accounting must speak with data, analyze and make decisions based on past and present data, summarize the present and foresee the future [9]. Decisions that cannot be made without data.

At the same time, with the rapid development of information technology, information system has also been more and more popular in enterprises. With the information platform, the value of management accounting is further excavated and amplified. Therefore, for management accounting talents, it has become a necessary ability to master certain information technology knowledge and skilled information system operation.

The relevant data of an enterprise is a gold mine. There is a lot of useful information for the enterprise, but if the enterprise does not pay attention to it, the useful information may be wasted. Therefore, management accounting must learn to use information systems and screen useful information from them.

Good communication skills

The communication ability mentioned here includes not only language communication ability, but also written communication ability and image communication ability. First, the ability to communicate and coordinate with and across departments; Secondly, the ability to write reports should be logical and clear, and can be directly reflected to the senior management of the headquarters [10]; Finally, they should have the ability to demonstrate, and be able to briefly and clearly introduce the specific actual situation of the whole business to the management personnel at the level of general manager of the headquarters, and communicate with them about the future forecast and analysis of the company.

5   Conclusion

What this paper studies is the research on the training path of management accounting talents under the background of the great wisdom moving to the cloud. Based on the characteristics of management accounting, the application of management accounting is not something that can be controlled by a single Department of the financial department, but requires the action of the whole company, especially the recognition and promotion of the management.

Therefore, as a management accounting talent, in addition to having the basic professional quality, it also needs to have the perspective of the manager, which is very important for the middle and senior management accounting personnel. Having a manager's perspective is conducive to team building and performance realization, because a lot of work of management accounting is to cooperate with all departments of the company. For example, in preparing the annual budget, we should first determine the annual strategic objectives, then decompose them to each department level by level, and then determine the specific plans of each department. In the whole process, the financial department should be responsible for communication and coordination, startup and promotion, etc., which require management accountants to be competent from the perspective of managers.

Acknowledgements. Accounting research project of Sichuan Provincial Department of Finance «Research on the innovation of management accounting talent training mechanism in Sichuan Province Based on the integration of industry and Finance», Item No: 2021-SCKJKT-025.

References

1. Tao, Y.: The explore and analysis to the training path of management accounting talents under big data environment. Guide Sci. Educ. (2019)
2. Cheng-Ai, L.I., Yue-Qin, G.E.: Research on the optimization of management accounting talents training model under the trend of financial sharing. J. Zhejiang Wanli Univ. (2019)
3. Gong, W., Huang, J., Xiao, Y.: A summary of research on the training path of applied big data talents under the background of new engineering. In: International Conference on Education, Economics and Information Management (ICEEIM 2019) (2020)
4. Zhong, R.: Research on the Reform Path of Southeast Asia Bilingual Talents Training Mode under the Background of the Belt and Road (2019)
5. Tang, X.: The training model of translation talents under the background of the "belt and road" initiative. In: Proceedings of the 3rd International Seminar on Education Innovation and Economic Management (SEIEM 2018) (2019)
6. Zhao, B., Yang, J., Ma, D., et al.: Exploration and Research on the Training Mode of New Engineering Talents Under the Background of Big Data (2), 1 (2018)
7. Liu, Y.: Research on the Influence of Artificial Intelligence on the Training of Accounting Talents and Strategy (2021)
8. Cultivation of International Accounting Talents under the Background of Informatization. J. Phys. Conf. Ser. **1852**(4), 042030 (7pp) (2021)
9. Sheng, S.: New Connotation of Financial Management under the Background of Internet + (2018)
10. Wei, Z., Huang, Y., Chen, F., et al.: Software Testing Talents Training Mode under the Background of Talent Training (12), 6 (2018)

The Value Added Promotion of Internet Technology to Tourism Economy

Yaojin Zhou[✉]

School of Tourism Management, Wuhan Business University, Wuhan 430056, Hubei, China
zhouyaojin@wbu.edu.cn

Abstract. "Internet+" is widely used in today's society. Simply speaking, it refers to the introduction of Internet technology into the traditional industrial structure by using the current information and communication technology and the major platforms of the Internet, so as to guide the development of the traditional industry into a new era. The characteristics of the "Internet+" era include online information, precise consumption and digital management. In the era of "Internet+", the discussion on the mutual integration of the tourism industry should clearly analyze why the tourism industry can develop with the Internet technology under the background of this era, and why the Internet technology can vigorously promote the reform of the tourism industry. Suggestions for promoting the integration of tourism industry are as follows: promulgate relevant policies and measures, improve the level of tourism informatization, strengthen the establishment of relevant laws and regulations, protect consumers' personal privacy, develop tourism education, and cultivate compound tourism talents.

Keywords: Tourism Industry Chain · "Internet plus" Technology · Integration of the Industrial Chain

1 Introduction

"Internet+" is widely used in today's society. To put it simply, it refers to the introduction of Internet technology into the traditional industrial structure by using the current information and communication technology and various platforms of the Internet, so as to guide the development of traditional industries into a new era. Although in our country, the Internet related technology is developed in recent years to gradually, but today, the Internet technology already has a mature system, and into the high speed development stage, according to the national network information center related according to the report, released in recent years, the scale of the Internet users in China has reached billion level, this information is showed in our country has formally entered the era of "Internet+".

In the era of "Internet+", the combination of the tourism industry and information technology can be made use of the Internet technology. From the perspective of supply, it is conducive to promoting the innovation of the tourism industry, the optimization of the structure of the tourism industry, the improvement of the competitiveness of the tourism industry and the optimal allocation of tourism resources [1]. From the perspective of demand, it is beneficial to meet the increasingly diversified, differentiated and personalized tourism consumption needs of tourism consumers. Therefore, it is a problem to combine the traditional tourism with the Internet so as to change the traditional tourism with the Internet technology.

2   Features of the "Internet Plus" Era

With the advent of the era of "Internet+", the Internet can be used as a new technology in people's life and production. And the traditional industry is also in this time of a new revolutionary change, and want to use the Internet to bring dividends, we must seize the characteristics of the Internet to combines with traditional industry chain, only the correct grasp to the characteristics of the Internet era, and the Internet into the traditional industrial chain, can maximize the strengths of the Internet technology (Fig. 1).

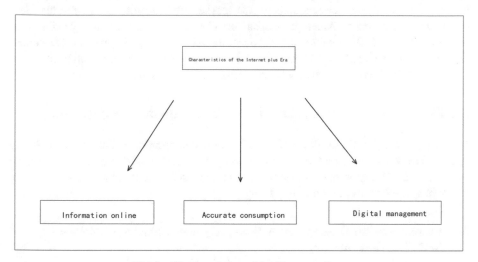

Fig. 1. Characteristics of the "Internet+" era

2.1   Information Online

In today's market, information asymmetry has caused many industries to fail to maximize profits, and the emergence of Internet technology can fundamentally solve this problem. The arrival of the "Internet+" era is also accompanied by the improvement of mobile communication network speed [2]. Consumers can access more and greater information through connecting to the network through mobile phones or computer terminals, and can use fast network communication to realize the rapid transmission of information, so as to realize the online information.

2.2   Precision Consumption

In the era of mass consumption, most consumers have a typical herd mentality, the market is relatively unified, and the goods can not meet the individual needs of consumers. With the advent of the era of "Internet+", combined with the level of the national economy continued to rise, consumers in consumption is also has a more and more strict requirements, and the use of the Internet technology can largely improve the quality of consumer spending, and for the individual needs of consumers using Internet technology to fast and accurate access to consumer's personal needs, to meet consumer demand of precision.

2.3   Digital Management

The wide application of Internet in today's society has opened an era of digital management. Digital management has advantages that traditional management mode does not have. Its advantages are mainly reflected in two aspects [3–5]. First, digital management mode can make the scarce resources in the market give full play to their value. Secondly, through digital management, consumers can be systematically analyzed according to their consumption habits and big data analysis, and reliable product recommendations can be provided for this analysis, and even customized products can be made for consumers to achieve high-quality consumption in the real sense.

3   The Dynamic Mechanism of Tourism Industry Integration

In the era of "Internet+", the discussion on the mutual integration of the tourism industry should clearly analyze why the tourism industry can develop with the Internet technology under the background of this era, and why the Internet technology can vigorously promote the reform of the tourism industry.

3.1   Prerequisite for Integration: A High Degree of Industrial Relevance in the Tourism Industry

As a tertiary industry, tourism industry has a strong correlation within the industry. Every node in the industry can be connected from top to bottom, which also forms the characteristics of tourism industry. High industrial correlation enables Internet technology to be given full play.

3.2   The Fundamental Driving Force of Integration: To Meet the Precise Needs of Consumers

In the tourism industry, consumers pay attention to the sense of experience, and to be able to quickly and accurately meet the needs of consumers has become the biggest problem. There are many problems in the tourism industry at the present stage, especially the poor experience of consumers [6]. Therefore, the use of Internet technology can intuitively present the tourist attractions and some attached products in front of consumers, so that consumers can choose to consume, and then improve the experience of consumers.

4 Obstacles Existing in China's Tourism Industry Integration

There is a huge gap in the integration of related industries, which makes the boundaries between different industries become particularly obvious, so that similar industries can not achieve unity. The use of Internet technology to integrate related industries can not only make the correlation between industries stronger, but also improve the interoperability of industries, so as to change the nature of the industrial chain and achieve a win-win situation. In the tourism industry, the basic conditions for the integration of industries have been mature, but what we can see is that in the tourism industry, industries have not formed a strong correlation, and it is still in the traditional development mode [7].This phenomenon shows that in the tourism industry, the integration of industries exists, but due to the limitations of various factors, it is impossible to truly achieve the integration of industries. From the perspective of the characteristics of the "Internet+" era, this paper summarizes the limiting factors of the integration of tourism industry chain into the following four main factors: 1.2. Restrictions on personal privacy security of tourism consumers; 3. The tourism management system is not perfect; 4. Lack of versatile talents (Fig. 2).

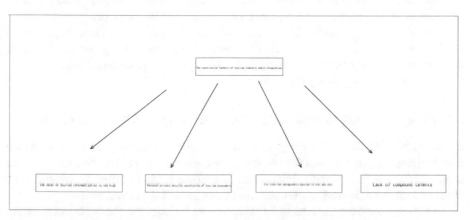

Fig. 2. Constraint factors of the mutual integration of tourism industry chain

5 Suggestions on Promoting the Integration of Tourism Industry

5.1 Introduce Relevant Policies and Measures to Improve the Level of Tourism Informatization

The characteristics of online information in the "Internet+" era determine to some extent the depth of the integration of the tourism industry chain. Whether the tourism industry chain can be integrated, the most important thing is whether the products derived from the tourism industry can meet the consumer demand. Whether the consumer demand can be satisfied depends on whether the integrated tourism industry chain has realized the online development of information [8, 9]. Therefore, relevant government departments should issue relevant policies to support the development of tourism informatization, improve Internet technology and actually apply it to the integration of the tourism industry, realize online management of consumer demand and meet personalized consumer demand of consumers.

Therefore, suggestions can be taken as follows: comprehensively guide and enhance the awareness of tourism informatization among the general public, and deepen this idea among the general public, so that every public can realize the importance of tourism informatization. For the integration of the tourism industry chain, the government should also give strong support, preferential treatment in the policy aspect, support in the coordination aspect, and then promote the perfect integration of the tourism industry chain. Talent training is also quite important. In terms of talent training, the existing traditional tourism practitioners should be promoted in information technology, and the training of new talents should also focus on information technology.

5.2 Strengthen the Establishment of Relevant Laws and Regulations to Protect Consumers' Personal Privacy

The arrival of the era of "Internet+" makes the application of mobile communication technology more mature, which provides a huge platform for the development of online tourism. However, we should also clearly understand that the continuous improvement of Internet technology also means that consumers' personal privacy has great security risks. In order to solve this problem, relevant legal systems should be established in time to protect consumers' privacy security.

5.3 Develop Tourism Education and Cultivate Compound Tourism Talents

In the era of "Internet+", the integration of the tourism industry is being promoted to a wider range, so the demand for compound tourism talents is also increasing. Therefore, it is the key to realize the integration of tourism industry chain to strengthen the development of tourism education and train more technical tourism talents.

In this regard, the measures to be taken are as follows: for the high-level tourism talents should be introduced vigorously, study first and then train. According to the current situation of China's tourism industry and talent needs, the introduction of the lack of high-level tourism talent, and to create a good development environment for them [10]. On the one hand, a reasonable reward and promotion system should be developed for high-level tourism talents to improve the attraction of the tourism industry to talents. On the other hand, it is necessary to strengthen the connection between tourism industry education and the tourism industry system, cultivate talents that can be applied to the current market demand, improve the education and training system, so as to form a comprehensive talent training mechanism.

6   Conclusion

The arrival of the "Internet+" era can truly integrate the traditional tourism-related industries, so as to create an industrial chain with the flavor of the new era. This is also a typical example of the practical application of Internet technology, and provides a foundation and reference for other industries to use this technology. As for the talent training after the integration of the tourism industry chain, information technology training should also be emphasized, and the training should be conducted in different levels and directions to adapt to the new tourism industry chain after the integration. In this paper, from the time characteristics of "Internet+", the integration of the tourism industry chain and the talent training after the integration, the application of this technology to the tourism industry chain is analyzed and interpreted in an all-round way and reasonable suggestions are given.

Acknowledgments. Research on Coordinated Development of Hotel Industry from the Perspective of Whole Area Tourism – A Case Study of Wuhan City (Philosophy and Social Science Research Project of Hubei Education Department, Project No.: 19G060).

References

1. Luo, M.: Principles of Tourism Economics. Fudan University Press (2004)
2. Chen, Z., Wu, X., Tang, S., et al.: Spatial difference of tourism economy development in Jiangsu Province economic geography **028**(006), 1064–1067 (2008)
3. Lu, L., Yu, F.: Analysis of spatial characteristics of China's tourism economic differences. Econ. Geogr. **25**(003), 406–410 (2005)
4. Chen, X., Wang, D., Zhang, Y., et al.: Analysis on the evolution of tourism economy in Liaoning Province in time and space economic geography **29**(001), 147–152 (2009)
5. Zhu, D., Lu, L., Yu, H.: J. of the role of tourist destinations in the Yangtze River Delta Metropolitan Area from the Perspective of Tourism Economic Network Economic Geography (04), 149–154 (2012)
6. Fang, Y.L., Huang, Z., Wang, K., et al.: Spatial-temporal difference analysis of tourism economy in China Province based on PCA—ESDA. Econ. Geogr. (08), 149–154 (2012)
7. Ning, Z.Q.: M. on Tourism Economy, Industry and Policy China Tourism Press (2005)
8. Late. Exploration of Tourism Economy for 20 Years of Reform and 20 Years of Reform and Opening up Guangdong Tourism Press (1998)
9. Li, F.: Research on vulnerability measurement of tourism economic system in China based on set pair analysis (SPA). In: Proceedings of the 16th National Symposium on Regional Tourism Development (2012)
10. Xiang, Y., Zheng, L., Wang, C.: A Study on Spatial Measurement of Factors of Growth in Tourism Economy. Econ. Geogr. **32**(006), 162–166 (2012)

Traffic Engineering Investment Estimation Method Based on Genetic Algorithm

Tao Wu[⊠] and Qun Zhou

Cccc First Highway Consultants Co., Ltd., Zhuozhou 150077, Hebei, China
wutao0515@foxmail.com

Abstract. The investment estimation of urban rail transit project is an important basis for investment decision-making of urban rail transit project construction. Accurate and rapid engineering investment estimation directly determines the correctness of investment decision-making. At the same time, it is also a decisive factor for the success of cost control in the later stage of the project. The traffic engineering investment estimation method based on genetic algorithm is a technology that can be used to estimate the traffic engineering investment cost. It is an evolutionary computation (EC), which uses the principle of evolution, also known as natural selection. In this method, the initial population of candidate solutions is generated by randomly selecting individuals from a set of possible solutions. This process will continue until all individuals are selected or there are no more candidates in the solution space. Then, the remaining populations will undergo a selection process to eliminate solutions that do not meet certain survival criteria.

Keywords: Genetic algorithm · Traffic engineering · Investment estimation

1 Introduction

In the early stage of urban rail transit construction, the estimation of project investment is very necessary. It is an important basis for investment decision-making of urban rail transit projects, and also directly determines the capital bearing capacity of the authorities. If the project investment estimation deviates greatly from the actual situation, either the estimation is greater than the actual situation, resulting in the project unable to pass the decision, or the decision is passed, but due to the high estimation, a lot of unnecessary waste is caused; Either the estimate is lower than the actual, and the actual investment results may also lead to the problem that the financial capacity of local governments is ultimately unbearable [1]. Many indicators will act on the final result of cost estimation. In the early stage of the project, the relationship between characteristic indicators and cost can be analyzed, and targeted technical measures can be taken to effectively control the total construction cost of the project.

At present, in the field of urban rail transit engineering investment, in addition to the application of analytic hierarchy process, fuzzy mathematics theory, grey system theory, regression analysis and other basic scientific achievements, the invention and application of various software (such as Guanglianda, Luban, sville, shenjimiaoxian,

J. C. Hung et al. (Eds.): IC 2023, LNEE 1045, pp. 570–576, 2023.
https://doi.org/10.1007/978-981-99-2287-1_82

Pinming, Hongye, Xindian, etc.) have greatly improved the efficiency of engineering investment and accelerated the development of engineering investment estimation [2]. However, the premise of using these software to calculate the quantities and further obtain the project investment is that the blueprint of the project to be evaluated must be prepared first. This method has relatively high requirements for the design in the decision-making stage, and may be applicable to some builders with rich experience in rail transit construction. However, if the technical capacity of the consulting unit or the design institute is limited, or the owner himself is unfamiliar with the construction of urban rail transit, it is easy to have large errors between the project blueprint and the actual situation, resulting in large errors in the results of project investment estimation, The accuracy is not high [3]. Because genetic algorithm can process nonlinear data quickly and accurately, it quickly becomes a fast and accurate estimation and prediction of complex data, and is widely used in the field of engineering investment. However, the research on the application of genetic algorithm to project investment estimation is currently mainly concentrated in the fields of road and bridge, real estate and other construction projects, and the application research applied to urban rail transit project investment estimation is relatively few [4]. Therefore, in this context, it is particularly urgent to give full play to the advantages of its method, apply genetic algorithm to the field of urban rail engineering investment and obtain more accurate and rapid cost estimation.

2   Related Work

2.1   Influencing Factors of Project Investment Estimation

In 2002, Hu Zhenyi analyzed and summarized the main relevant indicators of urban rail transit investment estimation based on the cases of urban rail transit projects organized by the state in which he participated, and put forward suggestions on unified and standardized investment estimation of urban rail transit projects.

In 2009, the "investment estimation index of Urban Rail Transit Project" was released, which is a unified specification of investment estimation index of urban rail transit project nationwide, which is conducive to the project investment estimation of new and expansion projects of urban rail transit in various parts of China according to the unified estimation index, meets the preparation requirements of project proposal and feasibility study report, and promotes the further effective control of project construction investment [5].

In 2012, after analyzing the main index factors affecting the project cost of urban rail transit, the full text Juan believed that the project cost could be effectively reduced scientifically and reasonably through five key elements and measures, such as doing a good job in the planning of urban rail transit network and paying attention to resource sharing [6].

In 2012, Wang Fangying summarized historical project cases, analyzed the factors that caused the differences in the comprehensive cost indicators of the project and affected the comprehensive cost indicators based on the composition of the project cost, and summarized the key index factors that can act on the final cost results. Thus, it provides an important basis for the investment estimation in the early stage of the project, the budget estimation in the design stage, and the project cost of the whole project cycle [7].

In 2013, from the perspective of the composition of urban rail transit construction project investment, Wang Guofu analyzed the cost investment of each system part according to the relevant completion samples and indicators, and determined the influencing factors affecting the urban rail transit project cost. Combined with these influencing factors, the optimization of scheme design, station structure and scale are put forward.

2.2 Application of Research Methods for Project Investment Estimation

In the early stage, the project investment estimation of the proposed project is carried out by using the project investment data of similar proposed projects and adjusting the relevant project index values. Although this method can predict the cost to a certain extent, the disadvantages of large error are also obvious. With the improvement of domestic economic level and the development of computer science, computer science has been used in various fields.

In the 1990s, some technology companies developed many advanced cost software. In 2002, the first domestic international symposium on software measurement was held at Tsinghua University. In October 2013, the specification for software R & D cost measurement was officially released, which is the first unified standard in the field of software measurement in China. Through continuous development, the function of cost software in China is becoming more and more perfect and widely used [8]. These softwares store the information of local quota and bill of quantities pricing specifications into the computer, so that the computer can have automatically available specification information. Then engineering technicians prepare blueprints and import these blueprints into the cost software. The cost software first calculates the quantities automatically, and then grabs the quota and specification information for project investment. This method takes more time and is highly resource dependent [9].

In recent years, scholars have also achieved fruitful results by using some new valuation models and artificial neural network technology to estimate project investment.

3   Traffic Engineering and Investment Estimation

3.1   Project Investment

Project investment is the construction cost of the project, and the construction price of the estimated or actual expenditure of the construction project. Engineering investment has two meanings.

The first meaning: engineering investment in a broad sense is the total cost of engineering construction and the sum of fixed assets, intangible assets and working capital spent on engineering projects. This meaning is from the perspective of investors and builders, including all the investment costs of the construction process from the early stage evaluation and decision-making of the project to the final completion acceptance.

The second meaning: project investment in a narrow sense is the contract price of construction projects formed on the basis of socialist commodity economy and recognized by both parties. From the perspective of the construction unit, this cost is the project contracting price. From the perspective of the construction unit, this cost is the contract price of the project.

It can be seen from the above definition that the project investment studied in this paper is the category of the first meaning. The total price estimation of urban rail transit project is the estimation of the investment amount in the early stage of urban rail transit project construction [10].

The investment of urban rail transit is divided into engineering costs, special costs, other costs and reserve funds. According to the professional category, the project cost can be divided into 16 items, including station, section, track, communication, signal, main substation, power supply, integrated monitoring, disaster prevention and alarm, environment and equipment monitoring, security and access control, ticket vending, escalator and elevator, platform door, operation control center, depot and comprehensive base, civil air defense engineering, etc.

3.2 Confirmation of Estimation Index System

According to the construction principle of the index system, the experts sorted out the table information based on the expert interview results of the characteristic factors of urban rail transit engineering. Combined with the research object of this paper, after research and discussion, they believed that the engineering characteristics of stations, sections and station spacing are important factors affecting the cost, which are largely determined by the landform, geological conditions and construction methods, but they are relatively difficult to quantify, Similar cases can be adopted to simplify the impact of these elements. At the same time, if the station spacing is determined by the line length and the number of stations, it should be deleted according to the principle of simplicity. At the same time, the line length and the number of stations can also be more comprehensive response indicators. Therefore, these indicators can be simplified into the number of stations and line length. The rail system is determined by vehicle type selection and marshalling. At present, the most commonly used metro system in China is the 6-car marshalling metro system. This research object focuses on the 6-car marshalling metro system rail transit, making the sample comparable. Due to the obvious homogeneity of communication and signal, the line length can be used for estimation, so that the project investment can be estimated on the premise of selecting the communication and signal with common configuration without spending a lot of energy in the early stage of the project. Integrated monitoring system, disaster prevention and alarm, environment and equipment monitoring, security and access control, automatic ticket selling, elevator, platform door and civil air defense engineering are mainly affected by the number of stations, depots, parking lots, main substations and traction substations in the underground rail transit system. Considering that the two indicators of station and line length have been adopted previously, these factors can be ignored according to the principle of independence. Depot, parking lot and main substation are greatly affected by their functions and have strong independence, which should be retained as the main influencing factor. Regional economic differences and the rise in the price of labor and materials are all affected by the region and the year of construction. According to the principle of comprehensiveness and independence, the year of construction and regional influencing factors are retained. Considering the difficulty of obtaining the number of samples, this paper plans to select the case as a case of similar projects. Therefore, in this case analysis, the regional influencing factors are simplified, which is convenient for the normal progress of the study without affecting the scientific and reasonable premise, and the rise of labor and material prices is not the main influencing factor.

4 Traffic Engineering Investment Estimation Method Based on Genetic Algorithm

4.1 Genetic Algorithm

Genetic algorithm (GA) is an American scholar pr John Holland and his students did computer simulation research on biological systems in nature. It was developed by pr Johnholland first proposed a kind of optimization algorithm based on Darwin's theory of biological evolution and Mendel's theory of genetic variation to simulate the process of biological evolution in 1975, which is a global optimization search algorithm with adaptive heuristics.

Genetic algorithm starts with a population representing the potential solution set of the problem, and a population is composed of a certain number of individuals encoded by genes. Each individual is actually an entity with characteristic chromosomes. Chromosome is the main carrier of genetic material, that is, the collection of multiple genes. Its internal performance genotype is the arrangement and combination of certain genes, which determines the external performance of individual morphology. For example, the feature of black hair is determined by a certain genome combination in the chromosome that controls this feature. Therefore, it is necessary to map and encode from phenotype to genotype at the beginning. Because the work of imitating gene coding is very complex, binary coding is generally used to simplify it. After the emergence of the first generation population, according to the principle of survival of the fittest and survival of the fittest, generation by generation evolution produces better and better solutions. In each generation, individuals are selected according to the fitness of individuals in the problem domain, and combined crossover and mutation are carried out with the help of genetic operators of natural genetics to produce a population representing the new solution set. This process will lead to the offspring population produced by the population like natural evolution, which is more adaptable to the environment than the original population. After decoding, the optimal individual in the last generation population can be used as the approximate optimal solution of the problem, as shown in Fig. 1.

4.2 Traffic Engineering Investment Estimation Based on Genetic Algorithm

Coding is the primary problem to be solved when using genetic algorithm optimization. The determination of coding method will affect the selection, crossover, mutation and other operators. Generally, binary coding or real coding can be used to code the research problem, and the coding methods of different problem types are different. Binary coding is simple and convenient, and has strong anti-interference ability. At present, it is widely used. Considering that the network weights and thresholds to be optimized in this paper are all real numbers, in order to ensure that the search space of the algorithm is sufficient,

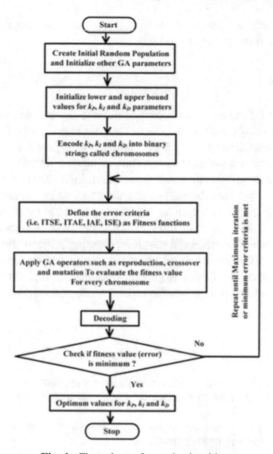

Fig. 1. Flow chart of genetic algorithm

the search speed and accuracy of the algorithm meet the requirements, and the coding length will not be affected by human factors, this paper adopts real number coding. Suppose the number of input neurons is n, the number of output layer neurons is k, the number of hidden layer neurons is m, and the coding length is l, then:

$$L = \begin{cases} \max, f(m) \\ s.t.n \in R \\ k \subseteq U \end{cases} \tag{1}$$

Using genetic algorithm model to estimate the investment of urban rail transit project, we can use the engineering characteristic data of completed projects as the input of genetic algorithm, take the engineering investment of these projects as the output of genetic algorithm model, carry out a lot of repeated training and learning, and constantly adjust and optimize network parameters (i.e. weights and thresholds) in the training, until the optimal parameters are finally achieved. Finally, taking the engineering characteristic data of the project to be built as the input, the output value obtained by running the above trained genetic algorithm model is the predicted value of engineering investment.

Therefore, the genetic algorithm can realize the accurate and rapid prediction of the project investment of the project to be built.

5 Conclusion

Project investment estimation is an important basis for investment decision-making of urban traffic engineering construction, and it is also the main influencing factor to control the project investment in the whole project stage in the project investment management project, which determines the success or failure of urban rail transit project investment. In the early stage of project construction, under the realistic situation of relatively limited design conditions, how to use the engineering features that are relatively easy to obtain, simplify the engineering investment estimation work, and at the same time, be able to obtain more accurate and fast estimation results, is one of the efforts of the engineering investment management of urban rail transit engineering projects. From the perspective of the owner, this study constructs an urban rail transit project investment estimation index system, which can be applied to genetic algorithm. Compared with the traditional index system, this index system has the characteristics of simplicity, practicality and good accessibility.

References

1. Han, X., Zhang, D.: Research on the training mode of applied talents in traffic engineering based on big data under the background of "new engineering." J. Phys. Conf. Ser. **1744**(4), 042050 (2021)
2. Han, X., Wang, D.: Research on the application of system theory in traffic engineering. J. Phys. Conf. Ser. **1972**(1), 012112 (2021)
3. Gwa, E., Jtbc, D., Lzbc, D., et al.: Research on rapid source term estimation in nuclear accident emergency decision for pressurized water reactor based on Bayesian network. Nucl. Eng. Technol. (2021)
4. Li, L., Lei, K.: Research on influence factors of bearing capacity of concrete-filled steel tubular arch for traffic tunnel. Symmetry **14**, 167 (2022)
5. Zhou, Y., Guo, F., Yang, W., et al.: Research on the effectiveness of a intersection risk warning system based on driving simulation experiment. Sci. Rep. **12**(1), 12261 (2022)
6. Pérez-Galarce, F., Candia-Véjar, A., Maculan, G., et al.: Improved robust shortest paths by penalized investments. RAIRO – Oper. Res. **55**(3), 1865–1883 (2021)
7. Dai, Y.: Research on the Design of Sports Injury Estimation Model based on Big Data (2021)
8. Chen, Y.: Research on R&D investment of new energy companies under the trend of green transformation. Technol. Invest. **13**(2), 11 (2022)
9. Paul, J., Woodside, A.G.: Five decades of research on foreign direct investment by MNEs: an overview and research agenda (2021)
10. Xiong, X., Zhang, S., Shen, Y.: Research on urban road traffic accident characteristics and countermeasures: a case study of ningbo city. J. Phys. Conf. Ser. **1910**(1), 012008 (8pp) (2021)

University Laboratory Safety Education System Based on Big Data Technology

Qun Liang and Hai Wang[✉]

College of Mechanical, Electrical and Quality Technology Engineering, Nanning University,
Nanning 530000, Guangxi, China
815291562@qq.com

Abstract. University laboratories are not only the place for teaching and scientific research, but also the cradle of talent training. The state attaches great importance to the construction of university laboratories and laboratory safety management. In the document of higher education teaching reform project, the state clearly proposed to establish a number of demonstrative and influential teaching and research bases and a number of demonstration centers competent for basic courses, including the vigorous construction and transformation of teaching and research laboratories in Colleges and universities. The research on university laboratory safety education system based on big data technology is a scientific research aimed at investigating the effectiveness of university laboratory safety education system based on big data technology. The researcher from China conducted the study with a team of researchers from Canada, Australia and other countries. This study aims to understand the effectiveness of university laboratory safety education system based on big data technology in preventing laboratory accidents. The study also hopes to find out whether there are any problems and concerns in the use of such systems.

Keywords: Big data technology · University Laboratory · Safety education

1 Introduction

University laboratory is not only the place of teaching and scientific research, but also the cradle of talent training. In the document of higher education teaching reform project, the state clearly proposed to establish a number of exemplary and influential teaching and research bases and build a number of demonstration centers competent for basic courses, including the vigorous construction and transformation of teaching and research laboratories in Colleges and universities, and also proposed to increase the construction of laboratories in the medium and long-term teaching reform development plan [1]. The document of the Ministry of education points out that we should have a comprehensive understanding of the laboratory work of undergraduate colleges and universities. In view of the current situation of laboratory work in Colleges and universities, such as laboratory conditions and environment, fund guarantee, management and operation, team construction and experimental teaching reform, we should focus on analyzing the

J. C. Hung et al. (Eds.): IC 2023, LNEE 1045, pp. 577–583, 2023.
https://doi.org/10.1007/978-981-99-2287-1_83

reasons for the problems and solutions, and study and formulate policies and measures to strengthen the laboratory work of colleges and universities in the new period [2]. At the same time, in order to Promote Colleges and universities to strengthen laboratory construction and management, make full use of university equipment, provide support for scientific research, train talents and serve the society, a research group on the policy and mechanism of laboratory work in undergraduate colleges and universities has been established to carry out multi-month investigation and research work with the laboratory work branch of the Chinese Association of higher education and Tsinghua University.

With the continuous reform of the education system, the development of colleges and universities has led to the continuous expansion of the scale of university laboratories, the increasing number of laboratory related personnel, the continuous rise in the number, types and value of experimental resources in school, showing a trend of mobility and openness, enhanced sharing, high-frequency use, heavy tasks and so on [3]. At the same time, the safety pressure is also increasing. There are many kinds of unsafe factors involved, causing greater and greater harm. The probability of laboratory safety accidents is greater, which is very likely to harm the lives of teachers and students, and cause losses to research results and state-owned property. Judging from the current safety accidents, most of the safety accidents are sudden and accidental, and the number of victims is small, mainly due to the impact of teaching and research order and the damage of property research data. If the safety evacuation is carried out in an orderly manner when the accident occurs, the possibility of malignant accidents is small. However, in the era of informatization, accidents are very easy to ferment on the network platform, causing the attention of the media and the masses [4]. The increase in the amount of attention has accelerated the spread speed, making the social influence great, and bringing unpredictable negative effects to the school. The school often bears great pressure, affecting the long-term harmonious development of the school and the country.

2   Related Work

2.1   Safety Management Concept

Safety management is a comprehensive system discipline, which is an important branch of management science. Safety management means that the operators and production managers of production and business units scientifically organize, command and coordinate all employees to carry out safety production activities in accordance with certain safety management principles in order to achieve safety production objectives.

Safety management mainly refers to labor safety management, which is an integral part of enterprise management. It is to carry out relevant decision-making, planning, organization and control activities for the purpose of safety. The main content of safety management is a series of organizational measures taken to implement the national guidelines, policies, laws and regulations on work safety and ensure safety in the production process [5]. For example, establish and improve the safety organization, formulate and improve the safety management system, prepare and implement the plan of safety technical measures, carry out safety publicity and education, organize safety inspections, carry out safety competitions, summary and evaluation, rewards and punishments, etc.

Its task is to find, analyze and eliminate various dangers in the production process, prevent accidents and occupational diseases, avoid various losses, and ensure the safety and health of employees, so as to promote the smooth development of enterprise production and serve to improve economic and social benefits. The scope of safety management includes safety production and labor protection.

The elements to measure the risk are the possibility of the accident and the severity of the consequences caused by the harmfulness. By analyzing the frequency and severity of accidents, we can evaluate the risks and formulate risk control measures, so as to manage, reduce or eliminate the risks [6]. On the other hand, the traditional safety management method focuses on controlling the frequency of accidents to reduce accidents, and has achieved some success. However, the severity of accident consequences can be assessed, predicted and controlled in some aspects. This principle emphasizes that under certain conditions, the possible accidents and their severity can be evaluated and predicted, the risk can be reduced by reducing the severity of the accident, and the occurrence of major malignant accidents can be prevented and controlled; Not only by reducing the frequency of accidents to reduce losses.

2.2   Big Data Informatization Improves Laboratory Safety Management Level

In the process of national development, the proportion of investment in education, especially in the construction of university laboratories, is increasing. The development and expansion of university laboratories in quantity or scale has promoted the number and mobility of laboratory personnel to grow day by day. The use of experimental equipment and environmental changes are difficult to control, so all kinds of safety accidents follow, which has brought great challenges to the traditional laboratory safety management mode. In the era of information technology, the state clearly pointed out in the relevant development plan that it is the general trend to improve the laboratory safety management level with information technology, and it is also one of the most effective methods to deal with the laboratory safety problems and improve the management level at present [7]. Therefore, in the context of national support for the construction of laboratories, the safety management of laboratories can carry the development tools of the times. It is in this opportunity that major colleges and universities have fully seized the opportunity of the times and have successively embedded informatization into the safety management process of school laboratories. The informatization construction of the laboratory operation management guarantee system of Tianjin University proposes to take the laboratory as the core, build a full life cycle platform for laboratory operation management with the idea of opening and sharing, and use various informatization means [8]. Xi'an Jiaotong University has made full use of advanced information technology to build a comprehensive laboratory safety management system based on various informatization platforms. Both have achieved certain results one after another, It plays a great role in the management efficiency of the laboratory.

After discussing the development form of school laboratory informatization security management, the school leaders analyzed and discussed the development thinking of laboratory informatization security management at many Party committee and principal working meetings in the first half of 2016 in combination with the reflected problems of laboratory informatization security management development. After many meetings, the development idea of laboratory informatization security management was clarified, that is, the development of laboratory informatization security management needs to formulate development goals and feasible specific measures, put them into practice one by one according to the development goals and measures, supervise the completion of development tasks on time, make the development form a certain scale, collect suggestions and opinions in the development process, and adjust the direction of development, Promote the formation of orderly and dynamic development [9]. At the same time, after discussion at the meeting, it was believed that the content of laboratory information security management was complex and professional, and the leading department needed to organize and coordinate all relevant departments to participate in the construction, open up the management process, concentrate management forces, and form a dynamic management mode of multiple departments, multiple personnel, and multiple channels.

3   College Laboratory Safety Education System Based on Big Data Technology

3.1   Current Situation of Laboratory Management

After communicating with relevant stakeholders of the laboratory, according to their description, the following problems still exist in the management of items and equipment in the laboratory:

(1) There is no platform for laboratory related knowledge, so that students and laboratory staff can browse and download laboratory related learning materials and training courses at will.

(2) At present, more traditional offline teaching is still used, and the traditional manual examination is used to check the students' and employees' mastery of laboratory safety knowledge. The efficiency and results are not satisfactory.

(3) At present, the notification methods such as email and blackboard are still used to release the news and announcements related to the laboratory, and there is no unified information release site.

(4) At present, the purchase, requisition, management and other related processes of laboratory chemicals and biological reagents are still paper documents, and the implementation efficiency of the process is not high. Some dangerous chemicals lack information records and effective monitoring and management.

(5) Complex work processes such as large-scale equipment procurement process and project demonstration process need to be countersigned by multiple parties. At present, there is no reliable and stable online process processing mechanism.

(6) The basic information, equipment information and annual reports of the laboratory are lack of input and automatic statistical system.

In order to solve the above problems, a laboratory information management system with reasonable design, simple use and convenient maintenance is needed to help laboratory personnel manage and daily work more efficiently, so as to promote the rapid development of laboratory construction.

Strengthening the concept construction of the harmfulness of safety accidents and the role of information safety management bears the brunt. The school should organize the propaganda department to cooperate with the new media institutions to comprehensively understand and investigate the past school laboratory safety accident cases and the implementation of information management by major universities or social units, analyze and summarize the occurrence process, causes and results of the accident cases, especially focus on the social public opinion caused by the safety accidents and the impact on the development of the school, The excellent experience of information management should be used for reference [10]. Then, in the offline publicity work, we should pay attention to posting warnings and publicity information of safety accident cases and excellent cases of information management on the school bulletin boards and bulletin boards, regularly place safety accident publicity display boards in the public areas of the school's laboratories, teaching buildings, teacher apartments, student apartments and other buildings, and also use new media communication tools online, Establish a public window for publicity and combine the offline publicity with joint publicity, so as to make the rapid circulation of warning and publicity information such as the harmfulness of school laboratory safety accidents and the advantages of informatization safety management, so that teachers and students can easily understand the latest information anytime and anywhere, and promote relevant personnel to form a cognition of the importance of laboratory informatization safety management.

3.2 College Laboratory Safety Education System

The objects of safety education and training are lower and higher grade undergraduates, postgraduates and other scientific researchers. The content of safety education and training adopts the education and training of ideas, safety common sense, safety skills and safety operation procedures.

(1) Classroom training. Classroom teaching is the most formal teaching method of safety education. Take the safety training course as the basic textbook for classroom training, and systematically explain the laboratory safety knowledge. Through this way of teaching, students can have a very systematic and comprehensive concept of the safety knowledge they have learned, and carry out deeper human learning and research.

(2) Hold lectures. Compared with classroom training, safety lectures are thematic. You can select a topic in the safety course for detailed explanation, such as taking hazardous chemicals as the theme, guiding how to deal with emergencies through case analysis, or taking fire safety as the theme, preaching relevant knowledge of fire risk avoidance and escape. This kind of education method is highly operable, focused, vivid and touching.

(3) Safety practice. Safety practice is a teaching mode of integrating theory with practice. Compared with the traditional teaching method, it is more conducive to enhance the enthusiasm and sense of achievement of teachers and students to participate, improve the effect of safety education activities, cultivate emergency response ability, and master the basic skills of escape, self-help and rescue in danger, so as to ensure that the experimental personnel are injured as little as possible in accidents.

Specific practical drills can simulate chemical leakage for personnel evacuation drill, simulate poisoning first-aid measures, master necessary skills such as cardiopulmonary resuscitation, and simulate fire scene to carry out fire equipment drill, including the correct use of fire extinguishers and fire hydrants.

The system deployment architecture is shown in Fig. 1. The servers of the whole system are divided into four types: website server, database server, file server and cache server. The website server is used to deploy system applications, and users usually mainly access this server. Install the database software on the database server. The file server is used to store the file resources required by various systems such as videos, documents, pictures, etc. When users need to watch videos online or download documents, they can access the file server. In this way, the pressure on the website server will be greatly reduced. An application cache cluster is set up on the cache server to store the cached data of the system. All servers are Gigabit Ethernet bandwidth, set up in the school computer center, and connected to each other through LAN.

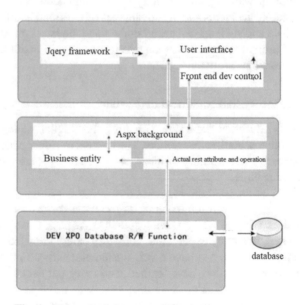

Fig. 1. University laboratory safety management system

4   Conclusion

University laboratory is an important base for talent training, and laboratory safety is the fundamental guarantee for the implementation of talent training and scientific research. Through the safety education in the laboratory, improve the safety awareness and master the necessary safety rules, so as to reduce the injury and damage to personnel and property in the process of experimental teaching and scientific research, put an end to the occurrence of safety accidents, create a harmonious environment for teachers and students, and enable the smooth development of experimental teaching and scientific research in Colleges and universities.

Acknowledgement. Professor Cultivation Project of Nanning University, No. 2021JSGC18; Research Foundation of Young and Middle-aged Teachers in Guangxi Universities, No. 2022KY1785.

References

1. Zheng, X., Miao, F., Chakpitak, N., et al.: Discussion of University Chemistry Laboratory Management Using DOSA Platform and Safety Education Based on Blockchain (2021)
2. Wei, C.: Research on university laboratory management and maintenance framework based on computer aided technology. Microprocess. Microsyst. **6**, 103617 (2020)
3. Gao, Z., Wu, B.: Research on the innovation system of university production and education integration based on computer big data. IOP Conf. Ser. Earth Environ. Sci. **692**(2), 022025 (15pp) (2021)
4. Yan, H., Yin, Q.: Research on open system of economic and management laboratory center based on cloud platform. J. Phys. Conf. Ser. **1550**(2), 022020 (6pp) (2020)
5. Ye, Y.: Research on university personnel management information system based on database. In: ICDEL 2020: 2020 the 5th International Conference on Distance Education and Learning (2020)
6. Zhang, H., Zhang, Y., Ma, Y., et al.: Dietary rumen-protected L-arginine or N-carbamylglutamate attenuated fetal hepatic inflammation in undernourished ewes suffering from intrauterine growth restriction. Anim. Nutr. **7**(4), 1095–1104 (2021)
7. Jump, R.C., Scavette, A.: The labor market effects of place-based policies: evidence from england's neighbourhood renewal fund. Working Paper (2022)
8. Hong, A.J.: Establishment of virtual-reality-based safety education and training system for safety engagement. Educ. Sci. **11**, 786 (2021)
9. Zhu, W.: Research on college english teaching system based on computer big data. JPhCS (2021)
10. Yang, B.J.: Research on teaching design of the course of fundamentals of college information technology based on big data. Sci-tech Innov. Prod. (2020)

University Teaching Quality Evaluation Technology Based on OLAP and SVM Algorithm

Miaomiao Xu[✉]

Anhui Xinhua University, Hefei 230088, Anhui, China
xumiaomiao232@163.com

Abstract. The focus of this research is to develop a new technology based on OLAP and SVM algorithm to evaluate the quality of university teaching. The main goal of this study is to find the best technology to measure the quality of teaching. This study will use descriptive statistics, correlation coefficient, regression analysis and cluster analysis and other data analysis techniques. The final results will be used to develop an effective evaluation system, which can measure the effectiveness of University Teachers' contributions to students' learning. In the research, we try to use OLAP and SVM algorithms to improve the quality of teaching evaluation. First, we used a large number of student data from different universities in China. We collect data from more than 20 universities every semester. Then, we build a large OLAP database based on this data set (about 1million records), which can be used as the input of SVM algorithm. Next, we designed three types of artificial neural network (ANN) models based on previous research. This method can be used for both undergraduate and graduate students. The results of this study will help universities evaluate their teaching quality, which will enable them to improve their teaching quality.

Keywords: College teaching · SVM algorithm · OLAP · Evaluation system

1 Introduction

Computer information technology has been applied to more and more colleges and universities and personal management business, which brings great convenience to various management of colleges and universities. At present, the efficiency of teaching evaluation management in most colleges and universities is very low, the management of information related to teaching evaluation cannot be effectively managed, and there are many flaws in a series of processes of teaching evaluation information management. Therefore, it is urgent to use the teaching evaluation system to assist colleges and universities to scientifically manage and control the information related to teaching evaluation, so as to improve the efficiency of teaching evaluation management [1], The purpose of scientific, accurate and convenient management teaching evaluation.

Teaching evaluation system is an indispensable part of a school. Its content is very important for school managers, so it should be able to provide users with sufficient information and fast query means. However, people have been using the traditional manual method to manage the relevant teaching evaluation information. This management method has many shortcomings, such as low efficiency and poor confidentiality. In addition, over time, a large number of documents and data will be generated, which brings a lot of difficulties to find, update and maintain.

It is required to investigate the teaching evaluation management platform and select the appropriate programming language and database management system. The whole system can be logically divided into data addition, deletion, modification and query functions.

The system manages different user roles according to their permissions. The main functions provided by the system for students' roles include personal information maintenance, teachers' teaching evaluation, and viewing their own evaluation information; It provides teachers with the ability to view the curriculum of their courses and their own evaluation information, evaluate students, view and manage class evaluation information; The functions provided for administrators (background users) include viewing and managing all departments, majors, courses, students, teachers' teaching evaluation information, class evaluation information management, and teachers' information. Through these management modules, we can effectively improve the management level of teaching evaluation. Based on this, this paper studies the evaluation technology of college teaching quality based on OLAP and SVM algorithm.

2 Related Work

2.1 SVM Algorithm

Support vector machine (SVM) is an important algorithm of statistical machine learning. It is a new machine learning method based on the statistical learning theory and the principle of structural risk minimization. It can effectively solve the problems of high dimension and nonlinearity, and effectively carry out classification and regression. Compared with other classifiers, SVM has better generalization [2].

The essence of SVM algorithm is to classify the collected two sample groups into two categories, which are expressed by the positive and negative values of kernel function respectively, and isolate the positive and negative categories. That is, the distance between heterogeneous vectors closest to the hyperplane is the largest. Figure 1 below shows the intuitive understanding of SVM algorithm.

In the figure, there are some two-dimensional data points and three straight lines belonging to two categories respectively. If the three straight lines represent three classifiers respectively, which classifier is better?

We should think that the answer is H3 by intuition. First of all, H1 can't separate categories, and this classifier is definitely not good; H2 is OK, but the separation line is only a small distance from the nearest data point. If the test data has some noise, it may be misclassified by H2 (that is, it is sensitive to noise and has weak generalization ability). H3 separates them at large intervals, so that it can tolerate some noise of the test data and classify correctly. It is a classifier with good generalization ability.

$$\min f(x_1, x_2, ..., x_n) = \frac{1}{m} \sum_{i=1}^{m} \left[Y_i^0 - Y_i \right]^2 \tag{1}$$

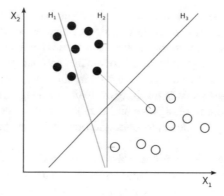

Fig. 1. Intuitive understanding of SVM algorithm

$$\Delta w(i, y) = -\eta \frac{\partial e}{\partial w(i, y)} \tag{2}$$

For support vector machine, if the data points are p-dimensional vectors, we use the hyperplane ° of P-1 dimension to separate these points. But there may be many hyperplanes that can classify data. A reasonable choice of the optimal hyperplane is the hyperplane that separates the two classes at the maximum interval. Therefore, SVM selects the hyperplane that can maximize the distance from the nearest data point to the hyperplane.

2.2 Teaching Evaluation

This paper analyzes the requirements of university subject core literacy on university classroom teaching, and based on the existing evaluation tools and expert interview results, constructs an index system. The index system is composed of six first-class indicators, which are "teaching objectives", "teaching content", "teaching organization", "teacher-student communication", "cognitive needs" and "literacy cultivation". Each index corresponds to three observation points, When establishing the weight, the analytic hierarchy process is used to assign values to each index. The evaluation index system constructed has fewer observation points and strong operability, but the effectiveness of evaluating core literacy is low. The three observation points of the "teaching organization" index are "whether the teaching language is clear and standardized", "whether the teaching methods are flexible and effective" and "whether the questioning and feedback are effective and timely" [3]. The determination of the observation points reflects the scientific concept, and there is no detailed description of how the index system reflects the core quality of disciplines in Colleges and universities.

The evaluation index system is completely constructed. In the process of construction, the requirements of the core quality of disciplines in Colleges and universities are combined, the core quality of disciplines in Colleges and universities is fully penetrated, and the classroom evaluation index system is transformed into the opinions of questionnaire survey experts on each index. After several rounds of expert opinion consultation, the correction of the index system is completed, and then the weight of each index system is constructed by logarithmic weighting method [4]. Finally, the developed evaluation index system is composed of two-level indicators and twenty evaluation standards. The evaluation index system constructed by this research institute comprehensively and accurately reflects the requirements of core literacy, as shown in Fig. 2:

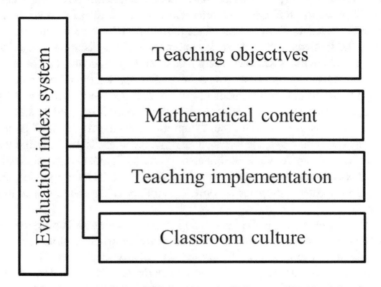

Fig. 2. Classroom evaluation system in Colleges and Universities

Through the multi-level evaluation form of the classroom evaluation scoring system (class) and the research of various scholars, it can be found that the classroom evaluation index system established by different scholars from different angles is different, but through the analysis, it can be found that the establishment of the index can be mainly divided into two perspectives. The first is from the perspective of teachers and classrooms: teachers should fully analyze the teaching content of this class before class to determine what abilities to develop students What qualities to cultivate students, etc. In classroom teaching, teachers should pay attention to the construction of students' thinking and communicate with students more. Second, from the perspective of students: what abilities students want to develop through this class, and whether students' abilities have been developed through the guidance of teachers [5]. But no matter from which angle, the ultimate goal is to implement students' core qualities and provide scientific guidance for teachers to adjust their teaching behavior through evaluation.

3 Research on University Teaching Quality Evaluation Technology Based on OLAP and SVM Algorithm

In the traditional teaching mode, only in the formal, open and strict teaching activities such as classroom teaching competitions, combined with student feedback, can the expert group give a more authoritative performance evaluation, so as to evaluate the teaching quality of a teacher or a class relatively accurately. However, in daily teaching activities, there is no condition to conduct a perfect evaluation of each teaching activity through an expert group, and a complete quality evaluation of a course from beginning to end naturally cannot be achieved. Teachers can understand the requirements of learning for teaching content through the teacher-side educational administration system, which can not only break the limitations of time and space, but also facilitate the school's management of teachers. Performance management performance management is actually to input the students' academic achievements of various disciplines in the teaching system [6]. There are many kinds of disciplines of college students, and the statistics of examination results is a very cumbersome work. The application of computer network technology into performance management can not only reduce the errors in the process of statistics, but also facilitate students to view it at the first time. When students have doubts, they can give feedback in time, which enhances the tightness of performance management. Evaluation and feedback college management is not only the unilateral management of teachers for students, but also the right of students to evaluate teachers. Of course, this evaluation refers to the evaluation of teachers' teaching content and quality during the semester. Computer network technology is also widely used in college management.

The learning methods of SVM include supervised learning and unsupervised learning. Supervised learning is called supervised learning when samples are given by input/output pairs. The example of input-output relationship is called training data. Input/output pairs usually reflect a functional relationship that maps input to output. When there is an intrinsic function from input to output, this function is called the objective function. The estimation of the objective function output by the learning algorithm becomes the learning solution. For classification problems, this function is also called decision function.

The above problems can be reasonably solved by combining the data mining method of support vector machine to deal with the data of teaching quality evaluation. Those data that deviate greatly from the actual results are often treated as dirty data or outliers in the process of data preprocessing, or when passing through the operation system, some dimensions or attributes correspond to a low weight, so they cannot have a fundamental impact on the final result, so a relatively accurate classification result is obtained [7]. In addition, SVM overcomes the shortcoming that the neural network structure depends on the designer's experience, solves the problems of high dimension, local minimum and small samples, and takes into account the advantages of neural network and gray model. Through the study of the existing classroom teaching quality evaluation samples, we can get the dependence between the classroom teaching quality and the influencing factors, and then we can make an accurate and objective judgment on the quality of a specific classroom teaching activity.

4 Simulation Analysis

Data preprocessing is an important step in the process of data mining, which refers to some processing of data before the main processing. When data mining, the amount of data is often very large, and it takes a long time to mine and analyze a small amount of data. Data reduction technology and other methods can be used to obtain the reduced representation of the data set, but it is still close to maintaining the integrity of the original data, and the result is the same or almost the same as that before reduction [8]. The original data sources include the online evaluation data submitted by students on the course in the existing teaching database and the data in the expert group evaluation report in the classroom teaching competition. Establish the teaching data warehouse of a school, unify and summarize the data format of the data source through extraction, conversion and other operations, divide it into daily teaching data and classroom teaching competition data, and load it into the data warehouse. Use the OLAP tool of data warehouse to drill and flip the data, which is convenient for the next step of data processing, and provide some parameters and preprocessed data for data mining [9]. The data preprocessing project code is shown in Fig. 3 below.

```python
def entropy(y_label):
    counter = Counter(y_label)
    ent = 0.0
    for num in counter.values():
        p = num / len(y_label)
        ent += -p * log(p)
    return ent
```

Fig. 3. Data preprocessing item code

SVM algorithm is used to train the sample data from the data set of classroom teaching competition, and the classifier is obtained. It is verified on the data of other classroom teaching competitions, and better classification results can be obtained. The classifier obtained by SVM algorithm through sample training is applied to new data, that is, it can evaluate daily teaching activities that have not been evaluated by experts. Provide the results to teaching managers and decision makers as the evaluation results of this daily teaching [10]. That is, through machine learning, using computers instead of experts to evaluate teaching, and get more accurate results at the same time.

5 Conclusion

This research aims to develop a teaching quality evaluation system based on OLAP and SVM algorithm. In order to achieve this goal, we first collect data from different university websites in China, including the following: (1) data collection forms; (2) Data analysis table; (3) Textbook database; (4) Textbook database; (5) University website information database. Then, we use these databases as input data for training and testing algorithms respectively.

References

1. Tantiwetchayanon, K., Vichianin, Y., Ekjeen, T., et al.: Comparison of the WEKA and SVM-light based on support vector machine in classifying Alzheimer's disease using structural features from brain MR imaging. J. Phys. Conf. Ser. **1248**(1), 012003 (6pp) (2019)
2. Gao, H., Ma, S.: Research on the third party evaluation index of higher vocational education based on IGSO-SVM. Bull. Sci. Technol. (2019)
3. Yang, S., Yang, Z., Zhang, S., et al.: Automatic grading method of tobacco leaves based on NIR technology and PSO-SVM algorithm. Guizhou Agric. Sci. (2018)
4. yangzhao. Research on the application of university teaching management evaluation system based on Apriori algorithm. J. Phys. Conf. Ser. **1883**(1), 012033 (6pp) (2021)
5. Gao, K.: Evaluation of college english teaching quality based on particle swarm optimization algorithm. In: CONF-CDS 2021: The 2nd International Conference on Computing and Data Science (2021)
6. Davardoost, F., Sangar, A.B., Majidzadeh, K.: An innovative model for extracting OLAP cubes from NOSQL database based on scalable Nave Bayes classifier. Math. Probl. Eng. (2022)
7. Wang, T.: Evaluation model of multimedia teaching effect about foreign language in university based on differential evolution algorithm. In: 2018 11th International Conference on Intelligent Computation Technology and Automation (ICICTA). IEEE Computer Society (2018)
8. Zhao, J., Wang, A.: Evaluation method and decision support of network education based on association rules. In: 2017 International Conference on Progress in Informatics and Computing (PIC) (2018)
9. Nie, D.X., Wei, W.K., Zhuang, Z.H., et al.: Water quality evaluation based on SVM optimized by the HPSOCS combined PSO and GA. Water Sci. Eng. Technol. (2019)
10. Yang, L.B.: Comprehensive evaluation of power quality based on improved PSO-SVM. Meas. Control Technol. (2018)

Video Quality Diagnosis System Based on Convolutional Neural Network

Hu Yi[✉] and Xiaodong Zhan

Training Center Beijing Polytechnic, Beijing 100176, China
huyi@bpi.edu.cn

Abstract. With the rapid development of modern society, people demand higher and higher performance of various products, there are many quality problems in the process of practical application. Therefore, in order to improve user experience and improve this situation this paper proposes a video quality diagnosis system based on convolutional neural network. The design includes various construction methods, several main framework structures and related databases. This paper takes the video quality during video conferencing as the research object, hopes to build a video quality diagnosis system using the theory of convolutional neural network.

Keyword: Convolutional Neural Network · Video Quality Diagnosis System

1 Introduction

Convolutional neural networks have achieved unprecedented success in the field of computer vision, many traditional computer vision algorithms have been replaced by deep learning convolutional neural networks. Meanwhile, with the development of network technology, video conferencing system has gradually become a mainstream and important way to disseminate multimedia information. In this paper, we use the convolutional neural network model to improve the video quality problems encountered in video conferencing, aim to establish a video quality diagnosis system based on convolutional neural network to improve the quality of video conferencing, so as to achieve the purpose of improving efficiency.

2 Convolutional Neural Networks in Brief

In recent years, convolutional neural networks have become more and more complex as the number of researchers in the field related to convolutional neural networks has increased and the technology is changing day by day. From the initial 5-layer, 16-layer, to the 152-layer ResNet proposed by MSRA [3] or even thousands of layers networks have become commonplace by a wide range of researchers and engineering practitioners.

A simple convolutional neural network is composed of various layers arranged in a sequence, each layer in the network uses a differentiable function to pass data from

J. C. Hung et al. (Eds.): IC 2023, LNEE 1045, pp. 591–597, 2023.
https://doi.org/10.1007/978-981-99-2287-1_85

one layer to the next. Convolutional neural networks consist of three main types of layers: convolutional layers, pooling layers, fully connected layers. By stacking these layers together, a complete convolutional neural network can be constructed. As shown in Fig. 1.

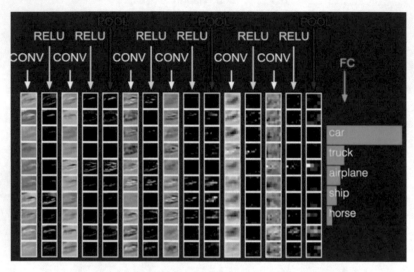

Fig. 1. Complete convolutional neural networks

3   Video Quality Diagnosis System

In recent years, deep learning convolutional neural networks have developed rapidly in the field of machine vision, which can build very complex models by simulating the human nervous system to analyze and interpret data, have powerful expression capabilities to handle complex practical application scenarios, especially convolutional neural networks, which are now widely used in the field of pattern recognition, have shown superior performance in various tasks of computer vision, their unique deep The unique deep structure can effectively learn the complex mapping between input and output. Therefore, the video quality diagnosis algorithm based on deep learning convolutional neural network can extract more features of abnormal video [4], adapt to more complex rules, detect and classify the abnormal types more accurately is completely feasible. Deep learning convolutional neural network algorithm is based on a large amount of data, so expect deep learning convolutional neural network algorithm applied to the field of video quality diagnosis, the following points need to be done (Fig. 2).

First, collect a comprehensive video quality database to form a deep learning convolutional neural network dataset, which is a prerequisite to ensure the generalization performance of the algorithm; second, the video database is organized into a dataset that conforms to the deep learning convolutional neural network model, then a set of schemes is developed, then each video in the video quality database is labeled with

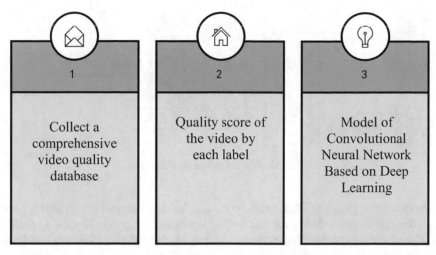

Fig. 2. Focus of building a convolutional neural network video diagnosis system

abnormal types, then the video is scored for quality according to each type of label, the process can be be called subjective evaluation of video; third, establish a deep learning convolutional neural network model based on which can accurately predict the video abnormality types and their video quality scores [5], the predicted results should be consistent with the manually labeled results, the process can be called an objective evaluation method of video quality.

4 Video Quality Diagnosis Algorithm Based on Convolutional Neural Network

4.1 Pool Layer of Convolutional Neural Network in Video Quality Diagnosis

Usually, a pooling layer is inserted periodically between successive convolutional layers. It serves to gradually reduce the spatial size of the data body, which in turn reduces the number of parameters in the network, making it less computationally resource intensive and also effective in controlling overfitting, as shown in Fig. 3. The pooling layer usually uses the MAX operation, which operates independently on each slice of the input data body to change its spatial dimensions. The most common form is to use a filter of size 2 × 2 to downsample each depth slice in steps of 2, discarding 75% of all activation information in it. Each MAX operation takes the maximum value from 4 numbers (i.e., in some 2 × 2 region of the depth slice) [6]. Note that the number of channels in the data body remains constant during the pooling process.

4.2 Convolutional Neural Network Video Diagnosis Algorithm in This Paper

The algorithm in this paper is a multi-task multi-label deep learning network model implemented based on the VGG-16 convolutional neural network as a prototype, using the convolutional network for feature map extraction of multi-frame images, which are

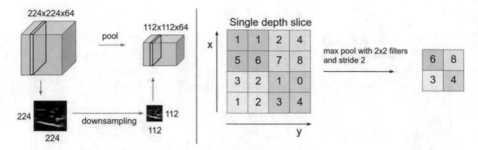

Fig. 3. Role of the pooling layer

then connected together to do anomaly type classification and quality scoring regression tasks. To extend the training set and testing environment, pyramid pooling (SPP) is introduced, the size of the input video images can be unrestricted. Different size images will get different size feature maps after convolutional layers. Since the number of parameters of the fully connected layer is fixed [7], it cannot be connected with the fully connected layer, the different size feature maps can be converted into feature vectors of the same size by means of pyramid pooling. The network structure block diagram is shown in Fig. 4.

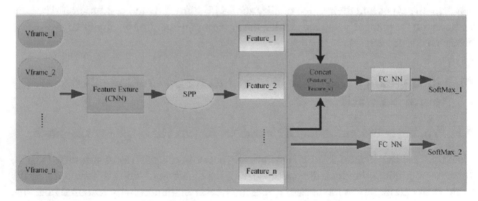

Fig. 4. Network structure block diagram

During the training process, the anomaly type classification is trained first, on this basis the quality score is trained based on the classification results. When using the trained model for prediction, video sequences of different sizes can be input, the abnormality type of the video can be obtained after model processing. If multiple consecutive frames have the same abnormality, it can be assumed that there is some kind of fault in the system, the quality score corresponding to each frame can be obtained by statistical analysis of the relative quality of each frame of the video.

5 Overall System Design Scheme

5.1 Overall System Architecture

The video quality diagnosis system adopts a modular design, including five modules: video frame interception module, OpenCV image processing module, image abnormality detection module, abnormality recording and display module, abnormality alarm module. The workflow of the video quality diagnosis system is mainly as follows: firstly, the required video frames are obtained from the stored surveillance video and saved as Mat entities: secondly. With the help of 0penCV image processing technology. Get the spatial domain structure information that can represent the image content: Finally, give the OpenCV processed image spatial domain structure information to the designed image abnormality detection algorithm for different faults to realize the automatic detection of image abnormality [8], the main functions of each module of the video quality diagnosis system are as follows.

(1) Intercept video frame module: The function of this module is mainly to intercept the stored video image frames, the acquired color image signal is color-separated, separately amplified and corrected to obtain RGB. And the basic image information and pixel data are encapsulated. as the image data to be detected.
(2) OpenCV image processing module: This module makes full use of the image processing library provided by OpenCV to pre-process the image to be detected, such as converting grayscale images, segmenting image channels, color clustering, image space conversion. And obtaining image pixel channel values and other required image structure information.
(3) Image anomaly detection module: this module is the core module of the whole video quality diagnosis system, the main function is to detect the image data information after OpenCV processing, determine whether the detected video frames have abnormal abnormalities including video signal lack of clarity, brightness noise, snow, streaks, color bias, screen freeze, PTZ motion out of control and other abnormalities each image anomaly detection is done by (1) independent algorithm class to complete the input set of video frames for sequential detection[9], return a variety of anomaly detection results.
(4) anomaly logging and display module: this module accepts the return value of the algorithm class of the image anomaly detection module, specifies the description and stores it in the log, so that it can be called when querying the detection results.
(5) abnormal alarm module: this module provides feedback on the detection results to detect abnormal images of the camera according to the abnormal type of mark to remind remind maintenance personnel to solve the problem in a timely manner.

5.2 Functional Module

Video quality diagnosis system mainly includes 6 modules.

(1) each quality diagnosis algorithm uses a single frame or two frames of video frame images at close moments to complete various diagnoses, which do not depend on

the background image, so there is no need for background modeling and the update of kenjing, reducing the false detection caused by unreasonable background model.

(2) Video quality diagnosis is completed by multiple servers, since each server is equally configured, the diagnosis task is equally distributed to each device. With the corpse only need to set up the pre-program of polling detection, LOTUS will start the task according to the start time set in the pre-program, without the need for manual intervention.

(3) Diagnostic results are stored for each recent detection, regardless of whether the camera is working properly or not. Users can query detection records by region [10], fault type.

(4) Fault information stores all the historical fault information of the problematic camera, also contains screenshots of the video frames when the fault is detected, which is easy for the user to view visually. The corpse can query the fault information records of a certain time period by camera, region, fault type. At the same time can be based on the stored video screenshots to determine whether the system is misdetected, can allow misdetected cameras to learn to reduce the probability of misdetection.

(5) In addition, users can count the number of failures and failure rate of cameras in different areas and brands in different time periods according to their own needs, which can be displayed in different forms to facilitate users to understand the operation status of cameras.

(6) In the pre-program management part, the user can set the inspection start time, detection items. In terms of algorithm parameter setting [11], at the early stage of system operation, the algorithm parameters of each camera are set according to the unified threshold; after the system runs for a period of time, the threshold of each detection algorithm of each camera will be adjusted due to various reasons such as equipment quality, service life, site environment and power supply, transmission line., that is, each camera has its own optimal algorithm threshold. In addition, users can set the thresholds of algorithm parameters applicable to special weather such as rain and snow to cope with these bad weather.

6   Conclusion

With the development of the times, video surveillance technology has been widely used, but in the process of its quality diagnosis and control, human operators are required to monitor it. The design of video image detection system based on convolutional neural network described in this paper can avoid the difficulties of traditional video diagnosis as much as possible, provide a new way for video quality diagnosis.

Acknowledgments. Application of neural network in human action recognition.

References

1. MLeCun, Y., et al.: Backpropagation applied to handwritten zip code recognition. Neural Comput. **1**(4), 541–551 (1989)

2. Krizhevsky, A., Sutskever, I., Hinton, G.E.: Imagenet classification with deep convolutional neural networks. In: Advances in Neural Information Processing Systems, pp. 1097–1105 (2012)
3. He, K., Zhang, X., Ren, S., Sun, J.: Deep residual learning for image recognition. In: CVPR (2016)
4. Simonyan, K., Zisserman, A.: Very deep convolutional networks for large-scale image recognition. arXiv preprint arXiv:1409.1556 (2014)
5. Szegedy, C., et al.: Going deeper with convolutions. In: Proceedings of the IEEE Conference on Computer Vision and Pattern Recognition, pp. 1–9 (2015)
6. Ioffe, S., Szegedy, C.: Batch normalization: accelerating deep network training by reducing internal covariate shift. In: ICML (2015)
7. Szegedy, C., Vanhoucke, V., Ioffe, S., Shlens, J., Wojna, Z.: Rethinking the inception architecture for computer vision. arXiv preprint arXiv:1512.00567 (2015)
8. Szegedy, C., Ioffe, S., Vanhoucke, V., Alemi, A.: Inceptionv4, inception-resnet and the impact of residual connections on learning. arXiv:1602.07261 (2016)
9. Huang, G., Liu, Z., Weinberger, K.Q., Maaten, L.: Densely connected convolutional networks. In: CVPR (2017)
10. Xie, S., Girshick, R., Dollar, P., Tu, Z., He, K.: Aggregated residual transformations for deep neural networks. In: CVPR (2017)
11. Hu, J., Shen, L., Sun, G.: Squeeze-and-Excitation Networks
12. Howard, A.G., et al. Mobilenets: Efficient convolutional neural networks for mobile vision applications. arXiv preprint arXiv:1704.04861 (2017)
13. Zhang, X., Zhou, X., Lin, M., Sun, J.: ShuffleNet: An Extremely Efficient Convolutional Neural Network for Mobile Devices

Virtual Reality Interior Home Design Based on Computer Animation Technology

Jinyang Zhou[✉]

College of Architecture and Art, Jiangxi Industry Polytechnic College, Nanchang 330096,
Jiangxi, China
zhou3231608@126.com

Abstract. Interior home design provides people with specific design ideas for better home design, but it is often inconsistent with the actual effect because the home design is not realistic enough. How to make better indoor home effect and reduce the deviation between home design and actual design has become a topic of great research significance. Virtual reality interior home design is a technology that uses computer animation to simulate the appearance and feeling of the real world environment using three-dimensional graphics. It allows you to experience the virtual world through the computer screen, which can be viewed on any modern flat-panel display (such as TV or monitor). Users wear special helmets with lenses to display images in front of their eyes. They can move in these virtual worlds as in the real world.

Keywords: Computer animation technology · Virtual reality · Interior home design

1 Introduction

Computer graphics technology provides us with an effective means to understand the world. With the help of graphics, we can express the objective world and subjective imagination of our thinking, and express the information we need with a strong visual effect. Compared with only using words, it is an important progress in computer application. After more than 30 years of development, computer graphics has been gradually mature and widely used in various fields of society. Today, whether you are from scientific computing, simulation, mechanical design, educational training and other scientific and educational activities, or in films, television, magazines and other media, you can see the existence of graphics [1]. Computer animation, as an important branch of graphics, has been particularly active in recent ten years. Compared with traditional graphics, which only uses static images to represent information, computer animation represents more information in the form of motion. It can be seen that computer animation will occupy an important position in the future information system with its unique charm [2].

With people's continuous pursuit of quality of life, home furnishing has received more and more attention. For home furnishings, people have changed from functional first to comfortable first, and the home style has also developed from simplicity to

© The Author(s), under exclusive license to Springer Nature Singapore Pte Ltd. 2023
J. C. Hung et al. (Eds.): IC 2023, LNEE 1045, pp. 598–603, 2023.
https://doi.org/10.1007/978-981-99-2287-1_86

elegance. Therefore, home design has become an industry. But in fact, when owners invite professional home designers to design, they will be restricted by the designer's design style, and the designer also takes the submission of plane renderings as the main way of expression. In this way, the effect of light and shadow and the difference of materials will lead to a big gap between the real object and the plane effect [3]. As a designer, I hope to have a better way to express my design intention; As a user, he wants to show his personality while getting the real design effect.

Facing this demand, it has become a new research field to apply VR based computer-aided industrial design to the field of home furnishing. It will integrate the subject and object of design, virtual and reality. Through three-dimensional computer graphics to achieve real, interactive and efficient goals, so that the design can be expressed in the computer before the real object, and it is convenient for further modification [4]. At the same time, with the help of the design idea and design method of computer-aided industrial design, we can fully consider the function, shape, structure, visual communication, amenity and other factors of furniture products, and carry out comprehensive creative design, so as to achieve the coordination and unity of human, furniture and environment. This mode will become a new development direction of the future home furnishing industry. In particular, the further improvement of computer software and hardware technology and the further popularization of Internet have also laid a solid technical foundation for this research.

2 Related Work

2.1 Virtual Reality Technology

At present, the industry has begun to use virtual reality technology to express the effect of interior design. So far, interior design has developed in a variety of forms, including hand-painted effect, two-dimensional software effect and three-dimensional effect. Through the application of new technology, it shows the new face of interior design. "Virtual reality, as an independent system, is developing rapidly in interior design. The advantage of using three-dimensional effect in virtual reality over traditional hand painting lies in the real size performance of space". Designers fully display their design ideas through virtual display technology design, which has the characteristics of design specialization, data technology, work efficiency and interactive humanization [5]. In the design process, it will be more efficient, realistic and intuitive, as shown in Fig. 1.

2.2 Problems and Prospects of VR Interior Design Application

At present, there are some problems in the application of VR due to some factors.

"First of all, the use effect of VR is far from expected, and the market reflects that there is a gap in the effect" °. When we search "VR interior design" on the Internet, we will find that many companies are involved in VR technology. However, the result is not as good as previously expected. After spending money to buy technology to build home VR, many home manufacturers found that although the effect of three-dimensional space is acceptable, the effect of users' actual experience is far from what they expected [6].

Fig. 1. Interior design based on virtual reality

Secondly, VR home experience is insufficient, and the technology is not yet mature. Home design is the most appropriate technology display for VR application. Whether it is house purchase or interior design and decoration, when users only see planar drawings and paper renderings, ordinary users cannot perceive the design space. At present, the hardware of VR technology is not perfect, and VR technology has not been popularized. Due to limitations and technical equipment, users will feel dizzy due to the unsmooth virtual sense when they reflect.

"The VR industry is an emerging industry and does not yet have perfect industry standards and norms" [7]. When developing VR technology, various enterprise manufacturers have no unified standards for device interfaces, algorithms, device compatibility, etc., which also affects the development speed of new technologies to a certain extent.

Restricted by hardware facilities, since China has not yet developed a chip for VR, the appearance, weight and display clarity of the product have not reached the expected goals, and various indicators need to be further improved.

"But in general, the prospect of VR technology application is very good. Last year, in some questionnaires for college students," virtual reality became the first, replacing the new energy vehicles and autonomous driving that students paid attention to the previous year. The reason for concern is that students want to get the first chance in their future work and entrepreneurship. The reason why relevant interior design majors pay attention to is the use of VR and AR for industry transformation. The industry transformation here is from plane display to three-dimensional visualization, and the second is user interaction. At present, VR technology is gradually moving from maturity to market. VR technology can bring a sense of existence, which can not be achieved by any previous technology [8]. More and more people will enjoy the immersive interactive experience provided by VR technology, and AR technology will promote the transfer of such experience, from desktop to pocket, and putting the world into pocket will not be limited by the screen. VR technology requires the unification of the virtual world and the real world. Each of us needs to establish a connection between the real world and the virtual world. Without a connection, these technologies are worthless.

3   Virtual Reality Interior Home Design Based On Computer Animation Technology

3.1   Virtual Reality Home Design Based on VRML

In VRML, the description, creation, interpretation and operation of objects are realized through instructions. Its basic working procedures include text description, remote transmission and object generation of local computer. In the part of text description, VRML describes the corresponding three-dimensional scene through a text markup language similar to HTML, which is similar to the programming language to a certain extent. In other words, VRML makes the corresponding description language standard for the text description of three-dimensional space [9].

Through the analysis of the needs of actual users, the virtual reality home design based on VRML should achieve the following three functions.

1) It can meet the personalized needs of users to the greatest extent. Users can query the specific information and quality of various furniture and furnishings in the home design through the virtual reality home design system, and can change the relevant home design, such as the change of home style, the adjustment of home color, the adjustment of furniture placement, etc.

2) Users can not only view the overall effect of indoor home design through the virtual reality home design system, but also move freely in various positions in the room and make virtual interaction. Users can make intuitive feelings about indoor functional areas.

3) In the experience of virtual reality interior home design system, users can intuitively and quickly understand the corresponding spatial scale, functional zoning and traffic lines, and can have a relatively accurate grasp of the overall effect after decoration. At the same time, users can put forward personalized improvement suggestions according to their actual feelings, which is convenient for designers to further improve before actual operation.

Through the demand analysis of virtual reality interior home design, combined with VRML, this paper summarizes the corresponding home design process. The working principle flow chart of VRML is shown in Fig. 2.

3.2   Setting Up the Scene Model

(1) Content of scene model construction

Virtual reality interior home design based on computer animation technology is different from traditional interior modeling. In this system, the design models of all furniture should be as detailed as possible, and the modeling should be carried out in strict accordance with the actual dimensions corresponding to the drawings. The content of modeling includes the overall appearance of furniture and the corresponding internal structure of furniture. At the same time, corresponding interactive operations should be added to facilitate users to have a complete roaming experience [10]. The main building objects of the scene model are shoe cabinets, wardrobe, cabinets, wall cabinets, windows,

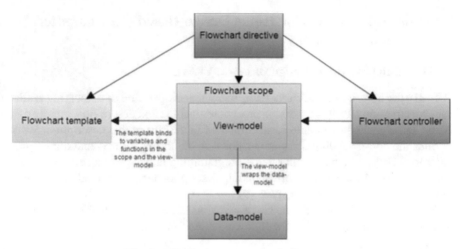

Fig. 2. VRML working principle flow chart

doors and other furniture. For this kind of home, we should make the corresponding independent max file and generate the corresponding WRL file at the same time, so as to realize more convenient modular operation and optimize the interaction in the virtual scene.

(2) Scene model optimization

In order to ensure that the virtual scene is more realistic, the corresponding requirements are put forward for the real-time rendering of the virtual scene system, which usually should be 12 frames per second and up. Therefore, in the process of making the model, the model can neither be very rough nor too fine, so it should be reasonably arranged according to the actual modeling requirements. Specifically, the scene model should be built in the form of polygon modeling as far as possible to ensure that it can not only maximize the satisfaction of user needs, but also avoid redundant model design. It can not only reflect the details of indoor furniture, but also reduce the number of corresponding surfaces of the model.

3.3   Realization of Interactive Animation

When users interact with virtual reality, VRML needs to provide corresponding interactive functions and give comfortable animation effects. This is the relatively attractive difference between VRML and other systems. The integrated scene is controlled by various interpolators and corresponding time sensors, so as to make the corresponding VRML animation. Specifically, in VRML animation, time sensor is defined as a clock generator, whose main function is to control various time parameters such as time interval, end time and start time, and then define the key corresponding to the expected nodes generated by various animations through interpolator. Generally speaking, the user's definition of animation is generated by the system using linear interpolation.

4   Conclusion

This paper designs interior home based on computer animation, and realizes three-dimensional modeling with the help of VRML language to realize the effect of high restoration interior home design. The design of this paper is more concise and convenient, and the investment cost is not high, which is conducive to practical application. The system can adjust the corresponding parameters and position of various home furnishings in the home design, so as to feel the layout effect of the actual space. At the same time, the comprehensive interactive function also provides great convenience for users to carry out better home design.

References

1. Chung, C.C., Cheng, Y.M., Lou, S.J.: Regarding the virtual reality environment design and evaluation based on STEAM learning (2021)
2. Valipoor, S., Ahrentzen, S., Srinivasan, R., et al.: The use of virtual reality to modify and personalize interior home features in Parkinson's disease. Exp. Gerontol. **159**, 111702 (2022)
3. Liu, X., Zhou, J.: Computer Visual Image Design Based on Virtual Reality Technology (2021)
4. Li, C.: Art image simulation design of craft products based on virtual reality and human-computer interactive processing. J. Amb. Intell. Human. Comput. **2021**(4) (2021)
5. Silva, T., Silva, P., Valenzuela, E., et al.: Serious game platform as a possibility for home-based telerehabilitation for individuals with cerebral palsy during COVID-19 quarantine – a cross-sectional pilot study. Front. Psychol. **12** (2021)
6. Silva, A.C.: Designing virtual reality environments through an authoring system based on CAD floor plans: a methodology and case study applied to electric power substations for supervision. Energies. 14 (2021)
7. Li, Y., Luo, H., Zhou, Y.: Design and implementation of virtual campus roaming system based on Unity3d. JPhCS (2022)
8. Xie, G., Chang, X.: Research on application of assembling parts method based on virtual reality technology in interior design. J. Phys. Conf. Ser. **2025**(1), 012078 (2021)
9. Ling, J., Zheng, Y., Chen, X., et al.: Virtual simulation technology for the design of the interior environment in an ultralong tunnel. IOP Conf. Ser. Earth Environ. Sci. **861**(7), 072025 (8p.) (2021)
10. Bai, Y.: Design and reconstruction of visual art based on virtual reality. Secur. Commun. Netw. **2021**(8), 1–9 (2021)

Visual Communication in New Media Art Design

Linlin Nong[1] and Biyue Long[2(✉)]

[1] Nanning University, Nanning City 530000, Guangxi Zhuang Autonomous Region, China
[2] Guangxi Commercial Technician College, Guilin 541000, Guangxi Zhuang Autonomous Region, China
nl1751221@sina.com.cn

Abstract. Art design is the link between life and art. We communicate countless information through vision every day. From ordinary traffic lights to all kinds of information we see are received through vision. It can be said that our life cannot be separated from visual communication design. The study of visual communication in new media art design is based on the analysis of various art works created by new media. The main focus is on the creation and presentation of images. In addition, other aspects such as color, texture and shape are also analyzed. In order to better understand this theme, it is necessary to have a deeper understanding of art history. This will help us understand how artists use color in their works and how they present color by selecting shapes or textures.

Keywords: New media · Visual communication · Art design

1 Introduction

This research topic was studied with my tutor, Mr. Zhou Yue. It is a small branch of Mr. Zhou's research direction. As a student majoring in visual communication, in the face of this ever-changing era, visual communication design is changing with the update of media and the progress of the times, and the designers of visual communication are also facing challenges. At the same time, challenges bring opportunities and elimination. Therefore, it is particularly important to summarize the characteristics of the current era and its impact on design. I think this topic is more practical and practical, So I chose this topic as my graduation thesis topic [1].

On the induction of the boundaries of visual communication, Nie Sen's "transformation and innovation of visual communication design in the digital age" puts forward that the concept of diversity also implies that the visual communication design mode in the new media era will completely break the boundaries of traditional design categories, and make art design a carrier that can include a variety of knowledge systems. In the 21st century, people's thinking has undergone tremendous reform, The connotation and extension of visual communication design have been completely different from the past[2]. Diversification is a significant change in today's visual communication design. On the prediction of the transformation trend of visual communication, Tan Xuhong's "when the new media Yishu Festival is over the visual communication design" clearly points out that communication and interaction are the focus of designers' attention in the design process, because connection and interaction are the most prominent features of new media art. First of all, it must be connected, not viewed from a distance, but can be realized through the keyboard, mouse, light and any sensing system, Or a more precise or even invisible "trigger" to produce an effect. In the exploration of media integration in visual communication, Chen Quan pointed out that the new media communication mode affects the cognitive habits of the audience [3]. Therefore, visual communication works must adapt to the changes in the way of experience and reception, and further enhance the comprehensive and interactive high-tech characteristics. In the digital era, the progress of science and technology is the first major factor affecting design. In Lu Xiangyi's study on establishing the artistic expression of animation in the era, it can be concluded that the influence of science and technology on animation is profound, which is hard to reach by any technical means in the past. Moreover, the influence of science and technology on design is deepening and accelerating. Science and technology enables designers to realize the imagination of animation artists without restriction, The audience can also enjoy more real and wonderful animation products [4]. In terms of how to consider the challenges of the new media era, Ren Gang's article "how to deal with the challenges of traditional media in the new media era" puts forward that science and technology makes information transmission more rapid, and the explosion of information also brings about problems of poor authority and insufficient credibility. These happen to be the advantages of traditional media. How to balance the advantages and disadvantages has become the biggest problem in the new road of traditional media [5].

2 Related Work

2.1 Visual Communication Design

The term "visual communication design" was popular at the world design conference held in Tokyo, Japan in 1960. Its contents include:

The visual design of newspapers and periodicals, the design of magazines, posters and other printed publicity materials, as well as the communication media such as films, television and electronic billboards, which convey the relevant contents to the eyes to carry out the expressive design of modeling, are collectively referred to as the visual communication design film posters. In short, the visual communication design is "the design for people to see and the design for information".

Nowadays, visual communication is no longer a new word. Frankly speaking, all things we can see with our eyes are called visual communication. Designers communicate with the public through the media. Images are pervasive in our lives. Film videos, web interfaces, magazines, packaging design, mid year reports, indoor and outdoor billboards, and movie posters are all media for information communication. The concept of visual communication is actually very easy to generalize. It is a design that takes "vision" as the form of expression [6]. As long as it is any design behavior that can be seen by human eyes, it can be called visual communication design. Visual communication is a practical design influenced by commerce, and commodities designed by visual communication are a bridge between merchants and customers. Visual communication design is mainly composed of three visual elements, namely, graphics, text and color. These three main visual elements are arranged and integrated according to the designer's ideas, so as to attract public attention and convey ideas [7].

Our life can only continue by relying on visual communication as a carrier. None of our learning, reading, writing, spelling and composition can be completed independently without vision. Moreover, pictures often perform an important mission, that is, to do what words cannot do. If you are a person who wants to make a difference in design, then you need to learn its concept. Light transmits images to our brain through the visual nerve [8]. The brain analyzes and defines the reflected things. What is interesting is that our visual cognition is connected with many concepts in our mind, including personal impression, education level, growth environment, etc. That is to say, our knowledge system filters the image information we see, which is obviously a complex process. The job of visual communication designers is to find the most moving things in these complicated clues, and then turn them into design works through artistic refinement and processing.

2.2   Concept of New Media Era

The word "new media" comes from P Goldmark got a development plan. After that, e Rosenberg used the term "new media" many times in his report submitted to President Nixon. Thus, the term "new media" became popular and spread to all parts of the world.

The development of the times has created new media. In the late 20th century, with the rapid development of science and technology, the so-called new media is relative to the traditional media, Professor Xiong Chengyu of Tsinghua University believes that "new media is an ever-changing concept. There are extensions on the basis of today's network, the problem of wireless mobile, and the emergence of other new media forms related to computers, which can be described as new media. For example, digital magazines, digital newspapers, digital broadcasting, mobile SMS, mobile TV, Internet, desktop windows, digital TV, digital movies, touch media, etc. in the United States Wired magazine has a relatively general new definition for new media: "communication from all people to all people." According to the data of Comcore Market Research Institute, the technology blog business insider recently listed the top 20 websites in the world. Among them, Sina ranks 19th, Sohu 16th, Taobao 13th, baidu 9th and Tencent 7th. There are five Chinese websites on the list. The more readers symbolize the more eyes, the stronger brand and the richer advertisements. It can be seen that Chinese websites have become full-fledged. It can be said that they have entered the network era, and it is also the prerequisite and necessary condition for the arrival of the new media era [9].

The signal of the arrival of the new media era is right around us. An online shopping mall was officially launched on the "double 11" in 2013. The group president of the "unexpectedly home" also said in the interview: "e-commerce is the trend of the times. Those who follow the trend will prosper, those who go against the trend will perish, and they can only adapt and cannot be stopped." Singles' day, a grass-roots festival that was born five years ago, has become the world's largest consumption Carnival in terms of sales volume and consumer interaction volume under the guidance of some leading commercial enterprises, According to the CCTV News survey data, in 2012, the 24-h sales of Alibaba's tmall mall reached 19.1 billion, 230 million independent Internet users, and 10 million consumers in the first minute; In 2013, that is, this year, this seemingly alarming figure has been refreshed. The sales volume of 19.1 billion in 24 h in 2012 reached within three hours in 2013. The sales volume in the first minute exceeded 100 million, and the sales volume in 6 min and 7 s exceeded 1 billion, exceeding the sales volume in September in Hong Kong. In 2009, it took 24 h to reach this figure, and five years later, it took about 6 min. The new media era has made online shopping a mainstream demand, This is a major trend for the whole people and the whole industry[10].

3   Analysis of Visual Communication Design Problems Under the New Media Trend

(1) Simplicity first

Since most of the current design fields focus on the requirements of customers, and a set of design schemes need to be modified many times, the designers have already had certain design templates and corresponding materials for backup. Such resources can make the design more efficient, but they will gradually form a fixed pattern, resulting in the design scheme gradually tending to be streamlined and uniform. Such design ideas are obviously not conducive to improving the design quality, If we don't pay enough attention to it and can't optimize it, it will inevitably make the designer's thinking become rigid gradually.

(2) Lack of intuitiveness

There is a great difference between the thoughts of many designers and those of the viewers. Therefore, the design scheme finally obtained by many designers will be more obscure, which is not conducive to the understanding of the viewers and will make the viewers unable to understand the main idea of the creator even after watching the design. This goes against the core of the visual communication concept, that is, the communication of information and the emphasis on interactivity. In short, if the viewer cannot clearly understand and understand the theme to be expressed in the design, the whole design process will fail and have no practical significance.

(3) Cannot actively fill content

During the design process, although the theme and part of the content are fixed, just like giving a framework, it is most important for the designer to use different elements to fill the framework, improve the design content, and integrate more relevant into the design scheme. Totems, murals, legends, history, poetry and ancient characters are inexhaustible. It is precisely for this reason that designers often feel that there is no place to start, so they will use some fixed materials, which leads to insufficient attraction of the design scheme. This is the key to enrich the design scheme, but it is also a problem that many designers are not aware of.

4   Methods of Visual Communication in New Media Art Design

(1) Use new media technology to achieve the purpose of work innovation

One problem that all designers need to consider is how to make use of new media technology to change visual communication works from "static" to "dynamic". First of all, designers must have innovative ideas, so as to design novel works. During the design planning, new media technology can be used to make the design content dynamic,

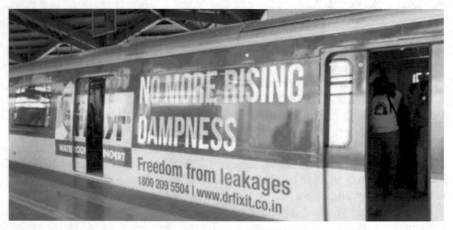
608   L. Nong and B. Long

express the theme of the work with new ideas, increase the visual communication effect and the artistry of the art work, and enrich the content. The "innovation" is not only expressed on the surface, but the key is to have innovative ideas. New media art takes people as the main body, so the works of visual communication have humanization.

The space of the high-speed railway is closed and the people flow is concentrated. Therefore, the high-speed railway station advertisement has a high degree of attention. The diversity of the high-speed railway media determines the richness of the high-speed railway station advertisement expression methods - the combination of dynamic and static. Generally, the continuous visual impression of the dynamic advertisement generates the story, and the information contained is easy to express. The impression generated by the people is more profound, and the visual effect is better than the static advertisement. With the continuous development of science and technology, advertisements on high-speed railway platforms will have better effects. As the high-speed railway platform advertisement has these characteristics, the high-speed railway platform advertisement needs more innovation, as shown in Fig. 1.

Fig. 1. Advertising design of high-speed railway platform

(2) Use new media technology to make works recognized

In order to solve the problem of insufficient acceptability of visual communication design works, it is necessary to fully apply new media to the design to ensure visual dynamics, and constantly check the design to make the works more acceptable to people. Different audiences have different requirements for the works and different degrees of acceptance of the works. Therefore, in the visual communication design, the actual investigation should be carried out according to the different audiences. According to their own experience, the real feelings of the audiences will be more clear, and the works can be continuously modified in the position of the audience. At the same time, because there are many ways for the audience to contact the new media visual communication design, the acceptance of the works is targeted to a certain extent. By using the understanding of the audience's real needs, the designer can delete and add the design content, so as to increase the acceptability of the works.

(3) Make use of new media technology to make the content of works more diverse

After the new media technology is added to the visual communication design, the content of the work is no longer single, and the audience can intuitively understand the meaning of the author. In the special environment, the new media technology enriches the design content and also makes the work full of vitality. For example, during the publicity of the TNT suspense drama "perception", in order to make more people accept it, the design works have been changed from different levels and angles in order to stimulate the audience's senses. Give full play to the design advantages of new media technology.

5 Conclusion

The new media art makes the visual communication design further develop, and the addition of digital technology makes the content of visual communication design more novel. Visual communication design cannot be constrained by graphic design, so it should have a broader development. The new media technology has made rapid progress, and the balance between science and technology and the media should be balanced. The interactive and high-tech characteristics of the new media can stimulate the innovative ideas of visual communication design, provide a platform for mutual communication for mass art, and promote the development of visual communication design.

Acknowledgements. This article is the 2019 New Century Guangxi Higher Education Teaching Re-form Pro-ject "Curriculum Reform and Practice of Visual Communication Design Majors in the Inheritance of Intangible Cultural Heritage in Guangxi and West Africa from the Perspective of New Media". Project No: One of the phased Achievements of Guijiao Higher Education (2019JGA395).

This article is one of the phased achievements of Guangxi Higher Education Higher Education Teaching Reform (2019JGZ163), a key project of Guangxi Higher Education Teaching Reform Project in 2019.

References

1. Yue, J., Liu, Z.: Research on the efficiency of visual communication in new media interface design. In: CIPAE 2021: 2021 2nd International Conference on Computers, Information Processing and Advanced Education (2021)
2. Wang, Y., Li, X.: Research on Automatic Layout Algorithm of Graphic Language in Visual Communication Design under New Media Context (2021)
3. Zhang, Z., Hu, W., Yang, Z.: Research on the innovation and development of visual communication design in the new media era. In: 2021 6th International Conference on Social Sciences and Economic Development (ICSSED 2021) (2021)
4. Li, Z.: Research on the spread path and evolution causes of oral language in the digital era. J. Math. **2022** (2022)

5. Zhang, M., Hou, K.: Research on the application of computer music making technology in new media environment. J. Phys. Conf. Ser. **1871**(1), 012142 (5p.) (2021)
6. Li, P., Tian, S.: Research on image communication of urban film and television advertisement based on complex embedded system. Microprocess. Microsyst. **2021**, 103996 (2021)
7. Wang, R.: Research on the communication mechanism of online rumors under the empowerment of new media technology-take the"nabobess having an affair with courier"rumor incident as an example. Psychol. Res. **12**(4), 10 (2022)
8. Chen, Y.: Research on the strategy of spreading socialist culture with Chinese characteristics in the era of new media. Open Access Libr. J. **9**(4), 5 (2022)
9. University of Bologna, ItalyUniversity of Bologna,Italy. Visual communication in research: a third space between science and art. Res. Educ. Med. **13**(2), 18–27 (2021)
10. Vanichvasin, P.: Effects of Visual Communication on Memory Enhancement of Thai Undergraduate Students, Kasetsart University (2021)

Water Conservancy Project Construction Supervision Quality Control Information Management System

Baihui Wang[✉]

Yunnan University of Bussiness Management, Yunnan, China
690233540@qq.com

Abstract. The quality of water conservancy project is related to the investment benefit, social benefit, environmental benefit and the safety of people's life and property. It directly affects the development of national economy and social stability. It is the key to the success or failure of project construction. In the construction supervision of water conservancy project, the project quality control is the core content of its work. It is not only the most important basis and standard to reflect the quality of supervision service, but also the basic requirement of the owner for supervision. Aiming at the construction of local water conservancy projects, this paper puts forward the necessity of developing the quality control information management system software of water conservancy project construction supervision. In the management of engineering construction quality evaluation, the fuzzy decision-making technology is applied to the comprehensive evaluation of engineering quality. Combined with the supervision practice, the evaluation grade is analyzed, and the four-level evaluation standard is established. Starting from the quality evaluation of the bottom unit project, the comprehensive evaluation of the engineering project is completed layer by layer through subdivisional works, divisional works and unit works.

Keywords: Hydraulic engineering · mis · supervisor · Quality Control

1 Introduction

Construction project supervision refers to a professional service activity in which an engineering supervision enterprise with corresponding qualifications accepts the entrustment of the construction unit, undertakes its project management work, and monitors the construction behavior of the contractor on behalf of the construction unit. In 1984, China first introduced the supervision mode in Lubuge Hydropower Station [1]. In 1990, the Ministry of water resources began to carry out construction supervision in the construction of water conservancy projects. By 1996, the construction supervision system was officially implemented in the water conservancy industry. Over the past 20 years, water conservancy project construction supervision has developed from scratch and gradually expanded. It has played an important role in water conservancy project construction quality, progress and investment control [2]. At the same time, it has promoted the reform of

© The Author(s), under exclusive license to Springer Nature Singapore Pte Ltd. 2023
J. C. Hung et al. (Eds.): IC 2023, LNEE 1045, pp. 611–617, 2023.
https://doi.org/10.1007/978-981-99-2287-1_88

water conservancy project construction management system, improved the management level of water conservancy project construction, and made remarkable achievements. In the process of implementing the project construction supervision system, the state and the Ministry of water resources have constantly revised and improved the laws, regulations, specifications and technical standards related to the construction supervision of water conservancy projects to meet the actual needs of water conservancy project construction [3]. At present, the construction supervision system has become a basic system that must be followed in the construction of water conservancy projects.

Project quality control is the core content of supervision work and the most important basis and standard to reflect the quality of supervision service; At the same time, it is also the basic requirement of the owner for supervision. The construction of engineering project involves many units such as design, construction, owner and supervisor. In the construction and implementation of the project, it is necessary to coordinate the relationship between all parties. In addition, the construction process of the project is affected by the natural environment such as geology, topography, meteorology and hydrology, the staffing and technical level of the construction unit, the capacity of construction machinery and equipment, construction organization and management, construction technology, technical measures and operation methods [4]. These factors not only directly affect the construction quality of the project, but also determine that the quality management of water conservancy project has the characteristics of large amount of information, high technical difficulty and strong comprehensiveness. The traditional project quality management information processing and quality evaluation mainly rely on the manual operation of supervisors, with heavy workload, long calculation cycle, poor reliability of results, non-standard use of quality evaluation form, and manual transmission of relevant data of project quality management among various units, which affects the timeliness, accuracy and comprehensiveness of information, which seriously affects the work of supervision engineers. Based on the above reasons, using advanced computer technology and system fuzzy decision-making technology, it is very necessary to research and develop the water conservancy project supervision quality control information management system.

2   Related Work

2.1   Contents of Supervision Quality Control in Project Construction Stage

The whole process of supervision quality control in the project construction stage can be divided into three links: pre control, in-process control and post control. The pre control supervision mainly includes: the quality control of the preparation work of the construction contractor, that is, the review of construction personnel, construction materials (raw materials, semi-finished products, components and accessories), construction machinery, construction methods and measures, environmental conditions necessary for construction, etc.; Do a good job in the prior quality assurance work that the supervision unit should do, such as being familiar with and mastering the technical basis of quality control, establishing and improving the quality monitoring system of the supervision department, making monitoring preparations, organizing the review of design drawings and issuing supervision engineer drawings, etc. [5]. In process control, use effective quality control methods, strictly inspect and inspect the process handover according to reasonable procedures, be responsible for the handling of quality accidents, and exercise the right of quality supervision. Post control, mainly responsible for reviewing the completion data, evaluating the quality status and level of the project, reviewing the quality inspection report and relevant technical documents provided by the construction contractor, sorting out the technical documents related to the quality of the project, cataloging and establishing archives; Organize relevant departments to make evaluation and conclusion on the project construction; Organize linkage test run, etc.

2.2 Research on the Application of Information Management System in Domestic Engineering Projects

In 1988, the Ministry of Construction issued the notice on carrying out construction supervision to implement the construction supervision system in China's construction field. In 1997, the state fully implemented the construction project supervision system. Since the implementation of the construction project supervision system, the construction project supervision system has played an important role in the project construction. Project construction management mainly controls the three objectives of project progress, investment amount and project quality. Among these three control objectives, quality control is the first [6]. As the core of project control, it determines the success or failure of project construction. Quality control runs through the whole process of the project, and the quality should be strictly controlled from planning, survey, design, construction and operation management after completion.

The quality control in the construction stage is particularly important. The construction stage is the process of finally turning the design blueprint into reality. The construction entity quality of the project is reflected in each process of the construction process. The quality control in the construction stage includes the whole process system quality control from the quality control of raw materials to the completion acceptance of the project. In order to improve the quality control and management level of engineering construction project construction, many software companies begin to study project management software [7]. At present, there are many kinds of project management software that have been developed. Because the construction quality management software involves a large number of construction standards and specifications, and there are differences in application regions and environments, many Chinese software companies have developed a number of engineering project quality management systems by combining foreign technologies and ideas with China's domestic management reality.

A set of calculation software with high reliability, high security and high expansibility is jointly developed by Henan water conservancy and hydropower project construction quality monitoring and supervision station and Henan Nochi Software Co., Ltd. The software is based on Windows operating system and Microsoft Net as the development platform, it adopts three-tier architecture and web service technology to carry out remote database transmission and quality control inspection through the Internet. It is a distributed and intelligent network information platform, covering the main processes of water conservancy and hydropower project quality management and 200 project quality evaluation forms. The system includes the correlation calculation of all forms, which greatly improves the work efficiency of staff, and also helps the management department standardize the specific filling in of forms [8]. The system carries out standardized management of quality management system control, project division, quality evaluation, project acceptance and other work through strict control of each process.

3 Quality Control in Advance in the Construction Stage of Project Supervision

3.1 Quality Control of Materials and Engineering Equipment Required for the Project

Raw materials, semi-finished products, components and fittings and permanent equipment required by the project will form an integral part of the permanent project in the future. Therefore, their quality will directly affect the quality of future engineering products, so the quality control is a very important link.

For the quality control of materials and engineering equipment, the whole process and overall control shall be carried out, that is, systematic supervision and control shall be carried out from the aspects of procurement, processing and manufacturing, transportation, mobilization, storage and use.

(1) Check whether the contractor has ordered and purchased materials and equipment according to the requirements of quality standards and project schedule. Before determining the order, the Contractor shall submit the quality report, supply capacity, price and test data of the supplier to the supervising engineer for review. The order can be placed only when the Supervising Engineer considers that it can meet the construction requirements of the project [9]. In case of lack of reliable data or doubts about materials, it is necessary to jointly investigate the production process, quality control and testing means, management mode, etc. of the supplier's factory, and order only after confirming that there are no problems.

(2) Check whether the storage and storage methods of materials and equipment after arrival are carried out in accordance with their performance, characteristics and operation and storage instructions, so as to ensure that materials and equipment will not deteriorate or change performance due to improper storage.

(3) Check whether the construction materials and equipment enter the site according to the requirements of the project construction schedule.

(4) The Contractor shall not change the supplier without authorization. If any change is necessary, it shall be reported to the supervising engineer for approval in advance.

(5) Various raw materials used for concrete and mortar shall be constructed according to the mix proportion approved by the supervising engineer after being confirmed and approved by the project supervisor. The mix proportion shall be designed by the supervision engineer's representative and the technical personnel of the construction unit together, and various data of multi scheme trial mixing, sample making and mixing shall be measured. The Contractor shall write a comprehensive report and submit it to the supervision project supervisor for selection and approval.

3.2 Construction Process Quality Control Method

(1) Establish project quality inspection system

The supervision organization shall establish various quality inspection systems in accordance with relevant laws, regulations, technical specifications and other provisions and in combination with the actual situation of the project. It is a mandatory and binding document that the supervisor and the contractor must abide by, such as the commencement application system; Quality inspection and evaluation system for process, unit, divisional and unit works; Quality acceptance and inspection system for concealed and key parts; Quality defect inspection and handling system; Quality accident investigation and handling system, etc.

(2) Formulate quality inspection procedures

After the completion of processes and unit works, the Contractor shall conduct self inspection on the construction quality according to the "three inspection system", make construction records, timely fill in the construction quality assessment form, and submit it to the supervisor and the on-site supervision engineer for review after passing the self-assessment. The quality level of processes and unit works shall be in accordance with DL/t5113 1-2005 standard for quality grade evaluation of unit works of water conservancy and hydropower capital construction projects (hereinafter referred to as the standard for evaluation of unit works) 136. For the quality acceptance of important works, important concealed works or key parts of the project, the construction unit shall first conduct self-assessment according to the quality acceptance standard. After passing the self-assessment, the owner, supervisor, designer and construction unit shall form a joint quality inspection team to check and verify the quality level. The quality evaluation of divisional works and unit works shall comply with the provisions of sl176-2007 code for construction quality inspection and evaluation of water conservancy and hydropower projects [10].

4 Water Conservancy Project Construction Supervision Quality Control Information Management System

4.1 System Functional Requirements Analysis

The problems existing in practice determine the functional requirements of the system, so the water conservancy project construction supervision quality control information management system should have the following functions:

(1) Information management function. For the quality data obtained from the test, the system shall have basic functions such as input, processing, query, statistics, analysis and output.

(2) Information sharing function. The project quality assessment involves five parties: the construction unit, the supervision unit, the construction unit, the design unit and the quality supervision organization. The quality information in the process of project construction should be transmitted and understood in time, and the database technology should be used to carry out systematic unified planning and design, so as to realize the timely exchange and sharing of quality information.

(3) Auxiliary decision-making function. The system shall have the functions of processing, statistics and analysis of the collected and entered data, which is conducive to giving correct decisions for project quality evaluation.

4.2 Overall System Function

According to the construction process quality control method, the structure diagram of water conservancy project construction supervision quality control information management system is established, as shown in Fig. 1.

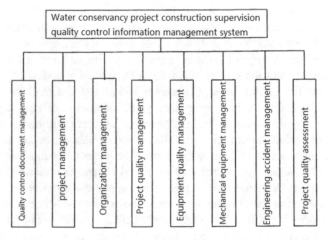

Fig. 1. Water conservancy project construction supervision quality control information management system

"Project division" is an important work completed by the construction unit (project legal person) with the participation of supervision, construction and design units. The project is divided into different levels according to the constituent units of the project. The first level of water conservancy project is divided into unit projects, which means that a water conservancy project is composed of several unit projects. The second level of water conservancy project is divided into divisional works, that is, a unit project is composed of several divisional works, and the third level of water conservancy project is divided into unit works, that is, a divisional project is composed of several unit works. In order to facilitate the quality supervision and evaluation in the construction process, each unit project, divisional project and unit project shall be numbered one by one. In order to determine the priority of each evaluation item in the quality inspection and evaluation, it is necessary to distinguish the main unit works, main divisional works, important concealed unit works and unit works at key parts. After the division of project composition is completed, it can be implemented only after it is confirmed by the corresponding engineering quality supervision organization. During the implementation of the project, if the project division needs to be adjusted, the project legal person shall re submit the adjusted project division to the project quality supervision organization for confirmation. For the partial adjustment of the project division that does not affect the unit works, main divisional works, important concealed unit works and unit works at key parts, the project legal person shall organize the supervisor, the design unit and the construction unit to make their own adjustment. Therefore, the project division management module includes three functions: project division table entry, modification and query.

5 Conclusion

The quality control of water conservancy project construction supervision is the core content of supervision work, and the information in the construction stage is a very important basis for quality control. This paper focuses on the water conservancy project construction supervision quality control information management system. After discussing and analyzing the research and application status of engineering project management at home and abroad and the engineering construction supervision system, this paper puts forward the necessity of developing the water conservancy project construction supervision quality control information management system software. The fuzzy identification model and level eigenvalue are used to evaluate the quality of the unit project, and on this basis, the whole project is comprehensively evaluated layer by layer. This method evaluates the quality of the project. By quantifying the qualitative indicators, the evaluation results can better reflect the primary and secondary importance of the system evaluation objectives.

References

1. Zhang, L., Tang, Z., Zhu, S., et al.: Strategic relationship between water conservancy project construction and environmentally sustainable development. Int. J. Environ. Technol. Manage. **2021**, 24 (2021)
2. Liu, X.: Research on water conservancy project construction and operation management based on cost management. E3S Web Conf. 276(9), 01034 (2021)
3. Ni, X.D., Hou, X.: Construction method of seepage control and leakage stoppage in water conservancy project construction. IOP Conf. Ser. Earth Environ. Sci. **768**(1), 012042 (2021)
4. Ni, X., Hou, X.: Application of big data technology in water conservancy project informatization construction. IOP Conf. Ser. Earth Environ. Sci. **768**(1), 012113 (2021)
5. Zhou, H., Li, M.: Research on application effect of construction technology in agricultural water conservancy project. IOP Conf. Ser. Earth Environ. Sci. **692**(4), 042030 (6p.) (2021)
6. Qiu, X.: Development in Chinese environmental water conservancy construction (2020)
7. Getu, A., Quezon, E.T., Genie, T.: Factors affecting supervision practice of public building construction projects in dire dawa administration. Am. J. Civil Eng. Architect. **9**(4), 134–141 (2021)
8. Tung, H.S., Tran, M.L., Hoang, H.V., et al.: External factors influencing bid/no-bid decision for supervision consultant service: a case of construction project in Hanoi. J. Asian Financ. Econ. Bus. **2020**, 7 (2020)
9. Song, L.L.: Key points of safety management in construction supervision of municipal engineering roads and bridges. Value Eng. (2020)
10. Su, W., Gao, X., Jiang, Y., et al.: Developing a construction safety standard system to enhance safety supervision efficiency in China: a theoretical simulation of the evolutionary game process. Sustainability. 13 (2021)

Choice Mechanism for Construction of University Enterprise Joint Cryptography Laboratory and Its Application in Hainan University

Yongheng Zhou[1,2], Lebing Huang[1,2], and Jun Ye[1,2(✉)]

[1] School of Cyberspace Security, Hainan University, Haikou, China
yejun@hainanu.edu.cn
[2] Key Laboratory of Internet Information Retrieval of Hainan Province, Haikou, China

Abstract. As the core technology of information security, the teaching of cryptography is paid more attention. Many schools have proposed a ' school - enterprise cooperation to build laboratories ' program. The main function of school-enterprise joint laboratory is to provide students with complete equipment and training environment. But for colleges and universities, how to match the most suitable company cooperation between schools and enterprises become the biggest problem, then establish a school-enterprise cooperation selection mechanism is very necessary. In this paper, the decision tree is used to simply screen all enterprises, and then the analytic hierarchy process is used to solve the weight from the arithmetic average method, the set average method and the eigenvalue method respectively. The final index weight is obtained by averaging the weight obtained by the three methods. Finally, the fuzzy comprehensive evaluation method is used to establish the selection mechanism of school enterprise cooperation. Through the case analysis of six enterprises in Hainan University, it is proved that the evaluation mechanism is effective, so as to provide objective standards for universities to select partners and jointly establish password laboratories. Keywords: First Keyword, Second Keyword, Third Keyword.

Keyword: Decision tree · Analytic hierarchy · Fuzzy comprehensive evaluation · Crypto School enterprise joint laboratory

1   Introduction

Cryptography is an important technology to ensure the three elements of information security, namely confidentiality, integrity and availability. As the core technology of information security, the teaching of cryptography is paid more attention. In the course of teaching, we should pay attention to the cultivation of students' ability, so that students have a solid theoretical foundation and strong application ability. Usually, universities should establish a cryptography laboratory for the practice of related courses. Many schools have proposed a 'school - enterprise cooperation to build laboratories' program.

J. C. Hung et al. (Eds.): IC 2023, LNEE 1045, pp. 618–628, 2023.
https://doi.org/10.1007/978-981-99-2287-1_89

The main function of the school-enterprise joint laboratory is to provide students with complete computer supporting equipment and training environment. After completion, the multi-source heterogeneous data collection, processing, analysis, storage, application and other operations can be carried out to meet the daily practice teaching, curriculum design, graduation design and so on. Second, improve teacher training, college students ' innovation training and competition training mechanism. But for colleges and universities, how to match the most suitable company cooperation between schools and enterprises become the biggest problem, then establish a school-enterprise cooperation selection mechanism is very necessary.

Liu Pengqi and others analyzed the three laboratory construction methods of school enterprise construction, joint construction with key disciplines and internal excavation, and proposed new ideas for the development of central laboratory [1]. Zhang Tao et al. Proposed a demand-oriented model of co-management laboratory construction of school enterprises [2]. Peng Zheng and others analyzed the significance of the construction of a laboratory by school enterprises to both parties with specific examples from this school [3]. Kang Wenbiao and Chen Ye introduced the contents and results of the construction of the laboratory of the school enterprise [4]. According to the analysis of previous research, there is a strong subjectivity in the choice of building crypto labs in cooperation with schools. There is no set of scientific selection mechanisms for schools to use, thus affecting the long-term development of the construction of crypto labs by schools. At present, there are also many applications of hierarchical analysis and vague comprehensive evaluation at home and abroad. A decision-making method based on qualitative and quantitative analysis of elements that are always relevant to selection mechanisms broken down into levels such as objectives, guidelines, schemes, etc. The relative weight of the indicators in the emergency management capability is determined by the hierarchical analysis method of Hou Fenglei, and the emergency capability of subway construction units is evaluated by the vague comprehensive evaluation method [5]. Xu Jianan and others used the AHP- vague comprehensive evaluation method to construct a children's street pedestrian space safety evaluation system, through the evaluation of the existing three street pedestrian spaces in Dalian [6]. In this article, the decision tree is used to perform simple screening of all enterprises, and then use the hierarchy analysis method and vague comprehensive evaluation method to establish the selection mechanism of school-enterprise cooperation. Thus, provide universities with objective criteria for selecting partners and jointly establishing password labs. Establishment of Evaluation Index for Construction of University -Enterprise Joint Cryptographic Laboratory.

2 Establishment of Evaluation Index for Construction of University -Enterprise Joint Cryptographic Laboratory

2.1 Selection of Evaluation Index

In the selection of evaluation indicators, the evaluation index system should have both indicators that reflect the level of development and indicators that reflect dynamic changes. The index should have a clear concept and a high correlation with the evaluation target to be achieved. It is important and meaningful to ensure the content of the index measurement for the evaluation target and the evaluation object. Combined with the basic principles of multi-dimensional, convenient access, and measurability of the evaluation system indicators, the selection model of this paper is obtained.

2.2 Establishing Two - Stage Multi - level Index System

The first stage needs to consider three main factors: enterprise research direction, enterprise operation situation and enterprise location. For example, in terms of operating conditions, for listed companies, for companies marked with ' ST or ST * ', the net profit of the audited two consecutive fiscal years is negative, or the net assets per share audited in the most recent year are lower than the current par value of the stock, then the risk of cooperation is huge, and these companies should be eliminated. In the second stage, objective multidimensional quantification is needed to comprehensively evaluate the companies left by the initial screening in the first stage. On the one hand, this stage needs to consider the basic re-sources of the enterprise, such as the ownership of high-level cryptographic technology and the situation of talents. On the other hand, the degree of matching between schools and businesses and the expectations of long-term cooperation need to be considered. After establishing the basic evaluation direction, through

Table 1. Selection mechanism index of university-enterprise joint crypto laboratory construction

Target layer	Criterion layer	Scheme layer
Enterprise selection mode (A)	Enterprise resource capability (B1)	Number of high-tech talents (C1)
		High-level cryptographic technology ownership(C2)
		Maintenance of national information security times (C3)
		Number of password talent reserve bases (C4)
	Basic business status (B2)	Financial situation (C5)
		Profitability (C6)
		Enterprise risk (C7)
		Corporate governance capacity(C8)
	Enterprise matching (B3)	Corporate research status (C9)
		School distance (C10)
		Enterprise cooperation willingness (C11)
		Internship students output (C12)

literature search and analysis, questionnaires, expert consultation and other methods, the selection mechanism indicators for the construction of university-enterprise joint cryptography laboratory were finally formed. The first-level indicators include enterprise resource capacity, enterprise basic situation, enterprise matching degree. There are three, and also 12 secondary indicators, as shown in Table 1.

3 Selection Mechanism of University- Enterprise Joint Cryptographic Laboratory Construction

3.1 Enterprise Selection Based on Decision Tree

Decision tree was first proposed to deal with decision problems. Decision tree has the advantages of simple structure, clear logic and good interpretability. The best decision tree is constructed by known 'prior data' to predict unknown data categories.[7] Using the idea of decision tree algorithm to extract the initial feature of some important indicators of enterprises, feature selection is to select the features with classification ability. If the classification result using a feature is not very different from the random classification result, this feature is said to have no classification ability. The indicators selected in this paper are whether the research direction of the enterprise is cryptography, whether the business situation is good, and whether the location of the enterprise is Hainan. The basic steps of constructing decision tree based on Gini value are as follows (Table 2):

Decision Tree Decision List and Calculation of Gini Value:
Two samples are randomly selected from the data set D, and the probability of inconsistent category labels is obtained. Therefore, the smaller the Gini (D) value, the higher the purity of the data set D:

$$\text{Gini}(D) = \sum_{k=1}^{|y|} \sum_{k' \neq k} p_k p_{k'} = 1 - \sum_{k=1}^{|y|} p_k^2 \tag{1}$$

Table 2. Decision Tree Decision List

Whether the research direction is cryptography ? (X)	Bussiness Status(Y)	Whether the location is Hainan? (Z)	Preliminary selection or not(E)
Yes	Good	Yes	Yes
No	Bad	No	No
Yes	Good	No	Yes
Yes	Bad	No	No
No	Good	No	No
No	Bad	Yes	No

Based on the decision list of the decision tree and the Gini value formula, the Gini values of the four indicators are as follows:

$$\text{Gini}(E) = 1 - \left(\frac{2}{6}\right)^2 - \left(\frac{4}{6}\right)^2 = 0.444444 \tag{2}$$

$$\text{Gini}\{X\} = 0.444444 - \frac{4}{6} \times 0.444444 - 0 = 0.148148 \tag{3}$$

$$\text{Gini}\{Y\} = 0.444444 - \frac{4}{6} \times 0 - \frac{2}{6} \times 0.48 = 0.284444 \tag{4}$$

$$\text{Gini}\{Z\} = 0.444444 - \frac{4}{6} \times 0.375 - \frac{2}{6} \times 0.5 = 0.0277773 \tag{5}$$

Construction of Decision Tree Model:
We then sorted the Gini values of the different indicators and build our decision tree to get the results shown in Fig. 1 below to facilitate the initial screening of enterprises.

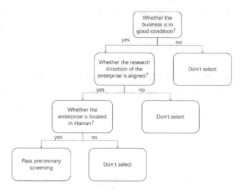

Fig. 1. Decision Tree Model

3.2 Determination of Evaluation Index Weight Based on Analytic Hierarchy Process

The comparison between many factors in the decision-making system often cannot be described in a quantitative way. At this time, semi-qualitative and semi-quantitative problems need to be transformed into quantitative calculation problems. Analytic Hierarchy Process is an effective method to solve such problems. The analytic hierarchy process layers the complex decision-making system, and provides a quantitative basis for analysis and final decision-making by comparing the importance of various related factors layer by layer. Because there are still few successful cases related to the school-enterprise joint cryptography laboratory, there is a lack of objective and accurate data. Therefore, the analytic hierarchy model is established in the form of consulting experts, teachers and questionnaires to empower the indicators. The specific steps are as follows:

Constructing the Judgment Matrix of Each Level Index:
The indicators are shown in Table 1 above. The original matrix A can be expressed as shown in (6), and then the judgment matrix is confirmed according to the classical nine-digit:

$$A = \begin{bmatrix} a_{11} & a_{12} & \cdots & a_{1n} \\ a_{21} & a_{22} & \cdots & a_{2n} \\ \vdots & \vdots & \ddots & \vdots \\ a_{n1} & a_{n2} & \cdots & a_{nn} \end{bmatrix} \tag{6}$$

Consistency Test of Judgment Matrix:
After constructing the judgment matrix, the consistency test is carried out to check whether the constructed matrix is too different from the consistency matrix:
    The 1 step: calculate the consistency index

$$CI = \frac{\lambda_{\max} - n}{n - 1} \tag{7}$$

Step 2: Find the corresponding average random consistency index RI.

n	1	2	3	4	5	6
RI	0	0	0.52	0.89	1.12	1.26
n	7	8	9	10	11	12
RI	1.36	1.41	1.46	1.49	1.52	1.54

Step 3: Calculate the consistency ratio CR.If CR < 0.1, it can be considered that the consistency of the judgment matrix is acceptable, otherwise the judgment matrix needs to be corrected until it passes the consistency test.

$$CR = \frac{CI}{RI} \tag{8}$$

Calculation of Index Weights:
In this paper, the arithmetic average method, the set average method and the eigenvalue method are used to solve the weights. Finally, the weights obtained by the three methods are averaged, so that the weights of the evaluation system are more accurate and reliable. The steps are as follows:
    The first step is to use the arithmetic average method. The judgment matrix is normalized according to the column (each element is divided by the sum of its columns),

and the normalized columns are added (summed by rows). Finally, each element in the obtained vector is divided by n to obtain the weight vector.

$$w_i = \frac{1}{n} \sum_{j=1}^{n} \frac{a_{ij}}{\sum_{k=1}^{n} a_{kj}} (i = 1, 2, \cdots, n) \tag{9}$$

The second step is to use the geometric average method for weight. The elements of A are multiplied by rows to get a new column direction. Each component of the new vector is multiplied by n times. Finally, the column vector is normalized to get the weight vector:

$$w_i = \frac{\left(\prod_{j=1}^{n} a_{ij}\right)^{\frac{1}{n}}}{\sum_{k=1}^{n} \left(\prod_{j=1}^{n} a_{ij}\right)^{\frac{1}{n}}} (i = 1, 2, \cdots n) \tag{10}$$

The third step, eigenvalue method for weight. Find the Maximum Eigenvalue of Matrix A and Its Corresponding Eigenvector.Normalize the obtained feature vector to get our weight. Finally, the weights of the three methods are averaged, that is, our weights are obtained.

3.3 Comprehensive Evaluation Model for Evaluation

Fuzzy comprehensive evaluation is a very effective multi-factor decision-making method to make a comprehensive evaluation of things affected by many factors. Its characteristic is that the evaluation results are not absolutely positive or negative, but represented by a fuzzy set [8].

The first step is to divide the factor set, because our criterion layer has three factors, so $U = \{U_1, U_2, U_3\}$ The second step is to determine the alternative set, $V = \{1, 2, \cdots, n\}$, representing n enterprises together for selection.

The third step is to establish the following single factor evaluation matrix, and then calculate the membership degree of each factor.

$$R_i = \begin{bmatrix} r_{11}^{(i)} & r_{12}^{(i)} & r_{13}^{(i)} & \cdots & r_{1n}^{(i)} \\ r_{22}^{(i)} & r_{22}^{(i)} & r_{23}^{(i)} & \cdots & r_{25}^{(i)} \\ r_{31}^{(i)} & r_{32}^{(i)} & r_{33}^{(i)} & \cdots & r_{35}^{(i)} \\ r_{41}^{(i)} & r_{42}^{(i)} & r_{43}^{(i)} & \cdots & r_{45}^{(i)} \end{bmatrix} \tag{11}$$

Note: For quantitative indicators, there are specific data of enterprises. We normalize these data to obtain single factor evaluation value; for qualitative indicators, we through market research, consumer satisfaction as a single factor evaluation.

The fourth step, we combine R_i and corresponding weight comprehensive evaluation, it is concluded that U_i for V membership:

$$B_i = W_{ij} \times R_i (i = 1, 2, 3) \tag{12}$$

The fifth step, combined with the entropy weight method to determine the weight of the first level factors $U = \{U_1, U_2, U_3\}$ to make a comprehensive judgment:

$$R = \begin{bmatrix} B_1 \\ B_2 \\ B_3 \end{bmatrix} \tag{13}$$

$$B = W \times R \tag{14}$$

Finally, the best enterprise is determined according to the principle of maximum membership.

4   Calculation and Application of Cases

Taking Hainan University as an example, this paper collects the indicators of six enterprises. The professional teachers and relevant experts of the school evaluate the index evaluation value of each company.

Determine according to the classic nines, and combine with expert scoring to finally obtain the judgment matrix Table 3.

Table 3. Judgment matrix of indicators at all levels

A	B1	B2	B3		B1	C1	C2	C3	C4
B1	1	2	2		**C1**	1	1	2	1
B2	1/2	1	1		**C2**	1	1	2	1
B3	1/2	1	1		**C3**	1/2	1/2	1	1/2
					C4	1	1	2	1
B2	C5	C6	C7	C8	**B3**	C9	C10	C11	C12
C5	1	1	3	3	**C5**	1	3	1/2	1
C6	1	1	3	3	**C6**	1/3	1	1/6	1/3
C7	1/3	1/3	1	1	**C7**	2	6	1	2
C8	1/3	1/3	1	1	**C8**	1	3	1/2	1

After transforming and normalizing the original data, it can be obtained according to the calculation steps of the above analytic hierarchy process. [9].

The weights corresponding to each index, and through the consistency test, the weights are shown in Table 4.[10].

Then obtain the single factor evaluation value of the six enterprises as shown in Table 5.

Then, the membership degree of each index is obtained respectively, as shown below.

$$B_1 = W_1 \times R_1 = [0.1314, 0.1429, 0.1929, 0.2043, 0.1357, 0.1514] \tag{15}$$

Table 4. Index Weight Table

Evaluation indicators	Weight
B1	0.2384
B2	0.3808
B3	0.3808
C1	0.2857
C2	0.2857
C3	0.1429
C4	0.2857
C5	0.3750
C6	0.3750
C7	0.1250
C8	0.1250
C9	0.2308
C10	0.0769
C11	0.4615
C12	0.2308

Table 5. Single factor evaluation table

Factors	1	2	3	4	5	6
C1	0.12	0.18	0.17	0.23	0.13	0.17
C2	0.15	0.13	0.18	0.25	0.12	0.17
C3	0.14	0.14	0.15	0.15	0.15	0.14
C4	0.12	0.12	0.25	0.16	0.15	0.12
C5	0.16	0.15	0.16	0.16	0.25	0.25
C6	0.13	0.16	0.23	0.14	0.16	0.16
C7	0.12	0.14	0.13	0.20	0.14	0.23
C8	0.18	0.20	0.14	0.12	0.12	0.15
C9	0.15	0.12	0.20	0.16	0.14	0.13
C10	0.25	0.18	0.12	0.13	0.20	0.14
C11	0.16	0.17	0.12	0.12	0.12	0.14
C12	0.23	0.16	0.18	0.18	0.18	0.20

$$B_2 = W_2 \times R_2 = [0.1463, 0.1588, 0.1800, 0.1525, 0.1863, 0.2012] \qquad (16)$$

$$B_3 = W_3 \times R_3 = [0.1808, 0.1569, 0.1523, 0.1438, 0.1446, 0.1515] \qquad (17)$$

$$R = \begin{pmatrix} B_1 \\ B_2 \\ B_3 \end{pmatrix} \qquad (18)$$

$$B = A \times R = [0.1559, 0.1543, 0.1725, 0.1616, 0.1583, 0.1704] \qquad (19)$$

From the above value of B, it can be concluded that the third enterprise has the highest degree of membership, so we should give priority to the third enterprise to achieve resource sharing and mutual benefit.

5 Conclusion

This paper studies the selection mechanism of school-enterprise joint construction laboratory, and puts forward a complete system for the current problem of how schools match the most suitable companies for cooperation, lacking a systematic and objective selection model. Combined with decision tree, analytic hierarchy process and fuzzy comprehensive evaluation method, the enterprises of cryptography laboratory co-constructed by school and enterprise are comprehensively analyzed and evaluated. Six enterprises taking Hainan University as an example are used to apply and analyze the cases, and the optimal cooperative enterprises are obtained, which overcomes the subjective and random selection methods and standardizes the selection mechanism. It provides a reference for the choice of joint venture between cryptology laboratory school and enterprise, and provides a basis for the development and construction of joint cryptology laboratory built by universities and enterprises to promote its promotion.

Acknowledgments. This work is partially supported by the Hainan Province Education and Teaching Research Found (Hnjg2022–23), the Science Project of Hainan University (KYQD(ZR)20021).

References

1. Liu, P., Xu, K., Zhang, Y.: A new path to joint development of the central laboratory. Lab. Res. Explor. **05**, 114–116 (1999)
2. Zhang Tao, G., Feng, D.P.: Exploring the model of building demand-oriented laboratory enterprises. Lab. Res. Explor. **34**(05), 228–230 (2015)
3. Peng, Z., Jia, H., Peng, J.: Discussion on the construction of laboratories between universities and enterprises. Univ. Lab. Work Res. **02** (2006)
4. Jiang, W., Chen, Y.: Implementing cooperative construction laboratory of schools and enterprises to strengthen practical teaching application. Lab. Res. Explor. **30**(11), 356–358+370 (2011)
5. Fenglai, H.: Research on comprehensive evaluation of emergency management capability based on hierarchy and hazy comprehensive evaluation method. Hyundai Urban Rail Transit. **09**, 87–92 (2022)

6. Jianan, X., Lianlian, L.: A study on the safety of pedestrian space in children's streets based on the vague comprehensive evaluation of AHP. City Build. **19**(18), 30–32 (2022). https://doi.org/10.19892/j.cnki.csjz.2022.18.10

7. Cheng, J., et al.: Emotion recognition from multi-channel eeg via deep forest. IEEE J. Biomed. Health Inform. (2020)

8. Shu, J.: Evaluation of urban emergency management capability based on AHP fuzzy comprehensive evaluation model

9. Yuan, C.: Evaluation and analysis of high quality economic development of Jiangsu Province based on entropy weight method

10. Wang, J.: Competitiveness evaluation of international consumption center cities based on entropy weight method. Food Indus. **41** 2), 8 (2020)

A Development Model of University Enterprise Joint Laboratory in School of Cryptography

Zhentao Li[1,2], Gantang Su[1,2], Jun Ye[1,2(✉)], and Zheng Xu[3]

[1] School of Cyberspace Security, Hainan University, Haikou, China
yejun@hainanu.edu.cn
[2] Key Laboratory of Internet Information Retrieval of Hainan Province, Haikou, China
[3] School of Computer and Information Engineering, Shanghai Polytechnic University, 2360 JinHai Road, Pudong District, Shanghai 201209, China

Abstract. University-enterprise joint cryptographic laboratory is an important measure to promote Hainan University to build a first-class university. We have the following three selection mechanisms to promote it after talking about. It is mainly realized in the following three directions: first, enterprises send people to teach technology, and enterprises can obtain talents in a directional manner. Second, different enterprises bid, guided by the technical problems of enterprises, to promote the growth of students' technical ability, corresponding to the graduation preferential treatment of enterprises. Third, the competition within the school to promote the growth and progress of students' technology. Keywords: First Keyword, Second Keyword, Third Keyword.

Keywords: School-enterprise cooperation · Directional screening · Dual system

1 Introduction

School-enterprise cooperation means that schools and enterprises make use of their own advantages, work together to complete a series of technological innovation behavior. Because of the improvement of the ability of labor force in the age of information technology, the cooperation between schools and enterprises is promoted, so that each takes what he needs and develops together.

School-enterprise cooperation was first born in Germany in the 19th century and has been popular in developed countries in Europe and the United States since the middle of the 20th century. The "dual system training" in Germany promoted the leap of the German economy at that time, while the "sandwich work-study system" in Britain, the "cooperative education" in the United States and the "school-base enterprise system" in Russia all played a positive role in the economic take-off of their country.

2 Introduction to Development Model

The university-enterprise collaborative development can be discussed in two forms. One is the collaborative development within the system. For example, in the context

J. C. Hung et al. (Eds.): IC 2023, LNEE 1045, pp. 629–637, 2023.
https://doi.org/10.1007/978-981-99-2287-1_90

of resources and environmental conditions, enterprises set up colleges and universities by themselves to realize their own expansion and drive the development of schools, and the development of schools also becomes an integral part of the development of enterprises. However, due to the macro management system and the limitations and constraints of the university and enterprise in organizational goals, motives, resources, etc., the enterprise-run universities have not yet formed the mainstream. The other is the synergetic development between the system, which is the synergetic formation between the enterprise and the school because they belong to different subjects, which is the research object of this paper. Combining with the above definition and analysis, we think to companies, universities as coordinated development between mode of network nodes, is in the "school-enterprise" under the framework of the internal market, with cooperative development as the goal, on the basis of system optimization to focus on enterprise and university development, adhere to the system open, pay attention to the internal market demand management and by means of organizing network management synergy, Through the coordination of organizational structure and process, we can realize the common development of the enterprise and the school, and achieve a harmonious and friendly model between the "school-enterprise" system and the external environment (see the following figure).

3 Predicament

3.1 Poor Operation Quality

First, the school-enterprise cooperation model is outdated. In China, school-enterprise cooperation in applied colleges and universities is still in the initial stage, and the relevant cooperation mode and cooperative management system still need to be improved. Haida for other schools temporarily cannot learn from the place.

At present, the school-enterprise cooperation of applied colleges and universities in China mainly focuses on scientific research and technology cooperation and the cultivation of talents, with single interest demands and single cooperation mode. At the same time, many enterprises still hold a wait-and-see attitude towards school-enterprise cooperation, while schools also show shortcomings in school-enterprise cooperation such as lack of broad thinking and low efficiency of resource integration.

Second, the depth of school-enterprise cooperation is insufficient. On the one hand, schools play a leading role in enterprise cooperation, but many applied colleges and universities have not found a suitable way for the development of applied education in their own schools.

On the other hand, in order to reduce the cost of talent reserve and company operation, enterprises often choose the cooperation mode with short period and quick effect, and avoid the cooperation mode with large investment, long period and high risky coefficient, which restricts the deep development of school-enterprise cooperation.

3.2 Serious Formalism

At present, there are significant differences in the attention paid to school-enterprise cooperation in various applied colleges and universities in China. On the one hand, many application-oriented colleges and universities supported by the state vigorously advocate school-enterprise cooperation, encourage teachers to strengthen exchanges and interactions with enterprises, and actively carry out the transformation of scientific research achievements.

Most of these colleges and universities have set up professional school-enterprise cooperation management departments to carry out detailed management of relevant cooperation matters and cooperation methods. On the other hand, due to the limited educational conditions and the lack of educational re-sources, most ordinary application-oriented colleges and universities have unclear cooperation intentions, and school-enterprise cooperation is often a mere formality, failing to truly promote education.

3.3 Absence of Government Management

The construction of school-enterprise cooperative education management system in application-oriented colleges and universities not only needs the mutual coordination between colleges and enterprises, but also needs the support and assistance from the government in all aspects. Firstly, application-oriented colleges and universities lack solid and operational laws and regulations in the process of school-enterprise cooperation, which leads to the lack of clear basis for school-enterprise cooperation management and little effect.

Secondly, when many application-oriented colleges and universities carry out school-enterprise cooperation management, the government does not strictly supervise the compliance of staff avoidance principle, which leads to the fact that some teachers are not only participants of school-enterprise cooperation projects, but also managers of school-enterprise cooperation.

This unclear division of labor reflects the inadequate management and the lack of corresponding responsibilities. Finally, the school-enterprise cooperation carried out by applied colleges in China is still in the threshold free cooperation stage, and the government lacks a macro-supervision and coordination system for school-enterprise cooperation.

3.4 Insufficient Enterprise Motivation

At present, the motivation of enterprise participation in school and enterprise cooperation is obviously insufficient, and the reasons are as follows: first, the lack of legal protection and interest protection system. The interests of enterprises cannot be effectively protected, leading to enterprises in school-enterprise cooperation.

Second, the enterprise's perception of university-enterprise cooperation is relatively one-sided, especially for university-enterprise cooperation may produce social benefit, economic effect of interest, such as lack of rational education achievements, the profound understanding, some corporate managers even believe that in the process of cooperation between colleges and enterprises is the capital both senders, school is a direct benefit, and bear the risk of major enterprises, This discourages many businesses from working with schools.

In general, the development of school-enterprise cooperation in running schools requires enterprises to improve the internal management system and talent team construction ideas, improve the management level and management ability.

4   Implementation of School Enterprise Cooperation

4.1   Overall Introduction

School-enterprise cooperation is a systematic project, which requires comprehensive cooperation from the aspects of talent training program formulation, embedded curriculum selection, joint construction of practice base, "double qualified" teacher training, quality supervision and control.

4.2   Specific Process

Establishing an Efficient School-Enterprise Cooperation Mechanism The university and the college attach great importance to the training of school-enterprise cooperation software service outsourcing talents. After several rounds of communication and field investigation, we visited several service outsourcing enterprises and combined with Hainan Software Park's R&D capability, project implementation ability and sustainable development ability of enterprises.

The major of cryptography finally decided to cooperate with Hainan Industrial Park in the development direction of JAVA big data of cryptography, and to cooperate with Software Park in the development direction of mobile Internet. The university and enterprise discussed, planned and coordinated various projects for many times. The Academic Affairs Office and Information Engineering College of the University, which undertook the construction, formed a service outsourcing project team, which was specifically engaged in the research, exploration and implementation of the project. The Steering Committee of Embedded Talents Training Project of School-enterprise Cooperative software Service Outsourcing has been set up, and the leading group of embedded talents training project of school-enterprise cooperative software service outsourcing has been set up.

4.3   Develop a Scientific "Embedded" Talent Training Program

Talent training program is the basis for the implementation of teaching. Both sides of the university and enterprise respect the law of higher education teaching, on the basis of full investigation, take the employment needs of the enterprise as the guide, combine the advantages of both sides of the university and enterprise, and work out the embedded talent training program in accordance with the process of knowledge formation and ability training.

Further, considering the rapid development of software technology and the continuous breeding of new models and forms of Internet+, both sides of the university and enterprise will survey the needs of enterprises for employment every year. Under the overall stability of the talent training program, some courses will be appropriately optimized and adjusted according to the relevant regulations of the university.

4.4 Building a Reasonable Curriculum System

According to the national quality standards for undergraduate talents training, the orientation of the university and the development plan of the major construction, the university and enterprise cooperation has constructed a reasonable professional curriculum system. The curriculum system consists of four major platforms: general education, professional education, concentrated practice, innovation and entrepreneurship, and quality improvement platform, reflecting the professional characteristics of "theory + practice". All four platforms offer compulsory courses and elective courses, emphasizing the cultivation of practical ability. The general education platform is mainly used to promote the all-round development of students.

The professional foundation platform mainly focuses on the cultivation of basic theories and knowledge such as mathematics, electronic technology, computer and software development. The professional application platform is divided into three parts: discipline foundation, professional compulsory courses and professional electives. The discipline foundation mainly focuses on training students' professional knowledge and skills in the field of software system integration development. The compulsory part focuses on the training of core professional knowledge and skills, mainly including object-oriented programming, database, cryptography and other core courses.

Specialized elective courses are divided into specialized elective courses and improvement elective courses, which introduce some mainstream software development technology of current industry enterprises, and cultivate students' application ability of emerging technologies such as mobile Internet application, big data and cloud computing. The centralized practice platform includes professional practice, curriculum design, comprehensive training and other links, reflecting the cultivation of students' skills. The innovation, entrepreneurship and quality promotion platform mainly cultivates students' innovation ability. Clear corporate responsibilities and obligations, participate in the whole process of the four-year training, each semester in different forms to participate in student training, practice teaching and professional direction teaching. Course system, a total of 174.5 credits. Among them, practical teaching accounted for 41.5%; Enterprise participation in the course reaches more than 25%.

4.5 Reasonably Embed Course Modules

Fully integrate into enterprise resources. Fully communicate with cooperative software enterprises, actively consult experts and scholars, deeply integrate the enterprise teaching resources and teachers throughout the four-year learning process of students. The three years in school are focused on 15 professional courses or professional quality training, and the one year outside school is focused on practical exercise. Cryptography major JAVA big data development direction and mobile Internet development direction of the main embedded courses.

In addition, the enterprise also undertakes the tasks of entrance education for students, team building, opening lectures on cutting-edge technology in the industry, and comprehensive training for primary majors.

4.6 Utilize the Superior Resources of Enterprises to Cultivate Practical Skills and Build a "TWo-Way Part-Time" Mode

On the one hand, it requires internal teachers to practice regularly in the enterprise, participate in the real project of the enterprise, improve the practical ability, and promote the construction of "double-qualified" team; On the other hand, enterprises are required to integrate real cases into the classroom and turn them into project cases that students can digest and absorb, so as to improve students' software skills.

4.7 Cooperative Construction Practice Base

With the strong support of the university and the college, the department of cryptography continues to cooperate with enterprises and sign off-campus practice bases, enriching students' practical training and practice fields and providing students with real practice scenes. In addition, Nanjing Dane-Suqian University Big Data Practice Base has been approved, and IFlytek Suqian branch has actively provided high-quality off-campus practice platform for software major students.

4.8 Sharing High-Quality Teaching Resources

Cryptography major and cooperative enterprises jointly build an online teaching resource library, and the platform provides functional modules such as "series of courses" and "technical Q&A".

Both cryptography professional and cooperative enterprise high levels of teachers, and between the two sides hand advanced teaching period and method, through teaching between colleges and technical project resources co-construction and sharing, improve the teaching quality and teaching level, improve the utilization rate of resources, to promote software engineering specialty construction and improve the quality of personnel training is of great significance.

4.9 School Enterprise Cooperation Helps Students Innovate and Start Businesses

Through the close and effective cooperation between schools and enterprises, the students' skill level, innovation and entrepreneurship awareness and ability are gradually improved through curriculum design, comprehensive training, professional practice, large-scale innovation projects, professional skills competition, and grade examination, etc., and the quality of talent training is gradually improved.

The main embodiment has the following aspects. Inspire morale, help students find the right direction; Enterprise elites will be invited to give lectures on the forefront of industry technology to help students grasp the new direction of the development of their major disciplines. To provide an environment to address student innovation; In addition to providing innovative and entrepreneurial places for students in the campus experimental center and library, it also makes full use of the real project scenarios of enterprises to provide on-site internship opportunities for students.

Borrowing teachers to help students improve their level, according to the plan, the "embedded" co-operative enterprise regularly arranges key technical personnel to the school to carry out "project" teaching for students, carries out group cooperation and other teaching method reform, and implements the case teaching of cryptography core course -- programming course, so as to improve students' practical programming skills and help improve their skill level.

Also practice, "also held joint cooperation enterprise software design competition, the challenge of artificial intelligence and other related professional skills contest, and require companies to arrange engineer Used the weekend to school counseling, adopts the teaching - online guide -" off-line "whereas", in the form of security to master new knowledge new technology or project experience.

4.10 Contract Guarantee

In order to further ensure the quality of embedded talents training and prevent the risks of school-enterprise cooperation, our school revised the school-enterprise cooperation agreement template.

The agreement fully considers the students and pays in installments according to the quality and quantity of students trained by enterprises. Enterprise experts and backbone are invited to the school to carry out cutting-edge lectures, and cooperate with enterprises to develop cases and projects for teaching. Arrange teachers to go to enterprises for exchange and study.

4.11 System Guarantee

The school issued some documents and regulations to form a sound professional teaching management system. Organize teachers (including enterprise engineers) to learn related quality standards and management system regularly, standardize the teaching process. Pay attention to quality control, strengthen and pay attention to the daily teaching process monitoring.

For the enterprise teaching engineers in the school to implement one-to-one pair of mutual assistance, for each enterprise teaching teachers in the school, a teacher in the same professional direction in the school is arranged to pair and help each other. Teachers in the school help enterprise teachers get familiar with school regulations, teaching requirements and students' learning conditions.

Enterprise engineers can guide pairs of teachers on new technologies, learn from each other, and jointly complete the task of school-enterprise cooperation in education. Each cooperative teaching class enterprise is equipped with a separate head teacher for easy management. Each time an enterprise comes to the school to teach or guide centralized practice, the enterprise will arrange technical engineers and administrative personnel to accompany each other, each performing their own duties. One is responsible for business guidance, and the other is responsible for class management.

4.12 Process Management

Enterprises establish archives for students in time. According to our enterprise embedded class's and grade's management system and process, as the embedded class new student and cooperative enterprises to establish a student information system for each student, in order to exchange and communication with management, in the late so as to realize each student in each stage of learning evaluation grades and often performance data real-time input system platform. Enterprises participate in the whole process of students' learning, from orientation activities to graduation practice. Every semester enterprises will come to the school to guide students in practical activities and carry out special activities. The enterprise dispatched project team members (class teachers, project managers, career development mentors, etc.) to participate in the orientation work, and organized a meeting for freshmen.

Through various kinds of orientation activities and interactive communication, new students can quickly adapt to the school learning and living environment, and achieve the ice-breaking effect between new students and project implementation members of cooperative enterprises, laying a foundation for the smooth implementation of the project in the future. Enterprises come to school to carry out industry cognition lectures.

The lecture content covers the cutting-edge fields and hot technologies of "Internet+" in the industry, so that students can have a preliminary understanding of the development prospect of "Internet+" industry and the cutting-edge technology field of IT, and look forward to the future career development. The enterprise offers career planning and innovation education courses for students, aiming to help students cultivate correct values and positive attitude, and ensure that students start their college life in a healthy and upward state.

At the same time, it helps students to establish career planning, develop the habit of team communication, stimulate learning interest, improve learning initiative, and provide a good foundation for the subsequent college study and the cultivation of professional quality. Continue to promote the project-based teaching process. The business backbone of the enterprise came to the school to implement the case teaching of the core course of cryptography -- programming course through teaching methods such as group coopera-tion, so as to improve students' practical skills of programming and further enhance their practical ability. School regularly visits the enterprise, puts forward reform suggestions on the training mode of the enterprise, and evaluates and supervises the teaching quality;

To the enterprise to carry out educational practice activities of the technical backbone according to the unified requirements of the school teachers and management. At the same time, the company assigns engineers to attend the teaching exchange in order to grasp the teaching rules of the school.

In the practice of school-enterprise cooperation and education, we constantly sum-marize experience, find problems, and then timely rectification, continuous progress, school-enterprise continue to deepen and efficient cooperation, week after week, take the road of "feedback - improvement" spiral progress.

5   Conclusion

The cooperation between the cryptography major of Hainan University and the enterprise will promote the cooperation and sharing relationship between the two sides, enhance the information exchange and internal cooperation between the two sides, and transform the technological advantages of the enterprise into technical reserves of students. On the one hand, it can increase the competitiveness of students and enhance the overall strength of cryptography major. On the other hand, through the close cooperation with enterprises, the strengthening of cooperation and sharing between the two sides will continue to promote the deepening of exchanges between the two sides, and the scientific and technological achievements will be transformed into sustained impetus. To achieve solid closed loop, and then cultivate effective development model and system.

Acknowledgments. This work is partially supported by the Hainan Province Education and Teaching Research Found (Hnjg2022-23), the Science Project of Hainan University (KYQD(ZR)20021). This work is supported by SSPU young talent project, China (EGD22QD02).

References

1. Wang, X.: Brand marketing strategy. Economic Management Press. 2 (2002)
2. Yu, J.: Marketing management. Southwestern University of Finance and Economics Press (1999)
3. Yang, T., Lu, R.: Exploration and research on the cultivation strategy of innovation ability of postgradu-ates of TCM specialty based on the model of "Co-creation of teachers and students". J. Higher Educ. **22**, 36–38+41 (2019)
4. Xie, J.: Study on the training model of modern apprenticeship in higher vocational colleges. Vocat. Educ. BBS **16**, 16–24 (2013)
5. Xiujin, H.: Research on the training model of modern apprenticeship system. J. Hebei Normal Univ. (Educ. Sci. Edn.) **3**, 97–103 (2009)
6. Wang, Q., Li, J., Wu, B., Dai, T., Fu, B., Song, C.: Discipline construction strategies in application-oriented universities under the back-ground of "Double First-class" Construction. Metallur. Educ. China
7. Wang, W., Li, N., Bai, W., et al.: Study on the mode of school-enterprise joint training of civil engineering applied talents. J. Anhui Univ. Technol. (Soc. Sci. Edn.) (2010)
8. Wang, W.S.: Reform and Practice of school-enterprise Joint education. China Vocat. Techn. Educ. **2**, 17–18 (2003). (in Chinese)
9. Xue, Z., Sanbo, C.: Thinking of integrating BIM technology into engineering management courses. Henan Sci. Technol. **9**, 269–280 (2013)
10. Lei, Y.: Research on the quality assurance mechanism of engineering applied talents training in Japan – based on the perspective of JABEE evaluation and certification. J. Yancheng Inst. Technol. Soc. Sci. Edn. **28**(1), 86–90 (2015)
11. Yu, H.: Strengthening practical teaching and improving college students' innovative practical ability. China Higher Educ. **51**(21), 23–25 (2010). (Ma, P., Wang, L., Hu, D.: Current situation analysis and coun-termeasure research of engineering practice teaching. Higher Eng. Educ. Res. **29**(1), 143–147 (2011)
12. Shaodong, L., Xin, Z., Zhaoqiang, Z.: Exploration and practice of school-enterprise joint talent train-ing mechanism in civil engineering. Higher Architect. Educ. **25**(02), 32–35 (2016)

Analysis of Instant Messaging Systems for Users Based on the Go Language

Shoulei Lu[1], Jun Ye[1,2(✉)], and Zheng Xu[3]

[1] School of Cyberspace Security (School of Cryptography), Hainan University, Haikou 570228, Hainan, China
yejun@hainanu.edu.com
[2] Key Laboratory of Internet Information Retrieval of Hainan Province, Haikou, China
[3] School of Computer and Information Engineering, Shanghai Polytechnic University, 2360 Jin-Hai Road, Pudong District, Shanghai 201209, China

Abstract. With the development of modernization, intelligence, and networking of enterprises, large-scale enterprises continue to emerge, and collaboration and communication between employees within an enterprise are increasingly important. It is time-consuming and labor-intensive, which directly leads to low work efficiency and efficiency. The current popular instant messaging technology provides a solution for this demand, and information and notifications can be released quickly and easily. A large number of instant messaging systems on the market can meet the basic needs of enterprises, but it is difficult to guarantee commercial security and low latency requirements. Therefore, it is necessary to develop an instant messaging system that meets the requirements of enterprises.

This article intends to analyze a set of instant messaging systems based on Go language according to actual needs. As a new language, the Go language has some features that many traditional programming languages do not have. It can make full use of the kernel of the device and make the system performance more stable and fast. Based on the above analysis, this paper uses the Go language to design a user instant messaging system.

Keywords: User instant messaging system · Go language

1 Introduction

With the development of Internet 5G, instant messaging applications have become more important in personal and corporate communications. Compared with the current instant messaging, the traditional short message service has high communication costs and can support less data types to be transmitted. At the same time, the infrastructure requirements for building SMS services are high, and it is difficult to achieve full coverage because the signal is not good and the communication quality is poor. Therefore, more and more people choose to use instant messaging to communicate. Using instant messaging software to communicate can reduce the cost of mass communication. Nowadays, the software that people mainly use to chat with friends in my country are instant messaging

© The Author(s), under exclusive license to Springer Nature Singapore Pte Ltd. 2023
J. C. Hung et al. (Eds.): IC 2023, LNEE 1045, pp. 638–646, 2023.
https://doi.org/10.1007/978-981-99-2287-1_91

tools such as WeChat and QQ. For personal communication, WeChat and QQ have been able to meet our basic needs. In today's mobile Internet era, enterprises need to have an instant communication tool that can meet the security of their own enterprise information and meet the needs of employees within the enterprise, so as to improve the work efficiency between employees. Due to the greater risk of enterprise data leakage and the greater loss of leakage, considering the data security of enterprises, many enterprises cannot use communication tools developed by third parties, especially between Internet companies that are not in the same camp, for the internal Important business information, if the use of third-party instant messaging software, will undoubtedly increase the risk of leaking the company's business secrets, and will bring certain crisis to the enterprise.

Instant messaging is one of the important factors that affect the efficiency of enterprise operation. The main functions of the communication system include user registration, login, sending messages, and displaying online users. The system can make the notification release more timely and the work arrangement more detailed, thereby improving the work efficiency of employees. How to design a convenient system is a difficult problem for many small and medium-sized enterprises, because they are limited in technology and do not have enough funds to support the design of the system. The survey found that existing systems can address the basic needs of businesses, although they have many problems. First of all, most programs are written in programming languages such as C, Java, and Python. The compilation speed of the system is relatively slow, and users often experience lag when using them. Second, they are less real-time and cannot better meet the requirements of enterprises. In addition, with the development of the times, the performance of enterprise servers and user devices has been greatly improved, but some systems do not support large concurrency very well, resulting in delay or failure of message sending. Based on the above analysis, we consider using Go language to design the system. Go language has the advantages of fast compilation and running of C language, as well as easy understanding of python language and convenient code writing. In addition, the Go language has some unique advantages, such as the ability to specify the number of cores to be used, and the automatic garbage collection function. The Go language does not allow the code to contain unused variables, function names, etc., which increases the system to a certain extent. Security.

With the development of 5G technology, information technology is also developing rapidly. With the impact of the epidemic, enterprises have to achieve informatization of offices. There is an urgent need for information-based office facilities. The application of a user-friendly instant messaging system can make the affairs of enterprises more convenient. The system is also a vital driving force in promoting the development of productive forces in enterprises. In this article, we mainly share the following:

(1) Through market research and analysis, this paper proposes a user instant messaging system based on the Go language to ensure efficient communication and quick message processing and forwarding capabilities.
(2) In the system's design, the Go language is used for development, and TCP socket programming is used to ensure the reliability of message transmission.

2 Related Work

Ye [1] uses HBuilder tool to develop client, uses MUI and H5Plus to develop mobile application client, realizes one-time development to generate mobile applications running on Android, iOS and applet. The main functions of the system are chat, address book, scan The functions of scanning, adding friends, uploading avatars and modifying personal information have certain reference for the construction of instant messaging systems. Ke [2] designed a distributed communication system, using a distributed architecture design, Linux Release as the operating system, and divided the system into a four-layer architecture of client, gateway, server, and distributed management, to achieve a large number of users. It can ensure the stable and efficient operation of the system. Li [3] designed a user communication system based on the Kylin system. The system is developed based on the Linux kernel, using the oracle database and RSA encryption technology, which can not only solve the problem of instant communication under the Kylin system, but also ensure the security of communication. Wang [4] designed an enterprise-oriented user communication system, considering both system security and system stability, and used the spring MVC framework to design the system. Section [5] analyzes synchronization and asynchrony, blocking and non-blocking in the development of network communication, and introduces the most widely used asynchronous and non-blocking network communication method. Zhang [6] designed a communication system based on WebRTC, analyzed the differences and commonalities between various signaling, and analyzed and designed audio and video processing. Enterprises should conform to the trend of informatization, keep pace with the development of the times, establish an intelligent enterprise platform, and improve enterprise efficiency. The conclusion shows that the user's instant messaging system is an important factor affecting the survival and development of enterprises.

3 Requirement Analysis

Through practical testing, this article summarizes the following functional requirements flowchart, as shown in Fig. 1.

3.1 The Problem with Traditional Workflow

Please note that the first paragraph of a section or subsection is not indented. The first paragraphs that follows a table, figure, equation etc. does not have an indent, either.

Subsequent paragraphs, however, are indented.

Traditional corporate communication has the following problems:

(1) Traditional corporate communication generally uses internal telephone calls or conferences, requiring employees to be in the workplace. At the same time, if the employee has something to delay, it cannot be carried out typically. Because of the need for negotiation, the whole process must last for a long time, which will seriously affect the working hours of other employees and hinder the company's production efficiency.

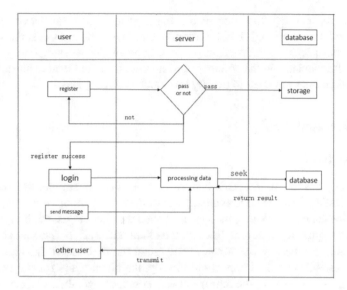

Fig. 1. Functional Requirements.

(2) Traditional methods need to collect employees' phone numbers, work content, etc., which leads to enterprises needing to manage and update employee information on schedule. Collecting requires dedicated personnel to maintain and some experience. An error in the work process may lead to severe impacts such as lack of notification or non-assignment of work.

(3) With the rapid development of 5G, our network speed has been increased, network latency has been significantly reduced, and the network coverage area is also expanding. Still, the traditional enterprise communication methods have not kept pace with the development of the times. First, informatization and intelligence have entered various fields, and enterprises' working mode is also changing. Secondly, the size of the enterprise is gradually getting more extensive, the number of employees is also increasing rapidly, and the use of traditional methods will significantly reduce the real-time and reliability of messages. Finally, with the improvement of the average level of education in our country, everyone will use computers to work to some extent.

3.2 Functional Requirement

Many researchers have recently proposed different technical solutions to implement user instant messaging systems. Combined with the actual needs of enterprises and the excellent ideas of other authors, we have summarized an online communication system with functions such as user registration, login, sending messages, displaying online user lists, and user withdrawal. When users use these functions, the system must ensure reliability in every message processing. In addition, to ensure the system's real-time, high reliability and high concurrency, it is also necessary to ensure the system's stability when multiple users are using it simultaneously. In this system, after the user has registered, they can log in, and then they can send a message and view the list of online users (employees who are working). Finally, after the employee finishes work, the system will exit. During this period, to improve the processing power, the working computer can run a terminal command to join the communication system, and the system has low memory requirements. It does not need to occupy too much storage space.

4 Requirement Analysis

4.1 Client Model

The client's task is to communicate between users, which is simple, flexible and reliable. The client's primary functions are registration, login, sending messages, displaying a list of online users, and exiting the system. When the client registers, first, it will send a connection request to the server. Second, the client will wait for the connection establishment result returned by the server. Third, send the registration information to the server after establishing the connection. Finally, the client waits for the server's processing result. If you receive the registration success message, then you can log in. The login and registration processes are similar. Both require a connection to be established and wait for the result of the server processing. While the login is successful, the user can use the other functions, and server processing is required when using each different function. When the user wants to exit the system, send an exit command and remove the connection established at the login time.

4.2 Server Model

The task of the server is to process the request submitted by the user and return the processing result to the client. It mainly handles registration requests, login requests, message forwarding processing, online user lists, and listening for connection requests. When the server starts, the port will always be in the listening state. When the user's connection request is heard, the server will authenticate the request. If the link is correct, the connection will be established. And then the next step is to process the request processing message from the client sent. If it is a registration message, it will access the database and return the processing result of the database to the client. A login request is also established, similar to a registration request. If a message forwarding request is made, the server will check the current online user and forward the message to the corresponding user. When the user wants to query the online user list, the server can directly send the user list maintained by itself to the user.

4.3 Database Model

The database is mainly used to store user data. When the user registers, it will check whether the registration information already exists in the database. Then send the query result to the server. When the user logs in, the user login information will be compared with the information in the database. The details of the system data sheet are shown in Tables 1, 2 and 3.

Table 1. User Info

Field name	Type
id	varchar
username	varchar
password	varchar
face-image	varchar

Table 2. Friend Information

Field name	Type
id	varchar
my-user-id	varchar
friends-id	varchar

Table 3. Chat message

Field name	Type
id	varchar
send-user-id	varchar
accep-user-id	varchar
mess	varchar

5 System Design

5.1 System Functional Design

The system is a user instant messaging system based on Go language. Initially, according to the different functions, the system can be divided into client-side and server-side. Then, each module is specifically divided, and the specific division of each module is shown in Fig. 2.

5.2 Register Design

This section focuses on the registration process. When each user uses the system for the first time, since the database does not have the user information, he needs to register. Generally, the registered account is the employee number, and the password is set by himself. Before user registration, the user must first send a connection establishment request to the server. After the connection is established successfully, the client can send the user registration information to the server. The server will compare the received

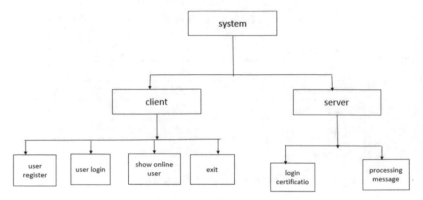

Fig. 2. System Function Design.

registration information with the information in the database. If there is no such informa-tion, it means that the registration is successful, and the server will return the successful registration result to the user. After the registration is successful, the user can proceed to the next step. The specific process is as shown in Fig. 3.

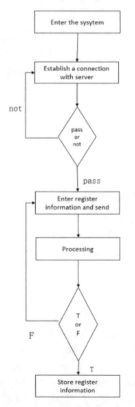

Fig. 3. Registration Process.

5.3 Login Design

This section mainly introduces the login process. Each user will use the previously registered account and password to log in. When a user logs in, a connection needs to be established first. After the establishment is successful, the login message will be sent to the server. The server will compare the login message with the account in the database. If the password is correct, the verification is successful, and the server will send the login success result. To the user, after a successful login, the information will be loaded into the cache, speeding up the system for other operations. The specific process is shown in Fig. 4.

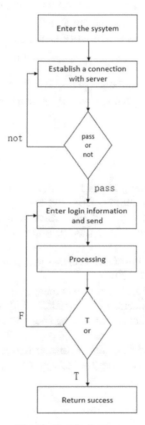

Fig. 4. Login Process.

5.4 Displays the Online User List Design

After the user logs in successfully, if you want to send a message, you can select the send message command and then send the message that you want to send to the server. After the server receives it, it will query the online user list and send the message.

5.5 Send Message Design

After the user logs in successfully, if you want to send a message, you can select the send message command, and then send the message you want to send to the server. After the server receives it, it will query the online user list and send the message.

6 Conclusion

This paper uses Go language and TCP socket for programming, and proposes an efficient and reliable design method. Compared with some existing designs, it has high concurrency, ensures work efficiency and process plasticity, and chooses a mature technical framework to greatly shorten the development time. After relevant discussions, the design is conducive to the collaborative work between employees of the enterprise and improves work efficiency.

Acknowledgments. This work is partially supported by the Science Project of Hainan University (KYQD(ZR)20021). This work is supported by SSPU young talent project, China (EGD22QD02).

References

1. Ye, W., Lin, S., Huang, L., Zhiming, X., Li, J.: Design and implementation of instant messaging system. Comput. Technol. Dev. **30**(02), 216–220 (2020)
2. Ke, Y.: Design and implementation of distributed instant messaging system. Huazhong University of Science and Technology (2020). https://doi.org/10.27157/d.cnki.ghzku.2020.007045
3. Li, Y.: Design and implementation of instant messaging system based on Kylin system. Autom. Technol. Appl. **39**(03), 51–55 (2020)
4. Wang, L.: Design and implementation of instant messaging system for enterprise users. Harbin Institute of Technology (2019). https://doi.org/10.27061/d.cnki.ghgdu.2019.001395
5. Nan, D.: Research on asynchronous non-blocking network communication technology. Modern Comput. **17,** 79–82 (2019)
6. Hang, Z.: Research and implementation of instant messaging system based on WebRTC. Huazhong University of Science and Technology (2019). https://doi.org/10.27157/d.cnki.ghzku.2019.003110

Research on the IoT and AI Under the Background of Blockchain

Dongfang Jia[1,2] and Longjuan Wang[1,2(✉)]

[1] School of Cyberspace Security, Hainan University, Haikou, China
wanglongjuan@hainanu.edu.cn
[2] Key Laboratory of Internet Information Retrieval of Hainan Province, Haikou, China

Abstract. As a new technology applied in computers, blockchain can encrypt computing, store data, transfer and consensus mechanism. Blockchain is a technology underlying Bitcoin with high security performance. The key of blockchain technology is to prevent the database information from being modified and destroyed. This technology can be divided into two types, mainly block and chain structure, which can reduce the probability of information theft or modification in the database and monitor the emergence of new blocks. Blockchain has a good effect in decentralizing the system, data traceability and tamper-proof, while the IoT is a platform for sharing resources and exchanging information between objects, and the value of data on it is constantly mined and revealed. Recently, a number of studies have explored the use of blockchain in the IoT. Under the background of blockchain, IoT technology and AI technology show an unstoppable development trend.

First, the basic concepts of blockchain technology are introduced, and secondly, based on the discussion of blockchain technology in the context of blockchain, the IoT is centered on the characteristics of IoT technology and the core of blockchain technology. Communication between the analysis of the problem of the network. Finally, the function of the blockchain is explained from the background of the blockchain. With regard to AI technology, the opportunities and challenges brought about by the integrated application of IoT and AI technologies and new technologies will be elaborated. At this stage, the combination of blockchain technology and the IoT has achieved initial results, showing its application and research value, but its further expansion still needs to rely on mutual integration with new research technologies.

Keywords: Block chain · IoT technology · AI

1 Introduction

At present, a new generation of information technology represented by AI and blockchain is becoming a new driving force for the development of China's digital economy. AI is an important guiding direction for a new round of industrial revolution and an important driving force for the development of the real economy [1]. Blockchain and AI are two technology trends, each unique and pioneering, with different priorities: blockchain

J. C. Hung et al. (Eds.): IC 2023, LNEE 1045, pp. 647–656, 2023.
https://doi.org/10.1007/978-981-99-2287-1_92

focuses on maintaining accuracy in recording, authentication, and enforcement, while AI facilitates decision making, evaluation, and understanding of certain patterns and datasets, leading to autonomous interactions. The academic definition of the IoT is "IoT". Its core system is built on the basis of the Internet, and gradually expanded and extended, forming a network in daily production and life [2]. Based on the background of blockchain technology, the IoT and AI technology have a broad application space, which can fundamentally improve the shortcomings of the Internet and improve the efficiency and security of information processing.

Right now, the two leading technologies in the tech industry are blockchain and AI. Through the discussion and research of these two technologies, blockchain technology basically uses consensus algorithm to generate and store data, uses smart contracts to operate data, and uses advanced encryption technology to prevent data theft, which can ensure the security of data. Information. The most important feature of blockchain technology is that it is decentralized and the data cannot be tampered with. And AI technology also has its unique advantages, through the analysis of the characteristics of these two technologies, it is believed that blockchain and AI can be integrated in data, algorithms, computing power and other aspects of the development, and with the in-depth integration, can be applied to more scenarios [3].

At present, with the development of science and technology, mutual network related technology is widely used in various fields of social science. In particular, the combination of the IoT and other related application equipment has promoted the development of technology in the direction of AI [4]. The traditional information processing methods fail to meet the expected expectations, and AI technology is needed to provide the conditions for information processing. Blockchain and AI technologies make up for the shortcomings of the IoT and can integrate and protect information and data.

2   Blockchain Technology Concept Elaboration

Blockchain technology is essentially a shared database for storing data or information, which is "unforgeable", "traceable", "decentralized" and other characteristics. Chain technology using the block chain validation data structure and data storage, using distributed node consensus algorithm to generate and update the data, the encryption technology to ensure the security of data transmission and access, automated script code intelligent use of contract to programming and operation of data, is a kind of new architecture based on distributed computing paradigm [5].

Block structure is a new node, with the network as the main form, the transaction in data as the performance of the structure. The block structure is mainly the block title and the block body. The block title contains the information in the database and its information, and the block function is to organize the time transaction information. The distribution structure of blockchain can store and distribute data information according to nodes. Different nodes also have the function of storing information data. After verification, the information is input into the block. The distribution structure will not be affected by other factors, and the database will not crash, so as to realize the optimization of the blockchain system [6].The classification of blockchain is shown in Fig. 1.

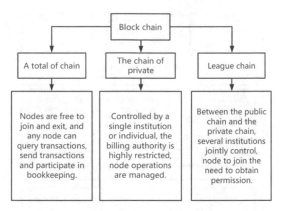

Fig. 1. Blockchain Classification

With the deepening of blockchain technology, the development of science and technology has provided a new model for our country, which has brought good results. Based on the data, algorithms and computing functions of AI, aiming at the pain points in the field of AI, the application architecture of AI technology based on blockchain is constructed. Blockchain technology was originally used in the financial industry, as the underlying technology and infrastructure applied to Bitcoin [7]. At present, its great application value has been expanded and extended from the financial industry to transportation, medical, media and other industries.

3 Blockchain Combined with IoT Technology

3.1 System Model

Based on RFID technology, objects (such as bitcoin, the initial application of blockchain) can be given a "private number", giving them the ability to "speak", so there is no confusion. AI can be broadly understood as the implantation of programs to make computers possess part of the "human brain" thinking ability, so as to complete a set of tasks with higher efficiency. Based on this, blockchain combined with the IoT architecture will further change human life.

The "decentralization" feature not only realizes the unique consensus mechanism of the blockchain, but also easily leads to the disclosure of user data privacy. With its in-depth application in various applications, the problem of data privacy disclosure is becoming more and more serious. A large number of data records will be generated in the transaction process of virtual currency, which brings opportunities to some people with improper intentions [8]. They will use the laws in the data to calculate the identity information and geographic location information of transaction participants, and then associate these information to steal the user's virtual currency. In energy application and other fields, if information leakage occurs, it will bring huge losses to the country. It can be seen that the data privacy security of blockchain technology is very important. The framework of blockchain is shown in Fig. 2.

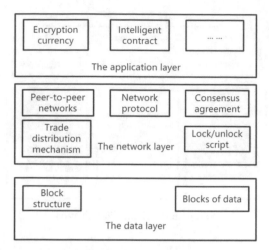

Fig. 2. Blockchain framework structure

The modern IoT technology is based on the computer Internet, using wireless data communication technology, radio frequency identification technology and so on, constitute the "big IoT" framework covering "everything" [9].

When modern scholars study the IoT, they often introduce the concept of "M2M", which is to realize the mutual communication and information transmission between Man and Machine, between people and machines. The framework of "M2M" has gradually expanded, providing a platform to apply more technologies. "Things" in IoT refers to conditions that would otherwise include data transmission paths, storage functions, central processing units, operating systems, applications used to implement specific functions, communication protocols, etc., all dependent on IoT, in The world's network has a unique identifiable number.

Blockchain can also be used as a basic framework, which can be subdivided into network layer, data layer, and application layer. The data layer generally gives the type and structure of data records of the blockchain. Data records are also called transactions, which are the evidence of specific interaction actions between nodes at a specific time.

3.2 Application of Blockchain and IoT

There are intelligent systems in the IoT system. After the integration of the IoT and AI technology, the services provided for users are more intelligent and humanized. The key link in the application of the IoT is control.

(1) Data privacy Protection

Iot of data in the whole from the life cycle of production, storage, management and sharing, data privacy and security protection of it involving nodes, iot devices and users access, access control and regulation of behavior such as process, the researchers on the stage, by using decentralized block chain, anonymity, data can be traced back and tamper-resistant features, The corresponding solutions are designed through blockchain technologies such as smart contracts and consensus mechanisms [10].

(2) Nodes are added

The nodes of the iot contain sensors, executing devices, and users who use the data, which are the basic elements of the underlying implementation of the iot. The identity authentication and privacy protection of nodes when accessing the network are the basic guarantee for secure communication in the IoT.

(3) Behavior supervision

The behavior regulation of iot plus blockchain covers both the behavior of data collected by devices and the behavior of data used by users. Researchers mostly focus on the use of smart contracts and consensus mechanisms to monitor and manage these behaviors [11].

(4) Application scenarios

Blockchain fills the security gap in the information exchange process of the IoT, further improves the availability of the IoT, and brings some new application scenarios.

(5) Fintech innovation

The application of blockchain plus iot in fintech focuses on supply chain finance. Supply chain is an important chain in economic activities, connecting suppliers, manufacturers, transporters, retailers and consumers [12]. In order to record and preserve all data of entities upstream and downstream of the supply chain, such as raw material supply, product manufacturing and processing, logistics transportation and sales, many researchers have explored the applicability of contract technology using blockchain technology and intelligence. These data are open, transparent, traceable and cannot be tampered with. The private data is encrypted to solve the trust issues, regulatory traceability issues and data privacy protection issues in the supply chain.

4   Analysis of Key Points of Blockchain and AI Integration

4.1   Blockchain Background AI Data Sharing

At present, there is a lack of unified and efficient sharing mechanism and management method for massive data in the AI industry [13]. The poor maintenance of open source data sets leads to uneven data quality and problems such as uncentralized and ununified data. In addition, a large amount of data required for AI training is mainly concentrated in the government and large companies [14]. Regulatory and commercial barriers and other conditions lead to poor data circulation and difficult access, which seriously restrict the pace of AI development in small and medium-sized enterprises [15]. In addition, when intelligent individuals make decisions, they need to obtain as much real-time data as possible as a reference. If there is not enough real-time data or data is not real-time enough, this kind of AI can only be limited intelligence. The application of AI technology is shown in Fig. 3.

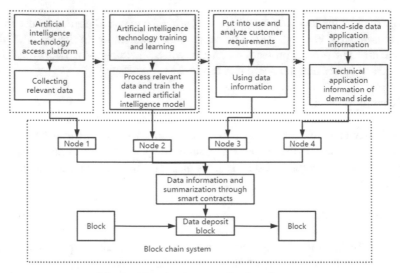

Fig. 3. AI technology application diagram

Blockchain distributed database can efficiently share data among nodes, so that every participant on the network can access data, which can provide more extensive data access and more effective data monetization mechanism for AI [16]. Firstly, relying on blockchain technology, a decentralized data sharing platform can be built. This platform is based on blockchain network information and data communication platform, and its update and information record are jointly completed by distributed subjects, rather than executed by an authority. Second, blockchain data validation will facilitate the establishment of cleaner and more organized personal data, which in turn will provide smoother data integration and the formation of new data markets. Finally, more open, shared, real-time data for analysis will enable machines to make more accurate predictions and assessments, generating more reliable algorithmic models, and thus advancing the overall state of the art of AI.

4.2   Blockchain Background AI Computing Power Supply

The cost of computing power is a big pain point in the AI industry. With the increase of geometric multiple of network data and the improvement of algorithm model complexity and accuracy, AI has reached the scale of one hundred billion parameters and one trillion training data sets, which undoubtedly requires larger and stronger computation [17]. Ordinary AI technology company needs high capital purchase of hardware resources such as the GPU computing center construction, it costs for most small and medium enterprises is extremely high, lead to "work force is not enough, the cost is expensive and difficult to obtain" the status quo of widespread, calculate the force has become the development of AI needs from one of the bottleneck of the cost benefit into consideration. The application architecture of AI technology is shown in Fig. 4.

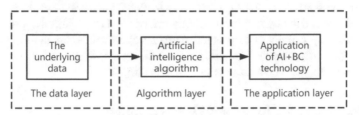

Fig. 4. AI technology application architecture

Most of the computing power of ordinary computers around the world is idle. If this power can be harnessed, the cost of AI modeling can be greatly reduced and the efficiency of resource utilization can be improved. The distributed characteristics of blockchain can make full use of the idle computing power distributed around the world, which is helpful to build a decentralized AI computing power infrastructure platform, change the traditional idea of constantly improving equipment performance to improve computing power and reduce the operation cost of enterprises. In addition, through the smart contract of blockchain, the network computing nodes are dynamically adjusted according to the amount of computation required by users, so as to provide elastic computing capacity to meet the computing needs of users [18]. AI research and development institutions make use of massive data and submit deep learning models to a blockchain-based computing power sharing platform, and allocate corresponding computing power according to the facility capacity of each node to train the AI model. The data sharing platform model is shown in Fig. 5.

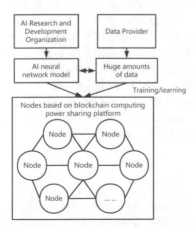

Fig. 5. Data sharing platform model

Blockchain distributed computing can use large GPU or FPGA server clusters, idle Gpus of small and medium-sized enterprises, and personal idle Gpus as computing nodes to share computing resources and provide computing power for AI [19]. In addition, if can be specially designed for mining machine block chain special integrated circuit ASIC chip part of the work force for AI, consensus in POW mechanism of AI is introduced into the ASIC chip friendly support matrix operations and convolution operation, can be used in ASIC chip can dig of distributed AI to accelerate the computation at the same time. Or adopt AI-friendly POW algorithm for ASIC chips, so that the miner can be used for AI acceleration after being idle or eliminated [20].

5 Difficulties and Challenges Opportunities

Blockchain has improved the functions of the IoT, but blockchain is not a panacea, and further improvement and innovation are needed. Breakthroughs in emerging technologies such as edge computing, big data and AI have brought new opportunities and challenges to blockchain + iot.

In this era of Internet development, there are tens of millions of devices in the Internet of Things, and the data generated is messy, complex and huge. The blockchain responsible for data maintenance and management needs to accommodate massive device users, track data generation and collection in real time, and process data transactions in the IoT in high concurrency [21]. However, the current blockchain solutions are constrained by ledger consistency, which makes it difficult to coordinate the contradiction between throughput and node number.

The "intelligence" of blockchain smart contract is reflected in the provision of Turing complete program execution [22]. In fact, it cannot support large-scale computation, and its input and output are also greatly limited. It is often only used to process preset simple business logic, so blockchain alone cannot effectively analyze and utilize large-scale IoT data. With the development of related AI technologies such as machine learning and knowledge graphs, the concept of intelligent Internet has been proposed. Through the IoT, information sources and knowledge can be obtained from the information Internet in the virtual world to realize the knowledge interconnection between knowledge entities [23]. To construct intelligent Internet, researchers need to automatically acquire knowledge from the emotionally mixed data of agents in the Internet on the basis of the IoT, cooperatively represent and transfer knowledge, and finally establish joint knowledge space to connect and cooperatively run knowledge [24]. Blockchain supports the collaborative operation of knowledge in intelligent networking to realize automatic operation and autonomous evolution in a distributed architecture.

6 Conclusion

In the context of blockchain, artificial intelligence and Internet of things technologies will break the existing application scope and bring greater convenience to human life and work. IoT is the upgrading of the Internet, through physical entity is linked together, the data in the IoT directly describes the basic attributes of the objective world and correlation, and block chain with distributed structure is fully reliable data, privacy, security and security interests, therefore block chain and the IoT will be future effective fusion of physical space and information space, realize the important development direction of technological change.

When related issues arise in the fields of AI and blockchain, complementary advantages can be gained by merging the two technologies. Currently, some application cases have fused the two together. Be worth what carry is, both in terms of technical indicators of the current block chain, or from the large data, the actual landing of AI, the real implementation technology integration and implementation of the ground is still facing many uncertainty factors, the fusion of the potential result is difficult to assess, so in the positive research of AI and block chain technology fusion as well as rational view, pay attention to practice, Organic combination, flexible innovation, truly realize the practice and exploration of the integration of AI and blockchain technology.

Acknowledgments. This work is partially supported by the Science Project of Hainan University (KYQD(ZR)20021).

References

1. Atlam, H.F., Azad, M.A., Alzahrani, A.G., et al.: A review of blockchain in IoT and AI. Big Data Cogn. Comput. **4**(4), 28 (2020)
2. Zhang, K., Zhu, Y., Maharjan, S., et al.: Edge intelligence and blockchain empowered 5G beyond for the industrial IoT. IEEE Netw. **33**(5), 12–19 (2019)
3. Alkhateeb, A., Catal, C., Kar, G., et al.: Hybrid blockchain platforms for the IoT (IoT): a systematic literature review. Sensors **22**(4), 1304 (2022)
4. Torky, M., Hassanein, A.E.: Integrating blockchain and the IoT in precision agriculture: Analysis, opportunities, and challenges. Comput. Electron. Agric. **178**, 105476 (2020)
5. Qiao, L., Dang, S., Shihada, B., et al.: Can blockchain link the future? Digit. Commun. Netw. 8, 687–694 (2021)
6. Li, D., Deng, L., Cai, Z., et al.: Blockchain as a service models in the IoT management: systematic review. Trans. Emerg. Telecommun. Technol. **33**(4), e4139 (2022)
7. Bublitz, M.F., Oetomo, A., Sahu, K.S., et al. Disruptive technologies for environment and health research: an overview of AI, blockchain, and IoT. Int. J. Environ. Res. Public Health **16**(20), 3847 (2019)
8. Kumar, S., Raut, R.D., Narkhede, B.E.: A proposed collaborative framework by using AI-IoT (AI-IoT) in COVID-19 pandemic situation for healthcare workers. Int. J. Healthc. Manag. **13**(4), 337–345 (2020)
9. Dwivedi, S.K., Roy, P., Karda, C., et al.: Blockchain-based IoT and industrial IoT: a comprehensive survey. Secur. Commun. Netw. **2021** (2021)
10. Ren, Q., Man, K.L., Li, M., et al.: Intelligent design and implementation of blockchain and IoT–based traffic system. Int. J. Distrib. Sens. Netw. **15**(8), 1550147719870653 (2019)
11. Pal, K.: Blockchain-integrated internet-of-Things architecture in privacy preserving for large-scale healthcare supply chain data. In: Blockchain Technology and Computational Excellence for Society 5.0. IGI Global, pp. 80–124 (2022)
12. Viriyasitavat, W., Da Xu, L., Bi, Z., et al.: New blockchain-based architecture for service interoperations in IoT. IEEE Trans. Comput. Soc. Syst. **6**(4), 739–748 (2019)
13. Alamri, M., Jhanjhi, N.Z., Humayun, M.: Blockchain for IoT (IoT) research issues challenges & future directions: a review. Int. J. Comput. Sci. Netw. Secur **19**(1), 244–258 (2019)

14. Daniels, J., Sargolzaei, S., Sargolzaei, A., et al.: The IoT, AI, blockchain, and professionalism. IT Prof. **20**(6), 15–19 (2018)

15. Eze K G, Akujuobi C M, Sadiku M N O, et al. IoT and blockchain integration: use cases and implementation challenges. In: International Conference on Business Information Systems, pp. 287–298. Springer, Cham (2019)

16. Sharma, D.K., Kaushik, A.K., Goel, A., et al.: IoT and blockchain: integration, need, challenges, applications, and future scope. In: Handbook of Research on Blockchain Technology, pp. 271–294. Academic Press (2020)

17. Gromovs, G., Lammi, K.: Blockchain and IoT require innovative approach to logistics education. Transp. Probl. **12** (2017)

18. Chen, F., Xiao, Z., Cui, L., et al.: Blockchain for IoT applications: a review and open issues. J. Netw. Comput. Appl. **172**, 102839 (2020)

19. Tan, L., Shi, N., Yu, K., et al.: A blockchain-empowered access control framework for smart devices in green IoT. ACM Trans. Internet Technol. (TOIT) **21**(3), 1–20 (2021)

20. Gadekallu, T.R., Pham, Q.V., Nguyen, D.C., et al.: Blockchain for edge of things: applications, opportunities, and challenges. IEEE IoT J. **9**(2), 964–988 (2021)

21. Gupta, R., Rakhra, A., Singh, A.: IoT Security using AI and blockchain. In: Machine Learning Approaches for Convergence of IoT and Blockchain, pp. 57–91 (2021)

22. Banafa, A.: Secure and Smart IoT (IoT). River Publishers (2018)

23. Ali, M.S., Vecchio, M., Pincheira, M., et al.: Applications of blockchains in the IoT: a comprehensive survey. IEEE Commun. Surv. Tutor. **21**(2), 1676–1717 (2018)

24. Rakovic, V., Karamachoski, J., Atanasovski, V., et al.: Blockchain paradigm and IoT. Wirel. Pers. Commun. **106**(1), 219–235 (2019)

Analysis of the Integration of Multimedia Technology and Dance Teaching in Colleges and Universities

Jingjing Wang[1], Dongfang Jia[2,3], and Ying Fan[1(✉)]

[1] School of Music and Dance, Hainan University, Haikou, China
4659626@qq.com
[2] School of Economics, Hainan University, Haikou, China
[3] Key Laboratory of Internet Information Retrieval of Hainan Province, Haikou, China

Abstract. Multimedia technology is the product of the new era and has been applied in various fields. Its appearance has brought a lot of convenience to our life, especially in education. Therefore, giving full play to the advantages of computer multimedia technology and improving the teaching quality of multimedia combined with dance has become an important measure of the development of The Times. This paper expounds the inevitability of the deep integration of computer multimedia technology and dance teaching in colleges and universities, and makes use of modern teaching media and network education resources to let teachers and students jointly participate in dance teaching practice.

The integration of dance teaching and the application of multimedia technology in Chinese colleges and universities are analyzed: the construction of cross-space resource sharing platform; The organic combination of traditional and modern multimedia teaching; And the teacher-led teaching mode of dance situation. This paper discusses the problems existing in the effective integration of the two, and tries to put forward effective teaching strategies: using multimedia technology to construct standardized dance textbooks; Integrating dance course resources in colleges and universities with network information; The development of wechat mini program online dance teaching platform. Aiming at the limitations and constraints of the inherent conditions of traditional dance teaching in colleges and universities, the teaching methods are constantly improved, so that college dance teaching must adapt to the requirements of the new era after class.

Keywords: Multimedia technology · Network education · College dance teaching · The organic integration

1 Introduction

Multimedia technology is a collection of text, image, audio, video, sound and other forms in one of the technology. It has these characteristics: digitalization, interactivity, complexity and integration. With the rapid development of computer network technology, the application of computer multimedia technology in college dance teaching is more and more extensive [1]. The use of new teaching models and teaching methods,

J. C. Hung et al. (Eds.): IC 2023, LNEE 1045, pp. 657–665, 2023.
https://doi.org/10.1007/978-981-99-2287-1_93

constantly tap the potential of students, students as the center, teachers as the leading, to achieve the goal of comprehensive ability training. Compared with traditional dance teaching methods in colleges and universities, multimedia has the outstanding advantages that ordinary teaching methods do not have. It can effectively integrate learning resources in a short time, clearly transfer dance knowledge to students, so as to maximize the enthusiasm of students and promote their mastery of dance professional knowledge [2]. Below is the computer multimedia technology and the innovation necessity and law of dance teaching in colleges and universities depth fusion, and analyzes how to change the dance teaching and multimedia technology related to application of the fusion in colleges and universities in China, finally, aimed at the problems existing in the computer multimedia fusion dance teaching put forward some related introduction, break through the limitations of traditional dance teaching and restriction, To improve the teaching methods. So that dance teaching stage in colleges and universities more in line with the new era of classroom requirements.

2 The Integration of Multimedia Technology and College Dance Teaching is Inevitable

Dance majors in colleges and universities shoulder the important task of training high-quality dance talents and carrying forward excellent dance culture, which is the training base of professional dance talents in China. Nowadays, the emergence of multimedia technology provides new teaching ideas for the reform of dance education in colleges and universities. Using computer network technology to cooperate with skill teaching can also choose to improve the quality of technical skills and movement finished products, and can obviously improve the efficiency of teaching, which is conducive to improve the decision-making ability of dance creation and cultivate students' good artistic quality [3]. In the teaching, through the application of computer motion capture technology, computer three-dimensional auxiliary system, motion editing technology and computer music editing technology, in the dance teaching classroom, text, sound, image and other materials can be fully integrated, enrich the classroom teaching content, and make contributions to the deepening and promotion of dance teaching.

It can greatly improve students' interest in learning, stimulate students' innovative thinking ability, so as to enhance students' learning enthusiasm, and finally achieve the expected teaching objectives. With the progress of multimedia technology, traditional teaching models are looking for greater opportunities for development, and all education systems will face unprecedented challenges. The integration of multimedia technology and dance teaching in colleges and universities will revolutionize dance education and inject new vitality into dance teaching.

3 Analysis of the Integration of Dance Teaching and Multimedia Technology in Colleges and Universities

3.1 Building a Platform for Resource Sharing Across Time and Space

Web-based multimedia technology has the characteristics of open, sharing and effective, which requires universities to establish the platform across time and space resources, according to the development trend of dance education in colleges and universities and their students' characteristics, highlight the advantages of talent training, positive and the integration of multimedia technology, the development of diversified education mode, and give full play to the advantages of multimedia teaching. For example, dance teaching in colleges and universities can be combined with multimedia technology to create a teaching hall of "please come in, go out". Teachers use multimedia network teaching software to teach dance artists in real time in the teaching process, and regularly set up special lectures by famous teachers in the online form to spread advanced knowledge of dance culture [4]. It can not only make up for the shortage of dance teaching in regional colleges and universities, but also realize the sharing of ancient and modern channels under the inspiration and infection of dance artists, improve students' exuberant thirst for knowledge, enrich teaching methods, and greatly improve the quality of teaching.

3.2 Organic Integration of Traditional and Modern Multimedia Teaching

The integration and application of multimedia technology in college dance teaching is not to abandon the traditional teaching methods and educational resources, but to build an education system that meets the requirements of college dance teaching through the effective combination of the two, so as to lay the foundation for improving the quality of college dance teaching. In demo dance moves, for example, teachers can make use of multimedia teaching software, the core of dance action is decomposed into several dynamic demonstration figure, and on the basis of the QQ, WeChat or PPT, will these images clearly show students, let students use plenty of time before class, group dancing in advance according to the pictures at the heart of the action, a preliminary understanding of the core.

In order to allow students to show each other in class, teachers take the initiative to collect and sort out students' self-taught problems before class, carry out interactive teaching, answer questions for students, and finally achieve seamless connection between before class and class [5]. However, the absorption degree of each of the students in the classroom, learning interest and learning ability is different, can't maintain full learning mood, some of the students in the class who were unable to complete the difficult dance moves, self-confidence, affect the study effect, which requires teachers in the classroom use of traditional teaching methods to guide students. As shown in Fig. 1, the integration means of traditional and modern multimedia teaching are illustrated.

In order to overcome the resistance of self-study before class, multimedia can be used as a link to narrow the distance between teachers and students. In addition, multimedia technology can be used to remotely correct students' homework after class, so as to further consolidate students' motor skills. Therefore, the organic combination of multimedia technology and traditional dance teaching in colleges and universities is an effective way to realize the interlinked teaching process before class, during class and after class.

3.3 Teacher-Led Dance Situational Teaching Mode

The traditional dance teaching mode undoubtedly limits students' creative thinking, while using multimedia technology to carry out dance scene teaching mode is conducive

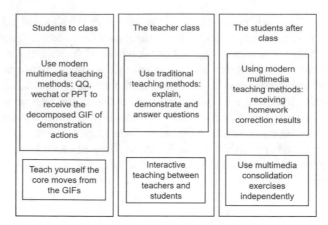

Fig. 1. Traditional and modern multimedia teaching fusion means

to students' deeper understanding of dance knowledge. Therefore, the teacher-led multimedia dance scene teaching model can start from three aspects: scene introduction, movement demonstration and learning feedback. Taking the teaching content of ethnic dance in colleges and universities as an example, that is, in the teaching of ethnic dance in colleges and universities, the cognitive management ability of most college students on ethnic traditional culture is still in a relatively weak stage. At the beginning, it is particularly important to introduce some scenes and reproduce them with multimedia technology [6].

Let college students have a more intuitive understanding of the national customs and customs, living environment and basic characteristics, so as to improve the students' dance performance ability. Secondly, teachers can use multimedia technology projection screen to show students the relevant combined movements of folk dance combination training in a complete and systematic way, and through slow motion analysis, press the pause button, replay and other ways in the courseware video, inspire and induce students to analyze the correlation and difference between combined movements and learning combination. Finally, learning feedback is an important performance to improve students' dancing ability. After guiding the students to analyze, imitate and practice the movements, the teacher can record the learning results of the students and watch them together. At this time, the expressions and gestures of the dance performance can be observed clearly at a slow speed, so that the teacher can reflect on the problems in the teaching process. Students can make targeted modifications to their own problems more intuitively, so as to improve the overall movement texture and learning effect.

4   Problem Analysis and Strategy Analysis

The use of audio, video and PPT has become an indispensable auxiliary means in college dance teaching, and the combination of virtual and real teaching online and offline has become a common practice in many disciplines. In the present stage, although some auxiliary means of computer multimedia technology are used in college dance classroom teaching, it is not mature enough to systematically apply computer multimedia technology to the knowledge structure and related courseware of college dance teaching [7]. Moreover, many college dance teachers have different levels in the production of multimedia courseware.

Teachers just play out the dance knowledge from the textbook through a few simple videos, and simply play the excellent dance performance videos. At this time, the video has become the so-called "movie", it is difficult to achieve the advantages of interactive and hypertext function, also can not fully reflect the significance of multimedia assisted teaching. In addition, it is not difficult to find that the ubiquitous phenomenon of multimedia equipment in colleges and universities, as well as the lack of awareness of modern multimedia teaching among school teachers, is still an important factor affecting the high degree of integration of multimedia touch control and dance teaching in colleges and universities. Without perfect multimedia teaching conditions, teachers do not realize the importance and necessity of developing and utilizing multimedia dance teaching.

To stay on the multimedia technology to consistent with the teaching content of the PPT courseware one-sided ideas, lead to teacher's information literacy and the multimedia teaching ability is limited, in the end it takes time and energy to complete the search of multimedia courseware, screening, sorting and production, increased the workload and pressure, affect the efficiency of dance teaching [8]. Colleges and universities dance teaching in order to improve the overall quality for the purpose, in addition to imparting professional dance skills, more emphasis on cultural heritage, which requires dance teacher through multimedia as auxiliary teaching means to supplement the output of the dance and rich cultural knowledge, such as: with the introduction of the related national cultural background, style characteristics of dance and movement of the thorough analysis, etc. At present, the teaching of these contents only depends on the teachers' own discipline quality, and there is no unified and standardized dance textbook, which leads to the difficulty in controlling the accuracy of the teaching content and uneven teaching quality.

4.1 Using Multimedia Technology to Construct Standardized Dance Teaching Materials

Dance in order to better regulate and unified teaching material, can use multimedia technology to the dance teaching content, make relevant courseware, in addition, the use of multimedia technology to build the dance teaching material not only confined to the theory of knowledge learning and skills training, should also be derived dance teaching demonstration, movement simulation training, interactive training, etc. In order to realize the sharing of remote resources, change the student-oriented independent learning and training mode, expand the scale of teaching and learning, promote the reform of dance education, improve the effect of dance training, and promote the informatization modernization process of dance education in colleges and universities [9]. With the application of multimedia technology and the innovation of multimedia dance education methods, dance textbooks constructed by multimedia technology have made a qualitative leap in technical design, material content, system integration, testing and evaluation, because there are more standardized requirements for the design and production of dance textbooks. Therefore, the rational use of multimedia technology to assist teaching can not only enrich the form of dance teaching in colleges and universities, standardize the teaching content, but also broaden the teaching space and improve the teaching effect, which is in line with the purpose of training dance professionals in colleges and universities in the new era.

4.2 Integrating Dance Curriculum Resources in Colleges and Universities by Using Network Information

With the help of network information, dance teachers in colleges and universities can collect a large number of rich content and various types of dance course resources, and develop a dance teaching resource library based on network carrier, which can effectively integrate the teaching content and related dance material resources involved in the teaching process. Firstly, the construction of university dance teaching resource library is divided into three modules. The first module is the dance teaching resource management module, which includes the collection and management of university dance fine course courseware, famous teachers' monographs, folk inheritors' interviews, dance case teaching database, courseware database, Q&A database and other teaching materials.

The second module is the resource system management module, which is used to manage the operation, safety and fault technology of the dance resource system. The third is the communication module used by teachers and students: teachers and students carry out real-time interactive discussion, online resource browsing and evaluation, and dance case teaching analysis through the university dance teaching resource library [10]. In addition, an online learning platform is built, virtual student behavior, dance, case teaching and classroom teachers are introduced into training courses in teaching, multimedia resources are rationally and effectively used to develop, excellent dance teachers are trained, doctoral education of dance in colleges and universities is improved, and the development of dance education is vigorously promoted. Finally, the use of network information to integrate dance course resources into the teaching activities of guarantee courses can help enhance the comprehensive, professional and interesting content of dance teaching, and help students expand their knowledge reserve, so that they can have a deeper and more comprehensive understanding of dance courses. As shown in Fig. 2, the functional structure of the university network dance course resource library.

4.3 Develop a Network Dance Teaching Platform for WeChat Applets

As a product of the development of The Times, wechat mini program has been widely used in daily life. With the development of network technology, many colleges and universities have set up wechat mini program courses as an extension of offline courses. The development of wechat mini program involves rich and colorful teaching content [11]. Based on WeChat small dance program to develop a network teaching platform, realize the dance professional communication and sharing of course materials and relevant learning materials, students in extra-curricular dance to receive information in network world, learning professional knowledge, in the long run can help cultivate students' dance artistic self-cultivation and dance comprehensive ability, fast development of dance education in colleges and universities.

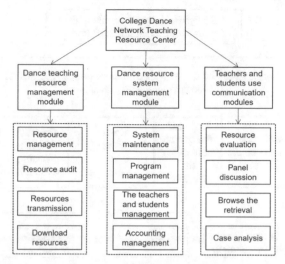

Fig. 2. Functional structure of college network dance course resource library

The school-based curriculum design of wechat mini program in colleges and universities includes the formulation of dance curriculum objectives, the arrangement of semester curriculum content, the selection of dance curriculum resources, the construction of learning curriculum evaluation, etc. Four basic principles must be taken into account when using wechat mini program to construct a curriculum plan closely related to the teaching objectives of dance in colleges and universities: the principle of starting from reality, basic principle, practical principle and practical principle. First of all, the course construction must conform to the reality of colleges and universities, and to the reality of teachers and students. Secondly, the course construction must pay attention to the basic knowledge, basic ability and basic thought of dance [12].

Third, the nature of dance curriculum determines that programming curriculum must emphasize practicality. Its content and various forms of practical links are very important. There are a lot of practical exercises arranged in the program, including demonstrations of movement details, entry of practice cards, etc., for students to enter the design forum with questions and study independently. Finally, because the applets connect people with various objects in various scenarios and provide various application services, the curriculum should be based on the actual needs in the field of dance learning. On the basis of guiding students to use small programs to facilitate the application of dance learning, through practical programs, let students master the knowledge and build real practical small programs. Figure 3 below: Schematic diagram of course design of dance teaching platform.

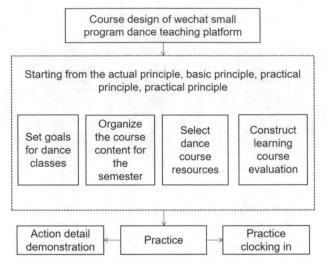

Fig. 3. Schematic diagram of course design of dance teaching platform

5 Conclusion

Multimedia technology is the combination of language, text, sound, image and other media information, which has been vividly used in education and teaching. Integrating computer multimedia technology into dance teaching in colleges and universities can not only create a good classroom atmosphere for dance learning, stimulate students' innovative spirit and practical ability of dance learning, but also provide students with more abundant dance teaching resources and flexible and open Internet learning environment [13].

It can also greatly improve the dance teachers' teaching and scientific research ability, realize the computer multimedia and organic synthetic analysis of colleges and universities dance teaching, trains the student to the core of the basic quality, at the same time to complete the Internet dance teaching reform and an important task of education in the new period, so as to effectively advance the construction of modern computer multimedia dance teaching system. In dance teaching in colleges and universities, therefore, the composition of multimedia technology to release some era is the need of developing professional, and how to use a variety of ways of integration of traditional and modern multimedia teaching dance, maximize the effectiveness of the teaching effect, promote the research and innovation of dance teaching in colleges and universities, colleges and universities under the new era has been exploring problems of dancing.

References

1. Wu, H., Leng, Y.: The Influence and analysis of multimedia interaction on the development of dance. J. Electr. Comput. Eng. **2022**, 1–10 (2022)
2. Wang, Y., Zheng, G.: Application of artificial intelligence in college dance teaching and its performance analysis. Int. J. Emerg. Technol. Learn. **15**(16), 178–190 (2020)

3. Liu, G.: The design and Implementation of sports dance teaching system based on digital media technology. In: The 1st EAI International Conference on Multimedia Technology and Enhanced Learning (2017)
4. Yang, X.: Analysis of the construction of dance teaching system based on digital media technology. J. Interconnect. Netw. **22**, 2147021 (2022)
5. Zhou, L.: Model construction of dance teaching system for college students under the background of information technology. Wirel. Commun. Mob. Comput. **2022** (2022)
6. Fernandes, C., Evola, V., Ribeiro, C.: Dance Data, Cognition, and Multimodal Communication. Taylor & Francis (2022)
7. Xu, Y.: Deep learning for dance teaching system. In: Hung, J.C., Chang, J.W., Pei, Y., Wu, W.C. (eds.) Innovative Computing, LNEE, vol. 791, pp. 1577–1581. Springer, Singapore (2022). https://doi.org/10.1007/978-981-16-4258-6_198
8. Rittershaus, D., Koch, A., deLahunta, S., et al.: Recording "effect": a case study in technical, practical, and critical perspectives on dance data creation. In: Dance Data, Cognition, and Multimodal Communication, pp. 71–88. Routledge (2022)
9. Wang, Z.: Modern social dance teaching approaches: studying creative and communicative components. Think. Skill Creat. **2021** (2021)
10. Xiaojun, W.: The application of multimedia and network technology on college physical education. In: 2018 4th International Conference on Education Technology, Management and Humanities Science (ETMHS 2018) (2018)
11. Hong, X.: Research on the online learning platform of multimedia teaching in higher vocational colleges. Wirel. Internet Technol. **1861** (2016)
12. Zhang, S.: Research on the infiltration and integration of modern dance elements in folk dance teaching. Arts Stud. Crit. **1**(3) (2020)
13. Parrish, M.: Technology in dance education. In: Bresler, L. (ed.) International Handbook of Research in Arts Education, vol. 16, pp. 1381–1397. Springer, Dordrecht (2007). https://doi.org/10.1007/978-1-4020-3052-9_94

A Strong Security Key Agreement Scheme for Underwater Acoustic Networks

Jinlong Wang[1], Shuai Zhang[1], Peijian Luo[1], and Xinwei Zhao[2(✉)]

[1] CETC Ocean Information Technology Research Institute Co., Ltd., LingShui 572400, China
[2] School of Computer Science and Technology, Hainan University, Haikou 570228, China
zxw1230927@163.com

Abstract. The underwater acoustic channel has the characteristics of broadcasting, and the frequent occurrence of various uncertain factors, which makes the underwater acoustic channel face many security threats. For this problem, starting from improving the security performance of the key, a strong security key agreement scheme based on the underwater acoustic channel is proposed. First, the Hash function is used to authenticate the identity of the legal node, and then, after the authentication is successful, the measured values of the two communicating parties are authenticated are obtained by error correcting coding, and the interleaving technology is used to eliminate the redundancy of the measured values and realize the safe transmission of the key in the open channel. Theoretical and experimental results show that the scheme can effectively reduce the bit error rate of the key, and can pass the eight tests of the NIST randomness test. Compared with existing scheme, the proposed method has significant advantages in terms of key security and robustness.

Keywords: Key generation · Security performance · Key agreement · Authenticate · Error correction coding

1 Introduction

In recent years, underwater acoustic (UWA) communication network has attracted extensive attention. Lots of countries have carried out many researches on UWA communication networks, so that UWA communication technology is developing rapidly. However, the interference of Marine environment makes the information easy to be stolen or even actively interfered by eavesdropping nodes in the process of transmission. Therefore, it is necessary to develop a highly confidential key agreement scheme for the security of the key, which allows communication nodes to negotiate the public key in the complex underwater environment, and provides integrity protection for the secure communication of legitimate nodes.

Key agreement is an important way for communication nodes to obtain shared keys. In recent years, there are lots of researches on key agreement, but the existing key agreement schemes are mainly based on wireless communication network. Different from wireless communication, the Marine environment is very complex, so simply applying

J. C. Hung et al. (Eds.): IC 2023, LNEE 1045, pp. 666–677, 2023.
https://doi.org/10.1007/978-981-99-2287-1_94

the key agreement scheme based on wireless communication network on land to under-water acoustic network will lead to higher bit error rate of key generation. Therefore, it is necessary to study the key agreement scheme for underwater acoustic channel to reduce the bit error rate of the key and improve the security performance of the key. Liu et al. proposed a double spread spectrum code structure and used zero sequence to provide interval protection for the transmitted data, which eliminated interference information in a large range and improved the consistency of the key. However, this method made the key transmission rate very low [1]. Shen et al. used local random pilot assisted channel detection protocol to study underwater acoustic channel, which enhanced the anti-interference ability of eavesdropping nodes in the nearby range, but the ability of active attack on defense eavesdropping nodes was weak [2]. D. Park et al. proposed an alternative quantum key coordination error correction method, which used 1/2 rate convolutional codes to describe the quantum key coordination process [3]. Jiang et al. designed a key agreement scheme for biometric identification, which can effectively resist the attack of eavesdropping nodes by constructing BCH coding [4].

However, most of the key agreement schemes are implemented under the assumption that they are in a relatively ideal situation. Some traditional key agreement schemes are applied to the underwater environment, but the complexity of the underwater environment is ignored. If the channel has no ability to resist the active interference of eavesdropping nodes, the security of underwater acoustic communication cannot be ensured. Among them, the literature [5] does not consider the influence of doppler effect and ignores the movement of objects, which increases many uncertain factors in the transmission process of communication nodes and affects the efficiency of key negotiation. Realistic application scenarios are generally random changes, such as underwater automatic vehicle communication. Doppler shift will destroy the orthogonality of the subcarrier in the system, so that the system cannot work normally, and the bit error rate of the generated key is very high. Literature [6] lacks discussion on the randomness of the key. The actual underwater acoustic physical layer key generation technology must ensure that the randomness of the generated key is strong enough, otherwise the eavesdropping node can guess the key according to the law of the key bit, which cannot guarantee the security performance of the key.

Through the above analysis, it can be seen that further research is needed to generate highly secure keys at the physical layer for complex underwater acoustic networks. Therefore, a strong security key agreement scheme is designed in this paper. The main tasks are as follows:

1) Construct a higher-order function with Hash function for identity authentication to realize the secure transmission of the initial key in the public channel.
2) Use the error correction coding technology of convolutional coding and interleaving technology to carry out key negotiation for the initial key.
3) The feasibility, correctness and security of the protocol are verified through experiments.

2 Characteristics of Underwater Acoustic Signals and System Model

2.1 Transmission Characteristics of Underwater Acoustic Signals

Underwater acoustic signal is easily affected by multipath effect and Doppler frequency shift when transmitted in the channel, which is analyzed in detail below.

The Multipath Effect. Complex multipath propagation is an important phenomenon in communication. Compared with wireless channel, the multipath effect of underwater acoustic channel is more serious. Multipath effect refers to that underwater acoustic signals are transmitted in multiple paths during transmission, and the signals coincide with each other, resulting in a higher bit error rate of the signals received by the receiver. Multipath effect influence the transmission performance of underwater acoustic channel, hindered the demodulation technology in the application of underwater acoustic channel. We use the underwater acoustic communication technology based on orthogonal frequency division multiplexing system to convert serial signals into parallel signal at low speed, which increases the signal transmission time, and promotes the signal transmission in complicated multipath environment.

Doppler Effect. Communication systems are generally affected by Doppler effect. In underwater acoustic channel, Doppler effect is hard because the speed of acoustic wave transmission is limited. Doppler effect is mainly due to the time-varying characteristics of underwater acoustic channel and the relative movement between transmitter nodes and receiver nodes. In real ocean environment, underwater node is dynamic, and the doppler effect causes great influence on the performance of the system, will lead to the deviation of the frequency of the signal extension. The length of the signal frame changes will seriously affect the SNR of the system, and make the system synchronization appear deviation, even directly lead to the interruption of communication. Therefore, Doppler compensation is required for each received underwater acoustic signal in channel estimation to balance the influence of object motion and medium instability.

2.2 Environmental Characteristics of Underwater Acoustic Signal

In underwater environment, underwater acoustic signal is affected by low propagation speed, high underwater noise and transmission loss, which leads to serious distortion of the receiver signal. The following is a detailed analysis of the environmental characteristics of underwater acoustic signals.

Low Propagation Speed. In underwater acoustic communication, we mainly use sound wave as the carrier to realize the information transmission over long distance. According to the theoretical analysis, the typical formula of the velocity of sound wave in water is

$$C = 1449 + 4.6T - 0.05T^2 + 0.00029T^3 + (1.34 - 0.012T)(S - 35) + 0.017Z \quad (1)$$

where, T is the temperature of seawater, S is the salinity, and Z is the depth of the ocean. The speed of sound propagation increases with the temperature of the water, the salinity of the ocean, and the depth of the ocean. The speed of sound waves at sea level is mainly affected by the temperature of the sea. With the deepening of the ocean depth, the influence of the change of water pressure on the speed of sound gradually increases. Traditional wireless channel key generation technology does not consider this environment characteristic. The underwater environment is complex, and many other factors lead to the low speed of sound waves, about 1500 m/s. If the communication distance is very close, large propagation delay also will be produced, which leads to communication on both sides of the measured value difference is big, the reciprocity of multipath underwater acoustic channel is damaged. Therefore, we measure the channel through the correlation between subcarriers in the channel state sensing stage to solve the problem of impaired reciprocity.

High Underwater Noise Level. Noise in the underwater environment is a big disturbing factor in the transmission of underwater acoustic signals. The cause of noise is very complex, and it is closely related to the geographical location of the ocean, the nearby climate and other factors. Underwater noise refers to the noise measured at the location of the system, which mainly includes artificial noise and noise generated by the natural environment, independent of the electronic noise of the system. Artificial noise mainly refers to the noise generated by ship operations, and environmental noise includes wave noise, white noise, eddy current noise, etc. Underwater noise model can be expressed as

$$n_s(f) = 40 + 20(G - 0.5) + 25log(f) - 55log(f + 0.03) \tag{2}$$

$$n_w(f) = 50 + 7\sqrt{w} + 20log(f) - 40log(f + 0.5) \tag{3}$$

$$n_{th}(f) = -16 + 20log(f) \tag{4}$$

$$n_t(f) = 15 - 30log(f) \tag{5}$$

where, n_s, n_w, n_{th}, n_t represents ship noise, wave noise, white noise and eddy current noise respectively, and f represents frequency in KHz. G is the factor of ship movement, and the value range is 0–1. w is the wind speed in m/s. The total noise composed of the above types of noise can be express ed as

$$n(f) = n_s(f) + n_w(f) + n_{th}(f) + n_t(f) \tag{6}$$

Unlike the noise level of the terrestrial wireless channel key generation scheme, the noise level of the underwater environment is higher, and there are many unsmooth points in the measured value after channel state sensing, which affects the consistency of the generated keys. In this paper, we use smooth filtering pre-processing to isolate the high frequency signals that exceed the set threshold value and reduce the influence of underwater noise.

Sound Propagation Loss. Sound waves in the marine environment will produce reflection, refraction, scattering, as well as the energy generated by the sound waves into other forms of energy, and absorbed by other materials and losses occur. Transmission 5 losses mainly include the loss caused by the outward expansion process and the loss caused by absorption, will change with the distance, temperature, frequency and so on. Expansion loss refers to the sound signal outward propagation gradually weakening characteristics, the size of the expansion loss and the physical structure of the medium. Absorption loss is caused by the conversion of sound energy into other energy, with the increase in distance and frequency and increase. Underwater sound propagation loss is a function of distance (L) and signal frequency (f), which can be expressed as

$$A(L,f) = L^a[\alpha(f)]^L \tag{7}$$

where, a represents the expansion factor, $\alpha(f)$ represents the absorption loss coefficient of seawater, and the propagation loss (TL) is represented in the form of decibel (dB) as

$$TL = 10LogA(L,f) = a * 10log(1000 * L) + L10log\alpha(f) \tag{8}$$

where, a is the expansion factor, the first part represents the outward expansion loss, and the second part represents the absorption loss. It can be seen that the propagation loss of sound waves in the underwater acoustic channel is related to distance, which leads to the bandwidth limitation in the underwater acoustic channel. Therefore, it is necessary to keep the frequency of the long-distance communication system below 10 kHz.

2.3 System Model

Due to the relative movement of communication nodes and the instability of the marine environment, there are many factors that affect the transmission of keys in the hydroacoustic channel, so this paper describes a relevant scenario in underwater networks based on the characteristics of hydroacoustic signals. The system configuration is shown in Fig. 1, a network model consisting of legitimate communication node A, node B and eavesdropping node E. The goal is for the legitimate node to establish a key, and node E is a passive listener to the legitimate communication node and wants to get the same key. Node B is the data collection center, Node A is an automatic underwater vehicle that collects data samples and sends them to Node B. Node E is an underwater robot that transmits information between the legitimate nodes by eavesdropping.

Node A and node B perform channel state sensing of the hydroacoustic channel by sending probe signals to each other. It is assumed that the eavesdropping node E can listen to all transmissions between two legitimate parties and can perform channel estimation on the probe signals sent by node A and node B. It also knows the key extraction scheme between node A and node B and the related parameters.

3 Scheme Design

In order to improve the security of the key and make the legitimate node A and node B obtain the exact same key by exposing the interactive negotiation information, we propose a key negotiation scheme with high security. It mainly consists of two parts: one is identity authentication and the other is negotiation error correction.

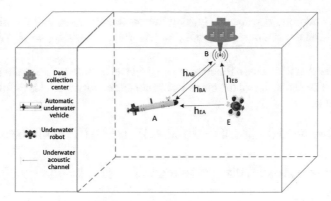

Fig. 1. System model diagram

3.1 Identity Authentication

The purpose of the authentication mechanism is to confirm the legitimate communication node by verifying the identity of the other node before negotiating error correction and avoiding key leakage to other eavesdropping nodes. The legitimate communication node knows the id number of the other node, but the eavesdropping node does not. We use the hash function to construct a unary higher order polynomial for identity verification, and the legitimate nodes after successful identity verification use the conversation key for negotiating error correction. As shown in Fig. 2, the specific steps are as follows.

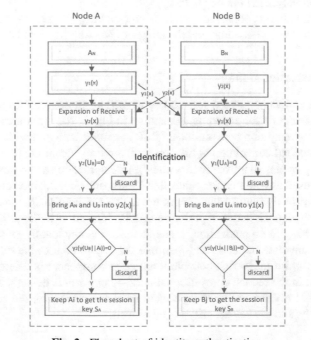

Fig. 2. Flowchart of identity authentication

(1) Node A and node B respectively send the bit sequence A_N and B_N obtained after quantization to each other, where $A_N = [A_1, A_2, \ldots, A_n]$, $B_N = [B_1, B_2, \ldots, B_n]$,

We verified that it uses hash function to construct univariate higher-order polynomials y1(x), y2(x). The expansion of the monadic higher-order polynomial is as follows, where x is the variable,

$$y_1(x) = [x - y(U_A||A_1)][x - y(U_A||A_2)]\ldots[x - y(U_A||A_n)][x - y(U_A)] \qquad (9)$$

$$y_2(x) = [x - y(U_B||B_1)][x - y(U_B||B_2)]\ldots[x - y(U_B||B_n)][x - y(U_B)] \qquad (10)$$

(2) Node A sends the expansion of $y_1(x)$ to node B. After receiving it, node B first verifies the identity of node A and calculates whether $y_1(U_A) = 0$ holds, if not, it proves that the sequence is not sent by node A and discards it directly. If it holds, then bring its own bit sequence B_N and U_A into the $y_1(x)$ expansion, if $y_1(y(U_A||B_j))$ $= 0$, then keep Bj, otherwise discard, and the final retained Bj in order to get S_B;

(3) Node B sends the expansion of $y_2(x)$ to node A. After receiving it, node A calculates whether $y_2(U_B) = 0$ holds. If it does not hold, it is discarded directly. If it is valid, it brings its own bit sequence A_N and B_N into the $y_2(x)$ expansion, and if $y_2(y(U_B||A_i))$ $= 0$, it keeps A_i, and then arranges the kept A_i in order to get S_A. $S_A = S_B$ is the negotiated key sequence.

(4) If the length of the key sequence S_A is less than n, repeat the above steps and insert the newly generated identical A_i and Bj into the negotiated key sequence S_A and S_B, respectively, until the length of $S_A \geq n$.

(5) Node A and node B obtain session keys M_A and M_B from key sequences S_A and S_B, respectively. We assume that the length of M_A and M_B is N bits, $N \geq n$.

3.2 Error Correction Code

The negotiated and authenticated session key is used as the source for secure encoding and convolutional code for encoding. As shown in Fig. 3, the input message is passed through a recursive system convolutional (RSC) encoder with and without interleaved blocks. The obtained convolutional bits are punctured to obtain the desired code rate. The controller provides adaptive security coding for interleaving and puncturing, and the session key is used as a source for controller 1 and controller 2, since only the session key is used to directly control the puncturer so that the legitimate node with the key can correctly distinguish between parity and parity bits. After the legitimate node gets the initial key bits after quantization, it has to make the key length equal to the frame length of the convolutional code and then use the key to control the puncturer. If the bit is 1, the parity bit in the first convolutional code component is deleted; if the bit is 0, the parity bit in the second convolutional code component is deleted. The parity bits are deleted randomly, and only the legitimate receiving node with the correct key can decode correctly, and the key security is guaranteed.

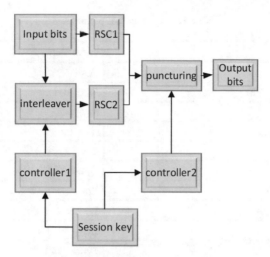

Fig. 3. Error correction coding process

Interleaving Techniques. Error correction coding techniques are usually combined with interleaving techniques. Many channels have sudden errors in the process of transmitting data, so that there is a strong coherence, and the range of errors are concentrated together, far beyond the error correction capability of error correction coding. Add data interleaver at the transmitter side, add the interleaver at the receiver side, to facilitate the sudden error branching of the channel, so that the channel becomes a separate arbitrary error channel, the full performance of the function of error correction coding. The role of interleaver is to disrupt the sequence of transmitted data, so that the coherence of the preceding and following sequences is lower. The interleaving and deinterleaving techniques are added to the channel, and the error correction performance of the system can be greatly improved.

Puncturing Technology. The puncturing technique is an algorithm that uses error correction codes to eliminate parity bits in the code word to improve the code rate and reduce redundant bits in the encoded data. The signal sequence passes through the interleaver to form a new sequence, which is transmitted with the new sequence to two component encoders, generating the sequences X^{p1} and X^{p2}. In order to improve the code rate, the sequence is passed through the puncturer, and some check bits are removed from the check sequence using the puncturing technique to form the check bit sequence X^p. The check bit sequence and the unencoded sequence are iteratively de-bugged to generate the convolutional code sequence.

4   Experiments and Results Analysis

We use BER and key randomness as evaluation metrics to evaluate the performance of convolutional coding and interleaving error correction coding. The experimental parameters are shown in Table 1.

Table 1. Experimental parameters

Items	Parameters
Coherence time	$T_c \approx 0.02$ s
Maximum Doppler shift	$F_d = 10$ Hz
Multipath number	$N_{ray} = 25$
Number of subcarriers	$N = 256$
Bandwidth	$B = 18$ kHz
Carrier frequency	$F_c = 10$ kHz
The length of the CP	128
OFDM symbol length	$T_{sym} = 0.018$ s
Propagation delay	$T_{delay} \approx 0.3$ s
Movement speed	$v = 5$ n mile/h
The communication distance	$d_{AB} = 500$ m

In this paper, we compare the performance of convolutional coding with that of uncoded and Hamming coding in order to verify the superiority of convolutional coding performance. As shown in Figs. 4 and 5, the BER of each error correction code generally decreases as the signal-to-noise ratio increases, but it is obvious that convolutional coding has better performance than uncoded and Hamming coding because in the coding process of convolutional codes, the input information is grouped first, and the coded signal of each group is not only related to the information bits of the current group, but also related to the signals of other groups. Similarly, in the decoding process of convolutional coding, not only the decoding information is obtained from the group received at the current moment, but also the coherent information is extracted from the related groups before and after. The correlation between the preceding and following groups is fully utilized in the coding process, which makes the convolutional code have a good performance advantage.

To verify the performance of interleaving, we compare convolutional coding, Hamming coding with interleaving and without interleaving. As shown in Fig. 6, the error correction coding of convolutional coding with interleaving has lower BER than that of convolutional coding without interleaving and Hamming coding with interleaving at the same signal-to-noise ratio, and it can be seen that the coding performance with interleaving is better than that without interleaving. In addition, we performed NIST tests on the interleaved keys. As shown in Table 2, in the current experimental setting, without interleaving, only a portion of the former NIST randomness test can be passed, while the keys generated by interleaving can pass almost all of the tests. The interleaving technique can effectively eliminate the redundancy of measurements and improve the randomness of keys.

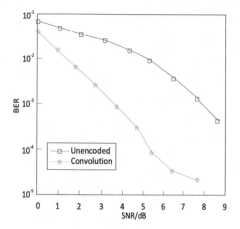

Fig. 4. Performance comparison between convolutional coding and uncoding

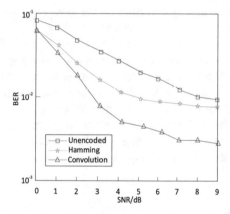

Fig. 5. Performance comparison of uncoding, convolutional coding and hamming coding

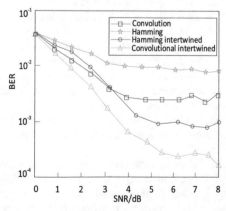

Fig. 6. Performance comparison between convolutional coding and hamming coding with interleaving

Table 2. NIST test

Items	Interwoven	Uninterwoven
Excursions	✓	✓
Block Excursions	✓	✓
Frequency	✓	✓
Block Frequency	✓	✓
Approximate Entropy	✓	×
Cusum	✓✓	× ×
Sequence	× ✓	× ×

5   Conclusion

In this paper, a strong security key agreement scheme is proposed for the key generation problem during the transmission of information in the hydroacoustic channel due to the complexity of the marine environment and the interference of the underwater eavesdropping nodes. The scheme constructs a one-dimensional high-order polynomial for authentication by hash function, and transmits the session key publicly under the open channel for the communication nodes after successful authentication. The initial key is negotiated and error corrected by convolutional coding plus interleaving error correction coding technique. Theoretical analysis and experiments show that the scheme effectively reduces the key error rate, improves the key randomness, and enhances the key security performance. The feasibility of this scheme in hydroacoustic networks is further demonstrated.

References

1. Liu, L.J., Jian-Fen, L.I., Zhou, L., et al.: An underwater acoustic direct sequence spread spectrum communication system using dual spread spectrum code. Front. Inf. Technol. Electr. Eng. **19**, 972–983 (2018)
2. Jing-Mei, L., Zhi-Wei, S., Qing-Qing, H., et al.: Underwater acoustic communication physical layer key generation scheme. J. Commun. **40**(2), 7 (2019)
3. Park, D., Heo, D., Kim, S., Hong, S.: Single trace attack on key reconciliation process for quantum key distribution. In: 2020 International Conference on Information and Communication Technology Convergence (ICTC), pp. 209–213 (2020)
4. Jiang, Q., Chen, Z., Ma, J., Ma, X., Shen, J., Wu, D.: Optimized fuzzy commitment based key agreement protocol for wireless body area network. IEEE Trans. Emerg. Topics Comput. **9**(2), 839–853 (2021)
5. Huang, Y., Zhou, S., Shi, Z., Lai, L.: Channel frequency response-based secret key generation in UWA systems. IEEE Trans. Wireless Commun. **15**(9), 5875–5888 (2016)
6. Luo, Y., Pu, L., Peng, Z., Shi, Z.: RSS-based secret key generation in UWA networks: advantages, challenges, and performance improvements. IEEE Commun. Mag. **54**(2), 32–38 (2016)

7. Choi, J.: A coding approach with key-channel randomization for physical-layer authentication. IEEE Trans. Inf. Forensics Secur. **14**(1), 175–185 (2019)
8. Junkai, L., Gangqiang, Z., Junqing, Z.: Key Generation technology based on multipath structure of UWA channel. In: 2021 OES China Ocean Acoustics (COA), pp. 636–641 (2021)
9. Zhang, Z., Li, G., Hu, A.: An adaptive information reconciliation protocol for physical-layer based secret key generation. In: 2019 IEEE 89th Vehicular Technology Conference (VTC2019-Spring), pp. 1–5 (2019)
10. Zhang, Q.K., Gan, Y., Wang, R.F., et al.: Inter-cluster asymmetric group key negotiation protocol. Comput. Res. Dev. **55**(12), 2651–2663 (2018)
11. Ali, S.T., Sivaraman, V., Ostry, D.: Zero reconciliation secret key generation for body-wornhealth monitoring devices In: ACM Conference on Security and Privacy in Wireless and Mobile Networks. ACM, pp. 39–50 (2012)
12. Nguyen, T.H.T., Barbot, J-P.: SATIE, ENS Cachan, CNRS, UniverSud, 61 Av. du Président Wilson 94235 CachanCedex, "Dynamic secret key generation for joint error control and security coding ". In: France IEEE WCNC 2014 Track 3 (Mobile and Wireless Networks)
13. Vogt, H., Awan, H.: Secret- key generation: Full-duplex versus half-duplex probing. IEEE Trans. Commun. **67**(1), 639–652 (2019)
14. Khiabani, Y.S., Wei, S.: A joint Shannon cipher andprivacy amplification approach to attaining exponentially decaying information leakage. Inf. Sci. **357**(3), 6–22 (2016)

BWA: Research on Adversarial Disturbance Space Based on Blind Watermarking and Color Space

Ziwei Xu[1(✉)], Chunyang Ye[1,2], and Shuaipeng Dong[2]

[1] School of Cyberspace Sevurity, Hainan University, Haikou 570228, China
zwx2642@163.com
[2] School of Computer Science and Technology, Hainan University, Haikou 570228, China

Abstract. Effective generation of adversarial examples can help to improve the training of neural models to avoid adversarial example attacks. Watermark-based adversarial example generation methods regard watermark as a meaningful noise to perturb the neural models. Therefore, the resulting adversarial examples are more similar to the original images yet more difficult to defend. Existing Watermark-based adversarial example generation methods adopt the visible watermarking technology. This however may reduce the success rate of the attacks because the adversarial examples with visible watermarks can be easily perceptible by humans. To address this issue, we propose a novel approach to generate adversarial examples based on the combination of frequency domain and color space perturbation. In particular, we use wavelet transform to hide the watermark, making it invisible and introducing noises to the frequency of the images. We then select the Lab color space Similarity as an optimization scheme for perturbations control. Experimental results show that under the same dataset, the maximum attack success rate of the adversarial example generated by our algorithm can reach 98.56%. In addition, the generated adversarial examples are highly portable, the successful attacks on VGG, Resnet101, and Inception-v3 can reach more than 95%, and the color space perturbation optimization achieves an average RGB channel similarity of 97.22%.

Keywords: Adversarial examples · neural networks · blind watermarks

1 Background

Deep Neural Networks show excellent performance in different fields [1], such as image classification [2, 3], text analysis [4], speech recognition [5], to name a few. However, deep learning models are often vulnerable to well-designed adversarial attacks, resulting in immeasurable security problems. For example, the existence of adversarial samples threatens [6] driving face recognition and road signs. In terms of voice, it causes instruction recognition errors, speaker recognition errors, information leakage, and even unintelligible problems. To improve the training of the neural models to avoid adversarial sample attacks, effective generation of adversarial examples is needed.

© The Author(s), under exclusive license to Springer Nature Singapore Pte Ltd. 2023
J. C. Hung et al. (Eds.): IC 2023, LNEE 1045, pp. 678–688, 2023.
https://doi.org/10.1007/978-981-99-2287-1_95

Currently, numbers methods have been came up to create adversarial examples [7] demonstrated that it is possible to add subtle perturbations to images that are imperceptible to humans, thus misleading deep neural network image classifiers to make wrong classifications. Goodfellow observed that in high-dimensional spaces, the linear behavior of deep neural networks can be utilized by adversarial examples, and proposed a Fast Gradient Sign Attack method to effectively calculate adversarial perturbations and generate adversarial examples [8]. The Deepfool algorithm [9] generates the smallest normative adversarial perturbation by iterative pushing the images to the classification boundary. By limiting L∞, L2 and the L0 norm makes the perturbation imperceptible, the optimization-based method C&W [10] search for adversarial examples with smaller perturbations amplitude. Typically, researchers use the L2 norm limit to evaluate distortion [11] (as a measure of perceptual similarity), because the attack strategy tricks the classifier by adding noise. L2 similarity has a distinct feature: it is highly sensitive to sample illumination and viewpoint changes [12], so this metric is not optimal. Different from other attacks, image watermarking is added to the original image as a meaningful noise without affecting people's recognition of the image. However, the disadvantage is that the adding of watermarks into the original image makes the difference between images large, and exposing the visible watermark information are more likely to raise suspicions about images with a high security factor.

To solve this problem, we put forward an adversarial examples generation method based on blind watermarking. To hide the watermark in the image, we use blind watermark to add peturbations in the frequency domain. In particular, in the frequency domain, add a blind watermark to the image with a random number in the transform domain. Then, the image is converted back to the original domain so that the difference between them cannot be identified by human eyes. We also add tiny color perturbations in the Lab color space to further optimize the image with a better attack effect and similarity with the original image. In this way, we can generate stable and strong general perturbations without a large amount of data.

The paper has the following outstanding contributions: First, we propose a method of making adversarial examples based on blind watermarking. Compared with the existing watermarking method, our method has stronger aggressiveness and better concealment. Second, by adding tiny color perturbations in the Lab color space, stable and aggressive perturbations can be generated without a large amount of data. This increases the adaptability and robustness of our approach. Third, we conduct extensive experiments to evaluate our proposal. Through experiments, the adversarial examples generated by our algorithm under the same dataset have a maximum attack success rate of 98.56%. In the test of attacking Vgg, it is 5.6% higher than the existing methods on average, and in the test of Resnet101 and Inception-v3. In addition, the generated adversarial examples are highly portable, and the successful attacks on VGG, Resnet101, and Inception-v3 can reach more than 95%, and the color space perturbation optimization achieves an average RGB channel similarity of 97.22%. We also analyze various factors of adding watermark, and discuss the influence of attack rate and image similarity.

The overall organizational structure of the rest of this article is as follows. Section 2 presents the background and related work on adversarial examples. Section 3 introduces the adversarial sample based on blind watermark and its improved idea of adding color perturbation, and the fourth part details the proposed adversarial method, including the comparison of some indicators. The last section summarizes our work and looks forward to future research directions.

Subsequent paragraphs, however, are indented.

2 Related Work

2.1 Adversarial Attack

Gradient-based attack methods include Carlini and Wagner attack (C&W) [13], Deepfool [9], JSMA [14]. Most of these attacks target image classification. However, with the deepening of research, adversarial examples not only attack image classification, but also become more and more popular in other computer vision tasks. [16, 17] Sharif proposed to estimate the prediction score of the model gradient using finite differences. These iterative attacks estimate the gradient by sampling from the noise distribution around the feature points. While this approach is successful, it requires a lot of model queries. Adversarial Transformation Network (ATN) [18] propose an autoencoder-based network to create adversarial examples. [19] Gragnaniello adopted a GAN network from which to create adversarial examples. However, since the adversarial samples have no direct correspondence with the original images, the perturbations may be very obvious and fail to deceive the human eye. The improved method proposed by [20] adopts the AdvGAN generator based on the auto-encoder to obtain the maximum range perturbation from the perspective of the original image. There are novel research algorithms such as one-pixel-attack [21], which modifies a single pixel point by random check and optimization to attack; there are also patch-based adversarial examples. [22–24] all paste patches on the original image, and only update the parameters of the patch by improving the loss backpropagation. The innovation lies in how the loss function is designed. However, online iterative attack strategies limit their application scenarios, in order to generate adversarial sequences, the only downside is that they always require access to the model's weights during the attack.

2.2 Visible Watermarking Methods

Digital watermarking is a kind of protection information embedded in the carrier file by applying computer algorithm. Digital watermarks can guarantee the security of information and protect the copyright of works. [25] proposed a blind technique based on fast Walsh-Hadamard transform, SVD, key mapping and coefficient sorting, but its effect against geometric attacks such as rotation and shearing is relatively weak. Difference. The above algorithms have caused varying degrees of changes to the image data. [26] proposed a robust watermarking scheme that exploits the multi-resolution and multi-scale properties of nonsub exampled wavelet transforms to analyze the orientation features of a given image. [7] Jia proposed a new optimization algorithm, when adversarial watermarks are generated using evolutionary algorithms (BHE) in a black-box attack environment, the location of the watermark and the watermark are highly correlated with the transparency attribute. In terms of the similar distance from the original image, the addition of visible watermarks will cause suspicion. In order to avoid this situation, we propose blind watermark perturbations. Considering the security issues of the Internet cannot be ignored, add an invisible watermark to protect their Copyright, which enables copyright protection of images while conducting adversarial attacks with better robustness.

3 Methodology

3.1 Research Ideas

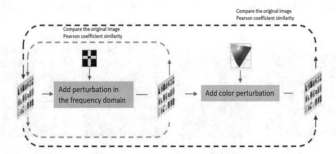

Fig. 1. The process of adversarial example making.

We add the watermark to the three color gamuts of RGB through wavelet transform and singular value decomposition, respectively, with the set random number seeds. In the transform domain algorithm, the multi-resolution characteristics of the transform domain and the inherent characteristics of the SVD singular value are fully utilized to enhance the invisibility and robustness of the watermark. Accordingly, we propose algorithm attack model (Fig. 1). We disguise the adversarial perturbation as a frequency-domain blind watermark. Our work is to generate adversarial images that cannot be classified correctly, the optimization or constraint of which is expressed by the following formula:

$$\begin{aligned} & minimize\ D(x, x + \theta) \\ & such\ that\ C(x + \delta) = t \\ & x + \delta \in [0, 1]^n \end{aligned} \tag{1}$$

where: $x \in Rm$ is a clean input, δ is the perturbation added, D is the distance metric, which measures the distance between the original image and the antagonistic sample, C is the classifier, t is the label of the misclassification of the adversarial example, $[0, 1]^n$ limits the perturbation between (0,1).

3.2 Algorithm Details

Using wavelet transform can improve the visual concealment and robustness of the watermark. Through the transformation, the features can be fully highlighted, localized analysis of temporal and spatial frequencies can be performed, and through a series of operations such as zooming, panning, etc., the purpose of refining the scale on the signal is achieved, and finally time subdivision is automatically realized at high frequencies, and frequency subdivision is automatically realized at low frequencies. Therefore, it will not miss every detail of the signal, and adding disturbances in the frequency domain can be more invisible. First complete the wavelet transform processing and singular value decomposition operations, we randomly add the watermark to the RGB color gamuts with the same random number seed, and add images to the wavelet frequency domain, and then improve the process of adding color disturbance (see Fig. 2). First of all, image scrambling is a mapping of two-dimensional space, and the function of scrambling is to change the arrangement and combination of the original images and the spatial correlation. For an image W of size N × N, use formula (1) to perform Arnold transform:

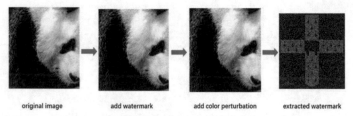

original image   add watermark   add color perturbation   extracted watermark

Fig. 2. From left to right, the images are the original image, the watermarked image, the image with added color disturbance, and the extracted watermark.

image W of size $N \times N$, use formula (1) to perform Arnold transform:

$$\begin{pmatrix} x' \\ y' \end{pmatrix} = \begin{pmatrix} 1 & 1 \\ k & k+1 \end{pmatrix} \begin{pmatrix} x \\ y \end{pmatrix} (modN) \tag{2}$$

In formula:

(x, y) is the pixel point of the original image, (x′, y′) is the pixel point of the new image after transformation, N is the image order, that is, the size of the image, generally considering a square image, k is an integer belonging to [1, N].

Denote the transformation matrix as W, thus, do the iterative procedure:

$$I_{xy}^{n+1} = WI_{xy}^n (modN), I_{xy}^n = (x, y), n = 0, 1, 2, ... \tag{3}$$

Singular value decomposition (SVD) in numerical analysis is a numerical algorithm that diagonalizes a matrix. From a linear algebra perspective, a grayscale image can be viewed as a non-negative matrix. The image W′ ∈ Rm × n, where R represents the real number domain:

$$W = U \Sigma V^T \tag{4}$$

Then perform first-level wavelet decomposition on the carrier image to obtain 3 high frequency subbands HH, LH, HL and 1 low frequency subband LL; divide the low frequency subband image into m blocks. Repeat the singular value decomposition and embed the watermark according to a certain intensity factor until all the watermark information is embedded, and then perform the inverse wavelet transform to obtain the watermarked image.

3.3 Loss Function

The parameter settings of adding watermark have different effects. When the depth is greater, the attack ability is stronger, but the picture changes will be blurred. The perturbation in the color space will not affect the perturbation in the frequency domain, and the watermark can still be extracted. Lab is designed based on people's perception of color, more specifically, it is perceptually uniform, if the number (the three channels of L, a, b) changes in the same magnitude, then it gives people a visual effect.

$$\Delta E_{00} = \sqrt{\left(\frac{\Delta L'}{k_L S_L}\right)^2 + \left(\frac{\Delta C'}{k_C S_C}\right)^2 + \left(\frac{\Delta H'}{k_H S_H}\right)^2 + \Delta R},$$

$$\Delta R = R_T \left(\frac{\Delta C'}{k_C S_C}\right)\left(\frac{\Delta H'}{k_H S_H}\right) \tag{5}$$

Fig. 3. The first column: original image. The second column: thermal map shown in the original image. The third column: thermal map of the image subjected to BWA attack. It shows that the feature regions that neural networks pay more attention to are not global.

4 Evaluation

4.1 Comparing the Results of Attacking Different Networks with Other Methods

Compared with the watermarking algorithm, our BWM algorithm has better conceal-ment. Figure 3 shows the comparison of the results of our method attacking different networks. Our method is compared with adv-watermark, one-pixel, FGSM, advGan methods respectively. The dataset adopts ImageNet datasets. We can see that the adv-gan method has the best effect and achieves the highest attack rate. In Resnet101 and Inception-v3, BWA'rate reaches the highest.

Table 1. Attack rate of different networks.

Adv-methods	VGG	Resnet101	Inception-v3
Adv-watermark	0.934	0.873	0.921
One-pixel	0.935	0.920	0.876
FGSM	0.827	0.881	0.955
AdvGan	0.973	0.923	0.970
BWA	0.954	0.952	0.976

The random number used for S_1 watermark encryption, S_2 is the random number for adding the watermark to the picture, mod_1, mod_2 are used for the divisor of the embedding algorithm. In theory, and the larger the divisor, the stronger the robustness, but the greater the distortion of the output image (see Table 1). Figure 4 shows the impact of different parameters on the attack power, the larger the mod_1, the greater the distortion of the picture, and the color space will have a certain blur and change.

Fig. 4. The parameters of the last three images are shown in Error! Reference source not found. With the increase of mod1, the color does not change much, and the texture is gradually blurred. It can be seen that the embedding depth of watermark first affects the texture features.

Table 2. The effect of different parameters.

Parameters	Example	Attack rate	Similarity (R G B)		
s1, s2	mod1,mod2		R	G	B
5539,3336	56,25	0.950	0.9665	0.9899	0.9800
5539,3336	79,25	0.954	0.9487	0.9844	0.9702
5539,3336	108,25	0.954	0.9265	0.9755	0.9526
5539,3336	175,25	0.967	0.9379	0.8945	0.9673

The same analysis can be analyzed as shown in Table 2. When S_1, S_2, and mod_2 are set to 8399, 5536, and 25, respectively, under the constant random number setting, as mod_1 becomes larger, the attack rate can also increase to a certain extent (see Fig. 5).

4.2   The Effect of Watermark Times

The number of watermark additions also affects the attack capability. The more times the watermark is added, the greater the disturbance added in the frequency domain space, and the greater the impact on the original image (see Table 4)

Table 3. The effect of different parameters.

Parameters	Example	Attack rate	Similarity (R G B)		
s1, s2	mod1,mod2		R	G	B
8539,5536	176,25	0.963	0.8607	0.9432	0.9036
8539,5536	258,25	0.958	0.7846	0.8949	0.8286
8539,5536	296,25	0.977	0.7108	0.8892	0.7814
8539,5536	336,25	0.983	0.6670	0.8497	0.7312

Fig. 5. The parameters of the next four pictures are shown in Error! Reference source not found. Slight changes in color and texture can be clearly seen, and high frequency features are destroyed, and the attack rate increases at the expense of the similarity of the images.

Table 4. The effect of watermark times ($s_1 = 8399$, $s_2 = 5536$, $mod_1 = 258$, $mod_2 = 25$).

Times	Similarity			Attack rate
	R	G	B	
1	0.8607	0.9432	0.9036	0.958
2	0.7846	0.8949	0.8226	0.962
3	0.7108	0.8892	0.7814	0.963
4	0.6670	0.8497	0.7321	0.963

Fig. 6. For the original image, you can refer to the picture on the left in the first row, the pictures are watermarked 2, 3, and 4 times in sequence.

4.3 The Effect of Color Space Perturbation on Watermark Extraction

Before and after the watermark is embedded, the human eye cannot directly perceive the existence of the watermark, which has good concealment of the watermark. We added color perturbation to make the adversarial example image more closer to the initial image (see Fig. 6). Compared with the original watermarking attack, adding color space perturbation has a higher similarity, the results of color perturbation are shown in Table 5. The extracted watermark is a grayscale image (see Figs. 7 and 8).

Fig. 7. The first column is the original picture, the middle column is the picture that has been watermarked once, the third column is the picture after adding color perturbation, and the fourth column is the confrontation sample of the Adwatermark algorithm.

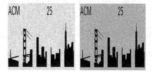

Fig. 8. The left picture: extracted watermark.

The right picture: watermark extracted after adding color disturbance.

Table 5. The effect of color space perturbation on watermark extraction (s1 = 8399, s2 = 5536 mod1 = 176, mod2 = 25).

Method	Similarity R	G	B	Attack rate
No color perturbation	0.8607	0.9432	0.9036	0.964
add color perturbation	0.9489	0.9824	0.9853	0.976

5   Conclusion

Our paper proposes a black-box attack, which adds perturbation in the form of blind water- mark in the wavelet frequency domain, and converts it to RGB space with a certain attack ability. We introduced the Lab color space for optimization, and added color disturbance in the lab color gamut. The results demonstrate that our method can successfully attack several networks. We hope that more researchers will pay attention to adversarial attacks and defenses in the field of neural networks in the future.

Acknowledgment. This work was supported in part by the Key Research and Development Program of Hainan Province under grant No. ZDYF2020008, ZDYF2020008, the Natural Science Foundation of Hainan Province under the grant No. 2019RCO88, 2019CXTD400, and grants from State Key Laboratory of Marine Resource Utilization in South China Sea and Key Laboratory of Big Data and Smart Services of Hainan Province.

References

1. He, K., Zhang, X., Ren, S., Sun, J.: Deep residual learning for image recognition. In: Proceedings of the IEEE Conference on Computer Vision and Pattern Recognition, pp. 770–778 (2016)
2. Krizhevsky, A., Sutskever, I., Hinton, G.E.: ImageNet classification with deep convolutional neural networks. Adv. Neural Inform. Process. Syst. **25**(2012)
3. Kurakin, A., Goodfellow, I., Bengio, S., et al.: Adversarial examples in the physical world. In: ICLR Workshop (2016)
4. Collobert, R., Weston, J.: A unified architecture for natural language processing: Deep neural networks with multitask learning. In: Proceedings of the 25th International Conference on Machine Learning, pp. 160–167 (2008)
5. Hinton, G., et al.: Deep neural networks for acoustic modeling in speech recognition: the shared views of four research groups. IEEE Signal Process. Mag. 29(6), 8297 (2012)
6. He, W., Wei, J., Chen, X., Carlini, N., Song, D.: Adversarial example defenses: ensembles of weak defenses are not strong (2017). https://arxiv.org/abs/1706.04701
7. Jia, X., Wei, X., Cao, X., Han, X.: Adv-watermark: a novel watermark perturbation for adversarial examples (2020). https://arxiv.org/abs/2008.01919
8. Goodfellow, I.J., Shlens, J., Szegedy, C.: Explaining and harnessing adversarial examples. In: 2017 IEEE Symposium on Security and Privacy (sp), pp. 3957. IEEE (2017).https://arxiv.org/abs/1412.6572
9. Moosavi-Dezfooli, S-M., Fawzi, A., Frossard, P.: Deepfool: a simple and accurate method to fool deep neural networks. In: 2016 IEEE Conference on Computer Vision and Pattern Recognition (CVPR), pp. 2574–2582 (2016). https://doi.org/10.1109/CVPR.2016.282
10. Papernot, N., McDaniel, P., Jha, S., Fredrikson, M., Berkay Celik, Z., Swami, A.: The limitations of deep learning in adversarial settings (2015). https://arxiv.org/abs/1511.07528
11. Gu, S., Rigazio, L.: Towards deep neural network architectures robust to adversarial examples. In: ICLR Computerence (2015)
12. Johnson, J., Alahi, A., Li, F-F.: Perceptual losses for real-time style transfer and super-resolution (2016). https://arxiv.org/abs/1603.08155
13. Carlini, N., Wagner, D.: Towards evaluating the robustness of neural networks. In: 2017 IEEE Symposium on Security and Privacy (SP)
14. Croce, F., Hein, M.: Sparse and imperceivable adversarial attacks. In: Proceedings of the IEEE/CVF International Conference on Computer Vision, pp. 4724–4732 (2019)
15. Madry, A., Makelov, A., Schmidt, L., Tsipras, D., Vladu, A.: Towards deep learning models resistant to adversarial attacks (2017). https://arxiv.org/abs/1706.06083
16. Engstrom, L., Tran, B., Tsipras, D., Schmidt, L., Madry, A.: Exploring the landscape of spatial robustness (2017). https://arxiv.org/abs/1712.02779
17. Sharif, M., Bauer, L., Reiter, M.K.: On the suitability of lp-norms for creating and preventing adversarial examples (2018). https://arxiv.org/abs/1802.09653
18. Eykholt, K., et al.: Robust physical-world attacks on deep learning models (2017). https://arxiv.org/abs/1707.08945

19. Gragnaniello, D., Marra, F., Poggi, G., Verdoliva, L.: Perceptual quality-preserving black-box attack against deep learning image classifiers (2019). https://arxiv.org/abs/1902.07776
20. Ren, S., He, K., Girshick, R., Sun, J.: Faster R-CNN: towards real-time object detection with region proposal networks (2015a). https://arxiv.org/abs/1506.01497
21. Su, J., Vargas, D.V., Sakurai, K.: One pixel attack for fooling deep neural networks. IEEE Trans. Evol. Comput. **23**(5), 828841 (2019). https://doi.org/10.1109/tevc.2019.2890858
22. Brown, T.B., Mane, D., Roy, A., Abadi, M., Gilmer, J.: Adversarial patch. arXiv preprint arXiv:1712.09665 (2017a)
23. Lee, M., Kolter, Z.: On physical adversarial patches for object detection. arXiv preprint arXiv: 1906.11897 (2019b)
24. Thys, S., Van Ranst, W., Goedeme, T.: Fooling automated surveillance cameras: adversarial patches to attack person detection (2019b). https://arxiv.org/abs/1904.08653
25. Khanam, T., Dhar, P.K., Kowsar, S., Kim, J-M.: SVD-based image watermarking using the fast walsh-hadamard transform, key mapping, and coefficient ordering for ownership protection. Symmetry **12**(1), 52, (2019). https://doi.org/10.3390/sym12010052
26. Zhao, J., Xu, W., Zhang, S., Fan, S., Zhang, W.: A strong robust zero-watermarking scheme based on shearlets high ability for capturing directional features. Math. Probl. Eng. **2016** (2016). https://doi.org/10.1155/2016/2643263
27. Jiang, F., Gao, T., Li, De.: A robust zero-watermarking algorithm for color image based on tensor mode expansion. Multim Tools Appl. **79**(11), 75997614 (2020). https://doi.org/10.1007/s11042-019-08459-3
28. Liu, X., Yang, H., Liu, Z., Song, L., Li, H., Chen, J.: Dpatch: an adversarial patch attack on object detectors. (2018a). https://arxiv.org/abs/1806.02299
29. Ye, M., Luo, J., Zheng, G., Xiao, C., Wang, T., Ma, F.: Medat- tacker: exploring black-box adversarial attacks on risk prediction models in healthcare (2021). https://arxiv.org/abs/2112.06063
30. Zheng, X., Fan, Y., Wu, B., Zhang, Y., Wang, J., Pan, S.: Robust physical-world attacks on face recognition (2021). https://arxiv.org/abs/2109.09320
31. Tram'er, F., Kurakin, A., Papernot, N., Goodfellow, I., Boneh, D., McDaniel, P.: Ensemble adversarial training: attacks and defenses (2017). https://arxiv.org/abs/1705.07204
32. Sharif, M., Bhagavatula, S., Bauer, L., Reiter, M.K.: A general frame work for adversarial examples with objectives. ACM Trans. Privacy Secur.**22**(3), 130 (2019b)

Blockchain Based Certificate Deposit System for Judicial Departments

Zhaoxing Jing[1,2], Chunjie Cao[1,2], Xiaoli Qin[1,2(✉)], and Hao Wu[1,2]

[1] School of Cyberspace Security, Hainan University, Haikou 570228, China
Xlqin@hainanu.edu.cn
[2] Key Laboratory of Internet Information Retrieval of Hainan Province, Haikou, China

Abstract. In the traditional judicial system, the public security, procuratorate, court and judicial bureau all involve private information, so the data current and sharing of each department will be greatly restricted. To solve these problems, a blockchain-based public security, procuratorial, and judicial evidence storage system is designed, which can, to a certain extent, solve the morass of data sharing among various departments of the judicial system, and can ensure the privacy of data. The proposed scheme uses the chain structure of the permissioned chain FISCO BCOS as the blockchain platform and combines the Merkle tree and hash function to access data. The structural storage of data can ensure that the uploaded data is traceable and without being tampered with. The public security, judicial and judicial evidence storage system designed with blockchain technology can realize database sharing among various departments of public security, procuratorate, court, and judicial bureau, eliminate the risk of leakage and tampering in the process of case evidence flow, and protect citizens' privacy and national information safety.

Keywords: Certificate Deposit System · Public Security · Procuratorial and Judicial Departments · Blockchain · Merkle Tree

1 Introduction

Traditionally, in the entire judicial enforcement project, the databases among the public security, procuratorate, courts, and judicial bureaus are independent of each other. The information systems of each department are in an independent and closed state. However, case acceptance, deadline control, document delivery, and other links require the connection of various departments. Consequently, case information cannot be transmitted in real-time, and complete data cannot be provided for higher-level departments' plans, which would largely affect case handling efficiency. On the other hand, various departments of the Public Security, Procuratorate and Judicial Department use their own credit to provide services such as depository, preservation, and witnessing of electronic data. However, there are many institutions for the circulation of case evidence, and the risk of being leaked and tampered with is high. Citizen privacy, once leaked or tampered with, the consequences will be unimaginable.

© The Author(s), under exclusive license to Springer Nature Singapore Pte Ltd. 2023
J. C. Hung et al. (Eds.): IC 2023, LNEE 1045, pp. 689–697, 2023.
https://doi.org/10.1007/978-981-99-2287-1_96

The emergence of blockchain technology has provided a new solution for the handover between traditional public security, procuratorial, and judicial departments. Through blockchain technology, the public security, procuratorate, court, and judicial bureau are placed in the blockchain ecosystem, and through the multi-level authority management of smart contracts and interfaces, government affairs processing and data sharing within a certain range are realized. Blockchain is a distributed ledger whose nodes involve data encryption, timestamping and consensus mechanisms. Due to the distributed storage of data, if one node is breached, it will not affect the overall data, and it is more difficult for the whole node to be breached. Securely handle sensitive information through smart contract authorization. At the same time, the data on the blockchain is non-tamperable and traceable, which can realize the immediate handling and accountability of data leakage incidents.

This project develops the integrated blockchain certificate deposit system for public security, procuratorial and judicial departments, which further improves the transparency, credibility and public satisfaction of judicial work in the advancement of smart city construction. It will provide a model for cross-departmental business collaboration and data sharing, which is indispensable to building an efficient and intelligent Smart City.

2   Related Work

Since the concept of blockchain was proposed, its implementation in various fields has been carried out.Bonomi et al. [1] proposed an improved monitoring chain (B-CoC) developed by Ethereum to automate the process of monitoring the chain to guarantee evidence integrity and traceability to the owner in the system. Ichikawa D et al. [2] developed and evaluated a trusted, auditable and tamper-proof mobile health system based on blockchain technology using a distributed network to address the problem of data management when mobile health information data is stored in a server. Chao Xie et al. [3] proposed a dual-chain architecture and proposed a data security storage scheme that is based on the puzzle that it is difficult to automatically store information on the chain when conducting blockchain traceability of agricultural products. The agricultural product quality data tracking in the blockchain ensures that the agricultural product data is not tampered with in the system. Yuqin Xu et al. [4] designed an education system for the problem that the current digital infrastructure for managing educational certificates cannot ensure the security of data and the trust of system, and most of the current blockchains rely on tokens, which cannot accurately and efficiently support certificate queries. The certificate manages the blockchain. It only takes a short time to realize the verification of the block, and at the same time, it can provide efficient on-chain transaction queries and historical transaction queries of on-chain accounts. Rui Q et al. [5] proposed a blockchain-based secure storage scheme for dynamic data in view of the possible tampering and forgery in the secure storage of dynamic data. Analyze the consistency between the local behavior of the consensus terminal to maximize its own interests and the overall goal of ensuring the overall security and effectiveness of the system through mathematical models, and design a consensus mechanism suitable for dynamic data security storage, data ownership state transition mechanism and storage

System architecture. Cebe et al. [6] constructed a permissioned blockchain scheme to tackle post-mortem analysis of traffic accidents in the Internet of Vehicles and put various sensory data collected by vehicle sensors on the blockchain, which can use the minimum storage space and handling overhead to enable post-incident analysis with traceable, trustless, and private information. Ryu et al. [7] proposed a digital forensics framework for IoT infrastructure based on blockchain technology, aiming at the problem that the current law enforcement agencies cannot meet the heterogeneity and distribution characteristics of digital forensics tools, investigation frameworks and processes in the IoT-related issues. Thus, the robustness of the existing depository data hosting process is improved. In their paper, Saraju P [8] et al. introduced the first-ever blockchain application in the IoT field that can tackle device limitations and data issues, in order to solve the scalability and latency. The consensus algorithm of this blockchain is better than Traditional PoW is 1000 times faster. Christian [9] studied the limitations of traditional encryption and access control models to solve security and privacy issues in the trend of transferring data and services in the healthcare field to the cloud. Nurzhan [10] studied the information security and privacy issues of transaction data in the smart grid field with interactive capabilities and combined blockchain technologies to overcome the security issues related to distributed smart grid power transactions in multiple ways. Miyachi Ken [11] et al. aimed at the problem of on-chain and off-chain collaboration in blockchain, explored the interaction between on-chain and off-chain storage and computing, and applied it to the medical industry, and proposed a modular hybrid privacy protection model, applied in three different reference frames, to protect medical privacy data.

3 Methods

3.1 Permissioned and Permissionless Chains

At present, the underlying platform of blockchain can be categorized into two types, namely permissioned chain and permissionless chain. This section will compare the two types of chains from multiple perspectives and evaluate the correct chain type selection.

In 2008, a scholar under the pseudonym Satoshi Nakamoto proposed Bitcoin, a decentralized digital currency payment system that does not require the endorsement of any authority. Later, it was discovered that the basic technology blockchain in Bitcoin, can also be employed in handling trust issues in information transmission between devices without trusting each other and without third-part intermediaries. To this end, a number of blockchain platforms, represented by Ethereum, have emerged to realize digital asset transactions. Any node can join/exit at any time without permission, so this type of blockchain is called a permissionless blockchain. The feature of permissionless chain that allows any node to enter and exit at will is obviously not suitable for enterprise-level applications. In a cross-institutional transaction scenario, multiple companies that cooperate with each other form an alliance, and only members of the alliance can join the blockchain and participate in transactions. A blockchain in which such nodes require permission to join is called a permissioned blockchain. Permissioned and permissionless blockchains target different application scenarios and solve different problem areas. The main differences between them are shown in Table 1.

Table 1. Comparison of Permissioned and Permissionless Chains

	Permissionless chain	Permissioned chain
Node admission	Nodes join freely	Nodes need permission to join
User management	Any user can join, the user identity is anonymous	Users need to be verified before they can join, and the user's identity is real-name
Decentralization	Deployed on a global scale to achieve complete decentralization	Deployed in the enterprise alliance to achieve multi-centralization
Consensus mechanism	Proof-based: PoW and POS	Voting-based: PBFT and Raft
Digital currency	Issue digital currency to motivate more nodes to participate in bookkeeping and operations	Built for inter-enterprise business without issuing digital currency and incentives
Transaction storage	Each node stores the entire network transaction data in full	Due to the trade secrets involved, each node consistently stores the hash of the transaction data related to its own business and the transaction data of other parties

As can be seen from the above table, due to the openness and completely decentralized structure of the non-licensed chain in terms of node access, user management, decentralization, consensus mechanism, etc., it is not suitable for the field of public security, procuratorial and judicial departments. Therefore, this system selects the permissioned chain as the available chain type. The mainstream underlying platforms of such chains are Hyperledger Fabric of the Linux Foundation, FISCO BCOS of the "Golden Chain Alliance", Coco of Microsoft, Enterprise Ethereum Alliance (EEA) and Corda of R3.

3.2 Merkle Tree

Merkle trees are a fundamental part of the blockchain. Merkle tree, also known as hash tree, is an important technical algorithm used in blockchain data storage. After the encryption process, the information extracted from the data stored in blocks will be stored in the node of a Merkel tree as a hash value. Hash trees can be used to authenticate data stored, processed, and transmitted in and between computers without considering the format of the data. The merits of the Merkle are that it can achieve a high level of security without losing the data transferring rate between devices and suffering from loss and tampering (Fig. 1).

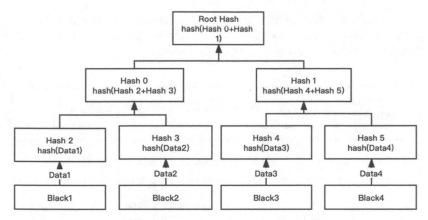

Fig. 1. Basic structure of Merkle tree

4 System

4.1 System Framework

The whole system framework design is mainly divided into four layers, namely data acquisition layer, data processing and chaining layer, data sharing layer, and application empowerment layer.

The data collection layer realizes the collection of the original data of various public security, procuratorial and judicial departments. To ensure the authenticity of the data before uploading to the blockchain, the identity of the data uploader is necessary to be identified and verified ahead, and the corresponding data to the data uploader in the organization through CA certification also need to be issued. The certificate ensures that the identity of the data uploader is authentic and credible. After identity verification, the data uploader can upload the relevant original data of the case.

The data processing and chaining layer mainly completes the following tasks: process the original case evidence, case-related documents and other information to ensure that the data to be uploaded to the chain is encrypted and protected, and after completing the encryption of the original data, receive data upload request, upload the encrypted and authoritative and reliable judicial evidence and other information to the blockchain for storage so that the uploaded data is traceable and cannot be tampered with.

The data sharing layer mainly completes the distributed sharing of judicial data privacy protection on the chain. At the same time, it cooperates with the control of regulatory power to protect the data security of all parties, and enables the relevant evidence and other data to flow in each judicial institution. Institutions and the public with supervisory authority can view the plaintext data of relevant case data and grasp the real situation.

The application empowerment layer is mainly oriented to business needs and builds a judicial evidence traceability system. All types of users can access the system after unified authentication through the user access interface. Different users set different query permissions, and users can search for relevant information through keyword searches. Case evidence is recorded on the chain (Fig. 2).

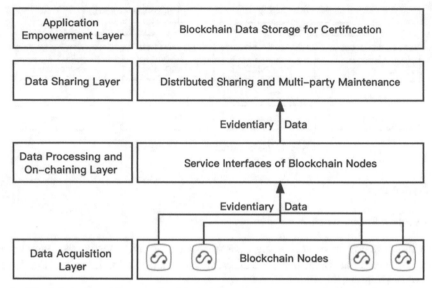

Fig. 2. Blockchain-based framework of the public security, procuratorial and judicial department certificate storage system

4.2 Block Structure

Blocks are the basic unit of a certain blockchain, normally it composes a block header and a block body. Version information is stored in the block header, such as the time when the block was generated (timestamp), the hash value of the subblock, and root node data of Markle tree which can summarize and quickly summarize all the data in the verification block. The block body uses the structure of Merkle tree to store data. Each node stores a hash value. Each leaf node at the bottom corresponds to a hash value of data information. Its parent node is two hash values again. Hash, and recursively get the final Merkle root hash (Fig. 3).

In the process of processing a case, a lot of documents and evidence will be generated. The uploader extracts keywords for the query according to the type and content of the data and generates an index from the keywords to describe the data, data abstract, data generation time and the original data is packaged into a file at the address stored in the local library. First, the abstract of the file is obtained, and the file is encrypted to generate a ciphertext. Utilizing his own private key, the uploader signs the index, file ciphertext, and file abstract and uploads it to the blockchain. After uploading all the evidence information in each link, the uploader also needs to generate a piece of information indicating the progress of the case, sign it with the private key and upload it to the blockchain. The

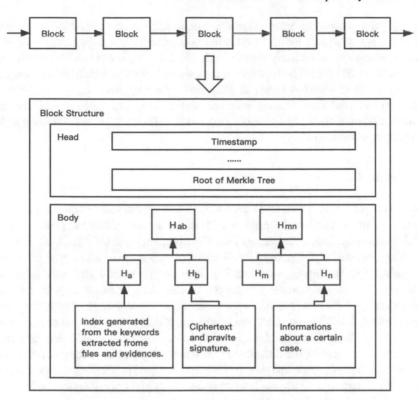

Fig. 3. Block structure

last leaf node of the Merkle tree of each block stores the information. It is the hash value of this piece of information, and the case information is stored in plaintext on the blockchain for the public query.

4.3 Performance

The emergence of electronic evidence has brought great changes to the judicial certification system. The development of electronic evidence has adapted to the tide of informatization, which is of great significance for improving judicial efficiency and reducing judicial costs; building a free trade port is a long-term and arduous task. In this process, legalization, transparency, and regulation are the objective requirements for building a high-level free trade port in the world. This system uses blockchain and other technologies to integrate and modernize the certificate deposit system and strengthen the application of the social credit system. Strengthen data security sharing and disclosure, improve government services and governance, and build a free trade port governance system with complete systems, scientific norms, and effective operation. Use zero-knowledge proof, verifiable secret sharing technology to protect data privacy and computational verifiability; use trusted input loading to ensure data authenticity. At the same time, the use of distributed ledger records ensures that the entire process service records of joint computing between entities can be verified and traceable. The improvement of the government's judicial credibility is significant to continuously promote the formation of high value-added industries in the pilot free trade zone and eventually move towards a free trade port. The blockchain depository has been widely piloted under the leadership of three Internet courts in Hangzhou, Beijing, and Guangzhou. With the assistance of blockchain technologies, it has accumulated and stored a large amount of electronic evidence, which has greatly improved the efficiency of case handling and has had a wide-ranging impact.

5   Conclusion

In view of the privacy and security issues and data sharing difficulty in the current judicial system, we employ blockchain technology to construct a certificate deposit system for the Public Security, Procuratorate and Law Division, and selected FISCO BCOS as the underlying platform of the permissioned chain of this system. The upper chain layer, data sharing layer and application enabling layer are the basic framework structure of the system. With the integrated advantages of the blockchain-based public security, procuratorial, judicial and judicial deposit system, data security sharing in the judicial system is realized. While ensuring the rapid sharing of data among various departments, it also ensures the data privacy and security of each department. And with the performance advantages of blockchain, the transparency, immutability and traceability of data on the chain can be achieved. After analysis, the blockchain-based public security, procuratorial and judicial department certificate storage system is of great significance for improving judicial efficiency and reducing judicial costs.

Acknowledgments. This work was supported in part by the Hainan Province Science and Technology Special Fund (ZDYF2020012, ZDYF2021GXJS216), the National Natural Science Foundation of China (62162020).

References

1. Bonomi, S., Casini, M., Ciccotelli, C.: B-coc: a blockchain-based chain of custody for evidences management in digital forensics. arXiv preprint arXiv:1807.10359 (2018)
2. Ichikawa, D., Kashiyama, M., Ueno, T.: Tamper-resistant mobile health using blockchain technology. Jmir Mhealth Uhealth 5(7), e7938 (2017)
3. Chao, X., Sun, Y., Luo, H.: Secured data storage scheme based on block chain for agricultural products tracking. In: 3rd International Conference on Big Data Computing and Communications (BIGCOM), IEEE, pp. 45–50 (2017)
4. Xu, Y., Shangli, Z., Lanju, K., Yongqing, Z., Shidong Z., Qingzhong, L.: ECBC: a high performance educational certificate blockchain with efficient query. In: International Colloquium on Theoretical Aspects of Computing, pp. 288–304 (2017)
5. Qiao, R., Dong, S., Wei, Q., Wang, Q.: Blockchain based secure storage scheme of dynamic data. Comput. Sci. **45**, 57–62 (2018)
6. Cebe, M., Erdin, E., Akkaya, K., et al.: Block4Forensic: an integrated lightweight blockchain framework for forensics applications of connected vehicles. IEEE Commun. Mag. **56**(10), 50–57 (2018)

7. Ryu, J.H., Sharma, P.K., Jo, J.H., Park, J.H.: A blockchain-based decentralized efficient investigation framework for IoT digital forensics. J. Supercomput. **75**(8), 4372–4387 (2019). https://doi.org/10.1007/s11227-019-02779-9

8. Mohanty, S.P., Venkata, P., Yanambaka, E.K., Deepak, P.: PUFchain: a hardware-assisted blockchain for sustainable simultaneous device and data security in the internet of everything (IoE). IEEE Consum. Electron. Mag. **9**(2), 8–16 (2020)

9. Aitzhan, N.Z., Svetinovic, D.: Security and privacy in decentralized energy trading through multi-signatures, blockchain and anonymous messaging streams. IEEE Trans. Depend. Secure Comput. **15**(5), 840–852 (2016)

10. Aitzhan, N.Z., Svetinovic, D.: Security and privacy in decentralized energy trading through multi-signatures, blockchain and anonymous messaging streams. IEEE Trans. Depend. Secure Comput. **15**(5), (2018)

11. Miyachi, K., Mackey, T.K.: hOCBS: a privacy-preserving blockchain framework for health-care data leveraging an on-chain and off-chain system design. Inf. Process. Manag. **58**(3), (2021)

Application Analysis of Blockchain Technology for 6G Network

Dongfang Jia[1,2] and Longjuan Wang[1,2](✉)

[1] School of Cyberspace Security, Hainan University, Haikou, China
wanglongjaun@hainanu.edu.cn
[2] Key Laboratory of Internet Information Retrieval of Hainan Province, Haikou, China

Abstract. In the 6G environment, society and information and communication technology are more and more closely combined. With the continuous improvement of communication rate, many data security problems will arise. Risk prevention will also be a focus of 6G technology that cannot be ignored.

Blockchain technology takes cryptography as the carrier, and the underlying supporting technology lays the foundation for 6G network. In the 6G era, blockchain will have more room for development. With the rapid development of information technology and the wide application of Internet technology, the accelerated development of global economic integration, and the increasingly frequent communication between countries, human society has entered a new historical period - the information age. In this process, the phenomenon of information leakage is becoming more and more serious.

Blockchain has the characteristics of decentralization, transparency and inviolability, and many technical improvements will provide a strong security guarantee for 6G. The latest development of 6G is described in detail, and the key performance of 6G and blockchain is discussed. Deeply analyze the technical basis of 6G blockchain, make a preliminary study on the application of blockchain technology in 6G, introduce blockchain related technologies, introduce the application of blockchain technology in spectrum management, analyze the challenges faced by the application of blockchain technology in 6G, and look forward to the future development prospects.

Keywords: Block chain · 6G · Data sharing · Information security

1 Introduction

1.1 A Subsection Sample

The pace of technological innovation in wireless communications has never been interrupted. 6G has higher transmission rate; Higher mobility; Greater security. In 6G network, people have higher and higher requirements on the service field [1]. There are a large number of edge devices and sensitive information of users, which need to be transmitted through traditional network infrastructure, thus bringing huge data security risks. Therefore, it is necessary to protect these sensitive data. By analyzing the security requirements of 6G network and the shortcomings of existing technologies, an information security solution based on blockchain technology is proposed to effectively cope with the challenges faced by 6G network security. Traditional network infrastructure is vulnerable to single point of attack, single point of failure and other factors, resulting in privacy leakage risk, which leads to serious data security problems [2].

Existing data privacy protection and centralized data processing methods require third-party trusted entities to serve them, and edge device data needs to be processed by third-party entities. None of these methods can guarantee that users' sensitive information will not be tampered with or deleted, nor can they meet the needs of distributed storage in cloud computing, so massive sensitive data will be abused, privacy leakage and other hidden dangers [3].

Blockchain is one of the key technologies to realize 6G, which can effectively solve the network security and privacy issues. The blockchain technology is introduced into 6G network, and the distributed collaboration mechanism of blockchain is used to support more secure and robust interaction between communication service nodes, so as to further improve the communication network coverage and communication capability.

2 6G Network and Blockchain Related Technology Overview

New basic communication technologies, including signal sampling and channel coding, support 6G network applications through innovations in coding mechanisms, spectrum sharing and spectrum balancing [4]. 6G era is based on artificial intelligence, with the performance of wireless communication technology, three-dimensional space integrated communication, wireless antenna network and other proprietary technologies, through the lowest level of wireless network, to break the limitations of the region and space, and finally achieve the purpose of transmission network.

6G mobile communication is bound to make new breakthroughs in technology, and has already realized industrial applications, but the key technology is the new spectrum communication technology.

With the rapid development of wireless communication services, the demand for spectrum is increasing. Due to the scarcity and non-renewable characteristics of spectrum, spectrum resources become very tight. At present, all countries in the world are facing a serious spectrum crisis [5]. Especially in China, the domestic spectrum resources are in short supply and demand is huge. When the traffic volume increases and the usage decreases sharply, the spectrum resources will become scarce.

6G trusted blockchain fragmentation system consists of blockchain layer and application layer. In the blockchain layer, the blockchain backbone is placed on the edge server, and the fragmented blockchain is deployed on the edge device. In order to deal with the large amount of data obtained from the edge network effectively, the blockchain sharding system is used to process massive transactions in parallel. This paper proposes a distributed transaction security solution based on blockchain fragment technology. Firstly, the public identification node of the blockchain is sliced and segmented, and separate blocks are established for different slices. In the slices, the consensus mechanism is used to verify the legitimacy and integrity of the block. On this basis, the hash table is used to store the association information between each block. The second is to reassemble the block established by each fragment, and get the final identification between fragments, and finally add a new block in the main chain [6].The blockchain hierarchy architecture is shown in Fig. 1.

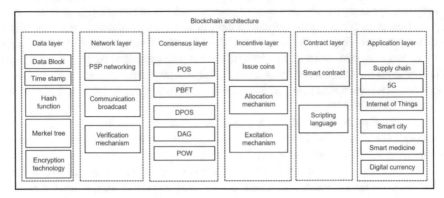

Fig. 1. Block chain hierarchy architecture diagram

Edge network can realize data sharing, data trading and security management. However, due to the lack of consensus mechanism for edge devices, the energy consumption of communication and computation is high, which affects the performance of edge devices [7]. In view of the deployment problem of blockchain fragmentation system, requesters of blockchain services can design contract combinations. In order to motivate edge devices to join the block consensus mechanism of blockchain, edge devices aim to maximize their own utility.

3   Manage 6G Spectrum Based on Blockchain

Limited spectrum resources have always plagued the development of mobile communication, green development, resource conservation and other concepts put forward higher requirements for spectrum co-construction and sharing. How to measure the sharing state of dynamic spectrum is the key to ensure the smooth operation of the whole network, which can achieve accurate processing, efficient and real-time settlement of dynamic spectrum sharing transactions [8]. The existing mainstream spectrum sharing methods have different degrees of shortcomings.

Blockchain spectrum sharing can provide users with flexible wireless services. This paper proposes a trust-based utility design, resource scheduling and negotiation mechanism, and a method to improve the efficiency of spectrum utilization across systems. For unreliable, decentralized, sensor nodes provide payment and other incentive mechanisms, smart contract for mobile operators, realize spectrum sensing and service functions, and has been successfully applied in a variety of occasions. Traditional wireless access networks are facing a rapid increase in the number of users and high real-time access requirements, resulting in slow allocation of spectrum resources, and it becomes more difficult to meet user requirements [9]. The entity relationship based on the blockchain spectrum management system is shown in Fig. 2.

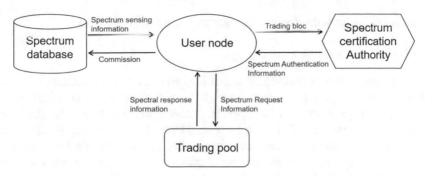

Fig. 2. Entity relationship of spectrum management system based on blockchain

The spectrum resource management and distributed spectrum sharing method based on blockchain has the characteristics of low spectrum utilization and good stability, which provides a scalable spectrum sharing and resource management solution for large-scale deployment in the future. The algorithm adopts asymmetric key agreement mechanism. Data distribution is secure, confidential and reliable, with good expansibility. The distributed blockchain network has simple structure, adopts point-to-point communication, has high security and low transmission cost, and effectively realizes spectrum sharing in wireless network through blockchain technology [10]. The advantages of blockchain applied to spectrum management are shown in Fig. 3.

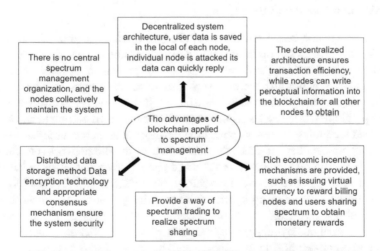

Fig. 3. Advantages of blockchain applied to spectrum management

In 6G network environment, digital encryption scheme based on distributed hash table cannot meet the demand of spectrum sharing. The distributed ledger algorithm based on decentralized structure can balance the contradiction between spectrum sharing and privacy protection well. At present, the research on the combination of blockchain and spectrum management has been carried out at home and abroad, but there are still many key problems to be solved.

4 Related Technologies of Blockchain and 6G Fusion

4.1 6G Network Enabling Technology

6G era will rely more on blockchain technology to achieve communication interconnection. As an emerging technology, blockchain technology has been widely used in various fields. Typical block-to-block, chain structure, or directed acyclic graph structure are used to store data. Cryptography is used to ensure the security of transmission and access, make data storage easy to tamper with, and prevent rejection [11]. In the low-cost competitive environment, this paper provides a new trust computing paradigm and cooperation mode of trust network.

With its unique trust establishment mechanism, blockchain is changing the application scenarios and operation rules of many industries. The higher the frequency, the higher the path loss, the smaller the coverage radius, the higher the cost. Regardless of the business requirements or cost effectiveness, network co-construction and sharing is likely to be the main direction of 6G network development in the future [12]. Blockchain is a new form of decentralized distributed ledger that is tamper-proof and traceable. In the scenario of network co-construction and sharing, trust interconnection based on blockchain is open and trusted for multiple operators under network co-construction, which can effectively track network quality, monitor network equipment, provide digital identity authentication and network roaming settlement services.

4.2 6G Hybrid Cloud and Blockchain

The 6G era brings new requirements and challenges to data centers. 6G has prominent technical characteristics such as large bandwidth and low latency, which can flexibly meet the specific needs of business, and then promote the improvement of social and economic efficiency and cost. In this process, how to coordinate the development of bandwidth and latency has become one of the research hotspots. Hybrid cloud is an effective method to solve this problem. Currently, there are many solutions, which complement each other. Hybrid cloud technology enables application developers and content service providers to seamlessly connect between the edge and center of the 6G network, enabling cloud computing [13]. Hybrid cloud is a complex and open system that needs to be managed by a trusted third party to ensure the security and integrity of user data. At present, there is no effective solution to hybrid cloud security problem. As a new distributed ledger technology, blockchain technology has been widely concerned and has broad prospects. In actual deployment, blockchain platforms or applications can be deployed on the hybrid cloud server side to support different application scenarios.

The hybrid cloud allows computing to take place at the edge of the mobile network. Hybrid clouds can be used to solve congestion and 6G network delays caused by mobile networks [14]. This paper proposes the combination of blockchain and hybrid cloud technology to construct the future 6G application scenarios under two different modes, and analyzes its service environment. Ensure the distributed deployment of computing resources, transaction data traceability, through the distributed deployment of hybrid cloud server, can be very convenient to achieve blockchain spectrum sharing.

4.3   6G Cybersecurity and Blockchain Technology

In order to deal with various complex privacy challenges, blockchain has brought possible solutions to 6G privacy protection, which can provide a strong guarantee for the construction of a distributed secure and trusted transaction environment [15]. Due to the complexity and imtamability of data information, traditional encryption methods are difficult to meet the demand of high access control for privacy, while blockchain technology can effectively solve this problem. Security is the most critical requirement and attribute of 6G network, and blockchain technology will play a decisive role in 6G network.

Traditional broadband wireless access technology based on fixed frequency can no longer meet the needs of users with rapid development in the future, and the demand for application speed and delay in the 6G era will continue to increase. New services also lead to increased requirements for lightweight and dynamic computing, so a communication network closely connected with information transmission and application needs is urgently needed to achieve full frequency domain, full scene effect, flexible adaptation of all services, and resource coordination [16].

Using the blockchain technology, combined with the decentralized and distributed 6G network deployment, the scalability of the blockchain is enhanced by reducing the processing overhead of data mining, so as to defend against various security attacks. Based on the consensus mechanism, the security of the system is enhanced. The use of consensus mechanism enhances transaction verification, prevents malicious attacks, tampering with information data and other behaviors, and reduces the risks of network attacks, malicious node spoofing and external environment attacks [17].

Network security technology is not mature, the future 6G era will have stronger computing power, but also faces the challenge of complex heterogeneous and distributed characteristics on computing power. Blockchain technology can be considered as a technological means to enhance the security of computing power and support the transaction of computing power.

5   The Fusion Application of Blockchain and 6G

With its unique key technology and distinctive characteristics, blockchain endows mobile communication with new connotation, enhances the security guarantee ability of mobile communication, and establishes a safe and reliable implementation mechanism.

6G is a new era of mobile communication, realizing the connection of intelligent artificial intelligence, the deep coverage of the Internet of things, holographic information interaction, all-weather ubiquitous connection and other ideas [18]. In this new era, people need more efficient and reliable network technology to support various business applications. It achieves high performance such as peak rate, high device strength and ultra-low latency, but still has data leakage and other security issues. Can achieve the purpose of protecting information, tamper-proof, traceability and other advantages, make up for security loopholes. The application prospect of blockchain in 6G is shown in Fig. 4.

Based on the basis of existing research, blockchain may play an important role in the 6G era:

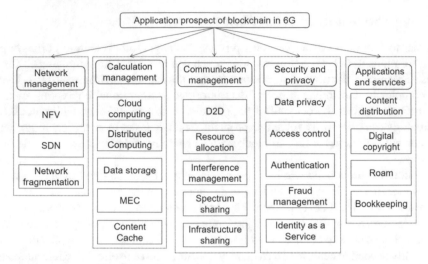

Fig. 4. Application prospect of blockchain in 6G

(1) Sharing spectrum, infrastructure and other resources. 6G can use the smarter, more distributed blockchain Dynamic Spectrum Sharing Access (DDS) technology to use blockchain to flexibly share network infrastructure and allocate it on demand to reduce operator costs and save energy.

(2) Authentication. With the increasing number of 6G network terminals, the centralized authentication may cause network bottlenecks and delays. In order to solve these problems, an identity authentication scheme based on blockchain is proposed, and its security is analyzed. The results show that the scheme has high security performance [19].

(3) Network security. With the advent of the 6G era, "Internet of everything" brings a large number of new business needs, but also challenges the next generation of Internet. In order to meet these new requirements, it is necessary to introduce security protection measures in the new generation network communication architecture. One of the most critical is information security.

The traditional "external" and "patch-type" network security protection mechanism can no longer adapt to the ubiquitous attacks that may exist in the future 6G networks, and the security risks are uncertain. With its unique communication technology, blockchain has become one of the hot spots in the future research of 6G network endogenous security strategy [20]. Blockchain will provide many security services for 6G, including access control, data integrity, identity verification and privacy protection.

6   Conclusion

Using cloud computing, edge computing, artificial intelligence and other emerging technologies, the blockchain is applied across the network, intelligent and efficient interconnection across fields. The integration of 6G and blockchain can improve user experience, reduce service costs, and improve communication channel utilization. At present, many countries around the world have put forward plans to promote the construction of network security, but the implementation effect is not optimistic [21]. Information security is the key to the success of 6G.

The application of blockchain technology in 6G is initially discussed. In the future, the development of blockchain will be promoted to the physical layer, and will develop in synergy with artificial intelligence and edge computing, which will significantly improve the computing, communication and storage performance of communication networks [22]. The deep integration of blockchain technology and 6G to realize intelligent consensus among distributed nodes will face new challenges such as new architecture and new technology, and also put forward new requirements for it.

In terms of blockchain technology itself, its basic theory and key technologies need to be further improved and breakthrough to lay the foundation for its integration into 6G. Combined with the characteristics and architecture of 6G, it is necessary to start with design and planning to form a blockchain technology suitable for 6G to support the needs of efficient, safe and convenient interconnection between humans and machines [23].

Block chain technology applied in the basic theory and key technology of 6G networks application was still in the stage of study, how to further in the next 6G network deployment, match and joint optimization performance index, security, stability, etc., the guiding ideology and target of the application of further integration specification, need further research, to ensure that the 6G block chain technology in health and sustainable development. Blockchain is both an idea and a technical approach, and further work is needed to evaluate the consistency of this idea with practical applications.

Acknowledgments. This work was partially supported by the Science Project of Hainan University (KYQD(ZR)20021).

References

1. Shen, X.S., Liu, D., Huang, C., et al.: Blockchain for transparent data management toward 6G. Engineering **8**, 74–85 (2022)
2. Jahid, A., Alsharif, M.H., Hall, T.J.: The convergence of Blockchain, IoT and 6G: potential, opportunities, challenges and research roadmap. arXiv preprint arXiv:2109.03184 (2021)
3. Shah, K., Chadotra, S., Tanwar, S., et al.: Blockchain for IoV in 6G environment: review solutions and challenges. Cluster Comput. **25**, 1–29 (2022)
4. Kumari, A., Gupta, R., Tanwar, S.: Amalgamation of blockchain and IoT for smart cities underlying 6G communication: a comprehensive review. Comput. Commun. **172**, 102–118 (2021)
5. Velliangiri, S., Manoharan, R., Ramachandran, S., et al.: Blockchain based privacy preserving framework for emerging 6G wireless communications. IEEE Trans. Industr. Inf. **18**(7), 4868–4874 (2021)
6. Faisal, T., Dohler, M., Mangiante, S., et al.: BEAT: Blockchain-enabled accountable and transparent network sharing in 6G. IEEE Commun. Mag. **60**(4), 52–56 (2022)
7. Wang, Z., Xu, Y., Liu, J., et al.: An efficient data sharing scheme for privacy protection based on blockchain and edge intelligence in 6G-VANET. Wirel. Commun. Mob. Comput. **2022**(2022)

8. Sekaran, R., Patan, R., Raveendran, A., et al.: Survival study on blockchain based 6G-enabled mobile edge computation for IoT automation. IEEE access **8**, 143453–143463 (2020)

9. Maksymyuk, T., Volosin, M., Gazda, J., et al.: Blockchain-based decentralized service provisioning in local 6G mobile networks. In: Proceedings of the 19th ACM Conference on Embedded Networked Sensor Systems, pp. 516–519 (2021)

10. Okon, A.A., Sholiyi, O.S., Elmirghani, J.M.H., et al.: Blockchain for spectrum management in 6G networks. In: Cao, B., Zhang, L., Peng, M., Ali, M. (eds.) Wireless Blockchain: Principles, Technologies and Applications, pp. 137–159 (2021)

11. Wright, S.A.: Blockchain-enabled decentralized network management in 6G. In: Dutta Borah, M., Singh, P., Deka, G.C. (eds.) AI and blockchain technology in 6G wireless network. Blockchain Technologies. Springer, Singapore (2022). https://doi.org/10.1007/978-981-19-2868-0_3

12. Garg, S., Goyal, S., Bhandari, A.: Role of blockchain in security of 6G networks. In: Challenges and Risks Involved in Deploying 6G and NextGen Networks, pp. 106–129. IGI Global (2022)

13. Aggarwal, S., Kumar, N., Tanwar, S.: Blockchain-envisioned UAV communication using 6G networks: Open issues, use cases, and future directions. IEEE Internet Things J. **8**(7), 5416–5441 (2020)

14. Kausar, F., Senan, F.M., Asif, H.M., et al.: 6G technology and taxonomy of attacks on blockchain technology. Alex. Eng. J. **61**(6), 4295–4306 (2022)

15. Li, B., Deng, S., Yan, X., et al.: The confluence of blockchain and 6G network: scenarios analysis and performance assessment. arXiv preprint arXiv:2207.04744 (2022)

16. Patel, F., Bhattacharya, P., Tanwar, S., et al.: Block6Tel: Blockchain-based spectrum allocation scheme in 6G-envisioned communications. In: 2021 International Wireless Communications and Mobile Computing (IWCMC), pp. 1823–1828. IEEE (2021)

17. Ni, Q., Linfeng, Z., Zhu, X., et al.: A novel design method of high throughput blockchain for 6G networks: performance analysis and optimization model. IEEE Internet of Things J. **9**, 25643–25659 (2022)

18. Zhang, H., Leng, S., Wu, F., et al.: A DAG Blockchain enhanced user-autonomy spectrum sharing framework for 6G-enabled IoT. IEEE Internet of Things J. **9**, 8012–8023 (2021)

19. Srivastava, V., Mahara, T., Yadav, P.: An analysis of the ethical challenges of blockchain-enabled E-healthcare applications in 6G networks. Int. J. Cogn. Comput. Eng. **2**, 171–179 (2021)

20. Jadav, N.K., Gupta, R., Tanwar, S.: Blockchain and Edge Intelligence-based secure and trusted V2V framework underlying 6G networks. In: IEEE INFOCOM 2022-IEEE Conference on Computer Communications Workshops (INFOCOM WKSHPS). IEEE, pp. 1–6 (2022)

21. Sun, W., Li, S., Zhang, Y.: Edge caching in blockchain empowered 6G. China Commun. **18**(1), 1–17 (2021)

22. Khan, A.H., Hassan, N.U.L., Yuen, C., et al.: Blockchain and 6G: the future of secure and ubiquitous communication. IEEE Wirel. Commun. **29**(1), 194–201 (2021)

23. Bhattacharya, P., Saraswat, D., Dave, A., et al.: Coalition of 6G and blockchain in AR/VR space: challenges and future directions. IEEE Access **9**, 168455–168484 (2021)

The 6th International Conference on Innovative Computing (IC 2023)

FCGSM: Fast Conjugate Gradient Sign Method for Adversarial Attack on Image Classification

Xiaoyan Xia[1], Wei Xue[1,2,3(✉)], Pengcheng Wan[4], Hui Zhang[1], Xinyu Wang[1], and Zhiting Zhang[1]

[1] School of Computer Science and Technology, Anhui University of Technology, Maanshan 243032, China
xuewei@ahut.edu.cn
[2] Anhui Engineering Research Center for Intelligent Applications and Security of Industrial Internet, Maanshan 243032, China
[3] Institute of Artificial Intelligence, Hefei Comprehensive National Science Center, Hefei 230088, China
[4] National Key Laboratory of Science and Technology on Automatic Target Recognition, National University of Defense Technology, Changsha 410073, China

Abstract. Deep neural network is sensitive to adversarial samples that crafted by adding imperceptible perturbations to original images, and many methods of generating adversarial samples have emerged. Although existing methods based on gradient direction have good attack performance, some ill-conditioned issues may reduce their performance on occasion. In this paper, we propose a novel attack method based on three-terms conjugate gradient direction, which is effectively for improving this limitation, and its is named as fast conjugate gradient sign method (FCGSM). The proposed method FCGSM can jump from the local maximum during the process of finding the maximum value of loss function, thus generating more adversarial samples than the SOTA methods APGD and ACG. Experiments conducted on two benchmark datasets show that the FCGSM works well in attacking deep neural network-based classification models.

Keywords: adversarial machine learning · deep learning · conjugate gradient · adversarial attack · adversarial training

1 Introduction

Deep neural networks (DNNs) have shown the tremendous capacity and ability in making a good progress in the filed of computer vision. However, it is also demonstrated that DNNs are highly vulnerable to adversarial samples [4,18], which are manufactured by adding small-imperceptible perturbations on input and make a model output incorrect classification. Plenty of methods in generating adversarial samples have been proposed since it helps to evaluate the vulnerability of models and enhance the robustness of various DNN algorithms by adversarial training [5,15]. Moreover, It is important for improving the robustness of models

J. C. Hung et al. (Eds.): IC 2023, LNEE 1045, pp. 709–716, 2023.
https://doi.org/10.1007/978-981-99-2287-1_98

by adversarial training to learn how to generate adversarial samples with better transferability [10, 19].

With full access to the knowledge structure of a model including the composition and parameters, most methods of generating adversarial samples can successfully attack the transparent model. This type of attack is known as a white-box attack, including optimization-based methods such as box-constrained L-BFGS [18], C&W [1], and gradient-based methods such as fast gradient sign method (FGSM) [4], iterative-FGSM (IFGSM) [9], projected gradient descent (PGD) [11], which use the steepest gradient to update the sign. In addition, some attack methods utilize the current and past gradient information to determine the next update, such as momentum iterative-FGSM (MI-FGSM) [3] and Auto-PGD (APGD) [2]. However, the steepest descent method may be inefficiently attack the deep learning models due to the fact that the convergence speed is relatively slow and the objective function of adversarial attack is highly nonconvex. To solve the above challenges, [20] applies the conjugate gradient (CG) method to generate adversarial samples. Although the traditional conjugate gradient method has some improvement in the accuracy of calculation and convergence of the objective function, for nonlinear objective functions, sometimes infinite cycling away from the optimal solution.

To this end, in this paper we propose a new adversarial attack algorithm, fast conjugate gradient sign method (FCGSM), based on a three-terms CG direction with adaptive stepsize selection strategy. In summary, we make the following contriutions:

1. We propose an effectual adversarial machine learning algorithm that based on the fast gradient sign method and auto CG attack method, which possesses the ability to search more diverse direction and generate more diverse adversarial samples;
2. We further use the obtained adversarial samples to execute the adversarial training to improve the robustness of the classification models that based on DNNs, and the corresponding experiments demonstrate adversarial training is an effective security defensive mechanism.

2   Proposed Method

In this section, we present the proposed adversarial attack method. In order to better describe our approach, we first give a brief review of the FGSM method and the CG method.

2.1   FGSM Method

FGSM is to generate an adversarial sample x^{adv} by stacking the original image x with variations that are consistent with the direction of gradients. Suppose that J is the object function, we can use it to compute the current gradient and then obtain the perturbation $\eta = \epsilon sign(\nabla_x J(x, y))$ constrained by $\| x^{adv} - x \|_\infty \leq \epsilon$, where $\epsilon > 0$ is an artificial parameter. Subsequently, the adversarial example generated by FGSM can be presented by $x^{adv} = x + \eta$.

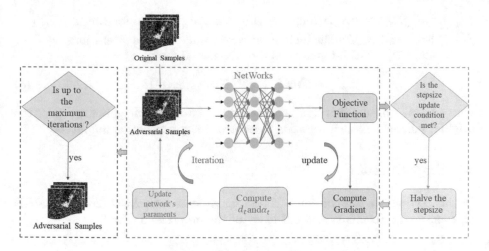

Fig. 1. Flowchart of FCGSM.

2.2 CG Method

CG method is a very efficient optimization algorithm. Consider the minimization problem $\min_{x \in R^n} f(x)$, where the objective function f is differentiable, then given an initial point x_0, a CG method generates a sequence $\{x_t\}$ by $x_{t+1} = x_t + \alpha_t d_t$, where α_t is the stepsize usually obtained by a line search, and the search direction d_t is computed by $d_k = -\nabla f(x_t) + \beta_t d_{t-1}$ with $d_0 = -\nabla f(x_0)$. Here, β_t is the CG update parameter. Some well-known formulas for β_t are β_t^{HS}, β_t^{PR}, $\beta_k t^{DY}$, et al., see [6] for details.

2.3 FCGSM Method

Based on the theories mentioned above, we now describe the proposed FCGSM method. The flowchart of FCGSM is shown in Fig. 1. Note that adversarial attack can be formulated as a maximization optimziation problem. Consider the problem $\max_{x \in R^n} f(x)$, where f is a continuous. Given the initial point x_0 and the initial search direction $d_0 = \nabla f(x_0)$, we then update x_t^{adv} (i.e., adversarial sample) at the t-th iteration with d_t as follows

$$x_{t+1}^{adv} = x_t^{adv} + \alpha_t \cdot sign(d_t), \tag{1}$$

and

$$d_t = \nabla f(x_t) - \beta_t d_{t-1} + \gamma_t y_{t-1}, \tag{2}$$

where d_t is the so-called three-terms CG direction. We set $\beta_t = \beta_t^{PR}(= \frac{(g_t)^T y_t}{(g_{t-1})^T g_{t-1}})$ and $\gamma_t = \frac{(g_t)^T d_{t-1}}{(g_{t-1})^T g_{t-1}}$, where $g_t = \nabla f(x_t)$, $y_t = g_t - g_{t-1}$.

For the stepsize α_t, we calculate it according to the following two conditions proposed in [2]: (1) $\sum_{i=w_{j-1}}^{w_j} N < \rho \cdot (w_j - w_{j-1})$, (2) $\eta^{(w_{j-1})} = \eta^{(w_j)}$ and

$f_{\max}^{(w_j-1)} = f_{\max}^{(w_j)}$, where N indicates the count of the cases for which $f(x_{t+1} > f(x_t))$ holds and $f_{\max}^{(w_j)}$ is the highest objective value in the w_j iterations.

In summary, FCGSM generates the adversarial samples by

$$x_{t+1}^{adv} = Cilp_{x,\epsilon} \left\{ x_t^{adv} + \alpha_t \cdot sign(d_t) \right\}, \tag{3}$$

where $Cilp_{x,\epsilon}$ means that x_t^{adv} has be cliped into the ϵ-neighbourhood of the original sample x at each iteration to control the perturbation amplitude.

3 Experiments

In this section, we present comparison experiments with ACG [20] and APGD [2] to show the feasibility and efficiency of the proposed method.

3.1 Experimental Setup

Datasets and Models. We choose six classification models (VGG-11, VGG-13, and VGG-16; ResNet-18, ResNet-34, and ResNet-50) and two benchmark datasets (MSTAR and CIFAR-10).

Hyperparameter setting. We set the maximum perturbation $\epsilon = 8/255$, the initial stepsize $\eta^{(0)} = 0.01$, the stepsize selection parameter $\rho = 0.75$, and the maximum number of iterations $T = 100$.

Choice of loss function. ACG and APGD use the CW loss proposed in [1] and DLR loss proposed in [2], respectively. For FCGSM, we choose the cross-entropy loss as the objective function.

Evaluation metrics. We adopt the evaluation metrics based on accuracy, attack success rate (ASR for short), and the diversity which is described as the Euclidean norm of two successive adversarial samples, where the ASR is defined as $ASR = \frac{\text{accuracy before attacking} - \text{accuracy after attacking}}{\text{accuracy before attacking}}$.

3.2 Analysis of Comparison Results

Table 1 reports the time of generating adversarial samples on MSTAR dataset for VGG-16 by using three attack methods. The time is calculated from the first sample started to be attacked to the end of the attack on the last example. We can see that the generating time of FCGSM is the lowest.

Table 1. Time of generating adversarial samples on MSTAR dataset.

CPU	RAM	GPU	VGG-16	APGD	ACG	FCGSM
Intel(R) Xeon(R)	24 GB	NVIDIA GeForce	ASR	95.46	98.15	98.48
Silver 4314 × 4		RTX 3090 × 2	time	5m46s	3m24s	3m22s

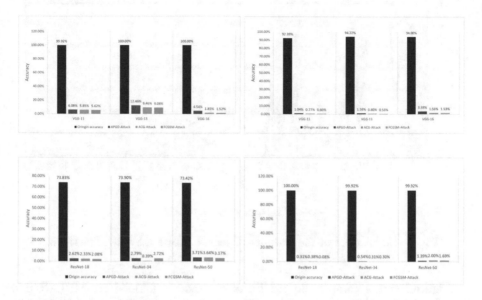

Fig. 2. Comparision results of accuracy on the original sample set and adversarial sample set. The left is on MSTAR, and the right is on CIFAR-10.

Table 2. The ASR of FCGSM, ACG and APGD for attacking the trained models. The highest ASR is in bold, and the second is underlined. diff is the difference between bold and underlined.

MSTAR	Attack success rate (%)			
Architecture	APGD	ACG	FCGSM	diff
VGG-11	93.92	_94.15_	**94.38**	0.23
VGG-13	87.54	_90.54_	**90.92**	0.38
VGG-16	95.46	_98.15_	**98.48**	0.33
ResNet-18	97.38	_97.67_	**97.92**	0.25
ResNet-34	97.21	**99.61**	_97.28_	2.33
ResNet-50	96.29	_96.36_	**96.83**	0.47
CIFAR-10	Attack Success Rate (%)			
Architecture	APGD	ACG	FCGSM	diff
VGG-11	98.96	_99.23_	**99.40**	0.17
VGG-13	98.84	_99.20_	**99.47**	0.27
VGG-16	96.62	_98.44_	**98.47**	0.03
ResNet-18	_99.69_	99.62	**99.92**	0.23
ResNet-34	99.46	_99.69_	**99.70**	0.01
ResNet-50	**98.61**	98.00	_98.31_	0.30

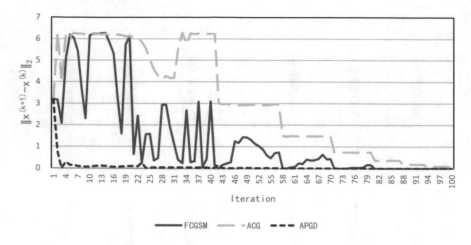

Fig. 3. Comparison of the diversity of search direction by three methods.

Figure 2 shows comparison results of six models trained on MSTAR and CIFAR-10 after being attacked by FCGSM as well as ACG and APGD. The highest is the original accuracy, and it can be seen that the accuracy value decreases significantly after attacking. Table 2 reports the ASR results, and overall, FCGSM has a higher ASR than other two methods in all scenarios.

We further examine the diversity of search direction by FCGSM. From the Fig. 3, we can see that the amount of perturbation between two points fluctuates widely, which indicates that FCGSM possesses the ability to search more diverse direction and generate more diverse adversarial samples.

3.3   Adversarial Training

In order to improve the robustness of the three classification models, we conduct the adversarial training by using the oabtined adversarial samples. Specifically, we divide the adversarial samples on VGG-16 generated by FCGSM into two parts. 70% of these are put into the training set for the adversarial training, and 30% of the adversarial samples generated from ACG and APDG are simultaneously chosen as test set to calculate the classification accuracy. We selecte VGG-16 as the adversarial training model and validate the effect on MSTAR and

Table 3. Adversarial training results on MSTAR.

MSTAR	Accuracy		
Architecture	APGD	ACG	FCGSM
VGG-16	34.53	21.70	33.76
RobustVGG-16	40.92	42.71	40.67

Table 4. Adversarial training results on CIFAR-10.

CIFAR-10	Accuracy		
Architecture	APGD	ACG	FCGSM
VGG-16	8.04	9.13	10.07
RobustVGG-16	44.82	46.56	46.02

CIFAR-10. Tables 3 and 4 show that the robustness of the model after adversarial training is significantly improved, while the model after adversarial training using the adversarial samples generated by FCGSM has good defense against the attacks of the adversarial samples generated by ACG and APGD.

4 Conclusion

In this paper, we proposed a three-terms conjugate gradient direction-based adversarial attack method, which has more diverse search ability to improve the attack performance. Experimental results verified the validity and feasibility of the proposed method, and in the future work we will apply this method to attack deep learning detection models.

Acknowledgements. This work was supported in part by the Anhui Provincial Natural Science Foundation (Grant No. 2208085MF168), the Program for Synergy Innovation in the Anhui Higher Education Institutions of China (Grant No. GXXT-2022-052), and the College Students' Innovation and Entrepreneurship Training Programs (Grant Nos. 202210360079, 202110360079, and S202110360291).

References

1. Carlini, N., Wagner, D.A.: Towards Evaluating the Robustness of Neural Networks. In: IEEE Symposium on Security and Privacy, pp. 39-57 (2017)
2. Croce, F., Hein, M.: Reliable evaluation of adversarial robustness with an ensemble of diverse parameter-free attacks. In: International Conference on Machine Learning, pp. 2206-2216 (2020)
3. Dong, Y., Liao, F., Pang, T., Su, H., Zhu, J., Hu, X., Li, J.: Boosting Adversarial Attacks with Momentum. In: IEEE Conference on Computer Vision and Pattern Recognition, pp. 9185–9193 (2018)
4. Goodfellow, I. J., Shlens, J., Szegedy, C.: Explaining and harnessing adversarial examples. In: International Conference on Learning Representations (2015)
5. Gowal, S., Qin, C., Uesato, J., Mann, T., Kohli, P.: Uncovering the limits of adversarial training against norm-bounded adversarial examples (2021). arXiv:2010.03593v3
6. Hager, W.W., Zhang, H.: Algorithm 851: CG_DESCENT, a conjugate gradient method with guaranteed descent. ACM Trans. Math. Softw. **32**, 113–137 (2006)
7. Ibrahim, A., Shareef, S.: Modified conjugate gradient method for training neural networks based on Logisting mapping. J. Univ. Duhok **22**(1), 45–51 (2019)

8. Krizhevsky, A., Sutskever, I., Hinton, G.: ImageNet classification with deep convolutional neural networks. Commun. ACM **60**(6), 84–90 (2017)
9. Kurakin, A., Goodfellow, I., Bengio, S.: Adversarial machine learning at scale. In: International Conference on Learning Representations, (2017)
10. Liu, Z., Liu, Q., Liu, T., Xu, N., Lin, X., Wang, Y., Wen, W.: Feature Distillation: DNN-oriented jpeg compression against adversarial examples. In: IEEE Conference on Computer Vision and Pattern Recognition, pp. 860-868 (2019)
11. Madry, A., Makelov, A., Schmidt, L., Tsipras, D., Vladu, A.: Towards deep learning models resistant to adversarial attacks. In: International Conference on Learning Representations (2018)
12. Ma, G., Lin, H., Jin, W., Han, D.: Two modified conjugate gradient methods for unconstrained optimization with applications in image restoration problems. J. Appl. Math. Comput. **68**, 4733–4758 (2022)
13. Papernot, N., Mcdaniel, P., Goodfellow, I., Jha, S., Celik, Z. B., Swami, A.: Practical black-box attacks against machine learning. In: ACM on Asia Conference on Computer and Communications Security, pp. 506-519 (2017)
14. Simonyan, K., Zisserman, A.: Very deep convolutional networks for large-scale image recognition. In: International Conference on Learning Representations (2015)
15. Song, C., He, K., Lin, J., Wang, L., Hopcroft, J. E.: Robust local features for improving the generalization of adversarial training. In: International Conference on Learning Representations (2020)
16. Sun, J., Zhang, J.: Global convergence of conjugate gradient methods without line search. Ann. Oper. Res. **103**, 161–173 (2001)
17. Szegedy, C., et al.: Going deeper with convolutions. In: IEEE Conference on Computer Vision and Pattern Recognition, pp. 1-9 (2015)
18. Szegedy, C., et al.: Intriguing properties of neural networks. In: International Conference on Learning Representations (2014)
19. Xie, C., Zhang, Z., Zhou, Y., Bai, S., Yuille, A. L.: Improving transferability of adversarial examples with input diversity. In: IEEE Conference on Computer Vision and Pattern Recognition, pp. 2730–2739 (2019)
20. Yamamura, K., et al.: Diversified adversarial attacks based on conjugate gradient method. In: International Conference on Machine Learning, pp. 24872–24894 (2022)
21. Zhang, L., Zhou, W., Li, D.-H.: A descent modified Polak-Ribiere-Polyak conjugate gradient method and its global convergence. IMA J. Numer. Anal. **26**(4), 629–640 (2006)

A Lightweight Network for Detecting Small Targets in the Air

Jiaxin Li[1], Hui Li[1], Ting Yong[2,3(✉)], and Xingyu Hou[4]

[1] Institute of System Engineering, Nanjing, China
[2] The Institute of North Electronic Equipment, Beijing, China
029588176@qq.com
[3] National Key Laboratory of Science and Technology on Information System Security, Beijing, China
[4] Zhengzhou Xinda Institute of Advanced Technology, Zhengzhou, China
25185703@qq.com

Abstract. Fast and accurate detection and identification of small airborne targets are of great importance, to security in the air. Unmanned aerial vehicle detection algorithms are mostly deployed on edge devices, and a yolov5-based aerial target lightweight detector is proposed by compressing channels and network cropping for the limited resource characteristics on edge devices. Firstly, the shallow cross-stage partial module is extended and optimized when designing the feature extraction network to maximize the use of shallow features. Secondly, the network is cropped to reduce the number of down-sampling, which makes the computation faster. Finally, the pyramid network used for feature fusion is simplified by modifying from two upsampling operations and two downsampling operations to only one upsampling operation. On the homemade dataset, the proposed Yolo-mini achieves 94.44% mean average accuracy on the test set and the Giga floating-point operations per second of the model is only 3.2, which achieves a better balance of accuracy and computation compared to other lightweight algorithms.

Keywords: Object Detection · UAV · Small object · Neural Networks · Deep Learning

1 Introduction

With the rapid development of the unmanned aerial vehicle (UAV) industrial industry, UAVs are widely used in various industries, such as industry, urban management, sports, peacekeeping, transportation, power cruising, agriculture plant protection, express delivery and disaster rescue, and other scenarios in which UAVs are used [1, 2]. Also in the military battlefield, drones are frequently seen as weapons [3, 4]. Despite attracting widespread attention in different civilian and commercial applications, there is no doubt that drones pose a threat to airspace security and may endanger people and property. Drones are also likely to be used for nefarious purposes, such as this collecting data from

© The Author(s), under exclusive license to Springer Nature Singapore Pte Ltd. 2023
J. C. Hung et al. (Eds.): IC 2023, LNEE 1045, pp. 717–727, 2023.
https://doi.org/10.1007/978-981-99-2287-1_99

private areas, tracking people alive vehicles as spies, remote bugging, carrying explosives for unpredictable terrorist attacks in public places, etc. Therefore, the development of drone countermeasure systems is crucial [5–7].

There are many methods for UAV detection based on video images, such as Faster-RCNN [8], SSD [9], YOLO [10], etc. These algorithms have achieved good results, but detection accuracy and detection speed are a pair of oxymorons, and these algorithms have their advantages and disadvantages, high detection accuracy means complex network structure, and complex network structure means limited detection speed.

To balance the detection speed of the model, a series of lightweight network structures have been proposed in the industry [11–14]. Widodo Budiharto et al. Constructed a detection model using Mobile Net and SSD to implement a fast detection algorithm with an accuracy that meets the practical requirements [15]. MobileNet is a lightweight network structure for edge devices. Sheng Yuan et al. proposed a lightweight network structure based on yolov5. They used the proposed CI network structure and then pruned the Neck structure. The method performs well on a bit of a stone detection task [16]. Haiying Liu et al. proposed an improved feature fusion method based on PANet and BiFPN, which effectively improves the detection of small objects [17]. These algorithms guarantee the model accuracy to meet the demand while ensuring the model size, and these models are relatively small in computation and suitable for deployment at the edge.

UAV countermeasure systems are mostly applied to edge devices. With the rapid development of UAVs, the detection accuracy and detection speed of UAV detection algorithms need to be further improved. Yolov5 is a more classical target detection algorithm with not bad detection accuracy and detection speed. In this paper, a lightweight network based on the yolov5 is proposed to detect small targets in the air. The backbone network is trimmed and optimized to improve the speed of the algorithm and reduce the model size without losing the accuracy of the algorithm, which is easy to deploy on embedded devices. The contributions of this paper are as follows.

1. A model tailoring idea for small targets is proposed.
2. The cross-stage partial (CSP) module is optimized for better expressiveness.
3. A detection algorithm for small targets in the air is proposed, which outperforms the original model in terms of speed and accuracy.
4. The model can identify the detected object in the images and mark the object's bounding box by joining the results across the regions.

2   Materials and Methods

2.1   Improving the YOLOV5s Network Structure

The aerial targets are generally acquired using ground equipment, and generally, the target in the image or video is small and belongs to small object detection. The original yolov5 backbone network contains a large number of down-sampling operations, and the feature map size after multiple down-sampling is small, which is not conducive to small object detection. In this paper, firstly, we reduce the number of down-sampling, which makes the computation faster. Secondly, the spatial pyramid pooling-fast (SPPF) module is added to the CSP module to improve the feature expression capability. Figure 1

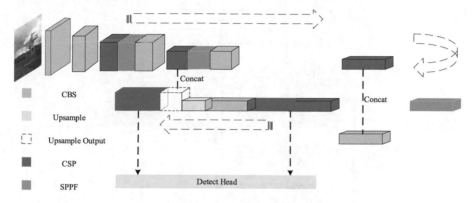

Fig. 1. Yolo-mini network

gives the network structure of the improved Yolo-mini network, which contains four down-sampling, one up-sampling, and two CONCAT operations, fully fusing features of different sizes and different channel numbers to achieve multiscale feature fusion. The overall network is built by the extended CSP module and CBS (Conv, batch-norming, and silo-activation) module, which does not contain complex operations and is easy to deploy.

As shown in Fig. 1, the network structure of Yolo-mini consists of three stages:

- Backbone (Down-sampling) stage: As shown in the first row of Fig. 1, the main role of this stage is to extract features and down-sample the images to reduce the computational effort. The CBS module represents the standard convolution, normalization, and activation function operations, and is mainly used for down-sampling. SPPF module contains four pooling sizes of $1 \times 1, 5 \times 5, 9 \times 9$, and 13×13, which are later fused by standard convolution. This stage expands the channel dimension of the feature map while down-sampling, as shown in Fig. 1. After the down-sampling stage, the feature map becomes slender from flat, and the feature information is concentrated in the channel dimension, which is easy to use for subsequent prediction.
- Neck (Up-sampling) stage: The main role of this stage is feature fusion, which fuses features of different dimension sizes to improve the expression of the features. As the second row of Fig. 1, after this stage, the feature map becomes flat from slender. This stage mainly uses modules such as the CSP module, CBS module, CONCAT module, and up-sampling module. The feature maps of different stages are stacked by CONCAT operation, and then the stacked feature maps are fused using the CSP module.
- Head stage: This stage is mainly for detection and classification based on the extracted feature maps. This part is consistent with the original yolov5.

2.1.1 CSP Extension Module

With the gradual deepening of the network level, the convolutional neural network can extract the semantic information of the high-level features better, but the resolution of the

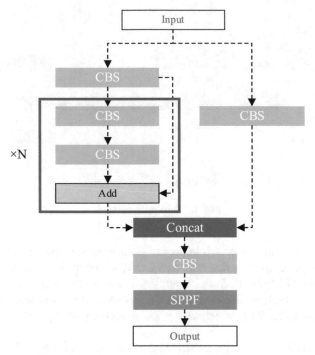

Fig. 2. Extended CSP module. The SPPF module is added after the last convolutional layer.

high-level feature maps is lower. In contrast, the resolution of the feature maps is higher at the shallow level, while the semantic information of features extracted from the shallow network is weaker. For an object with fewer and weaker features in the image, deep convolution can lead to difficult extraction or even loss of object features. To maximize the extraction of features that facilitate the detection of a weak object in the airborne, it is necessary to make full use of the high-resolution features of the convolutional neural network at the shallow layer. Therefore, in the feature extraction stage, we extend the thickness of the CSP module in the shallow feature extraction process. Through stepwise feedback iteration, the object features in the feature map can be fully extracted, and multi-feature extraction from shallow to deep layers can be achieved. Moreover, in deepening the CSP module in the whole feature extraction network by controlling the width and depth factors, we only extend the thickness of the CSP module to extract shallow features. This enhances the ability to extract shallow feature information without increasing the size of the network model and the complexity of the algorithm, which facilitates the detection of a weak object in images. In addition, the CSP structure divides the feature mapping into two branches for extracting features and then merges them, which can achieve a richer combination of gradients while reducing the computational effort.

The structure of the extended CSP module is given in Fig. 2. The red box is the dynamic expansion port, which will repeat N operations. Different CSPs have different N values. In general, a larger N value can increase the network depth and extract better features, but it will increase the computational effort. In the original yolov5, the N values of first these three CSPs in the backbone stage are [1–3], which gives less attention to the shallow features and more attention to the deep features. To give more attention to the shallow features, we modify the N values of the first three CSP modules in the backbone to [1–3].

The SPPF module in yolov5 is an improved version of the SPP module, which draws on the idea of spatial pyramids and enables the fusion of local and global features through the SPP module, enriching the expressiveness of the feature map and facilitating the detection of large differences in target size in the image to be detected, so it has a great improvement on the accuracy of detection. In the original yolov5 network structure, only an SPPF module is finally employed in the backbone network. To improve the expression of shallow features, as in Fig. 2, we add an SPPF module after each CSP module to enhance feature extraction.

2.1.2 Network Trimming

The original backbone network of the yolov5 network contains a large number of down-sampling operations, and the feature map size after multiple downsampling is small, which is not good for small object detection. As in Fig. 1, the improved backbone network is given. Firstly, the 5 downsampling operations are reduced to 4 downsampling operations, which prevents the features from being too small and unfavorable to small object detection. The size of the input image is 640×640, then the minimum feature map of the original yolov5 backbone network is 20×20, and the minimum feature map of the cropped backbone network is 40×40.

The pyramid network structure was also modified. Replacing the original two up-samples operations with only one upsampling operation, only retaining the 40×40 and 80×80 scales of output while adjusting to two anchors and two outputs.

3 Results

3.1 Introduction to the Data Set

The machine learning model should be trained on a set of annotated images with markers to detect and identify small targets in the air. There are three main sources of the dataset, one source is a public dataset, one source is public images collected online, and one source is real data. Among them, the public dataset is from [18], and the original data format is.mat format, which is converted to the applicable Yolo format after script processing is applied. The dataset covers several types of UAVs, civil aircraft, helicopters, and several species of birds of prey. Factors affecting the object detection results, such as foreground occlusion, smoke, target size, imaging angle of view, and color, were also considered. The dataset was scaled to the video source ratio to ensure the best detection results. The dataset was divided into a training set, test set, and validation set according to the ratio of 6:3:1. Then the sample data in the dataset were labeled using the image labeling software called LabelImg. There was a total of 218,173 images in the dataset, and each image containing at least one target frame.

A sample dataset is given in Fig. 3. The whole dataset is all aerial targets and the target size is small. Since it is an aerial target, the background is mostly the sky. Some

Fig. 3. Sample images from the dataset

of the targets have a white surface, which is close to the color of the clouds and poses a great challenge to detect. In addition, due to the difference in flight altitude of the four categories of targets: airplane, bird, UAV, and helicopter, the dataset does not contain multiple categories in the same image, in other words, a picture contains only one category of targets.

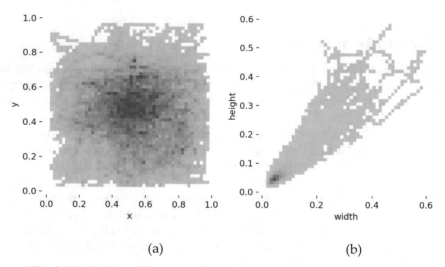

(a)   (b)

Fig. 4. Distribution of the sizes and locations of targets in images of the dataset

In the Yolo series network, the target is described using the four dimensions of the target box, which correspond to [x, y, width, height]. [X, y] is the coordinates of the center point of the target box, and [w, h] is the width and height of the target box. Figure 4 depicts the size and position distribution of the targets in the dataset image, where [x, y, width, height] are all normalized to between [0,1]. Figure 4(a) depicts the position statistics of the centroids of all target boxes in the dataset. It can be seen that the target distribution covers all positions of the image, and most of the targets are concentrated in the middle position of the image. The analysis concludes that the target position distribution in the dataset is close to the normal distribution and is relatively comprehensive. Figure 4(b) depicts the width and height statistics of all target frames in the dataset. It can be seen that the width-to-height ratio of the target boxes is relatively balanced and close to square, and the width-to-height distribution of most of the targets is between [0.0, 0.1], indicating that most of the targets in the dataset belong to small targets.

3.2 Experiment Introduction

The computer configuration for the experiments is as follows: 8 GB NVIDIA RTX3090 graphics processing unit (GPU), 16 GB main memory, 1.297 GHz CPU, and SSD hard disk. We used the original yolov5 model weights on the coco dataset (keeping only the uncropped part) to accelerate the training. To run the Yolo-mini process on this GPU, training was performed using cuda11.1 and cudnn8.0. The code was implemented in torch, using torch version 1.7.1. All experiments were trained for 300 epochs, and the first 3 epochs were hot started. We used various data enhancement techniques and set parameters (e.g., rotation, translation, scaling, and other parameters) to enable the model to generate various images from a single image to enrich the given dataset.

The training loss plot is given in Fig. 5, and it can be seen that the loss decreases relatively fast in the first 10 Epochs, after which the trend of train/box_loss and Val/box_loss also slowly decrease, indicating that the training is effective. Train/obj_loss and Val/obj_ The trend lines of loss almost overlap, which indicates that the model can discriminate the background and target well. Train/cls_loss keeps decreasing, but Val/cls_loss even shows an increasing trend, but the overall level remains low.

The detection results are given in Fig. 6, with several typical targets selected for display. To facilitate the display, all images are scaled to 480*640 size. As in the second row of Fig. 6, the model can detect well for some very small targets, which shows that our model is effective for small targets. Most of the target frames in Fig. 6 are close to the outer rectangle of the target, which indicates that the regression of our target frames is very good and close to the ideal effect. Figure 6 contains two types of images, infrared and visible, which indicates that our model can support the detection of both types of images.

3.3 Ablation Experiments

To investigate the impact of our improvements on the model, we have separately investigated the N values for the first three CSP modules of the backbone network in Fig. 1. The first three N values are [1–3] by default, and the three N values are [1–3] after

Loss Trend

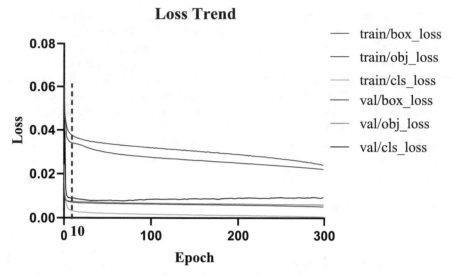

Fig. 5. Training loss trend.

Fig. 6. The detection results on the test dataset

taking the inverse. Plus whether to add the SPPF module after the CSP module, one has four combinations. Withsppf_inver is our improved model with N value taking inverse and CSP module adding SPPF two improvements. Withsppf and Withoutsppf_inver are controlled experiments, which add only N values taking inverse or CSP module adding SPPF. Withoutsppf is the original model without any improvements added.

As in Fig. 7, the results of the four comparison experiments are given. The horizontal coordinate is the number of epochs trained and the vertical coordinate is mAP@.5:.95.

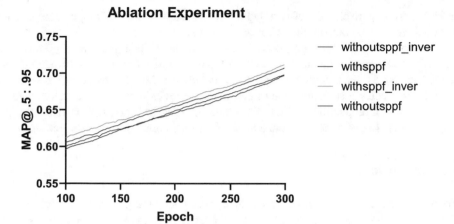

Fig. 7. Ablation experiment results

To facilitate the presentation of the results, we have taken, the results from 100 to 300 Epochs. From Fig. 7, we can see that Withsppf_inver achieves the best results, which shows that our improvement is effective. In addition, Withsppf and Withoutsppf_inver also work better than Withoutsppf (the original version), respectively, indicating that individual improvement is also useful for the model. We analyze the reasons for the usefulness of the improvements: since the backbone network of the model is continuously down-sampling, the N value is taken inverse to make the model focus more on shallow features, which improves the detection of small targets. On the other hand, since the backbone network in Fig. 1 is constantly down-sampled, we inverse the N value, which improves the computational effort. The SPPF module can integrate features with different granularity, which improves the feature representation, so it also has a role in map improvement.

3.4   Comparison with Classical Lightweight Object Detection Models

Table 1. Comparison with other models.

Network	Recall/%	Precision/%	mAp.5/%	mAP_0.5:0.95%	Gflops
Mobilenet [19]	92.98	86.03	89.52	46.93	6.4
Yolov7n [20]	96.35	94.59	94.98	58.25	13.2
Shffule [21]	91.96	81.59	85.12	40.30	1.6
YOLO-mini	95.57	93.06	94.44	71.17	3.2

In the field of object detection, there are many classic lightweight models, and to verify the effectiveness of our proposed method. We have selected three lightweight models for comparison. All network inputs were used with 640*640 inputs and trained with 300 epochs to compare the effect on the test set.

As shown in Table 1, the proposed Yolo-mini achieved the highest mAP_0.5:0.95% by 71.71 compared to the other models. Compared to Mobilenet, all accuracy performance achieved a lead, while the model size is half smaller. Compared to the yolov7n model, the precision is comparable to that of the yolov7n model with a model size of 1/4 of its size. Compared to Shffule, the model is twice as large, but the performance improvement is large enough to make these additional computations worthwhile.

4 Conclusions

To overcome the shortcomings of image detection of small targets, a lightweight detection model Yolo-mini is proposed for small air targets such as UAVs, flying birds, helicopters, and planes. The mAP of the model reaches 94.4%, and the Gflops of the model is only 3.2. In this paper, the network structure is firstly cropped to detect small targets, and only the feature maps with the larger resolution are retained, and then a series of optimizations such as order adjustment and module expansion is carried out for the backbone network. Through a series of comparative experiments, it is found that our model has advantages in the same volume model structure.

References

1. Azar, A.T., et al.: Drone deep reinforcement learning: a review. Electronics **10**, 999 (2021)
2. Sandino, J., Vanegas, F., Maire, F., Caccetta, P., Sanderson, C., Gonzalez, F.: UAV framework for autonomous onboard navigation and people/object detection in cluttered indoor environments. Remote Sensing **12**, 3386 (2020)
3. Udeanu, G., Dobrescu, A., Oltean, M.: Unmanned aerial vehicle in military operations. Sci. Res. Educ. Air Force **18**, 199–206 (2016)
4. Pedrozo, S.: Swiss military drones and the border space: a critical study of the surveillance exercised by border guards. Geographica Helvetica **72**, 97–107 (2017)
5. Restas, A.: others drone applications for supporting disaster management. World J. Eng. Technol. **3**, 316 (2015)
6. Gallacher, D.: Drone applications for environmental management in urban spaces: a review. Int. J. Sustain. Land Use Urban Plann. **3** (2016)
7. Lee, S., Choi, Y.: Reviews of unmanned aerial vehicle (drone) technology trends and its applications in the mining industry. Geosyst. Eng. **19**, 197–204 (2016)
8. Ren, S., He, K., Girshick, R., Sun, J.: Faster R-Cnn: towards real-time object detection with region proposal networks. Adv. Neural Inf. Process. Syst. **28** (2015)
9. Liu, W., et al.: SSD: Single Shot Multibox Detector. In: Proceedings of the European Conference on Computer Vision, Springer, Cham, pp. 21–37 (2016)
10. Redmon, J., Divvala, S., Girshick, R., Farhadi, A.: You only look once: unified, real-time object detection. In: Proceedings of the Proceedings of the IEEE Conference on Computer Vision And Pattern recognition, pp. 779–788 (2016)
11. Liu, Y., Zhang, X.-Y., Bian, J.-W., Zhang, L., Cheng, M.-M.: SAMNet: stereoscopically attentive multi-scale network for lightweight salient object detection. IEEE Trans. Image Process. **30**, 3804–3814 (2021)

12. Wieczorek, M., Si\lka, J., Woźniak, M., Garg, S., Hassan, M.M.: Lightweight convolutional neural network model for human face detection in risk situations. IEEE Trans. Indust. Inform. **8**, 4820–4829 (2021)
13. Du, X., Song, L., Lv, Y., Qiu, S.: A lightweight military target detection algorithm based on improved YOLOv5. Electronics **11**, 3263 (2022). https://doi.org/10.3390/electronics1120 3263
14. Yu, J., Zhou, G., Zhou, S., Qin, M.: A fast and lightweight detection network for multi-scale sar ship detection under complex backgrounds. Remote Sensing **14**, 31 (2021)
15. Budiharto, W., Gunawan, A.A., Suroso, J.S., Chowanda, A., Patrik, A., Utama, G.: Fast object detection for quadcopter drone using deep learning. In: Proceedings of the 2018 3rd International Conference On Computer and Communication Systems (ICCCS), IEEE, pp. 192–195 (2018)
16. Yuan, S., Du, Y., Liu, M., Yue, S., Li, B., Zhang, H.: YOLOv5-Ytiny: A miniature aggregate detection and classification model. Electronics **11**, 1743 (2022). https://doi.org/10.3390/ele ctronics11111743
17. Liu, H., Sun, F., Gu, J., Deng, L.: SF-YOLOv5: a lightweight small object detection algorithm based on improved feature fusion mode. Sensors **22**, 5817 (2022)
18. Svanström, F., Alonso-Fernandez, F., Englund, C.: A dataset for multi-sensor drone detection. Data Brief **39**, 107521 (2021)
19. Howard, A.G., et al.: Mobilenets: efficient convolutional neural networks for mobile vision applications. arXiv preprint arXiv:1704.04861 (2017)
20. Wang, C.-Y., Bochkovskiy, A., Liao, H.-Y.M. YOLOv7: Trainable bag-of-freebies sets new state-of-the-art for real-time object detectors. arXiv preprint arXiv:2207.02696 **(2022)**
21. Zhang, X., Zhou, X., Lin, M., Sun, J.: ShuffleNet: An extremely efficient convolutional neural network for mobile devices. In: Proceedings of the IEEE Conference on Computer Vision and Pattern Recognition, pp. 6848–6856 (2018)

Applying 5PKC-Based Skeleton Partition Strategy into Spatio-Temporal Graph Convolution Networks for Fitness Action Recognition

Jia-Wei Chang and Hao-Ran Liu[✉]

National Taichung University of Science and Technology, Taichung City, Taiwan
kn880701@gmail.com

Abstract. With the rise of health awareness, people's demand for fitness has gradually increased. However, improper exercise may easily cause damage to the body. It would be possible to avoid wrong actions if automatic action recognition can detect and judge the human motion of exercises. Therefore, we aim to grasp the user's fitness status through human action recognition. However, most human action recognition mostly uses CNN-based models to process images, which may introduce unnecessary noise other than the human body from the background. To address this problem, we use the Spatio-Temporal Graph Convolutional Network (ST-GCN) as the backbone and take skeleton data as input to learn skeleton relationships. To further improve the accuracy, we propose a novel partition strategy based on Five Primary Kinetic Chains (5PKC) to explore the skeleton partition status and then enrich the skeleton relationships. Finally, the proposed method with 9 ST-GCN blocks that integrated the proposed partition strategy achieved 99.5% of accuracy which outperforms the model using 9 ST-GCN blocks with 84.5%.

Keywords: Action recognition · Fitness · ST-GCN · Five primary kinetic chains

1 Introduction

In recent years, fitness has become increasingly popular around the world, and gyms have gradually increased. However, when exercising, improper exercise can easily cause damage to the body. Although there are fitness trainers who can assist, there are still some dangers that may be overlooked due to environmental or human factors. For gym owners, it is very important to improve the safety of the gym and reduce operating costs. Therefore, if automatic human action recognition can be used to continuously detect and judge the user's motion, it can not only effectively avoid human negligence and improve safety, also effectively reduce personnel costs. Even the user can use it at home, allowing the user to analyze whether the movement is qualified or not when exercising at home. Human action recognition has been widely used in multimedia computing, such as intelligent surveillance, virtual reality, and human–machine interaction. Although there

© The Author(s), under exclusive license to Springer Nature Singapore Pte Ltd. 2023
J. C. Hung et al. (Eds.): IC 2023, LNEE 1045, pp. 728–737, 2023.
https://doi.org/10.1007/978-981-99-2287-1_100

have been many advances in the research of human action recognition in recent years, the high complexity and variability of human motion make the recognition accuracy and efficiency still have much room for improvement.

Most of the existing human action recognition models are based on images and consider the background. However, the same action will show completely different results in different illumination, viewing angles, and backgrounds. Most of the existing models are based on images and take into account factors such as background, which makes it easy to introduce unnecessary noise when performing action recognition. In order to deal with these noises, these models need to improve ability of modeling the change of background, but the processing will also increase burden on the models. Although some people reduce the impact of background noise by converting the image into a depth, thermal view, the effect of removing background noise is still limited. And the human body is a deformable object with a high degree of freedom, rather than a fixed shape. This makes it difficult to capture human body.

In the related works of human action recognition, deep neural networks [1] have become the main tool for this task. In recent years, it has been proposed to use skeleton-based temporal CNN or RNN for action recognition. By rearranging structured data, the human skeleton data is represented as a vector sequence to adapt to the neural network. Representative works include [2–10]. However, since the skeleton is essentially a non-Euclidean graph, directly sending the coordinates of the skeleton to the network cannot effectively analyze the structural information of the skeleton data [11].

To address this problem, Graph Convolutional Networks (GCN) [10, 12, 13] have been applied to skeleton-based action recognition because it can efficiently analyze the structural information of skeleton data. Spatio-temporal Graph Convolutional Network (ST-GCN) [14] is one of the representative works in skeleton-based action recognition. By using joints as nodes and connections between joints as edges, an undirected spatio-temporal graph is constructed, and a partition strategy is designed according to distance and spatial configuration, and the graph convolution operation is performed based on this. This method has also been proven to effectively improve the effect of human action recognition. Many variants derived from ST-GCN [14] have achieved excellent results [15–19], and ST-GCN [14] has also become the One of the most used frameworks for tasks.

In this paper, we propose a spatio-temporal GCN-based skeleton classification and scoring network. Based on the spatio-temporal graph convolutional network (ST-GCN) [14], it can effectively explore the distribution relationship between joints and joints through spatial configuration. The angle extends the relationship between the joints into successive frames. Then multiple ST-GCN layers are stacked to jointly transfer joints information in space and time. Finally, the action is classified by the score from the loss function.

2   Related Works

2.1   Skeleton-Based Action Recognition

Human action recognition is based on the human body, which means that human skeleton is the most important basis for action recognition. Analysis of bones can also reduce the complexity of action recognition. Past methods usually rearrange the skeleton data into a grid-like structure or sequence of coordinate vectors and send it to CNN [4, 6, 7, 9] or RNN [1, 2, 8] architecture. However, as stated in [11], the skeleton is a non-Euclidean graph, and the spatial subdivision of the skeleton cannot be effectively analyzed using CNN and RNN. With the development of GCN, GCN has been widely used in skeleton-based action recognition [14, 15, 20, 21], because the spatial subdivision of skeleton can be effectively analyzed through GCN. The ST-GCN proposed by Yan et al. [14] first applied GCN to skeleton-based action recognition. It not only captures the relationship of joints in space, but also extends the relationship between joints to the concept of time, thereby capturing the relationship between joints in space and time between each consecutive frame.

2.2   Graph Convolutional Networks

Graph Convolutional Networks (GCN) generalize convolution to graph-structured data, applying them to irregular data such as interpersonal social networks and biological data. Graph convolutional networks can be divided into two types, namely in the spectral domain [10, 13, 22–24] or the spatial domain [20, 25–29] to transfer node features. The former considers graph convolution from a spectral point of view, using the Fourier transform of the graph to operate in the spectral domain, while the latter obtains the information of its neighboring nodes directly on the node. To improve the performance of GCN, someone introduced attention mechanism in GCN [28, 29].

3   Methodology

3.1   Human Skeleton Graph Construction

In action recognition, the skeleton graph uses joint as nodes and bones as edges to form a human body topology. There are 18 nodes in the human body topology, as shown in Fig. 1, which are left hip, right hip, left knee, right knee, left ankle, right ankle, left foot, right foot, left shoulder, right shoulder, left elbow, right elbow, left wrist, right wrist, the left eye, the right eye, the head, and a dynamic center of gravity. Dynamic center is calculated by adding and averaging the coordinates of all nodes. The spatio-temporal skeleton is based on the input video, and each frame is converted into a skeleton and stacked in time series.

Suppose a given skeleton sequence has T frames, each with N joints. According to the human skeleton diagram in Fig. 1, we construct an undirected graph and define it as $G = (V, E)$, where the joints are defined as $V = \{v_i \in R^C | i = 1, \cdots, N\}$, N represents the number of joints, C is the dimension of joint features, all adjacent joints of joint v_i are denoted as $N(v_i)$, and bones are defined as $E = \{v_i v_j | (i, j) \in e\}$, and e represents a set of edges formed by every two adjacent joints.

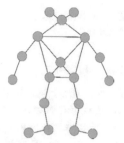

Fig. 1. Human skeleton graph

3.2 Spatial Graph Convolution

To aggregate joint information from adjacent joints into each joint using graph convolution operation, as shown in Fig. 2, we define the joint adjacency matrix $A \in R^{N \times N}$ according to the skeleton graph, if e is Existing $A_{ij} = 1$, otherwise $A_{ij} = 0$. Since the human skeleton graph is an undirected graph, A is a symmetric matrix.

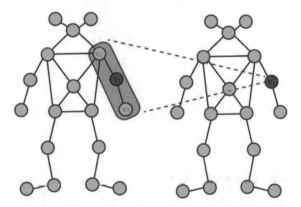

Fig. 2. Aggregate Adjacent Joint Information to each Joint

Next, for the graph convolution operation, we denote the adjacent region with 1-distance from v_i as $N(v_i)$, and the entire skeleton feature is defined as $V \in R^{N \times C}$. The formula for taking V and A as the input of the spatial GCN is as follows:

$$V^{l+1} = \sigma \left(\underline{A} V^l W^l \right) \tag{1}$$

$V^l \in R^{N \times C_l}$ is the skeleton feature of the l-th layer, $V^{l+1} \in R^{N \times C_{l+1}}$ is the skeleton feature of the $1 + 1$th layer, C_l and C_{l+1} are the channel numbers of the l-th layer and the $1 + 1$th layer, respectively. $W^l \in R^{C_l \times C_{l+1}}$ is the training weight of the l-th layer. $\underline{A} = \tilde{D}^{-\frac{1}{2}} \tilde{A} \tilde{D}^{-\frac{1}{2}}$ is the normalized adjacency matrix, and $\tilde{A} = A + I$ is the adjacency matrix that increases the identity matrix to keep the senior features. $\tilde{D} \in R^{N \times N}$ is the number of nodes matrix. Equation (1) updates the features of each node according to the weighted average of the adjacent node features, and transforms the number of channels in each layer separately by the weights.

3.3 Spatio-Temporal Graph Convolution

The above introduction only considers the space in the graph convolutional network, which is only suitable for static images, but because we want to take a movie as input, we convert the movie into a spatio-temporal skeleton, as shown in Fig. 3. According to [6], we extend the spatial GCN to the spatio-temporal GCN by redefining the adjacent positions of nodes. This means that the i-th node-adjacent node range $N(v_i)$ not only spatially adjacent nodes, but also contains the same joints on consecutive T frames. To this end, we redefine the skeleton feature V to the original space plus the dimension of time T as $V \in R^{N \times C \times T}$.

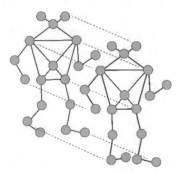

Fig. 3. Spatio-temporal graph for human skeleton

3.4 Skeleton Partition Strategy with Five Primary Kinetic Chains

We follow the spatial connection configuration of the skeleton in [14] and add other different connection methods. In the human skeleton, the human body is a structure composed of interconnected joints. In the process of exerting force, these joints that are responsible for transmitting force along the way are connected to each other, and they are connected together like a "chain" to form a kinetic chain, and the kinetic chain is the path of power output. Therefore, we refer to the Five Primary Kinetic Chains (5PKC) systems mentioned by Joseph in [30]. The functions by different joints are different, and they follow a certain logical distribution. Different kinetic chains exist as a single entity but also depend on each other to create a balanced and efficient movement, so we added it to the spatial connection configuration to extract more useful information from the human skeleton.

In a normal connection, the adjacent area of a node can be defined as Fig. 2. The neighbors of node v_1 are $\{v_2, v_3\}$. And our redefined connection configuration is shown in Fig. 4. The spatial connection configuration of Fig. 4(a) is to divide each node into centrifugal groups that are farther from the center of gravity in each joint itself, according to the distance of other nodes in the adjacent area of each node and our custom dynamic center of gravity. And the centripetal groups in the adjacent regions that are closer to the center of gravity. Then there are the five primary kinetic chains systems mentioned by Joseph in [30], in which only the other four except the posterior oblique sling (POS) are used because the skeleton does not distinguish between front and rear. The deep longitudinal sling (DLS) Fig. 4(b), the anterior oblique sling (AOS) Fig. 4(c), the lateral sling (LS) Fig. 4(d), the intrinsic (IS) Fig. 4(e), through these four connection configurations, the potential information of each joint in the skeleton can be analyzed more effectively.

And we also re-divide the original adjacent matrix A into four sub-matrices according to the above definition, and redefine Eq. (1) as Eq. (2):

$$V^{l+1} = \sigma\left(\sum\nolimits_{I=1}^{4} \underline{A_i} V^l W_i^l\right) \tag{2}$$

Here i is the index of the sub-matrix, $\underline{A_i}$ is the i-th sub-matrix separated from the adjacent matrix A, and W_i^l is the trainable weight of the i-th sub-matrix.

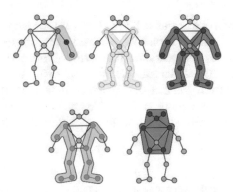

Fig. 4. Skeleton Partition Strategy

3.5 Network Architecture

Our network architecture is shown in Fig. 5. We use a spatio-temporal graph convolution network based on ST-GCN [14], which takes the human skeleton converted from the video as input V^0, and then passes V^0 through Multiple stacked ST-GCN modules process the spatio-temporal relationship of joints, and each ST-CGN is composed of GCN and TCN. Then use the Cross Entropy loss function to obtain the loss score, such as Eq. (3):

$$L = -\frac{1}{M} \sum_i \sum_{m=1}^{M} log\left(y_{im} log\left(P(X_i)\right)\right) \tag{3}$$

M is the number of categories, $y_{im} \in \{0, 1\}$ is the sign function, which is used to indicate whether the category of the input sample i is equal to the real category m, if they are equal, it is equal to 1, otherwise it is 0. $P(X_i)$ represents the probability that the input sample X_i belongs to the real sample Y_m. The loss score can be obtained by Eq. (3).

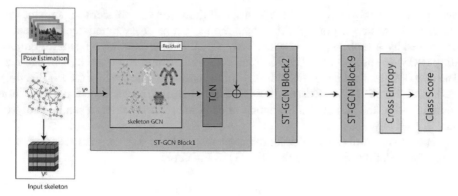

Fig. 5. Network architecture of ST-GCN with 5PKC

4   Experimental Results

The dataset used in our experiments is InfiniteRep [31], which is suitable for detecting fitness and extracting human skeletons. We compare our proposed model with ST-GCN and judge the accuracy for classification. All experiments are performed on the PyTorch, a deep learning package. The initial learning rate is set to 0.1 and running for 50 epochs, the learning rate is reduced by a factor of 10 every 10 epochs. Batch size is set to 64. Each input consists of a multi-frame skeleton. The model consists of 9 ST-GCN layers. The number of channels is 64, 64, 64, 128, 128, 128, 256, 256, 256, respectively.

4.1   Datasets

InfiniteRep [31] is a synthetic dataset for fitness and physiotherapy (PT), which mainly consists of performing everyday fitness activities and repeating them multiple times, which are then converted into 3D joint points. This data set has ten categories, arm raise, bird dog, curl, fly, leg raise, overhead press, push up, squat, bicycle crunch, and superman. A total of there are 1,000 action data, and each will repeat the action 5 to 10 times in 7 indoor scenes.

4.2   Testing of ST-GCN Blocks

We conduct an ablation study on the choice of the number of ST-GCN blocks, and we test the effect of one to nine blocks on the accuracy with ST-GCN plus dynamic center of gravity and five kinetic chains. The output channels of the 11 ST-GCN are 64, 64, 64, 128, 128, 128, 256, 256, 256, 512, and 512, respectively. We sequentially increase the number of ST-GCN, and the results are shown in Fig. 6. As the number of ST-GCN blocks increases, the classification accuracy gradually increases. The best accuracy of 99.5% is achieved when the number of ST-GCN block is 9, and then the accuracy starts to decrease as the number of stacks increases. It shows that stacking too many ST-GCN blocks may lead to overfitting and lower accuracy.

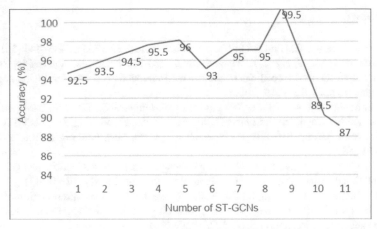

Fig. 6. Performance on ST-GCN blocks.

4.3 Partition Strategy Tests

In this section, we compare our model with the original ST-GCN model on InfiniteRep, and the results are shown in Table 1. The results show that adding a dynamic center of gravity to the skeleton can improve the accuracy compared to the original ST-GCN. Compared with ST-GCN, adding a dynamic center of gravity to the head can better capture the changes in motion. After adding five primary kinetic chains (5PKC) to the connection configuration, the accuracy has been improved, which proves that adding 5PKC can effectively transmit useful information between joints.

Table 1. Ablation comparisons of ST-GCN models

Models with different partition strategy	Accuracy
ST-GCN with 9 blocks	84.5%
ST-GCN with 9 blocks + Dynamic Center	93.5%
ST-GCN with 9 blocks + Dynamic Center + 5PKC	99.5%

5 Conclusions

Based on ST-GCN, this paper adds a dynamic center of gravity in the skeleton and five primary kinetic chains in the connection configuration. The proposed model can achieve abnormal action recognition, effectively process spatial relationships of the skeleton, and better capture the changes of motion. Experiments on the fitness action dataset show that the ST-GCN with the proposed partition strategy achieves classification accuracy of 99.5%, better than the original ST-GCN of 84.5%, proving the effectiveness of the proposed method.

References

1. LeCun, Y., Bengio, Y., Hinton, G.: Deep learning. Nature **521**(7553), 436–444 (2015)
2. Shahroudy, A, Liu, J, Ng, T-T, Wang, G.: Ntu rgb+ d: a large scale dataset for 3d human activity analysis. In: Proceedings of the IEEE Conference on Computer Vision and Pattern Recognition, pp. 1010–1019 (2016)
3. Liu, J., Shahroudy, A., Xu, D., Wang, G.: Spatio-temporal LSTM with trust gates for 3d human action recognition. In: Leibe, B., Matas, J., Sebe, N., Welling, M. (eds.) ECCV 2016. LNCS, vol. 9907, pp. 816–833. Springer, Cham (2016). https://doi.org/10.1007/978-3-319-46487-9_50
4. Li, C., Hou, Y., Wang, P., Li, W.: Joint distance maps based action recognition with convolutional neural networks. IEEE Signal Process. Lett. **24**, 624–628 (2017)
5. Ke, Q., Bennamoun, M., An, S., Sohel, F., Boussaid, F.: A new representation of skeleton sequences for 3D action recognition. In: Proceedings of the IEEE Conference on Computer Vision and Pattern Recognition, pp. 3288–3297 (2017)
6. Li, C., Zhong, Q., Xie, D., Pu, S.: Co-occurrence feature learning from skeleton data for action recognition and detection with hierarchical aggregation. In: IJCAI, 2018, pp. 1–8 (2018)
7. Liu, M., Liu, H., Chen, C.: Enhanced skeleton visualization for view invariant human action recognition. Pattern Recogn. **68**, 346–362 (2017)
8. Song, S., Lan, C., Xing, J., Zeng, W., Liu, J.: An end-to-end spatio-temporal attention model for human action recognition from skeleton data. In: Proceedings of the AAAI conference on artificial intelligence, vol. 1 (2017)
9. Zhang, P., Lan, C., Xing, J., Zeng, W., Xue, J., Zheng, N.: View adaptive recurrent neural networks for high performance human action recognition from skeleton data. In: Proceedings of the IEEE International Conference on Computer Vision, pp. 2117–2126 (2017)
10. Kim, T.S., Reiter, A.: Interpretable 3D human action analysis with temporal convolutional networks. In: 2017 IEEE Conference on Computer Vision and Pattern Recognition Workshops (CVPRW), pp. 1623–1631 (2017)
11. Li, M., Chen, S., Chen, X., Zhang, Y., Wang, Y., Tian, Q.: Actional-structural graph convolutional networks for skeleton-based action recognition. In: CVPR (2019)
12. Liu, K., Gao, L., Khan, N.M., Qi, L., Guan, L.: A Vertex-edge graph convolutional network for skeleton-based action recognition. In: 2020 IEEE International Symposium on Circuits and Systems (ISCAS), pp. 1–5 (2020)
13. Kipf, T.N., Welling, M.: Semi-supervised classification with graph convolutional networks. arXiv preprint arXiv:1609.02907 (2016)
14. Yan, S., Xiong, Y., Lin, D.: Spatial temporal graph convolutional networks for skeleton-based action recognition. In: Thirty-Second AAAI Conference on Artificial Intelligence (2018)
15. Defferrard, M., Bresson, X., Vandergheynst, P.: Convolutional neural networks on graphs with fast localized spectral filtering (2016)
16. Li, C., Cui, Z., Zheng, W., Xu, C., Yang, J.: Spatio-temporal graph convolution for skeleton based action recognition. In: Proceedings of AAAI, pp. 3482–3489 (2018)
17. Li, B., Li, X., Zhang, Z., Wu, F.: Spatio-temporal graph routing for skeleton-based action recognition. In: Proceedings of the AAAI Conference on Artificial Intelligence, pp. 8561–8568 (2019)
18. Gao, X., Hu, W., Tang, J., Liu, J., Guo, Z.: Optimized skeleton-based action recognition via sparsified graph regression. In: Proceedings of the 27th ACM International Conference on Multimedia, pp. 601–610 (2019)
19. Shi, L., Zhang, Y., Cheng, J., Lu, H.: Two-stream adaptive graph convolutional net-works for skeleton-based action recognition. In: CVPR (2018)

20. Monti, F., Boscaini, D., Masci, J., Rodolà, E., Svoboda, J., Bronstein, M.M.: Geometric deep learning on graphs and manifolds using mixture model CNNs. In: CVPR (2016)
21. Peng, W., Hong, X., Zhao, G.: Tripool: Graph triplet pooling for 3D skeleton-based action recognition. Pattern Recogn. **115** (2021)
22. Peng, W., Shi, J., Zhao, G.: Spatial temporal graph deconvolutional network for skeleton-based human action recognition. IEEE Signal Process. Lett. pp. 244–248 (2021)
23. Henaff, M., Bruna, J., LeCun, Y.: Deep Convolutional networks on graph-structured data (2015)
24. Duvenaud, D., et al.: Convolutional networks on graphs for learning molecular fingerprints. In: Advances in Neural Information Processing Systems 28 (NIPS 2015) (2015)
25. Li, Y., Tarlow, D., Brockschmidt, M., Zemel, R.: Gated graph sequence neural net-works (2015)
26. Bruna, J., Zaremba, W., Szlam, A., LeCun, Y.: Spectral networks and locally connected networks on graphs. In: International Conference on Learning Representations (ICLR2014), CBLS, April 2014 (2013)
27. Zoph, B., Le, Q.V.: Neural architecture search with reinforcement learning (2016)
28. Veličković, P., Cucurull, G., Casanova, A., Romero, A., Liò, P., Bengio, Y.: Graph attention networks. (2017)
29. Vaswani, A., et al.: attention is all you need (2017)
30. Schwartz, J.: The 5 primary kinetic chains (5PKC). https://dna-assessment.com/the-master-template/ (2016). Accessed 20 Aug 2016
31. Weitz, A., Colucci, L., Primas, S., Bent, B.: InfiniteForm: a synthetic, minimal bias dataset for fitness applications (2021)

A Skeletal Sequence-Based Method for Assessing Motor Coordination in Children

Zitong Pei[1], Wenai Song[1], Nanbing Zhao[1], Zhiyu Chen[1], Wenbo Cui[1], Yi Lei[2], Yanjie Chen[3(✉)], and Qing Wang[4,5(✉)]

[1] School of Software Engineering, North University of China, Taiyuan 038507, China
[2] Faculty of Information Technology, Beijing University of Technology, Beijing 100124, China
[3] Department of Children's Health Care Centre, Beijing Children's Hospital, Capital Medical University, Beijing 100045, China
chenyanjie@bch.com
[4] Department of Automation, Tsinghua University, Beijing 100084, China
qing.wang@tsinghua.edu.cn
[5] Pharmacovigilance Research Center for Information Technology and Data Science, Cross-Strait Tsinghua Research Institute, Xiamen 361015, China

Abstract. Children's motor coordination is an important component of physical fitness test for young children. The development of children's motor coordination occurs throughout children's motor development, and it is not only limited by the maturity of children's neurodevelopment, but also plays an important role in promoting children's neurodevelopment. This study proposes an automatic assessment method based on deep learning to improve assessment efficiency and reduce costs. The method combines human posture estimation, similarity calculation and time series feature extraction for the assessment of children's movements. The results showed that the accuracy rate and redundancy rate of the fine action coin toss keyframes finding algorithm are 89.8% and 7.5%; the accuracy rate of the dynamic action standing long jump keyframes finding algorithm is 74.3%; the accuracy rate, precision rate and recall rate of the fine action coin toss assessment algorithm are 74.4%, 72.5% and 90.0%; the accuracy rate, precision rate and recall rate of the static action single-leg balance assessment algorithm are 87.1%, 73.5% and 87.1%; the accuracy rate, precision rate and recall rate of the dynamic action standing long jump assessment algorithm are 71.6%, 73.5% and 71.4%; the results of the three actions generally matched the expert assessment results. The method provides a good auxiliary tool for determining the motor development level of young children, and provides a good technical support for achieving the goal of "promoting the early comprehensive development of young children through actions".

Keywords: Human Posture Estimation · Similarity · Action Assessment · Time Series Features

1 Introduction

The development of children's movement is very important in early life [1]. The development of children's movements is not only the achievement of milestone projects, but also the identification of development risks behind children through movement development, and the earlier the risk is identified, the higher the value is. Many children with developmental disorders have certain problems in their motor development, such as children with autism will have somatization movements, children with developmental retardation will have body use disorders, children with developmental coordination disorders will have clumsy movements, and so on [2, 3]. Developmental coordination disorder is mainly characterized by clumsy movement and poor physical coordination. At present, the incidence is about 6%~8% [4].

At present, MABC-2 [5] test is used to evaluate children's motor coordination. The test method is an international test method for children's sports coordination development level, which is suitable for children aged 3~6 years. At present, the assessment of the development of children's movement in China mainly adopts the form of scale, mainly based on standardized guidelines, toolbox, etc. The assessment standard is interpreted manually, which may produce certain deviation. There will be certain differences in the judgment of children by different personnel, different institutions and different environments. The assessment of intervention effect will also have a certain impact.

The automatic assessment of the development of children's movement using artificial intelligence is mainly based on key technologies such as human target detection, human keypoints recognition, motion capture, pose estimation and fine statistical measurement. These technologies are currently mostly used in film entertainment and sports video analysis. According to the report released by Fior Markets in 2018, the global motion capture market is expected to grow from $163.2 million in 2018 to $261.7 million in 2026, the CAGR is 8.13% in the forecast period 2019–2026. With the development of global health care and sports industry and the improvement of people's attention to sports health, the demand and scale of motion capture related industries are growing driven by market applications in many fields such as health and sports, and the market share is also increasing [6].

This study attempts to implement the research results in the related fields of human pose estimation and movement assessment into the diagnosis system of children's developmental dyskinesia. If the system can be applied to the actual auxiliary examination, it can improve the diagnosis and treatment ability of areas with insufficient medical ability, timely intervene and treat children with the disease, so as not to affect their normal development. Based on the existing manual design features and depth features, this paper explores the assessment method to optimize motion assessment, and puts forward a motion assessment method for developmental coordination disorders.

2 Methods

The motion assessment method uses the following methods: keypoints detection, keyframes detection and similarity calculation [7]. The process is based on cropping and formatting the video, keypoints detection and output of the skeletal sequences for the test video, data processing and extraction of the keyframes, similarity calculation between the skeletal sequences of the test action keyframes and the skeletal sequences of the standard action keyframes, and finally a composite score where the scored parts of each action are weighted and summed up using different weights.

2.1 Keypoints Detection

In this paper, we use MediaPipe algorithm to detect keypoints. In the cross platform artificial intelligence work pipeline framework MediaPipe [8], Google launched the body pose attention function BlazePose with the latest technology, which can instantly and accurately locate the keypoints of body pose on the mobile phone. Now the standard model for focusing on pose is based on the COCO topology, but it mainly depends on the powerful computing power of desktop computers. The human pose sensing method BlazePose released by Google now uses machine learning to infer 33 2D feature points of human body, as shown in Fig. 1. In addition to being more accurate than the coco topology, BlazePose can use the CPU of the mobile device to make real-time speculation. In addition to pose, BlazePose can also pay attention to facial expression and hand pose at the same time.

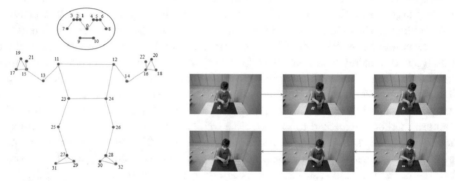

Fig. 1. Topological graph.   **Fig. 2.** Coin action keyframes.

Google showed the application scenarios of BlazePose, including squats and push up. The application can automatically count user data, verify technology and train quality. Therefore, the human pose estimation based on MediaPipe algorithm proposed in this paper can meet the basic needs of children's pose estimation.

2.2 Assessment Methods

After obtaining the preprocessed skeleton sequence, we can compare it with a preset template motion skeleton sequence, obtain the assessment results of motion quality through the comparison with the standard motion, then calculate the similarity of keypoints between the corresponding coordinate points to measure the matching error between the two skeleton sequences, and then score the quality assessment score.

OKS (Object Keypoints Similarity). In the human body keypoints assessment task, the quality of the keypoints obtained by the network is not calculated only by simple Euclidean distance, but by adding a certain scale to calculate the similarity between the two points. This indicator is mainly used in multi-person pose estimation tasks. However, the original calculation method does not take the importance of each keypoints into account. We introduce a penalty factor σ. Different penalty factors are set according to the importance of the keypoints. The greater the penalty factor, the greater the impact caused by the deviation of the keypoints.

$$OKS_p = \frac{\sum_i e^{\frac{-d_{p\sigma}^2}{2S_p^2 k_i^2}} \delta(v_{p^i} > 0)}{\sum_i \delta(v_{p^i} > 0)} \tag{1}$$

In the above formula: The p represents the ID of a person in the growth truth. The p^i indicates the ID of the keypoints. δ is 1 at $v_{p^i} > 0$ and 0 otherwise, meaning that only visible keypoints are calculated. The S_p represents the square root of the area occupied by this person, which is calculated according to the box of people in the ground truth. The k_i represents the normalization factor of the i-th bone point, which is obtained by calculating the standard deviation of all ground truth in the existing data set, reflecting the impact of the current bone point on the whole. The larger the value, the worse the standard effect of this point in China in the whole data set; The smaller the value, the better the annotation effect of this point in the whole data set. The d_p represents the Euclidean distance between the detection keypoints and the standard keypoints.

According to the clinician's instructions, each subject's movement is judged as a composite, so that each movement is weighted and summed using different weights, and when the weighted sum of the scores is greater than 0.5, the movement is considered normal. The composite score is calculated by means of Eq. (2).

$$S = \sum_{i=1}^{n} We_i * S_i \tag{2}$$

Of which, n is the number of scored parts, We_i is the weight of the i-th score, and S_i is the score of the ith score.

Fine Movements. For the fine movements, a coin toss (6 tosses with the right hand as the dominant hand) is selected as a case study. The coin action is subdivided into four parts: scoring the similarity of keyframes, scoring the duration, scoring the standard deviation of the right hand keypoints coordinates of keyframes and scoring the complexity of the right hand keypoints coordinates waveform graph time series. The keyframes are extracted by setting the prominence of the peaks, the minimum height of the peaks, and the minimum horizontal distance between adjacent peaks of the right hand keypoints coordinate time series waveform graph [9]. The keyframes are shown in Fig. 2, and Fig. 3 shows the number of keyframes for all data. According to Fig. 4, we can observe that the peak wave tip is for the action of throwing in the coin. The similarity scoring of keyframes is based on the similarity calculation between the keypoints coordinates of the keyframes of the test action and the keyframes of the standard action according to Eq. (1). According to the specific situation of collecting the coin throwing action, avoiding the inaccuracy of keypoints identification caused by the obscuration of the table and the judging criteria of the coin action, we adopt the upper itself keypoints for the similarity calculation. Time duration scoring is based on the Movement Assessment Battery for Children-Second Edition (MABC-2). Time series complexity is scored by calculating the degree of confusion between the front and back parts of the keypoints coordinate waveform. The final assessment is based on Eq. (2) and the scoring results for all data are shown in Fig. 5.

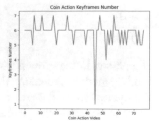

Fig. 3. The number of keyframes for coin action.

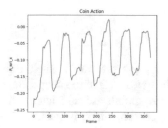

Fig. 4. X coordinate of the right hand side of the coin action.

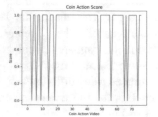

Fig. 5. Coin action scoring results.

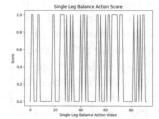

Fig. 6. Single leg balance action scoring results.

Static movements. For the static movements, a single-leg balance movement is chosen as a case study. The single-leg balancing movement is subdivided into three parts: the similarity score of keyframes, the duration score and the standard deviation of keypoints coordinates. According to the assessment rules, every frame in the standard movement video is immovable and identical, so we choose one frame from the standard movement video as the standard frame. The scoring of keyframes is a similarity calculation between the keypoints sequence of each frame of the test action video and the keypoints sequence of the standard frame based on Eq. (1). Duration scoring is based on the Movement Assessment Battery for Children-Second Edition (MABC-2). The standard deviation score is calculated by calculating the standard deviation of the coordinates of each keypoints and the resulting standard deviation data is used to indicate the stability of the single-leg balance movement. The final assessment is based on Eq. (2) and the scoring results for all data are shown in Fig. 6.

Dynamic Movements. For the dynamic movements, the standing long jump is chosen as a case study. The vertical jump is refined into five keyframes for the similarity calculation to be assessed. Based on the clinician's guidance, we select five keyframes of the standing long jump movement by combining the angle of the joint, the coordinates of the keypoints and the positions of the keyframes that have been derived, as shown in Fig. 7. The five keyframes of the test movement and the five keyframes of the standard movement are calculated according to Eq. (1). The final assessment is based on Eq. (2) and the scoring results for all data are shown in Fig. 8.

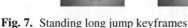

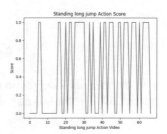

Fig. 7. Standing long jump keyframes

Fig. 8. Standing long jump action scoring results

3 Experimental Results and Discussion

The medical action video data used in this paper is not a public dataset and is provided by the collaborators - the Digital Medical Health Engineering Research Centre of Tsinghua University Institute of Information Technology and Beijing Children's Hospital. Medical data is sensitive and is used for scientific research to strictly protect patient privacy. Once the collection is complete, the clinician will do a diagnostic assessment of the subjects, complete the labelling of the data to separate the normal samples from the abnormal samples, and label the movement of the subject samples as positive abnormal as well.

We designed experiments to compare the effectiveness of OpenPose and MediaPipe in detecting the three movements of coin, single leg balance and standing long jump, as well as the effectiveness of the three movement assessment methods mentioned above [10–12].

3.1 Keypoints Detection

In order to try to avoid the problem of loss of detection values due to overlap of some keypoints of the human body in 2D images, the data acquisition criteria are developed with requirements on the angle of the shot to ensure the integrity of the action but also to avoid overlap as much as possible and to minimize the loss of some joint points caused by overlap, which could affect subsequent analysis and processing. We used OpenPose and MediaPipe to estimate human posture for the three movements of coin toss, single-leg balance and standing long jump respectively, as shown in Fig. 9. And the recognition rate calculation of each action by OpenPose and MediaPipe is derived by Eq. (3).

   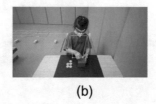

(a)   (b)

Fig. 9. Keypoints detection chart. (a)OpenPose Detection(b) MediaPipe Detection

$$RecognitionRatio = \frac{X}{X + Y} \qquad (3)$$

In the formula RecallRatio is the recognition degree, X is the number of keypoints detected and Y is the number of keypoints undetected.

Table 1. OpenPose and MediaPipe recognition effects.

Recognition algorithms	Coin toss		Single leg balance		Standing long jump		Average	
	Velocity(fms)	Recognition ratio (%)	Velocity(fms)	Recognition ratio (%)	Velocity(fms)	Recognition ratio (%)	Velocity(fms)	Recognition ratio (%)
OpenPose	12	76.8	12	98.5	12	74.7	12	83.3
MediaPipe	8	99.9	7	100	7	98.8	7.3	99.5

Table 1 shows that the recognition rate of MediaPipe is higher than that of OpenPose. The MediaPipe model is slow due to its top-down detection method, while the Open-Pose model is fast due to its bottom-up detection method. OpenPose and MediaPipe are comparable in terms of single-person recognition accuracy, MediaPipe outperforms OpenPose overall due to the high level of timeliness and accuracy required by the human motion recognition algorithm. The MediaPipe model is chosen for skeletal keypoints detection based on relevant performance comparisons and usage context.

3.2 Action Assessment

A control group is made based on the clinician's artificially selected keyframes images and assessment results. The data collected for the three movements of coin toss, single leg balance and standing long jump are processed separately using specific keyframes extraction algorithms and assessment algorithms, and the results obtained are analyzed in combination with the control group. The redundancy rate, accuracy rate, precision rate and check-all rate are used as assessment metrics to test the keyframes and assess the method performance.

$$Fa = \frac{n}{m} \qquad (4)$$

$$Fr = \frac{M - n}{M} \qquad (5)$$

The formula Fa is the accuracy, n is the accuracy value and m is the number of frames extracted by the clinician; Fr is the redundancy and M is the number of frames extracted.

$$AccuracyRatio = \frac{TP + TN}{TP + FP + TN + FN} \qquad (6)$$

Table 2. Keyframes extraction results for each action.

Type of action	Coin toss	Standing long jump
Extracting the correct number of frames	418	299
Extracting frames	452	402
Specified number of frames to be extracted	462	402
Accuracy (%)	89.8	74.3
Redundancy (%)	7.5	–

$$PrecisionRatio = \frac{TP}{TP + FP} \tag{7}$$

$$RecallRatio = \frac{TP}{TP + FN} \tag{8}$$

In the formula, AccuracyRatio is the accuracy ratio, PrecisionRatio is the precision ratio and RecallRatio is the recall ratio; TP is a positive sample with a positive prediction result; TN is a negative sample with a negative prediction result; FP is a positive sample with a negative prediction result; and FN is a negative sample with a positive prediction result.

Table 3. Analysis of action assessment results.

Type of action	Coin	Single leg balance	Standing long jump
Recall ratio (%)	90.0	87.1	71.4
Precision ratio (%)	72.5	73.5	73.5
Accuracy ratio (%)	74.4	87.1	71.6

According to Table 2 respectively, the keyframes extraction accuracy for the standing long jump movement is 74.3%, and the keyframes extraction accuracy for the coin movement is 89.8%, with a redundancy of 7.5%, which achieved a relatively good keyframes extraction effect. According to Table 3, the recall ratio, precision ratio and accuracy ratio for the coin action assessment algorithm are 90%, 72.5% and 74.4%; the recall ratio, precision ratio and accuracy ratio for the single leg balance action assessment algorithm are 87.1%, 73.5% and 87.1%; the recall ratio, precision ratio and accuracy ratio for the standing long jump action assessment algorithm are 71.4%, 73.5% and 71.6%. The assessment methods for all three movements were generally consistent with the physician's assessment.

4 Conclusion

In this paper, a combination of human pose estimation, similarity calculation and time series feature extraction methods is used to achieve movement assessment of children.

Different methods are used for data pre-processing, keyframes extraction and movement assessment for the coin, single leg balance and standing long jump movements, and the results of keyframes selection and movement assessment are approximately the same as our manual judgement results. At a later stage, when more and more data are available, deep learning can be used to train models for relevant movement assessment, which can be more accurate and convenient and will make a great contribution to the development of movement assessment for children.

Acknowledgements. This work was supported by the National Key R&D Program of China (2020YFC2006702, 2020YFC2005503).

References

1. Chen, Y.J., Wang, H., Liang, A.M.: Study on the relationship between children's social skills and developmental coordination disorder. Chin. J. Reprod. Health **32**(04), 311–314 (2021)
2. Dong, Y.G.: Study on the relationship between motor development and physical health level of children in grades 1–3 in. Capital Institute of Physical Education, Beijing (2021)
3. Chen, Y.J., Wang, H., Liang, A.M.: Study on the relationship between children's physical evaluation and coordination disorder evaluation index. Chin. J. Child Health Care **29**(05), 542–544 + 549 (2021)
4. Wu, D., Tang, J.L.: Diagnosis and treatment of developmental motor coordination disorder. Chin. J. Rehabil. Med. **35**, 513–516 (2020)
5. Hua, J., Wu, Z.C., Meng, W., et al.: Preliminary analysis on the application validity of child developmental coordination disorder assessment tool in China. Chin. J. Child Health Care 28–31 (2010)
6. Zhu, Z.P.: Analysis on Age Characteristics of Standing and Jumping Learning of Preschool Children. Overseas Chinese University (2020)
7. Wang, J., Qiu, K., Peng, H., et al.: AI coach: deep human pose estimation and analysis for personalized athletic training assistance. In: Proceedings of the 27th ACM International Conference on Multimedia, pp. 374–382 (2019)
8. Bazarevsky, V., Grishchenko, I., Raveendran, K., et al.: BlazePose: on-device real-time body pose tracking. arXiv preprint arXiv:2006. 10204 (2020)
9. Ziyi, W.: Research on Time Series Classification Methods Based on Feature Extraction. Nanjing University, Nanjing (2019)
10. Guan, C.: Realtime multi-person 2d pose estimation using shufflenet. In: 2019 14th International Conference on Computer Science & Education (ICCSE), pp. 17–21. IEEE (2019)
11. Chen, K.: Sitting posture recognition based on OpenPose. In: IOP Conference Series: Materials Science and Engineering, vol. 677, no. 3, p. 032057 (2019)
12. Okugawa, Y., Kubo, M., Sato, H., et al.: Evaluation for the synchronization of the parade with OpenPose. J. Robot. Networking Artif. Life **6**(3), 162 (2019)

An Artificial Intelligence Camera System to Check Worker Personal Protective Equipment Before Entering Risk Areas

Watthanaphong Muanme[✉], Sawat Pararach, and Phisan Kaeprapha

Thammasat School of Engineering, Thammasat University, Rangsit Campus Khlong Luang, Bangkok 12120, Pathum Thani, Thailand
Wattanaphong@outlook.co.th

Abstract. Factories overlooking the need to check personal protective equipment (PPE) before workers enter risk areas is a factor contributing to injuries. To assist in resolving such issues, we are investigating and designing a safety management system (SMS) using artificial intelligence (AI) camera technology to detect PPE devices with a You Only Look Once (Yolo) deep learning algorithm. The technology checks and displays real-time operator PPE equipment inspections. The AI camera system tabulates and processes data from the extant database for accurate detection. The camera can detect and notify about the following PPE devices: reflective clothing, helmets, safety goggles, and safety gloves. A warning will be displayed on the screen when PPE devices are not worn to check that each worker is wearing the correct PPE devices before entering risk areas. The model AI camera system can also operate in conjunction with automatic doors to prevent entry to a risk area and has been designed for use in industrial plant or job site safety management. Additionally, tests have shown that using this equipment increases the incidence of wearing PPE on entry to risk areas.

Keywords: Artificial intelligence (AI) · Safety management (SMS) · System design

1 Introduction

Researchers studied the root causes of problems arising from the neglect of joint safety checks for workers who do not wear PPE [1]. The designed solution has been combined with modern industrial technology to develop a safety management system to make workers aware of the importance of the dangers that will arise during work that do not wear protective equipment. Some factories have neglected to check workers readiness checks, such as checking equipment, workers safety, and availability. Currently, some factories have a poor safety culture and underestimate the importance of less safe work practices, such as unsafe work due to employees neglecting to wear PPE and a lack of PPE checks before entering the facility. Working with risks that can cause accidents every time, such as an accident from hand cut workpieces from not wearing safety gloves, fire

J. C. Hung et al. (Eds.): IC 2023, LNEE 1045, pp. 747–755, 2023.
https://doi.org/10.1007/978-981-99-2287-1_102

splashes in the eyes from not wearing safety glasses [2], or pieces of workpieces on high ground falling on the head when not wearing safety helmets, etc.

The researcher aim is to design a working system for detecting each type of device and to classify each device as follows: reflective vests, helmets, safety glasses, safety gloves. This detection uses a camera to detect PPE devices in real time [3, 4], checking workers before entering work, This research uses safety management principles and AI technology systems, and Image Processing pro-working with The Yolov3 algorithms to help improve the process [5] of PPE checks for industrial facilities [5] or on-site to help reduce the problem of neglect of personal protective equipment inspections and to reduce accidents caused by not wearing protective equipment [7]. This work builds on and improves upon works that have gone before, improvements have come from technological and hardware improvements allowing more accurate detection of more objects.

2   Related Works

2.1   Computer Vision and Machine Learning

Image processing and computer vision [8] can be applied management instead of humans without bias. It depends on the training information they receive. They can provide accurate and fast approval of entry into the safe zone. In the future, the computer system may extend to the reporting of violation. Image Processing [9] or Computer Vision is the processing of a learning algorithm and verifying the results to improve the outcome such that for a given task it can become better at making decisions and analyzing results. Images are processed using a set of algorithms known as YOLO that clearly defines the area to be searched for objects of potential interest to be classified. The system is trained on a set of images to classify target images [10]. In this system the targets are items of PPE: - Safety glasses, Hard hat, Reflective vest, and Gloves. The greater the number of images of PPE the more accurate the system should become.

3   Proposed System

3.1   The Process of Problem Analysis

Researcher uses the concept of finding the cause of safe work using the fishbone diagram analysis technique, which will be used to analyze [11] the true cause of the problem with the main idea of the heading of fish and any factors that contribute to the important issues that need to be determined to determine the root cause. (See Fig. 1).

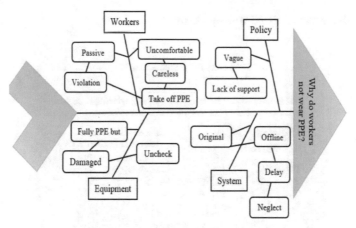

Fig. 1. Fishbone diagram analysis.

3.2 System of PPE Detection Algorithms

From Object Detection with Image Processing and learning, PPE Detection by Yolov3, the system uses a database of images of selected PPE to test images taken from the camera against. The more images and conditions that the systems test the greater the accuracy, according to the learning principles. An area of future work may be to attempt to improve training by using human assist training (Figs. 2 and 3).

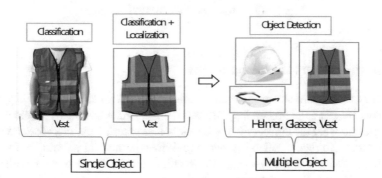

Fig. 2. Algorithms for object detection.

3.3   PPE Detections Algorithms by Yolov3

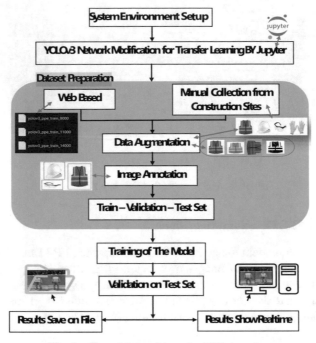

Fig. 3. Chart AI algorithms for PPE detection.

4   System Design and Experiment

In this study, the researchers propose an AI camera system for checking PPE by combining the concept of image processing with safety management. to solve problems arising from neglect to wear the correct PPE or check before entering the risk work area. Using a digital camera to record video images in real time on the central computer screen, to show whether the workers is wearing PPE or not, rough digital processing with a theory called "Object Detection," the system can detect PPE devices and identify the type of safety device.

4.1   The Design of Systems Work

The work will include a central computer for analyzing the processed data and sending commands through the camera, and the AI camera system will work continuously in real time, with the door opening and closing in response to the specified program's commands. Before going to work, for workers to check by categorizing each type of equipment as follows: reflective vest, safety helmet, safety goggles, and safety gloves, The researcher has designed a system so that PPE equipment detection can be divided into 2 cases as follows this below.

- Case 1: The worker is wearing all required PPE, entry is allowed.
  Case 2: Some or all the of the PPE is missing or not worn correctly, entry is not allowed (Table 1).

Table 1. In case of detection PPE

Object	Vest	Helmet	Glasses	Gloves
Present	Vest	Helmet	Glasses	Gloves
Not Present	No-Vest	No-Helmet	No-Glasses	No-Gloves

The system design of the camera-based detection system to check PPE, to make it easier to understand. (see in Fig. 4, 5).

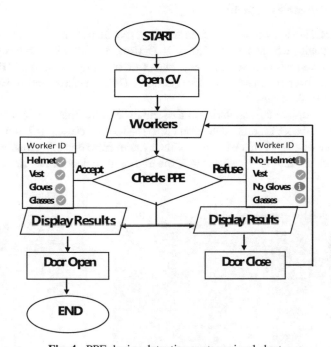

Fig. 4. PPE device detection system visual chart.

The researcher uses a central computer to process the images and execute the commands. The program was designed to carry out the analysis, process and operate the cameras and doors. It is essential to use a central computer that is detached from the normal computer because it must have a relatively high processing power specification for real time performance to be achieved. (As in Fig. 6, 7). The camera receives commands from the central computer to record detections in real time, detects PPE devices and transmits the resulting data values to the display. The door either opens or closes

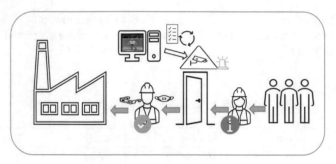

Fig. 5. PPE device detection system model.

according to the main program command. It opens if all required PPE detected otherwise remains close and wait for the next command from the central computer.

4.2 PPE Detection System Trials

A test of the PPE detection system when wearing five devices is (as shown in Fig. 3). A split-level test using a database with 5,000, 10,000 and 15,000 image training algorithms (Fig. 4) is used in real-time worker image detection training, compared to datasets with each level of training for Intersection Over Union (IOU) measurements of PPE detection performance (Table 2 and 3).

The researchers set the standard value of the probability that each device object can be detected to have an accurate detection (IOU) value of between 0.7 and 1. If there is an (IOU) value below 0.7 this will prevent the system from detecting that type of device likewise, the closer the (IOU) value is to 1, the higher the detection accuracy. (See Fig. 6, 7).

- Detection experiment when not wearing PPE

Fig. 6. The result is detected when not wearing PPE.

From training 5,000, 10,000 and 15,000 images in order for the camera system to learn and remember to detect PPE, there will be differences in the IOU values as a measure of detection assessment. In our testing, 15,000 image trained detection AI camera systems yielded a better detection IOU than 5,000 and 10,000image training, training with more than 10,000 or more images should be sufficient for this camera system to detect PPE accurately.

Table 2. Image training for PPE detection an accuracy (IOU) when not wearing PPE.

Image train	Accuracy Intersetion Over Union (IOU)					
	No_Helmet	No_Glasses	No_Vest	No_Gloves(L)	No_Gloves(R)	AVE.
5,000	0.8453	0.8611	0.7454	0	0.7301	0.63638
10,000	0.8868	0.8911	0.9808	0.7188	0.8208	0.85966
15,000	0.9874	0.9825	0.878	0.861	0.9558	0.93294

- Detection experiment when wearing PPE

Fig. 7. The result is detected when wearing PPE.

Table 3. Image training for PPE detection an accuracy (IOU) when wearing PPE.

Image train	Accuracy Intersetion Over Union (IOU)					
	Helmet	Glasses	Vest	Gloves(L)	Gloves(R)	AVE.
5,000	0.9973	0.7098	0.9982	0	0.888	0.71866
10,000	0.9976	0.8083	0.9473	0.8516	0.9552	0.912
15,000	0.9991	0.7868	0.9918	0.9967	0.9808	0.95104

The experiment confirmed that the AI camera system was able to detect PPE devices and was able to identify each type of personal protective equipment (PPE). Each area will have different protection tools, so the AI camera system can be adjusted to the work of that area.

5   Conclusion

Constraints on this system; currently the system will accurately detect PPE if the target is standing in a fixed pose at less than 150 cm from the camera.

At a Test site with 78 employees the results obtained with the system were as follows:

This research developed an AI camera system for monitoring personal safety. It distinguishes the security of each type of device in real time, eliminating problems caused by a lack of monitoring. Inspect each area within the factory or on the job site

Table 4. Test site PPE results

	No PPE Worn	%	Partial PPE worn	%	Full PPE Worn	%
Prior to installation	22	28.21	30	38.46	26	33.33
After Installation	0	0	6	7.69	72	92.31

before entering work to speed up batch inspections. The Table 4 shows that the system can detect and differentiate each PPE, being worn by employees in real time detection through the camera system.

6 Further Improvement

The real time training assisted by human operator to provide potential improvement in either accuracy or speed of detection.

Possibility of improvements in either accuracy or speed of detection can be obtained by having the candidate at variable distances and in fixed or free poses in front of the camera, this may limit the use of the system in some modes.

Peripheral connection of the system to door controls, ID card reader or Facial recognition to be able to report workers who exceed a pre-determined number of exceptions.

Trial the system for use with workers in high-risk areas who may be at risk of removing items of safety equipment.

References

1. Balakreshnan, B., Richards, G., Nanda, G., Mao, H., Athinarayanan, R., Zaccaria, J.: PPE compliance detection usingartificial intelligence in learning factories. In: 10th Conference on Learning Factories, CLF2020, pp. 2–5 (2020)
2. Zhou, L., Zhao, H., Leng, J.: MTCNet: multi-task collaboration network for rotation-invariance face detection. Pattern Recognit. 3–10 (2021)
3. Yang, X., Yu, Y., Shirowzhan, S., Sepasgozar, S., Li, H.: Automated PPE-tool pair check system for construction safety using smart IoT. J. Build. Eng. **32**, 2–10 (2020)
4. Mollah, A.F., Majumder, N., Basu, S., Nasipuri, M.: Design of an optical character recognition system for camera based handheld devices. IJCSI Int. J. Comput. Sci. Issues **8** (2011)
5. Gonzalez, R.C., Woods, R.E.: A textbook on "Digital Image Processing", 2nd edn, Publications of Pearson, New York (2002)
6. Dzyubachyk, O., Niessen, W., Meijering, E.: Advanced level – set based multiple -cell segmentation and tracking in time - lapse fluorescence microscopy images. In: Olivo- Marin, J.C., Bloch, I., Laine, A. (eds.) IEEE International Symposium on Biomedical Imaging: From Nano to Macro, Piscataway, NJ, pp. 185–188. IEEE (2008)
7. Tran, Q.-H., Le, T.-L., Hoang, S-H.: A fully automated vision-based system for real-time personal protective detection and monitoring. Research Gate 2–6 (2020)

8. Torres, L.: Is there any hope for face recognition? In: Proceedings of the 5th International Workshop on Image Analysis for Multimedia Interactive Services (WIAMIS 2004), Lisboa, Portugal (2004)
9. Delhi, V.S.K., Sankarlal, R., Thomas, A.: Detection of personal protective equipment (PPE) compliance on construction site using computer vision based deep learning techniques. Front. J. **6**, 1–8 (2020)
10. Srinivasan, Dr., G.N., Shobha, G.: Segmentation techniques for target recognition. Int. J. Comput. Commun. **3**(1), 313–333 (2007)
11. Kuruvilla, J., Sankar, A., Sukumaran, D.: A study on image analysis of Myristica fragrans for automatic harvesting. IOSR J. Comput. Eng. (IOSR-JCE) ISSN: 2278–0661.p-ISSN: 2278–8727PP50–55

The International Workshop on Big-Data, IoT, Cloud Computing Technologies and Applications (BICTA2023)

A Comparative Study of Female Image in "Eouyadam" and "Yojaejii"

She Shaoshuo[1], Young-Hoon An[1(✉)], and Hwa-Young Jeong[2]

[1] Department of Korean Language and Literature, Kyung Hee University, 26, Kyungheedae-Ro, Dongdaemun-Gu, Seoul 02447, Republic of Korea
sheshaoshuo@naver.com, yhnahn@khu.ac.kr
[2] Humanitas College, Kyung Hee University, 26, Kyungheedae-Ro, Dongdaemun-Gu, Seoul 02447, Republic of Korea
hyjeong@khu.ac.kr

Abstract. In the new era, women are becoming more colorful in society and at home as women pursue equal rights and status with men and pay more and more attention to unique styles and attractions, a sign of self-awareness that has taken a step further in women's social history development. The report shows the diversity and progress of women's images in the 17th and 18th centuries, focusing on the female stories of Korean writer Yoo Mong-in's "Eouyadam" and Chinese writer Pu Song-ling's "Yojaejii".

Keywords: Eouyadam · Yojaejii · the female image · comparative study

1 Introduction

In the 21st century, many people value mental needs and human rights have been the focus in Korea. In particular, women with low social status are trying to have proper rights and equal social status with men because they have been oppressed by feudalism since ancient times. It can be felt that the female aspects of modern society are becoming rich and colorful. Not only Korea, but also China. This may show progress in the times. However, such progress did not appear suddenly, but the promotion of history and the change of society are inevitable products that have worked and appeared together. Therefore, when understanding the aspects of women in ancient times, the character and aspects of modern women are helpful in understanding the overall change and progress. It recognizes the aspects of women in ancient times, and through the specific characters of classical literature, it is easier to understand the aspects of women at that time, and the situation of women of that time can be seen from the overall perspective. Therefore, the text will study various female aspects by organizing and comparing female characters that appeared in the book, focusing on Yoo Mong-in's "Eouyadam" and Pyo Song-ryeong's "Yojaejii" in the early Qing Dynasty of China.

Yoo Mong-in is a literary man in the middle of the Joseon Dynasty, and the representative work is "Eouyadam", and Posongryeong is a famous literary man in the early Qing Dynasty of China, and the representative work is "Yojaejii." The two writers have

J. C. Hung et al. (Eds.): IC 2023, LNEE 1045, pp. 759–766, 2023.
https://doi.org/10.1007/978-981-99-2287-1_103

similar backgrounds, who live at the end of feudal society and are deeply involved in Confucianism and Cheng Zhu's Philosophy the same era. In the male perspective of this social background, these psychological activities are well represented through female characters who appeared in literary works, so the female aspects of the two works, "Eouyadam" and "Yojaejii", have something in common. Although the social environment is similar, there are differences in the female aspects created because the personal experiences of Yoo Mong-in and Posong-ryeong are different and their social status is different.

The text organizes and compares information on Yoo Mong-in and the conscription decree in a table so that personal history can be seen more clearly. In addition, information on the two works "Eooyadam" and "Yojaejii" was simply organized, and classified and organized according to the aspects of female characters that appeared in the work. The female aspect created by Yoo Mong-in and Posong-ryeong can be found through information contrast that summarizes the causes and reasons that have commonalities and differences. In addition, it is hoped that we can understand the aspects of modern women by looking at the appearance and aspects of women living in feudal society.

2 Related Works

Kim Jin-sun revealed that women are in a desperate social position by exploring the relationship between women and women in "Eouyadam" and analyzing women's self-awareness and efforts to escape secular ideas. Hyun Hye-kyung tried to identify the characteristics and meaning of the various and rich shapes of women's life in "The Shape and Meaning of the Shape of Women's Life" in "Eouyadam". Gao Hai-lui revealed that women's social status is underground and they have to rely on men even after death according to the analysis of the aspects of the maiden ghost in "Eouyadam", Ko Sook-hee briefly analyzed the status and aspects of women's marriage in the traditional feudalism in the work and simply divided them into two types: "traditional women" and "progressive women.". Lee Soo-yeon studied and analyzed the types of female characters in the love story in "Yojaejii", and explored and studied four types of women according to their characteristics: "Current wife-in-law model," "Affectionatement-seeking type," "Chongmyeongjae daughter type," and "Great martial arts type". Jeon Soon-nam classified women who appeared mainly in the feudal ethics system into three types: "feudal female shape," "anti-feudal female shape," and "complex female shape," and examined women's psychology and survival under the feudal system. Wang Meng "Comparison of Night Talk in the Late Joseon Dynasty and the Women's Talk in the Novels of the Qing Dynasty" in the paper systematically sorted out and analyzed the women's talks introduced in the five late Korean night talk collections and the Qing Dynasty Notes Novel Collection. In the above study, a meaningful and rewarding study was conducted on the female aspects of "Eouyadam" and "Yojaejii". However, no detailed comparative study was conducted on the two works. Therefore, the text used big data and conducted a simple comparative study of "Eouyadam" and "Yojaejii", so that the commonalities and differences between the two works could be more intuitive and detailed.

3 Comparison of Yoo Mong-In and Po Song-ryeong's Personal History

Since literature originates from life and the writer's career has a great influence on the literary work he creates, it is necessary to understand the author's personal career first to understand the literary work.

First, let's look at the personal history of Yoo Mong-in and Po Song-ryeong (Table 1).

Table 1. A list of personal details of Yoo Mong-in and Po Song-ryeong

	Yoo Mong-in	Po Song-ryeong
life time	1559-1623	1640 - 1715
pseudonym	Ngmun, Eudang, Ganjae, Mukhoja	Yuseon, Yucheon, Yucheon Geosa

Personal History

Age	Year	Event	Year	Event
1	1559	Born in Goheung, Jeollanam-do (noble family)	1640	Born in Shandongseong jinan Chicheon (merchant family)
18			1657	Married Miss Liu
19			1658	He passed the childbirth testand became a doctor's student. a disciple of Si Yun-jang
21			1660	Fail the imperial examination
23			1662	The first son was born.
24	1582	Pass the entrance examination for a student	1663	Fail the imperial examination
26			1665	Work as a governess
30			1670	Work as personal secretary
31	1589	The first place in the imperial examination	1671	Resign from secretarial position The second son was born.

(*continued*)

Table 1. (*continued*)

32			1672	Visited Laoshan Mountain Fail the imperial examination
34	1592	War Responsible for Diplomatic Operations		
35	1593	Became the prince's teacher.	1675	Fail the imperial examination The third child was born.
40			1679	A Preliminary Completion of "Yojaejii"
41			1680	Mother's dead.
48			1687	Fail the imperial examination
51	1609	Third Mission to the Ming Dynasty	1690	Fail the imperial examination
54	1612	Served as Yejochampan and Ijochampan.		
55	1622	A compilation of "Eouyadam"		
56	1623	Resign from office, Be sentenced to death		
63			1702	Fail the imperial examination
72			1711	Took the imperial examination and became a preparatory student.
74			1713	Wife is dead.
76			1715	Dead.

Through this table, more clearly, the personal history of Yoo Mong-in and Posong-ryeong is compared. As you know, passing past exams and serving in government posts in ancient times was the only way for writers to participate, so both Yoo Mong-in and Po Song-ryeong took the past exams. However, because of this, the fate of Yoo Mong-in and Posong-ryeong unfolded completely differently. Yoo Mong-in passed the examination for the first birthplace and Jinsashi at the age of 24, and passed the examination as a manor in the department of Jeunggwang Literature at the age of 31. It was followed by a smooth entry into government office, which became important. On the other hand, at the age of 19, Fosongryeong, a Cheongdae literate, took the examination for his younger brother's poem and received the first prize, became a doctor's student and a disciple of Si Yoon-jang. For a 19-year-old child, this was a very high evaluation and good result, but after that, he took the exam eight times in the past for more than 40 years, but failed and passed the examination until the age of 72 and became a craftsman.

As mentioned earlier, in ancient times, the past was the only way for writers. Yoo Mong-in is from an aristocratic family and is a great-great-grandchild of Yu Yi, and his grandfather is Yu Chung-gwan, Sagan. And my father is Yu Taeng, a housewife, and my mother is the daughter of Cham Bong-min. Yoo Mong-in, a native of an aristocratic family, passed the examination in the past and became a government official, and his job was smooth, so he had no worries about living.

However, the conscription is the opposite of Yoo Mong-in. Fosongryeong was from a scholar's family, and both her great-grandfather passed the floodgates, but her grandfather failed to pass the floodgates, so her family began to decline, and her father also failed to pass the floodgates, so she made a lot of money by doing business. However, in the middle age, as he believed in Buddhism and stopped doing business, the situation gradually became difficult. Po Song-ryeong was born into this family. When the family situation became difficult, Fosongryeong's father took the role of a teacher and taught Fosongryeong's knowledge, and from an early age, Fosongryeong listened to his father's merchant thoughts and Buddhist ideas.

Due to poor family circumstances, Posongryeong had no worries about living in her childhood, but because she devoted her body and mind to creating literature in the past, she had to worry about her livelihood because she had no income source, passed the exam several times, and had to feed her wife and children, and eventually became a Seodang teacher.

In addition, this table is prepared based on age comparison, and if you look at the table, the difference in fate between Yoo Mong-in and Posong-ryeong may seem more intuitive. First of all, Yoo Mong-in and Posong-ryeong both took past tests at the age of 24, but the results were completely different, and Yoo Mong-in passed the test, but Posong-ryeong failed. At the age of 35, Yoo Mong-in had already become a government official, but Posong-ryeong failed again in the past. In addition, at the age of 51, Yoo Mong-in was dispatched to the Ming Dynasty as a Seongjeolsa and a private teacher, but Posongryeong still failed while fighting the past system.

Through this clear contrast, it can be seen that the difference between the fate of Yumongin and Posongryeong is very large. The difference in fate has become one of the many reasons why the two expressed their thoughts through the female aspects that appeared in literary works.

On the other hand, as shown in the conclusion of the Yumongin, Yumongin was a noble and had no financial worries, but he was involved in partisan battles and had a lot of heartache and despondency. Posongryeong suffered from difficulties in life as she failed the past exams several times, and her mind was filled with disappointment and resentment for the world. Based on this, it is possible to understand that the thoughts and emotions contained in the two literary works have many things in common.

4   Comparison of "Eouyadam" and "Yojaejii"

In this part, information on the two works is organized and contrasted from various angles, such as the period of creation, the background of the times, and mainstream social ideas. In addition, the aspects of women appearing in the work are classified and prepared by organizing the number of copies. Through this table, you can examine the characteristics of women in the mid-Joseon Dynasty and early Qing Dynasty, and feel the diversity of ancient women in the late feudal society and the progressiveness of the early implementation of women (Table 2).

The creation period of "Eouyadam" and "Yojaejii" was in the 17th and 18th centuries, the last feudal period in the history of both Korea and China, and social economy, ideology, and culture changed significantly at the end of feudal society. Social-led ideas were still studying abroad, but Jeongjuri and Yangmyeonghak gradually became mainstream ideas.

At a time when social ideas clashed, new changes began to appear in the lives of women who were weighed down by the feudal social system, and self-consciousness became more awakened and women's patterns diversified.

Yoo Mong-in and Posong-ryeong depicted various female aspects in the work, and according to their characteristics, women in the work were classified into types with several social representations. In other words, the human part contains Hyunbuckam, Wisdom Story, Yeolnyeo Story, Heopnyeo Story, Destruction Story, and Akcheom Story, and Sinseondam, Fox, and Ghost Story in the second-class part. Through various female aspects, the diversity and abundance of female aspects of society at the time can be seen. This can also be said to be a common feature of the female aspects created by Yoo Mong-in and Posong-ryeong.

Another common feature is that looking at this table, it can be seen that wisdom accounts for a large proportion of the human female aspect. This is deeply related to the times. At the end of the feudal period, Yoo Mong-in and Posong-ryeong, who were greatly influenced by Confucian ideas and Jeong Ju-ri ideas, represented the character and aspects that women should have as women in the eyes of most men at that time. In feudal society, men's thoughts that women are pretty, nice, wise, good at housework, managing large families, and always being able to help their husbands with good strategies when needed were well expressed through women of Wisdom.

In addition to the traditional female aspects such as virtuous woman, clever woman, and woman of strong character, there were also female aspects such as chivalrous woman, defile one's chastity, and bad wife, which deviate from the traditional female aspect and rebel against the oppression of women's human rights and nature in the feudal society. This aspect of women can be seen as a progressive awakening of women's will to traditional feudal society.

Table 2. Female Characters in "Eouyadam" and "Yojaejii"

Work	Eouyadam	Yojaejii
the year of creation	1618-1622	About 1672-1710
Time	the mid-Joseon Period	the early Qing Dynasty
mainstream social thought	Confucianism, Jeongjuri, Jeongjuri	
bibliography	Wan Zongqi Edition, Stone Pillow Publishing, 2006.	Chinese bookshop, 2015.
Total number of articles	522	491
Number of female content	68	181

Eouyadam									
	human beings						inhumanity		total
female image	virtuous woman	clever woman	woman of strong character	chivalrous woman	defile one's chastity	bad wife	Other types	fairy	Foxes and ghosts.	
Number	1	16	6	4	12	1	20	1	7	68
specific gravity	1%	24%	9%	6%	18%	1%	29%	1%	11%	100%

Yojaejii									
Number	11	15	11	4	14	8	22	22	70	181
specific gravity	6%	8%	9%	2%	7%	4%	12%	12%	38%	100%

On the other hand, this part of Foxes and ghosts stories deserves attention. Foxes and ghosts stories accounted for 11% in Eouyadam, but 38% in "Yojaejii". This is also the biggest difference that can be found through this table. Exploring the cause is related to the writer's origin and personal career. It is more urgent and necessary to express feelings of dissatisfaction, anger, and disappointment about reality that have long been built through surreal characters such as foxes and ghosts, even though they were born in poverty and failed in the past.

5   Conclusion

In this text, Yoo Mong-in and Posong-ryeong also worked on the women's aspects in "Eouyadam" and "Yojaejii" which were composed of two works, briefly organized and studied in a table using digital. In addition, the commonalities and differences of female aspects were analyzed by combining the author's personal history and the background of the times. In particular, in Chapter 3, the personal history of Yoo Mong-in and Po song-ryeong, the creative background of the work, and the female images of the works were organized in a table using big data, so that comparative research on the female aspects of both Korea and China could be conducted more coherently and intuitively. In addition, after calculating the number and ratio by clearly listing various female aspects shown in "Eooyadam" and "Yojaeji," you can further highlight the diversity and abundance of female aspects and feel the progressiveness of women's times.

References

1. Mong-in, Y.: Eooh Yadam, stone pillow (2006)
2. Posongryeong, E.R.: Yojaejii. China Bookstore, Beijing (2015)
3. Sook-hee, K.: The Study of Women in Yojaejii, Master's thesis at Sookmyung Women's University (1995)
4. Soo-yeon, L.: A Study on the Types of Female Characters in Yojaejii Love Story, Kyunghee University's Master's Degree thesis (2005)
5. Soon-nam, J.: The Study of Women in Yojaejii, Master's thesis at Yeungnam University (2011)
6. Jin-sun, K.: A Study on the Existence of Women in Yadamjip, A Master's thesis at Kyung Hee University (2006)
7. Hai-lui, G.: The Descriptive Patterns and Features of Ghosts in Eouyadam'', Master's thesis at Pusan National University (2020)
8. Mong, W.: Comparison of Yadam in the Late Joseon Dynasty and Women's Stories in Written Fiction in Qing Dynasty, Korea University's Ph.D. thesis (2017)
9. Hyun, H.-K.: The Pattern and Meaning of the Shape of Women's Life in Eouyadam, Korean Classical Women's Literature Research, Korean Classical Women's Literature Society (2004)

Smart Farm Management System Using Humidity Meter

Yuseung Shin[1] and Jaeyun Jeong[2(✉)]

[1] Kyunghee High School, Seoul 02447, Korea
[2] Hankuk University of Foreign Studies, Seoul 02450, Korea
3500jjy@gmail.com

Abstract. With the recent development of IoT technology, farmers can enjoy convenient and practical lives with smart farms created by combining agriculture and IoT technology. In this paper, we introduce the characteristics of plants and explain the direction of Beacon devices and smart devices through AAP. When managing smart farm moisture using a hygrometer, it is useful for promoting plant growth as well as saving water.

Keywords: Smart Farm · IoT · ICT

1 Introduction

The development of the Internet of Things (IoT) has made people enjoy a more comfortable life. Since then, the combination of agriculture and the Internet of Things has allowed farmers to enjoy a convenient and practical life with smart farms.

Although cultivation kits are being released as personal smart farms, there is a limit that plants are not compatible with various pots, and the types of plants are limited to vegetables, and temperature and humidity light should be correlated in the actual environment. To solve this problem, moisture can be measured with existing moisture sensors and weight sensors, but measurement errors and weight sensors are unstable due to plant growth, so an automated humidity control algorithm is needed with development of humidity sensors and beacons. In this paper, we propose a method to provide the appropriate humidity of plants using a humidity sensor.

2 Related

This section introduces existing smart farm and beacon technologies and explains the characteristics of plants and the plants introduced by Korea.

J. C. Hung et al. (Eds.): IC 2023, LNEE 1045, pp. 767–773, 2023.
https://doi.org/10.1007/978-981-99-2287-1_104

2.1 Definition of Beacon

Beacon is a Bluetooth protocol-based NFC device. Beacon's wireless communication has recently been in the spotlight as a near-field communication technology due to many advances such as low power, miniaturization, life extension, and increased reception distance, without the pairing process that had to be done to connect between the two devices using Bluetooth. In addition, the maximum communication distance is relatively long at about 50m, and sophisticated location can be identified indoors. Beacons classify certain objects with beacons as UUID values and transmit signals to users without a separate pairing procedure for each close-range section using RSSI (Received Signal Strength Indicator) to individuals with smartphones at low cost. The beacon transmitter periodically signals its UUID and RSSI values, and when a person with a smartphone comes within the reach of this signal, the smartphone recognizes it and sends signal information to the server [3].

2.2 Implementing Beacons

There are Starbucks siren orders, hospital appointments, and mobile payments for medical expenses using APP, but the service is not working well in some places in the hospital due to battery consumption problems, but the problem is expected to be solved in the future. There is also a disadvantage of weak security.

2.3 Smart Farm

It is a system created by the fusion of precision agriculture and ICT technology that emerged in the 1980s, and a system that collects data on plant growth and environment and helps decision-making is called a smart farm. It uses crop data collected through satellites, weather information, and environmental information collected using various sensors [1].

2.4 Smart Farm Trends

According to the Korea Institute for Science and Technology Jobs, industrial trends by smart farm country are spreading to areas such as distribution and consumption of smart farms in Korea, but so far, agricultural production has been the core. It is believed that it is concentrated in the monitoring and control stages, and developing optimized algorithms using big data and automation technologies related to robots are currently in the R&D stage Currently, the smart farm system applied to our farms remains at the level of opening and closing of cultivation facilities (insulation cover, ceiling, curtain, ventilator, sprinkler, fluid, hot air, etc.) through smart media based on environmental information (temperature, humidity, CO_2, illumination, etc.). In the future, it is required to develop a growth optimal environment setting model for precise crop management by growth stage based on cultivation growth information and to develop a specialized model for diagnosis of crop physiological disorders and pests.

The Netherlands is a representative smart farm-using country, and although its land area is only 1/2 of that of Korea, it has become the world's second-largest exporter of agricultural products through the introduction of ICT. The Netherlands is a representative horticultural country, and 99% of all greenhouses are glass greenhouses, and various sensors and control solutions have been developed based on decades of accumulated big data and experience optimized for the cultivation environment. Through these agricultural ICT technologies, production and quality optimization will be planned, and Priva, a leading Dutch company, is producing the world's best greenhouse environment control system and exporting it to countries around the world.

The U.S. is attempting to use not only IoT but also nanotechnology and robot technology for agriculture in earnest. In the case of Google, it is trying to develop an artificial intelligence decision support system technology that helps spread seeds, fertilizers, and pesticides by collecting big data on soil, moisture, and crop health.

In Japan, companies such as IBM, NEC, Fujitsu, and NTT provide various services by incorporating ICT technology into the agricultural field.

Examples of Japan are IBM's agricultural product history tracking service, NEC's M2M-based growth environment monitoring and logistics service, and Fujitsu's agricultural management cloud service system.

Israel is a leader in monitoring the growing environment and automatically measures crop growth information such as crop size, stem change, and leaf temperature, and predicts accurate yields by automatically adjusting water supply cycles and water supply, especially, the development of crop stress sensors has increased production by more than 40% [2].

2.5 Plants

The current status of inflow-oriented plants in Korea and their generative characteristics the distribution of origin of 114 species of inflow-oriented plants is shown in Fig. 1. There were 17 species of plants native to North and South America, accounting for 14.9% of the total. Next, 15 species of plants native to Africa and Asia each accounted for 13.3%. In addition, there were 14 species of plants native to North America and 11 species in South America, 42 species native to North and South America, accounting for 36.8% of the total. Therefore, thorough quarantine should be carried out because seeds of imported plants are most likely to be mixed or adhered to agricultural products imported from North and South America, Africa, and Asia. And there were nine species of plants native to the Mediterranean coast. Therefore, the nine species were distributed on three continents: Europe, Africa, and Asia. It was included in the top 100 malignant weeds designated by IUCN and was designated as an introductory plant in Korea, but some of them are native to tropical regions, so they cannot survive even if they enter Korea. Although it is judged that plants of this inflow should be excluded, even if some tropical regions are native, Jeju Island has a tropical climate due to global warming, suggesting the possibility of survival.

3   Smart Farm Management System Using Humidity Meter

This section presents prior research and the direction in which Beacon devices and smart devices configure smart farm systems through (APP) apps.

3.1   System Configuration Diagram

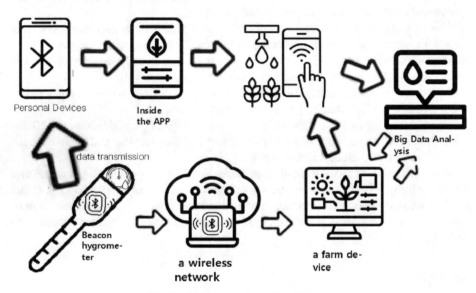

Fig. 1. System Configuration Diagram

After connecting the Beacon device built into the hygrometer and the smart device (smartphone) through the (APP) app, farm use is presented at startup. Users can choose plant types by presenting a list of plants, register photos and names, and finish setting up Wi-Fi after connecting the mobile device and the humidity sensor using a beacon in the process of adding them. For farms, help connect the farm device to the sensor.

The hygrometer settings are as follows (Figs. 2 and 3).

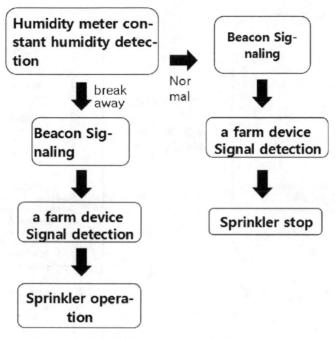

Fig. 2. The hygrometer settings

3.2 APP Internal

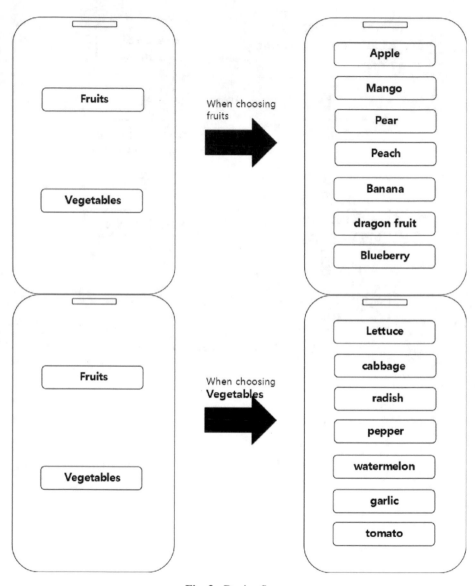

Fig. 3. Device Screen

After connecting the personal device and the hygrometer through APP, fruits, vegetables, and fruits are presented, and when the user selects fruits, the fruit type is presented, and even if the vegetables are selected, the vegetable type is presented. When the user selects the type of fruit or vegetable, set the appropriate humidity on the hygrometer.

There is a 'farm type' installation method so that the humidity controller can be applied in various places.

4 Conclusion

In this paper, we limited the humidity measurement system using soil humidity sensors that secure the limitations of plant types applied to existing smart farms and increase utilization and efficiency in smart farms. The system may expand the scope of application of existing smart farms such as various types of flower pots, vinyl houses, and open fields using various materials. Also, due to global warming, fruit production in Korea is changing little by little by little. It can also be applied to tropical fruits and plants such as mangoes and apple mangoes, which are tropical fruits grown on Jeju Island, suggesting higher viability. By implementing a humidity meter using Beacon and implementing an (APP) app, it presents a direction to grow various types of plants and fruits beyond smart farms, where the types of plants are currently limited to vegetables.

References

1. Park, S., Cho, J., Eom, S.H.: Smart Farm water management system using weight sensors. In: Proceedings of the Korean Information Science Society Conference, pp. 1731–1733 (2022)
2. https://www.bioin.or.kr/InnoDS/data/upload/tech/CF87C038-7017-11A5-6A54-5581930C2 D2F.pdf
3. Park, J., Jungminwoo, S.L., Dae-Young, K.: Fire evacuation system using beacon for hard-of-hearning people. In: JKICS, pp. 319–330 (2022)

A Study of OSMU for Henan Seolheon's Works

Zhao Wenxuan[1], Young-Hoon An[1], and Hwa-Young Jeong[2](✉)

[1] Department of Korean Language and Literature, Kyung Hee University, 26, Kyungheedae-Ro, Dongdaemun-Gu, Seoul 02447, Republic of Korea
munseon981011@naver.com, yhnahn@khu.ac.kr
[2] Humanitas College, Kyung Hee University, 26, Kyungheedae-Ro, Dongdaemun-Gu, Seoul 02447, Republic of Korea
hyjeong@khu.ac.kr

Abstract. The purpose of this paper was to study the contents of the work theory of the famous Korean ancient female writer Henan Seol-heon. When introducing classical works or classical poets to modern people, they are looking for content such as more fun and easier to receive, and storytelling. When referring to the keywords of classical literature and female poets, the name Henan Seol-heon appears a lot. It needs to know the works of female poets who have a great position in classical literature. During the Joseon Dynasty, when Heo Nan-seol-heon lived, most of the creators of literary works were male writers, claiming Confucian ideas. Due to Confucianism, women of that era were restricted in various fields such as status, recognition, freedom, and study. In this paper, we investigate the contents using the current works of Heo Nan-seol-heon and investigate the big direction of how to promote Heo Nan-seol-heon and how to proceed with the contents. In this era, almost everything is related to the database, but classical works seem to be difficult to relate to the database. In addition, it seems that to achieve this, we must support technology on many levels.

Keywords: Heo Nanseolheon · content using classical literature · works by Heo Nanseolheon

1 Introduction

Heo Nan-seol-heon was a female poet, painter, writer, and government official in the mid-Joseon Period. In an era when women had no name, Henan Seol-heon made his own name. Her real name is Heo Cho-hee, and it is passed down as Heo Ok-hye. The pen name is Nan Seolheon, and the ruler is Gyeongbeon. Both fathers and children of Heo Nan-seol-heon's family were excellent in writing, and people in the world called Heo's five sentences (Heo-yeop, Heo-sung, Heo-bong, Heo Nan-seol-heon, and Heo Gyun), but considering the Confucian society at the time, they were relatively generous to women, and they were able to study Chinese characters. Figure 1 is the standard image of Henan Seol-heon. This painting was created by artist Son Yeon-chil in 1997 [1]. The painting is now in the collection of the National Museum of Modern and Contemporary Art in Korea.

J. C. Hung et al. (Eds.): IC 2023, LNEE 1045, pp. 774–782, 2023.
https://doi.org/10.1007/978-981-99-2287-1_105

When of Heo Nan-seol-heon was 8 years old, she was called a prodigy and famous among scholars after building the Gwanghanjeonbaek Okru Sangryangmun. At the age of 15, she married Kim Seong-rip of Andong Kim's family. The people who influenced Heo Nan-seol-heon the most while learning writing was her brother Ha-gok and her teacher Son-gok. While learning poetry from Songok, Heo Nan-seol-heon even accepted it as his person. Songok's dissatisfaction, sense of defiance, arrogance, and rudeness in knowing the world would have been in line with Gyosan and Nanseolheon. Misfortunes are continuously encountered in Henan Seol-heon's family while her marriage is not smooth and her relationship with her mother-in-law is not good. Her father Heo Yeop died in 1580, and she had a son and a daughter as children, both of whom died at a young age due to an epidemic. Henan Seol-heon says that she died in 1589 at the young age of 27 because her family declined in the middle, and her father, brother, and her children died one after another, and she was under a lot of pressure and stress from her mother-in-law.

Fig. 1. Henan Seol-heon, the picture if from Namuwiki.com

The 16th century was a time when Confucian ethics were strictly applied throughout politics, economy, and culture. At that time, a society in which people demanded strict moral ethics and closed allowed women to admire the unreal world and find desires that could not be solved in a fictional world.

Although the period when Heo Nan-seol-heon lived was a period of great development in the literary and artistic aspects, the political turmoil of the Joseon Dynasty was at that time. From the 15th century to the 17th century, data on women's songs in the early and mid-Joseon Periods are mainly concentrated on sijo, and the writers' class is also centered on the kisaeng class. In comparison, not only are there very few female writers in the upper class but there is also a problem with the credibility of the author. However, the literary activities of the female class were active in the late Joseon Dynasty, centering on the Gyubang lyrics [2]. In the late 16th century, especially during the reign of King Seonjo, the literary atmosphere was so strong that it was called "Mure-ungseongse," and Seongrihak had Hwang Jin-deok, Song Soon, Imje, Jeong Cheol, Park In-no, and Sinheum, Jang Yu, Lee Jeong-gu, Seo Yang Sa-eon, and Hanho [3]. It can be said that the period when Heonnan Seolheon lived was the most prosperous period of Joseon literature.

2   Related Works

Until now, papers on poet Heo Nan-seol-heon can be largely divided into parts. First, the study of life and poetry in poets, i.e., work theory and writer theory, the second is the study of the poet's work and domestic and international poet's comparison, the third is the study of the problem of work belonging to works, the fourth is the study of translation problems of poetry by Heo Nan-seol-heon, and the fifth is the study of women's ideas in poet's works.

There are only a few content and storytelling papers on Henan Seol-heon's work. The main contents are as follows.

Lee Hyuk-jin and Shin Ae-kyung presented A Study on the Directions of Utilization for Cultural Tourism Contents of Gangneung City in Gangwon Province - Focused on Specific Historical Figures and Places - [4]. Among the papers, Gangneung-si, Gangwon-do, was presented as a case area, and the direction of exploration and use of cultural tourism contents centered on geography, tourism resource status, and historical figures. In addition, the purpose was to promote Gangneung-si through historical figures such as Kim Si-seup, Sin Saimdang, Yulgok Yi-i, Heo Gyun and Heo Nan-seol-heon of the Joseon Dynasty, and related places.

In Kang Myung-ye's 'A method of reality correspondence and storytelling of Heo Nanseolheon and Yoon Heesoon' [5]. In preparation for Heo Nan-seol-heon and Yoon Hee-soon, who are believed to have many things in common because they are marginalized and foreigners in the background of the times, especially local (Gangwon-do), they reviewed their world view, self-response, and writing patterns, and even briefly promoted the storytelling as an appendix.

In the thesis of Kim Hee-sook and Jang Woo-kwon's 'A Study on the Content and Composition of Digital Character Archive in Works and Subjects: Female Writers in the mid of the Joseon Dynasty [6]' The purpose of this study was to explore the contents and composition of works and subject-type digital character archives for the works of Shin Saimdang, Heo Nan-seol-heon, and Song Deok-bong among female writers in the mid-Joseon Dynasty.

Park Yong-jae's 'A Study on the Extensiveness of Cultural Contents in Hernanseulhen Poetry [7]' The paper studied the cultural background and storytelling method of the creation of the play "Dream Journey to the Peach Blossom" through the medium of Henan Seolheon, and the expansion of the poem into cultural contents.

Shin Soo-yeon's 'Analysis of storytelling elements of the memorial spaces for Korean female artists [8]' The thesis focused on the feminist perspective, which has recently become a hot topic in the cultural world. Among them, Heo Nan-seol-heon's example was seen as a change in the perception of oppression imposed on women in history.

3   Contents Related to the Research and Work of Henan Seol-Heon

Looking at Korean domestic papers, the contents of alternative studies are shown in the following table for the study of Henan Seol-heon's works. Among the DBPIA.co.kr papers, it is written focusing on the results that come out by setting the keyword 'Henan Seol-heon' (Table 1).

Table 1. Previous research papers related to Henan Seol-heon

Number	Year	author and thesis name	Major Research Directions
1	2021	LIM MIJUNG, Reconsideration on the Materials of Heo Nanseolheon's Poems	complementary work
2	2021	Lee Cheol-hui, A critical investigation into the authorship of two proses in Nanseolheonsijip, allegedly written by Heo Nanseolheon	the question of quieting one's work
3	2018	Jeong Soyeon, Diglossia of Literature in the Middle Ages and Literacy Education: -Hwang Jini and Heo Nanseolheon in the 16th century-	a contrast study
4	2017	Yunhyeji, The Depressed Mood in Poetry by Female Writers from Ancient Korea and China - focusing on Huh-Nanseolhun and Wang-Fengxian	a contrast study
5	2017	Yun Inhuyn, Heonanseolheon's Consciousness through her Chinese Poems	the theory of works
6	2016	Lee, Hwa-hyung, A Study on the Consciousness of "Subject and Liberty" in HuhNanseolHeon's Life and Literature	the theory of works
7	2016	Park Hyun Kyu, Study on the Selected Edition of Heo Nanseolheon's Nanseol sihan Compiled by Heo Gyun in 1597	the problem of ripening/distribution

(continued)

Table 1. (*continued*)

Number	Year	author and thesis name	Major Research Directions
8	2016	Kang-myeonghye, Heo Nanseolheon's Yousun Poems and the Poems' Color Aesthetics - Focusing on Comparing Characteristics Color aesthetic, with other Youson poems, China and Joseon dynasty	a work theory/comparative study
9	2015	Yu Yukrye,A Study on Nanseolheon Heo Romantic Love and Yearning Feeling Poems	the theory of works
10	2015	Han Seonggeum, Speculation on Chinese Poetry Written by Women from Noble Families in the 16th Century and the Expressive Aspects Used - Chinese Poetry by Song Duk-Bong and Huh Nanseolhun-	Comparative Research & Theory of Works
11	2014	Son Aenghwa, The study of Unfortunate consciousness that appears in Yuseonsa by Heonanseolheon - On the basis of poetic-word statistics and analysis	the theory of works
12	2014	Kang Minkyoung, The study on the time images in Yusun literature of Heonanseolheon	the theory of works
13	2030	Yi Dongha, Fictionalization of the Noble Women's Life during the Chosun Dynasty	writer's theory
14	1990	Lee Sanglan, A Comparative Study of HuLansulhun and Emily Dickinson -A Long Night Journey to the "Mother's space"-	comparative study
15	1980	Jangjin, A Study on Heo Nan-seol-heon's poems	The Theory of Writers and Works

In the table above, we can see the research history of Heo Nan-seol-heon in Korea. Most of them write papers on writer theory, work theory, comparative research, and work acquisition problems. In addition, many books about Henan Seol-heon's works are now included in Korea. He always studies the works of Heo Nan-seol-heon, focusing on "Nansol-heon Poetry," which Heo Gyun, Heo Nan-seol-heon"s younger brother, edited. "Nanseolheonsi" was edited by Heo Gyun in 1608 and contained 210 poems in total. Until now, Heo Nan-seol-heon's content paintings have been produced in the form of dance, music drama, ballet, and musicals, but there are only a few works except for special performances related to the Pyeongchang-dong Mirror Olympics. The format used as the content is shown in the following table (Table 2).

Table 2. Utilization of content by Heo Nan-seol-heon

Time	content/work	Content utilized
2014.02.24	About the life and work of Heo Nanseolheon	documentary drama/Gangneung MBC
2016.08.20	The works "Kyuwon" and "Gamwoo"	Chamjak Dance (Gangneung Wonju University, Haerang Cultural Center)
2016.12.23	About the life and work of Heo Nanseolheon	Music Drama (Gangneung Wonju University, Haerang Cultural Center)
2017.05.05	"Gamwoo", "Dream Journey to the Peach Blossom Land"	Ballet (CJ Towol Theater, Arts Center)
2018 PyeongChang Winter Olympics	About the life and work of Heo Nanseolheon	special performance
–	"Dream Journey to the Peach Blossom Land"	Musical
–	–	Heo Gyun and Heo Nanseolheon Memorial Hall (Gangneung)

Fig. 2. MBC documentary drama "Henan Seol-heon". Koo Hye-sun, a picture that appeared in an article titled "The 24th Broadcast," which released a still of the documentary drama "Heonan Seol-heon."

Figure 2 is in 2014, actress Koo Hye-sun filmed, made, and acted in the MBC documentary drama "Henan Seol-heon". Heo Nan-seol-heon, played by actress Koo

Hye-sun in 2014 [9], is a female literary scholar who is easier and more understood by the public. If you watch the video rather than the book, you can learn about Heo Nan-seol-heon's life more simply and interestingly.

Fig. 3. "Heonan Seolheon's Musical poster by Naver.com

Figure 3 is poster of a musical play about Henan Seol-heon [10] and Fig. 4 is a music drama promotion book about Heo Nanseolheon [11]. In this format, it is introduced to the public by Henan Seol-heon. Since the number of spectators is also large, the method of combining classical and contemporary content through this can say a successful word.

Fig. 4. 'Heonan Seolheon' Memorial Hall image picture by Naver.com

Figure 5 is Ballet created by Heo Nan-seol-heon, using "Gamwoo" and "Dream Journey to the Peach Blossom Land" [12]. Kang Hyo-hyung will present 55-min works under the themes of Heo Nan-seol-heon's poems "Gamwoo" and "Mongyu Gwangsang-shansi." In the first half, he expressed Henan Seol-heon's warm and happy time through "Gamwoo," and in the second half, he expressed his painful and sad later life through "Mongyu Gwangsangshansi."

Heo Nan-seol-heon's life can be largely divided into two parts. The first half was when she was at home, and the days before the breakup were favored by her family, and unlike the women of the time, she was a woman from a prestigious family who could learn letters or literature. Before marriage, Heo Nan-seol-heon was a girl who lived without any worries. On the other hand, Heo Nan-seol-heon's poetry changed greatly

Fig. 5. Heona Seolheon Ballet

Fig. 6. Back door of Gangneung, Heo Gyun, and Heo Nanseolheon Memorial Hall

Fig. 7. Front door of Gangneung, Heo Gyun, and Heo Nanseolheon Memorial Hall

due to her unhappy life after marriage, conflicts with her husband, her son's early death, family misfortune, and these causes.

Figure 6 [13]and Figure 7 [14] is Gangneung, Heo Gyun, and Heo Nanseolheon Memorial Hall.

4 Conclusion

Only those who are interested in classical literature and scholars who have studied classical literature have read a lot. Most modern people are familiar with classical works. In addition, classical works are recorded in Chinese characters, not in Korean, and young people who use only Hangul today use longer to understand the works.

Following the contents of Heo Nan-seol-heon's work mentioned in Chapter 3, it is possible to create various contents by re-interpreting and quoting the original work of Heo Nan-seol-heon and receiving them more easily from young people and foreigners. Classical literature works are easily received by modern people by mixing content or storytelling methods. It seems that it is the current trend to create more interesting content after using the media than reading the original text. Even if this is not easy to realize, the combination of classical literature and content, and the combination of classical literature and database can be said to be a future trend.

References

1. Figure 1 is the standard image of Henan Seol-heon. This painting was created by artist Son Yeon-chil in 1997, by (Korean website) naver website, Namuwiki - Henan Seolheon
2. Lee Hye-soon, A Study on Women in Korean Classical Literature, Taehaksa, p. 122 (1999)
3. Kim Myung-hee, Literature by Heo Nan-seol-heon, p. 9 (1987)
4. Lee Hyuk-jin and Shin Ae-kyung presented A Study on the Directions of Utilization for Cultural Tourism Contents of Gangneung City in Gangwon Province - Focused on Specific Historical Figures and Places- Journal of Photographic Geography **28**(1), 63–78 (2018))
5. Kang Myung-ye: A method of reality correspondence and storytelling of Heo Nanseolheon and Yoon Heesoon Onji Nonchong **30**, 137–172 (2012)
6. Kim Hee-sook and Jang Woo-kwon: A Study on the Content and Composition of Digital Character Archive in Works and Subjects: Female Writers in the mid of the Joseon Dynasty. Journal of the Korean Literature and Information Society **54**(1), 145–174 (2020)
7. Park Yong-jae: A Study on the Extensiveness of Cultural Contents in Hernanseulhen Poetry. Ph.d paper of Graduate School of Dankook University (2020)
8. Shin Soo-yeon: Analysis of storytelling elements of the memorial spaces for Korean female artists. Ph.d paper of Konkuk University Graduate School (2020)
9. Figure 2, Segye Ilbo, 2014.02.24 Koo Hye-sun, a picture that appeared in an article titled "The 24th Broadcast," which released a still of the documentary drama "Heonan Seol-heon."
10. Figure 3, NAVER's photo (by Korean website in naver). Heonan Seolheon's Music Drama
11. Figure 4, NAVER's photo (by Korean website in naver). Heonan Seolheon's Music Drama
12. Figure 5, NAVER's 'Heona Seolheon Ballet' photo (by Korean website in naver)
13. Figure 6, NAVER's 'Heonan Seolheon' Memorial Hall image picture (by Korean website in naver)
14. Figure 7, NAVER's 'Heonan Seolheon' Memorial Hall image picture (by Korean website in naver)

Author Index

© The Editor(s) (if applicable) and The Author(s), under exclusive license
to Springer Nature Singapore Pte Ltd. 2023
J. C. Hung et al. (Eds.): IC 2023, LNEE 1045, pp. 783–785, 2023.
https://doi.org/10.1007/978-981-99-2287-1

Printed in the United States
by Baker & Taylor Publisher Services